Embeddings in Manifolds

Embeddings in Manifolds

Robert J. Daverman
Gerard A. Venema

Graduate Studies in Mathematics
Volume 106

American Mathematical Society
Providence, Rhode Island

2000 *Mathematics Subject Classification*. Primary 57N35, 57N30, 57N45, 57N40, 57N60, 57N75, 57N37, 57N70, 57Q35, 57Q30, 57Q45, 57Q40, 57Q60, 57Q55, 57Q37, 57P05.

For additional information and updates on this book, visit
www.ams.org/bookpages/gsm-106

Library of Congress Cataloging-in-Publication Data

Daverman, Robert J.
Embeddings in manifolds / Robert J. Daverman, Gerard A. Venema.
p. cm. — (Graduate studies in mathematics ; v. 106)
Includes bibliographical references and index.
ISBN 978-0-8218-3697-2 (alk. paper)
1. Embeddings (Mathematics) 2. Manifolds (Mathematics) I. Venema, Gerard. II. Title.

QA564.D38 2009
514—dc22 2009009836

∞ The paper used in this book is acid-free and falls within the guidelines
established to ensure permanence and durability.
Visit the AMS home page at `http://www.ams.org/`

10 9 8 7 6 5 4 3 2 1 14 13 12 11 10 09

To Lana and Patricia
for persistent support and patient tolerance
despite their doubts this project would ever end.

Contents

Acknowledgments xi

Introduction: The Main Problem xiii

Chapter 0. Prequel 1

§0.1. More definitions and notation 1

§0.2. The Seifert-van Kampen Theorem 3

§0.3. The ultimate duality theorem 6

§0.4. The Vietoris-Begle Theorem 9

§0.5. Higher homotopy groups 9

§0.6. Absolute neighborhood retracts 12

§0.7. Dimension theory 14

§0.8. The Hurewicz Isomorphism Theorem and its localization 16

§0.9. The Whitehead Theorem 17

§0.10. Acyclic complexes and contractible manifolds 17

§0.11. The 2-dimensional PL Schönflies Theorem 19

Chapter 1. Tame and Knotted Embeddings 23

§1.1. Knotted and flat piecewise linear embeddings 23

§1.2. Tame and locally flat topological embeddings 29

§1.3. Local co-connectedness properties 30

§1.4. Suspending and spinning 35

Chapter 2. Wild and Flat Embeddings 41
§2.1. Antoine's necklace and Alexander's horned sphere 41
§2.2. Function spaces 53
§2.3. Shrinkable decompositions and the Bing shrinking criterion 55
§2.4. Cellular sets and the Generalized Schönflies Theorem 59
§2.5. The Klee trick 66
§2.6. The product of $\mathbb{R}^1$ with an arc decomposition 67
§2.7. Everywhere wild cells and spheres 73
§2.8. Miscellaneous examples of wild embeddings 75
§2.9. Embeddings that are piecewise linear modulo one point 91

Chapter 3. Engulfing, Cellularity, and Embedding Dimension 97
§3.1. Engulfing without control 98
§3.2. Application: The cellularity criterion 106
§3.3. Engulfing with control 114
§3.4. Application: Embedding dimension 123
§3.5. Embeddings of Menger continua 130
§3.6. Embedding dimension and Hausdorff dimension 136

Chapter 4. Trivial-range Embeddings 145
§4.1. Unknotting PL embeddings of polyhedra 146
§4.2. Spaces of embeddings and taming of polyhedra 152
§4.3. Taming 1-LCC embeddings of polyhedra 159
§4.4. Unknotting 1-LCC embeddings of compacta 161
§4.5. Chart-by-chart analysis of topological manifolds 162
§4.6. Detecting 1-LCC embeddings 163
§4.7. More wild embeddings 169
§4.8. Even more wild embeddings 172

Chapter 5. Codimension-three Embeddings 181
§5.1. Constructing PL embeddings of polyhedra 182
§5.2. Constructing PL embeddings of manifolds 190
§5.3. Unknotting PL embeddings of manifolds 196
§5.4. Unknotting PL embeddings of polyhedra 211
§5.5. 1-LCC approximation of embeddings of compacta 227
§5.6. PL approximation of embeddings of manifolds 243
§5.7. Taming 1-LCC embeddings of polyhedra 261
§5.8. PL approximation of embeddings of polyhedra 263

Chapter 6. Codimension-two Embeddings 273
§6.1. Piecewise linear knotting and algebraic unknotting 274
§6.2. Topological flattening and algebraic knotting 283
§6.3. Local flatness and local homotopy properties 293
§6.4. The homology of an infinite cyclic cover 298
§6.5. Properties of the Alexander polynomial 311
§6.6. A topological embedding that cannot be approximated by PL embeddings 325
§6.7. A homotopy equivalence that is not homotopic to an embedding 334
§6.8. Disk bundle neighborhoods and taming 346

Chapter 7. Codimension-one Embeddings 349
§7.1. Codimension-one separation properties 350
§7.2. The 1-LCC characterization of local flatness for collared embeddings 353
§7.3. Unknotting close 1-LCC embeddings of manifolds 358
§7.4. The Cell-like Approximation Theorem 366
§7.5. Determining n-cells by embeddings of M_n^{n-1} in S^n 371
§7.6. The 1-LCC characterization of local flatness 377
§7.7. Locally flat approximations 382
§7.8. Kirby-Siebenmann obstruction theory 402
§7.9. Detecting 1-LCC embeddings 403
§7.10. Sewings of crumpled n-cubes 410
§7.11. Wild examples and mapping cylinder neighborhoods 417

Chapter 8. Codimension-zero Embeddings 425
§8.1. Manifold characterizations 425
§8.2. The α- and β-Approximation Theorems 427
§8.3. Ends of manifolds 427
§8.4. Ends of maps 432
§8.5. Quinn's obstruction and the topological characterization of manifolds 435
§8.6. Exotic homology manifolds 437
§8.7. Homotopy tori are tori 439
§8.8. Approximating stable homeomorphisms of $\mathbb{R}^n$ by PL homeomorphisms 439

§8.9. Rigidity: Homotopy equivalence implies homeomorphism 442
§8.10. Simplicial triangulations 443

Bibliography 445

Selected Symbols and Abbreviations 461

Index 463

Acknowledgments

Thanks to David Koop for converting early versions of parts of the manuscript from *WordPerfect®* to LaTeX and for creating the basic database from which the bibliography was built. Thanks also to Daniel Venema for making many of the diagrams, especially those in Chapters 2 and 6. Hearty thanks to Helaman Ferguson for providing photographs of his marvelous sculptures, which were inspired by some of our favorite wild embeddings, and for generously allowing their reproduction here.

The following colleagues read parts of the manuscript and offered valuable comments: John Bryant, Craig Guilbault, John Engbers, Denise Halverson, Sergey Melikhov, David Snyder, Fred Tinsley, Paul Thurston and Violeta Vasilevska.

Both authors owe a great debt to Benny Rushing. His book *Topological Embeddings* spurred developments in the subject and played a major role in our thinking about topological embeddings. While there was considerable flux in perspectives when he put his book together, now the topic has attained a mature stability, which we have attempted to organize and elucidate. He was advisor, friend and collaborator, and we miss him.

Robert J. Daverman
Gerard A. Venema
February, 2009

Introduction: The Main Problem

Prerequisites. What background is needed for reading this text? Chiefly, a knowledge of piecewise linear topology. For many years the standard reference in that area has been the text *Introduction to Piecewise-Linear Topology*, by C. P. Rourke and B. J. Sanderson (1972), and we assume familiarity with much of their book. To be honest, that book presumes extensive understanding of both general and algebraic topology; as a consequence we implicitly are assuming those subjects as well. In an attempt to limit our presumptions, we specifically shall take as granted the results from two fairly standard texts on general and algebraic topology, both by J. R. Munkres—namely, his *Topology: Second Edition* (2000) and *Elements of Algebraic Topology* (1984), each of which can be treated quite effectively in a year-long graduate course.

Unfortunately, even those three texts turn out to be insufficient for all our needs. The purpose of the initial Chapter 0, the Prequel, is to correct that deficiency.

Basic Terminology. The notation laid out in this subsection should be familiar to those who have read Rourke and Sanderson's text. Nevertheless, we spell out the essentials needed to fully understand the forthcoming discussion of the primary issues addressed in this book.

Here $\mathbb{R}$ denotes the set of real numbers and $\mathbb{R}^n$ denotes n-dimensional Euclidean space, the Cartesian product of n copies of $\mathbb{R}$. For $1 \leq k < n$ we regard $\mathbb{R}^k$ as included in $\mathbb{R}^n$ in the obvious way, as the subset containing all points whose final $(n-k)$-coordinates are all equal to zero.

We use B^n to denote the standard n-ball (or n-cell) in $\mathbb{R}^n$, $\operatorname{Int} B^n$ to denote its interior, and S^{n-1} to denote the standard $(n-1)$-sphere, the boundary, ∂B^n, of B^n. Specifically,

$$B^n = \{\langle x_1, x_2, \ldots, x_n\rangle \in \mathbb{R}^n \mid x_1^2 + x_2^2 + \cdots + x_n^2 \leq 1\},$$
$$\operatorname{Int} B^n = \{\langle x_1, x_2, \ldots, x_n\rangle \in \mathbb{R}^n \mid x_1^2 + x_2^2 + \cdots + x_n^2 < 1\}, \text{ and}$$
$$S^{n-1} = \partial B^n = \{\langle x_1, x_2, \ldots, x_n\rangle \in \mathbb{R}^n \mid x_1^2 + x_2^2 + \cdots + x_n^2 = 1\}.$$

We call any space homeomorphic to B^n or S^{n-1} an *n-cell* or an *$(n-1)$-sphere*, respectively. The k-ball B^k is defined as a subset of $\mathbb{R}^k$, but for each $k < n$ the inclusion $\mathbb{R}^k \subset \mathbb{R}^n$ determines a standard k-ball B^k and a standard $(k-1)$-sphere S^{k-1} in $\mathbb{R}^n$ as well.

All simplicial complexes and CW complexes are assumed to be locally finite. A *polyhedron* is the underlying space of a simplicial complex. While a simplicial complex K and the underlying polyhedron $|K|$ are two different things, we will not always maintain this distinction in our terminology. Piecewise linear is abbreviated PL.

An *n-dimensional (topological) manifold* is a separable metric space in which each point has a neighborhood that is homeomorphic to $\mathbb{R}^n$. Such a neighborhood is called a *coordinate neighborhood* of the point.

The Main Problem. The central topic in this text is topological embeddings. Formally, an *embedding* of one topological space X in another space Y is nothing more than a homeomorphism of X onto a subspace of Y. The domain X is called the *embedded space* and the target Y is called the *ambient space*. Two embeddings $\lambda, \lambda' : X \to Y$ are *equivalent* if there exists a (topological) homeomorphism Θ of Y onto itself such that $\Theta \circ \lambda = \lambda'$. The main problem in the study of topological embeddings is:

Main Problem. *Which embeddings of X in Y are equivalent?*

In extremely rare circumstances all pairs of embeddings are equivalent. For instance, if X is just a point, the equivalence question for an arbitrary pair of embeddings of X in a given space Y amounts to the question of homogeneity of Y, which has an affirmative answer whenever, for example, Y is a connected manifold.

Ordinarily, then, our interest will turn to conditions under which embeddings are equivalent, and we will limit attention to reasonably well-behaved spaces X and Y. Specifically, in this book the embedded space X will ordinarily be a compact polyhedron[1] and the ambient space Y will always be a manifold, usually a piecewise linear (abbreviated PL) manifold. If there are embeddings of the polyhedron X in the PL manifold Y that are homotopic

[1]A major exception is the study of embeddings of the Cantor set.

but not equivalent, then X is said to *knot* in Y. For given polyhedra X and Y, it is often possible to identify a distinguished class of PL embeddings of X in Y that are considered to be *unknotted*; any PL embedding that is not equivalent to an unknotted embedding is then said to be *knotted.*

While we do place limitations on the spaces considered, we intentionally include the most general kinds of topological embeddings in the discussion. Let X be a polyhedron and let Y be a PL manifold. An embedding $X \to Y$ is said to be a *tame embedding* if it is equivalent to a PL embedding; the others are called *wild.* For embeddings of polyhedra the Main Problem splits off two fundamental special cases, one called the Taming Problem and the other the (PL) Unknotting Problem.

Taming Problem. *Which topological embeddings of X in Y are equivalent to PL embeddings?*

Unknotting Problem. *Which PL embeddings of X in Y are equivalent?*

The point is, for tame embeddings the Main Problem reduces to the Unknotting Problem, and PL methods provide effective – occasionally complete – answers to the latter. As we shall see, local homotopy properties give very precise answers to the Taming Problem. This also means that local homotopy properties make detection of wildness quite easy. There are related crude measures that adequately differentiate certain types of wildness, but the category of wild embeddings is highly chaotic. In fact, at the time of this writing very little effort had been devoted to classifying in any systematic way the wild embeddings of polyhedra in manifolds.

A closed subset X of a PL manifold N is said to be *tame* (or, *tame as a subspace*) if there exists a homeomorphism h of N onto itself such that $h(X)$ is a subpolyhedron; X itself is *wild* if it is homeomorphic to a simplicial complex but is not tame. Here the focus is more on the subspace X than on a particular embedding. One can provide a direct connection, of course: a closed subset X of a PL manifold N is tame as a subspace if and only if there exist a polyhedron K and a homeomorphism $g : K \to X$ such that $\lambda = \text{inclusion} \circ g : K \to N$ is a tame embedding.

We say that a k-cell or $(k-1)$-sphere X in $\mathbb{R}^n$ is *flat* if there exists a homeomorphism h of $\mathbb{R}^n$ such that $h(X)$ is the standard object of its type. Generally, whenever we have some standard object $S \subset \mathbb{R}^n$ and a subset X of $\mathbb{R}^n$ homeomorphic to S, we will say that X is *flat* if there is a homeomorphism $h : \mathbb{R}^n \to \mathbb{R}^n$ such that $h(X) = S$. In other words, S represents the preferred copy in $\mathbb{R}^n$, and another copy in $\mathbb{R}^n$ is flat if it is ambiently equivalent (setwise) to S.

Flatness Problem. *Under what conditions is a cell or sphere in $\mathbb{R}^n$ flat?*

The problems listed above are the main ones that will occupy attention in this text. They all can be viewed as uniqueness questions in the sense that they ask whether given embeddings are equivalent. There are also existence questions for embeddings, which will be studied alongside the uniqueness questions. We identify two such: one global, the other local.

Existence Problem. *Given a map $f : X \to Y$, is f homotopic to a topological embedding or a PL embedding?*

Approximation Problem. *Which topological embeddings of X in Y can be approximated by PL embeddings?*

The flatness concept has a local version. A topological embedding $e : M \to N$ of a k-dimensional manifold M into an n-dimensional manifold N is *locally flat at $x \in M$* if there exists a neighborhood U of $e(x)$ in N such that $(U, U \cap e(M)) \cong (\mathbb{R}^n, \mathbb{R}^k)$. An embedding is said to be *locally flat* if it is locally flat at each point x of its domain. The last two problems have local variations: for example, one can ask whether a map of manifolds is homotopic to a locally flat embedding or whether a topological embedding of manifolds can be approximated by locally flat embeddings.

When considering an embedding $e : X \to Y$, the dimension of Y is called the *ambient dimension.* Almost all of the examples and theorems in this book involve embeddings in manifolds of ambient dimension three or more. We skip dimension two because classical results like the famous Schönflies theorem (Theorem 0.11.1) imply that no nonstandard local phenomena arise in conjunction with embeddings into manifolds of that dimension.

While isolated examples of wild embeddings were discovered earlier, the work of R. H. Bing in the 1950s and 1960s revealed the pervasiveness of wildness in dimensions three and higher. His pioneering work led to a proliferation of embedding results, first concentrating on dimension three, but soon expanding to include higher dimensions as well. The subject of topological embeddings is now a mature branch of geometric topology, and this book is meant to be a summary and exposition of the fundamental results in the area.

Organization. As mentioned earlier, the initial Chapter 0 addresses background matters. The real beginning, Chapter 1, treats knottedness, tameness and local flatness; it provides examples of knotted, PL codimension-two sphere pairs in all sufficiently large dimensions, and it delves into the local homotopy properties of nicely embedded objects. Chapter 2 presents the basic examples that motivate the study and offers context for theorems to come later; it also includes several flatness theorems that can be proved without the use of engulfing. Engulfing – the fundamental technical tool for the subject – is introduced and carefully examined in Chapter 3. The remaining chapters strive to systematically investigate the central embedding problems. That investigation is organized by codimension. The *codimension* of an embedding $e : X \to Y$ is defined by $\operatorname{codim}(e) = \dim Y - \dim X$, the difference between the ambient dimension and the dimension of the embedded space. Generally speaking, the greater the codimension the easier it is to prove positive theorems about embeddings. Chapter 4 treats the *trivial range*, the range in which the codimension of the embedded space exceeds its dimension, where the most general theorems hold. Next, Chapter 5 moves on to codimension three, to which many trivial-range theorems extend with appropriate modifications. However, very few of the codimension-three theorems extend to codimension two, so Chapter 6 is largely devoted to the construction of codimension-two counterexamples. In codimension one the situation changes once more, and again there are many positive results, which form the subject of Chapter 7. The book concludes in Chapter 8 with a quick description of some codimension-zero results.

Chapter 0

Prequel

This Prequel sets forth – with references, but with few proofs – important background results covered by neither Rourke and Sanderson nor Munkres. Readers may want to briefly familiarize themselves with the contents of this chapter and then begin their serious study with Chapter 1. Chapter 0 can be used as a reference for topics that arise later and consulted as needed. The prerequisites covered in this chapter should be enough to carry the reader through the first five chapters of the book. Beyond that point, additional deep material occasionally will be interwoven, without proof, to present a complete picture of current developments.

0.1. More definitions and notation

The *n-cube* I^n is the n-fold product $[-1,1]^n$. Following Rourke and Sanderson (1972, page 4), we consistently use I^1 to denote the interval $[-1,1]$, but sometimes use I to denote the interval $[0,1]$. Whether I denotes $[0,1]$ or $[-1,1]$ should be clear from the context. Of course I^n is homeomorphic to B^n, so the n-cube is an n-cell. In some contexts a k-cell will also be called a *k-disk* and will be denoted D^k.

Let X and Y be two spaces with base points x_0 and y_0, respectively. The *wedge* (or *wedge sum*) of X and Y is the quotient of the disjoint union $X \sqcup Y$ obtained by identifying x_0 and y_0. The wedge of X and Y is denoted $X \vee Y$ and might also be called the *one-point union* of X and Y; the wedge of a finite number of circles is often called a *bouquet of circles* (Figure 0.1).

Upper half space $\mathbb{R}^n_+$ consists of all the points in $\mathbb{R}^n$ whose last coordinate is nonnegative; i.e.,

$$\mathbb{R}^n_+ = \{\langle x_1, x_2, \ldots, x_n\rangle \mid \text{ each } x_i \text{ is a real number and } x_n \geq 0\}.$$

Figure 0.1. A bouquet of six circles

Note that $\mathbb{R}^k \subset \mathbb{R}^n_+$ if $k < n$.

An *n-dimensional ∂-manifold* (read "boundary manifold") is a separable metric space in which each point has a neighborhood[1] that is homeomorphic to $\mathbb{R}^n_+$. We will use superscripts to denote the dimension of a manifold or a ∂-manifold. Thus the statement "M^n is a manifold" is to be interpreted to mean that M is an n-dimensional manifold.

Let M be an n-dimensional ∂-manifold. The *interior of M* (denoted $\operatorname{Int} M$) consists of all points $x \in M$ such that x has a neighborhood that is homeomorphic to $\mathbb{R}^n$. The *boundary of M* (denoted ∂M) is defined by $\partial M = M \smallsetminus \operatorname{Int} M$.

Remark. Our use of the term *∂-manifold* is somewhat nonstandard, but we prefer it to the more awkward *manifold-with-boundary.* The use of the term ∂-manifold allows us to be consistent in our use of the word *manifold*: in this book a manifold always has empty boundary.

A *closed manifold* is a manifold that is compact and has empty boundary. Since all our manifolds have empty boundary (by definition), there is no difference between a closed manifold and a compact manifold.

The Invariance of Domain Theorem (Munkres, 1984, Theorem 4-36.5) should be used to work several of the following exercises.

Exercises

0.1.1. The dimension of a manifold is well defined: two manifolds of different dimensions cannot be homeomorphic.

0.1.2. The dimension of a ∂-manifold is well defined.

0.1.3. Let M be an n-dimensional ∂-manifold and let y be a point in M. If the last coordinate of $h(y)$ is zero for one pair (U, h) in which U

[1]A neighborhood is not necessarily an open set. A *neighborhood* of the point x in the space X is any subset U of X such that x is contained in the topological interior of U.

is a neighborhood of y and $h : U \to \mathbb{R}^n_+$ is a homeomorphism, then the last coordinate of $h(y)$ is zero for every such pair (U, h).

0.1.4. The interior and boundary of a ∂-manifold are well defined. Specifically, if M and N are ∂-manifolds and $h : M \to N$ is a topological homeomorphism, then $h(\partial M) = \partial N$ and $h(\operatorname{Int} M) = \operatorname{Int} N$.

0.1.5. If M is an n-dimensional ∂-manifold, then ∂M is an $(n-1)$-dimensional manifold (without boundary).

0.1.6. If M and N are ∂-manifolds, then $M \times N$ is a ∂-manifold and

$$\partial(M \times N) = (\partial M \times N) \cup (M \times \partial N).$$

0.2. The Seifert-van Kampen Theorem

We will assume familiarity with the fundamental group and the theory of covering spaces. The Seifert-van Kampen Theorem relates the fundamental group of the union of two spaces to the fundamental groups of the two constituent pieces. The setting for the theorem posits the following data: U_1 and U_2 are pathwise connected, open subsets of a space X such that $X = U_1 \cup U_2$ and $U_0 = U_1 \cap U_2$ are pathwise connected, $x \in U_0$, $\phi_i : \pi_1(U_0, x) \to \pi_1(U_i, x)$, $i \in \{1, 2\}$, and $\psi_i : \pi_1(U_i, x) \to \pi_1(X, x)$, $i \in \{0, 1, 2\}$, are the inclusion-induced homomorphisms.

Theorem 0.2.1 (Seifert-van Kampen). *If H is a group and $\rho_i : \pi_1(U_i, x) \to H$ are any homomorphisms for $i = 0, 1, 2$ such that the diagram*

$$\begin{array}{ccccc} & & \pi_1(U_1, x) & & \\ & \overset{\phi_1}{\nearrow} & & \overset{\rho_1}{\searrow} & \\ \pi_1(U_0, x) & & \xrightarrow{\quad \rho_0 \quad} & & H \\ & \underset{\phi_2}{\searrow} & & \underset{\rho_2}{\nearrow} & \\ & & \pi_1(U_2, x) & & \end{array}$$

is commutative, then there exists a unique homomorphism $\sigma : \pi_1(X, x) \to H$ such that $\sigma\psi_i = \rho_i$ for $i = 0, 1, 2$; that is, σ renders the following diagram commutative:

$$\begin{array}{ccccc} & & \pi_1(U_1, x) & & \\ & \overset{\phi_1}{\nearrow} & \downarrow \psi_1 & \overset{\rho_1}{\searrow} & \\ \pi_1(U_0, x) & \xrightarrow{\psi_0} & \pi_1(X, x) & \xrightarrow{\sigma} & H \\ & \underset{\phi_2}{\searrow} & \uparrow \psi_2 & \underset{\rho_2}{\nearrow} & \\ & & \pi_1(U_2, x) & & \end{array}$$

The full theorem and proof are presented in (Munkres, 2000, §70). There is also a thorough exposition of the theorem in (Massey, 1967, pp. 113–122) (or (Massey, 1991, pp. 86–96)). Later in the chapter (Theorem 0.11.5) we will prove the following addendum to 0.2.1: *if* $\phi_i : \pi_1(U_0, x) \to \pi_1(U_i, x)$ *is one-to-one for* $i = 1, 2$, *then* $\psi_i : \pi_1(U_i, x) \to \pi_1(X, x)$ *is also one-to-one for* $i = 0, 1, 2$.

As the next two examples illustrate, the Seifert-van Kampen Theorem can often be used to gain useful information about a fundamental group without explicitly computing the group itself.

Example 0.2.2. Examples abound where $\pi_1(X, x)$ is trivial despite nontriviality of all $\pi_1(U_i, x)$. But the theorem immediately gives nontriviality of $\pi_1(X, x)$ when one can locate a group H and pair of homomorphisms $\rho_i : \pi_1(U_i, x) \to H$ $(i = 1, 2)$ satisfying the commutativity relationship in the statement with either ρ_1 or ρ_2 nontrivial. □

Example 0.2.3. The Seifert-van Kampen Theorem can also be exploited to detect nonabelian fundamental groups. We illustrate with a simple example, in which each $\pi_1(U_i, x)$ is infinite cyclic, ϕ_1 amounts to multiplication by 2, and ϕ_2 to multiplication by 3. To see why $\pi_1(X, x)$ can be nonabelian, simply consider $H = S_3$, the symmetric group on the symbols $\{a, b, c\}$, define $\rho_1 : \pi_1(U_1, x) \cong \mathbb{Z} \to S_3$ by defining $\rho_1(1)$ to be the transposition (ab) and $\rho_2 : \pi_1(U_2, x) \cong \mathbb{Z} \to S_3$ by defining $\rho_2(1)$ to be the 3-cycle (abc). Then the trivial homomorphism ρ_0 fleshes out the commutative diagram, and the presence of noncommuting elements (ab), (abc) in $\sigma(\pi_1(X, x)) \subset S_3$ indicates $\pi_1(X, x)$ is nonabelian. A similar argument can be provided whenever ϕ_1, ϕ_2 amount to multiplication by relatively prime integers greater than 1. The Seifert-vanKampen Theorem will be used this way in the next chapter to prove that torus knots are truly knotted. □

When the theorem is used to compute $\pi_1(X, x)$, it is important to bear in mind that $\pi_1(X, x)$ is generated by the images of ψ_1 and ψ_2 (Munkres, 2000, Theorem 9-59.1). With this additional piece of information it is relatively easy to see that the version of the Seifert-van Kampen Theorem stated above implies the classical version of the theorem. The classical version asserts that

$$\pi_1(X, x) \cong \pi_1(U_1, x) * \pi_1(U_2, x)/N,$$

where $\pi_1(U_1, x) * \pi_1(U_2, x)$ is the free product and N is the normal subgroup generated by elements of the form $\phi_1(g)(\phi_2(g))^{-1}$ with $g \in \pi_1(U_0, x)$.

Example 0.2.4. One of the simplest, but most useful, applications of the classical Seifert-van Kampen Theorem is to the computation of the fundamental group of a 2-dimensional CW complex. Let X be a CW complex that consists of one 0-cell $\{x_0\}$ with 1-cells $a_1, a_2, \ldots, a_n$ attached. Then X is a bouquet of circles (Figure 0.1). Decompose X into open sets $U_1, U_2, \ldots, U_n$, each homotopy equivalent to a circle, any two of which intersect in a fixed contractible neighborhood of x_0, and apply the theorem inductively to see that $\pi_1(X, x_0)$ is a free group with one generator for each of the 1-cells in X. To be specific, let α_i be the loop that goes once around a_i. Then $\pi_1(X, x_0)$ is the free group generated by $\{\alpha_1, \alpha_2, \ldots, \alpha_n\}$.

Now attach 2-cells $b_1, b_2, \ldots, b_k$ to X to form a 2-dimensional CW complex Y. Each 2-cell b_i is attached via a map $f_i : \partial I^2 \to X$. We can think of f_i as representing an element of $\pi_1(X, x_0)$, and $[f_i]$ can be written as a word β_i in $\alpha_1, \alpha_2, \ldots, \alpha_n$. Applying the classical Seifert-van Kampen Theorem inductively yields that $\pi_1(Y, x_0)$ is isomorphic to the group with presentation $\langle \alpha_1, \alpha_2, \ldots, \alpha_n : \beta_1, \beta_2, \ldots, \beta_k \rangle$. □

Often we will encounter subsets of manifolds that do not have the homotopy type of finite CW complexes. Such sets, in general, can have more complicated fundamental groups than those computable via the Seifert-vanKampen Theorem.

Example 0.2.5. Let c_n be the circle of radius $1/n$ centered at the point $\langle 1/n, 0 \rangle$ in $\mathbb{R}^2$. Each c_n passes through the base point $z_0 = \langle 0, 0 \rangle$. The *Hawaiian earring* is the compact set $Z = \cup_{n=1}^{\infty} c_n$ (see Figure 0.2). For each n there is a loop γ_n that wraps once around c_n. Even though Z looks superficially like a straightforward generalization of the X in the previous example, the group $\pi_1(Z, z_0)$ is not generated by $\{\gamma_n\}$. To see this, note that the loop $\beta : [0, 1] \to Z$ that wraps the subinterval $[1/(n+1), 1/n]$ once around c_n defines an element of $\pi_1(Z, x_0)$ that cannot be written as a finite product of γ_n's. The structure of $\pi_1(Z, z_0)$ is surprisingly large and complex (and interesting); a detailed description of the group can be found in (Cannon and Conner, 2000). □

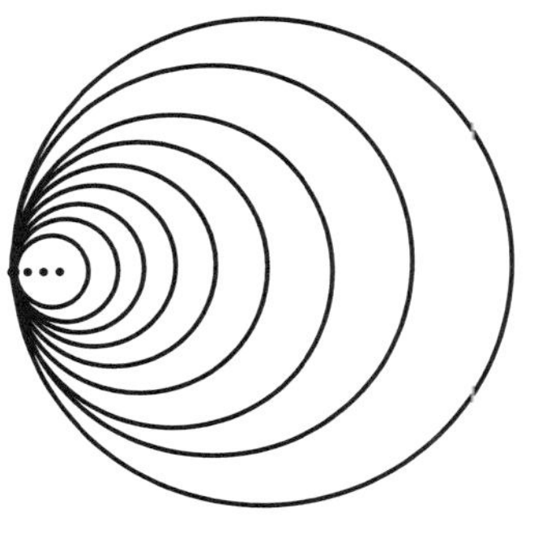

Figure 0.2. The Hawaiian earring

0.3. The ultimate duality theorem

Homology and cohomology are powerful tools in the study of embeddings. One reason for their usefulness is the many duality theorems they satisfy that relate the homology/cohomology of an embedded object to the cohomology/homology of its complement. The duality theorems found in (Munkres, 1984), which apply to triangulable homology manifolds, are not quite general enough for our purposes, so we state the theorems to be employed. The following basic duality theorem can be found on page 342 of (Spanier, 1966).

Theorem 0.3.1 (Duality). *Let G be an abelian group, and let (A, B) be a pair of closed subsets of the G-orientable n-manifold M. Then for all $p \geq 0$ there exists a natural isomorphism*

$$H_{n-p}(M \smallsetminus B, M \smallsetminus A; G) \xrightarrow{\cong} \check{H}^p_c(A, B; G).$$

The group on the right is called the pth Alexander (relative) cohomology group of (A, B) with G-coefficients and compact supports. (Throughout this section G will denote an Abelian group.) When A and B are compact, $\check{H}^p_c(A, B; G)$ equals the more usual pth Alexander cohomology of the pair (again with G-coefficients), and when, in addition, both A and B are themselves complexes, manifolds, or absolute neighborhood retracts (to be defined in §0.6), it equals the usual singular cohomology of the pair. A connected n-manifold M is *G-orientable* if $H^n_c(M; G) \cong G$; an arbitrary n-manifold is *G-orientable* if each of its components is. Every n-manifold is $\mathbb{Z}_2$-orientable; those that contain a copy of $MB \times I^{n-2}$, where MB denotes the Möbius Band, fail to be $\mathbb{Z}$-orientable.

Say that a ∂-manifold W is G-orientable if $\operatorname{Int} W$ is.

Corollary 0.3.2 (Poincaré-Lefschetz Duality). *If W is a G-orientable n-dimensional ∂-manifold, then $H_{n-p}(W; G) \cong H^p_c(W, \partial W; G)$ for every p.*

Proof. It follows from Collaring Theorem 2.4.10, to be proved later, that the manifold $W' = W \cup \partial W \times (-1, 0]$, obtained from the disjoint union of W and $\partial W \times (-1, 0]$ by attaching the product $\partial W \times (-1, 0]$ to $\partial W \subset W$ along $\partial W \times \{0\}$ in the obvious way, is homeomorphic to $\operatorname{Int} W$ and, thus, is G-orientable. Apply the duality theorem in W' with $A = W$ and $B = \partial W$ to obtain

$$\begin{aligned} H^p_c(W, \partial W; G) &\cong H_{n-p}(W' \smallsetminus \partial W, W' \smallsetminus W; G) \\ &= H_{n-p}(\operatorname{Int} W \cup \partial W \times (-1, 0), \partial W \times (-1, 0); G) \\ &\cong H_{n-p}(\operatorname{Int} W; G) \\ &\cong H_{n-p}(W; G). \end{aligned}$$

□

For a compact pair (A, B), there is a fundamental relationship between the cohomology with compact supports of $A \setminus B$ and the unrestricted cohomology of that pair (Spanier, 1966, p. 321):

Theorem 0.3.3. *For any compact Hausdorff pair (A, B),*

$$\check{H}^q_c(A \setminus B; G) \cong \check{H}^q(A, B; G).$$

Corollary 0.3.4. *A compact, connected, n-dimensional ∂-manifold M is G-orientable if and only if $H^n(M, \partial M; G) \cong G$.*

Corollary 0.3.5. *If A is a locally compact Hausdorff space and B is a closed subset of A, then*

$$\check{H}^q_c(A \setminus B; G) \cong \check{H}^q_c(A, B; G).$$

Proof. Let A^+, B^+ denote the one-point compactifications of A, B, respectively, with ∞ the ideal point. Then

$$\check{H}^q_c(A; G) \cong \check{H}^q(A^+, \{\infty\}; G) \cong \tilde{\check{H}}^q(A^+; G)$$

and similarly for B. The Five Lemma and another application of Theorem 0.3.3 yield

$$\check{H}^q_c(A, B; G) \cong \check{H}^q(A^+, B^+; G) \cong \check{H}^q_c(A \setminus B : G). \qquad \square$$

Corollary 0.3.6. *Every open subset of a G-orientable manifold is G-orientable.*

Proof. If U is a connected open subset of the G-orientable manifold M^n, then $G \cong H_0(U; G) \cong \check{H}^n_c(M, M \setminus U; G)$ by duality, and $\check{H}^n_c(M, M \setminus U; G) \cong H^n_c(U; G)$ by 0.3.5. $\square$

Corollary 0.3.7. *If S is a locally compact Hausdorff space and $q \geq 2$, then*

$$\check{H}^q_c(S \times \mathbb{R}; G) \cong \check{H}^{q-1}_c(S; G).$$

Proof. Let S^+ denote the one-point compactification of S, with ∞ the ideal point. A Mayer-Vietoris argument shows that $\check{H}^q(\text{Suspension of } S^+; G) \cong \check{H}^{q-1}(S^+; G)$ for $q \geq 2$. Thus,

$$\begin{aligned}
\check{H}^q_c(S \times \mathbb{R}; G) &\cong \check{H}^q(\text{Suspension of } S^+, \text{Suspension of } \infty; G) \\
&\cong \check{H}^q(\text{Suspension of } S^+; G) \cong H^{q-1}(S^+; G) \\
&\cong \check{H}^{q-1}(S^+, \{\infty\}; G) \\
&\cong \check{H}^{q-1}_c(S; G). \qquad \square
\end{aligned}$$

Corollary 0.3.8. *A manifold M is G-orientable if and only if $M \times \mathbb{R}$ is G-orientable.*

The equivalence of Alexander-Spanier cohomology with Čech cohomology on compact spaces (Spanier, 1966, p. 319) and computation with the latter yields:

Theorem 0.3.9 (Dimension). *$H^i_c(S) \cong 0$ whenever $i > \dim S$ and S is locally compact.*

Corollary 0.3.10 (Local Duality). *Let M be a G-orientable n-manifold, X a closed, k-dimensional subset of M, and $x \in X$. If V is a coordinate neighborhood of x, then $\widetilde{H}_p(V \smallsetminus X; G) = 0$ for $0 \le p \le n - k - 2$.*

Proof. Let V be a coordinate neighborhood of x. Then

$$\check{H}^{n-p-1}_c(X \cap V; G) \cong H_{p+1}(V, V \smallsetminus X; G) \cong \widetilde{H}_p(V \smallsetminus X; G),$$

the latter isomorphism coming from the long exact homology sequence of the pair $(V, V \smallsetminus X)$. All groups in this line are trivial, since $n - p - 1 > k$ and $k = \dim(X) \ge \dim(X \cap V)$. Thus $\widetilde{H}_p(V \smallsetminus X; G) \cong 0$. □

One of the delicate issues that will require attention later in the book is the question of when the homology connectivity of Corollary 0.3.10 can be promoted to connectivity in the sense of homotopy. Dimension one is special in that regard.

Corollary 0.3.11. *If X is a k-dimensional closed subset of a connected n-manifold M, $k \le n - 2$, then $M \smallsetminus X$ is pathwise connected. Indeed, each $x \in X$ has arbitrarily small neighborhoods U such that $U \smallsetminus X$ is path connected.*

Exercises

0.3.1. If $\Sigma \subset S^n$ and Σ is homeomorphic to S^k, $k < n$, then

$$\widetilde{H}_i(S^n \smallsetminus \Sigma; \mathbb{Z}) \cong \begin{cases} \mathbb{Z} & \text{if } i = n - k - 1, \\ 0 & \text{otherwise.} \end{cases}$$

0.3.2. (Jordan Separation Theorem) If $\Sigma \subset S^n$ and Σ is homeomorphic to S^{n-1}, then $S^n \smallsetminus \Sigma$ has exactly two components and Σ is the topological frontier of each.

0.3.3. If X is a k-dimensional closed subset of a ∂-manifold M^n, $k \le n-2$, then $M \smallsetminus X$ is connected.

0.3.4. No manifold M^n that contains a copy of $MB \times I^{n-2}$ (MB = Möbius band) is $\mathbb{Z}$-orientable.

0.4. The Vietoris-Begle Theorem

Another algebraic result needed much later on is the Vietoris-Begle Mapping Theorem. It implies that a closed, surjective mapping for which all point preimages are cohomologically acyclic induces isomorphisms of Čech or Alexander cohomology. See (Spanier, 1966, p. 344) for the proof.

Theorem 0.4.1. *Let $f\colon X' \to X$ be a closed surjective map between paracompact Hausdorff spaces, and let G be an abelian group. Assume that there is some $m \geq 0$ such that $\check{H}^q(f^{-1}(x); G) = 0$ for all $x \in X$ and $q < m$. Then*

$$f^* : \check{H}^q(X; G) \to \check{H}^q(X'; G)$$

is an isomorphism for $q < m$ and a monomorphism for $q = m$.

0.5. Higher homotopy groups

Let X be a space with basepoint $x_0 \in X$. For each integer $n \geq 0$ there is a group $\pi_n(X, x_0)$, called the *nth homotopy group of X with basepoint x_0*. The definition of $\pi_n(X, x_0)$ closely parallels that of the fundamental group, only with the role of the unit interval taken over by the unit n-cube. For that reason, throughout this section I^n will be used to denote $[0, 1]^n$.

Let F^n denote the set of all maps $(I^n, \partial I^n) \to (X, x_0)$. Define two maps in F^n to be *equivalent* if they are homotopic relative to ∂I^n; i.e., $f, f' \in F^n$ are equivalent if there exists a homotopy $\Psi_t : I^n \to X$ between f and f' with $\Psi_t(\partial I^n) = \{x_0\}$ for all $t \in I$. As a set, $\pi_n(X, x_0)$ is just the set of equivalence classes.

Define the group operation on F^n, $n \geq 1$, as follows: for $f, g \in F^n$ their sum $f + g \in F^n$ is the function determined by

$$(f+g)(t_1, t_2, \ldots, t_n) = \begin{cases} f(2t_1, t_2, \ldots, t_n) & \text{if } 0 \leq t_1 \leq 1/2 \\ g(2t_1 - 1, t_2, \ldots, t_n) & \text{if } 1/2 \leq t_1 \leq 1. \end{cases}$$

Since the homotopy class $[f + g]$ depends only on the classes $[f]$ and $[g]$, addition in F^n induces a well-defined addition in $\pi_n(X, x_0)$ via $[f] + [g] = [f + g]$. One can easily verify that this makes $\pi_n(X, x_0)$ a group; its identity element is the class of the (only possible) constant map.

The 0-dimensional cube I^0 is just a point and $\partial I^0 = \emptyset$. With this understanding the definition of the set $\pi_0(X, x_0)$ makes sense (it consists of the set of path components of X), but there is no group structure defined on it. Even though $\pi_0(X, x_0)$ does not have a group structure, we still designate the equivalence class of the map $I^0 \to x_0$ as the identity element. This allows us to include $\pi_0(X, x_0)$ in an important exact sequence to be described below.

A map $\psi : (X, x_0) \to (Y, y_0)$ leads to an *induced homomorphism* $\psi_\# : \pi_n(X, x_0) \to \pi_n(Y, y_0)$ defined by $\psi_\#([f]) = [\psi \circ f]$. The nth homotopy group has the expected functoriality properties with respect to such base-point-preserving maps of spaces. Unlike the fundamental group, however, $\pi_n(X, x_0)$ is always abelian when $n > 1$.

Application of lifting theorems such as (Munkres, 2000, Lemma 79.1) to a covering map $\Theta : X' \to X$ quickly yields that $\Theta_\# : \pi_n(X', x) \to \pi_n(X, \Theta(x))$ is an isomorphism for $n > 1$. Consequently, $\pi_n(X, x_0) = 0$ for all $n > 1$ if the universal covering space of X is contractible.

Note that X is path connected if and only if $\pi_0(X, x_0) = 0$ and X is simply connected if and only if both $\pi_0(X, x_0)$ and $\pi_1(X, x_0)$ are trivial. There is a corresponding kind of connectivity in every dimension and it is detected by the higher homotopy groups.

Definition. A space X is said to be *k-connected*, $k \geq 0$, if each map $\partial I^{n+1} \to X$ can be extended to a map $I^{n+1} \to X$ for $n = 0, 1, \ldots, k$.

Theorem 0.5.1. *Let X be a pathwise connected space with basepoint x_0. For each $k \geq 0$ the following are equivalent.*

(1) *X is k-connected.*

(2) *$\pi_n(X, x_0) = 0$ for $n \leq k$.*

(3) *If K is a k-dimensional polyhedron and L is a closed subpolyhedron of K, then any map $(K, L) \to (X, x_0)$ is homotopic, rel L, to the constant map $K \to x_0$.*

Proof. Exercise 0.5.2. □

There is also a local version of connectivity, the usefulness of which will be illustrated in the next two sections.

Definition. A space X is said to be *locally connected in dimension k* (abbreviated k-LC) if for every $x \in X$ and for every neighborhood U of x in X there exists a neighborhood V of x such that each map $S^k \to V$ extends to a map $B^{k+1} \to U$. The space X is said to be *locally k-connected* (abbreviated LC^k) if X is n-LC for $0 \leq n \leq k$.

We now define a *relative homotopy group* $\pi_n(X, A, x_0)$, which is associated with a space X, a nonempty subset A of X, and a basepoint $x_0 \in A$. In order to define the relative group, it is convenient to identify I^{n-1} with the face $I^{n-1} \times \{0\} \subset \partial I^n$ and to use J^{n-1} to denote the union of the remaining faces of I^n (so $\partial I^n = I^{n-1} \cup J^{n-1}$ and $I^{n-1} \cap J^{n-1} = \partial I^{n-1}$). An element of $\pi_n(X, A, x_0)$ is represented by a map of triples $(I^n, I^{n-1}, J^{n-1}) \to (X, A, x_0)$. Two such maps are equivalent if they are homotopic via a homotopy of triples.

The group operation is defined exactly as in the absolute case. The only catch is that the coordinate t_n plays a special role in the definition of the relative homotopy group, and the definition of the sum of two equivalence classes uses the coordinate t_1 in an independent way. As a result, the definition of the group operation only makes sense if $n \geq 2$; $\pi_1(X, A, x_0)$ is simply considered to be a set of equivalence classes with no natural group structure (just like $\pi_0(X, x_0)$). As in the absolute case, a continuous map $\psi : (X, A, x_0) \to (Y, B, y_0)$ induces a homomorphism $\psi_* : \pi_n(X, A, x_0) \to \pi_n(Y, B, y_0)$ defined by $\psi_*([f]) = [\psi \circ f]$.

We may identify the absolute group $\pi_n(X, x_0)$ with the relative group $\pi_n(X, \{x_0\}, x_0)$, so there is a natural (inclusion-induced) homomorphism $j_* : \pi_n(X, x_0) \to \pi_n(X, A, x_0)$. There is also an inclusion-induced homomorphism $i_* : \pi_n(A, x_0) \to \pi_n(X, x_0)$. An element of $\pi_{n+1}(X, A, x_0)$ is represented by a map $f : I^{n+1} \to X$ such that $f(I^n) \subset A$ and $f(J^n) = \{x_0\}$. In particular, $f(\partial I^n) = \{x_0\}$, so $f|I^n$ may be viewed as representing an element of $\pi_n(A, x_0)$. This correspondence induces a natural homomorphism $\partial : \pi_{n+1}(X, A, x_0) \to \pi_n(A, x_0)$. The homomorphism ∂ is called the *boundary operator* because $f|J^n$ is constant and so $f|I^n$ is essentially $f|\partial I^{n+1}$. The three homomorphisms just described combine to form an extremely valuable long exact sequence, the *homotopy sequence of the pair* (X, A) *with base point* x_0:

$$\cdots \to \pi_{n+1}(X, A, x_0) \xrightarrow{\partial} \pi_n(A, x_0) \xrightarrow{i_*} \pi_n(X, x_0) \xrightarrow{j_*} \pi_n(X, A, x_0) \to \cdots$$

A proof of the exactness of the homotopy sequence of a pair can be found on pages 344 and 345 of (Hatcher, 2002).

Definition. A topological pair (X, A) is said to be *k-connected*, $k \geq 0$, if each map $(I^n \times \{0\}, \partial I^n \times \{0\}) \to (X, A)$ can be extended to a map $(I^{n+1}, J^n) \to (X, A)$ for $n = 0, 1, \ldots, k$.

The final theorem in the section will be applied in the proof of the basic engulfing theorem of Chapter 3.

Theorem 0.5.2. *Let X be a pathwise connected space with subset A and basepoint $x_0 \in A$. For each $k \geq 0$ the following are equivalent.*

(1) *The pair (X, A) is k-connected.*
(2) $\pi_n(X, A, x_0) = 0$ *for* $n \leq k$.
(3) *If K is a k-dimensional polyhedron and L is a closed subpolyhedron of K, then any map $(K, L) \to (X, A)$ is homotopic, rel L, to a map $K \to A$.*

Proof. Exercise 0.5.4. □

Exercises

0.5.1. S^k is $(k-1)$-connected.

0.5.2. Prove Theorem 0.5.1.

0.5.3. If X is k-connected and A is $(k-1)$-connected, then (X, A) is a k-connected pair. More generally, if the inclusion-induced homomorphism $\pi_i(A) \to \pi_i(X)$ is an epimorphism for $i = k$ and an isomorphism for $i < k$, then (X, A) is a k-connected pair.

0.5.4. Prove Theorem 0.5.2.

0.5.5. $\pi_n(X, x_0)$ is abelian provided $n \geq 2$.

0.5.6. $\pi_n(X, A, x_0)$ is abelian provided $n \geq 3$. Show by example that $\pi_2(X, A, x_0)$ need not be abelian.

0.6. Absolute neighborhood retracts

Although many spaces considered in this text are polyhedra, there are some situations in which more general kinds of spaces are the appropriate ones to use. The class of ANRs is particularly relevant.

Definition. A metric space Y is an *absolute retract* (abbreviated AR) if for every metric space X and for every embedding $e : Y \to X$ such that $e(Y)$ is a closed subset of X, there is a retraction $r : X \to e(Y)$. We say that Y is an *absolute neighborhood retract* (abbreviated ANR) if for every metric space X and for every embedding $e : Y \to X$ such that $e(Y)$ is a closed subset of X, there exist a neighborhood U of $e(Y)$ in X and a retraction $r : U \to e(Y)$.

Finite-dimensional ANRs can be characterized in terms of local connectivity and finite-dimensional ARs can be characterized in terms of both connectivity and local connectivity. Detailed proofs of the next two theorems may be found in (Hu, 1965, Chapter V).

Theorem 0.6.1. *Let Y be a separable metric space of dimension $k < \infty$. Then the following are equivalent.*

(1) *Y is an ANR.*

(2) *Y is locally contractible.*

(3) *Y is locally k-connected.*

Theorem 0.6.2. *Let Y be a separable metric space of dimension $k < \infty$. Then the following are equivalent.*

(1) *Y is an AR.*

(2) *Y is contractible and locally contractible.*

(3) *Y is k-connected and locally k-connected.*

It follows easily from Theorem 0.6.1 that every finite-dimensional polyhedron or CW complex is an ANR. The Hawaiian earring is not an ANR; neither is the sine $(1/x)$ curve (Figure 2.13).

The next theorem identifies an especially useful property of ANRs.

Theorem 0.6.3. *Let Y be a compact ANR and ϵ a positive number. There exists $\delta > 0$ such that for any two maps $f_0, f_1 : S \to Y$ of an arbitrary space S to Y with $d(f_0, f_1) < \delta$, there is a homotopy $H : S \times I \to Y$ such that $H_0 = f_0$, $H_1 = f_1$, and $\operatorname{diam} H(\{s\} \times I) < \epsilon$ for all $s \in S$.*

Sketch of Proof. Embed Y in the Hilbert cube $I^\infty = \Pi_{n=1}^\infty [0, 1/n]$, where it has a neighborhood U that retracts to the embedded copy Y' of Y. Choose δ small enough that the straight line homotopy from f_0 to f_1 stays in U and still has diameter less than ϵ after being retracted into Y'. □

Theorem 0.6.4 (Estimated Homotopy Extension Theorem). *Let Y be an ANR, X a normal space, $f : X \to Y$ a map, $b : Y \to (0, \infty]$ another map, A a closed subset of X, U a neighborhood of A in X, and $\mu : A \times I \to Y$ a homotopy such that $\mu_0 = f|A$ and $\operatorname{diam} \mu(\{a\} \times I) < b(\mu(a, t))$ for all $a \in A$ and $t \in I$. Then there exists a homotopy $H : X \times I \to Y$ such that $H_0 = f$, $H|A \times I = \mu$, $H(\{x\} \times I) = f(x)$ for all $x \in X \setminus U$, and $\operatorname{diam} H(\{x\} \times I) < b(H(x, t))$ for all $x \in X$ and $t \in I$.*

Proof. Define a map F' on $Z = (X \times \{0\}) \cup (A \times I) \subset X \times I$ as f on $X \times \{0\}$ and μ on $A \times I$. Since Y is an ANR, F' has an extension $F : O \to Y$ over some neighborhood O of Z in $X \times I$. Find an open subset V of X, $A \subset V \subset U$, such that $V \times I \subset O$ and $\operatorname{diam} F(\{v\} \times I) < b(F(v, t))$ for all $v \in V$ and $t \in I$. Apply Urysohn's Lemma to obtain a map $\eta : X \to [0, 1]$ for which $\eta(X \setminus V) = 0$ and $\eta(A) = 1$. Finally, define $H : X \times I \to Y$ as $H(x, t) = F(x, \eta(x)t)$. □

Corollary 0.6.5. *Suppose $R : Y \to A$ is a retraction of an ANR Y to a compact subset A, and suppose $\eta : A \times I \to Y$ is a homotopy between $\eta_0 = incl_A$ and an embedding $\lambda = \eta_1$. Then Y retracts to $\lambda(A)$; moreover, if λ moves points less than $c > 0$ and $\operatorname{diam} R\eta(\{a\} \times I) < b$ for all $a \in A$, then there is a retraction $R' : Y \to \lambda(A)$ such that $d(R'(y), R(y)) < b + 2c$ for all $y \in Y$.*

Proof. The homotopy $\eta' = \lambda R \eta(\lambda^{-1} \times \mathrm{Id}) : \lambda(A) \times I \to \lambda(A)$ satisfies $\eta'_1 = \lambda R|\lambda(A)$ and $\eta'_0 = \mathrm{Id}$; obviously η'_1 extends to $\lambda R : Y \to \lambda(A)$. □

Exercises

0.6.1. Every n-cell is an AR.

0.6.2. Every ∂-manifold is an ANR.

0.6.3. Every polyhedron is an ANR.

0.6.4. Give an example of an ANR that is not homeomorphic to a polyhedron.

0.7. Dimension theory

Munkres (2000, §50) (or (1975, §7-9)) contains most, but not quite all, of what the reader needs to know about dimension theory. This section fills in the necessary background. All topological spaces considered in this book are separable metric spaces, so some of the technical complications of dimension theory for more general spaces can be avoided.

Let $\mathcal{U}$ be an open cover of the space X. Say $\text{order}(\mathcal{U}) \leq k+1$ if no point of X lies in more than $k+1$ of the elements of $\mathcal{U}$. The statement $\dim X \leq k$ means that for every open cover $\mathcal{U}$ of X there exists an open cover $\mathcal{V}$ of X such that $\mathcal{V}$ is a refinement of $\mathcal{U}$ and $\text{order}(\mathcal{V}) \leq k+1$.

A function $f : X \to Y$ defined on a metric space is an *ϵ-mapping* if $\operatorname{diam} f^{-1}(y) < \epsilon$ for every $y \in Y$. When $\dim X \leq k$ and $\epsilon > 0$ is given, it is a fairly simple matter to use the nerve of a cover to construct an ϵ-mapping of X into a k-dimensional polyhedron (Hurewicz and Wallman, 1948, pages 67–70). Alexandroff proved that, by modifying the polyhedron, this mapping can be made onto and that the existence of ϵ-mappings into k-dimensional polyhedra characterizes compacta of dimension $\leq k$. The following result appears as Theorem V10 in (Hurewicz and Wallman, 1948).

Theorem 0.7.1 (Alexandroff). *A compact metric space X has dimension $\leq k$ if and only if for every $\epsilon > 0$ there exists an ϵ-mapping of X onto a polyhedron of dimension $\leq k$.*

Sketch of proof. When X admits an ϵ-mapping ϕ onto a k-dimensional polyhedron P, one can compute $\eta > 0$ such that $\operatorname{diam} \phi^{-1}(x) < \epsilon$ for all $x \in P$. By subdividing P and carefully thickening the interiors of the various simplices of the subdivision, one can produce an open cover $\mathcal{V}$ of P with $\text{order}(\mathcal{V}) \leq k+1$ and with $\operatorname{diam} V < \eta$ for all $V \in \mathcal{V}$. Then

$$\mathcal{U} = \{\phi^{-1}(V) \mid V \in \mathcal{V}\}$$

is a small mesh open cover of X of order $k+1$. □

If X is a subset of a PL manifold, the polyhedron can be realized as a subcomplex of the manifold. The following corollary illustrates how we will make use of the dimension of an arbitrary compactum and the proof is an early indication of how local connectivity will be used in this book.

Corollary 0.7.2. *Let X be a compact subset of the PL manifold W and let $k = \dim X$. For every $\epsilon > 0$ there exists a k-dimensional polyhedron $K \subset W$ and an onto map $g : X \to K$ such that $d(x, g(x)) < \epsilon$ for every $x \in X$.*

Proof. Since X is compact, we can replace W with a compact ∂-manifold W' that contains X in its interior. Now W' is an ANR, so it is n-LC at x for every n and for every x. The compactness of W' together with a Lebesgue number argument establishes the following uniform version of local connectivity: for each nonnegative integer n and for each $\epsilon > 0$ there exists $\delta > 0$ such that any map of ∂I^n into a δ-subset of W' extends to a map of I^n into an ϵ-subset of W'.

Let $\epsilon > 0$ be given. Choose δ_k such that any map of ∂I^k into a δ_k-subset of W' extends to a map of I^k into an $(\epsilon/2)$-subset of W'. Then recursively choose $\delta_{k-1}, \delta_{k-2}, \ldots, \delta_1$ such that any map of ∂I^j into a δ_j-subset of W' extends to a map of I^j into a $(\delta_{j+1}/2)$-subset of W'. Set $\delta = \delta_1$.

By the Alexandroff Theorem there is a δ-mapping $f : X \to L$, where L is a compact polyhedron of dimension $\leq k$. The idea is to use the local connectivity of W' to construct a map $h : L \to W'$ that is an approximate inverse for f. Then we can define $K = h(L)$ and $g = h \circ f$.

If $d(x, x') \geq \delta$, then $f(x) \neq f(x')$. Compactness gives a positive number η such that $d(x, x') \geq \delta$ implies that the distance from $f(x)$ to $f(x')$ is $\geq \eta$ in L. Let T be a triangulation of L such that each simplex in T has diameter $< \eta$; then $\operatorname{diam} f^{-1}(\sigma) < \delta$ for every $\sigma \in T$.

For each vertex $v \in T$, define $h(v)$ to be some point in $f^{-1}(v)$. Let σ be a 1-simplex in T. We have already defined $h|\partial\sigma$ and the choice of δ_1 allows us to extend h to σ in such a way that $\operatorname{diam} h(\sigma) < \delta_2/2$. Now consider a 2-simplex $\tau \in T$. Note that $h|\partial\tau$ is already defined and $\operatorname{diam} h(\partial\tau) < \delta_2$. Hence the choice of δ_2 allows us to extend h to τ in such a way that $\operatorname{diam} h(\tau) < \delta_3/2$. This process is continued inductively and results in a map $h : L \to W'$ such that $\operatorname{diam} h(\sigma) < \epsilon/2$ for every $\sigma \in T$. We may assume that h is a PL map in general position.

Define $K = h(L)$ and define $g : X \to K$ by $g(x) = h(f(x))$. Fix $x \in X$. We must check that $d(x, g(x)) < \epsilon$. Locate a simplex $\sigma \in T$ such that $f(x) \in \sigma$ and choose a vertex v of σ. Then

$$d(x, g(x)) \leq d(x, h(v)) + d(h(v), g(x)) = d(x, h(v)) + d(h(v), h(f(x))).$$

Now x and $h(v)$ are both in $f^{-1}(\sigma)$, so $d(x, h(v)) < \delta < \epsilon/2$. In addition, both $h(v)$ and $h(f(x))$ are in $h(\sigma)$, so $d(h(v), h(f(x))) < \epsilon/2$. Thus $d(x, g(x)) < \epsilon$. □

For years the classic monograph by W. Hurewicz and H. Wallman (1948) stood as the standard dimension theory reference. The more recent book by J. van Mill (1989) is an excellent alternative.

0.8. The Hurewicz Isomorphism Theorem and its localization

The Hurewicz Theorem relates homotopy and homology groups. The following statement appears on page 394 of Spanier (1966) and on page 366 of Hatcher (2002).

Theorem 0.8.1 (Hurewicz Isomorphism). *Let X be a $(k-1)$-connected space, $k \geq 2$, with $x_0 \in X$. Then there is a natural isomorphism $\pi_k(X, x_0) \to H_k(X)$.*

Corollary 0.8.2. *If X is 1-connected and $H_i(X) \cong 0$ for $1 \leq i \leq k$, then $\pi_i(X, x_0) \cong 0$ for $i \leq k$.*

There is also a useful local version of the theorem that does not appear in any of the standard references on algebraic topology.

Theorem 0.8.3 (Local Hurewicz). *Suppose $V \subset U_0 \subset \cdots \subset U_k$, $k \geq 2$, are open sets such that $H_k(V) \to H_k(U_0)$ is trivial and $\pi_q(U_q) \to \pi_q(U_{q+1})$ is trivial for $0 \leq q \leq k-1$. Then $\pi_k(V) \to \pi_k(U_k)$ is trivial.*

Proof. Consider any map $\alpha : S^k \to V$. As $[\alpha] = 0$ in $H_k(U_0)$, there exist a subdivision L of S^k and a singular $(k+1)$-chain $c = \Sigma_j n_j \sigma_j$ carried by U_0 such that $\Sigma_i \alpha_{\#}(\tau_i) = \partial c$, where $\{\tau_i\}$ denotes a collection of 1–1 simplicial maps $\Delta^k \to L$, one for each k-simplex of L, determined by some ordering of the vertices. Let K denote a geometric realization of the finite, singular complex determined by the $\{\sigma_j\}$; here K contains L as a subcomplex, and $\alpha : L \to V \subset U_0$ has a natural extension $\beta : |K| \to U_0$. Let K' be the union of K and the cone on its $(k-1)$-skeleton. Since $\pi_q(U_q) \to \pi_q(U_{q+1})$ is trivial, we can extend β over successive skeleta to a map $\beta' : |K'| \to U_k$. Now $[S^k]$ is zero in $H_k(K)$ and hence in $H_k(K')$. One can easily check that K' is $(k-1)$-connected. By the Hurewicz Isomorphism Theorem, $[S^k] = 0$ in $\pi_k(K')$. Application of β confirms that $[\alpha] = 0$ in $\pi_k(U_k)$. □

Theorem 0.8.3 is also known as the *Eventual Hurewicz Theorem*—see (Ferry, 1979, Proposition 3.1) and (Quinn, 1979, Theorem 5.2).

Several applications require a relative version of the Hurewicz Theorem. A complete statement of the relative Hurewicz Theorem must take account of the action of π_1 on the higher homotopy groups. In order to avoid that technicality we state the relative theorem only in the simply connected case.

The full version can be found in (Spanier, 1966, page 397) or (Hatcher, 2002, page 371).

Theorem 0.8.4 (Relative Hurewicz Isomorphism). *Let (X, A) be a $(k-1)$-connected pair, $k \geq 2$, such that A is nonempty and simply connected. Then for each $x_0 \in A$ there is a natural isomorphism $\pi_k(X, A, x_0) \to H_k(X, A)$.*

0.9. The Whitehead Theorem

The Whitehead Theorem allows one to detect algebraically that a map of complexes is a homotopy equivalence.

Theorem 0.9.1 (Whitehead). *A map $f : K \to L$ between simplicial complexes (or CW complexes) is a homotopy equivalence if and only if $f_* : \pi_n(K) \to \pi_n(L)$ is an isomorphism for every n.*

This particular statement of the theorem can be found in (Hatcher, 2002, page 346), for example. Here is a related result that is often easier to apply, and which follows from the Whitehead Theorem, the Relative Hurewicz Isomorphism Theorem, and a mapping cylinder construction.

Theorem 0.9.2. *A map $f : K \to L$ between 1-connected simplicial complexes (or CW complexes) is a homotopy equivalence if and only if $f_* : H_*(K; \mathbb{Z}) \to H_*(L; \mathbb{Z})$ is an isomorphism.*

0.10. Acyclic complexes and contractible manifolds

As an application of the theorems in the last few sections we briefly consider acyclic and contractible spaces.

Definition. A space X is *acyclic* if $\widetilde{H}_*(X; \mathbb{Z}) \cong 0$.

The following is an immediate consequence of the Hurewicz and Whitehead Theorems.

Theorem 0.10.1. *A 1-connected complex K is contractible if and only if it is acyclic.*

Example 0.10.2. *There exists a compact 2-dimensional CW complex that is acyclic but not contractible.*

Proof. The classic example is the CW complex Y that has one 0-cell, two 1-cells a and b, and two 2-cells attached to the loops a^5b^{-3} and $b^3(ab)^{-2}$. The cellular chain complex for Y has the form $\cdots \to 0 \to \mathbb{Z}^2 \xrightarrow{\partial} \mathbb{Z}^2 \to \mathbb{Z}$, where ∂ is represented by the matrix $A = \begin{bmatrix} 5 & -3 \\ -2 & 1 \end{bmatrix}$. Since $\det A = -1$, ∂ is an isomorphism; hence $\widetilde{H}_*(Y; \mathbb{Z}) \cong 0$ and Y is acyclic.

By Example 0.2.4, $\pi_1(Y)$ has presentation $\langle \alpha, \beta : \alpha^5\beta^{-3}, \beta^3(\alpha\beta)^{-2}\rangle$. Thus $\pi_1(Y)$ has two generators α and β and two relations $\alpha^5 = \beta^3 = (\alpha\beta)^2$. This is the well-known *binary icosahedral group*, so called because there is an order-two homomorphism of $\pi_1(Y)$ onto the group G of rigid motions of the icosahedron.

In order to complete the proof we need to show that $\pi_1(Y)$ is nontrivial. We will do that by exhibiting a homomorphism of $\pi_1(Y)$ onto a nonabelian subgroup of G. To determine such a homomorphism, send α to the counterclockwise rotation through $2\pi/5$ radians about the point x indicated in Figure 0.3 and send β to the counterclockwise rotation through $2\pi/3$ radians about the point y. Then $\alpha\beta$ is the rotation through π radians about the point z, so all three of α^5, β^3, and $(\alpha\beta)^2$ represent the identity motion. As a result, this assignment extends to a homomorphism of $\pi_1(Y)$ to G. Since $\beta\alpha$ is rotation through an angle of π radians about z', we see that $\alpha\beta \neq \beta\alpha$. $\square$

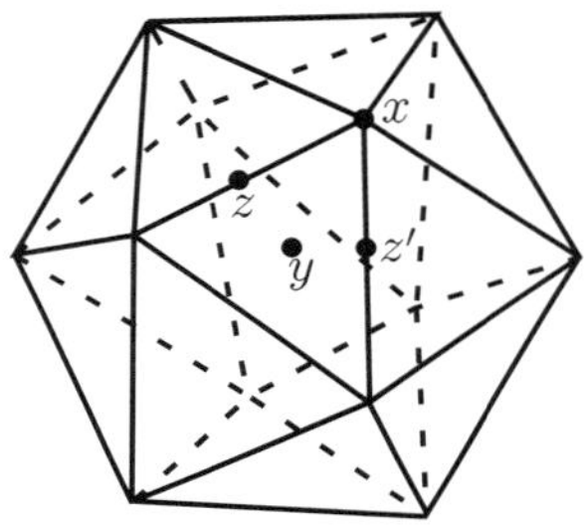

Figure 0.3. The icosahedron

In the preceding argument it was not necessary to compute $\pi_1(Y)$ explicitly to determine its non-triviality, but it is known that G has 60 elements, and that the order of $\pi_1(Y)$ is 120.

Example 0.10.3. *There exists a compact, contractible n-dimensional ∂-manifold in S^n, $n \geq 5$, that is not a ball.*

Proof. Start with an acyclic 2-complex P (such as that in the preceding example) and embed it in S^n, $n \geq 5$. Name a regular neighborhood N of P and set $M = S^n \setminus \operatorname{Int} N$. General position considerations yield that $S^n \setminus P$ is 1-connected; the same holds for M, which is a (deformation) retract of $S^n \setminus P$, since $N \setminus P \cong \partial N \times [0, 1)$. Like P, N is acyclic; more importantly, so is M, by duality or a simple Mayer-Vietoris argument (it helps to know $\partial N = \partial M$ is orientable, due to §0.3). Hence M is contractible. Note that ∂M need not be a sphere. In particular, ∂M is 1-connected if and only if P is, for general position implies that the arrow in the line below,

$$\pi_1(\partial M = \partial N) \cong \pi_1(\partial N \times [0, 1)) \cong \pi_1(N \setminus P) \to \pi_1(N) \cong \pi_1(P),$$

represents an isomorphism. □

Historical Notes. The contractible ∂-manifold M constructed in Example 0.10.3 is called a *Newman contractible manifold*. It is named after M. H. A. Newman (1948), who was the first to describe contractible ∂-manifolds this way. The same construction can also be carried out (topologically) in S^4, but the 4-dimensional case requires more care to ensure that M is simply connected (see (Lickorish, 2003)). No such example is possible in S^3.

Exercise

0.10.1. The homomorphism described in the last paragraph of the proof of Example 0.10.2 is an epimorphism.

0.11. The 2-dimensional PL Schönflies Theorem

A *simple closed curve* in a space X is the image of a continuous one-to-one function $f : S^1 \to X$. The classical Jordan Curve Theorem states that any simple closed curve in $\mathbb{R}^2$ separates $\mathbb{R}^2$ into exactly two components, with the simple closed curve being the topological frontier of each of the complementary domains. In this form the theorem generalizes to high dimensions—see Exercise 0.3.2. There is another, stronger form of the Jordan Curve Theorem, called the Schönflies Theorem, unique to two dimensions, which provides context for many of the results in this book. In effect, the Schönflies Theorem shows that most of the unusual phenomena to be studied in this book are high-dimensional and do not occur in dimension two. This section offers a short review of this key result.

Theorem 0.11.1 (Topological Schönflies). *For any two simple closed curves P_1 and P_2 in $\mathbb{R}^2$, there is a (compactly supported) topological homeomorphism $\Theta : \mathbb{R}^2 \to \mathbb{R}^2$ such that $\Theta(P_1) = P_2$.*

We will not prove the topological Schönflies Theorem. A thorough exposition of the proof can be found in (Moise, 1977, Chapter 9). The original version appeared in (Schönflies, 1908).

The tools of (Rourke and Sanderson, 1972) come very close to proving the PL variant of the theorem; we complete the proof in that special case.

Theorem 0.11.2 (PL Schönflies). *For any two polygonal simple closed curves P_1 and P_2 in $\mathbb{R}^2$, there is a (compactly supported) PL homeomorphism $\Theta : \mathbb{R}^2 \to \mathbb{R}^2$ such that $\Theta(P_1) = P_2$.*

Lemma 0.11.3. *Each polygonal simple closed curve P in $\mathbb{R}^2$ with at least 4 vertices has a pair of nonadjacent vertices v and w for which the interior of the segment vw misses P.*

Proof. Identify consecutive vertices v_1, v_2, v_3 of P, and form the triangle Δ determined by them. Impose coordinates on $\mathbb{R}^2$ with v_1, v_3 on the x-axis and v_2 below it. Let K denote the complement in P of the two segments v_1v_2 and v_2v_3. If no point of K meets Δ, then the segment determined by $v = v_1$ and $w = v_3$ obviously has the desired property. Otherwise, choose w to have least y-coordinate among the finitely many vertices of $K \smallsetminus \{v_2\}$ touching the 2-cell bounded by Δ (see Figure 0.4); in this case the segment formed by $v = v_2$ and w works.[2] □

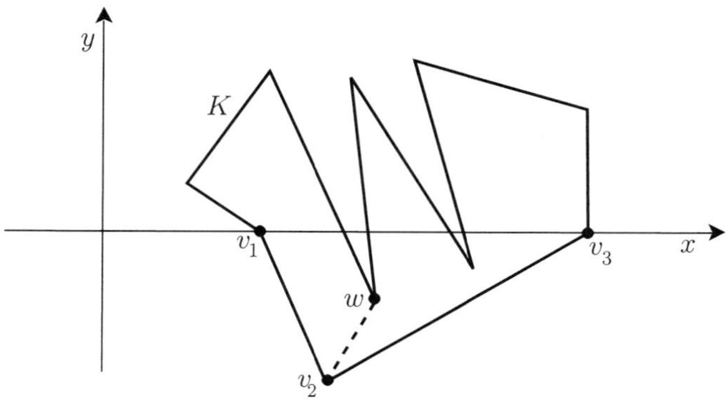

Figure 0.4. The interior of segment v_2w misses P

Theorem 0.11.4. *Every polygonal simple closed curve P in $\mathbb{R}^2$ bounds a PL 2-cell.*

Proof. By induction on the number n of vertices in P, with the initial case $n = 3$ being trivial. Inductively assume the result for all polygons of fewer than n vertices, and consider a polygon P having n vertices. Apply Lemma 0.11.3 to form two polygonal simple closed curves J_1, J_2, each having fewer than n vertices, with $J_1 \cup J_2 = P \cup vw$ and $J_1 \cap J_2 = vw$; by induction, each J_i bounds a PL 2-cell D_i $(i = 1, 2)$. If $D_1 \cap D_2 = vw$, then $D_1 \cup D_2$ is a disk and $\partial(D_1 \cup D_2) = P$. Otherwise, one of the J_i misses the interior of the other 2-cell D_j, for if $J_1 \cap \operatorname{Int} D_2 \neq \emptyset$, then $J_2 \cap \operatorname{Int} D_1 = \emptyset$, as $D_1 \subset D_2$ (see Figure 0.5). Let us say $J_1 \cap \operatorname{Int} D_2 = \emptyset$ for definiteness. Let C denote the closure of the bounded component of $\mathbb{R}^2 \smallsetminus P$. It is left to the reader to check that there is an elementary shelling (see Rourke and Sanderson (1972), p. 40) of D_1 to C across D_2. Hence, $C \cong D_1$ is a 2-cell. □

Proof of Theorem 0.11.2. The theorem is an immediate corollary of Theorem 0.11.4 and the Disc Theorem of Rourke and Sanderson (1972, Theorem 3.34). □

[2] J. W. Cannon deserves credit for this elegant argument.

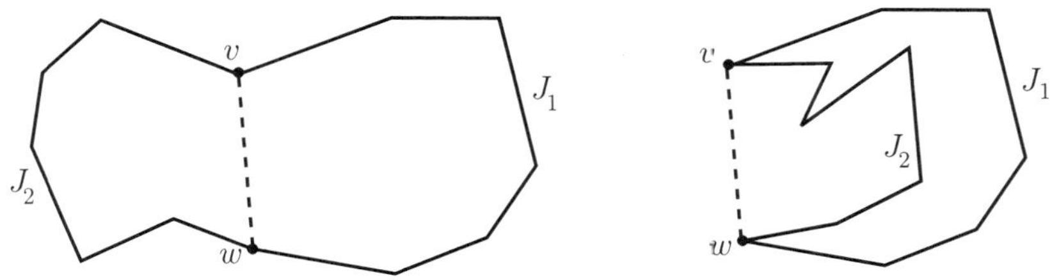

Figure 0.5. Two possibilities for $J_1 \cup J_2$

We close the chapter with an application of Theorem 0.11.4 that furnishes a promised technical improvement to the Seifert-van Kampen Theorem.

Theorem 0.11.5. *In the setting of Seifert-van Kampen Theorem 0.2.1, if both* $\phi_i : \pi_1(U_0, x) \to \pi_1(U_i, x)$, $i \in \{1, 2\}$, *are 1-1, then so are* $\psi_i : \pi_1(U_i, x) \to \pi_1(X, x)$, $i \in \{1, 2\}$.

Proof. Consider a loop $\alpha : \partial I^2 \to U_1$ representing an element of $\pi_1(U_1, x)$. It suffices to show that if α is inessential in X, then α is inessential in U_1. Let $A_1 = X \smallsetminus U_2$ and $A_2 = X \smallsetminus U_1$, and note that $A_1 \cap A_2 = \emptyset$. Given a path homotopy $h : I^2 \to X$ between α and the constant path, choose $\delta > 0$ less than the distance in I^2 between $h^{-1}(A_2)$ and $\partial I^2 \cup h^{-1}(A_1)$. Build a compact PL 2-dimensional ∂-manifold $M \subset I^2$ satisfying

$$h^{-1}(A_2) \subset \operatorname{Int} M \subset M \subset N(h^{-1}(A_2); \delta) \subset \operatorname{Int} I^2 \smallsetminus h^{-1}(A_1).$$

List the components $J_1, J_2, \ldots, J_k$ of ∂M. Each J_j is a PL simple closed curve; by Theorem 0.11.4, J_j bounds a disk $D_j \subset \operatorname{Int} I^2$. Order the J_j's with the innermost ones listed first – specifically, order so that $s > j$ implies $J_s \cap D_j = \emptyset$. Since J_1 is an innermost curve, $\partial M \cap \operatorname{Int} D_1 = \emptyset$, which yields that either $D_1 \subset M$ or $D_1 \cap M = J_1$; consequently, either $h(D_1) \cap A_1 = \emptyset$ or $h(D_1) \cap A_2 = \emptyset$. This means $J_1 \subset h^{-1}(U_0)$ is mapped to a nullhomotopic loop in either U_2 or U_1, and by hypothesis then $h(J_1)$ must be nullhomotopic in U_0. Hence, $h = h_0$ can be modified to produce a map $h_1 : I^2 \to X$ satisfying

(1) $h_1 | I^2 \smallsetminus \operatorname{Int} D_1 = h | I^2 \smallsetminus \operatorname{Int} D_1$
(2) $h_1(D_1) \subset U_0$.

Repeating, we recursively produce additional maps $h_2, \ldots, h_k : I^2 \to X$ such that for $t = 1, 2, \ldots, k$

(3) $h_t | I^2 \smallsetminus \operatorname{Int} D_t = h_{t-1} | I^2 \smallsetminus \operatorname{Int} D_t$
(4) $h_t(\cup_{i=1}^t D_i) \subset U_0$.

In particular, condition (3) insures that $h_t | \partial I^2 = h | \partial I^2$. Assuming $h_0, h_1, \ldots, h_t$ satisfy the preceding, we observe that $\operatorname{Int} D_{t+1} \smallsetminus \cup_{i=1}^t D_i$ either misses

M or lives in Int M, and exactly as in the initial case, either $h_t(D_{t+1}) \subset U_1$ or $h_t(D_{t+1}) \subset U_2$; in either event J_t is mapped by h_t to a loop nullhomotopic in U_0, implying the existence of the required map h_{t+1}. As $\cup_{i=1}^k(\text{Int } D_i) \supset h^{-1}(A_2)$, the final map $h_k : I^2 \to X$ has image in $U_1 = X \smallsetminus A_2$, so it provides the desired path homotopy in U_1 between α and the constant path. $\square$

Chapter 1

Tame and Knotted Embeddings

This chapter offers an introductory exploration of tameness, knottedness, and local flatness. A classic theorem in piecewise linear topology assures that any cell or sphere PL embedded in S^n is flat, in the PL category, provided its codimension is greater than two. Thus, every manifold M^m PL embedded in a PL manifold N^n is locally flat, provided $n - m > 2$; indeed, even when $m = n - 1$, M^{n-1} is locally flat in the topological sense. §1.4 lays out two fundamental methods, suspending and spinning, that, among other effects, promote a knotted codimension-k object in S^n to a knotted codimension-k object in S^{n+1}. The suspension operator can introduce local knotting; it almost instantly gives examples of PL codimension-two spheres in S^n that fail to be locally flat. In contrast, the spinning operator circumvents the introduction of local knotting; it easily leads to examples of PL codimension-two spheres in S^n that are locally flat but not flat.

Section 1.3 sets forth definitions of local homotopy conditions on the complement of an embedded object. It systematically analyzes which of them are possessed by tame and locally flat objects. Certain of these conditions turn out to play an absolutely essential role much later on in the characterization of tameness and local flatness.

1.1. Knotted and flat piecewise linear embeddings

We begin with a review of some simple examples that illustrate various ways in which PL embeddings of polyhedra can be knotted. Later in the section we will observe that PL embeddings of cells and spheres are flat, provided the codimension is different from two.

Let X denote the disjoint union of S^{n-1} and a point. If $X_1 \supset S^{n-1}$ is a copy of X in $\mathbb{R}^n$ with the special point lying in the bounded component of $\mathbb{R}^n \setminus S^{n-1}$ and $X_2 \supset S^{n-1}$ is another copy in $\mathbb{R}^n$ with the special point lying in the unbounded component of $\mathbb{R}^n \setminus S^{n-1}$, then X_1, X_2 are inequivalently embedded in $\mathbb{R}^n$. To obtain a connected polyhedron with inequivalent copies in $\mathbb{R}^n$, do something similar to the wedge of S^{n-1} and a line segment (see Figure 1.1).

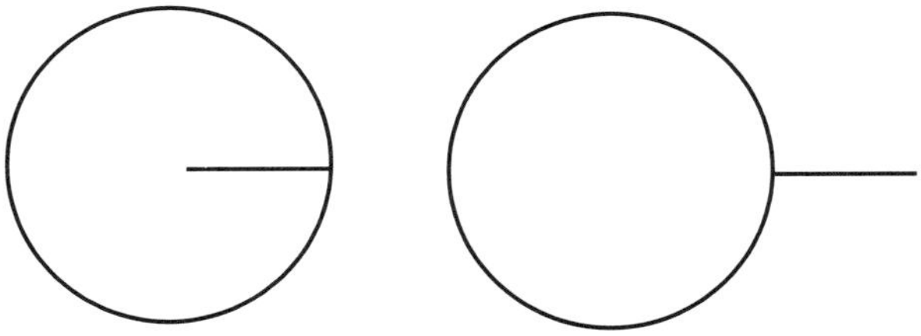

Figure 1.1. Inequivalent embeddings of $S^1 \vee I$ in $\mathbb{R}^2$

Links of spheres represent another basic class of examples. Regard S^{p+q+1} as the join of S^p and S^q (Rourke and Sanderson, 1972, Proposition 2.23), and let L denote the natural copies of S^p and S^q forming this join. By elementary properties of the join, the inclusion $S^p \hookrightarrow S^{p+q+1} \setminus S^q$ is a homotopy equivalence, so it cannot be homotopic to a constant. Now specify disjointly embedded $(p+1)$- and $(q+1)$-cells B^{p+1}, B^{q+1} in S^{p+q+1} and let $L' = \partial B^{p+1} \cup \partial B^{q+1}$. This time each component of L' is null-homotopic in the complement of the other component, so L and L' are inequivalent embeddings of $S^p \sqcup S^q$. The linked and unlinked embeddings are pictured in Figure 1.2.

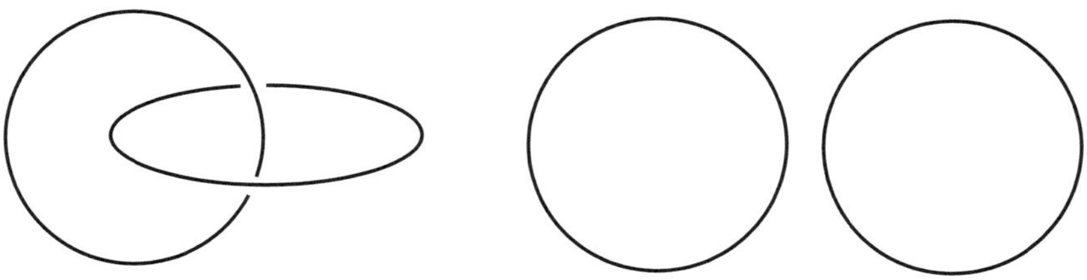

Figure 1.2. Linked and unlinked 1-spheres in S^3

When $p = q$ the links constructed in the previous paragraph all have dimension approximately half the ambient dimension. It is necessary for the dimension of the embedded space to be at least this large because PL embeddings of manifolds into $\mathbb{R}^n$ or S^n unknot in the trivial range (Rourke and Sanderson, 1972, Corollary 5.9). In Chapter 4 we will generalize that result to embeddings of arbitrary polyhedra (see Theorem 4.1.1).

Standard algebraic devices, like homology and homotopy groups, provide imperfect but efficacious invariants for measuring inequivalence. If λ, λ' are embeddings of L in K and ξ represents an element of, say, $H_k(L)$ for which $\lambda_*(\xi)$ and $\lambda'_*(\xi)$ are intrinsically different elements of $H_k(K)$, different in the sense that no automorphism of $H_k(K)$ carries one to the other, then λ,λ' must be inequivalent. For instance, when λ carries $L = S^1$ homeomorphically onto $S^1 \times \{\text{origin}\}$ in $K = S^1 \times \mathbb{R}^2$, then any embedding $\lambda' : L \to K$ equivalent to λ must send a generator of $H_1(L) \cong \mathbb{Z}$ to a generator of $H_1(K) \cong \mathbb{Z}$.

It is obvious from the definition that equivalent embeddings have homeomorphic complements. Thus another way to establish inequivalence is to employ the standard algebraic invariants to distinguish the topological types of the complements. Knots are the classic examples. These are 1-spheres PL embedded in S^3, typically in nonstandard fashion. A deep result about 3-manifolds, the famous Loop Theorem of Papakyriakopoulos (1957), implies that a knot K is flat if and only if $\pi_1(S^3 \smallsetminus K) \cong \mathbb{Z}$. Compared with proving the Loop Theorem, finding examples of nontrivial (= nonflat) knots is quite easy. For example, as in (Massey, 1967), for any (p, q)-torus knot $K_{p,q}$, where p, q are relatively prime and $p > 1 < q$, $\pi_1(S^3 \smallsetminus K_{p,q})$ is a nonabelian group. (Torus knots are those equivalent to a 1-sphere on the boundary of a solid torus standardly embedded in S^3.) The method of Example 0.2.3 can be used to prove that the group is not abelian (Exercise 1.1.3), thereby confirming that $K_{p,q}$ is not flat. The knot $K_{2,3}$ is shown in Figure 1.3.

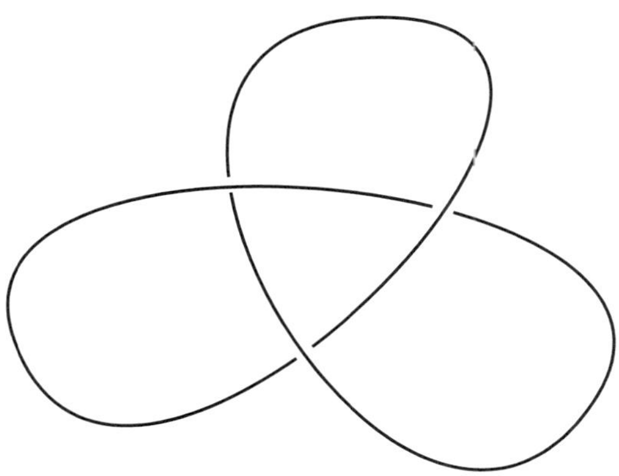

Figure 1.3. The trefoil knot $K_{2,3}$

Knotted $(n-2)$-spheres in S^n, $n \geq 4$, can be constructed from the 1-dimensional examples by suspending and spinning. Those constructions will be treated in §1.4. When applied to PL objects, suspending and spinning produce PL embeddings; hence there are knotted PL $(n-2)$-spheres in S^n for every $n \geq 3$.

Two is the only codimension in which PL sphere pairs can be knotted in the topological category. The following is a minor extension of the Unknotting of Sphere Pairs Theorem (Rourke and Sanderson, 1972, Theorem 7.1).

Theorem 1.1.1. *For $k \leq n-3$, every PL embedded k-cell or k-sphere in $\mathbb{R}^n$ or S^n is unknotted and thus flat.*

Comments on the proof. Since S^n is the one-point compactification of $\mathbb{R}^n$, any PL embedding of a sphere into $\mathbb{R}^n$ naturally defines a PL embedding into S^n. The ambient homeomorphism that unknots a PL k-sphere in S^n can be chosen to fix one point, so it induces an unknotting homeomorphism in $\mathbb{R}^n$. Thus any PL k-sphere, $k \leq n-3$, in $\mathbb{R}^n$ is PL flat as well.

The proof for embeddings of cells is left as an exercise. □

Corollary 1.1.2. *If $k \leq n-3$, then every tamely embedded k-cell or k-sphere in $\mathbb{R}^n$ or S^n is flat.*

The issue of whether an embedded $(n-1)$-sphere $\Sigma \subset S^n$ is flat is known as the *Schönflies Problem* due to its relationship with the Schönflies Theorem (Theorem 0.11.1). The complement $S^n \smallsetminus \Sigma$ has two components by the Jordan Separation Theorem (Exercise 0.3.2). The components of $S^n \smallsetminus \Sigma$ are called *complementary domains* of Σ. It is relatively easy to check that Σ is flat if and only if the closure of each of the complementary domains is an n-cell, so the real question is whether those complementary domain closures are n-cells.

Assume Σ is PL embedded and consider a complementary domain U of Σ. There are two parts to the issue of PL flatness: (1) is $\overline{U}$ a PL ∂-manifold and (2), if so, is it a PL n-cell? If (1) has an affirmative answer, then for $x \in U$ the Weak Schönflies Theorem (Rourke and Sanderson, 1972, Theorem 3.38) promises that $\overline{U} \smallsetminus \{x\}$ is PL homeomorphic to $S^{n-1} \times [0,1)$. It follows automatically that $\overline{U}$ is a topological n-cell, but it does not follow that $\overline{U}$ is a PL n-cell. Indeed, the following conjecture is still open for $n \geq 4$.

PL Schönflies Conjecture. *If Σ is a PL $(n-1)$-sphere in S^n, then Σ bounds two PL n-cells in S^n.*

The PL Schönflies Conjecture is known to be true in dimension three. It is also known, in all dimensions, that the closure of one complementary domain U is a PL n-cell if the other one is (Rourke and Sanderson, 1972, Corollary 3.13). If $\overline{U}$ is a PL ∂-manifold, then attaching a PL n-cell B to $\overline{U}$ along $\Sigma = \partial B$ produces a PL n-manifold M^n which has the homotopy type of S^n. When $n \geq 5$, M^n will be PL homeomorphic to S^n by the PL Poincaré Theorem (Rourke and Sanderson, 1972, p. 9); being the complement in $M = S^n$ of the PL n-cell B, $\overline{U}$ also will be a PL n-cell. With these results in hand, examination of the link of a vertex shows that the PL Schönflies conjecture in dimension $n-1$ implies that the closures of both complementary domains are PL ∂-manifolds. As a consequence, a proof of the PL Schönflies conjecture in dimension 4 would imply that the conjecture holds in all higher dimensions as well (Rourke and Sanderson, 1972, page 47). At present, however, the 4-dimensional case is unresolved.

In §2.4 we will prove a topological Schönflies theorem. One of its corollaries attests that an $(n-1)$-sphere in S^n is flat if the closures of its two complementary domains are (topological) ∂-manifolds. Temporarily assuming this corollary, we derive topological flatness results for PL embedded cells and spheres in codimension one.

Theorem 1.1.3. *Every PL embedded $(n-1)$-sphere in $\mathbb{R}^n$ or S^n is topologically flat.*

Proof. By induction on n. Assume the result holds for dimension $k-1$, and consider a PL embedded $(k-1)$-sphere Σ in S^k. Since all link pairs are topologically standard, by induction, Σ is locally flat. Thus, the closure of each component of $S^k \setminus \Sigma$ is a ∂-manifold. Hence, by Corollary 2.4.11, Σ is flat, and the inductive step is established. □

A similar argument shows that any PL embedded $(n-1)$-cell in S^n is locally flat.

Theorem 1.1.4. *Every locally flat $(n-1)$-cell in $\mathbb{R}^n$ is flat.*

This relies upon the following, the proof of which is left as an exercise.

Lemma 1.1.5. *Suppose $\lambda : I^{n-1} \to \mathbb{R}^n$ is an embedding such that both $\lambda(I^{n-2} \times [0, 2/3])$ and $\lambda(I^{n-2} \times [1/3, 1])$ are flat. Then $\lambda(I^{n-1})$ is flat.*

Corollary 1.1.6. *Every PL embedded $(n-1)$-cell in $\mathbb{R}^n$ or S^n is topologically flat.*

Historical Notes. The PL Schönflies Conjecture is known to hold for $n \leq 3$. For $n = 2$ this is Theorem 0.11.2 and for $n = 3$ this is a theorem due to J. W. Alexander (1924a). The Alexander-Newman Theorem states that the closure of one complementary domain is a PL n-cell if the other one is (Alexander, 1930), (Newman, 1926). The proof that the PL Schönflies Conjecture in dimension four implies the conjecture in higher dimensions relies on the high-dimensional PL Poincaré Theorem of Smale (1961). The proof of the Poincaré Theorem in dimension 5, which Rourke and Sanderson leave untreated, depends on the theorem that every PL n-manifold with the homology of S^n, $n \geq 5$, bounds some compact, contractible, PL, (n+1)-dimensional ∂-manifold (Kervaire, 1969).

Exercises

1.1.1. The two embeddings of $S^1 \vee I$ shown in Figure 1.1 are equivalent in S^2.

1.1.2. Let L be the link of two circles in S^3 that is shown in the left half of Figure 1.2 and let L' be the link in the right half of the same figure. (We will refer to L as the *Hopf link* and to L' as the *unlink*.) Prove that $S^3 \smallsetminus L$ has the homotopy type of the torus $S^1 \times S^1$ and that $S^3 \smallsetminus L'$ has the homotopy type of $S^1 \vee S^1 \vee S^2$. Conclude that $H_1(S^3 \smallsetminus L) \cong H_1(S^3 \smallsetminus L')$ but that $\pi_1(S^3 \smallsetminus L) \ncong \pi_1(S^3 \smallsetminus L')$.

1.1.3. Let $T^2 = S^1 \times S^1$ be the 2-torus. The *standard embedding* of the torus in S^3 is the one shown in Figure 1.4; it is a tame embedding that extends to an embedding $S^1 \times B^2 \to S^3$. A circle of the form $\{\text{point}\} \times S^1$ is called a *meridian* and a circle of the form $S^1 \times \{\text{point}\}$ is called a *longitude* of the torus. Let p and q be two relatively prime positive integers.
   (a) Prove that there is a simple closed curve $K_{p,q}$ on T^2 that is homologous to p times a longitude plus q times a meridian. The curve $K_{p,q} \subset T^2 \subset S^3$ is called the *(p,q)-torus knot.*
   (b) Prove that $T^2 \smallsetminus K_{p,q}$ is an open annulus; i.e., $T^2 \smallsetminus K_{p,q} \cong S^1 \times (-1, 1)$.
   (c) Use the Seifert-vanKampen Theorem to prove that $\pi_1(S^3 \smallsetminus K_{p,q})$ has presentation $\langle x, y : x^p = y^q \rangle$.
   (d) Prove that $K_{1,q}$ and $K_{p,1}$ are unknotted for every p and q.
   (e) Use the method of Example 0.2.3 to prove that $\pi_1(S^3 \smallsetminus K_{p,q})$ is not abelian if $p \geq 2$ and $q \geq 2$.

1.1.4. Every tame k-cell in S^n or $\mathbb{R}^n$ is flat provided $k \leq n - 3$.

1.1.5. Every tame arc in $\mathbb{R}^3$ is flat.

1.1.6. A PL embedding $h : S^{n-1} \to S^n$ is PL unknotted if and only if the closure of each complementary domain is a PL n-cell.

1.1.7. A topological $(n-1)$-sphere $\Sigma \subset S^n$ is flat if and only if the closure of each complementary domain is a topological n-cell.

1.1.8. Prove Lemma 1.1.5.

1.1.9. Prove Theorem 1.1.4.

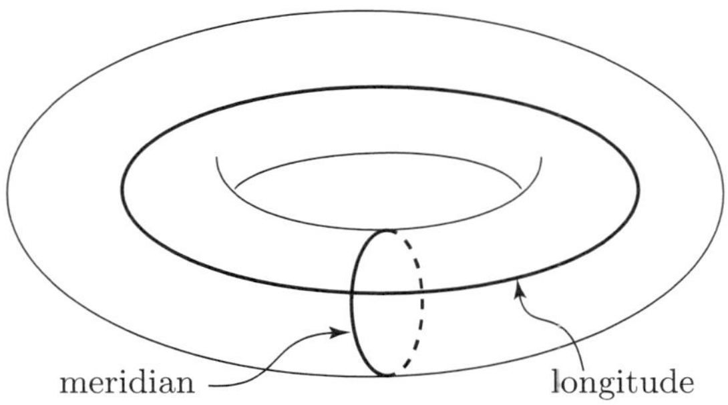

Figure 1.4. The standard embedding of the torus

1.2. Tame and locally flat topological embeddings

We now turn our attention to local questions and show that, in most codimensions, a PL embedding of one PL manifold in another is locally flat. The exception is codimension two, in which PL embeddings may be locally as well as globally knotted. Theorems 1.1.1 and 1.2.1 illustrate a general principle of the subject: global flatness results for spheres and cells often serve as prototypes for local flatness results regarding embeddings of manifolds.

Theorem 1.2.1. *If M^m is a PL m-manifold tamely embedded in a PL n-manifold N^n, then M^m is locally flat in N^n at each $x \in M^m$ provided $n - k \neq 2$.*

Proof. We may assume that M^m is a PL submanifold of N^n. In the case when $m \leq n-3$, the desired conclusion follows from (Rourke and Sanderson, 1972, Corollary 7.2). We must examine the inductive structure of that proof in order to see how to extend it to codimension one. The link of a simplex σ from M^m in (N^n, M^m) is a sphere pair of codimension $n-m \geq 3$ but ambient dimension less than n. This sphere pair is unknotted and the unknotting homeomorphism can be extended across the join structure of the star of σ to obtain a PL homeomorphism that flattens the associated star pair.

The proof breaks down in codimensions one and two because sphere pairs in these codimensions are not necessarily flat in the PL sense. But, as observed in the proof of Theorem 1.1.1, a codimension-one PL sphere pair is topologically flat. Hence the inductive proof works in codimension one as well, but gives only a (weaker) topological conclusion. □

Since tameness implies local flatness for embeddings of manifolds in all codimensions except two, we will say that an embedding $e : M \to N$ of a k-dimensional PL manifold M into an n-dimensional PL manifold N, $k \neq n-2$, is *wild at* $e(x)$ when $e(M)$ fails to be locally flat at $e(x)$.

As we will see in §1.4, PL embeddings can be locally knotted in codimension two. It follows that a codimension-two PL embedding can fail to be locally flat at some points. However, if N is a PL manifold and X is a subcomplex of some triangulation, then X obviously is locally flat at each point $x \in X$ belonging to the interior of a top-dimensional simplex; hence every tamely embedded PL manifold X is locally flat at a dense set of points, even in codimension two.

On the other hand, it follows from work of Kirby and Siebenmann (1977), to be treated in more detail much later, that local flatness does not imply tameness, not even for embedded manifolds. This is because manifolds can fail to support PL structures and because locally flat embeddings might not be approximable by PL embeddings. The distinctions between the concepts of local flatness and tameness are inherently derived from the categories to be considered – the concept of tameness belongs to the PL category and that of flatness to the topological category.

Exercises

1.2.1. Every locally flat cell in $\mathbb{R}^n$ is flat.

1.2.2. Every convex n-cell in $\mathbb{R}^n$ is flat.

1.2.3. Every tame 1-sphere in $\mathbb{R}^3$ is locally flat.

1.2.4. (Transitivity of local flatness). If $N_3 \subset N_2 \subset N_1$ are manifolds with N_{i+1} locally flat in N_i $(i = 1, 2)$, then N_3 is locally flat in N_1.

1.2.5. If M and N are PL manifolds of dimensions $n-2$ and n, respectively, and $e : M \to N$ is a PL embedding, then the set of points at which e is not locally flat is contained in the $(n-4)$-skeleton of M. It follows that the dimension of the set of nonlocally flat points is at most $n-4$ and every tame embedding of PL manifolds is locally flat at a dense set of points.

1.3. Local co-connectedness properties

This section presents definitions of local homotopy conditions on the complement of an embedding. Heavy emphasis will be placed on these conditions throughout our treatment of embeddings because of their prominent role in the characterization of local flatness and tameness. We simply observe here that the various local homotopy conditions are necessary for local flatness; an overarching goal of later chapters in this book will be to prove that these conditions are also sufficient to guarantee local flatness or tameness. The definitions of 1-LCC, 1-alg and locally flat will be used in the next section and then extensively in Chapter 2; the rest of this section may be postponed until later, and consulted when needed.

Definition. Let A denote a closed subset of the metric space X. Say that A is *locally k-co-connected* in X *at* $a \in A$ (abbreviated A is k-LCC at a) if each neighborhood U of a in X contains a smaller neighborhood V of a such that each map $\partial I^{k+1} \to V \smallsetminus A$ extends to a map $I^{k+1} \to U \smallsetminus A$.

The "co" in this definition points to the complement of A.

When X is an n-manifold and $A \subset X$ is a finite set, A is k-LCC in X for $k < n-1$ but not for $k = n-1$.

In general, the $k = 0$ case is readily characterized: A is 0-LCC at a if for each neighborhood U of $a \in A$ there is another neighborhood $V \subset U$ of a such that $V \smallsetminus A$ is contained in a path component of $U \smallsetminus A$. As indicated in Chapter 0, duality ensures that each closed subset A of an n-manifold N with $\dim A + 2 \leq n$ is 0-LCC in N. Unexpectedly, perhaps, even the $k = -1$ case has content: A is (-1)-LCC in X if and only if $X \smallsetminus A$ is dense in X. The k-LCC concept is meaningless for $k < -1$.

Definition. We say that A is *locally k-co-connected* (in X), written k-LCC, if A is k-LCC at a for every $a \in A$. We write A is LCC^k (in X) if it is i-LCC for $i = -1, 0, 1, \ldots, k$.

The case $\dim X = \dim A + 2$ is exceptional because then one cannot expect A to be 1-LCC in a manifold X. The following definitions are the appropriate ones for codimension-two manifold pairs.

Definition. Let A be a closed subset of X. We say that A is *1-alg* at $a \in A$ if each neighborhood U of a contains a smaller neighborhood V of a such that the inclusion-induced homomorphism $\pi_1(V \smallsetminus A) \to \pi_1(U \smallsetminus A)$ has abelian image. Say that A is *locally homotopically unknotted* (in X) *at* $a \in A$ if A is locally 1-alg at a and A is k-LCC at a for every $k \neq 1$. Finally, say A is *locally homotopically unknotted* if A is locally homotopically unknotted at every $a \in A$.

The *alg* in the definition above stands for "abelian local groups." In the case of a codimension-two submanifold, the abelian image will be an infinite cyclic subgroup.

Immediately we see that these k-LCC properties give necessary conditions for local flatness. One of the major goals of later chapters in this book is to show that they also provide sufficient conditions. Indeed, we shall learn that almost invariably the 1-LCC condition alone implies local flatness of embedded submanifolds and local tameness of embedded polyhedra. The exception is codimension two, where the full force of the locally homotopically unknotted condition is needed for the submanifold case.

Proposition 1.3.1. *Suppose the manifold $M^m \subset N^n$ is locally flatly embedded in the n-manifold N^n.*

If $m < n-2$, then M^m is LCC^{n-m-2} in N^n.

If $m = n-1$, M^{n-1} is k-LCC in N^n for all $k \geq 1$ but not for $k = 0$.

If $m = n-2$, M^{n-2} is locally homotopically unknotted in N^n.

Proof. Local flatness of M^m at $z \in M^m$ ensures that z has arbitrarily small neighborhoods U with $(U, U \cap M^m)$ pairwise homeomorphic to $(\mathbb{R}^n, \mathbb{R}^m)$. Hence,

$$U \setminus M^m \approx \mathbb{R}^n \setminus \mathbb{R}^m \approx \mathbb{R}^m \times (\mathbb{R}^{n-m} \setminus \{\text{origin}\}),$$

implying that $U \setminus M^m$ has the same homotopy type as S^{n-m-1}. □

In the Introduction we defined local flatness for embeddings of manifolds; now we extend that definition to embeddings of ∂-manifolds.

Definition. Let e denote an embedding of an m-dimensional ∂-manifold M^m into an n-manifold N^n. We say that e is *locally flat* at $x \in M^m$ (and that $e(M^m)$ is *locally flat at* $e(x)$) if there exist a neighborhood U of $e(x)$ in N^n and a homeomorphism h of U onto $\mathbb{R}^n$ such that

(1) $h(U \cap e(M^m)) = \mathbb{R}^m \subset \mathbb{R}^n$ when $x \in \operatorname{Int} M^m$ or

(2) $h(U \cap e(M^m)) = \mathbb{R}^m_+ \subset \mathbb{R}^n$ when $x \in \partial M^m$.

Taking the opposite tack, one says that $e(M^m)$, $m \neq n-2$, is *wild at* $e(x)$ when it fails to be locally flat there.

For points in the interior of the ∂-manifold this definition of local flatness agrees with the earlier one, and then Proposition 1.3.1 applies; for points on the boundary we have the following complementary result.

Proposition 1.3.2. *Let M^m be a ∂-manifold locally flatly embedded in an n-manifold N^n and $z \in \partial M^m$. Then M^m is k-LCC in N^n at z for all $k \geq 0$.*

Proof. With U carefully chosen,

$$U \setminus M^m \approx \mathbb{R}^n \setminus \mathbb{R}^m_+ \approx \mathbb{R}^{m-1} \times (\mathbb{R}^{n-m+1} \setminus \mathbb{R}^1_+),$$

and both factors in the latter are contractible. ($\mathbb{R}^{n-m+1} \setminus \mathbb{R}^1_+$ contracts to any point of $\mathbb{R}^1 \setminus \mathbb{R}^1_+$ via a straight line homotopy.) □

Definition. Let P be a polyhedron and let $e : P \to N^n$ be an embedding of P in a topological n-manifold. The embedding e is said to be *locally tame* if for every $x \in P$ there exist a PL neighborhood U of $e(x)$ in N^n and a homeomorphism $h_x : U \to \mathbb{R}^n$ such that $h_x \circ e|e^{-1}(U)$ is PL.

In the preceding definition there should be one fixed triangulation of P and each of the maps $h_x \circ e$ should be PL relative to that triangulation. A straightforward general position argument yields the following result.

Proposition 1.3.3. *If a p-dimensional polyhedron P is locally tamely embedded in an n-manifold N^n, then P is LCC^{n-p-2} in N^n.*

Proof. Exercise 1.3.2. □

The next result discloses the value of the Local Hurewicz Theorem.

Proposition 1.3.4. *If a closed subset X of an n-manifold N^n is 1-LCC in N^n, then X is LCC^k in N^n, where $k = n - \dim X - 2$.*

Proof. The argument relies on the consequence of Duality Theorem 0.3.1 that, for any orientable open subset W of M and any $q \le k$, $H_q(W \smallsetminus X;\mathbb{Z}) \to H_q(W;\mathbb{Z})$ is an isomorphism, since $H_{q+1}(W, W \smallsetminus X;\mathbb{Z}) \cong H_c^{n-q-1}(X \cap W;\mathbb{Z}) \cong 0$ and, similarly, $H_q(W, W \smallsetminus X;\mathbb{Z}) \cong 0$. Hence, for any coordinate chart W, $H_q(W \smallsetminus X;\mathbb{Z}) \cong 0$.

Fix a neighborhood U of $x \in X$. By Corollary 0.3.10, hypothesis, and induction on k, one can obtain neighborhoods

$$V \subset U_0 \subset U_1 \subset \cdots \subset U_{k-1} \subset U_k = U$$

of x, where the inclusion-induced homomorphisms $H_k(V \smallsetminus X) \to H_k(U_0 \smallsetminus X)$ and $\pi_q(U_q \smallsetminus X) \to \pi_q(U_{q+1} \smallsetminus X)$ $(q = 0, 1, \ldots, k-1)$ are all trivial. That $\pi_k(V \smallsetminus X) \to \pi_k(U \smallsetminus X)$ also is trivial is then ensured by Local Hurewicz Theorem 0.8.3. □

There is a related result for complements of codimension one manifolds S. The proof retraces the inductive argument of 1.3.4. To obtain the triviality of $H_k(V \smallsetminus S) \to H_k(U_0 \smallsetminus S)$, interpolate both a coordinate chart and a smaller open set W that intersects S in a copy of $\mathbb{R}^{n-1}$, since then

$$H_k(W, W \smallsetminus S;\mathbb{Z}) \cong H_c^{n-k}(W \cap S;\mathbb{Z}) \cong H_c^{n-k}(\mathbb{R}^{n-1}) \cong 0 \quad (k > 1).$$

The restriction of attention to LCC properties of S in $\overline{U}$, rather than in N^n itself, is simply to assure 0-LCC.

Proposition 1.3.5. *Let N^n be a connected n-manifold and S a connected (n-1)-manifold embedded in N^n as a closed subset such that $N^n \smallsetminus S$ has two components, U and V, and suppose S is 1-LCC in $\overline{U} = U \cup S$. Then S is LCC^k in $\overline{U}$ for all $k \ge 0$.*

There is also a stability result for local co-connectedness.

Proposition 1.3.6. *Suppose Y is a locally contractible space and $A \subset X$. Then A is k-LCC in X (respectively, is locally homotopically unknotted in X) if and only if $A \times Y$ is k-LCC in $X \times Y$ (respectively, is locally homotopically unknotted in $X \times Y$).*

Proof. Exercise 1.3.3. □

Proposition 1.3.7. *Let X be a locally compact, locally contractible space and Y a compact subset which is LCC^k in X. Then for each $\epsilon > 0$ and each map $f : (K, L) \to (X, X \smallsetminus Y)$ defined on a pair of simplicial complexes with dim $K \leq k$, there exists a map $g : K \to X \smallsetminus Y$ such that $g|L = f|L$ and $\rho(g, f) < \epsilon$.*

Proof. We treat only those A having a compact neighborhood N in X; the general case is similar but requires more care. Given such a compact neighborhood N, for each $\delta > 0$ there exists $\eta > 0$ such that every map of ∂I^i, $i \leq k$, into an η-subset of $N \smallsetminus A$ extends to a map of I^i into a δ-subset of $X \smallsetminus A$.

Find another neighborhood N' of A, $N' \subset \overline{N'} \subset \operatorname{Int} N$, and limit ϵ, if necessary, so $\epsilon < \min\{d(N', X \smallsetminus N), d(A, X \smallsetminus N')\}$. Produce successive positive numbers $\epsilon = \delta_{k+1} > \delta_k > \cdots > \delta_1 > 0$ such that each map of ∂I^i into a δ_i-subset of $N \smallsetminus A$ extends to a map of I^i into a $\delta_{i+1}/3$-subset of $X \smallsetminus A$. Subdivide K using a triangulation T of such small mesh that all images of simplices of T have diameter less than $\delta_1/3$, and let T_A denote the subcomplex consisting of all simplices whose images miss A. Note that $|T_A| \supset |L|$.

We define g to coincide with f on $|T_A|$, and we extend g over the simplices of $T \smallsetminus T_A$ in order of increasing dimension. For each vertex $v \in T \smallsetminus T_A$, we choose $g(v) \in X \smallsetminus A$ within $\delta_1/3$ of $f(v)$. For each 1-simplex τ of $T \smallsetminus T_A$, $g|\partial\tau$ has been determined by this rule and its image is a δ_1-subset of $N \smallsetminus A$, so g can be extended over τ so $g(\tau)$ is a $\delta_2/3$-subset of $X \smallsetminus A$. Generally, for each i-simplex σ of $T \smallsetminus T_A$, $g|\partial\sigma$ will be determined with image a δ_i-subset of $N \smallsetminus A$, and then g can be extended over σ so its image is a $\delta_{i+1}/3$-subset of $X \smallsetminus A$. The resulting map g sends K to $X \smallsetminus A$ and is a satisfactory approximation to f. □

Exercises

1.3.1. Let C be a nowhere dense subset of a metric space X and let c be a point in C such that C is 0-LCC at c. Construct a map $\mu : [0, 1] \to X$ with $\mu(0) = c$ and $C \cap \mu((0, 1]) = \emptyset$.

1.3.2. Prove Proposition 1.3.3.

1.3.3. Prove Proposition 1.3.6 (stability of local co-connectedness).

1.4. Suspending and spinning

There are two operations, known as suspending and spinning, that transform embeddings of polyhedra into related higher-dimensional embeddings. In most cases they transform PL embeddings into PL embeddings and they transform wild or nonflat embeddings into new embeddings that are still wild or nonflat. These operations allow us to construct examples of PL $(n-2)$-spheres in S^n that are either locally or globally knotted.

Suspending. Let K be a compact polyhedron (usually a sphere). By the *suspension of* K, written $\mathrm{Susp}(K)$, we mean the (external) join $K * S^0$ of K with the 0-sphere $S^0 = \{-1, 1\}$. The join is the union of all line segments that join a point of S^0 to a point of K (Rourke and Sanderson, 1972, pages 22–24). The points of S^0 are called the *suspension points*. For any $A \subset K$, $K * S^0$ naturally contains a copy of the suspension of A, and $\mathrm{Susp}(A)$ is a subcomplex of $\mathrm{Susp}(K)$ when A is a subcomplex of K.

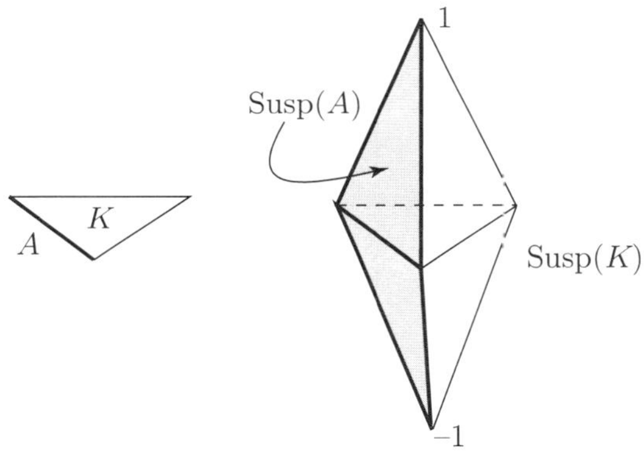

Figure 1.5. Susp(A) is a subcomplex of Susp(K)

Note that $S^{n+1} = \mathrm{Susp}(S^n)$, so the suspension of a subset of S^n is a subset of S^{n+1}. This observation allows the definition of suspension to be extended to spaces that are not polyhedra: If X is any subset of S^n, define $\mathrm{Susp}(X)$ to be the union of all the straight line segments in $\mathrm{Susp}(S^n)$ that join a point of X to one of the suspension points. Using this definition we can speak, for example, of the suspension of a Cantor set.

Suspending converts any embedding $e : K \to S^n$ into an embedding $\mathrm{Susp}(e) : \mathrm{Susp}(K) \to S^{n+1}$. If e is a PL embedding, then $\mathrm{Susp}(e)$ is a PL embedding; if e is a topological embedding, then $\mathrm{Susp}(e)$ is a topological embedding.

Lemma 1.4.1. *Let D be a subset of S^n.*

(1) *If D is a d-cell, then* $\operatorname{Susp}(D)$ *is a* $(d+1)$*-cell.*

(2) *If D is a d-sphere, then* $\operatorname{Susp}(D)$ *is a* $(d+1)$*-sphere.*

(3) $S^{n+1} \smallsetminus \operatorname{Susp}(D)$ *and* $S^n \smallsetminus D$ *have the same homotopy type.*

(4) *If D fails to be 1-LCC at $x \in D$, then* $\operatorname{Susp}(D)$ *fails to be 1-LCC at all points in the interior of the arc* $\operatorname{Susp}(\{x\})$.

Proof. The first two parts of the lemma follow from the traditional suspension rules (Rourke and Sanderson, 1972, Proposition 2.23). Part (3) follows from the observation that $\operatorname{Susp}(S^n) \smallsetminus \operatorname{Susp}(D) \cong (S^n \smallsetminus D) \times (-1,1)$. Use U to denote the complement in $\operatorname{Susp}(S^n)$ of the two suspension points. Since $(U, U \cap \operatorname{Susp}(D))$ is equivalent to $(S^n \times (-1,1), D \times (-1,1))$, the last part follows from Proposition 1.3.6. □

Suspension preserves tameness but not necessarily local flatness, so it can be used to construct tame embeddings that are not locally flat. Of course such embeddings must have codimension two since Section 1.2 indicates that, for manifolds, tameness implies local flatness in all other codimensions.

Example 1.4.2. *For every $n \geq 4$ there exist embeddings $S^{n-2} \to S^n$ that are piecewise linear but not locally flat.*

Proof. Let K be a PL 1-sphere in S^3 whose complement has nonabelian fundamental group. For example, K could be one of the torus knots $K_{p,q}$ in S^3 that were constructed in §1.1. Observe that $\operatorname{Susp}(K)$ is a PL 2-sphere in S^4. If B is any small convex 4-ball centered at one of the suspension points, then $B \smallsetminus \operatorname{Susp}(K) \cong (S^3 \smallsetminus K) \times [0,1)$. Hence $\pi_1(B \smallsetminus \operatorname{Susp}(K))$ is nonabelian. Furthermore, if $B_2 \subset \operatorname{Int} B_1$ are two such balls, then the inclusion-induced map $\pi_1(B_2 \smallsetminus \operatorname{Susp}(K)) \to \pi_1(B_1 \smallsetminus \operatorname{Susp}(K))$ is an isomorphism. It follows that $\operatorname{Susp}(K)$ is not 1-alg at the suspension point and is, therefore, not locally flat at the suspension point (by Proposition 1.3.1). Examples in higher dimensions are constructed by iterating the suspension operation. □

The suspension operation, when applied to a globally knotted sphere, produces a sphere that is locally knotted. Therefore suspension cannot be used to produce higher-dimensional examples of locally flat but knotted spheres. The second operation described in this section will rectify the failure to preserve local flatness that is inherent in the suspension operation.

Spinning. The perspective to take for spinning is that of generalized cylindrical coordinates, with points of $\mathbb{R}^{k+1}$ represented in terms of distance from the origin and direction. Explicitly, regard $\mathbb{R}^{k+1}$ as the quotient space

$$\mathbb{R}^{k+1} = ([0,\infty) \times S^k)/G_k = (\mathbb{R}^1_+ \times S^k)/G_k,$$

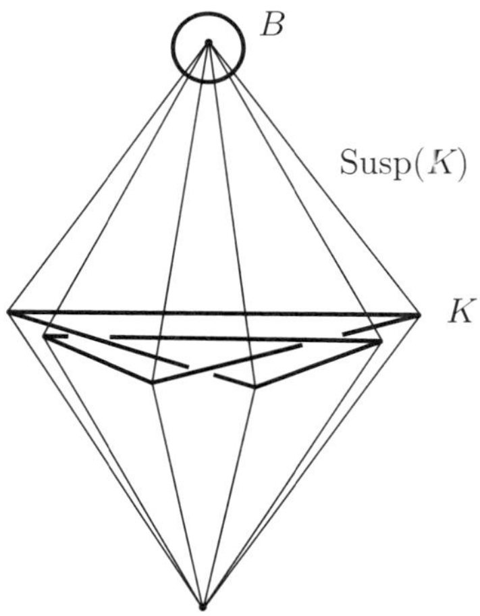

Figure 1.6. ∂B contains a small copy of K and $B \cap \text{Susp}(K)$ is a cone

where G_k denotes the decomposition whose only nondegenerate element is $0 \times S^k$. (See §2.3 for a general review of decomposition spaces.) Taking products with $\mathbb{R}^{n-k-1}$, one recovers $\mathbb{R}^n$ as

$$\mathbb{R}^n = (\mathbb{R}^{n-k-1} \times \mathbb{R}^1_+ \times S^k)/G_k = \mathbb{R}^{n-k}_+ \times S^k/R,$$

where R now denotes the decomposition of $\mathbb{R}^{n-k}_+ \times S^k$ whose nondegenerate elements are the sets $\{x\} \times S^k$, $x \in \mathbb{R}^{n-k-1} \times \{0\}$. This is easiest to visualize when $k = 1$, for then one can imagine sweeping out $\mathbb{R}^n$ by rotating $\mathbb{R}^{n-1}_+$ about its "edge" $\mathbb{R}^{n-2}$; in case $n = 3$, this is the familiar process of sweeping out $\mathbb{R}^3$ by rotating a half-plane about its boundary line; see Figure 1.7.

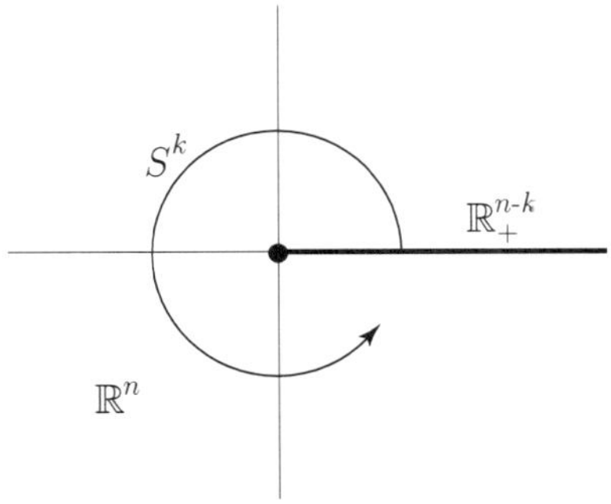

Figure 1.7. Viewing $\mathbb{R}^n$ as the k-spin of $\mathbb{R}^{n-k}_+$

Similarly, upon taking one-point compactifications, one can view S^n as the quotient space $S^n = (B^{n-k} \times S^k)/T_k$, where T_k denotes the decomposition of $B^{n-k} \times S^k$ into points and the k-spheres $\{x\} \times S^k$, $x \in \partial B^{n-k}$. Here one should imagine holding ∂B^{n-k} fixed and revolving $\operatorname{Int} B^{n-k}$ around S^k to sweep out B^n; see Figure 1.8.

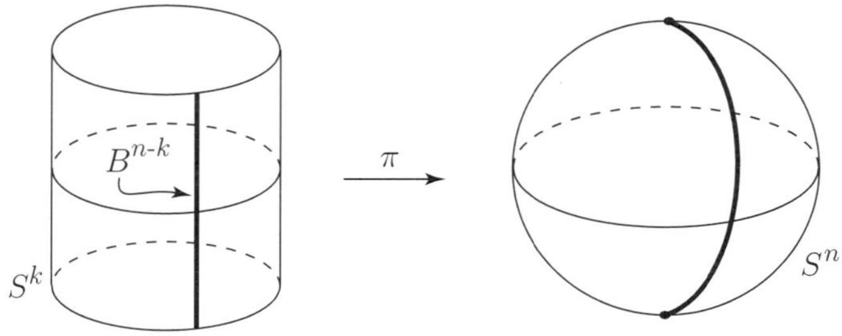

Figure 1.8. Viewing S^n as the k-spin of B^{n-k}

Let $\pi : B^{n-k} \times S^k \to S^n = (B^{n-k} \times S^k)/T_k$ denote the quotient map. Given a subset A of B^{n-k}, let $\operatorname{Spin}^k(A)$ denote the image $\pi(A \times S^k)$. Sometimes it pays to identify the induced map $\chi_k : S^n \to B^{n-k}$, defined to produce commutativity of the diagram below.

$$\begin{array}{ccc} B^{n-k} \times S^k & \xrightarrow{\ \pi\ } & S^n \\ \big\downarrow & \swarrow{\scriptstyle \chi_k = (\text{proj})\circ\pi^{-1}} & \\ B^{n-k}. & & \end{array}$$

Definition. A map $f : M \to N$ between ∂-manifolds is said to be *faithful* if $f^{-1}(\partial N) = \partial M$.

Lemma 1.4.3. *If D is a d-cell faithfully embedded in B^{n-k}, then* $\operatorname{Spin}^k(D)$ *is a $(d+k)$-sphere in S^n. Moreover, if D is locally flat in B^{n-k}, then* $\operatorname{Spin}^k(D)$ *is locally flat in S^n. Conversely, if D fails to be 1-LCC at $y \in$* $\operatorname{Int} D$, *then* $\operatorname{Spin}^k(D)$ *fails to be 1-LCC at each point of* $\operatorname{Spin}^k(\{y\})$.

Proof. The fact that $\operatorname{Spin}^k(D)$ is a sphere is clear from the spinning construction. At points of $\operatorname{Spin}^k(\operatorname{Int} D)$, local flatness is immediate since

$$(\operatorname{Spin}^k(\operatorname{Int} B^{n-k}), \operatorname{Spin}^k(\operatorname{Int} D)) \cong (\mathbb{R}^{n-k} \times S^k, \operatorname{Int} D \times S^k).$$

At points of $\operatorname{Spin}^k(\partial D)$ local flatness is a little more involved and is left as an exercise.

To establish the converse, take $x = \pi(y, z) \in \operatorname{Spin}^k(\{y\})$, equate B^{n-k} with $\pi(B^{n-k} \times \{z\})$, and fix a neighborhood U of y in B^{n-k}. If $\operatorname{Spin}^k(D)$ were 1-LCC at x, we could find a neighborhood W of x in $\operatorname{Spin}^k(U)$ such that every loop in $W \smallsetminus \operatorname{Spin}^k(D)$ would contract in $\operatorname{Spin}^k(U \smallsetminus D) = \operatorname{Spin}^k(U) \smallsetminus \operatorname{Spin}^k(D)$. Choose a neighborhood V of y in B^{n-k} with $V \times \{z\} \subset W$. Then any given loop γ in $V \smallsetminus D$ would contract in $\operatorname{Spin}^k(U \smallsetminus D)$ via some map f_t, and $\chi_k f_t$ would show that γ contracts in $U \smallsetminus D$, a contradiction. $\square$

Example 1.4.4. *For every $n \geq 3$ there exist knotted embeddings $S^{n-2} \to S^n$ that are both piecewise linear and locally flat.*

Proof. Focus on $n > 3$. Let (B^3, D) be a PL $(3,1)$-cell pair with $\pi_1(B^3 \smallsetminus D)$ nonabelian. Then by Lemma 1.4.3, $\Sigma = \text{Spin}^{n-3}(D)$ is an $(n-2)$-sphere locally flatly embedded in $S^n = \text{Spin}^{n-3}(B^3)$, and the restriction of χ_k determines a "retraction" $S^n \smallsetminus \Sigma \to B^3 \smallsetminus D$ certifying that $\pi_1(S^n \smallsetminus \Sigma)$ is also nonabelian. □

Remark. The spheres constructed in Example 1.4.4 are seen to be knotted because the fundamental group of the complement is nonabelian. A PL 1-sphere in S^3 is unknotted if and only if the fundamental group of its complement is abelian, but in higher dimensions more subtle forms of knotting are possible. In Chapter 6 we will construct locally flat PL $(n-2)$-spheres in S^n whose complements have abelian fundamental groups, but have nontrivial higher homotopy groups.

Historical Notes. The construction of higher-dimensional knots by suspending and spinning originated with E. Artin (1925). D. Rolfsen (1990) has a lucid exposition of spinning and of a generalization known as "twist spinning" that is due to E. C. Zeeman (1965).

Exercises

1.4.1. For $A \subset S^k$, $\text{Susp}(A)$ is 1-LCC at a suspension point $p \in S^0$ if and only if $\pi_1(S^k \smallsetminus A) \cong \{1\}$. If A is a $(k-2)$-sphere and $\pi_1(S^k \smallsetminus A)$ is abelian, then $\text{Susp}(A)$ is 1-alg at each suspension point; if $\pi_1(S^k \smallsetminus A)$ is nonabelian, then $\text{Susp}(A)$ is not locally flat at either suspension point.

1.4.2. Let K be a locally flat PL $(m-2)$-sphere in S^m, $m \geq 3$, such that $\pi_1(S^m \smallsetminus K)$ is nonabelian and suspend K $(n-m)$ times to define a PL embedding of S^{n-2} into S^n. Prove that the set of nonlocally flat points of this embedding is an $(n-m-1)$-sphere. Conclude that the dimension of the set of nonlocally flat points of a PL embedding $M^{n-2} \to N^n$ of PL manifolds can be any integer in the range $-1, \ldots, n-4$. (Compare Exercise 1.2.5.)

1.4.3. If D is a d-cell faithfully and locally flatly embedded in B^{n-k}, then $\text{Spin}^k(D)$ is locally flat at points of $\text{Spin}^k(\partial D)$.

1.4.4. If A is an open, connected subset of B^{n-1} such that $A \cap \partial B^{n-1} \neq \emptyset$, then $\pi_1(\text{Spin}^1(A)) \cong \pi_1(A)$.

Chapter 2

Wild and Flat Embeddings

To truly appreciate results about flatness, one must be keenly aware of the existence of wildness. In this chapter we set forth a wealth of examples of wild embeddings, beginning with two classics discovered by Antoine and Alexander in the 1920s. Then we describe a technique involving decomposition spaces by which wild arcs in S^n are transmuted into wild arcs in S^{n+1}. Toward the end of the chapter we introduce additional examples of wildly embedded 1-, 2- and 3-cells in S^3; some of them were discovered in the 1940s by R. H. Fox and E. Artin while others were discovered in the 1960s by R. H. Bing and his followers. The net effect is to make available examples of wild embeddings in all possible dimensions and codimensions.

Partly for contrast, we also present several results about flat embeddings. All are derived by elementary methods, independent of the engulfing techniques to be developed in subsequent chapters. The results include the Generalized Schönflies Theorem of M. Brown confirming the flatness of any codimension-one sphere in S^n that is locally flat, a version of work by J. C. Cantrell assuring the flatness of a codimension-one sphere in $\mathbb{R}^n$, $n > 3$, that is locally flat everywhere except possibly one point and a result of V. Klee attesting to the flatness of an arc in $\mathbb{R}^n$ that lies in a hyperplane.

2.1. Antoine's necklace and Alexander's horned sphere

Here we reproduce two fundamental, historically important examples of wild embeddings in $\mathbb{R}^3$. The first is a wild embedding of the Cantor set and the second is a wild embedding of S^2. For each of them the invariant used to detect wildness is the fundamental group of the complement. Related high-dimensional examples are constructed by suspension.

Example 2.1.1. *There exists a wild Cantor set in* $\mathbb{R}^3$.

A *Cantor set* in a manifold is an embedded copy of the familiar middle-thirds Cantor set. Cantor sets are characterized as the compact, totally disconnected metric spaces that are perfect, meaning that they have no isolated points. The example we construct is known as *Antoine's necklace.*

Remark. It is a mild but common and traditional misnomer to speak of a Cantor set as being "wild"; strictly speaking, a Cantor set cannot be wild because it is not an embedded polyhedron. There is a standard copy of the Cantor set in $[0,1] \subset \mathbb{R}^1 \subset \mathbb{R}^n$, so what we really mean when we call a Cantor set in $\mathbb{R}^n$ wild is that it is not flat. In the same spirit, when we speak of "tame" Cantor sets in $\mathbb{R}^n$, what we really mean is that they are flat.

A *solid torus* is a ∂-manifold homeomorphic to $S^1 \times B^2$. Let T be a solid torus standardly positioned in $\mathbb{R}^3$, and let T_1, T_2, T_3, and T_4 be solid tori embedded in $\operatorname{Int} T$ as shown in Figure 2.1. (Note that $(\mathbb{R}^3, T)$ and $(\mathbb{R}^3, T_i)$ are pairwise homeomorphic.) Set $A_0 = T$ and $A_1 = \cup_{i=1}^4 T_i$. In each T_i let $T_{i1}, T_{i2}, T_{i3}, T_{i4}$ be solid tori embedded there exactly as the T_i are placed in T. Replicate infinitely often, so that at the k-th step we have a ∂-manifold A_k, the union of 4^k (pairwise disjoint) solid tori, where each component τ of A_k contains exactly 4 components of A_{k+1}, and where there exists a homeomorphism of the triples $(\mathbb{R}^3, \tau, \tau \cap A_{k+1})$ and $(\mathbb{R}^3, T, A_1)$. Arrange these pieces so that each component τ of A_k has diameter at most ϵ_k, where $\epsilon_k \to 0$ as $k \to \infty$. (It is permissible to let the number of components of A_k be larger than 4^k in order to achieve small size.)

Set $A = \cap_k A_k$. Then A is a compact, totally disconnected metric space with no isolated points. Hence, A is homeomorphic to the standard middle-thirds Cantor set C in $[0,1] \subset \mathbb{R}^1 \subset \mathbb{R}^3$. (If one wants to secure a homeomorphism between A and C directly, without appeal to the topological characterization of the Cantor set, one easily can show, based on the construction, that $A \cong \Pi_{i=1}^{\infty} X_i$, where $X_i = \{1,2,3,4\}$ is endowed with the discrete topology, and exploit the related, more familiar $C \cong \Pi_{i=1}^{\infty} S_i$, where each S_i is a two-point set with the discrete topology.)

Proposition 2.1.2. $\pi_1(\mathbb{R}^3 \smallsetminus A) \neq \{1\}$.

Since $\pi_1(\mathbb{R}^3 \smallsetminus C) \cong \{1\}$, this proposition will confirm that A is wild. The argument will be based on the following pair of technical facts.

Lemma 2.1.3. *The inclusion-induced* $\phi_\# : \pi_1(\partial T_i) \to \pi_1(\mathbb{R}^3 \smallsetminus \operatorname{Int} A_1)$ *is one-to-one.*

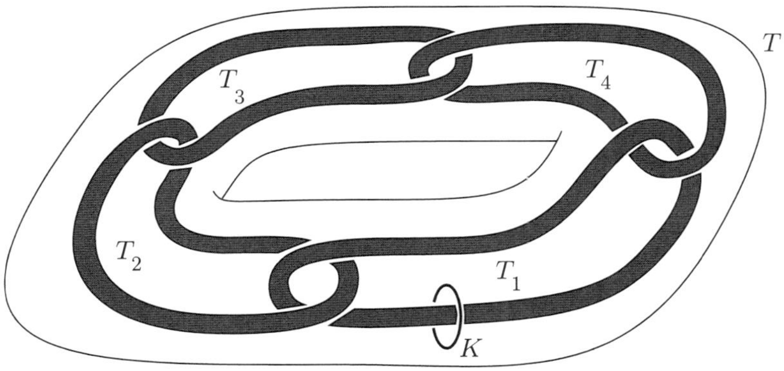

Figure 2.1. The first two stages in the construction of Antoine's necklace

Lemma 2.1.4. *The inclusion-induced* $\phi'_{\#} : \pi_1(\partial T) \to \pi_1(T \smallsetminus \operatorname{Int} A_1)$ *is one-to-one.*

Assuming Lemmas 2.1.3 and 2.1.4 for the moment, we complete the proof of Proposition 2.1.2. Thicken $\mathbb{R}^3 \smallsetminus \operatorname{Int} A_0$ to an open set W_0 that admits a strong deformation retraction to $\mathbb{R}^3 \smallsetminus \operatorname{Int} A_0$. Similarly, for $k \geq 1$ thicken $A_{k-1} \smallsetminus \operatorname{Int} A_k$ to an open set W_k that admits a strong deformation retraction to $A_{k-1} \smallsetminus \operatorname{Int} A_k$. Impose control on these thickenings to ensure that $W_{k-1} \cap W_k$ is naturally homeomorphic to $\partial A_k \times (-1, 1)$. Add the components of W_k to $\cup_{i=0}^{k-1} W_i$ one at a time and apply Lemma 2.1.3 and 2.1.4 in conjunction with Theorem 0.11.5 to establish that $\pi_1(\cup_{i=0}^{k-1} W_i) \to \pi_1(\cup_{i=0}^{k} W_i)$ is 1-1. It follows that $\mathbb{Z} \cong \pi_1(W_0) \to \pi_1(\mathbb{R}^3 \smallsetminus A = \cup_{i=0}^{\infty} W_i)$ is 1-1. Thus, $\pi_1(\mathbb{R}^3 \smallsetminus A) \neq \{1\}$.

We now turn our attention to Lemmas 2.1.3 and 2.1.4. Their proofs are based on the following claims.

Claim 2.1.5. *Let J and C denote linked circles in $\mathbb{R}^3$, as shown in Figure 2.2. Then J is a retract of $\mathbb{R}^3 \smallsetminus C$.*

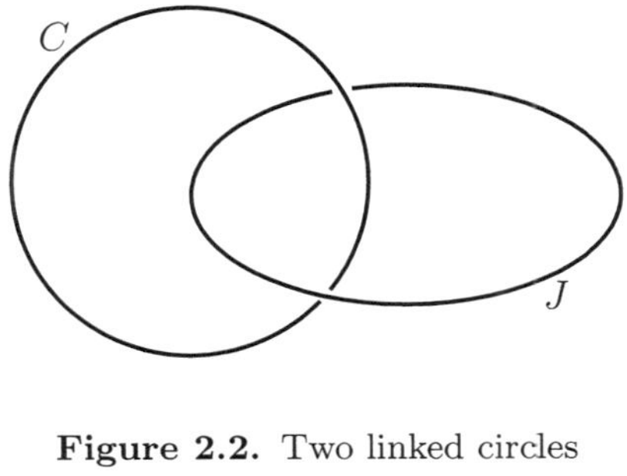

Figure 2.2. Two linked circles

Proof. Build a (round) 3-cell B containing $J \cup C$, and split it with a 2-cell into two hemispherical balls B_L and B_R such that $B_R \supset J$ and B_L intersects C in a standard spanning arc. Find retractions:

(1) of $\mathbb{R}^3 \smallsetminus C$ to $B \smallsetminus C$,

(2) of $B_L \smallsetminus C$ to $\partial B_L \smallsetminus C$ and, by extension, of $B \smallsetminus C$ to $(B_R \cup \partial B_L) \smallsetminus C$,

(3) of the latter to $B_R \smallsetminus C$, and

(4) of $B_R \smallsetminus C$ to J. □

A similar argument yields:

Claim 2.1.6. *Let C and C' denote circles in $\mathbb{R}^3$ and E a planar disk in $\mathbb{R}^3$, as shown in Figure 2.3. Then $E \smallsetminus (C \cup C')$ is a retract of $\mathbb{R}^3 \smallsetminus (C \cup C')$.*

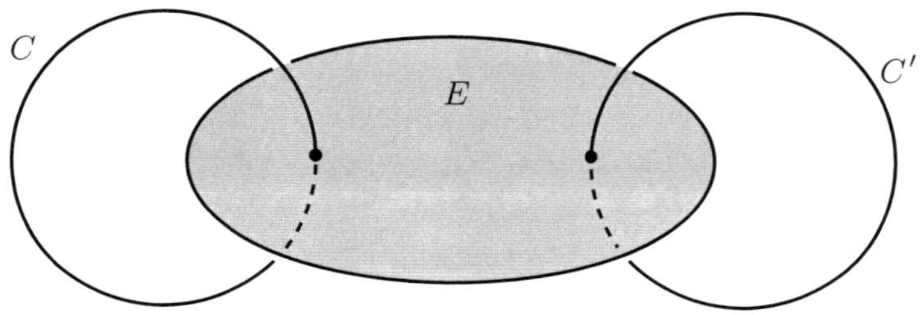

Figure 2.3. Three linked circles

Proof of Lemma 2.1.3. The proof of Lemma 2.1.3 is a relatively straightforward application of Claim 2.1.5 and is left as an exercise. □

Call a simple closed curve C a *center line* of a solid torus T if there exists a homeomorphism of $S^1 \times B^2$ onto T carrying $S^1 \times \{0\}$ onto C.

Proof of Lemma 2.1.4. Let C_1 and C_3 denote center lines of T_1 and T_3, respectively. Find disks E_2 and E_4 in $\operatorname{Int} T$ as shown in Figure 2.4, where $E_j \cap \partial A_1 = \partial E_j \subset \partial T_j$. Since $C_1 \cup E_2 \cup C_3 \cup E_4$ contains a center line of T, there exists a retraction

$$\rho : T \smallsetminus (C_1 \cup E_2 \cup C_3 \cup E_4) \to \partial T.$$

Thicken each E_j ($j = 2, 4$) to a 3-cell B_j such that, among other things, B_j meets ∂T_j in an annulus and meets each of C_1 and C_3 in a standard arc spanning B_j. Then E_j splits B_j into two 3-cells and the closure of $\partial B_j \smallsetminus T_j$ consists of two parallel copies, E_j^+ and E_j^-, of E_j.

Suppose $f : I^2 \to T \smallsetminus A_1$ with $f(\partial I^2) \subset \partial T$. Find a 2-dimensional PL ∂-manifold M in $\operatorname{Int} I^2$ such that

$$f^{-1}(E_2 \cup E_4) \subset \operatorname{Int} M \subset M \subset f^{-1}(B_2 \cup B_4).$$

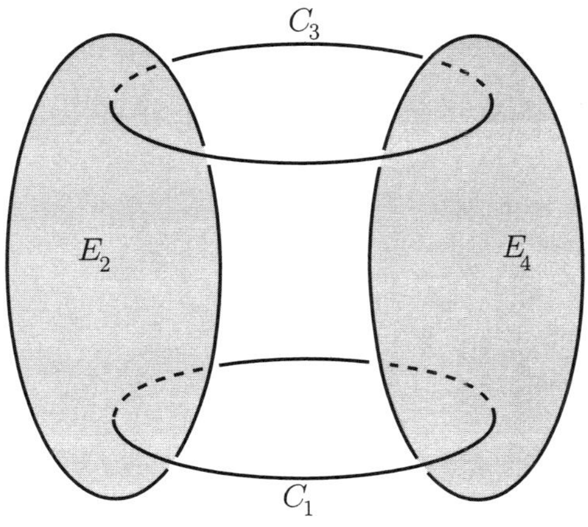

Figure 2.4. Four linked circles

Identify the component P of $I^2 \setminus \operatorname{Int} M$ containing ∂I^2, then name the simple closed curves J_i of $P \cap M$ as well as the 2-cells $D_i \subset \operatorname{Int} I^2$ bounded by J_i $(i = 1, \ldots, k)$.

Fix i. There exists a j (either 2 or 4) such that $f(J_i) \subset B_j \setminus E_j$, so $f(J_i)$ may be homotoped to lie entirely in either E_j^+ or E_j^-. This homotopy takes place inside of $B_j \setminus E_j$, so it may be extended to all of I^2 and we may assume that $f(J_i)$ is contained in either E_j^+ or E_j^-; to be specific let us say $f(J_i) \subset E_j^+$. The map $f|D_i$ shows that $f|J_i$ is nullhomotopic in $\mathbb{R}^3 \setminus (T_1 \cup T_3)$, and Claim 2.1.6 implies that f can be redefined on D_i so that its image lies in $E_j^+ \setminus (C_1 \cup C_3)$.

If this process is carried out for each i, then f will be replaced by a map $F : I^2 \to T \setminus (C_1 \cup E_2 \cup C_3 \cup E_4)$ with $F|\partial I^2 = f|\partial I^2$. Now ρF reveals that $f|\partial I^2$ is nullhomotopic in ∂T. □

This completes the construction of Antoine's necklace. As mentioned earlier, some variation is allowed in the number of solid tori used at each stage of the construction: each solid torus at one stage may be replaced by more than four solid tori at the next stage, and it is even permissible for the number of solid tori to vary from one stage to another and from one link to another. For this reason Antoine's necklace can be regarded as one specific member of a whole class of Antoine Cantor sets. There are two conditions that must be satisfied by the construction of the objects in this class. First, each component τ of the k-th stage ∂-manifold A_k must be an unknotted solid torus and all the next-stage solid tori in τ must be simply linked in a chain that winds exactly once around τ. This will ensure that $\pi_1(\mathbb{R}^3 \setminus A) \neq \{1\}$. The second condition is that each solid torus at one stage must be replaced by enough solid tori at the next stage so that the diameters of the components of A_k approach 0 as $k \to \infty$. This second condition ensures that $\cap_{k=1}^{\infty} A_k$ is totally disconnected and therefore a Cantor set. Figure 2.5 shows three consecutive stages in a typical construction.

Figure 2.5. An Antoine Cantor set

Different Antoine Cantor sets may be inequivalently embedded. In fact, varying the number of links in the Antoine construction results in an uncountable number of different equivalence classes of embeddings of the Cantor set in $\mathbb{R}^3$ (Sher, 1968).

We now turn our attention to the construction of wild spheres.

Example 2.1.7. *There exist wild 2-spheres in* $\mathbb{R}^3$.

We will describe two different examples. The first is based on Antoine's necklace. Start with a round 3-cell F_0 in $\mathbb{R}^3$ that is disjoint from A. Add a tube F_1 that connects the first 3-cell to the solid torus T. The tube is solid, so $F_0 \cup F_1$ is another 3-cell. Add thin tubes in T to connect $F_1 \cap T$ to the four components of A_1; then do the same at later stages, adding 4^k tubes at stage $k+1$. There should be four disjoint tubes in each component of A_k as indicated in Figure 2.6. Define the *Antoine 3-cell* to be F, the union of all the tubes together with Antoine's necklace A, and define the *Antoine sphere* to be ∂F.

It is not difficult to construct a homeomorphism from a 3-cell to F. We will not explicitly describe that construction, although we will give some indication of how a similar homeomorphism is constructed when we describe Alexander's horned sphere, below. The Antoine sphere bounds a topological

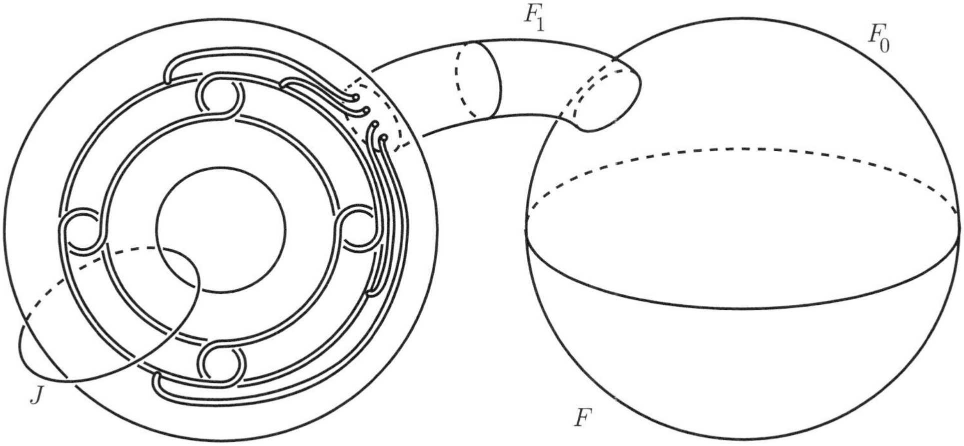

Figure 2.6. The Antoine sphere

3-cell on the inside, but the exterior is not simply connected. In order to see that the exterior is not simply connected, observe that F can be constructed so that it does not intersect the loop $J \subset \mathbb{R}^3 \setminus F$ shown in Figure 2.6. Since J is homotopically essential in the complement of Antoine's necklace, it is homotopically essential in $\mathbb{R}^3 \setminus \partial F$ as well.

Note that we have not only constructed a wild 2-sphere, but have also constructed a wild 3-cell F. Any arc in F that contains Antoine's necklace must be wild since the loop J represents a nontrivial loop in the complement. Similarly any 2-cell in F that contains A must be wild. Hence we have the following.

Example 2.1.8. *There exist wild cells of dimension 1, 2, and 3 in* $\mathbb{R}^3$.

We now construct a second wild 2-sphere in $\mathbb{R}^3$, the famous *Alexander horned sphere.* The construction relies on a certain pillbox replacement procedure.

Definition. A *pillbox* is a cylindrical 3-cell C with top disk τ and bottom disk β containing simply linked solid tori T_1 and T_2, with $T_1 \cap \partial C = \tau$ and $T_2 \cap \partial C = \beta$. (See Figure 2.7.)

Lemma 2.1.9. *Let C be a pillbox, let X be a closed subset of $\mathbb{R}^3$ such that $X \cap C = \tau \cup \beta$, and let J be a 1-sphere in $\mathbb{R}^3 \setminus (X \cup C)$ as shown in Figure 2.8. If $\pi_1(J) \to \pi_1(\mathbb{R}^3 \setminus (X \cup C))$ is one-to-one, then $\pi_1(\mathbb{R}^3 \setminus (X \cup C)) \to \pi_1(\mathbb{R}^3 \setminus (X \cup T_1 \cup T_2))$ is also one-to-one.*

Proof. Use Claim 2.1.5 and the technique of proof of Lemma 2.1.4 to show that if J is null-homotopic in $\mathbb{R}^3 \setminus (X \cup T_1 \cup T_2)$, then J is null-homotopic in $\mathbb{R}^3 \setminus (X \cup C)$. □

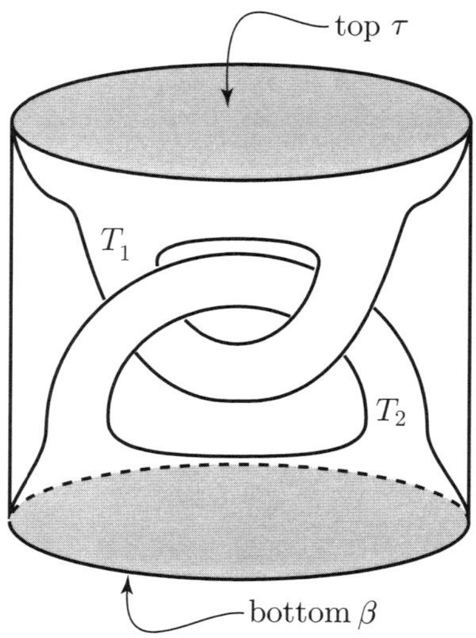

Figure 2.7. A pillbox

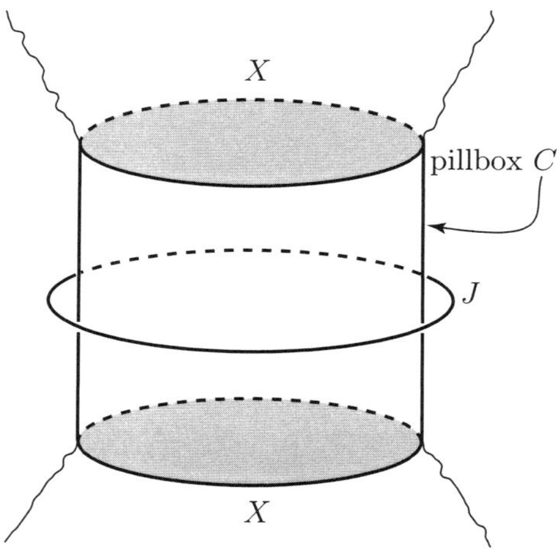

Figure 2.8. Lemma 2.1.9

To begin the construction of the horned sphere, let S_1 be an unknotted solid torus in $\mathbb{R}^3$. We will refer to this solid torus as the first stage in the construction. It is obvious that $\pi_1(\mathbb{R}^3 \setminus S_1) \neq \{1\}$; in fact, the loop J shown in Figure 2.9 represents a nontrivial element of $\pi_1(\mathbb{R}^3 \setminus S_1)$. Inside S_1 identify a pillbox C_1 as indicated in Figure 2.9. Let D_1 denote the complementary 3-cell in S_1. Then $S_1 = C_1 \cup D_1$ and $C_1 \cap D_1 = \tau_1 \cup \beta_1$, the top and bottom of the pillbox.

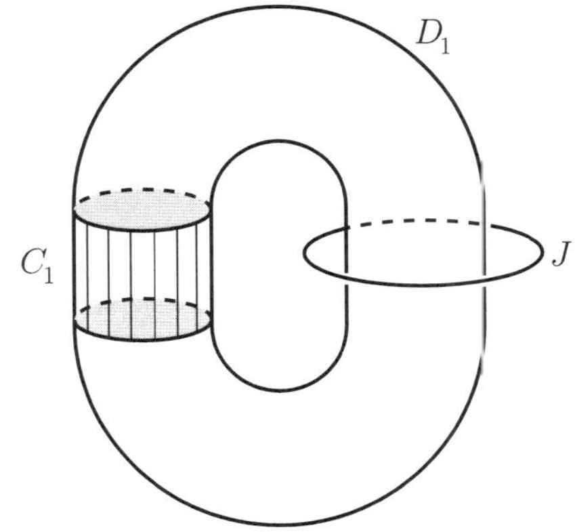

Figure 2.9. The first stage in the construction

Let T_{11} and T_{12} be the two distinguished solid tori in the pillbox C_1. Define the second stage of the construction by $S_2 = (S_1 \smallsetminus C_1) \cup (T_{11} \cup T_{12})$. It follows from Lemma 2.1.9 that the loop J represents a nontrivial element of $\pi_1(\mathbb{R}^3 \smallsetminus S_2)$.

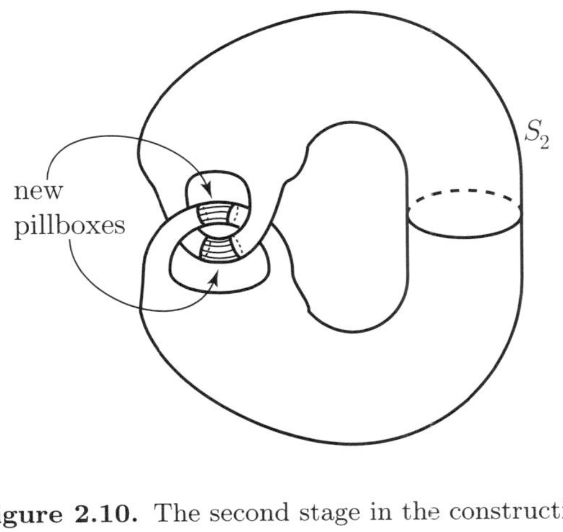

Figure 2.10. The second stage in the construction

Inside T_{11} and T_{12} identify two new pillboxes C_{11} and C_{12} as indicated in Figure 2.10. Inside each of those two pillboxes we can identify two distinguished solid tori. Define the third stage S_3 to be the solid object obtained from S_2 by removing the two new pillboxes and replacing them with the four solid tori just described. This construction is continued inductively, with arrangements to ensure that the diameters of pillboxes at stage $k \geq 2$ is bounded by 2^{-k}. The process results in a nested sequence of compact 3-dimensional solids $S_1 \supset S_2 \supset S_3 \supset \ldots$. Define the *Alexander 3-cell* to be the compact set $B = \cap_{i=1}^{\infty} S_i$ and define the *Alexander horned sphere* to be the boundary of B. Figure 2.11 shows a drawing of Alexander's horned sphere. Color Plates 2–4 display photographs of physical models of the first few stages in the construction.

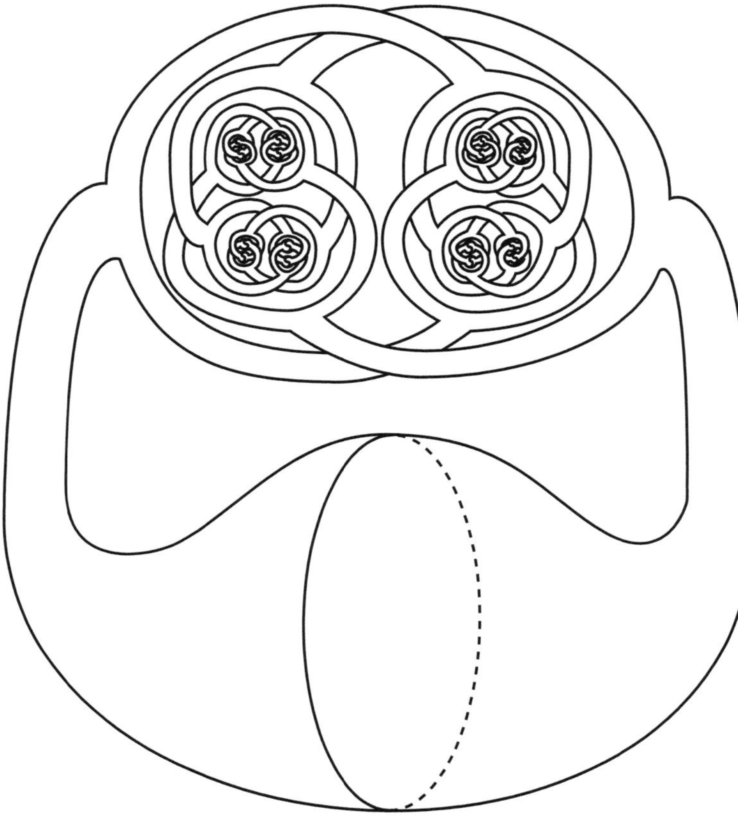

Figure 2.11. The Alexander horned sphere

In order to complete the proof that the horned sphere has the stated properties, we must show two things: first, B is a topological 3-cell and second, $\pi_1(\mathbb{R}^3 \smallsetminus B) \neq \{1\}$. If the loop J shown in Figure 2.9 were inessential in $\mathbb{R}^3 \smallsetminus B$, then compactness of the track of the shrinking homotopy would provide an n such that J is inessential in $\mathbb{R}^3 \smallsetminus S_n$. But induction and Lemma 2.1.9 show that J is essential in $\mathbb{R}^3 \smallsetminus S_n$ for every n. Hence J is essential in $\mathbb{R}^3 \smallsetminus B$.

To see that B is a 3-cell, it helps to think of it as a union rather than an intersection. At the nth stage of the construction we have 2^{n-1} pillboxes, each of which contains two distinguished solid tori. Each of these 2^n solid tori is then divided into a pillbox and a complementary 3-cell (the 3-cell D_1 at the first stage). Let D_n denote the union of the 2^{n-1} complementary

3-cells at the nth stage. Inductively define $B_1 = D_1$ and $B_n = B_{n-1} \cup D_n$. Observe that

$$B = \overline{\cup_{n=1}^{\infty} B_n}.$$

It is relatively simple to use the B_n to construct a homeomorphism from a 3-cell to B. The construction is indicated in Figure 2.12, which shows the domain of the homeomorphism. Map the large region at the bottom to D_1, map the union of the next two regions to D_2, map the union of the next four regions to D_3, etc. Note that $B \smallsetminus \cup_{n=1}^{\infty} B_n$ is a Cantor set. We will call this Cantor set the *Alexander Cantor set.* This completes the construction of the Alexander horned sphere.

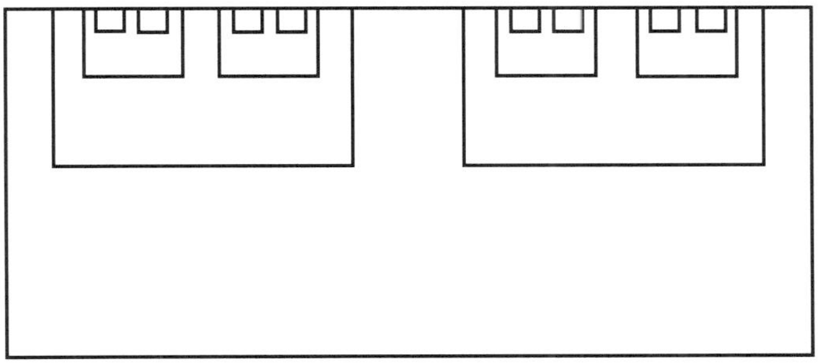

Figure 2.12. Construction of a homeomorphism from a 3-cell to B

Remark. It is interesting to compare the wildness of the two embeddings of S^2 that were constructed in this section. The Antoine sphere and the Alexander horned sphere are alike in that each has one complementary domain whose closure is a 3-cell while the other complementary domain fails to be simply connected. The two embeddings are also alike in that each of them is locally flat except at the points of a Cantor set. There is, however, a significant qualitative difference in the wildness exhibited by the two embeddings. In each case the Cantor set of wild points can be considered either as a subset of the 2-sphere itself or as a subset of $\mathbb{R}^3$. Antoine's necklace is flat when considered as a subset of the Antoine sphere, but it is wild when considered as a subset of $\mathbb{R}^3$. By contrast, the Alexander Cantor set is twice flat in the sense that it is flat both as a subset of the Alexander 2-sphere and as a subset of $\mathbb{R}^3$ (Exercise 2.1.3).

It is clear from the definition that local flatness is an open condition, so the set of points at which an embedding is wild is always a closed set. We will refer to this set as the *wild set* of the embedding and a point in this set is called a *wild point* of the embedding. The wild set of each of the spheres constructed in this section is a Cantor set. Later in the chapter we will construct wild embeddings of the 2-sphere in $\mathbb{R}^3$ whose wild sets are as small as a single point or as large as the entire sphere.

High-dimensional examples are constructed by suspension.

Example 2.1.10. *There exist wild cells and spheres in S^n for all $n \geq 3$.*

Proof. For $n > 3$, iteration of the suspension operator applied to the examples constructed earlier in the section produces examples of nonflat codimension-two and codimension-one spheres in S^n, as well as nonflat cells in codimensions 0, 1, and 2. By Lemma 1.4.1, cells whose complements have nontrivial fundamental groups suspend to cells with the same property and the same codimension. As a result, the existence of wild cells and spheres in all dimensions $n \geq 3$ follows immediately from the 3-dimensional examples. □

The wild cells constructed in Example 2.1.10 have dimensions n, $n-1$, and $n-2$ and are locally flat except at the points of the iterated suspension of a Cantor set. Later in the chapter we will use other methods to construct everywhere wild cells in $\mathbb{R}^n$ of all codimensions.

Historical Notes. Antoine's necklace and the Alexander horned sphere are named for their inventors, L. Antoine and J. W. Alexander, respectively. The discovery of these two examples dates back to the 1920s; see (Antoine, 1921) and (Alexander, 1924b). Alexander pointed out (1924c) that Antoine's construction of a wild Cantor set could also be used to construct a wild 2-sphere, as detailed in this section.

Exercises

2.1.1. Prove Lemma 2.1.3.

2.1.2. Let H be a compact, 2-dimensional ∂-manifold in $\mathbb{R}^2$. Show that for each map $f : H \to T \smallsetminus A_1$ with $f(\partial H) \subset \partial T$ there exists a map $F : H \to \partial T$ with $F|\partial H = f|\partial H$. [Hints: Show that every loop in ∂T is null-homotopic in $\mathbb{R}^3 \smallsetminus (T_i \cup T_{i+1} \cup T_{i+2})$; then show that there exists a map $f' : H \to T \smallsetminus A_1$ such that $f'|\partial H = f|\partial H$ and $f'(H) \cap (E_2 \cup E_4) = \emptyset$.]

2.1.3. The Alexander Cantor set $\mathcal{A}$ is tame in $\mathbb{R}^3$. [Hint: Alter the embedding of $\mathcal{A}$ so linear projection to the axis perpendicular to the plane of the page in Figure 2.11 restricts to an embedding on $\mathcal{A}$.]

2.1.4. Let C be a Cantor set in a connected n-manifold M. Construct an arc α satisfying $C \subset \alpha \subset M$.

2.1.5. Let C be a Cantor set in a connected n-manifold M, $n > 2$. Construct an n-cell B satisfying $C \subset \partial B \subset B \subset M$, with C flat as a subset of ∂B.

2.1.6. Construct a 2-sphere in $\mathbb{R}^3$ such that neither complementary domain is simply connected.

2.1.7. Every compact, totally disconnected subset C of an n-manifold M has a neighborhood $U \supset C$ that can be embedded in $\mathbb{R}^n$.

2.2. Function spaces

Several subsets of the function space $C(X,Y)$—the space of all continuous functions of X to Y—will prove useful. For this discussion, one should assume that Y admits a complete and bounded[1] metric d and that $C(X,Y)$ is endowed with the complete metric ρ defined by

$$\rho(f,g) = \mathrm{lub}\{d(f(x), g(x)) \mid x \in X\}.$$

We will be interested in the following subsets of $C(X,Y)$:

Surj(X,Y) = the set of all mappings of X onto Y (the surjections);

Emb(X,Y) = the set of all embeddings of X in Y; and

Homeo(X,Y) = the set of all homeomorphisms of X onto Y.

In case X and Y both are simplicial complexes, we will use $C_{\mathrm{PL}}(X,Y)$, $\mathrm{Emb}_{\mathrm{PL}}(X,Y)$, $\mathrm{Homeo}_{\mathrm{PL}}(X,Y)$ and $\mathrm{Surj}_{\mathrm{PL}}(X,Y)$ to denote the collection of all PL mappings of the specified type. For the Main Problem to be non-vacuous, in this notation we must have that Emb(X,Y) is nonempty. Correspondingly, to solve the Taming Problem, we must determine which elements of Emb(X,Y) are equivalent to elements of $\mathrm{Emb}_{\mathrm{PL}}(X,Y)$, whereas to answer the PL Unknotting Problem, we must decide which elements of $\mathrm{Emb}_{\mathrm{PL}}(X,Y)$ are equivalent.

Lemma 2.2.1. *Let (X, d_X) be a compact metric space, (Y, d_Y) a complete metric space, and $C(X,Y)$ the space of all continuous functions of X to Y with metric ρ, as above. Then* Surj(X,Y) *is a closed subset of $C(X,Y)$. Moreover,* Emb(X,Y) *and* Homeo(X,Y) *are G_δ-subsets of $C(X,Y)$.*

Proof. Showing that the complement of Surj(X,Y) is open in $C(X,Y)$ is straightforward (even when X is non-metrizable). In order to confirm that Emb(X,Y) is a G_δ-subset, consider the set of $(1/k)$-mappings

$$A_k = \{f \in C(X,Y) \mid \mathrm{diam}\, f^{-1}f(x) < 1/k \text{ for each } x \in X\}.$$

One can prove that A_k is open in the function space $C(X,Y)$ by producing, for any $f \in A_k$, a corresponding $\eta = \eta(f) > 0$ such that $d_Y(f(x_1), f(x_2)) > \eta$ whenever $x_1, x_2 \in X$ satisfy $d_X(x_1, x_2) \geq 1/k$. Each $g \in C(X,Y)$ with $\rho(g,f) < \eta/2$ belongs to A_k, since then $d_Y(g(x_1), g(x_2)) > 0$ whenever

[1] Recall that if d' is an arbitrary complete metric on Y, then the rule $d(y,y') = \min\{1, d'(y,y')\}$ defines a complete and bounded metric on Y that is equivalent to the original in the sense that they induce the same topology.

$d_X(x_1, x_2) \geq 1/k$. It follows that $\cap_{k=1}^{\infty} A_k = \text{Emb}(X, Y)$ is a G_δ-set in $C(X, Y)$. The G_δ-property also holds for $\text{Homeo}(X, Y)$ because

$$\text{Homeo}(X, Y) = \text{Surj}(X, Y) \cap \text{Emb}(X, Y). \qquad \square$$

The point of Lemma 2.2.1, of course, is that these subsets all admit complete metrics and, therefore, have the Baire property.

One should observe that ordinarily $\text{Homeo}(X, Y)$ fails to be closed in $C(X, Y)$. Consequently, an arbitrary Cauchy sequence $\{h_k\}$ in $\text{Homeo}(X, Y)$ need not converge to a homeomorphism, although it will always converge in $C(X, Y)$ to a surjection. Later we will want to know conditions under which a Cauchy sequence of homeomorphisms *does* converge to a homeomorphism, and, conveniently, one can recover appropriate conditions from the proof of Lemma 2.2.1. Here is an all-important philosophical perspective that should be extracted: when constructing a sequence of homeomorphisms $h_k : X \to Y$ recursively, if for each k one can impose control limiting $\rho(h_{k+1}, h_k)$ that is specified after $h_1, \ldots, h_{k-1}$ and h_k have all been determined, then one can construct the entire sequence $\{h_i\}$ so that it converges to a homeomorphism. The principle is embodied in the next Proposition.

Proposition 2.2.2. *Let (X, d_X) be a compact metric space and (Y, d_Y) a complete metric space. Suppose $\{h_k \mid k = 1, 2, \ldots\}$ is a sequence of embeddings of X in Y and $\{\epsilon_k \mid k = 0, 1, 2, \ldots\}$ is a sequence of positive numbers such that for $k > 0$*

(a) $\epsilon_k < \epsilon_{k-1}/2$;

(b) $d_Y(h_k(x_1), h_k(x_2)) \geq 4\epsilon_k$ *for all* $x_1, x_2 \in X$ *with* $d_X(x_1, x_2) \geq 1/k$, *and*

(c) $\rho(h_{k+1}, h_k) < \epsilon_k$.

Then $\{h_k\}$ converges in $C(X, Y)$ to an embedding $h_\infty : X \to Y$. Moreover, if each h_k is a homeomorphism, then so is h_∞.

Proof. Exercise 2.2.1. $\square$

Remark. Conditions (a) and (c) assure that $\{h_k\}$ forms a Cauchy sequence in $C(X, Y)$. When (b) is added to the mix, the Conditions mimic similar ones appearing in the proof of Lemma 2.2.1 that force all successive h_{k+i} to belong to the open subset A_k of $C(X, Y)$.

Theorem 2.2.3. *If X is a compact metric space of dimension at most k, then* $\text{Emb}(X, \mathbb{R}^{2k+1})$ *is a dense G_δ-subset of* $C(X, \mathbb{R}^{2k+1})$.

Proof. The fact that $\text{Emb}(X, \mathbb{R}^{2k+1})$ is a G_δ-set follows from Lemma 2.2.1. See (Munkres, 1975, Theorem 7.9.6) for a proof of density. The full theorem may also be found on page 56 of (Hurewicz and Wallman, 1948). $\square$

Theorem 2.2.4. *If K is a finite k-complex and M is a PL m-manifold, $2k < m$, then* $\mathrm{Emb}_{PL}(K, M)$ *is dense in* $C(K, M)$.

Proof. Munkres's argument, which establishes density of $\mathrm{Emb}_{\mathrm{PL}}(K, \mathbb{R}^m)$ in $C(K, \mathbb{R}^m)$, can be applied in one chart at a time to give the result—see (Rourke and Sanderson, 1972, Theorem 5.4). □

Remark. Theorem 2.2.3 is sharp. For $k = 1$ there are famous examples, reproduced in Munkres, of finite 1-complexes that do not embed in $\mathbb{R}^2$. In Chapter 5 we will prove the more general result that the k-skeleton of a $(2k + 2)$-simplex cannot be embedded in $\mathbb{R}^{2k}$.

Exercise

2.2.1. Prove Proposition 2.2.2.

2.3. Shrinkable decompositions and the Bing shrinking criterion

Many wild embeddings arise from decompositions: a tame embedding into a manifold is followed by a quotient of the ambient manifold. It becomes important then to have tools available for detecting when the quotient space is a manifold. In this section we develop tools for that purpose.

We begin with a quick review of some basic definitions. A *decomposition* G of a space X is simply a partition of X (ordinarily into closed sets). The *decomposition space* (= quotient space) is the space X/G whose points are the elements of G. There is a natural quotient map $\pi : X \to X/G$ and X/G is assigned the quotient topology. (A subset U of X/G is defined to be open if $\pi^{-1}(U)$ is open in X.) An open set $V \subset X$ is said to be *G-saturated* if it is the union of elements of G; thus, $U \subset X/G$ is open if and only if it is the image of a G-saturated open subset of X.

During the 1950s R. H. Bing introduced and exploited several forms of a remarkable condition now called the *Bing shrinkability criterion* or *Bing shrinking criterion.* It prompted a major change in decomposition theory, shifting the focus from the decomposition space back to the source. The need for a fresh point of view arose from the study of decomposition maps $q : S^3 \to Q$ because, even when it appeared certain that Q had to be homeomorphic to S^3, one then had no effective characterization of S^3 to exploit for establishing the topological equivalence. The shrinkability criterion aimed at realizing Q as the homeomorphic image of the known source space, a realization achieved as the end result of manipulations in the source on the decomposition elements.

In its most general form, the criterion is expressed as follows: a partition G of a space X is *shrinkable* if and only if the following condition is satisfied.

Shrinkability criterion. For each G-saturated open cover $\mathcal{U}$ of X and each arbitrary open cover $\mathcal{V}$ of X there is a homeomorphism h of X onto itself satisfying

(a) for each $x \in X$ there exists $U \in \mathcal{U}$ such that $x, h(x) \in U$, and

(b) for each $g \in G$ there exists $V \in \mathcal{V}$ such that $h(g) \subset V$.

In other words, the homeomorphism h must shrink elements of G to small size, where "small" is determined by $\mathcal{V}$, under an action that is limited by $\mathcal{U}$.

Experience suggests that the decomposition space associated with a shrinkable decomposition is often homeomorphic to the source space S. To guarantee that this is true, additional restrictions, like local compactness or complete metrizability, must be imposed on S. This section explores some relatively coarse aspects of those restrictions. A good starting point is the compact metric case.

Definition. Let ρ denote a complete metric on $C(X, Y)$, where X and Y are compact metric spaces. A surjection $f : X \to Y$ is a *near-homeomorphism* if for each $\epsilon > 0$ there exists $h \in \text{Homeo}(X, Y)$ such that $\rho(h, f) < \epsilon$.

Lemma 2.3.1. *Let X and Y be compact metric spaces. If $f \in \text{Surj}(X, Y)$ and $h \in \text{Homeo}(X, X)$, then $\rho(f, fh) = \rho(f, fh^{-1})$*

Proof. $\rho(f, fh^{-1}) = \rho(fhh^{-1}, fh^{-1}) = \rho(fh, f)$. □

Theorem 2.3.2 (Shrinkability criterion in the compact metric case). *Let X, Y be compact metric spaces and ρ a metric on $C(X, Y)$. Then $f \in \text{Surj}(X, Y)$ is a near-homeomorphism if and only if for each $\epsilon > 0$ there exists $h \in \text{Homeo}(X, X)$ satisfying:*

(a) $\rho(f, fh) < \epsilon$, *and*

(b) $\operatorname{diam} h(f^{-1}(y)) < \epsilon$ *for each* $y \in Y$.

Proof. The forward implication is the easier. Fix a near-homeomorphism f and $\epsilon > 0$. By hypothesis there exists $F \in \text{Homeo}(X, Y)$ with $\rho(F, f) < \epsilon/2$. Uniform continuity of F^{-1} provides $\delta > 0$ such that the image under F^{-1} of each δ-subset of Y has diameter less than ϵ. Again, there exists $F^* \in \text{Homeo}(X, Y)$ with $\rho(F^*, f) < \min\{\epsilon/2, \delta/2\}$. For each $y \in Y$, $F^*(f^{-1}(y))$ lies in the $(\delta/2)$-neighborhood of y, implying that $\operatorname{diam} F^*(f^{-1}(y)) < \delta$. Define $h \in \text{Homeo}(X, X)$ as $F^{-1}F^*$. The choice of δ guarantees that h satisfies condition (b). To see that h satisfies condition (a) as well, note that

$$\begin{aligned}\rho(f, fh) &\le \rho(f, F^*) + \rho(F^*, fF^{-1}F^*)\\ &< \epsilon/2 + \rho(F(F^{-1}F^*), f(F^{-1}F^*))\\ &= \epsilon/2 + \rho(F, f)\\ &< \epsilon/2 + \epsilon/2 = \epsilon.\end{aligned}$$

To prove the reverse implication, fix $f \in \mathrm{Surj}(X, Y)$ satisfying shrinkability conditions (a) and (b) and let A denote the closure in $C(X, Y)$ of the subset consisting of all maps fh^{-1}, where $h \in \mathrm{Homeo}(X, X)$. For $n = 1, 2, \ldots$ define

$$A_n = \{\varphi \in A \mid \operatorname{diam} \varphi^{-1}(y) < 1/n \text{ for each } y \in Y\}.$$

The claim is that each A_n is open and dense in A. Openness follows exactly as in the proof of Lemma 2.2.1. To prove denseness, we start with $\varphi \in A$ and $\eta > 0$ and produce $\varphi^* \in A_n$ such that $\rho(\varphi, \varphi^*) < \eta$. To do so, first obtain fh^{-1}, $h \in \mathrm{Homeo}(X, X)$, such that $\rho(\varphi, fh^{-1}) < \eta/2$ and then apply uniform continuity of h and the shrinkability criterion to obtain another $H \in \mathrm{Homeo}(X, X)$ for which $\rho(f, fH) < \eta/2$ and $\operatorname{diam} Hf^{-1}(y)$ is so small that $\operatorname{diam} hHf^{-1}(y) < 1/n$. Clearly the map $\varphi^* = fH^{-1}h^{-1}$ satisfies $\operatorname{diam}(\varphi^*)^{-1}(y) < 1/n$ for each $y \in Y$. Moreover,

$$\begin{aligned}\rho(\varphi, \varphi^*) &= \rho(\varphi, fH^{-1}h^{-1})\\ &\le \rho(\varphi, fh^{-1}) + \rho(fh^{-1}, fH^{-1}h^{-1})\\ &\le \rho(\varphi, fh^{-1}) + \rho(f, fH^{-1})\\ &= \rho(\varphi, fh^{-1}) + \rho(f, fH) \qquad \text{(by Lemma 2.3.1)}\\ &< \eta/2 + \eta/2 = \eta.\end{aligned}$$

To conclude the argument, observe that A itself is complete, being a closed subset of the complete metric space $\mathrm{Surj}(X, Y)$. By the Baire Category Theorem, $\cap_n A_n$ is dense in A, and $\cap_n A_n \subset \mathrm{Homeo}(X, Y)$, as before. Thus, $f \in A$ can be approximated by homeomorphisms $F \in \cap_n A_n$. □

Theorem 2.3.3. *Let X be a compact metric space and $f \in \mathrm{Surj}(X, Y)$. Then f is a near-homeomorphism if and only if, for each $\epsilon > 0$, there exists $\mu \in \mathrm{Surj}(X, X)$ such that $\{f^{-1}(y) \mid y \in Y\} = \{\mu^{-1}(x) \mid x \in X\}$ and $\rho(f, f\mu) < \epsilon$.*

Proof. First assume $\mu \in \mathrm{Surj}(X, X)$ satisfies $\{f^{-1}(y) \mid y \in Y\} = \{\mu^{-1}(x) \mid x \in X\}$ and $\rho(f, f\mu) < \epsilon < 1$. Then $F = f\mu^{-1}$ defines a homeomorphism of

X onto Y. Moreover, for each $x \in X$ there exists $x^* \in \mu^{-1}(x)$ and

$$\begin{aligned}\rho(f(x), F(x)) &= \rho(f(x), f\mu^{-1}(x)) \\ &= \rho(f(x), f(x^*)) \\ &= \rho(f\mu(x^*), f(x^*)) \\ &\leq \rho(f\mu, f) < \epsilon.\end{aligned}$$

Thus, $\rho(f, F) < \epsilon$ and f is a near-homeomorphism.

Conversely, assume f is a near-homeomorphism. Given ϵ, $0 < \epsilon < 1$, identify $F \in \text{Homeo}(X, Y)$ satisfying $\rho(f, F) < \epsilon$, and define μ as $\mu = F^{-1}f$. Clearly then $\{f^{-1}(y) \mid y \in Y\} = \{\mu^{-1}(x) \mid x \in X\}$, and

$$\rho(f, f\mu) = \rho(f, fF^{-1}f) = \rho(FF^{-1}f, fF^{-1}f) = \rho(F, f) < \epsilon,$$

as required. □

Technical needs make it advantageous to impose further controls on the shrinking process. To that end, given $f \in \text{Surj}(X, Y)$ let N_f denote the *nondegeneracy set* of f, defined by

$$N_f = \left\{x \in X \mid f^{-1}f(x) \neq \{x\}\right\}.$$

Furthermore, given a closed subset C of X missing N_f, say that the induced partition $G_f = \{f^{-1}(y) \mid y \in Y\}$ of X is *shrinkable fixing* C if shrinking homeomorphisms $h : X \to X$ fulfilling the shrinkability criterion can be obtained that keep each point of C fixed, and say that G_f is *strongly shrinkable* if, for every closed set $C \subset X$ with $C \cap N_f = \emptyset$, G_f is shrinkable fixing C.

By restricting the action on C, one can readily adapt the proof given for Theorem 2.3.2 to establish the following, which lends itself to quick application of shrinkability in the locally compact metric case.

Theorem 2.3.4. *Suppose X is a compact metric space, $f \in \text{Surj}(X, Y)$, and C is a closed subset of X with $C \cap N_f = \emptyset$. Then f can be approximated by homeomorphisms agreeing with f on C if and only if G_f is shrinkable fixing C.*

A mapping $f : X \to Y$ is *proper* if, for each compact subset C of Y, $f^{-1}(C)$ is compact. Several key results concerning near-homeomorphisms between compact metric spaces have analogs pertaining to proper mappings between locally compact metric spaces.

Theorem 2.3.5. *Suppose (X, d_X) and (Y, d_Y) are locally compact metric spaces. Then a proper, surjective mapping $f : X \to Y$ can be approximated (in the space of maps $X \to Y$ endowed with the compact-open topology) by homeomorphisms if for each compact subset C of Y and each $\epsilon > 0$ there exists a homeomorphism $h : X \to X$ satisfying*

(a) $d_Y(f(x), fh(x)) < \epsilon$ *for each* $x \in f^{-1}(C) \cup h^{-1}f^{-1}(C)$, *and*

(b) $\operatorname{diam} hf^{-1}(c) < \epsilon$ *for each* $c \in C$.

Proof. Let X^* and Y^* denote the one-point compactifications of X and Y, respectively, and $f^* : X^* \to Y^*$ the obvious extension of f. Properness of f is equivalent to continuity of f^*. Since X and Y are locally compact and second countable, X^* and Y^* are compact metric spaces. The point is that f can be approximated (in the compact-open topology) by homeomorphisms if f^* can be approximated in $C(X^*, Y^*)$ by homeomorphisms preserving the points at infinity, which reduces Theorem 2.3.5 to Theorem 2.3.4. □

Historical Notes. The shrinking criterion is a profound insight of R. H. Bing. It appeared implicitly in (Bing, 1952) and developed over time into a general method; see (Bing, 1957a), for example, or (van Mill, 1989, §6.1).

Exercises

2.3.1. Every proper continuous mapping $f : X \to Y$ between metric spaces is a closed mapping.

2.3.2. Let X, Y be locally compact metric (or even Hausdorff) spaces and $X^* = X \cup \{\infty\}$, $Y^* = Y \cup \{\infty'\}$ their one-point compactifications. Then $f \in C(X, Y)$ is proper if and only if the obvious extension $f^* : X^* \to Y^*$ (where $f^*(\infty) = \infty'$) is continuous.

2.3.3. Let C denote the Cantor set. Show that each $f \in \operatorname{Surj}(C, C)$ is a near-homeomorphism. [Hint: any subset $X \subset C$ that is both open and closed in C is homeomorphic to C.]

2.4. Cellular sets and the Generalized Schönflies Theorem

Next we identify a crucial property possessed by the point preimages of a near-homeomorphism of manifolds. The first application, later in the section, will be the proof of a Generalized Schönflies Theorem. Historically this argument was an early signal of the crucial relationship between topological embeddings and decompositions of manifolds.

Definition. A subset X of $\mathbb{R}^n$ (or, more generally, of an n-manifold) is said to be *cellular* if there exists a sequence $\{B_i\}$ of n-cells in $\mathbb{R}^n$ such that $B_{i+1} \subset \operatorname{Int} B_i$ and $X = \cap B_i$. Alternatively, a compact $X \subset \mathbb{R}^n$ is cellular if each neighborhood U of X contains an n-cell B such that $X \subset \operatorname{Int} B \subset B \subset U$. As yet another possibility, a compact $X \subset \mathbb{R}^n$ is cellular if and only if it has arbitrarily small neighborhoods homeomorphic to $\mathbb{R}^n$.

Cellular sets are compact and connected, but they need not be locally connected. (Consider the $\sin(1/x)$-continuum in $\mathbb{R}^2$, Figure 2.13.)

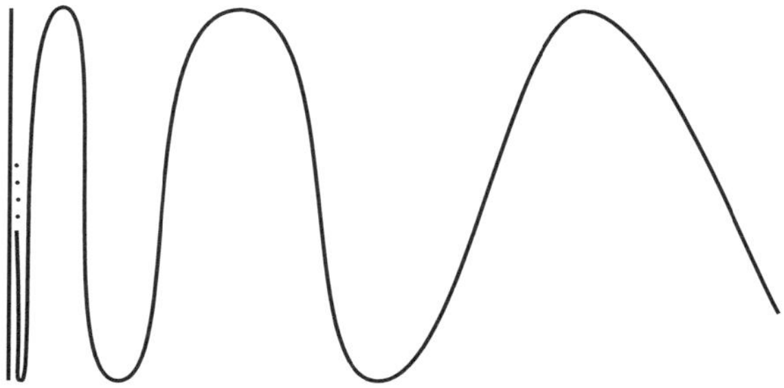

Figure 2.13. The $\sin(1/x)$-continuum

Definition. A map $f : M \to Y$ defined on an n-manifold M is said to be a *cellular map* if $f^{-1}(y)$ is a nonempty cellular set in M for every $y \in Y$.

We use $\mathrm{Cell}(M, Y)$ to denote the set of all cellular maps. Note that

$$\mathrm{Cell}(M, Y) \subset \mathrm{Surj}(M, Y) \subset C(M, Y).$$

Cellular maps defined on manifolds and near-homeomorphisms are closely linked; in fact, we will see that under various special conditions the two kinds of maps are the same. The next theorem asserts that cellularity of point preimages is a necessary condition for a map defined on an n-manifold to be a near-homeomorphism. It is not, however, a sufficient condition in general: the quotient map defining the famous dogbone space (Bing, 1957b) is a counterexample, but that example is too specialized for treatment here.

Proposition 2.4.1. *If $f \in \mathrm{Surj}(X, M^n)$ is a near-homeomorphism, M^n is a compact n-manifold, and $z \in M^n$, then $f^{-1}(z)$ is cellular.*

Proof. Let U be any neighborhood in X of $f^{-1}(z)$, find an n-cell B in M^n satisfying $z \in \mathrm{Int}\, B \subset B \subset M^n \smallsetminus f(X \smallsetminus U)$, and choose $\epsilon > 0$ smaller than both $d(x, M^n \smallsetminus B)$ and $d(B, M^n \smallsetminus f(X \smallsetminus U))$. By hypothesis there exists $F \in \mathrm{Homeo}(X, M^n)$ with $\rho(F, f) < \epsilon/2$. Then $F^{-1}(B)$ is an n-cell in X, and a routine check indicates that $f^{-1}(z) \subset \mathrm{Int}\, F^{-1}(B) \subset F^{-1}(B) \subset U$, so $f^{-1}(z)$ is cellular. □

Corollary 2.4.2. *If $f \in \mathrm{Surj}(M^n, Y)$ is a near-homeomorphism and $y \in Y$, then $f^{-1}(y)$ is cellular.*

A closed subset X of a space M determines a decomposition whose only nondegenerate element is X. We use M/X to denote the associated decomposition space. In this special case cellularity is sufficient to imply that the quotient map is a near-homeomorphism.

Proposition 2.4.3. *If X is a cellular subset of an n-manifold M and Q is the quotient space M/X, then the quotient map $q : M \to Q$ is a near-homeomorphism.*

Proof. Given $\epsilon > 0$, let U denote the $(\epsilon/2)$-neighborhood of $q(X)$ in Q. Apply cellularity of X to obtain an n-cell B such that $X \subset \operatorname{Int} B \subset B \subset q^{-1}(U)$. Equate B with B^n, the standard n-cell; interior to $B = B^n$ construct another round n-cell $B' \supset X$ centered at the origin of $B = B^n$; radially compress B' very near the origin, keeping ∂B pointwise fixed, via a homeomorphism $h^* : B \to B$ such that $\operatorname{diam} h^*(X) < \epsilon$. By Theorem 2.3.2 or 2.3.5 the extension of h^* across $M \setminus B$ via the identity to $h \in \operatorname{Homeo}(M, M)$ shows that q is a near-homeomorphism. □

Definition. An *inverse set* of a map $f : X \to Y$ is a nondegenerate point preimage of f; i.e., an inverse set is a set of the form $f^{-1}(y)$ that contains more than one point.

Corollary 2.4.4. *If U is an open subset of an n-manifold and f is a closed map of $\overline{U}$ onto an n-cell B for which the only inverse set under f is a cellular subset X of U, then $\overline{U}$ is an n-cell.*

The topic of cellularity leads to one of the major themes of this book: the intimate connections between decomposition theory and taming theory. J. W. Cannon probably was the first to stress this theme explicitly, but the connections themselves have been, or should have been, visible from the outset, in the work dating back to the 1950s of R. H. Bing, E. E. Moise, and M. Brown. Brown's important Generalized Schönflies Theorem (1960), one of the first and perhaps the most elegant flatness theorem, displays an aspect of that connection through its dependence on decomposition methods. As noted in §1.1, an $(n-1)$-sphere Σ in S^n is flat if and only if it bounds two n-cells. It is this observation that allows us to make the connection between flatness of $(n-1)$-spheres in S^n and certain decompositions of S^n.

Proposition 2.4.5. *Let Q be an n-cell and let X be a compact subset of* $\operatorname{Int} Q$. *If $f \in C(Q, S^n)$ has X as its only inverse set and $f(\operatorname{Int} Q)$ is open, then X is cellular in Q.*

Proof. Since f is one-to-one on ∂Q, $f(\partial Q)$ is an $(n-1)$-sphere. The inverse set does not touch ∂Q, so the connected set $f(\operatorname{Int} Q)$ must be contained in one of the two complementary domains of $f(\partial Q)$; in particular, f is not onto. Choose a point $z \in S^n \setminus f(Q)$. Then $S^n \setminus \{z\} \cong \mathbb{R}^n$, so $S^n \setminus \{z\}$ has a radial structure centered at the point $f(X)$.

Let U denote an open subset of $\operatorname{Int} Q$ containing X. Then $f(U) = f(\operatorname{Int} Q) \setminus f(Q \setminus U)$ is an open subset of S^n. Use the radial structure of $S^n \setminus \{z\}$ to construct a homeomorphism $\theta : S^n \to S^n$, fixed on some neighborhood V of $f(X)$ and a neighborhood of z, such that $\theta(f(Q)) \subset f(U)$. Define $F : Q \to U$ as the identity on $f^{-1}(V)$ and as $f^{-1}\theta f$ on $Q \setminus X$. Note that F is well defined, continuous, and one-to-one. Thus F is an embedding and $F(Q)$ is an n-cell in U that contains X in its interior. □

Proposition 2.4.6. *If $\psi \in \mathrm{Surj}(S^n, S^n)$ has exactly two inverse sets, then each of them is cellular.*

Proof. Let A and B denote the inverse sets of ψ. We will show that B is cellular. Let Q be an n-cell in S^n containing $A \cup B$ in its interior. Then $\psi(\mathrm{Int}\, Q)$ is open and contains an open set U for which $\psi(A) \in U$ but $\psi(B) \notin U$. Use the structure of S^n as the union of two n-cells to find $\theta \in \mathrm{Homeo}(S^n, S^n)$ such that $\theta(\psi(Q)) \subset U$ and θ fixes some neighborhood V of $\psi(A)$. Define $f \in C(Q, S^n)$ as the identity on $\psi^{-1}(V)$ and as $\psi^{-1}\theta\psi$ on $Q \setminus A$. Then $f(\mathrm{Int}\, Q)$ is open and B is the only inverse set of f. By Proposition 2.4.5, B is cellular. □

A similar proof establishes the following generalization.

Proposition 2.4.7. *If $\psi \in \mathrm{Surj}(S^n, S^n)$ has only a finite number of inverse sets, then $f \in \mathrm{Cell}(S^n, S^n)$.*

The next theorem is the main theorem of the section.

Theorem 2.4.8 (Generalized Schönflies). *If h is an embedding of $S^{n-1} \times [-1, 1]$ in S^n, then $h(S^{n-1} \times \{0\})$ is flat. In particular, the closure of each component of $S^n \setminus h(S^{n-1} \times \{0\})$ is an n-cell.*

Proof. Let A denote the closure of the component of $S^n \setminus h(S^{n-1} \times \{1\})$ that does not contain $\Sigma = h(S^{n-1} \times \{0\})$ and B the closure of the component of $S^n \setminus h(S^{n-1} \times \{-1\})$ that does not contain Σ (see Figure 2.14). Furthermore, let D_A (respectively D_B) denote the closure of that component of $S^n \setminus \Sigma$ containing A (respectively B).

Let $q : S^{n-1} \times [-1, 1] \to Q$ denote the quotient mapping to the quotient space obtained from $S^{n-1} \times [-1, 1]$ by identifying the spheres $S^{n-1} \times \{\pm 1\}$ to (separate) points. As Q is the suspension of S^{n-1}, there exists $\lambda \in \mathrm{Homeo}(Q, S^n)$ sending the image of $S^{n-1} \times \{0\}$ to the standard $S^{n-1} \subset S^n$. Extend the map $\lambda q h^{-1}$ from $h(S^{n-1} \times [-1, 1])$ onto S^n to $f \in \mathrm{Surj}(S^n, S^n)$ by defining $f(A) = \lambda q h^{-1}(h(S^{n-1} \times \{1\}))$ and $f(B) = \lambda q h^{-1}(h(S^{n-1} \times \{1\}))$. Each of A and B is cellular (Proposition 2.4.6) and, therefore, D_A and D_B are n-cells by Corollary 2.4.4. □

An $(n-1)$-manifold Σ contained in an n-manifold M is said to be *bicollared* if there exists an embedding $h : \Sigma \times [-1, 1] \to M$ such that

Photograph by Claire Ferguson

PLATE 1. Tame sphere, Inner Mongolian black granite, 16" diameter, by Helaman Ferguson

Photograph by Claire Ferguson

PLATE 2. Alexander horned wild sphere, bronze, by Helaman Ferguson

Photograph by Claire Ferguson

PLATE 3. Alexander horned wild sphere, patina bronze, 9" diameter, by Helaman Ferguson

Photograph by Claire Ferguson

PLATE 4. Incised torus wild sphere, polished bronze, 9" diameter, by Helaman Ferguson

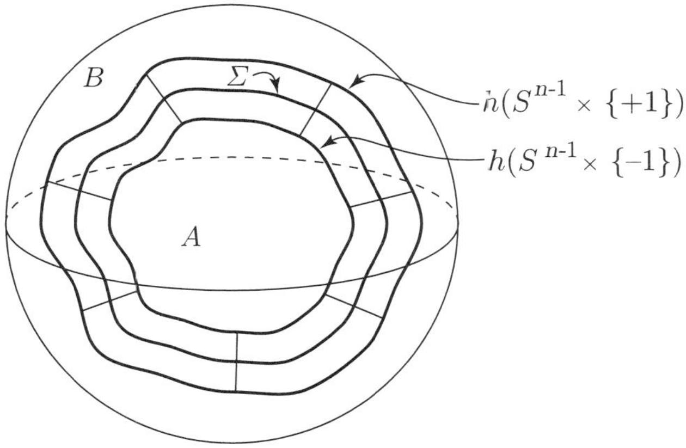

Figure 2.14. Proof of the Generalized Schönflies Theorem

$h(\Sigma \times \{0\}) = \Sigma$. The Generalized Schönflies Theorem can be simply paraphrased using this terminology: every bicollared $(n-1)$-sphere in S^n is flat.

It should be clear from the examples described earlier in the chapter that the bicollar hypothesis is necessary in the Generalized Schönflies Theorem. The complement of any $(n-1)$-sphere embedded in S^n will always have exactly two connected components, but the closure of these complementary domains need not be n-cells. In each of the wild examples constructed earlier, one of their complementary domains was not simply connected. Later in the chapter we will see that the closure of a complementary domain may fail to be an n-cell even if the complementary domain itself is homeomorphic to the interior of an n-cell.

Application of the techniques used in the proof of the Generalized Schönflies Theorem leads to a simple manifold structure theorem.

Proposition 2.4.9. *Any compact n-manifold that can be expressed as the union of two open n-cells is homeomorphic to S^n.*

Proof. Suppose M can be expressed as the union of open sets U and V, each homeomorphic to $\mathbb{R}^n$. Name a homeomorphism $f : V \to \mathbb{R}^n$, and regard the target $\mathbb{R}^n$ as $S^n \smallsetminus \{p\}$. Then f extends to $F \in \mathrm{Surj}(M, S^n)$ by setting $F(M \smallsetminus V) = \{p\}$, and F has $X = M \smallsetminus V$ as its only inverse set. Since X is contained in the interior of some n-cell $Q \subset U$, Proposition 2.4.5 implies that X is cellular. Finally, by Proposition 2.4.3, F is a near-homeomorphism, implying that M is an n-sphere. □

To complete the coverage of the Generalized Schönflies Theorem we show that every locally flat codimension-one sphere is bicollared. It is convenient to work with one-sided collars.

Definition. A subset C of a space X is said to be *collared in* X provided there exists an embedding λ of $C \times [0,1)$ onto an open subset of X such that $\lambda(c,0) = c$ for all $c \in C$, and it is said to be *locally collared* if it can be covered by a collection of open sets (relative to C), each of which is collared in X. The image of λ is called a *collar on* C.

Theorem 2.4.10 (Collaring). *The boundary ∂M of a ∂-manifold M is collared in M.*

Proof. Form a new ∂-manifold M' from $M \cup (\partial M \times [-1,0])$ by identifying each $p \in \partial M$ with $\langle p, 0\rangle \in \partial M \times [-1,0]$. It has the advantage that $\partial M'$, which corresponds to $\partial M \times \{-1\}$, is clearly collared in M'.

We treat only compact ∂M. Cover ∂M by finitely many open subsets $\{W_i\}$, each collared in M, and let V_i denote a collar on W_i in M. Inductively build collars on $\cup_{i=1}^{k} W_i$; the general case quickly reduces to the case $k = 2$. Find $C_i \subset W_i$, closed in $W_1 \cup W_2$, such that $C_1 \cup C_2 = W_1 \cup W_2$. Name a continuous $\gamma_1 : W_1 \cup W_2 \to [-1,0]$ with $\gamma_1(C_1) = -1$ and $\gamma_1(W_2 \smallsetminus W_1) = 0$. After parametrizing $V_1 \cup (W_1 \times [-1,0])$ as $W_1 \times [-1,1)$ in the natural way, define an embedding $\psi_1 : M \to M'$ by declaring $\psi_1 \mid M \smallsetminus V_1 = \text{incl}$, next specifying (for $w \in W_1$)

$$\langle w, 0\rangle \to \langle w, \gamma_1(w)\rangle \text{ and } \langle w, t\rangle \to \langle w, t\rangle \text{ for } t \in [1/2, 1),$$

and then extending linearly to prescribe correspondences between the various intervals $\{w\} \times [0, 1/2]$ and $\{w\} \times [\gamma_1(w), 1/2]$. A similar construction with the constant function $\gamma_2 : W_1 \cup W_2 \to \{-1\}$ gives an embedding $\psi_2 : \text{image}\, \psi_1 \to M'$ for which the composite $\psi_2 \cdot \psi_1$ sends M homeomorphically onto $M \cup (W_1 \cup W_2) \times [-1,0]$. The inverse of $\psi_2 \cdot \psi_1$ exposes a collar on $W_1 \cup W_2$. □

A related argument shows that a closed subset C of a metric space X is collared in X if and only if C is locally collared in X.

Corollary 2.4.11. *An $(n-1)$-sphere Σ in S^n is bicollared, and hence flat, if and only if the closure of each component of $S^n \smallsetminus \Sigma$ is a ∂-manifold.*

Corollary 2.4.12. *Every compact ∂-manifold in S^n bounded by an $(n-1)$-sphere is an n-cell.*

The following corollary of Theorems 2.4.8 and 2.4.10 is often called the Generalized Schönflies Theorem.

Corollary 2.4.13. *Every locally flat $(n-1)$-sphere in S^n is flat.*

Corollary 2.4.14. *The boundary of every G-orientable ∂-manifold M is G-orientable.*

Proof. Now we know M contains a copy of $\partial M \times \mathbb{R}$. Corollary 0.3.6 assures that the latter is G-orientable, and Corollary 0.3.8 does the same for ∂M. □

Corollary 2.4.15. *Let M be a ∂-manifold and $\phi_t : \partial M \to \partial M$ an isotopy such that $\phi_0 = \mathrm{Id}_{\partial M}$. Then, for each neighborhood U of ∂M, ϕ_t extends to an ambient isotopy Φ_t of M supported in U such that $\Phi_0 = \mathrm{Id}_M$.*

Proof. Produce a collar $\lambda : \partial M \times [0,1] \to M$ on ∂M with image in U, where $\lambda_0 = \mathrm{incl}_{\partial M}$. Then define $\Phi_1 : M \to M$ as the identity on $M \smallsetminus \lambda(\partial M \times [0,1])$ and as $\lambda(\phi_{1-t}(x), t)$ for $\lambda(x,t) \in \lambda(\partial M \times [0,1])$. Specification of an isotopy Φ_t extending ϕ_t and running from $\Phi_0 = \mathrm{Id}_M$ to Φ_1 is left to the reader. □

Historical Notes. The generalized Schönflies theorem was first proved by M. Brown (1960), who developed the elegant method of shrinking cellular sets used in the proof. Earlier B. Mazur (1959) had proved the theorem with an additional technical hypothesis, and eventually M. Morse (1960) showed how to remove that condition to provide an alternative proof of the theorem.

Cellularity, as an important concept, not the term itself, appeared in the 1920s with the analysis by R. L. Moore (1925) of cellular decompositions of 2-manifolds.

Collaring Theorem 2.4.10 is also due to Brown (1960). The argument here is taken from R. Connelly (1971), who conceived the simplification of appending an abstract collar.

Exercises

2.4.1. The three definitions of cellular set given at the beginning of the section are equivalent.

2.4.2. A compact set X in S^n is cellular if and only if $S^n \smallsetminus X \cong \mathbb{R}^n$.

2.4.3. Every arc $\alpha \subset \mathbb{R}^n$ that is locally polyhedral modulo one point is cellular.

2.4.4. (A one-sided Schönflies theorem.) Let $\Sigma \subset S^n$ be an embedded $(n-1)$-sphere and let U be one of its complementary domains. If $\overline{U}$ is a ∂-manifold, then $\overline{U}$ is an n-cell.

2.4.5. Let Σ_1 and Σ_2 be two disjoint $(n-1)$-spheres in S^n, let U_1 be the complementary domain of Σ_1 that contains Σ_2, and let U_2 be the complementary domain of Σ_2 that contains Σ_1. Define $A = U_1 \cap U_2$. Prove that $\overline{A} \smallsetminus \Sigma_i \cong S^{n-1} \times [0,1)$.[2]

[2] In a later chapter we will make use of the annulus theorem, which asserts that $\overline{A} \cong S^{n-1} \times [0,1]$.

2.5. The Klee trick

A simple, elegant application of the Tietze Extension Theorem leads to an unknotting result for embeddings into hyperplanes.

Theorem 2.5.1. *Suppose* $\lambda : C \to \mathbb{R}^n$ *and* $\lambda' : C \to \mathbb{R}^m$ *are embeddings of a compact metric space* C*. Then the associated embeddings* $e, e' : C \to \mathbb{R}^n \times \mathbb{R}^m$*, where* $e(c) = \langle \lambda(c), 0 \rangle$ *and* $e'(c) = \langle 0, \lambda'(c) \rangle$*, are each equivalent to the diagonal embedding* $d = \lambda \times \lambda' : C \to \mathbb{R}^n \times \mathbb{R}^m$*.*

Proof. It suffices to show that e is equivalent to d. Since $\mathbb{R}^m$ has the universal extension property (Munkres, 1975, page 216), the map $\lambda'\lambda^{-1} : \lambda(C) \to \mathbb{R}^m$ can be extended to a map $\psi : \mathbb{R}^n \to \mathbb{R}^m$. Define $\Psi : \mathbb{R}^n \times \mathbb{R}^m \to \mathbb{R}^n \times \mathbb{R}^m$ as $\Psi(\langle x, y \rangle) = \langle x, y + \psi(x) \rangle$. Clearly Ψ is continuous; indeed, it is a homeomorphism, for the map $\langle x, y \rangle \to \langle x, y - \psi(x) \rangle$ acts as its inverse. Furthermore,

$$\Psi e(c) = \Psi(\langle \lambda(c), 0 \rangle) = \langle \lambda(c), \psi(\lambda(c)) \rangle = \langle \lambda(c), \lambda'(c) \rangle = d(c),$$

as required. □

Corollary 2.5.2. *Any two embeddings* λ, λ' *of a compact metric space into* $\mathbb{R}^n$ *are equivalent when considered as embeddings to their images in* $\mathbb{R}^n \times \{0\} \subset \mathbb{R}^n \times \mathbb{R}^n = \mathbb{R}^{2n}$*.*

Corollary 2.5.3. *Every arc in* $\mathbb{R}^n = \mathbb{R}^n \times \{0\} \subset \mathbb{R}^{n+1}$ *is flat in* $\mathbb{R}^{n+1}$*.*

Another corollary could be listed—that every k-cell in $\mathbb{R}^n = \mathbb{R}^n \times \{0\} \subset \mathbb{R}^{n+k}$ is flat in $\mathbb{R}^{n+k}$—but for $k > 1$ this is far from best possible. In later chapters we shall learn that all k-cells in $\mathbb{R}^n = \mathbb{R}^n \times \{0\} \subset \mathbb{R}^{n+1}$ are flat in $\mathbb{R}^{n+1}$.

Historical Notes. Theorem 2.5.1 is due to V. Klee (1955).

Exercises

2.5.1. Show that for every arc $A \subset \mathbb{R}^n$, $A \times [-1, 1]$ is a cellular subset of $\mathbb{R}^n \times \mathbb{R}^1$.

2.5.2. Suppose $X \times \mathbb{R}^1$ is a manifold. Show that each arc of the form $\{x\} \times [-1, 1]$ is cellular in $X \times \mathbb{R}^1$.

2.5.3. Let X be a compact subset of $\mathbb{R}^n$ and let $f : X \to \mathbb{R}^m$ be continuous. Show that $X \times \{0\}$ and the graph of f are equivalently embedded in $\mathbb{R}^n \times \mathbb{R}^m$.

2.5.4. Any arc $\alpha \subset \mathbb{R}^n$, $n > 3$, that is a countable union of points and segments is flat. [Hint: Find a line L such that any line parallel to L intersects α in at most one point.]

2.6. The product of $\mathbb{R}^1$ with an arc decomposition

Next we turn to the construction of everywhere wild embeddings in all dimensions and all codimensions. The examples of wild embeddings constructed in §1.4 all have relatively low codimension, and a new technique is required to produce examples in codimensions greater than two. The idea is this: start with an arc in S^n, suspend it to produce a 2-cell in S^{n+1}, and then shrink out the arcs in the levels of the suspension to produce a new arc in S^{n+1}. The necessary shrinking theorem is proved in this section and the examples will be constructed in the following section.

Let A be an arc in $\mathbb{R}^n$ and $q : \mathbb{R}^n \to \mathbb{R}^n/A$ the quotient map. Urysohn's Metrization Theorem assures that $\mathbb{R}^n/A$ is a locally compact metric space.

Theorem 2.6.1. *If A is an arc in $\mathbb{R}^n$, then $(\mathbb{R}^n/A) \times \mathbb{R}^1 \approx \mathbb{R}^{n+1}$.*

Note that $(\mathbb{R}^n/A) \times \mathbb{R}^1$ is the same as $(\mathbb{R}^n \times \mathbb{R}^1)/\{A \times \{t\} \mid t \in \mathbb{R}\}$. We intend to prove that the decomposition of $\mathbb{R}^n \times \mathbb{R}^1$ into points and the arcs $A \times \{t\}$, $t \in \mathbb{R}^1$, is shrinkable. To that end, name a homeomorphism $\alpha : [0,1] \to A$, and fix $\epsilon > 0$. Partition $[0,1]$ by points $\{t_i\}$ with $0 = t_0 < t_1 < \cdots < t_{m+1} = 1$ such that $\operatorname{diam} \alpha([t_{i-1}, t_{i+3}]) < \epsilon$ for $i \in \{1, 2, \ldots, m-2\}$. Expand each $\alpha([t_{i-1}, t_i])$ slightly to an open subset U_i of $\mathbb{R}^n$, where

$$U_i \cap U_j \neq \emptyset \text{ if and only if } |i - j| \leq 1, \text{ and}$$
$$\operatorname{diam}(U_i \cup U_{i+1} \cup U_{i+2} \cup U_{i+3}) < \epsilon \quad (i = 1, 2, \ldots, m-2).$$

These U_i's will supply motion controls on $h \in \operatorname{Homeo}(\mathbb{R}^{n+1}, \mathbb{R}^{n+1})$ for the $\mathbb{R}^n$ factor; to maintain control in the $\mathbb{R}^1$ direction, we identify some intervals and related sets: for $i \in \{1, 2, \ldots, m-1\}$ let $J_i = [i, 2m - i]$ and then let $L_i = J_i \smallsetminus \operatorname{Int} J_{i+1}$ $(i < m - 1)$.

According to Corollary 2.5.3, each level arc $A \times \{t\}$ is flat, so any one of them can be shrunk to small size. To support our aim of shrinking all level arcs simultaneously, Lemma 2.6.2 shows how to combine a vertical compression with the pinching of one level arc to achieve shrinking of the product of J_{i+1} and a subarc of A. This basic move is applied finitely often in Lemma 2.6.3 to achieve a partial shrinking of certain blocks, and these block moves, carefully arranged, achieve the desired shrinking of the entire family of arcs.

Lemma 2.6.2. *Let V_i be a neighborhood of $\alpha([0, t_i])$ in $\mathbb{R}^n$. Then there exists $h_i \in \operatorname{Homeo}(\mathbb{R}^{n+1}, \mathbb{R}^{n+1})$ satisfying:*

(1) $h_i \mid \mathbb{R}^{n+1} \smallsetminus (V_i \times J_i) = \operatorname{Id}$,

(2) $h_i \mid \alpha([t_i, 1]) \times \mathbb{R}^1 = \operatorname{Id}$, *and*

(3) $h_i(\alpha([0, t_i]) \times J_{i+1}) \subset U_{i+1} \times J_i$.

Proof. By Corollary 2.5.3, $A \times \{i+1\}$ is flat. One can shrink the subarc $\alpha([0,t_i]) \times \{i+1\}$ near the point $\alpha(t_i) \times \{i+1\}$ via $\mu \in \text{Homeo}(\mathbb{R}^{n+1}, \mathbb{R}^{n+1})$ such that

$$\mu \mid \mathbb{R}^{n+1} \setminus (V_i \times J_i) = \text{Id},$$
$$\mu \mid \alpha([t_i, 1]) \times \mathbb{R}^1 = \text{Id}, \text{ and}$$
$$\mu(\alpha([0,t_i]) \times \{i+1\}) \subset U_{i+1} \times J_i.$$

It follows that $\mu^{-1}(U_{i+1} \times \text{Int } J_i) \supset (\alpha([0,t_i]) \times \{i+1\}) \cup (\alpha(t_i) \times J_{i+1})$, so certainly there exists $\delta > 0$ such that

$$\mu^{-1}(U_{i+1} \times \text{Int } J_i) \supset \alpha([t_i - \delta, t_i]) \times J_{i+1}).$$

Now one can produce $\upsilon \in \text{Homeo}(\mathbb{R}^{n+1}, \mathbb{R}^{n+1})$, which compresses points of $\alpha([0,t_i]) \times J_{i+1}$ into $\mu^{-1}(U_{i+1} \times \text{Int } J_i)$ and changes only the $\mathbb{R}^1$ coordinates, subject to the restrictions

$$\upsilon \mid \mathbb{R}^{n+1} \setminus (V_i \times J_i) = \text{Id},$$
$$\upsilon \mid \alpha([t_i, 1]) \times \mathbb{R}^1 = \text{Id}, \text{ and}$$
$$\upsilon(\alpha([0,t_i]) \times J_{i+1}) \subset \mu^{-1}(U_{i+1} \times \text{Int } J_i).$$

To produce υ more explicitly, name $d \in (0,1)$ for which

$$\mu^{-1}(U_{i+1} \times \text{Int } J_i) \supset \alpha([0,t_i]) \times [i+1, i+1+d];$$

use Urysohn's Lemma to define a map $s : \mathbb{R}^n \to [i+1+d, 2m-i-1]$ sending $(\mathbb{R}^n \setminus V_i) \cup \alpha([t_i, 1])$ to $\{2m-i-1\}$ while sending $\alpha([0, t_i \setminus \delta])$ to $\{i+1+d\}$. Finally, define $\upsilon \in \text{Homeo}(\mathbb{R}^{n+1}, \mathbb{R}^{n+1})$ as the identity above $R^n \times \{2m-i\}$ and below $\mathbb{R}^n \times \{i+1\}$, with $\upsilon(\langle p, 2m-i-1\rangle) = \langle p, s(p)\rangle$ for each $p \in \mathbb{R}^n$, and with υ acting as the obvious linear homeomorphism on all vertical intervals $\{p\} \times J_{i+1}$ and $\{p\} \times [2m-i-1, 2m-i]$. The effect of υ is illustrated in Figure 2.15. Note that υ is the identity in a neighborhood of the shaded region.

Now simply define h_i as $\mu\upsilon$. Then

$$h_i(\alpha([0,t_i]) \times J_{i+1}) = \mu\upsilon(\alpha([0,t_i]) \times J_{i+1}) \subset \mu\mu^{-1}(U_{i+1} \times J_i) = U_{i+1} \times J_i,$$

as desired. The other requirements of Lemma 2.6.2 are easily confirmed. □

Lemma 2.6.3. *There exists* $\lambda \in \text{Homeo}(\mathbb{R}^{n+1}, \mathbb{R}^{n+1})$ *satisfying:*

(1) $\lambda \mid \mathbb{R}^{n+1} \setminus \cup_{i=1}^{m-2}(U_i \times J_i) = \text{Id}$,

(2) $\lambda(\alpha([0,t_{i+1}]) \times L_i) \subset (U_i \cup U_{i+1}) \times J_{i-1}$ *for* $i \in \{1, 2, \ldots, m-2\}$,

(3) $\lambda(\alpha([0,t_m]) \times J_{m-1}) \subset (U_{m-1} \cup U_m) \times J_{m-2} \subset (U_{m-1} \cup U_m) \times J_1$.

Proof. Here λ will arise as a composition $h_1 h_2 \cdots h_{m-2}$ of homeomorphisms from Lemma 2.6.2. To get started, obtain h_1 from 2.6.2 for the neighborhood $V_1 = U_1$ of $\alpha([0,t_1])$.

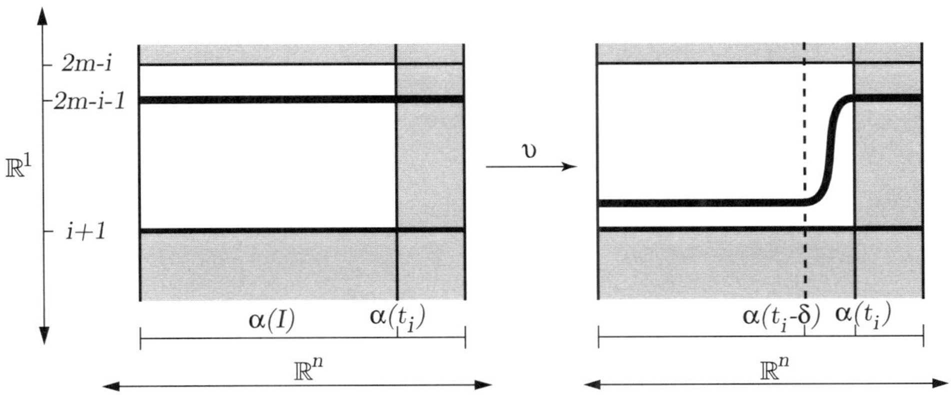

Figure 2.15. The homeomorphism υ

Since h_1 acts as the identity on $\alpha([t_1, t_2]) \times \mathbb{R}^1$ and carries $\alpha([0, t_1]) \times J_2$ into $U_2 \times J_1$, there exists a neighborhood $V_2 \subset U_1 \cup U_2$ of $\alpha([0, t_2])$ such that $h_1(V_2 \times J_2) \subset U_2 \times J_1$. Apply Lemma 2.6.2 again with this neighborhood V_2 to obtain h_2.

The iterative step repeats the pattern of the second step. After h_{i-1} has been obtained, subject to the conditions

$$h_{i-1} \mid \alpha([t_{i-1}, 1]) \times J_{i-1} = \text{Id and}$$
$$h_{i-1}(\alpha([0, t_{i-1}]) \times J_i) \subset U_i \times J_{i-1},$$

determine a neighborhood V_i of $\alpha([0, t_i])$ in $U_1 \cup \cdots \cup U_i$ such that $h_{i-1}(V_i \times J_i) \subset U_i \times J_{i-1}$ and then apply Lemma 2.6.2 with this neighborhood V_i to obtain h_i.

The composition $\lambda = h_1 h_2 \cdots h_{m-2}$ is shown in Figure 2.16. In a neighborhood of the shaded region, λ is the identity.

It should be obvious from the choices of $V_i \subset U_1 \cup \cdots \cup U_i$ and conclusion (1) of Lemma 2.6.2 that $\lambda = h_1 h_2 \ldots h_{m-2}$ satisfies conclusion (1) above. In analyzing conclusions (2) and (3), it is useful to keep in mind that $U_1 \cup \cdots \cup U_i$ and $U_{i+2} \cup \cdots \cup U_{m+1}$ are disjoint. Due to the choices of V_i, conclusion (1) of Lemma 2.6.2 then yields, for $j \geq i$,

$$(*) \qquad \begin{aligned} &h_i \mid (U_{j+2} \cup \cdots \cup U_{m+1}) \times \mathbb{R}^1 = \text{Id and} \\ &h_i \mid \mathbb{R}^n \times (\mathbb{R}^1 \setminus J_j) = \text{Id}. \end{aligned}$$

Since $L_i \subset \mathbb{R}^1 \setminus \text{Int}\, J_{i+1}$, the latter implies

$$(**) \qquad h_j \mid \mathbb{R}^n \times L_i = \text{Id whenever } i < j.$$

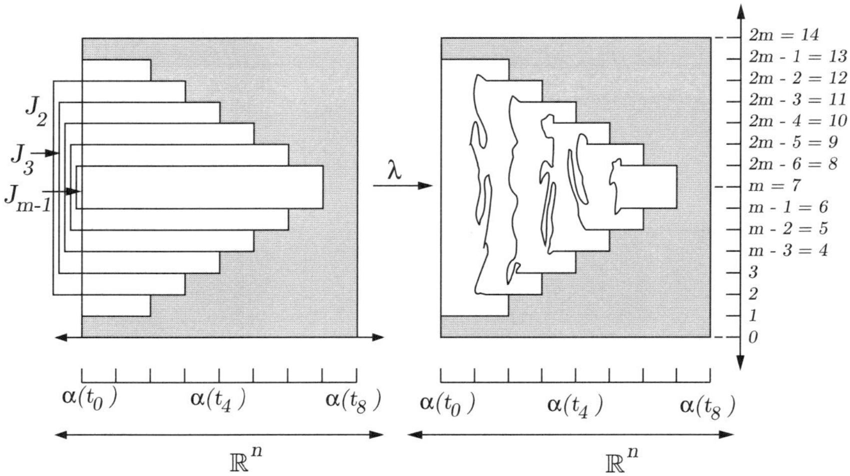

Figure 2.16

To see why conclusion (3) holds, note that

$$\begin{aligned}\lambda(\alpha([0,t_{m-2}])\times J_{m-1}) &= h_1h_2\cdots h_{m-2}(\alpha([0,t_{m-2}])\times J_{m-1})\\ &\subset h_1h_2\cdots h_{m-3}(U_{m-1}\times J_{m-2})\\ &\subset U_{m-1}\times J_{m-2}\end{aligned}$$

by (3) of Lemma 2.6.2 and (*). In addition, by conclusion (2) of the Lemma,

$$\lambda(\alpha([t_{m-2},t_m])\times J_{m-1}) = \alpha([t_{m-2},t_m])\times J_{m-1}\subset (U_{m-1}\cup U_m)\times J_{m-1},$$

and these two inclusions quickly combine to yield (3).

To verify conclusion (2), first observe that $h_i(\alpha([0,t_i])\times J_i)\subset h_i(V_i\times J_i) = V_i\times J_i$ by conclusion (1) of Lemma 2.6.2. Then

$$\begin{aligned}\lambda(\alpha([0,t_{i+1}])\times L_i) &= h_1h_2\cdots h_{m-2}(\alpha([0,t_{i+1}])\times L_i)\\ &= h_1h_2\cdots h_i(\alpha([0,t_{i+1}])\times L_i) \qquad \text{by } (**)\\ &\subset h_1h_2\cdots h_i((\alpha([0,t_i])\times L_i)\cup(\alpha([t_i,t_{i+1}])\times J_i))\\ &\subset h_1h_2\cdots h_i(\alpha([0,t_i])\times J_i)\cup(U_{i+1}\times J_i)\\ &\qquad\qquad \text{by (2) of Lemma 2.6.2}\\ &\subset h_1h_2\cdots h_{i-1}(V_i\times J_i)\cup(U_{i+1}\times J_i) \qquad \text{as above}\\ &\subset h_1h_2\cdots h_{i-2}(U_i\times J_{i-1})\cup(U_{i+1}\times J_i) \text{ by choice of } V_i\\ &= (U_i\times J_{i-1})\cup(U_{i+1}\times J_i) \qquad \text{by } (*)\\ &\subset (U_i\cup U_{i+1})\times J_{i-1}.\end{aligned}$$

Why the conclusion also holds for $i = 1$ should be evident to anyone who understands the preceding lines. □

Proof of Theorem 2.6.1. Let $q' : \mathbb{R}^{n+1} = \mathbb{R}^n \times \mathbb{R}^1 \to (\mathbb{R}^n/A) \times \mathbb{R}^1$ be the map $q \times \mathrm{Id}$. Our intention is to show that q' is a near homeomorphism, which will follow from Theorem 2.3.5 almost instantly, once we construct $h \in \mathrm{Homeo}(\mathbb{R}^{n+1}, \mathbb{R}^{n+1})$ satisfying

$$h \mid (\mathbb{R}^n \setminus N(A;\epsilon)) \times \mathbb{R}^1 = \mathrm{Id},$$
$$h(\mathbb{R}^n \times t) \subset \mathbb{R}^n \times [t-\epsilon, t+\epsilon], \text{ and}$$
$$\operatorname{diam} h(A \times t) < 3\epsilon \text{ for each } t \in \mathbb{R}^1.$$

This will be accomplished by exploiting the structures named for Lemmas 2.6.2 and 2.6.3, carefully pieced together.

Formally, let k range over the integers and set

$$D_k = \cup_{i=1}^{m-2} U_i \times [2mk+i, 2mk+2m-i] \text{ and}$$
$$D'_k = \cup_{i=1}^{m-2} U_{m+2-i} \times [2mk+m+i, 2mk+3m-i].$$

In addition, let $D = \cup_k D_k$ and $D' = \cup_k D'_k$; D and D' are mirror images of each other, and act as supports for homeomorphisms obtained from Lemma 2.6.3. A key feature is $D \cap D' = \emptyset$. For details in a particular instance, consider $\langle x, t\rangle \in D_0$, where $t \le m$. Choose the least integer i such that $x \in U_i$; then $t \ge i$, by definition of D_0. The only possible D'_k that might contain $\langle x,t\rangle$ is D'_{-1}. If that were the case, note that $x \in U_{m+2-j}$ for $j = m+2-i$, so $x \in U_{m+2-j}$ can hold only for $j \in \{m+2-i, m+1-i\}$. In either situation, the definition of D'_{-1} forces

$$t \le m - (m+1-i) = i - 1 < i,$$

a contradiction.

For $k \in Z$ and $i \in \{1, 2, \ldots, m-1\}$, define

$$J'_i = \cup_k [2mk+i, 2mk+2m-i]$$
$$P'_i = \cup_k [2mk+m+i, 2mk+3m-i]$$

and for $i < m-1$ let $L'_i = J'_i \setminus \mathrm{Int}\, J'_{i+1}$ and $Q'_i = P'_i \setminus \mathrm{Int}\, P'_{i+1}$. According to Lemma 2.6.3, there exist a homeomorphism λ_R for the J'_i and L'_i and another homeomorphism λ_L for the P'_i and Q'_i, each involving translates of the Lemma 2.6.3 homeomorphism λ to the appropriate levels, satisfying

(1) $\lambda_R \mid \mathbb{R}^{n+1} \setminus D = \mathrm{Id}$ and $\lambda_L \mid \mathbb{R}^{n+1} \setminus D' = \mathrm{Id}$;

(2) $\lambda_R(\alpha([0, t_{i+1}]) \times L'_i) \subset (U_i \cup U_{i+1}) \times J'_i$ and $\lambda_L(\alpha([t_{m-i}, 1]) \times Q'_i) \subset (U_{m-i+1} \cup U_{m-i+2}) \times P'_i$;

(3) $\lambda_R(\alpha([0, t_m]) \times J'_{m-1}) \subset (U_{m-1} \cup U_m) \times J'_1$ and $\lambda_L(\alpha([t_1, 1]) \times P'_{m-1}) \subset (U_2 \cup U_3) \times P'_1$.

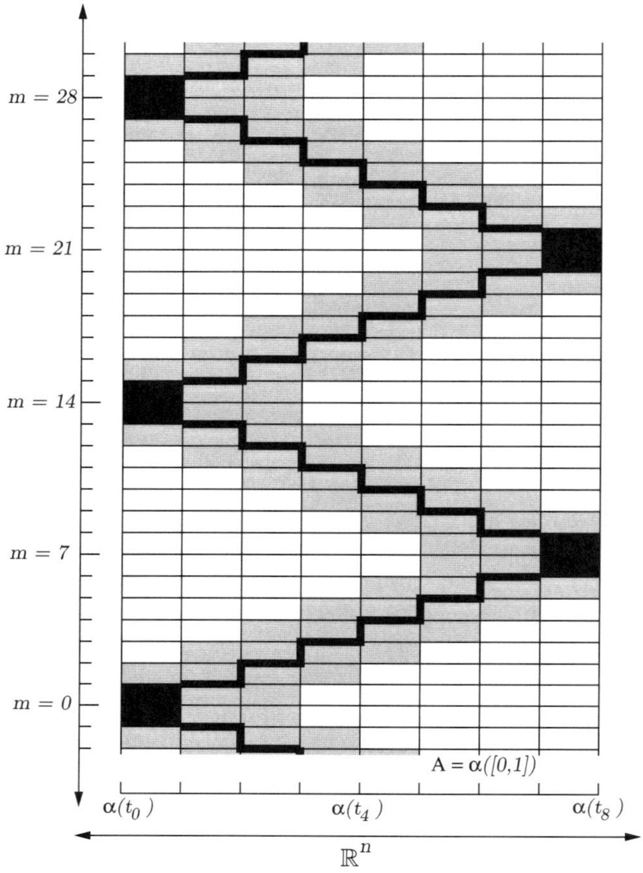

Figure 2.17

Now define h as $\lambda_R\lambda_L$. The resultant shrinking depends on this explicit juxtaposition of left and right stacks. For example, if $t \in \operatorname{Int} L_i'$ then $t \in Q_j'$, where $j = m - 1 - i$. Thus, by (2),

$$\begin{aligned} h(A \times t) &\subset \lambda_R\lambda_L(\alpha([0, t_{i+1}]) \times L_i') \cup \lambda_R\lambda_L(\alpha([t_{i+1}, 1]) \times Q_j') \\ &\subset \lambda_R(\alpha([0, t_{i+1}]) \times L_i') \cup \lambda_R((U_{i+3} \cup U_{m-i+3}) \times P_i') \\ &\subset [(U_i \cup U_{i+1}) \times \mathbb{R}^1] \cup [(U_{i+2} \cup U_{i+3}) \times \mathbb{R}^1]. \end{aligned}$$

Of course, neither λ_R nor λ_L moves points vertically more than $2m$, and due to the disjointness of D and D', this gives both

$$\begin{gathered} h(A \times t) \subset (U_i \cup U_{i+1} \cup U_{i+2} \cup U_{i+3}) \times [t - 2m, t + 2m] \text{ and} \\ h(\mathbb{R}^n \times t) \subset \mathbb{R}^n \times [t - 2m, t + 2m]. \end{gathered}$$

Recall that initially the U_i's were chosen so that the diameters of any four consecutive ones were small. To complete this proof, rescale the $\mathbb{R}^1$-coordinate with $2m < \epsilon$. □

Corollary 2.6.4. *For each arc A in S^n, the suspension of the quotient space S^n/A is homeomorphic to S^{n+1}.*

Historical Notes. Theorem 2.6.1 is due to J. J. Andrews and M. L. Curtis (1962). The dogbone space constructed by R. H. Bing (1959a) was the first example of a non-manifold decomposition space X such that $X \times \mathbb{R}^1$ is a manifold.

2.7. Everywhere wild cells and spheres

Straightforward application of the product arc-shrinking theorem from §2.6 leads to embeddings that are wild at every point. An embedding of a manifold or ∂-manifold that is not locally flat at any point is called *everywhere wild.*

Example 2.7.1. *For each $n \geq 3$ and $0 < k < n$, S^n contains an everywhere wild k-cell.*

Lemma 2.7.2. *For each $n \geq 3$, S^n contains a wild arc α for which $S^n \setminus \alpha$ fails to be simply connected.*

Proof. Such an arc in S^3 was described in Example 2.1.8. Given an arc A in S^{n-1}, $n > 3$, with nonsimply connected complement, Corollary 2.6.4 allows us to identify S^n with the suspension of S^{n-1}/A. Let α be the arc in S^n that corresponds to the suspension of the special point in the quotient space S^{n-1}/A. Then $S^n \setminus \alpha$ is topologically equivalent to $(S^{n-1} \setminus A) \times (-1, 1)$, which is not simply connected. □

The arcs α provided in the preceding lemma are everywhere wild, starting in dimension four. In order to prove this, we need pinpoint information about the way in which the complement fails to be simply connected: we need to know that the complement of α contains loops that are very close to α but essential in the complement, and the following condition—known as the "cellularity criterion" because it implies cellularity for certain subsets of manifolds, a result to be proved in the next chapter—paves the way. The cellularity criterion is a global version of the 1-LCC condition.

Definition. A compact set X in an m-manifold M is said to satisfy the *cellularity criterion* if for every open neighborhood U of X in M there exists an open set V such that $X \subset V \subset U$ and every map $\partial I^2 \to V \setminus X$ extends to a map $I^2 \to U \setminus X$.

Lemma 2.7.3. *The arc α in Lemma 2.7.2 fails to satisfy the cellularity criterion.*

Proof. Start with $n = 3$; let A be the wild arc in Example 2.1.8, which contains a copy of Antoine's necklace. The simple closed curve J shown in Figure 2.6 is essential in the complement of A and there are related curves just like J that link later stages of the construction. Hence every neighborhood V of A contains a simple closed curve that is essential in $S^3 \setminus A$.

We now show that if A is an arc in S^{n-1} such that $S^{n-1} \setminus A$ is non-simply connected, then the arc α constructed from it (as in the proof of Lemma 2.7.2) fails to satisfy the cellularity criterion. Suppose there exists a neighborhood of α such that every loop in $V \setminus \alpha$ is null-homotopic in $S^n \setminus \alpha$. A loop in $S^n \setminus \alpha$ can be pushed up the product structure on $S^n \setminus \alpha \cong (S^{n-1} \setminus A) \times (-1, 1)$ into V and so the existence of V would mean that $S^n \setminus \alpha$ is simply connected. This contradicts the conclusion of Lemma 2.7.2. □

The proof of the following lemma is left as an exercise.

Lemma 2.7.4. *A compact set $X \subset S^n$ satisfies the cellularity criterion in S^n if and only if the arc corresponding to the suspension of X in* $\mathrm{Susp}(S^n/X)$ *is 1-LCC at each interior point.*

Proof of Example 2.7.1. Consider, first, the case $k = 1$ and $n \geq 4$. By Lemma 2.7.4, the arcs constructed in Lemmas 2.7.2 and 2.7.3 fail to be 1-LCC at all interior points. Proposition 1.3.1 and Exercise 2.7.1 imply that these arcs are everywhere wild.

Now assume that $k > 1$ and $n - k > 2$. By the previous paragraph, there is an arc in S^{n-k+1} that fails to be 1-LCC at each interior point. The $(k-1)$-fold suspension is a k-cell in S^n that is everywhere wild because, by Lemma 1.4.1, it fails to be 1-LCC at each interior point.

Finally, consider the cases $k = n - 2$ and $k = n - 1$. Example 2.1.10 provides wild cells in those codimensions, but they are not everywhere wild, since the basic examples of wild arcs and disks in $\mathbb{R}^3$ on which they are based are not everywhere wild. To address this issue, in the next section we will produce examples of everywhere wild arcs and disks in S^3. Once those examples are in place, multiple suspension to S^n yields everywhere wild cells of dimensions $n - 2$ and $n - 1$. □

Historical Notes. The idea of exploiting the Andrews-Curtis Theorem to produce everywhere wild embeddings is due to Brown (1967). Earlier, W. A. Blankinship (1951) devised wild embeddings in all dimensions and codimensions, based on his construction of wild Cantor sets in $\mathbb{R}^n$, $n > 3$; his Cantor set construction will be set forth in §4.7.

Exercises

2.7.1. Let $C \subset M$ be a k-cell topologically embedded in an n-manifold. If C is nonlocally flat at every interior point, then C is nonlocally flat at every boundary point as well.

2.7.2. Prove Lemma 2.7.4.

2.7.3. Every cellular subset of an n-manifold ($n > 2$) satisfies the cellularity criterion.

2.7.4. If $\alpha \subset S^n$ is an arc that satisfies the cellularity criterion, then $S^n \setminus \alpha$ is contractible.

2.8. Miscellaneous examples of wild embeddings

This section offers more examples of wild embeddings in $\mathbb{R}^3$. These new examples exhibit wildness that is qualitatively different from that of the examples presented earlier in the chapter. The two original examples of wildly embedded 2-spheres in $\mathbb{R}^3$, the Antoine sphere and the Alexander horned sphere, share one property: each of them contains a Cantor set such that the embedding is wild at every point of the Cantor set and is locally flat at all other points. The examples in the section show that a variety of wild sets are possible; the first examples to be presented are wild at just one point while the later examples are wild at positive-dimensional sets. In particular, among the later ones are some everywhere wild codimension-one and -two cells in $\mathbb{R}^3$, which fill a gap in the proof of Example 2.7.1. The section contains an outline of the proofs that the examples have the properties indicated, but many details are left as exercises.

2.8.1. The Fox-Artin arc. The first example is an arc whose wildness is minimal in the sense that the arc is locally flat at every point except one and the complement of the arc is the same as that of a flat arc. The construction begins with the basic building block shown in Figure 2.18. The building block consists of three arcs A, B, and C embedded in a 3-cell as indicted in the figure.

Put an infinite sequence of these building blocks together in such a way that they converge to a point p. Include the point p and delete the first copy of B to form the arc α pictured in Figure 2.19. This arc, known as the *Fox-Artin arc*, is wild because it fails to be locally flat at the endpoint p.

The arc α is obviously locally flat and PL at every point other than p. In order to see that α is not locally flat at p one must prove that α is not

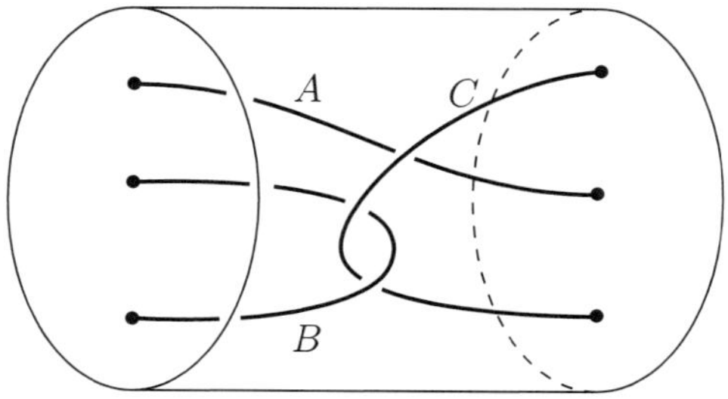

Figure 2.18. The basic Fox-Artin building block

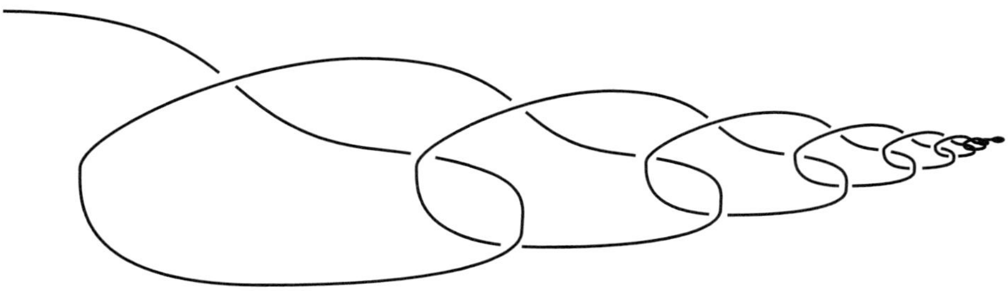

Figure 2.19. The Fox-Artin arc

1-LCC at p. (A tame arc is 1-LCC at each of its endpoints.) The Fox-Artin arc is not 1-LCC at p because for any small neighborhood U of p there exist loops in $U \setminus \alpha$ that cannot be shrunk to a point in a small subset of $S^3 - \alpha$; in fact, they cannot be shrunk to a point without going all the way over the other end of α. A proof is sketched in the exercise below.

Notice that α is cellular (Exercise 2.4.3) and thus the complement of α in S^3 is an open 3-cell. In particular, $S^3 \setminus \alpha$ is simply connected, which means that the wildness of the Fox-Artin arc is more subtle than that of the wild arcs studied earlier, which were known to be wild because their complements were not simply connected. One can obtain wild but cellular embeddings in higher dimensions by suspending the Fox-Artin arc.

This example is unique to S^3 in the sense that three is the only ambient dimension in which an arc can fail to be locally flat at just a single point (Exercise 2.5.4).

Exercise 2.8.1. This exercise contains an outline of the proof that α is not 1-LCC at p. The problem is to fill in the details in the argument. First we need some notation. Choose a sequence $D_1, D_2, D_3, \ldots$ of 3-cells such that $D_{i+1} \subset \operatorname{Int} D_i$ for each i, $\cap_{i=1}^{\infty} D_i = \{p\}$, and D_i intersects α as indicated in Figure 2.20. Let A_1, B_1, and C_1 be the arcs in $D_1 \setminus \operatorname{Int} D_2$ that correspond to A, B, and C, respectively, and let E_1 be a flat disk in $D_1 \setminus \operatorname{Int} D_2$ such that $C_1 \subset \partial E_1$ and $\partial E_1 \setminus C_1 \subset \partial D_2$ (Figure 2.20).

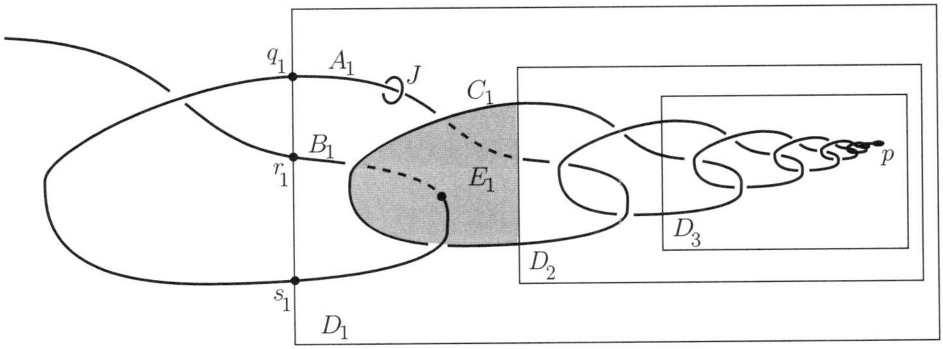

Figure 2.20. A sequence of 3-cells and 2-cells

(a) Let q_1, r_1, and s_1 be the three points at which α intersects ∂D_1 and view D_1 as the cone on its boundary. Show that
$$D_1 \setminus (A_1 \cup D_2 \cup E_1 \cup B_1) \cong D_1 \setminus \operatorname{Cone}(\{q_1, r_1, s_1\})$$
via a homeomorphism that is the identity on the boundary.

(b) Use an argument like that in the proof of Lemma 2.1.4 to prove that the inclusion induced homomorphism $\pi_1(D_1 \setminus (A_1 \cup D_2 \cup E_1 \cup B_1)) \to \pi_1(D_1 \setminus (A_1 \cup D_2 \cup C_1 \cup B_1))$ is one-to-one.

(c) Combine the preceding results to show that $\pi_1(\partial D_1 \setminus \{q_1, r_1, s_1\}) \to \pi_1(D_1 \setminus (\alpha \cup D_2))$ is one-to-one.

(d) Use an argument like that in the proof of Theorem 0.11.5 to show that $\pi_1(\partial D_1 \setminus \{q_1, r_1, s_1\}) \to \pi_1(D_1 \setminus \alpha)$ is one-to-one. Conclude, in particular, that the loop J shown in Figure 2.20 is essential in $D_1 \setminus \alpha$.

(e) Observe that each D_i contains a loop that is homotopic to J in $D_1 \setminus \alpha$ and use this observation to prove that the inclusion induced homomorphism $\pi_1(D_i \setminus \alpha) \to \pi_1(D_1 \setminus \alpha)$ is nontrivial for every i.

(f) Prove that α is not 1-LCC at p.

2.8.2. Double Fox-Artin arcs. Variations on the Fox-Artin arc can have interesting properties. Two that are worthy of mention are the "double Fox-Artin arcs" shown in Figures 2.21 and 2.22.

The double Fox-Artin arc in Figure 2.21 is constructed from a doubly infinite sequence of copies of the basic Fox-Artin building block. It is a wild arc because it fails to be 1-LCC at both endpoints. Its complement is not simply connected.

Figure 2.21. A double Fox-Artin arc with nonsimply connected complement

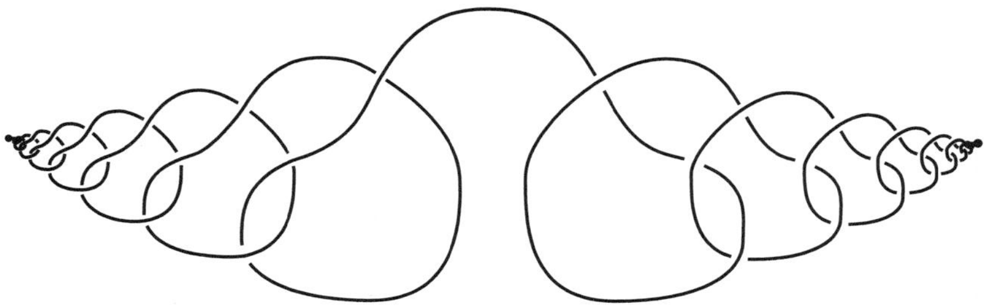

Figure 2.22. A double Fox-Artin arc with simply connected complement

The double Fox-Artin arc shown in Figure 2.22 is the wedge of two copies of α. The unusual feature of this second double Fox-Artin arc is that its complement is simply connected but is not an open 3-cell. In other words, the complement of this arc is simply connected but the arc is not cellular, because it does not satisfy the cellularity criterion. Notice that the only difference between the two arcs is that one of the two crossings in the center of Figure 2.21 has been changed to produce Figure 2.22.

Exercise 2.8.2. Use the techniques of Exercise 2.8.1 to prove the following.

(a) The complement of the arc in Figure 2.21 is not simply connected.

(b) The complement of the arc in Figure 2.22 is simply connected.

(c) The arc in Figure 2.22 does not satisfy the cellularity criterion.

2.8.3. Fox-Artin spheres. The Fox-Artin arc can be used to construct wild embeddings of spheres in S^3. To do so, start with the round 1-sphere or 2-sphere and add a feeler that follows the Fox-Artin arc. This construction is indicated in Figures 2.23 and 2.24. These embeddings are examples of what are called weakly flat spheres. An embedding $e : S^k \to S^n$ is *weakly flat* if $S^n \setminus e(S^k) \cong S^n \setminus S^k$. Neither example is flat since each contains the nonflat arc α. In particular, neither sphere is 1-LCC at the endpoint of the feeler.

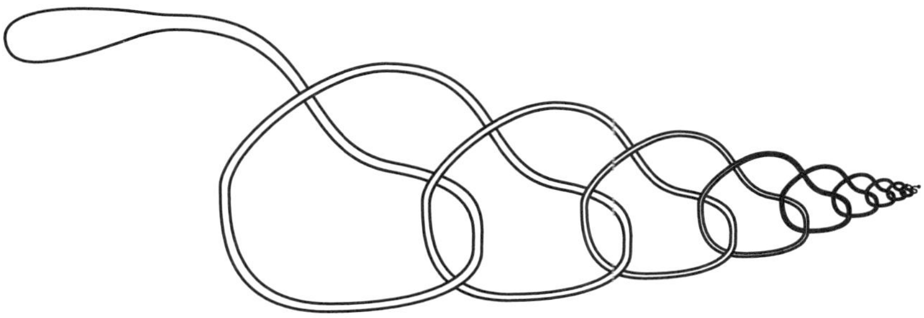

Figure 2.23. The Fox-Artin 1-sphere

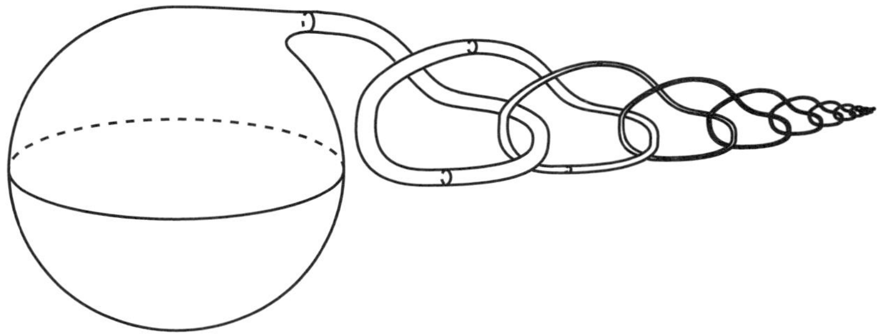

Figure 2.24. The Fox-Artin 2-sphere

Exercise 2.8.3.

(a) The Fox-Artin 1-sphere is not 1-alg at the exceptional point.

(b) The Fox-Artin 2-sphere is not 1-LCC at the exceptional point.

(c) The Fox-Artin 1-sphere is weakly flat.

(d) The Fox-Artin 2-sphere is weakly flat.

2.8.4. Mildly wild arcs. Not every arc that is formed by concatenating an infinite converging sequence of polygonal blocks is wild. In fact, the arc shown in Figure 2.25 is a tame arc.

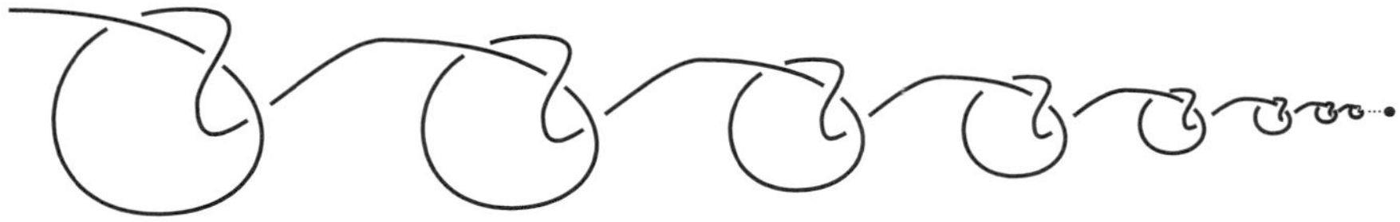

Figure 2.25. A tame arc

Interestingly, if two such arcs are joined end-to-end, the resulting arc is wild. An arc in S^3 is said to be *mildly wild* if it is wild but can be written as the union of two flat arcs. Figure 2.26 shows an example of a mildly wild arc.

Figure 2.26. A mildly wild arc

Exercise 2.8.4.

(a) Prove that the arc in Figure 2.25 is flat. [Hint: Find a nested sequence $B_1 \supset B_2 \supset \dots$ of 3-cells such that $\cap_{i=1}^{\infty} B_i$ is the endpoint of the arc and each B_i intersects the arc in a single point. For each i there is an ambient homeomorphism that is the identity on $(S^3 \smallsetminus \operatorname{Int} B_i) \cup B_{i+1}$ and that straightens out $\alpha \cap (B_i \smallsetminus \operatorname{Int} B_{i+1})$. The flattening homeomorphism is a limit of a composition of such homeomorphisms.]

(b) Prove that the arc in Figure 2.26 is not flat. [Hint: Use the Seifert-van Kampen Theorem to prove that the arc is not 1-alg at the wedge point.]

2.8.5. The Bing sling. The Bing sling is an example of an everywhere wild 1-sphere $\Sigma \subset \mathbb{R}^3$. Moreover, any arc in Σ is an everywhere wild 1-cell.

The construction begins with the basic building block shown in Figure 2.27. The building block consists of three arcs embedded in a cylindrical 3-cell; it is nearly identical to the one used in the Fox-Artin construction, but for historical accuracy we use this variation.

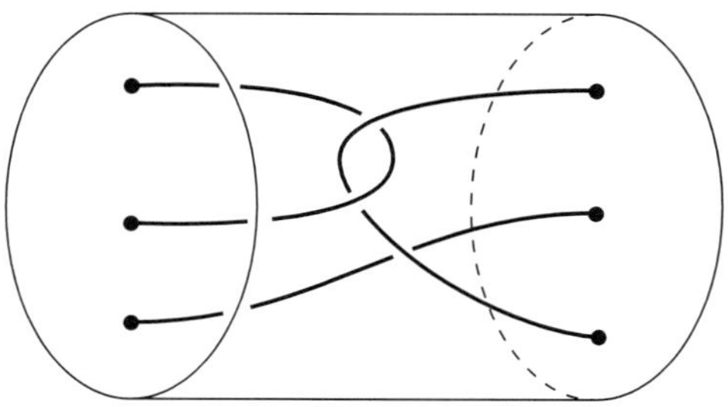

Figure 2.27. The basic building block for the Bing sling

The Bing sling arises as the intersection of a nested sequence of solid tori $T_1 \supset T_2 \supset \cdots$. The first solid torus T_1 is formed from six copies of the basic building block fit together end-to-end in a cycle as shown in Figure 2.28. The core of T_1 is a circle J_1. Inside T_1 there is a distinguished simple closed curve J_2 formed by the union of the subarcs of the six blocks that constitute T_1. This simple closed curve is the center line of a second solid torus, T_2, which is composed of many copies of the basic building block placed end-to-end along J_2. Just a few of those blocks are indicated in Figure 2.28. The subarcs of the blocks that make up T_2 combine to form a simple closed curve J_3, which is the centerline of a third solid torus T_3. The construction is continued recursively and Σ is defined by

$$\Sigma = \cap_{i=1}^{\infty} T_i.$$

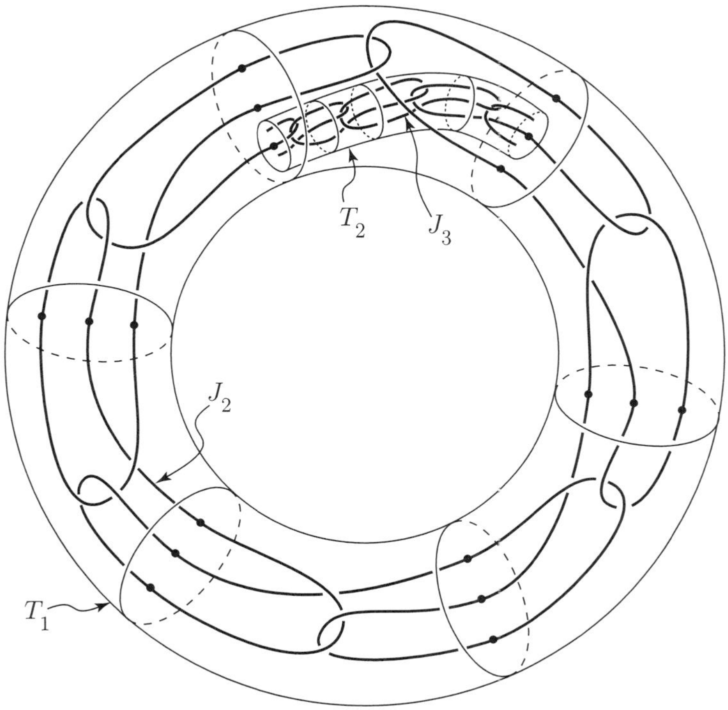

Figure 2.28. The Bing sling

At first glance it might appear that the intersection of the solid tori will be a complicated continuum, but it is, in fact, a simple closed curve. To verify this, observe that there is a homeomorphism h_i from J_i to J_{i+1}, and h_i can be kept as close to the identity as we wish by inserting multiple copies of the basic building block into the ith stage of the construction. Here h_i can be specified so as to send the portion of J_i in a block B from T_i into $J_{i+1} \cap (B \cup B')$, where B' is one of the two blocks from T_i touching B. Thus Proposition 2.2.2 shows that we can perform the construction in such a way that the composition of these homeomorphisms converges to an embedding $e : J_1 \to \mathbb{R}^3$. It is not hard to see that $e(J_1) = \Sigma$ (Exercise 2.8.5(a)).

To prove that Σ is everywhere wild, we show that Σ fails to be 1-alg at every point. Fix a point $x \in \Sigma$ and a neighborhood U of x. Assume that U is contained in the union of two of the building blocks in T_1 so that homotopies in U cannot go all the way around T_1. For any smaller neighborhood V of x there is an index i and one of the building blocks B_0 that make up T_i such that $x \in B_0 \subset V$. Consider the loop K shown in Figure 2.29. It is clear that K is null-homologous in $B_0 \setminus J_{i+1}$, as it bounds an orientable surface there; thus, K represents a commutator in $\pi_1(V \setminus \Sigma)$. If Σ were 1-alg at x, we would be able to choose V small enough so that K is inessential in $U \setminus \Sigma$. In the next two paragraphs we will show that, to the contrary, K is essential in $U \setminus \Sigma$, so no such V exists and we can conclude that Σ is not 1-alg at x.

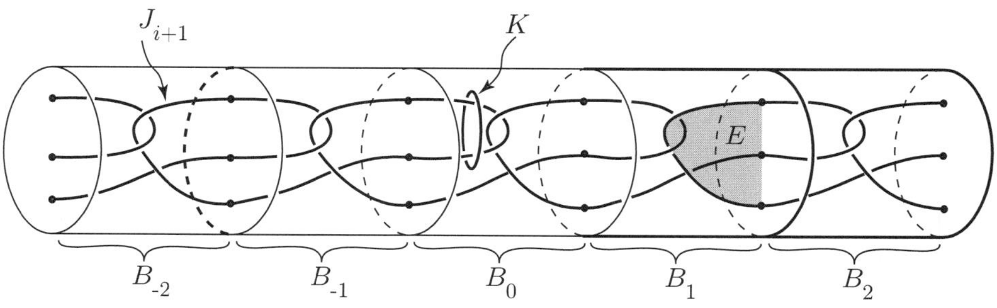

Figure 2.29. The loop K is linked around J_{i+1}

Suppose an embedding $S^1 \to K$ extends to a map $g : B^2 \to U \setminus \Sigma$. The choice of U implies that $g(B^2)$ will miss at least one of the blocks in T_i, so we can find a sequence of consecutive blocks $B_{-n}, \ldots, B_n$ such that $g(B^2) \cap T_i$ is contained in the 3-cell $A = B_{-n} \cup \cdots \cup B_n$ and that $g(B^2)$ does not intersect either end of A (see Figure 2.29). Put g in general position relative to ∂T_i. Then $g^{-1}(\partial T_i)$ will consist of a finite number of disjoint simple closed curves. Consider one such simple closed curve C that is innermost in the sense that no other curve is in its interior relative to B^2. The interior of C (in B^2) is mapped by g either to $\mathbb{R}^3 \setminus J_i$ or to $J_i \setminus \Sigma$. In either case, it follows that $g(C)$ does not link J_i homologically and thus is an inessential curve on the annulus $\partial A \cap \partial T_i$. Hence we can modify g so that it maps the interior of C into $\partial A \cap \partial T_i$ and then push the image to one side to eliminate C from $g(B^2) \cap \partial T_i$. This process can be continued inductively and results in a new map g with the property that $g(B^2) \cap \partial T_i = \emptyset$, which means that $g(B^2) \subset A \setminus \Sigma$.

The previous paragraph shows that if K is inessential in $U \setminus \Sigma$, then K is inessential in $A \setminus \Sigma$. In this paragraph we show that K is essential in $A \setminus \Sigma$, which completes the proof that Σ is not 1-alg at x. An argument like that in the construction of Antoine's necklace (see Exercise 2.8.5(b)) establishes that K is essential in $A \setminus J_{i+1}$. Assume that $S^1 \to K$ extends to a map $g : B^2 \to A \setminus \Sigma$. Put g in general position with respect to ∂T_{i+1}. Then $g^{-1}(\partial T_{i+1})$ will consist of a finite number of simple closed curves. Since the images of these curves do not go all the way around T_{i+1}, each of them represents some multiple of the meridian of T_{i+1}. Let C be one of the curves. If $g(C)$ is inessential on ∂T_{i+1}, then g can be modified (as in the previous paragraph) to eliminate that curve of intersection. Thus there must be at least one of these curves C whose image is a nonzero multiple of the meridian of T_{i+1}. But then $g(C)$ homologically links Σ and so $g(B^2) \cap \Sigma \neq \emptyset$.

Exercise 2.8.5.

(a) Prove that the Bing sling Σ is a simple closed curve by verifying that the embedding $e : J_1 \to \mathbb{R}^3$ described above satisfies $e(J_1) = \Sigma$.

(b) Prove that the loop K shown in Figure 2.29 is essential in $A \setminus J_{i+1}$, where $A = B_{-n} \cup \cdots \cup B_n$. [Hint: First observe that K is essential in $B_0 \setminus J_{i+1}$ by results established earlier in the chapter. Then consider the inclusion of K into $(B_0 \cup B_1) \setminus J_{i+1}$. Use an argument like that in the proof of Lemma 2.1.4 to show that if K is inessential in $(B_0 \cup B_1) \setminus J_{i+1}$ then K is inessential in $(B_0 \cup B_1) \setminus (J_{i+1} \cup E)$, where E is the disk shown in Figure 2.29. Check that the embedding of J_{i+1} in $(B_0 \cup B_1) \setminus E$ is the same as the embedding of J_{i+1} in B_0. Next add in B_{-1} and proceed inductively.]

(c) Let $f : S^1 \to \mathbb{R}^3 \setminus \Sigma$ be a map such that $f(S^1)$ homologically links Σ and let $F : B^2 \to \mathbb{R}^3$ be an extension of f. Prove that $F^{-1}(\Sigma)$ contains a Cantor set. Use this fact to give an alternative proof that Σ is everywhere wild.

(d) Prove that Σ is homogeneously embedded; i.e., for every pair of points $x, y \in \Sigma$ there exists a homeomorphism $h : \mathbb{R}^3 \to \mathbb{R}^3$ such that $h(\Sigma) = \Sigma$ and $h(x) = y$.

2.8.6. Bing's hooked rug. Bing's hooked rug is an example of an everywhere wild 2-sphere in $\mathbb{R}^3$. The wild sets of the Alexander and Antoine spheres are Cantor sets while the wildness of the Fox-Artin sphere is concentrated at a single point. By contrast, the wildness of the hooked rug is totally diffused: the embedding is wild at every point. Nevertheless, each arc in the 2-sphere is tame. The complement of Bing's hooked rug is not simply connected. To the contrary, near any point of the 2-sphere one can find a loop that is essential in the complement.

Like the previous examples, the construction of Bing's hooked rug is described in two different ways. The hooked rug can be understood as the boundary of the intersection of a nested sequence of compact ∂-manifolds; this view is useful in proving that the example fails to be 1-LCC (and is therefore wild). The example can also be realized as the limit of a sequence of embeddings of the 2-sphere; this view is useful in proving that it is a topologically embedded sphere.

The construction begins with a round 3-cell F_0. Cover the surface of F_0 with a sequence $E_1, E_2, \ldots, E_n$ of disks that have disjoint interiors and such that $E_i \cap E_{i+1}$ is an arc in the boundary of each. (Count cyclically so that $E_n \cap E_1$ is also an edge of each.) Attach to each E_i a tube with a solid torus at the end. The union of the tube and solid torus is called an *eyebolt*—see Figure 2.30.

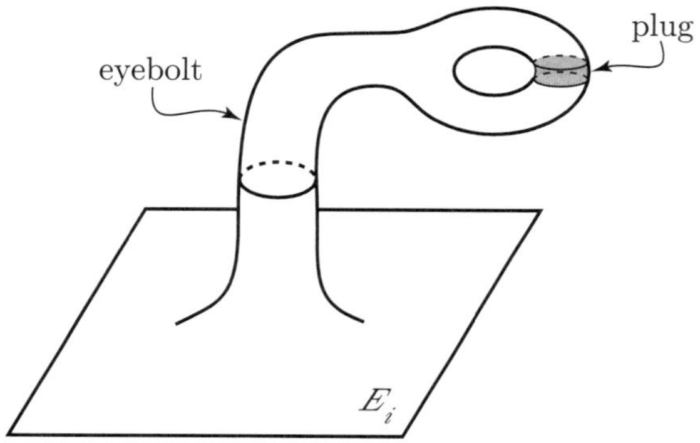

Figure 2.30. A disk E_i with an eyebolt attached

Hook the eyebolt on E_i to the base of the eyebolt on E_{i+1} and the eyebolt on E_n to the base of that on E_1 in a cyclic pattern as indicated in Figure 2.31. The original ball F_0 together with the union of all the eyebolts forms a solid 3-dimensional object H_1. Note that H_1 consists of a 3-cell with eyebolts attached, so H_1 is a cube with handles.[3] Shrink F_0 slightly before attaching the eyebolts so that H_1 is contained in the interior of F_0.

A *plug* for an eyebolt is a copy of $B^2 \times (0, 1)$ that cuts off the eyebolt as shown in Figure 2.30. Remove a plug from each of the eyebolts in H_1; the resulting solid is a 3-cell F_1. The 2-sphere ∂F_1 is the first approximation to Bing's hooked rug. There is an obvious homeomorphism $F_0 \to F_1$. The distance any point is moved by this homeomorphism is at most twice the maximum diameter of the disks $E_1, E_2, \ldots, E_n$, so we can control the size of this homeomorphism by controlling the number and size of the disks E_i.

[3]A *cube with handles* is the regular neighborhood in $\mathbb{R}^3$ of a 1-dimensional polyhedron.

Figure 2.31. H_1, the first stage in the hooked rug construction

The surface of F_1 is covered by disks $E_1', E_2', \ldots, E_n'$ that have the same boundaries as the original disks $E_1, E_2, \ldots, E_n$. Cover each E_i' with a sequence of 15 (or more) disks and erect a new, smaller eyebolt on each of the disks. Hook the eyebolts on E_i' together in a circular pattern as indicated in Figure 2.32. Define H_2 to be the union of F_1 and all the second-stage eyebolts. Then H_2 is another cube with handles. We again shrink F_1 slightly before attaching the second-stage eyebolts so that $H_2 \subset \text{Int}\, H_1$.

Figures 2.31 and 2.32 provide drawings of stage one and stage two, respectively, of the hooked rug construction; Color Plates 5–6 display photographs of physical models of those same stages.

Remove plugs from each of the second-stage eyebolts to form a 3-cell F_2. Note that again there is a homeomorphism $F_1 \to F_2$ and that the distance any point is moved by this homeomorphism is at most twice the maximum diameter of any of the disks used at the second stage. Thus we can make the homeomorphism $F_1 \to F_2$ close to the identity by simply subdividing into more second stage disks and making the corresponding eyebolts small.

The construction is continued inductively to produce a nested sequence $H_1 \supset H_2 \supset H_3 \supset \ldots$ of cubes with handles and a sequence $F_0, F_1, F_2, \ldots$ of 3-cells. Define

$$H = \cap_{i=1}^{\infty} H_i$$

and define Bing's hooked rug to be the boundary of H. We claim that H is a topological 3-cell and that ∂H is an everywhere wild 2-sphere.

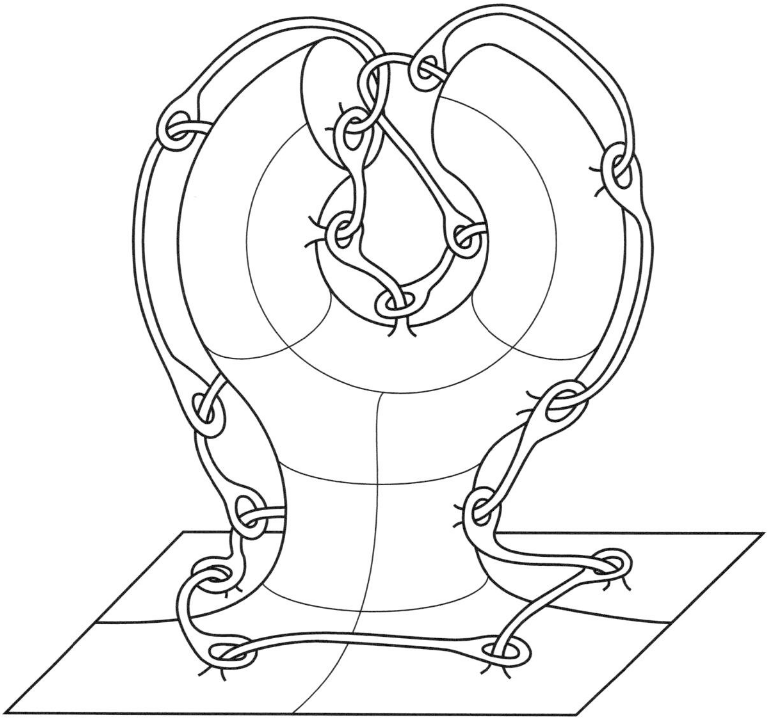

Figure 2.32. The second stage in the hooked rug construction

To prove that H is a 3-cell, we observe that the construction can be done in such a way that the homeomorphism $F_{i-1} \to F_i$ is close to the identity. While Proposition 2.2.2 does not quite apply, the same kind of proof as was used for the Alexander horned sphere shows that the composition of these homeomorphisms converges to an embedding $h : F_0 \to \mathbb{R}^3$. It is not difficult to see that h maps F_0 onto H (Exercise 2.8.6 (a)).

We prove that ∂H is everywhere wild by showing that H fails to be 1-LCC at each point of ∂H. Specifically, we prove that a loop in the complement of H that circles the base of one of the eyebolts at stage i is essential in the complement at each subsequent stage and therefore will be essential in the complement of the intersection H. The small eyebolts at a later stage are spread densely over the sphere, so there is such a loop near every point on the limiting sphere. Thus H fails to be 1-LCC at any point of ∂H.

In order to prove the claims in the preceding paragraph we break down the transition from H_{i-1} to H_i into three steps. Start with H_{i-1}. Remove a plug from each of the eyebolts in H_{i-1} and replace it with a pillbox (see definition on page 47). Call the new ∂-manifold H_i'. Note that H_i' is a 3-cell with two solid handles attached for each eyebolt in H_{i-1}. Now shrink

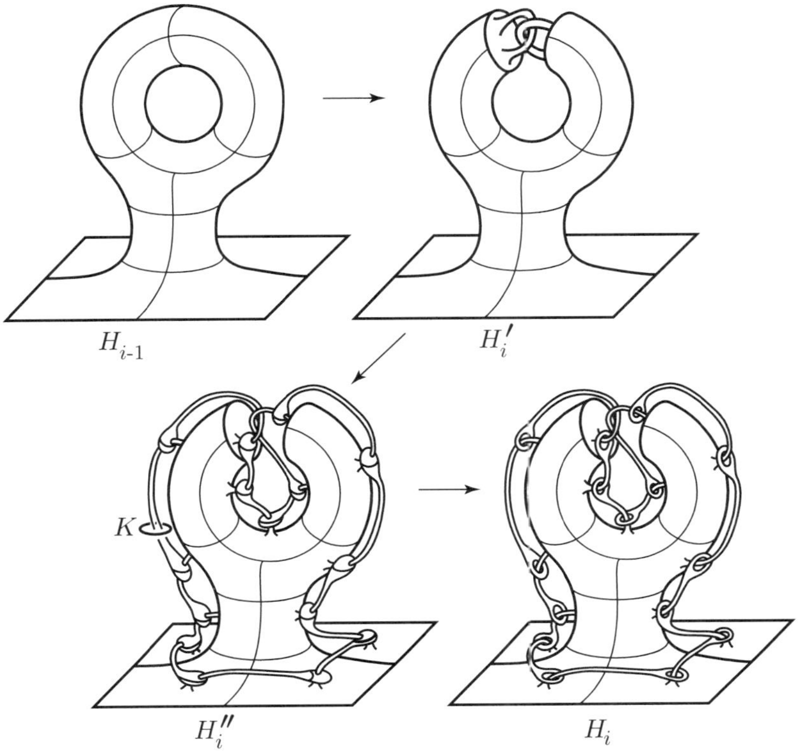

Figure 2.33. The three-step transition from H_{i-1} to H_i

H'_i slightly and add additional simple unlinked handles until there is one handle for each eyebolt in H_i. These handles can be isotoped to resemble the eyebolts that we want to add at this stage. More specifically, each of them can be deformed to look like a tube with a bulb at the end and with the base of one handle going straight through the bulb of the previous handle. The ∂-manifold formed in this way is called H''_i and it is illustrated in Figure 2.33. Finally, convert the handles to eyebolts by drilling a tunnel through the bulb in each handle to create a hole for the next handle to pass through. The resulting ∂-manifold is H_i.

Repeated application of the following lemma shows that $\pi_1(\mathbb{R}^3 \smallsetminus H''_i) \to \pi_1(\mathbb{R}^3 \smallsetminus H_i)$ is one-to-one.

Lemma 2.8.1. *Let C be a 3-cell in $\mathbb{R}^3$, let B_1, B_2, and B_3 be three disjoint disks on ∂C, let T be a solid torus in C such that $T \cap \partial C = B_1$, and let S be a 3-cell in C such that $S \cap \partial C = B_1 \cup B_2$. Assume T and S are linked as indicated in Figure 2.34. Let X be a closed subset of $\mathbb{R}^3$ such that $X \cap C = B_1 \cup B_2 \cup B_3$. If $\pi_1(\partial C \smallsetminus (B_1 \cup B_2 \cup B_3)) \to \pi_1(\mathbb{R}^3 \smallsetminus (X \cup \operatorname{Int} C))$ is one-to-one, then $\pi_1(\mathbb{R}^3 \smallsetminus (X \cup C)) \to \pi_1(\mathbb{R}^3 \smallsetminus (X \cup S \cup T))$ is one-to-one.*

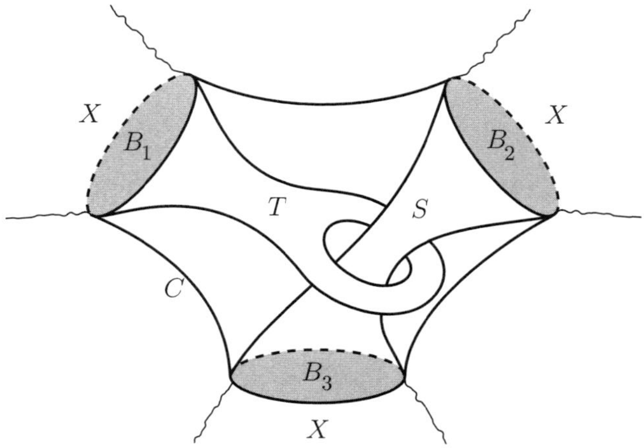

Figure 2.34. Detail of $H_i'' \to H_i$

Proof. Exercise 2.8.6 (b). □

It is clear that a loop K that goes around one of the small handles of H_i'' is essential in $\mathbb{R}^3 \smallsetminus H_i''$ (see Figure 2.33). By Lemma 2.8.1, K is also essential in $\mathbb{R}^3 \smallsetminus H_i$. But then K is essential in $\mathbb{R}^3 \smallsetminus H_{i+1}'$ by Lemma 2.1.9. It is clear that any loop that is essential in $\mathbb{R}^3 \smallsetminus H_{i+1}'$ is also essential in $\mathbb{R}^3 \smallsetminus H_{i+1}''$. Several applications of Lemma 2.8.1 show that K is also essential in $\mathbb{R}^3 \smallsetminus H_{i+1}$. Since this argument can be continued inductively, we see that K is essential in $\mathbb{R}^3 \smallsetminus H$. This completes the proof that H fails to be 1-LCC at every point of ∂H.

As asserted earlier, every arc in ∂H is tame in $\mathbb{R}^3$. We will not prove this, but in a later chapter we will develop tools that could be used to show that every arc in ∂H is 1-alg. A complete proof that arcs in ∂H are tame may be found in (Bing, 1961a).

Exercise 2.8.6.

(a) Prove that Bing's hooked rug is a topological sphere by verifying that there is an embedding $h : F_0 \to \mathbb{R}^3$ such that $h(F_0) = H$.

(b) Prove Lemma 2.8.1.

2.8.7. The Alford sphere. Our final example is a 2-sphere in $\mathbb{R}^3$ whose wild set is an arc. Its construction involves a retooling of the one just given for Bing's hooked rug. Instead of having the eyebolts wander all over the sphere so as to be dense in the limit, we will erect eyebolts just along an arc in the 2-sphere.

Start with a round 3-cell. On its boundary identify a narrow rectangle R with centerline L. Subdivide R into a large number of squares and then attach an eyebolt to each of the squares. Two consecutive eyebolts should be hooked together, but the last one should be left dangling as shown in Figure 2.35.

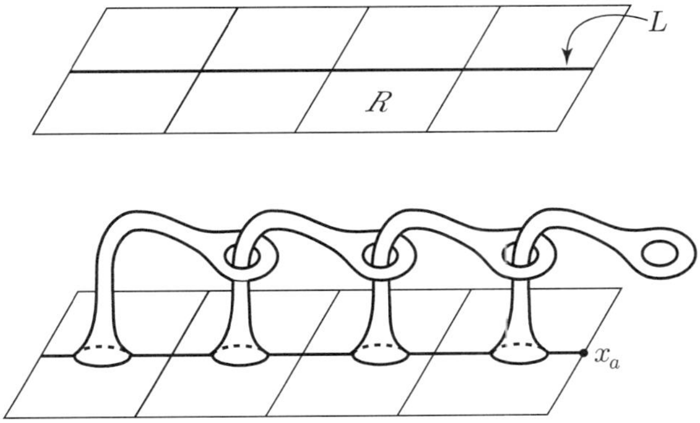

Figure 2.35. The first stage in the construction of the Alford sphere

Now remove a plug from each eyebolt. The resulting solid is a 3-cell and there is a new segment L on its boundary that goes up and over each cut eyebolt. Cover L with a sequence of much smaller squares and erect a new sequence of smaller eyebolts, one in each of the small squares. Again hook consecutive eyebolts together in a linear chain and leave the last one dangling. More specifically, the second-stage eyebolts associated with the ith disk from the first stage should be hooked together as indicated in Figure 2.36; the last second-stage eyebolt for the ith disk should be hooked to the first second-stage eyebolt associated with the $(i+1)$st disk and the very last second-stage eyebolt should be left dangling.

The process is continued inductively and the limit is a 3-cell. The boundary of this 3-cell is the Alford 2-sphere S_A. It is clear that S_A is locally flat at each point not on the limit arc. The limit arc γ_A is called the Alford arc, and x_A is used to denote the endpoint of γ_A near which the eyebolts are left dangling. We prove that S_A fails to be 1-LCC at each point of $\gamma_A \setminus x_A$, so the wild set of S_A is exactly γ_A. In fact, the proof shows that the Alford arc itself fails to be 1-alg at points of $\gamma_A \setminus x_A$, so γ_A is a new example of an everywhere wild arc.

In order to demonstrate that S_A fails to be 1-LCC at points of $\gamma_A \setminus x_A$, we identify a small loop near most points of γ_A that cannot be shrunk to a point in a small set without hitting the Alford sphere. Specifically, let K be a small loop in the complement of S_A that circles the base of one of the ith stage eyebolts as shown in Figure 2.37. Add a short arc α that connects the

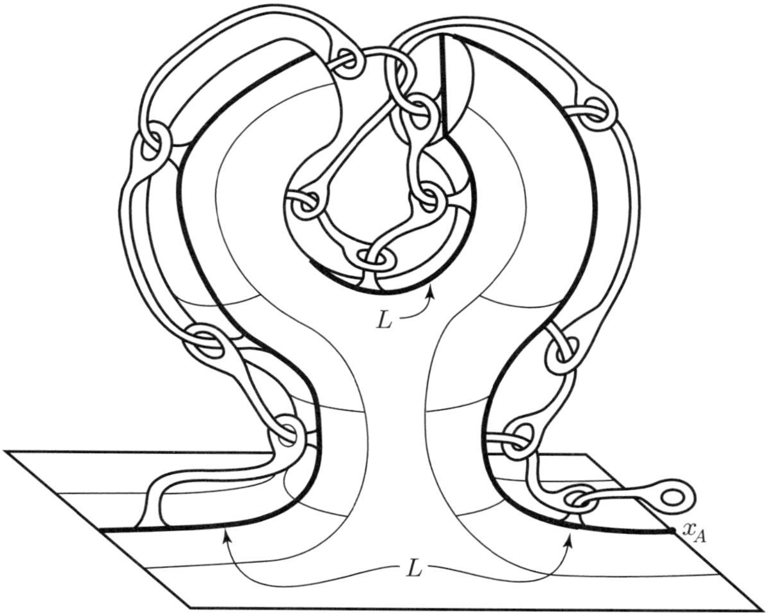

Figure 2.36. The linking of the second stage eyebolts

end of the dangling ith-stage eyebolt to the end of L. By Lemma 2.8.1, K is essential in the complement of the ith stage with α added. Any homotopy that shrinks K to a point in the complement of the ith stage must therefore intersect α and cannot be confined to a small neighborhood of a point on $\gamma_A \setminus x_A$.

Figure 2.37. The arc α

In fact, K cannot be shrunk to a point in a small subset of the complement of any subsequent stage either. In order to see this, let us say that the construction is done in such a way that α touches the end of the dangling eyebolt at each subsequent stage of the construction. Then the techniques of the preceding section can be used to show that K is essential in the complement of each stage with a subarc of α added. It follows that K cannot be shrunk to a point in a small subset of the complement of S_A.

Exercise 2.8.7.

(a) Prove that the construction described above can be carried out in such a way that the limit is an embedded 3-cell.

(b) Fill in the details of the proof that the Alford sphere is wild at each point of the Alford arc. Prove that the Alford arc itself is an everywhere wild embedding of $[0,1]$ in $\mathbb{R}^3$.

(c) Prove that the Alford arc is 1-LCC at one of its endpoints. [This shows that an embedding can be 1-LCC at a wild point.]

(d) Prove that the Alford arc is cellular.

(e) Prove that the Alford construction can be modified to produce an example of an embedding of S^2 in $\mathbb{R}^3$ whose wild set is homeomorphic to any finite tree. Prove that this construction can be done in such a way that the wild set is cellular.

(f) Prove that there are uncountably many inequivalent embeddings of S^2 in $\mathbb{R}^3$ by producing embeddings whose wild sets are 1-dimensional compacta that are limits of trees.

Historical Notes. The Fox-Artin arc is one of many examples of wild embeddings discovered by R. H. Fox and E. Artin (1948). The mildly wild arc shown in Figure 2.26 was described by R. H. Fox and O. G. Harrold (1962); they named such arcs Wilder arcs after R. L. Wilder, who was the first to consider them. The Bing sling was described in (Bing, 1956). Bing's hooked rug appeared in (Bing, 1961a). D. Gillman (1964) revised the hooked rug technology to produce an everywhere wild 2-sphere in $\mathbb{R}^3$ that bounds a cellular 3-cell. W. R. Alford (1962) capitalized on the work of both Bing and Gillman in developing the Alford sphere.

2.9. Embeddings that are piecewise linear modulo one point

We conclude the chapter with a flattening theorem for codimension-one spheres in S^n. It assures that any $(n-1)$-sphere in S^n, $n \geq 4$, that is piecewise linear modulo one point is flat. This contrasts with the situation in ambient dimension three, where the Fox-Artin sphere is locally PL modulo one point but still wild.

The promised result stands among many flattening theorems to be proved in the text. We include it here in this preliminary chapter because its proof stems from a marvelous argument, one that does not rely on the more elaborate techniques to be developed later, and because it serves as an early indication of the fact that high-dimensional embedding phenomena differ from those encountered in dimension three.

To briefly describe the generalizations to be proved later, we need one additional bit of terminology. Let Σ be an $(n-1)$-sphere topologically embedded in S^n. A subset K of Σ that is homeomorphic to either a cell or a sphere is said to be *twice flat* provided it is flat when considered as a subset of the sphere Σ and also flat when considered as a subset of S^n. In Chapter 7 we will generalize Theorem 2.9.3 in two different ways, showing that Σ is flat if it is locally flat modulo a twice flat Cantor set or if it is locally flat modulo a twice flat cell of dimension not equal to $n-3$.

We make no attempt to state the theorems in this section in their ultimate generality since we plan to improve them later. Instead we state them with hypotheses strong enough to minimize proof technicalities, in order to more easily expose the pivotal ideas.

Definition. Let K be a finite simplicial complex with $p \in |K|$. A map $f : |K| \to M$ into a piecewise linear manifold M is said to be *piecewise linear modulo p* if there exists a locally finite triangulation K' of the noncompact polyhedron $|K| \smallsetminus \{p\}$ such that each simplex of K' is contained in a simplex of K and f is linear on each simplex of K'.

We treat a special case first. The hypothesis $n \geq 4$ is already needed in this special case.

Proposition 2.9.1. *Let C^n be an n-simplex. If $e : C^n \to S^n$, $n \geq 4$, is an embedding that is piecewise linear modulo one vertex, then $S^n \smallsetminus \operatorname{Int} e(C^n)$ is a topological n-cell.*

Definition. Let (A, B) be a pair of closed subsets of the space X. Define $G_{A,B}$ to be the decomposition of $A \times [0,1]$ whose nondegenerate elements are the arcs $\{x\} \times [0,1]$ with $x \in B$. A *collar of A pinched at B* is an embedding $h : A \times [0,1]/G_{A,B} \to X$ such that $h(x,0) = x$ for every $x \in A$ and $h(A \times [0,1]/G)$ is a neighborhood of $A \smallsetminus B$. In case the subset B is clear from the context, we will simply refer to h as a *pinched collar.*

Lemma 2.9.2. *Let (A, B) be a pair of closed subsets of the space X. If A is locally collared in X at each point of $A \smallsetminus B$, then there is a collar of A pinched at B.*

Proof. Exercise 2.9.1. □

Proof of Proposition 2.9.1. Let C^n be an n-simplex and let $e : C^n \to S^n$ be an embedding that is piecewise linear modulo the vertex $v \in C^n$. Let D^n be a second n-simplex such that $C^n \subset D^n$ and $C^n \cap \partial D^n = \{v\}$. Let E^n be a third n-simplex such that $E^n \subset \operatorname{Int} C^n$. Pick a vertex w of E^n and let α be the straight line segment from w to v. We may assume that $\alpha \cap E^n = \{w\}$. (See Figure 2.38.)

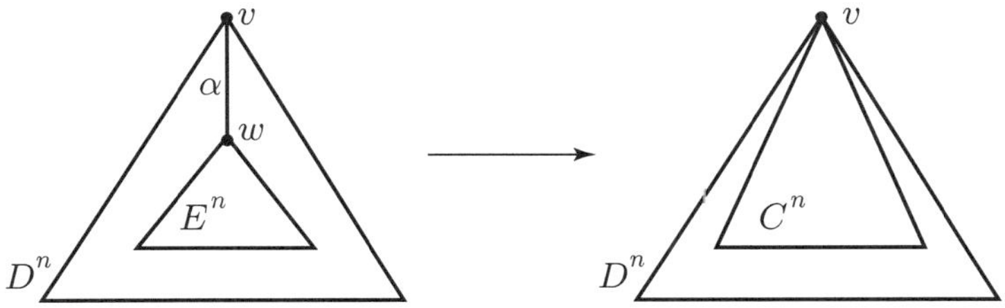

Figure 2.38. Shrink α to a point to map (D^n, E^n) to (D^n, C^n)

By Lemma 2.9.2, e can be extended to a topological embedding $h : D^n \to S^n$. Define

$$G = \overline{S^n \setminus h(C^n)} \quad \text{and} \quad F = \overline{S^n \setminus h(E^n)}.$$

Generalized Schönflies Theorem 2.4.8 guarantees that F is a topological n-cell. We will prove that G is also an n-cell by proving that G is homeomorphic to F.

Note that there is a map $(D^n, E^n) \to (D^n, C^n)$ that is the identity on ∂D^n and whose only nondegenerate inverse set is α. This map of D^n induces a map from $F \cap h(D^n)$ to $G \cap h(D^n)$ that is the identity on $h(\partial D^n)$. Extending via the identity produces a continuous map $g : F \to G$ whose only nondegenerate inverse set is $h(\alpha)$.

Now $h(\alpha)$ is a locally flat arc by Exercise 2.5.4. Hence there is a continuous function $f : F \to F$ whose only nondegenerate inverse set is $h(\alpha)$. It is easy to check that $f \circ g^{-1}$ is a well-defined homeomorphism from G to F. □

Definition. Let $\Sigma \subset S^n$ be a topologically embedded $(n-1)$-sphere and let $p \in \Sigma$. A *bicollar of Σ pinched at p* is an embedding

$$c : S^{n-1} \times [-1, 1]/\{v\} \times [-1, 1] \to S^n$$

such that $c(S^{n-1} \times \{0\}) = \Sigma$ and $c(v, 0) = p$. (Here v is a point in S^{n-1} and $S^{n-1} \times [-1, 1]/\{v\} \times [-1, 1]$ is the quotient space of $S^{n-1} \times [-1, 1]$ formed by shrinking $\{v\} \times [-1, 1]$ to a point.)

The following result is the main theorem in the section.

Theorem 2.9.3. *If $\Sigma \subset S^n$, $n \geq 4$, is an embedded $(n-1)$-sphere and Σ has a bicollar pinched at $p \in \Sigma$, where the bicollar is piecewise linear modulo the preimage of p, then Σ is flat.*

Proof. To simplify the notation, we denote $S^{n-1} \times [-1, 1]/\{v\} \times [-1, 1]$ by Q and use v^* to denote the point in Q corresponding to $\{v\} \times [-1, 1]$. By hypothesis, there exists an embedding $c : Q \to S^n$ such that $c(v^*) = p$

and c is piecewise linear modulo v^*. Define H to be the closure of the complementary domain of $c(S^{n-1}\times\{-1\})$ that does not contain Σ and define K to be the closure of the complementary domain of $c(S^{n-1}\times\{1\})$ that does not contain Σ. We will prove that $c(S^{n-1}\times[-1,0]/\{v\}\times[-1,0])\cup H$ and $c(S^{n-1}\times[0,1]/\{v\}\times[0,1])\cup K$ are both n-cells.

Let C^n be an n-simplex in $\operatorname{Int} Q\cup\{v^*\}$ such that v^* is a vertex of C^n and $C^n\cap(S^{n-1}\times\{0\})$ is a flat disk; define $Q' = Q\smallsetminus\operatorname{Int} C^n$. Figure 2.39 shows two views of the pinched bicollar with Q' shaded in each. In the first view, all of Q is shown with $S^{n-1}\times\{0\}$ as its core. In the second view, ∂C^n has been turned inside out so that only Q' is visible. We can choose C^n so that Q' is PL homeomorphic to an n-simplex Δ with the interiors of two PL n-cells removed. Those two n-cells meet at v^* and are otherwise in the interior of Δ. Their boundaries are $S^{n-1}\times\{-1\}$ and $S^{n-1}\times\{1\}$. In addition, $S^{n-1}\times\{0\}$ separates Q' into two pinched collars, Q_- and Q_+, so that Q_- contains $S^{n-1}\times\{-1\}$ and Q_+ contains $S^{n-1}\times\{1\}$. The left half of Figure 2.39 shows that we can choose C^n so that Q_- is naturally homeomorphic to $S^{n-1}\times[-1,0]/\{v\}\times[-1,0]$ and Q_+ is naturally homeomorphic to $S^{n-1}\times[0,1]/\{v\}\times[0,1]$. Thus we can complete the proof by showing that $c(Q_-)\cup H$ and $c(Q_+)\cup K$ are both n-cells.

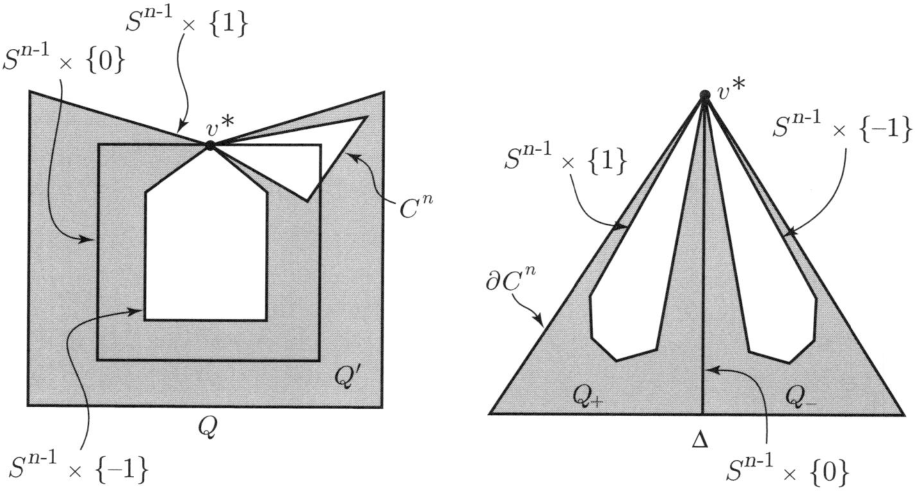

Figure 2.39. Two different views of the bicollar

Proposition 2.9.1 implies that the closure of the complement of $c(C^n)$ is an n-cell. But $S^n\smallsetminus\operatorname{Int} c(C^n) = c(Q')\cup H\cup K$, so $c(Q')\cup H\cup K$ is an n-cell. That is to say, sewing H and K to Q' along $S^{n-1}\times\{-1\}$ and $S^{n-1}\times\{1\}$, respectively, results in an n-cell; more specifically, if h and k are the maps defined by $h = c^{-1}|\operatorname{Fr} H$ and $k = c^{-1}|\operatorname{Fr} K$, then $Q'\cup_h H\cup_k K$ is an n-cell. We will use an infinite construction to show that $Q_-\cup_h H$ and $Q_+\cup_k K$ are also n-cells.

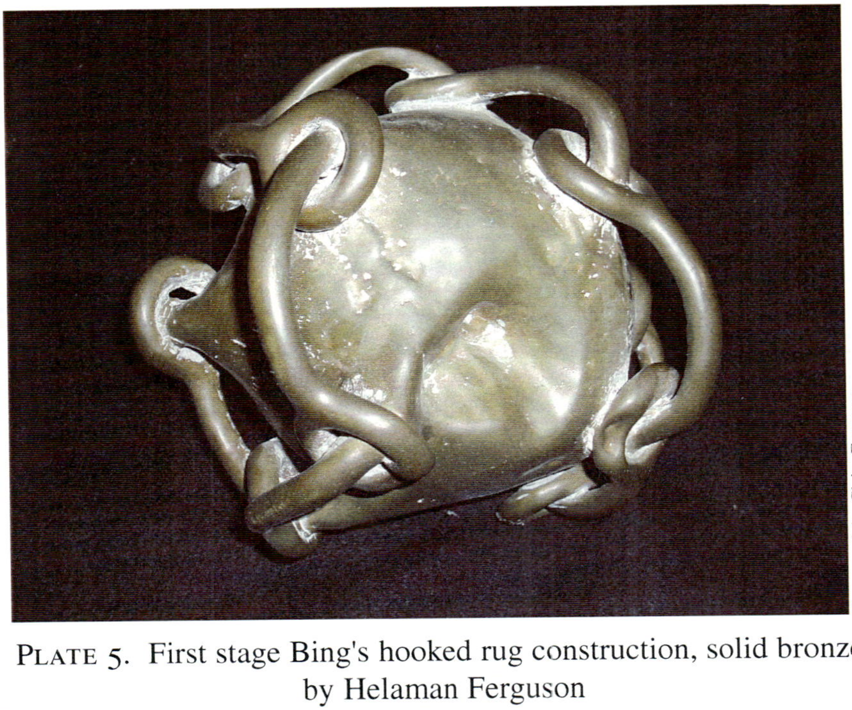

Photograph by Claire Ferguson

PLATE 5. First stage Bing's hooked rug construction, solid bronze, by Helaman Ferguson

Photograph by Claire Ferguson

PLATE 6. Second stage Bing's hooked rug construction, solid bronze, by Helaman Ferguson

Let $B_1, B_2, B_3, \ldots$ be a sequence of n-cells such that, for each i, $B_i \cap B_{i+1}$ is a flat $(n-1)$-cell standardly embedded in both ∂B_i and ∂B_{i+1}, $\cup_{i=1}^{\infty} B_i$ is an n-cell, and there is a point q such that $B_i \cap B_j = \{q\}$ for $|i-j| > 1$. Figure 2.40 shows one way to construct such a sequence by starting with a round n-cell B^n and subdividing it via hyperplanes, any two of which intersect B at a point $q \in \partial B^n$.

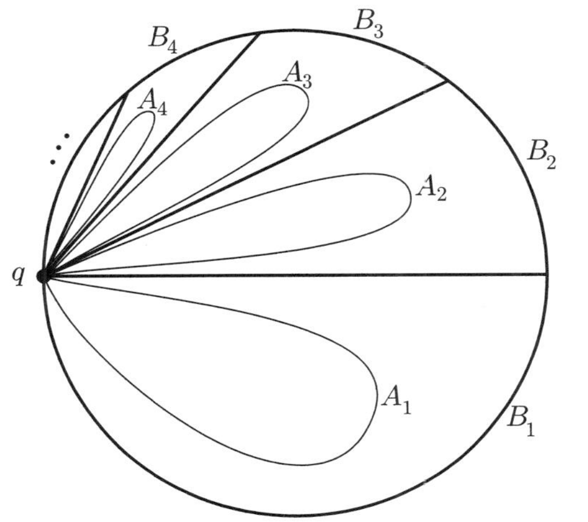

Figure 2.40. A sequence of cells whose union is a cell

For each i, choose a flat n-cell $A_i \subset B_i$ such that $A_i \cap \partial B_i = \{q\}$. If i is odd, define B_i' to be $(B_i \setminus \operatorname{Int} A_i) \cup_{h'} H$; if i is even, define B_i' to be $(B_i \setminus \operatorname{Int} A_i) \cup_{k'} K$. The maps h' and k' are appropriate modifications of h and k, respectively. As demonstrated above, $B_i' \cup B_{i+1}'$ is an n-cell for every odd integer i. Thus $\cup_{i=1}^{\infty} B_i'$ is an n-cell. On the other hand, there is a homeomorphism of Q' to itself that interchanges $S^{n-1} \times \{-1\}$ and $S^{n-1} \times \{1\}$ and is the identity on the other component of $\partial Q'$. [This does not look right in the 2-dimensional figure, but such a self homeomorphism exists as long as $n \geq 3$.] Therefore $B_i' \cup B_{i+1}'$ is also an n-cell for i even and so $\cup_{i=2}^{\infty} B_i'$ is an n-cell. It follows that B_1' is an n-cell. A similar argument shows that B_2' is an n-cell, so the proof is complete. □

Historical Notes. Theorem 2.9.3 is due to J. C. Cantrell. The statement appeared in (Cantrell, 1963a) and the proof is contained in (Cantrell, 1963b) and (Cantrell and Edwards, 1963). The technique of pairing off the infinite sequence of cells in two different ways is often called the "Mazur swindle." Mazur (1959) (1961b) first used the technique to prove the special case of the Generalized Schönflies Theorem in Exercise 2.9.5. Other applications of the technique are described in (Mazur, 1964b) and (Mazur, 1966). The result in Exercise 2.9.4 was first proved by P. H. Doyle and J. G. Hocking (1960).

Exercises

2.9.1. Prove Lemma 2.9.2. [Hint: The open subset $A \setminus B$ has an ordinary collar by Theorem 2.4.10. Carefully trim this collar back to a pinched collar.]

2.9.2. For each $n \geq 3$ there exists a wild $(n-1)$-sphere $\Sigma \subset S^n$ whose wild set is a twice flat $(n-3)$-cell.

2.9.3. For each $n \geq 3$ there exists a wild $(n-1)$-sphere $\Sigma \subset S^n$ whose wild set is an $(n-2)$-cell that is tame in Σ.

2.9.4. If $\Sigma \subset S^3$ is a 2-sphere that is locally flat modulo a point p and there is an arc $A \subset \Sigma$ passing through p that is flat in S^3, then Σ is flat.

2.9.5. Use the technique of proof of Theorem 2.9.3 to give a new proof of the following special case of the Generalized Schönflies Theorem: If $e : S^{n-1} \to S^n$ is a locally flat topological embedding such that $e|U$ is PL for some open subset U of S^{n-1}, then e is flat.

Chapter 3

Engulfing, Cellularity, and Embedding Dimension

Engulfing, a powerful piecewise linear method, has proved itself to be an effective tool for unlocking mysteries of high-dimensional-manifolds. It is a process by which an open subset of a manifold is adjusted, typically via ambient isotopy, to absorb a predetermined polyhedron. Although engulfing receives brief mention in Rourke and Sanderson (1972), its full power is not developed there.

There are several distinct versions of engulfing, connected by a common thread. This chapter presents two basic types of engulfing theorems, Stallings engulfing and Bryant-Seebeck engulfing. In the first of them the hypotheses are global in nature and in the second the conditions are more local. In the first form of engulfing one makes no effort to limit the amount of the motion necessitated by the engulfing process, whereas in the second form one subjects the engulfing isotopy to strict motion control. The two theorems presented do not cover every application of engulfing—far from it—but they illustrate the essence of the method and their statements are sufficiently general to address many applications of interest.

The chapter also contains sample applications that demonstrate the power and utility of the two types of engulfing theorems. The Stallings method is employed to characterize the cellular sets in a PL manifold while the more controlled Bryant-Seebeck method is used to explicate a notion of dimension for compact subsets of a manifold. In both situations, engulfing allows us to see that certain local fundamental group conditions detect the well-behaved embeddings. Specifically, the cellularity criterion detects cellular embeddings of contractible compacta and the 1-LCC condition detects those embedded compacta whose topological dimension equals their embedding dimension. The last two sections of the chapter offer a brief exploration of fractals, Menger continua and Hausdorff dimension, as well as of their connections to embedding dimension.

3.1. Engulfing without control

This section contains a statement and proof of a basic uncontrolled engulfing theorem. Before presenting the result itself, we strive to motivate our study by illustrating how the ability to engulf certain polyhedra can bring insight into the structure of the surrounding manifold.

Since engulfing is rooted in PL topology, we use the notation and terminology of Rourke and Sanderson (1972) and begin with a quick review. Let K be a simplicial complex and L a subcomplex of K. The *simplicial complement* $C(L, K)$ of L in K is the subcomplex consisting of all simplices in K disjoint from $|L|$. A subcomplex L of K is called a *full subcomplex* if $\sigma \in K$ and $\partial\sigma \subset L$ imply $\sigma \in L$. Note that if a given subcomplex L is not already full, it can be made full by simply taking a derived subdivision near L (Rourke and Sanderson, 1972, page 32). When L is full in K, then each simplex in K is the join of a simplex in L and a simplex in $C(L, K)$.

Consider an n-dimensional PL ∂-manifold M with triangulation T. Define L to be the k-dimensional skeleton of T and define $P = C(L', T')$, where T' is a derived subdivision of T and L' is the corresponding subdivision of L. It is easy to check that $\dim P = n - k - 1$. The complex P is called the *dual* $(n - k - 1)$-skeleton. Every simplex in T' is the join of a simplex in the k-skeleton and one in the dual $(n - k - 1)$-skeleton.

The following simple lemma signals how to exploit this join structure to stretch an open set containing one join factor to cover the complement of an open set containing the other join factor. Once the lemma is proved, we give an immediate application showing how the stretching process, in turn, helps characterize the n-sphere.

Lemma 3.1.1 (Stretching across the join structure). *Suppose K is a triangulation of a PL w-manifold W, L is a finite, full subcomplex of K, and $C(L, K)$ is the simplicial complement of L in K. Let U and V be neighborhoods of $|L|$ and $|C(L, K)|$, respectively. Then there exists an ambient isotopy of W, starting at Id_W and ending at $h : W \to W$, such that*

$h(U) \cup V = W$; moreover, the isotopy can be required to fix $|L \cup C(L, K)|$, to have compact support, and to move points less than mesh(K).

Proof. Choose derived neighborhoods N_1 and N_2 of $|L|$ with $N_1 \subset U$ and $N_2 \supset W \setminus V$. (The join structure mentioned above ensures that N_2 exists.) The canonical isotopy between derived neighborhoods (Rourke and Sanderson, 1972, Theorem 3.24) carries N_1 onto N_2 and has the required features. In particular,

$$h(U) \cup V \supset h(N_1) \cup V = N_2 \cup V \supset (W \setminus V) \cup V = W. \quad \square$$

Proposition 3.1.2. *If M^m is a compact PL m-manifold such that every finite k-dimensional polyhedron in M^m, $k \leq m/2$, is contained in the interior of an m-ball, then M^m is (topologically) homeomorphic to S^m.*

Proof. Choose a triangulation T of M^m. Let p be the greatest integer less than or equal to $m/2$, denoted $[m/2]$. Let L denote the p-skeleton of T and $C(L, T')$ the simplicial complement of L in T', a subdivision of T derived near L. The crucial point is that both L and $C(L, T')$ have dimension at most p, so the hypothesis promises that each lives inside some m-cell. Applying Lemma 3.1.1 to stretch one of these cells across the join structure of T', we see that M^m can be covered by the interiors of two open m-cells. The result follows from an application of the Generalized Schönflies Theorem (see Proposition 2.4.9). $\square$

Engulfing homeomorphisms typically are guided by (a finite collection of) homotopies. The main reason there are so many different forms of engulfing is that there are many different ways in which these homotopies may be specified. The assumption in the next result about r-connectedness of the pair (W, U) permits use of Theorem 0.5.2 to construct the necessary homotopies.

Theorem 3.1.3 (Stallings Engulfing). *Let W be a w-dimensional PL ∂-manifold, U an open subset of W, K a complex in W of dimension $k \leq w-3$ such that $|K|$ is closed in W and $|K| \cap \partial W \subset U$, and L a subcomplex of K such that $|L| \subset U$ and $K \setminus L$ is covered by a finite r-subcomplex R of K. If (W, U) is r-connected, then $|K|$ can be engulfed by U keeping $|L|$ fixed; i.e., there exists a compactly supported PL ambient isotopy ψ_t of W such that $\psi_0 = \mathrm{Id}_W$, $\psi_1(U) \supset |K|$ and ψ_t is the identity on a neighborhood of $|L| \cup \partial W$.*

Before taking up the proof, we look at an important application: the topological version of the Poincaré Theorem in high dimensions.

Theorem 3.1.4 (Weak Poincaré). *Any closed PL m-manifold M^m with the homotopy type of S^m, $m \geq 5$, is homeomorphic to S^m.*

Proof. One can easily check that, for any open m-cell U in M^m, (M, U) is $(m-1)$-connected (Exercises 0.5.1 and 0.5.3). The restriction to $m \geq 5$ yields $[m/2] \leq m-3$, so the Stallings Engulfing Theorem promises that each finite $[m/2]$-complex K in M can be engulfed in an open m-cell. Proposition 3.1.2 does the rest. $\square$

Remark. The weakness of 3.1.4 is its merely topological conclusion, despite the PL data in the hypothesis. We cannot prove the existence of a PL homeomorphism with the present techniques due to the reliance upon the Generalized Schönflies Theorem in this proof of Proposition 3.1.2.

We use the notation $K \searrow L$ to indicate that K collapses to L.

Lemma 3.1.5 (Shadow Building). *Given subpolyhedra L and Σ of a polyhedron K such that $K \searrow L$, there exists another subpolyhedron L' of K such that*

$\Sigma \subset L'$,

$K \searrow L' \searrow L$, *and*

$\dim(L' \setminus L) \leq \dim \Sigma + 1$.

Definition. The polyhedron L' is called the *shadow* of Σ under the collapse $K \searrow L$.

Proof. The existence of L' is proved in item (5) on page 40 of Rourke and Sanderson (1972). The proof is by induction on the number of elementary collapses in the collapse $K \searrow L$. In the case of one elementary collapse, the shadow can be regarded rather literally as the shadow of Σ under the projection of the n-cube onto the bottom codimension-one face (see (Rourke and Sanderson, 1972), pp. 39–40). Rourke and Sanderson reserve the term "shadow" for that special case and use the term "trail" for the general case. $\square$

Recall (Rourke and Sanderson, 1972, page 60) that the *singular set* of a piecewise linear map f is defined as

$$S(f) = \mathrm{Cl}(\{x \text{ in the domain of } f \mid f^{-1}f(x) \neq \{x\}\}),$$

where Cl denotes closure. It should be clear that the singular set of a PL map is a subpolyhedron of the domain.

Corollary 3.1.6. *If K is a compact polyhedron in a PL n-manifold M^n and the inclusion map* incl : $K \to M^n$ *is null-homotopic, then there exist polyhedra P and Q in M^n such that $K \subset P \searrow Q$,* $\dim P \leq \dim K + 1$, *and* $\dim Q \leq 2\dim K - n + 3$.

Proof. Let $c(K)$ be the cone on K. The inclusion $K \hookrightarrow M^n$ extends to a map $f : c(K) \to M$. Shift f into general position and define $\Sigma = S(f)$. Then $\dim S(f) \leq 2(\dim K + 1) - n$. Let L be the shadow of Σ under the collapse $c(K) \searrow$ point. Define $P = f(c(K))$ and $Q = f(L)$. Since $f|c(K) \smallsetminus S(f)$ is one-to-one, we have $P \searrow Q$ as needed. □

Proof of Stallings Engulfing Theorem 3.1.3. The proof proceeds by induction on r. The result is obviously true for $r = -1$. Less trivially, the case $r = 0$ quickly reduces to the special case in which R consists of a single vertex v not in L. We may assume $v \in \operatorname{Int} W$. The 0-connectedness of (W^n, U) and Theorem 0.5.2 give a map $g : [0,1] \to W$ with $g(0) = v$ and $g(1) \in U$. By general position, g can be approximated by a PL embedding $G : [0,1] \to \operatorname{Int} W$ such that $G([0,1]) \cap L = \emptyset$, $G(0) = v$, and $G(1) \in U$. A regular neighborhood of $G(1)$ in $U \smallsetminus L$ can be expanded to a regular neighborhood of $G([0,1])$ via an ambient isotopy ψ_t supported in a compact subset of $\operatorname{Int} W \smallsetminus L$. The extension of ψ_t over the rest of W via the identity has the desired effect.

Now assume that the engulfing theorem holds for $r = 0, 1, \ldots, i-1$, and consider $r = i \leq w - 3$. It suffices to establish the result for the case in which R consists of a single i-simplex Δ, for by induction we can engulf the $(i-1)$-skeleton of any finite i-complex R and then engulf any number of i-simplices, one at a time. Add to L any part of R previously engulfed before turning to the next i-simplex.

Identify Δ with $\Delta \times \{0\} \subset \Delta \times [0,1]$. The i-connectedness of (W, U) together with Theorem 0.5.2 ensures that the inclusion $(\Delta, \partial\Delta) \hookrightarrow (W, U)$ extends to a map

$$g : (\Delta \times [0,1], (\partial\Delta \times [0,1]) \cup (\Delta \times \{1\})) \to (W, U).$$

Extend g over $|L|$ via the identity. Assume g is PL and in general position as a map, thus obtaining a triangulation T of the domain with respect to which g is simplicial, is one-to-one on the simplices, and satisfies

$$\dim(g(\sigma) \cap g(\tau)) \leq \dim(\sigma) + \dim(\tau) - w$$

for every pair of simplices $\sigma, \tau \in T$.

Observe that $L \cup (\Delta \times [0,1])$ collapses to $L \cup (\partial\Delta \times [0,1]) \cup \Delta \times \{1\}$. Modify T so that the collapse can be treated as simplicial with respect to T (any barycentric subdivision of a collapsible complex is collapsible). Then enumerate certain simplices $A_1, A_2, \ldots, A_s$ from the triangulation T so that for

$$S_{j-1} = |L| \cup (\partial\Delta \times [0,1]) \cup (\Delta \times \{1\}) \cup A_1 \cup A_2 \cup \ldots A_{j-1},$$

$S_{j-1} \cap A_j = v_j * \partial B_j$ where $A_j = v_j * B_j$, and S_j collapses to S_{j-1}. This enumeration lists only half the simplices from $T \smallsetminus L$, namely, the top-dimensional simplices appearing in the sequence of simplicial collapses, not the faces across which those collapses occur. Here each S_j admits an elementary collapse to S_{j-1}; the collapse is across A_j, from B_j onto $v_j * \partial B_j$.

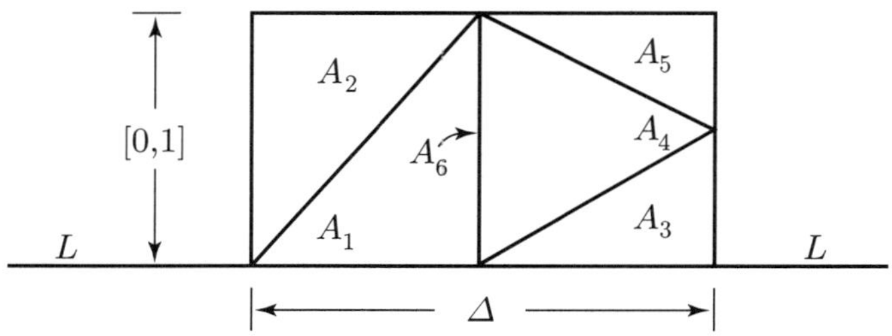

Figure 3.1. The collapse across $\Delta \times [0,1]$

The proof that we can engulf $g(S_j)$ is another inductive argument; this time it proceeds by induction on s. Let's first restrict slightly to the case $i \leq w-4$. Clearly we have already engulfed $g(S_0) = g(|L| \cup (\partial\Delta \times [0,1]) \cup (\Delta \times \{1\}))$ by U. Inductively assume that $g(S_{j-1})$ has been engulfed by U and let $\Sigma_j = S(g) \cap A_j$; then

$$\dim(\Sigma_j) \leq 2(i+1) - w \leq (i+1) + (w-3) - w = i-2.$$

The Shadow Building Lemma (3.1.5) promises an $(i-1)$-subcomplex L' of A_j containing Σ_j such that $A_j \searrow L' \searrow B_j$. Based on the inductive hypothesis, we can engulf $g(L' \cup S_{j-1})$ keeping $g(S_{j-1})$ fixed. Since $g|A_j$ is one-to-one, $g(S_j)$ collapses to $g(L' \cup S_{j-1})$, implying that the adjusted U can be expanded to engulf $g(S_j)$ keeping the points of $g(L' \cup A_{j-1})$ fixed. As this completes the induction, we see we can engulf $g(S_s) \supset g(\Delta \times \{0\}) = \Delta$ with U.

When $i = w-3$, we must be more careful, because $\dim(\Sigma_j)$ is only limited by $i-1$ and the inductive hypothesis does not apply to the i-complex L' obtained from the Shadow Building Lemma. Instead, in this case we let D_{j-1} be the part of the $(w-3)$-skeleton of T contained in S_{j-1} and let $\Sigma_j = A_j \cap S(g \mid A_j \cup D_{j-1})$. Now

$$\dim(\Sigma_j) \leq (w-2) + (w-3) \leq i-2,$$

so the inductive hypothesis applies, and the rest of the proof proceeds as above, furnishing an engulfing of, first, $g(\Sigma_j)$ and then $g(A_j)$, keeping $g(D_j)$ fixed. Ultimately, when $j = s$, we secure the desired engulfing of $|K| \subset g(D_s \cup A_s)$. □

Remark. In codimension four it is possible to engulf not only the polyhedron K but also the entire track of a homotopy that pulls K into U. In codimension three we must settle for less. In the inductive step of the proof, the entire image of an A_j is engulfed by U. But it is not necessarily possible for U to hold on to that entire set at later stages in the construction. Instead we keep $g(\partial A_j)$ in U as the engulfing progresses, which is good enough.

This subtlety is illustrated in Figures 3.2 through 3.4. None of the figures is dimensionally accurate, but the three figures do indicate what will actually happen. In all three the function g is suppressed. In Figure 3.2 the first isotopy engulfs the entire track of B_1 and the second isotopy engulfs the track of B_2. Part of the track of B_1 is exposed by the second isotopy, but B_1 itself remains in the image of U.

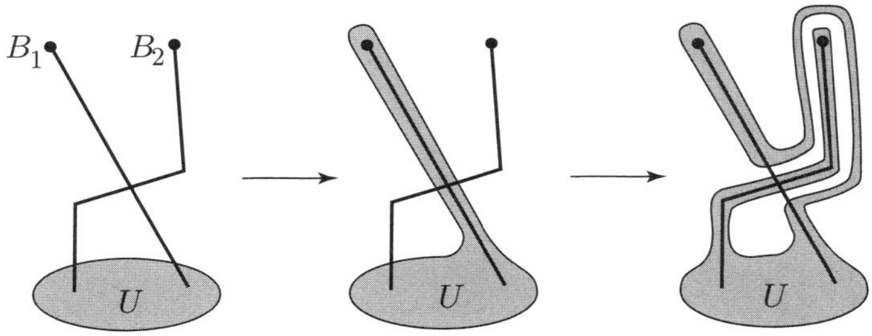

Figure 3.2. The second isotopy uncovers part of the track of the first isotopy

Figure 3.3 shows how it is possible to uncover part of A_j while keeping ∂A_j in U. In the figure it is possible to engulf all of A_1 and A_2, but when A_3 is engulfed, part of the interior of A_1 will be uncovered.

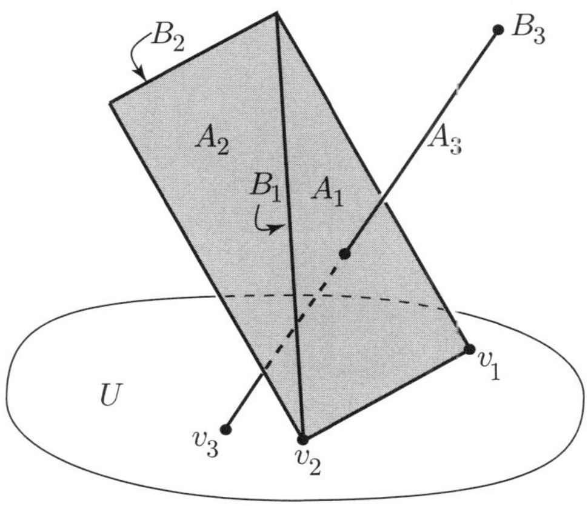

Figure 3.3. The third isotopy can uncover part of Int A_1, but will hold on to ∂A_1

Figure 3.4 is an attempt to show how both the B_j's could have boundary. In the diagram, imagine that the lower portion of U is first pushed straight up to engulf B_1 and later the left-hand portion of U is pushed left-to-right to engulf B_i. The vertical isotopy will stretch U out to cover the entire vertical slab below B_1, but part of that slab will be uncovered by the later horizontal isotopy.

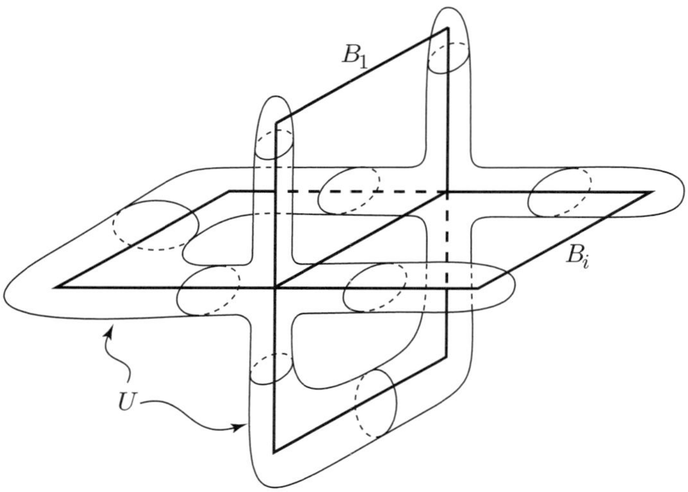

Figure 3.4. A later isotopy must uncover part of Int A_1 but can hold on to B_1

The power of engulfing lies in the technique, not in any particular formulation of an engulfing theorem. The theorem stated here illustrates the technique and applies in a fairly wide variety of settings. However, it will not apply perfectly in all the situations where we want to use it; in particular, there will not always be a single open set U such that (W, U) is highly connected. That connectedness hypothesis guarantees the existence of homotopies that pull certain polyhedra into U, and those homotopies are the crucial ingredient for the proof. It should be noted, though, that in order to engulf a given polyhedron K it is not enough to have one homotopy that pulls K into U; instead, it is necessary to pull many different polyhedra into U. The inductive structure of the proof starts with a homotopy that pulls a polyhedron like K into U. The image of that homotopy contains a shadow that must also be homotoped into U. This second homotopy contains a second shadow, etc. The number of additional layers of homotopies needed is bounded by the dimension of the polyhedron to be engulfed. This observation is made precise in the following variation of 3.1.3.

Theorem 3.1.7 (Modified Stallings Engulfing). *Let M^n be an n-dimensional PL ∂-manifold, $U_0 \subset W_0$ open subsets of M, K a complex in W_0 of dimension $k \leq n-3$ such that $|K|$ is closed in M and $|K| \cap \partial M \subset U_0$, and L a subcomplex of K such that $|L| \subset U_0$ and $K \smallsetminus L$ is covered by a finite*

r-subcomplex R of K. Suppose

$$(W_0, U_0) \subset (W_1, U_1) \subset \cdots \subset (W_{k+1}, U_{k+1})$$

are pairs of open sets in M such that $(|R|, |R \cap L|) \subset (W_0, U_0)$ and the inclusion-induced homomorphism $\pi_j(W_i, U_i) \to \pi_j(W_{i+1}, U_{i+1})$ is trivial for each $i \leq r$ and each $j \leq r$. Then K can be engulfed by U_{r+1} keeping L fixed via a PL isotopy that is supported on a compact subset of W_{r+1}.

We end the section with another application.

Theorem 3.1.8 (Weak h-Cobordism). *If (W, M_0, M_1) is a compact PL h-Cobordism and* $\dim W \geq 5$, *then $W \setminus M_1$ is PL homeomorphic to $M_0 \times [0, 1)$.*

Proof. Let $C_0' \subset C_0$ be PL collars on M_0 in W (Theorem 2.4.10). It suffices to show that any compact subset C of $W \setminus M_1$ can be engulfed by C_0 keeping C_0' fixed. Once this is established, it is a simple matter to piece together a countable collection of closed PL collars to form the open collar that covers all of $W \setminus M_1$.

Let $C_1' \subset C_1$ be PL collars on M_1 in $W \setminus C$, and let U_i denote the interior of a regular neighborhood of C_i' in C_i. Fix a triangulation T of $\overline{W \setminus (C_0' \cup C_1')}$ and identify the codimension-3 skeleton L of T and the dual 2-skeleton L' in T'. Since (W, U_i) is j-connected for all j, Stallings Engulfing Theorem promises that L can be engulfed by U_0 and L' can be engulfed by U_1. That is, there exist PL ambient isotopies of W fixing ∂W and ending in g, γ such that $g(U_0) \supset L$ and $\gamma(U_1) \supset L'$, where we require g and γ to fix C_0' and C_1' respectively. Now by stretching across the join structure of T', we can obtain another PL ambient isotopy of W fixing ∂W and ending in h, where

$$hg(U_0) \cup \gamma(U_1) = W.$$

Applying γ^{-1} we obtain

$$\gamma^{-1} hg(U_0) \cup U_1 = \gamma^{-1}(W) = W.$$

Since $C \cap U_1 = \emptyset$, it is clear that $C \subset \gamma^{-1} hg(U_0)$, which is part of the collar $\gamma^{-1} hg(C_0)$ on M_0 as required. □

Historical Notes. The concept of engulfing originated with E. C. Zeeman, who formulated engulfing theorems in a somewhat different way than we have. Zeeman constructed one polyhedron that contains the original complex K and collapses to a subset of U—see Zeeman (1963a), for example. Later authors thought in terms of an isotopy that stretches U out to engulf K. Engulfing Theorem 3.1.3 is due to J. Stallings (1962b, Theorem 3.1), and its applications presented in the section are found in the same paper by Stallings. In particular, Stallings first proved the Weak Poincaré Theorem. The strong form of the theorem, in which both the hypothesis and conclusion are PL, is due to S. Smale (1961).

Exercises

3.1.1. If W^w is a contractible PL manifold, $w > 1$, and $S \subset W$ is a compact PL $(w-1)$-manifold, then all but one component of $W \smallsetminus S$ has compact closure.

3.1.2. A noncompact space W is said to be *simply connected at* ∞ if for every compact set $C_1 \subset W$ there is a larger compact set C_2 such that every loop in $W \smallsetminus C_2$ is null-homotopic in $W \smallsetminus C_1$. Prove that every contractible PL n-manifold, $n \geq 5$, that is simply connected at ∞ is homeomorphic to $\mathbb{R}^n$. [Hint: Proceed as in the proof of the Weak h-Cobordism Theorem; fill the entire manifold with a PL n-cell plus a sequence of collars.]

3.1.3. Two compact PL n-manifolds M_0 and M_1, $n \geq 4$, are h-cobordant if and only if $M_0 \times \mathbb{R}$ and $M_1 \times \mathbb{R}$ are PL homeomorphic.

3.2. Application: The cellularity criterion

This section is devoted to another valuable application of Stallings engulfing. There will be many other applications later, but this one serves as a model illustrating the power of the technique in the study of topological embeddings. Previously we encountered examples indicating that some wildly embedded arcs and cells are cellular while others are not. The main theorem in this section confirms that the fundamental group condition introduced in Chapter 2 distinguishes the cellular embeddings from the others.

We have already seen that the cellularity criterion, defined in §2.7, is necessary for cellularity (Exercise 2.7.3); now we will show that it is also sufficient. Before doing that, we must address the question of which spaces potentially could be embedded as cellular subsets of a manifold.

Lemma 3.2.1. *If X is a compact, contractible subset of a PL w-manifold W and U is a neighborhood of X in W, then there exists a neighborhood V of X such that $X \subset V \subset U$ and the inclusion map $V \hookrightarrow U$ is null-homotopic.*

Proof. Name a contraction $\xi : X \times I \to X$, where $\xi_0 = \mathrm{Id}_X$ and $\xi_1(X) = x_0 \in X$. Define a map $f : A \to U$ on $A = (X \times I) \cup (U \times \{0,1\}) \subset U \times I$ as

$$f(a) = \begin{cases} u & \text{if } a = \langle u, 0\rangle \\ \xi_t(x) & \text{if } a = \langle x, t\rangle, \\ x_0 & \text{if } a = \langle u, 1\rangle. \end{cases} \quad \text{and}$$

Then f extends to a map $F : Z \to U$ defined on some neighborhood Z of A in $U \times I$. Invoke compactness of X to choose a neighborhood V of X in W such that $X \times I \subset V \times I \subset Z$. The restriction of F to $V \times I$ gives the required homotopy. □

Corollary 3.2.2. *If X is a compact, contractible subset of a PL w-manifold W, then there exists a sequence $\{Q_i\}_{i=1}^{\infty}$ of w-dimensional, compact PL ∂-manifolds in W such that $Q_{i+1} \subset Q_i$, $Q_{i+1} \hookrightarrow Q_i$ is null-homotopic and $X = \cap_{i=1}^{\infty} Q_i$.*

This is a useful property, so we give it a name. We will learn that any compact subset of a manifold satisfying the conclusion of Lemma 3.2.1 is enough like a cell that it admits an embedding as a cellular subset of some Euclidean space.

Definition. A compact set X in an ANR Y is *cell-like* if for each neighborhood U of X in Y there exists an open subset V of Y such that $X \subset V \subset U$ and the inclusion $V \hookrightarrow U$ is null-homotopic.

Lemma 3.2.1 certifies that contractible sets are cell-like. Familiar examples like the topologist's sine curve (Figure 2.13) show that cell-likeness is more general than contractibility. Obviously cellular subsets of manifolds are cell-like, but the converse sometimes fails: wild arcs in S^n with nonsimply connected complements are cell-like but not cellular. Thus cell-likeness is also more general than cellularity.

Although cellularity and cell-likeness both appear to depend on the embedding of X in W, the specific embedding really is irrelevant for the latter; Lemma 3.2.1 makes this plain for embeddings of contractible objects in manifolds and a similar argument disposes of the general case (Exercise 3.2.1). Hence cell-likeness of the compactum is invariant under embeddings into ANRs.

We can now state the main theorem of the section.

Theorem 3.2.3 (Cellularity Criterion). *Let X be a compact, cell-like subset of a PL w-manifold W, $w \geq 5$. Then X is cellular if and only if X satisfies the cellularity criterion.*

The proof relies on two lemmas that cover the two halves of the engulfing argument.

Lemma 3.2.4. *Let $\{Q_i\}$ denote a sequence of compact, n-dimensional PL ∂-manifolds such that $Q_{i+1} \subset Q_i$ and each inclusion $Q_{i+1} \hookrightarrow Q_i$ is null-homotopic. Let K^k be a polyhedron in Q_{i+k+1}, where $k \leq n-3$. Then there exists a PL n-cell B^n such that $K^k \subset \operatorname{Int} B^n \subset B^n \subset Q_i$.*

Proof. This follows immediately from the Modified Stallings Engulfing Theorem (Theorem 3.1.7). □

Lemma 3.2.5. *If W is a PL w-manifold, $w > 2$, and X is a cell-like subset of W that satisfies the cellularity criterion, then $(W, W \smallsetminus X)$ is 2-connected.*

Proof. Let $p : \widetilde{W} \to W$ be the universal cover of W and let $X_1 = p^{-1}(X)$. The fact that X is cell-like implies that incl : $X \hookrightarrow W$ can be lifted to $\widetilde{W}$ and so each component of X_1 is homeomorphic to X. That means that the components of both X_1 and its one-point compactification X_1^* are acyclic. Hence, $\check{H}^i(X_1^*) \cong 0$ for $i > 0$. It follows from 0.3.3 that $\check{H}^i_c(X_1) \cong 0$ for $i > 0$. It is not difficult to see that the cellularity criterion implies $\widetilde{W} \smallsetminus X_1$ is simply connected. Using this one proves the lemma by an application of the Relative Hurewicz Theorem and Duality Theorem 0.3.1:

$$\pi_k(W, W \smallsetminus X) \cong \pi_k(\widetilde{W}, \widetilde{W} \smallsetminus X_1) \cong H_k(\widetilde{W}, \widetilde{W} \smallsetminus X_1) \cong \check{H}^{w-k}_c(X_1).$$

The last group is trivial as long as $k < w$. □

Proof of Cellularity Criterion Theorem 3.2.3. The necessity of the cellularity criterion is Exercise 2.7.3.

For sufficiency, consider any neighborhood U of X in W. Apply Corollary 3.2.2 to obtain a sequence $\{Q_i \mid i = 1, \dots w - 2\}$ of w-dimensional, compact PL ∂-manifolds such that $Q_{i+1} \subset \operatorname{Int} Q_i \subset Q_i \subset U$ and each inclusion $Q_{i+1} \hookrightarrow Q_i$ is null-homotopic. Let K^{w-3} denote the $(w-3)$-skeleton of Q_{w-2} and let L denote the dual 2-skeleton. By Lemma 3.2.4, Q_1 contains a PL n-cell B with $\operatorname{Int} B \supset K^{w-3}$. Apply Lemma 3.2.5 and Stallings Engulfing Theorem 3.1.3 to obtain a PL homeomorphism g of W such that g acts as the identity on $Q_1 \smallsetminus Q_{w-2}$ and $g(W \smallsetminus X) \supset L$. Stretch across the join structure of this triangulation of Q_{w-2} to further adjust g so that

$$\operatorname{Int} B \cup g(W \smallsetminus X) = W.$$

Apply g^{-1} to obtain

$$g^{-1}(\operatorname{Int} B) \cup (W \smallsetminus X) = g^{-1}(W) = W.$$

Now clearly $X \subset g^{-1}(\operatorname{Int} B)$. Since $B \subset Q_1$ and $g(Q_1) = Q_1$, we have $g^{-1}(B) \subset Q_1 \subset U$. □

Corollary 3.2.6. *If X is any cell-like subset of a PL w-manifold W, $w \geq 4$, then $X \times \{0\}$ is a cellular subset of $W \times \mathbb{R}^1$.*

Proof. That $X \times \{0\}$ satisfies the cellularity criterion in $W \times \mathbb{R}^1$ is Exercise 3.2.4. □

Corollary 3.2.7. *A finite-dimensional compact metric space X is cell-like if and only if X can be embedded as a cellular subset of $\mathbb{R}^n$ for some n.*

Corollary 3.2.6 promises that every cell-like subset X of $\mathbb{R}^n$ is stably cellular in the sense that X is cellular when considered as a subset of any higher-dimensional Euclidean space. Restated more simply, every cell-like subset of $\mathbb{R}^n$ admits a cellular embedding in $\mathbb{R}^{n+1}$. In general, one cannot improve on this to obtain a cellular embedding in $\mathbb{R}^n$. For example, a Newman contractible ∂-manifold M^n (Example 0.10.3) admits no cellular embedding e in $\mathbb{R}^n$: $e(\partial M^n)$ would be 0-LCC in $\mathbb{R}^n \smallsetminus e(\operatorname{Int} M^n)$ and, coupled with the cellularity criterion, that would imply the triviality of $\pi_1(\partial M^n)$.

One aspect of Theorem 3.2.3 worth special notice is the fact that the proof delivers a little more than the statement advertises. The cells constructed in the proof are piecewise linear cells. We have defined a compact subset of a manifold to be cellular if it is the intersection of a nested sequence of topological cells. But any such set will satisfy the cellularity criterion and thus the proof of Theorem 3.2.3 shows that it will be the intersection of a nested sequence of PL cells. Consequently, the concepts "cellular by topological cells" and "cellular by PL cells" are equivalent for subsets of PL manifolds.

Corollary 3.2.8. *Let X be a finite-dimensional cell-like set and G an Abelian group. Then $\check{H}^q(X;G) \cong 0$ for all $q > 0$.*

Proof. Regard X as a cellular subset of (some) S^n. Given $q > 0$ and a relative $(n-q)$-cycle z from $(S^n, S^n \smallsetminus X)$, S^n contains an n-cell $C \supset X$, where C lies in the complement of a carrier for ∂z. Then z is null-homologous in $(S^n, S^n \smallsetminus C)$, so it follows that $H_{n-q}(S^n, S^n \smallsetminus X; G) \cong 0$. Duality gives the cohomology conclusion. □

Definition. A map $f : Y \to Z$ is said to be *cell-like* if each $f^{-1}(z), z \in Z$, is a cell-like set.

By definition, cell-like maps are surjective. Moreover, they preserve Čech cohomology.

Proposition 3.2.9. *Any proper, cell-like mapping $f : X \to Y$ between paracompact Hausdorff spaces induces isomorphisms $f^* : \check{H}^k(Y;G) \to \check{H}^k(X;G)$ for all $k \geq 0$ and all G.*

Proof. This is an immediate consequence of the Vietoris-Begle Theorem (0.4.1), as cell-like sets have trivial Čech cohomology. □

Cell-like maps also enjoy a valuable approximate lifting property.

Proposition 3.2.10 (Approximate lifting of cell-like maps). *Let $f : Y \to Z$ be a proper cell-like mapping from a locally compact ANR to a metric space Z, and let $\mu : K \to Z$ be a map defined on a finite k-complex K. For each*

$\epsilon > 0$ there exists a map $\tilde{\mu} : K \to Y$ such that $\rho(f\tilde{\mu}, \mu) < \epsilon$; moreover, if L is a subcomplex of K and $\mu_L : L \to Y$ is a map for which $f\mu_L = \mu|L$, then $\tilde{\mu}$ can be obtained satisfying $\tilde{\mu}|L = \mu_L$.

Proof. Restrict attention to a compact neighborhood $Z' \subset Z$ of $f(K)$ and the compact neighborhood $Y' = f^{-1}(Z') \subset Y$ of $f^{-1}(A)$. Identify an open cover $\mathcal{U}_k$ of $f(K)$ in Z' by sets of diameter less than $\epsilon/2$. Use the hypothesis that f is cell-like to find open covers $\mathcal{U}_{k-1}, \ldots, \mathcal{U}_1, \mathcal{U}_0$ of $f(K)$ such that, for $i = 0, 1, \ldots, k-1$, $\mathcal{U}_i$ refines $\mathcal{U}_{i+1}$ and for all $U_i \in \mathcal{U}_i$, $f^{-1}(\text{star}(U_i; \mathcal{U}_i))$ is null-homotopic in some element of $\mathcal{U}_{i+1}$, where

$$\text{star}(U_i; \mathcal{U}_i) = \{\cup U' \in \mathcal{U}_i \mid U' \cap U_i \neq \emptyset\}.$$

Choose a triangulation T of K with small enough mesh that $\{f(\sigma) \mid \sigma \in T\}$ refines $\mathcal{U}_0$. We shall produce successive extensions $\mu_i : L_i \to Y'$ of μ_L, where $L_i = L \cup T^{(i)}$, such that $\{f\mu_i(\tau) \mid \tau \in T \text{ in the domain of } \mu_i\}$ refines $\mathcal{U}_i$.

Extend μ_L over the vertices of T not in L by choosing $\mu_0(v) \in f^{-1}(v)$. Given a 1-simplex σ of T, we are assured of the existence of $U_\sigma \in \mathcal{U}_0$ such that $f(\sigma) \subset U_\sigma$, and $f^{-1}(U_\sigma) \supset \mu_o(\partial\sigma)$ is null-homotopic in some $f^{-1}(U_1), U_1 \in \mathcal{U}_1$, so $\mu_0|\partial\sigma$ can be extended to μ_1 with $f\mu_1(\sigma) \subset U_1$. Suppose inductively that μ_i has been defined as required. Consider an $(i+1)$-simplex $\tau \in T$. Pick an i-dimensional face γ of τ, and find $U_\gamma \in \mathcal{U}_i$ such that $f\mu_i(\gamma) \subset U_\gamma$. Check that $f\mu_i(\partial\tau) \subset \text{star}(U_\gamma; \mathcal{U}_i)$. The prearrangements in place assure that μ_i can be extended to μ_{i+1} so $\mu_{i+l}(\tau)$ is contained in some $f^{-1}(U_\tau), U_\tau \in \mathcal{U}_{i+1}$, for all $(i+1)$-simplices $\tau \in T$.

Set $\tilde{\mu} = \mu_k$. To see that $f\tilde{\mu}$ is close to μ, look at $x \in K$ and then at a $\sigma_x \in T$ for which $x \in \sigma_x$. Each of $f\tilde{\mu}(\sigma_x)$ and $\mu(\sigma_x)$ is contained in some element of $\mathcal{U}_k$, and the two elements must intersect, since the two images of vertices of σ_x are identical. Hence, $\rho(f\tilde{\mu}(x), \mu(x)) < 2 \cdot \text{mesh } \mathcal{U}_k < \epsilon$. □

Approximate lifting implies that any cell-like mapping f between ANRs induces an isomorphism of all homotopy groups (see Exercise 3.2.8). As a result, the Whitehead Theorem yields that cell-like mappings between simplicial complexes are homotopy equivalences; the same is true for cell-like mappings between locally compact ANRs, since the latter all have the homotopy type of some simplicial complex (West, 1975). Cell-like mappings carry an even stronger property. A mapping $f : X \to Y$ is a *fine homotopy equivalence* if it has a homotopy inverse $g : Y \to X$ such that, for each open cover $\mathcal{V}$ of Y, fg is $\mathcal{V}$-homotopic to Id_Y and gf is $f^{-1}(\mathcal{V})$-homotopic to Id_X; in other words, there is a homotopy $\Psi_t : Y \to Y$ between fg and Id_Y and another homotopy $\Phi_t : X \to X$ between gf and Id_X such that $\{\Psi(y \times I) \mid y \in Y\}$ refines $\mathcal{V}$ and $\{\Phi(x \times I) \mid x \in X\}$ refines $f^{-1}(\mathcal{V})$. Approximate lifting also implies that cell-like mappings between ANRs are fine homotopy equivalences. When both the domain and target are polyhedra, this follows almost immediately from Proposition 3.2.10 (Exercise 3.2.9); in general, one interpolates "approximations" to the ANR domain and target by polyhedra and does similar approximate lifting of maps and homotopies, but with more elaborate bookkeeping to account for the interpolations. Since the result is not used here, we omit details.

Proposition 3.2.11. *Every cell-like map $f : X \to Y$ between locally compact ANRs is a fine homotopy equivalence.*

Proposition 3.2.12. *Let $f : Y \to Z$ be a proper cell-like mapping between locally compact ANRs. Then a compact subset A of Z satisfies the cellularity criterion in Z if and only if $f^{-1}(A)$ satisfies the cellularity criterion in Y.*

Proof. Assume A satisfies the cellularity criterion in Z. Given a neighborhood U of $f^{-1}(A)$, first produce a neighborhood U' of A such that $f^{-1}(U') \subset U$ and next a smaller neighborhood V' of A such that loops in $V' \setminus A$ are null-homotopic in $U' \setminus A$. Then the image under f of any loop in $V = f^{-1}(V')$ bounds a singular disk in $U' \setminus A$. The approximate lifting result assures that the original loop bounds a singular disk in $U \setminus f^{-1}(A)$.

The proof of the other implication is left to the reader. □

Corollary 3.2.13. *Every proper cell-like mapping $f : M \to M'$ between PL n-manifolds, $n \geq 5$, is a cellular mapping.*

We conclude this section by describing an intrinsically interesting and historically important example of a cell-like continuum in S^3. Its construction clearly illustrates how a continuum can have neighborhoods that contract as in Corollary 3.2.2 without the continuum itself being contractible.

The Whitehead continuum. Begin with an unknotted solid torus $V_0 \subset S^3$. Inside V_0 embed a second solid torus V_1 as indicated in Figure 3.5. There is a homeomorphism $h : V_0 \to V_1$; define V_i, $i \geq 2$, recursively by $V_i = h(V_{i-1})$. The *Whitehead continuum* is the compact, connected set X defined by

$$X = \bigcap_{i=0}^{\infty} V_i.$$

Since $V_i \hookrightarrow V_{i-1}$ is null-homotopic, X definitely is cell-like. As we shall see, X does not satisfy the cellularity criterion in S^3, so X is not cellular in S^3. Using the natural inclusion $S^3 \subset S^4$, we can consider X to be a subset of S^4, where it is cellular (Exercise 3.2.11), although Corollary 3.2.6 does not apply.

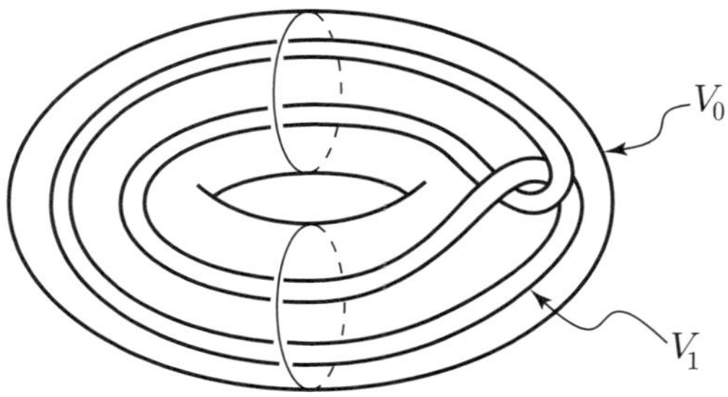

Figure 3.5. The first two stages in the construction of the Whitehead continuum

The complement $W = S^3 \smallsetminus X$ of the Whitehead continuum is known as the *Whitehead manifold.* The Whitehead manifold is interesting because, although contractible, it is not homeomorphic to $\mathbb{R}^3$. The assertion that X does not satisfy the cellularity criterion is equivalent to the statement that W is not simply connected at infinity (see Exercise 3.1.2 for the definition), so W cannot be homeomorphic to $\mathbb{R}^3$.

To verify that W is contractible, write $W = \cup_{i=0}^{\infty} W_i$, where $W_i = S^3 \smallsetminus \operatorname{Int} V_i$. Since each V_i is an unknotted solid torus, W_i is also an unknotted solid torus. The key observation is that the embedding of W_{i-1} in W_i is exactly the same as the embedding of V_i in V_{i-1}. To confirm this, note that W_0 is embedded in W_1, the complement of V_1, as indicated in Figure 3.6. The link in Figure 3.6 is known as the *Whitehead link* and it is symmetric in the sense that there is a homeomorphism of $\mathbb{R}^3$ interchanging the two components. (This can easily be verified by building a model out of string.) It follows that W_0 is null-homotopic in W_1 and therefore each W_{i-1} is null-homotopic in W_i. Since the image of any map $S^k \to W$ is contained in W_i for some i, $\pi_k(W) = 0$ for every k. The Whitehead Theorem (0.9.1) implies W is contractible.

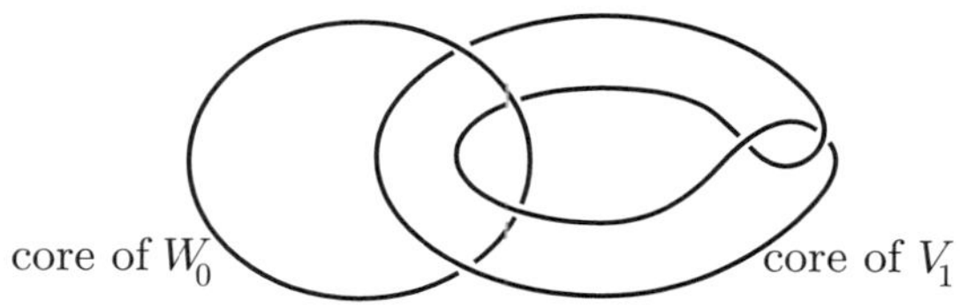

Figure 3.6. The core of W_0 and the core of V_1 form a Whitehead link

Finally, we explain why the Whitehead continuum X does not satisfy the cellularity criterion in S^3. The proof is based on the following claim.

Claim 3.2.14. *For $i = 0, 1$ the inclusion-induced homomorphism $\pi_1(\partial V_i) \to \pi_1(V_0 \smallsetminus \operatorname{Int} V_1)$ is one-to-one.*

Proof. Due to the symmetry of the Whitehead link, the two inclusions are essentially the same, so it is enough to prove this for $i = 0$. Compare Figures 2.1 and 3.5. There is a four-to-one covering map from a slightly twisted version of (T, A_1) in Figure 2.1 to (V_0, V_1) in Figure 3.5. If a loop on ∂V_0 is null-homotopic in $V_0 \smallsetminus \operatorname{Int} V_1$, then the null-homotopy can be lifted to the covering space. By the proof of Lemma 2.1.4, the lifted loop must be trivial on ∂T and hence the original loop must be null-homotopic on ∂V_0. Thus $\pi_1(\partial V_0) \to \pi_1(V_0 \smallsetminus \operatorname{Int} V_1)$ is one-to-one. □

Application of Theorem 0.11.5 inductively yields the following generalization.

Claim 3.2.15. *For every pair of nonnegative integers i and j, $i < j$, the inclusion-induced homomorphisms $\pi_1(\partial V_i) \to \pi_1(V_i \smallsetminus \operatorname{Int} V_j)$ and $\pi_1(\partial V_j) \to \pi_1(V_i \smallsetminus \operatorname{Int} V_j)$ are one-to-one.*

Further applications of Theorem 0.11.5 and compactness of I^2 images give that for every pair of nonnegative integers i and j, $i \leq j$, the inclusion-induced homomorphism $\pi_1(\partial V_j) \to \pi_1(V_i \smallsetminus X)$ is one-to-one. Clearly this last statement implies that X does not satisfy the cellularity criterion.

Historical Notes. Cellularity Criterion Theorem 3.2.3 is mainly due to D. R. McMillan, Jr. (1964). The concept of cell-likeness had not yet entered the nomenclature when McMillan proved his result, so he stated his theorem only for absolute retracts, but his proof actually works for cell-like sets. The theorem is also known to hold in dimension four by work of M. H. Freedman (1982). The 3-dimensional case of Theorem 3.2.3 follows from (McMillan, 1964, Theorem 1′) and Perelman's recent solution to the Poincaré Conjecture.

The definition of *cell-like* is due to R. C. Lacher (1969), although earlier McMillan singled out the contractibility of neighborhoods as the feature essential to his proof of 3.2.3. Lacher has written an extensive survey of cell-like sets and cell-like mappings (Lacher, 1977). W. J. R. Mitchell and D. Repovš (1988) have a more recent survey analyzing the impact of examples constructed by A. N. Dranishnikov (1989) of cell-like, dimension-raising mappings.

Theorem 3.2.11 about cell-like maps between ANRs being fine homotopy equivalences is due to G. Kozlowski (unpublished manuscript) and W. Haver (1975).

The Whitehead manifold was first described in (Whitehead, 1935); Whitehead used it to give a counterexample to his own purported proof of the Poincaré Conjecture.

Exercises

3.2.1. (Topological invariance of cell-likeness) Suppose X is a cell-like subset of an ANR Y and e is an embedding of X in another ANR Y'. Then $e(X)$ is cell-like in Y'.

3.2.2. Let X be a compact subset of the Hilbert cube Q. Then X is cell-like if and only if each map $f : X \to Y$ to an ANR Y is null-homotopic.

3.2.3. A compact subset X of a manifold M is cell-like if and only if for each neighborhood U of X there exists a neighborhood V of X such that the images of $\pi_1(V)$ in $\pi_1(U)$ and of $H_*(V;\mathbb{Z})$ in $H_*(U;\mathbb{Z})$ are trivial.

3.2.4. If X is any cell-like subset of a manifold W, $\dim W > 1$, then $X \times \{0\}$ satisfies the cellularity criterion in $W \times \mathbb{R}^1$.

3.2.5. Prove Corollary 3.2.7.

3.2.6. If $e : S^{n-1} \to S^n$, $n \geq 5$, is a topological embedding such that $e(S^{n-1})$ satisfies the cellularity criterion, then e is weakly flat. (Refer to definition of weakly flat on page 78.)

3.2.7. Let $e : S^{n-1} \to S^n$ be a topological embedding. If $e(S^{n-1})$ is 1-LCC, then $e(S^{n-1})$ satisfies the cellularity criterion. Show by example that the converse is false.

3.2.8. If $f : Y \to Z$ is a cell-like mapping between ANRs, then f induces an isomorphism $f_* : \pi_k(X, x_0) \to \pi_k(Y, f(x_0))$ for all k.

3.2.9. Every cell-like map $f : K \to L$ between finite polyhedra is a fine homotopy equivalence.

3.2.10. If the map $f : X \to Y$ between metric compacta is a fine homotopy equivalence, then $\dim Y \leq \dim X$.

3.2.11. Prove that the Whitehead continuum X is cellular when considered as a subset of S^4.

3.2.12. Find a cellular embedding of the Whitehead continuum in S^3.

3.3. Engulfing with control

In establishing that the Cellularity Criterion implies cellularity, we exercised a partial motion control: the motion associated with the engulfing activity was restricted to a predetermined neighborhood of the cell-like set X, but the distance points moved was unrestricted. Next we turn to the matter of imposing control on the amount of motion in an engulfing isotopy. The statement of a general controlled engulfing theorem can be quite complicated, so we focus initially on a special situation in which the necessary hypotheses are relatively simple to state. This special case, known as Bryant-Seebeck engulfing, will be extremely useful in the remainder of the chapter. In an appendix to this section we treat a more general theorem that can be proved with the techniques developed here. Situations will arise much later in the text in which this more general (and more complicated) statement is needed.

Throughout this section and the next we will use d to denote the metric on the ambient space and ρ to denote the sup-norm on the function space. We will use $B(A;\epsilon)$ to denote the ϵ-neighborhood of A:

$$B(A;\epsilon) = \{x' \mid d(x,x') < \epsilon \text{ for some } x \in A\}.$$

Definition. Let A be a subset of a complete metric space Y and let ϵ be a positive number. An *ϵ-push of* (Y,A) is a homeomorphism $\psi \in \text{Homeo}(Y,Y)$ for which there is an isotopy $\Psi_t : Y \to Y$ such that $\Psi_0 = \text{Id}_Y$, $\Psi_1 = \psi$, $\rho(\Psi_t, \text{Id}_Y) < \epsilon$ for all $t \in [0,1]$, and Ψ_t is supported in $B(A;\epsilon)$. (The last condition means that Ψ_t is the identity outside a closed subset of $B(A;\epsilon)$.) The isotopy Ψ_t is called the *supporting isotopy* of the push ψ.

Observation 3.3.1. If ψ is an ϵ-push of (Y,A), then ψ^{-1} is an ϵ-push of (Y,A) and a 2ϵ-push of $(Y,\psi(A))$.

Theorem 3.3.2 (Bryant-Seebeck Engulfing). *Let W^w be a PL w-manifold and $X \subset W^w$ a compact, x-dimensional, 1-LCC subset of W^w, $x \leq w-3$. For each closed polyhedral subset K^k of W^w with $k \leq \min\{w-x-1, w-3\}$ and each $\epsilon > 0$, there exists a PL ϵ-push ψ of (W,X) such that $\psi(W \smallsetminus X) \supset K$.*

As stated, the theorem asserts that k-dimensional polyhedra can be engulfed from $W \smallsetminus X$ in a controlled way. But ψ^{-1} is an ϵ-push of (W,X) such that $\psi^{-1}(K) \cap X = \emptyset$, so the theorem also shows that k-dimensional polyhedra can be pushed off X in a controlled way. That is the form in which we usually will apply it. The statement does not include the usual engulfing feature about fixing points of K that are already in $W \smallsetminus X$ since the positive number ϵ regulating the push can be reduced to achieve this.

Before addressing the proof of Theorem 3.3.2, we develop some crucial properties of LCC sets.

Lemma 3.3.3. *Suppose X is a closed LCC^{k-1} subset of a locally compact ANR Y, $f : K \to Y$ is a map defined on a finite k-complex K, L is a subcomplex of K with $f(L) \subset Y \smallsetminus X$, and $\epsilon > 0$. Then there exists a map $F : K \to Y$ satisfying (i) $F(K) \subset Y \smallsetminus X$, (ii) $F|L = f|L$, and (iii) $\rho(F,f) < \epsilon$. Moreover, F can be obtained so that there exists an ϵ-homotopy $h_t : K \to Y$ such that $h_0 = f$, $h_1 = F$, $h_t|L = f|L$ for every $t \in I$, and* $\text{diam}\, h(q \times I) < \epsilon$ *for all $q \in K$.*

Proof. Local compactness of Y assures the existence of a compact neighborhood C of $f(K)$ in Y. A standard Lebesgue number argument shows that the following uniform variant of the i-LCC properties holds near C: to each $\eta > 0$ and $i \leq k$ corresponds a positive number $\delta = \delta(\eta, i) > 0$ such that each map of ∂I^i into a δ-subset of $B(C;\delta) \smallsetminus X$ extends to a map of I^i into an η-subset of $Y \smallsetminus X$. (Note that $\delta(\eta, i) \leq \eta$.)

Fix $\epsilon > 0$, set $\delta_k = \delta(\epsilon/3, k)$, and then recursively determine $\delta_{k-j} = \delta(\delta_{k-j+1}/3, k-j)$ for $j = 1, \ldots, k-1$. Subdivide K via a triangulation T so that $\operatorname{diam} f(\sigma) < \delta_1/3$ for every $\sigma \in T$. Let L' denote the union of all $\sigma \in T$ with $f(\sigma) \subset Y \smallsetminus X$ (clearly L' contains a subdivision of L), and define $F|L' = f|L'$. Extend F on vertices v of $T \smallsetminus L'$ so $F(v)$ is a point in $Y \smallsetminus X$ within $\delta_1/3$ of $f(v)$. For 1-simplices τ of $T \smallsetminus L'$, check that $\operatorname{diam} F(\partial\tau) < \delta_1$ and $F(\partial\tau) \subset B(C;\delta_1)$, so F extends to a map (still called F) sending τ into a $(\delta_2/3)$-subset of $Y \smallsetminus X$. Assuming F defined on the $(i-1)$-skeleton of T, $i > 1$, so as to send each $(i-1)$-simplex into a $(\delta_i/3)$-subset of $Y \smallsetminus X$, we check that $\operatorname{diam} F(\partial\sigma) < \delta_i$ and $F(\partial\sigma) \subset B(C;\delta_i)$ for every i-simplex $\sigma \in T$ and exploit the prearranged choice of δ_i to extend F to a map sending σ into a $(\delta_{i+1}/3)$-subset of $Y \smallsetminus X$. Ultimately, when $i = k$, this yields $F : K \to Y$ sending k-simplices of T into $(\epsilon/3)$-subsets of $Y \smallsetminus X$.

To estimate $\rho(F, f)$, fix $q \in K$ and choose $\sigma \in T$ with $q \in \sigma$, as well as a vertex v of σ. Then

$$\begin{aligned} d(F(q), f(q)) &< d(F(q), F(v)) + d(F(v), f(v)) + d(f(v), f(q)) \\ &\leq \operatorname{diam} F(\sigma) + d(F(v), f(v)) + \operatorname{diam} f(\sigma) \\ &< \epsilon/3 + \delta_1/3 + \delta_1/3 \leq \epsilon. \end{aligned}$$

The map F satisfies (i), (ii), and (iii). In order to achieve the additional condition in the last sentence of the theorem statement, first apply Theorem 0.6.3 to find a number $\epsilon' \leq \min\{\epsilon, d(f(K), Y \smallsetminus C)\}$ so that ϵ'-close maps of K into C are ϵ-homotopic in Y; then apply the preceding argument to find an F satisfying (i), (ii), and (iii) with ϵ replaced by ϵ'. □

Lemma 3.3.4 (π_i-negligibility). *If X is a closed, LCC^{k-1} subset of a locally compact ANR Y, then the inclusion-induced homomorphism $\pi_i(Y \smallsetminus X) \to \pi_i(Y)$ is an isomorphism for $i < k$ and an epimorphism for $i = k$.*

Proof. Any $\alpha \in \pi_i(Y \smallsetminus X)$ can be represented by a map $f : \partial I^{i+1} \to Y \smallsetminus X$. If α is trivial in $\pi_i(Y)$, then f extends to $f : I^{i+1} \to Y$. By Lemma 3.3.3 there is a map $F : I^{i+1} \to Y \smallsetminus X$ such that $F|\partial I^{i+1} f|\partial I^{i+1}$, and thus α is trivial in $\pi_i(Y \smallsetminus X)$. In a similar way $\beta \in \pi_k(Y)$ can be represented by a map $g : S^k \to Y$. Apply the last part of the statement of Lemma 3.3.3 to see that g is homotopic to $G : S^k \to Y \smallsetminus X$. It follows that $\pi_k(Y \smallsetminus X) \to \pi_k(Y)$ is an epimorphism. □

Lemma 3.3.5. *Let W^w be a w-manifold and let X be a k-dimensional closed subset of W^w, $k \leq w-2$, that is 1-LCC in W^w. Then X is LCC^{w-k-2}.*

Proof. Consider a point $x \in X$ and the interior V of a small open ball containing x. By Lemma 3.3.4, $V \setminus X$ is simply connected, and by Corollary 0.3.10, $H_i(V \setminus X) = 0$ whenever $i \leq w-k-2$. The Hurewicz Isomorphism Theorem certifies that $\pi_i(V \setminus X) = 0$ for $i \leq w-k-2$. □

Proof of Theorem 3.3.2. Inductively assume that the theorem holds for complexes of dimension less than k. Then, in treating the k-complex K, subdivide so all simplices have diameter less than ϵ and apply the inductive hypothesis to move the $(k-1)$-skeleton off X with a small push ϕ. We suppress ϕ, though, and simply presume that the $(k-1)$-skeleton already misses X.

List the k-simplices $\Delta_1, \ldots, \Delta_m$ of K that meet X. For $i = 1, \ldots, m$, identify a slightly smaller k-simplex Δ_i' such that $\Delta_i \cap X \subset \operatorname{Int} \Delta_i' \subset \Delta_i' \subset \operatorname{Int} \Delta_i$. Invoke Lemma 3.3.5 and the hypothesis $x \leq w-k-1$ to confirm that X is LCC^{k-1}. Use Lemma 3.3.3 with a very small number $\eta < \epsilon/2$ to find homotopies $h^i : \Delta_i' \times I \to W$ such that $h_0^i =$ inclusion, $h_1^i(\Delta_i') \subset W \setminus X$, $h^i(\Delta_i' \times I) \cap X \subset \operatorname{Int} \Delta_i'$, $\operatorname{diam} h^i(\Delta_i' \times I) < \epsilon$, and $h^i(\Delta_i' \times I) \cap h^j(\Delta_j' \times I) = \emptyset$ for $i \neq j$. After a PL approximation, the homotopy images are polyhedra $S_i \supset \Delta_i' \cup L_i$, where $L_i = h_1^i(\Delta_i') \subset W \setminus X$.

Focus on the case $k < w-3$, which is a bit easier than the $k = w-3$ case, since then (by application of Shadow Building Lemma 3.1.5 exactly as in the proof of Theorem 3.1.3) each S_i has a subcomplex $L_i' \supset L_i$ such that $S_i \searrow L_i'$ and $\dim(L_i' \setminus L_i) < k$. Here L_i' is the image under h^i of a shadow containing the singularities of the general position map h^i. Induction again promises an $(\epsilon/2)$-push ψ of (W, X) such that $\psi(W \setminus X) \supset L_i'$; moreover, ψ can be constrained so strictly that it fixes all points of $K \setminus (\cup_i \Delta_i')$. Enclose the various S_i's in pairwise disjoint sets U_i, each of diameter less than $\epsilon/2$. By the proof of the regular neighborhood theorem, we can construct a regular neighborhood N of $[K \setminus (\cup_i \Delta_i')] \cup (\cup_i L_i')$ in $\psi(W \setminus X)$, a regular neighborhood N' of $[K \setminus (\cup_i \Delta_i')] \cup (\cup_i S_i) = K \cup (\cup_i S_i)$, and a sequence of intermediate simplicial neighborhoods $N = N_0 \subset N_1 \subset \cdots \subset N_q = N'$ such that the closure of each $N_j \setminus N_{j-1}$ is a w-ball B_j in some U_i, where B_j meets ∂N_{j-1} in a $(w-1)$-face. Hence there is an ambient isotopy carrying N to N' and supported in $\cup_i U_i$ near the balls covering $N' \setminus N$. The end of that isotopy is another $(\epsilon/2)$-push ψ' of (W, X), and $\psi' \circ \psi(W \setminus X) \supset K \cup (\cup_i S_i) \supset K$. Hence, $\psi'\psi$ has all the features required in the theorem.

Finally, in case $k = w-3$, analogs of the small sets S_i and U_i provide the motion control. The uncontrolled codimension-three engulfing argument given in 3.1.3 works equally well here. One repeatedly identifies more limited singular sets, just as before, engulfs the shadows L_i' of those singularities, and then uses regular neighborhoods to cover the inverses of related collapses. As the resulting isotopies are all supported in $\cup_i U_i$, the outcome is an ϵ-push. □

Appendix on Generalized Controlled Engulfing

Techniques related to those developed in the proof of Theorem 3.3.2 can be used to prove other, even more powerful, controlled engulfing theorems. We conclude this section with a theorem general enough for use in several applications to come in Chapters 5 and 7. It seems efficient to deal with the result immediately, while the proof of 3.3.2 is close at hand, rather than to derive it later, when it is needed. We set it off in an appendix, in case the reader prefers to move forward toward applications of 3.3.2.

The main ingredient required for controlled engulfing is a collection of short homotopies that pull various polyhedra into an open set. In the proof of Bryant-Seebeck engulfing, those homotopies are constructed using the 1-LCC condition and duality (Lemma 3.3.5). The controlled engulfing theorem of this appendix will explicitly assume the existence of the short homotopies. Except for the motion size controls in the conclusion, the statement exactly parallels that of Modified Stallings Engulfing Theorem 3.1.7.

The following special definition simplifies and shortens the statement of the controlled engulfing theorem.

Definition. Let M be a PL manifold, $(W, U) \subset (W', U')$ two pairs of open subsets of M, and $\delta > 0$. We say that *r-dimensional relative polyhedra in (W, U) can be δ-homotoped to U' in W'* if for every pair $(K, L) \subset (W, U)$ of compact polyhedra with $\dim K \leq r$ there exists a homotopy $g : K \times [0, 1] \to W'$ such that

(1) $g_0 = \mathrm{incl}_K : K \to W'$,

(2) $g_1(K) \subset U'$,

(3) $g_t|L = \mathrm{incl}_L$ for every t, and

(4) $\mathrm{diam}\, g(\{x\} \times [0, 1]) < \delta$ for every $x \in K$.

Due to the inductive structure of an engulfing proof, the final engulfing isotopy will move points many times farther than any one of the individual homotopies guiding the construction of the isotopy. An important observation regarding the proof of Theorem 3.3.2 is that the relationship between the size of the final engulfing isotopy and the size of the homotopies used to construct it is mainly determined by the number of homotopy layers built up in the course of the inductive construction. Thus, in the presence of a uniform bound on the motion of the homotopies, the motion control is independent of both the ambient manifold itself and the particular polyhedron to be engulfed; it is dependent solely on the dimension of that polyhedron.

Lemma 3.3.6. *Let σ be an r-simplex and $p_\sigma : \sigma \times I \to \sigma$ the obvious projection. For any two i-simplices $\tau, \tau' \subset \sigma \times I$, $i < r$, such that* $\operatorname{Int}\tau \cap \operatorname{Int}\tau' = \emptyset$ *and $p_\sigma(\tau) = p_\sigma(\tau')$ and for any $\epsilon > 0$, there exists a PL ϵ-homeomorphism $\theta : \sigma \times I \to \sigma \times I$ fixing $\tau \cup \partial\tau'$ and supported near τ' such that $p_\sigma\theta(\tau) \cap p_\sigma\theta(\tau') = p_\sigma(\partial\tau)$.*

Proof. Figure 3.7 suggests how to adjust τ' to achieve this, simply by shifting the barycenter of τ' off $p_\sigma(\tau) \times I$. □

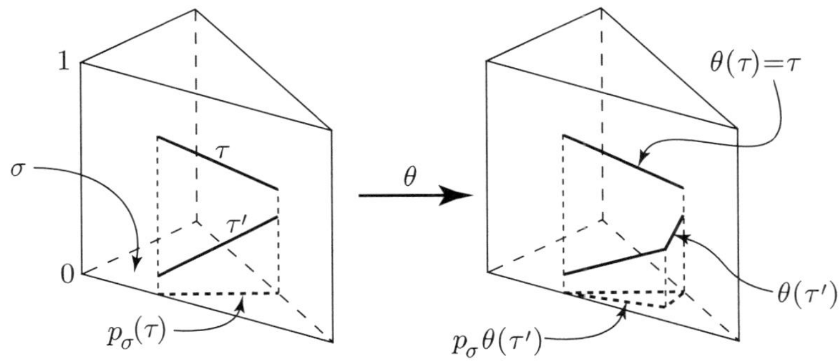

Figure 3.7. Modifications to eliminate coincident projections

Theorem 3.3.7 (Generalized Controlled Engulfing). *Let δ denote a fixed positive number, M an n-dimensional PL ∂-manifold, (W_0, U_0) a pair of open subsets of M, K a complex in W_0 of dimension $k \leq n-3$ such that $|K|$ is closed in M^n and $|K| \cap \partial M \subset U_0$, and L a subcomplex of K such that $|L| \subset U_0$ and $K \smallsetminus L$ is covered by a finite r-subcomplex R of K. Suppose*

$$(W_0, U_0) \subset (W_1, U_1) \subset \cdots \subset (W_{r+1}, U_{r+1})$$

are pairs of open sets in M such that r-dimensional relative polyhedra in (W_i, U_i) can be δ-homotoped to U_{i+1} in W_{i+1}, $0 \leq i \leq r$, and $(|R|, |R \cap L|) \subset (W_0, U_0)$. Then K can be engulfed by U_{r+1}, keeping L fixed, via a PL ϵ-push of (M, U_{r+1}) compactly supported in $W_{r+1} \cap \operatorname{Int} M$, where $\epsilon = (r+2)\delta$.

Proof. This is proved by induction on r. We can assume no simplex of R meets ∂M, by first subdividing K so all simplices that do touch ∂M actually lie in U_0, then deleting those simplices from R while adding both them and all their faces to L. From here on we treat M as a boundaryless manifold, with the understanding that compactly supported pushes there extend via the identity over ∂M when ∂M is restored.

First, in all cases except $r = n-3$, just as in the proof of Theorem 3.1.3, we show that the entire image of an appropriate homotopy $g : R\times[0,1] \to W_1$ pulling R into U_1 can be engulfed by U_{i+r+1} via a PL ϵ'-push compactly supported in W_{i+r+1} and fixing L, where $\epsilon' = (r+1)\delta$. The initial $r = 0$ step is readily obtained. Assume this controlled engulfing can always be accomplished for complexes of dimension less than $s \leq n-4$, and consider a finite s-complex R with $(|R \cup L|, |L|) \subset (W_i, U_i)$, $0 \leq i \leq r-s$. The hypothesis promises a δ-homotopy $g : R \to W_{i+1}$ such that $g_0 = \text{incl}_R$, g_t fixes $R \cap L$ and $g_1(R) \subset U_{i+1}$. Put g in general position (rel $R \cap L$), both internally and with respect to how its image intersects L, so that S_g, its singular set, and $g^{-1}(L) \smallsetminus ((R \cap L) \times I)$ have dimension at most

$$(s+1) + (n-3) - n \leq s-2.$$

Name a triangulation T of R such that (1) $\operatorname{diam} g(\gamma \times I) < \delta$ for all simplices $\gamma \in T$ and (2) $S_g \cup g^{-1}(L)$ is contained in $T^{(s-2)} \times I$, where, as usual, $T^{(s-2)}$ denotes the $(s-2)$-skeleton of T. Induction, applied to $g(T^{(s-2)} \times I) \subset W_{i+1}$, yields an $s\delta$-push ψ, fixing $L \cup g_1(K)$ and compactly supported in W_{i+s+1}, such that

$$\psi(U_{i+s+1}) \supset g(T^{(s-2)} \times I) \cup g_0(R).$$

Let C denote the complement of $T^{(s-2)}$ in $R \smallsetminus L$. Here g embeds $C \times I$, since $(C \times I) \cap S_g = \emptyset$; moreover, by choice of $T^{(s-2)}$, $g(C \times I)$ misses $L \cup g(T^{(s-2)} \times I)$. Hence, by (Rourke and Sanderson, 1972, Lemma 3.25) one can produce another push ψ' in M, compactly supported in W_{i+1} and fixing $L \cup g((R \times \{1\}) \cup (T^{(s-2)} \times I))$, such that

$$\psi'\psi(U_{i+s+1}) \supset g(T \times I) = g(R \times I).$$

This ψ' can be realized as the composition of two δ-pushes, the first supported in pairwise disjoint δ-sets near $\cup_\tau g(\tau \times I)$, indexed by the $(s-1)$-simplices $\tau \in T$, and the second in pairwise disjoint δ-sets near $\cup_\sigma g(\sigma \times I)$, indexed by s-simplices $\sigma \in T$. The first portion of the push ψ' expands $\psi(U_{i+s+1})$ to cover $g(T^{(s-1)} \times I)$ and the second portion expands further to cover the entire $g(R \times I)$. With extra care during the expansion associated with $(s-1)$-simplices τ, one can force ψ' to be a δ-push, not merely a 2δ-push, by insisting that the motion near $g(\tau \times I)$ be close to all $g(\sigma \times I)$ for which τ is a face of σ. Then $\psi'\psi$ is an $(s+1)\delta$-push that engulfs $g(R \times I)$ subject to all the required inductive constraints. This completes the proof of the inductive step ($r \leq n-4$).

Remark. The extra care to assure that ψ' is a δ-push, not merely a 2δ-push, is not ultimately crucial, provided one is willing to tolerate essentially double the motion of the composite push $\psi'\psi$.

Now we turn to the $r = n-3$ case, which stands apart from the others, in that we cannot engulf the entire image of the homotopy, for the usual reasons. Consider a finite relative $(n-3)$-complex $(|R \cup L|, |L|)$ in (W_0, U_0). Find a δ-homotopy $g : R \times I \to W_1$, as before, such that $g_0 = \mathrm{incl}_R$, g_t fixes $R \cap L$ and $g_1(R) \subset U_1$.

The key to successful completion of the argument involves exceedingly meticulous general position considerations imposed on g. Put g in general position so that its singular set S_g has dimension at most

$$(n-2) + (n-2) - n \leq n-4$$

and the preimage S_{TP} under g of all triple points of $g((R \setminus L) \times I)$—points having at least three preimages—has dimension at most $3(n-2)-2n \leq n-6$. In addition, adjust to locate an $(n-5)$-polyhedron $P \subset K \times I$ such that the projection of $S_g \setminus P \subset K \times I$ to R is 1-1. Again name a small mesh triangulation T of R so diam $g(\gamma \times I) < \delta$ for all $\gamma \in T$ and

$$P \cup S_{TP} \cup S'_g \cup (g^{-1}(L) \setminus (R \cap L) \times I) \subset T^{(n-5)} \times I,$$

where S'_g is the reduced singular set consisting of all preimages of singularities arising from $g(\sigma), g(\sigma')$ for which $\dim \sigma + \dim \sigma' < 2(n-2)$.

We modify the $(n-4)$-skeleton of T slightly to obtain a collection of small PL $(n-4)$-cells τ in R, with each $\partial \tau$ covered by $T^{(n-5)}$, each diam $g(\tau \times I) < \delta$, and $g|(\cup_\tau \operatorname{Int} \tau) \times I$ an embedding. The modifications occur only for those $(n-4)$-simplices $\gamma \in T$ for which $(\operatorname{Int} \gamma \times I) \cap S_g \neq \emptyset$; then γ is replaced by two $(n-4)$-cells τ, τ', where $\partial \gamma = \partial \tau = \partial \tau'$, $g|(\operatorname{Int} \tau \cup \operatorname{Int} \tau') \times I$ is an embedding and $\tau \cup \tau'$ bounds a small $(n-3)$-cell C_γ containing γ such that $\operatorname{diam} g(C_\gamma \times I) < \delta$. Two such $C_\gamma, C_{\gamma'}$ are allowed to intersect only at $\partial \gamma \cap \partial \gamma'$. Furthermore, for $(n-3)$-simplices $\sigma \in T$, the closure of each $\sigma \setminus \cup_\gamma C_\gamma$ should be a PL $(n-3)$-cell. As a result, the closures $C_1, C_2, \ldots, C_t$ of components of $R \setminus (T^{(n-5)} \cup (\cup_\tau \tau))$ are small PL $(n-3)$-cells such that, for each i, diam $g(C_i \times I) < \delta$ and $g(\operatorname{Int} C_i \times I)$ meets at most one other $g(\operatorname{Int} C_j \times I)$, $j \neq i$.

Now application of the $n-4$ case to $g(T^{(n-5)} \times I) \subset W_1$ yields an $(n-3)\delta$-push ψ, fixing $L \cup g_1(R)$ and compactly supported in W_{n-2}, such that $\psi(U_{n-2}) \supset g(T^{(n-5)} \times I) \cup g_1(R)$. We can ambiently expand $\psi(U_{n-2})$, fixing $L \cup g(T^{(n-5)} \times I) \cup g_1(R)$, along the various $g(\tau \times I)$ via a δ-push ψ' to achieve

$$\psi' \psi(U_{n-2}) \supset g((T^{(n-5)} \cup (\cup_\tau \tau)) \times I) \cup g_1(R)$$

and follow that with further expansions (still denoted as ψ') along $g(C_i \times I)$ to achieve

$$\psi' \psi(U_{n-2}) \supset g((T^{(n-5)} \cup (\cup_\tau \tau)) \times I) \cup g_0(R) \supset R.$$

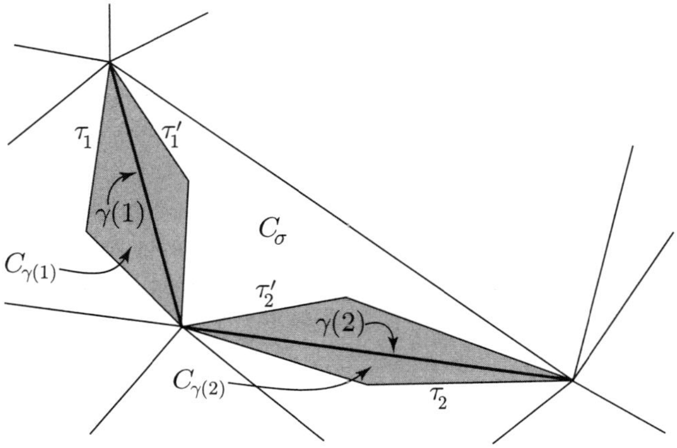

Figure 3.8. Modifications of the triangulation T

As $g(R \times I) \subset W_1$, ψ' can be obtained with compact support in W_1. The collection of further expansions along the C_i might change things by nearly 2δ, since any given point could be moved near the union of at most two intersecting $g(C_i \times I)$. The bound on that motion cannot be reduced to δ itself. However, by insisting that the support of the initial adjustments along $g(\tau \times I)$ be very close to all $g(C_i \times I)$ where $\tau \subset \partial C_i$, just as was done in the cases $r \leq n-4$, one can obtain ψ' as a 2δ-push instead of a 3δ-push. Then $\psi'\psi$ is a $(n-1)\delta$-push having compact support in W_{n-2}. □

Historical Notes. Bryant-Seebeck engulfing and its applications were developed by Bryant and Seebeck in a series of papers written in the late 1960s (see Bryant and Seebeck (1968a), (1968b), (1968c), and (1970)) as well as Bryant (1969).

There are other forms of controlled engulfing. The theorem in the appendix imposes the condition that no point moves more than a fixed amount under the relevant homotopies; another way in which to impose control is to specify in advance a collection of tracks along which points may move. A "radial" engulfing theorem is one in which the tracks of certain homotopies are identified and then an engulfing isotopy is constructed having the property that the movement of each point is confined to a small neighborhood of a limited number of those tracks. Bryant-Seebeck engulfing is a particular case in which the specified homotopies have short tracks.

The first radial engulfing theorem is due to E. H. Connell (1963), who used the technique to approximate stable homeomorphisms of $\mathbb{R}^n$; stable homeomorphisms are touched upon in §8.8. In Connell's theorem the tracks literally are radial. Connell proved a codimension-four theorem, which Bing

(1963) later improved to codimension three. Radial engulfing theorems are surveyed in Bing (1968) and (1975).

Exercises

3.3.1. Verify Observation 3.3.1.

3.3.2. Every cell-like LCC[1] subset X of an ANR Y satisfies the cellularity criterion.

3.3.3. Let X be a compact subset of a locally compact ANR Y. Then $\{F \in C(I^k, Y) \mid F(I^k) \cap X = \emptyset\}$ is dense in $C(I^k, Y)$ if and only if X is LCC^k in Y.

3.3.4. Let X be a compact LCC^k subset of a locally compact ANR Y, $f : K \to Y$ a map defined on a k-complex K, U an open subset of K, and $\epsilon > 0$. Then there exists a map $F : K \to Y$ such that (i) $F|K \smallsetminus U = f|K \smallsetminus U$, (ii) $\rho(F, f) < \epsilon$ and (iii) $F(U) \cap X = \emptyset$.

3.4. Application: Embedding dimension

A major theme of this book is that π_1 conditions detect many useful properties of embeddings. We have already seen, earlier in this chapter, that a π_1 condition on neighborhoods distinguishes cellular embeddings from noncellular ones. Next we use the Bryant-Seebeck engulfing to prove that a local π_1 condition implies that a compactum has neighborhoods with a simple local PL structure. The ultimate goal is to prove that 1-LCC embeddings of polyhedra are tame and that 1-LCC embeddings of compacta behave in many ways like PL embeddings of polyhedra. The results in this section represent the first step in that direction and also serve to illustrate how the controlled engulfing theorem is used.

In order to define embedding dimension we must review some definitions and constructions related to mapping cylinders.

Definition. Let $f : A \to B$ be a map. The *mapping cylinder of f*, denoted $\mathrm{Map}(f)$, is the quotient space obtained from the disjoint union $(A \times [0,1]) \sqcup B$ under identification of each point $\langle a, 1\rangle \in A \times [0,1]$ with $f(a) \in B$. Standard practice identifies A and B with the images of $A \times 0$ and B, respectively, in $\mathrm{Map}(f)$.

In case K and L are polyhedra and $f : K \to L$ is a PL map, the mapping cylinder $\mathrm{Map}(f)$ is a polyhedron. Care must be taken, though, with how it is triangulated. In general it is *not* possible to triangulate $K \times [0,1]$ in such a way that the quotient map $K \times [0,1] \sqcup L \to \mathrm{Map}(f)$ is PL. The impossibility is evident whenever f is degenerate on some simplex; there is

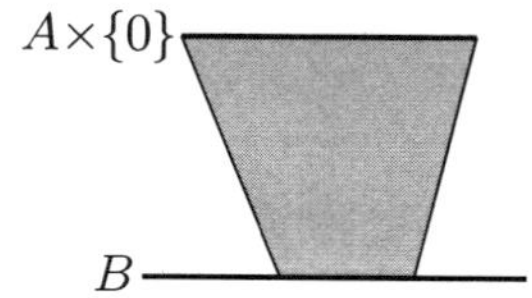

Figure 3.9. The mapping cylinder of $f : A \to B$

no impediment to producing such a triangulation when K is compact and f is injective.

Rather than triangulating $K \times [0,1]$ and L separately and pasting them together, we will define a simplicial version of the complete mapping cylinder. Start with triangulations T of K and S of L such that f is simplicial relative to T and S. The *simplicial mapping cylinder* associated with T and S is a simplicial complex whose underlying space is homeomorphic to $\mathrm{Map}(f)$; it contains S and the first barycentric subdivision of T as subcomplexes. For each vertex $v \in T$, triangulate $\mathrm{Map}(f|v) = \{v\} \times [0,1] \cup L$ so that $\{v\} \times [0,1]$ is a 1-simplex. Given a simplex $\sigma \in T$, we may assume inductively that the simplicial mapping cylinder structure on $\mathrm{Map}(f|\partial\sigma)$ has already been defined. Triangulate $\mathrm{Map}(f|\sigma)$ as the cone on $\mathrm{Map}(f|\partial\sigma) \cup f(\sigma)$ from the barycenter of σ as illustrated in Figure 3.10. The simplicial mapping cylinder of f consists of S plus the union over all simplices in T of these partial mapping cylinders. Whenever $f : K \to L$ is PL we will assume that $\mathrm{Map}(f)$ is triangulated as a simplicial mapping cylinder.

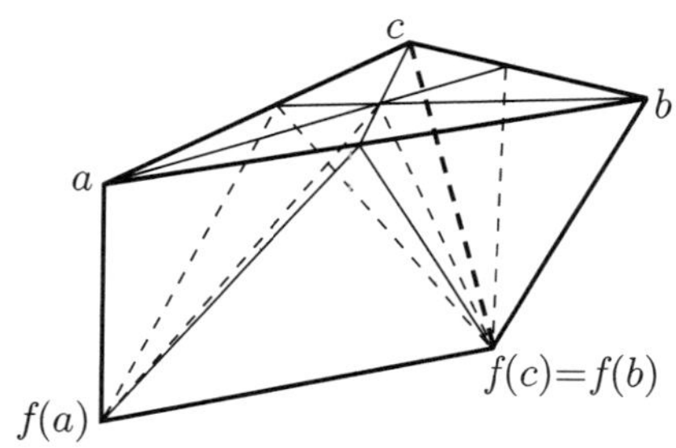

Figure 3.10. The simplicial mapping cylinder of a linear map from a 2-simplex onto a 1-simplex

Definition. Let N be a regular neighborhood of the polyhedron K in the PL manifold M. Say that N is an *ϵ-regular neighborhood* of K if there exist a PL map $r : \partial N \to K$ and a PL homeomorphism $H : \mathrm{Map}(r) \to N$ such that $H(z,0) = z$ for every $z \in \partial N$, $H|K = \mathrm{Id}$, and $\operatorname{diam} H(\{z\} \times [0,1]) < \epsilon$ for every $z \in \partial N$.

The regular neighborhoods constructed by Rourke and Sanderson (1972) are ϵ-regular neighborhoods. Since they neither explicitly state nor prove that fact, we include a proof here.

Lemma 3.4.1. *If K is a compact polyhedron in the PL manifold M, then K has an ϵ-regular neighborhood for every $\epsilon > 0$.*

Proof. Take a triangulation T of M that includes a triangulation of K as a full subcomplex. Each simplex in the simplicial neighborhood $N(K,T)$ is the join of a simplex in K with one that is disjoint from K. Thus each point $z \in \operatorname{Int}|N(K,T)| \smallsetminus K$ can be written uniquely as $z = tx + (1-t)y$, where $x \in K$, $y \in \operatorname{Fr} N$, and $0 < t < 1$. The map $r : \operatorname{Int}|N(K,T)| \to K$ defined by $r(z) = x$ if $z \notin K$ and $r(z) = z$ if $z \in K$ is a retraction. Let $N = N(K,T')$ be the simplicial neighborhood of K in a first derived subdivision T' of T. Then N is a regular neighborhood of K by the Simplicial Neighborhood Theorem (Rourke and Sanderson, 1972, Theorem 3.11). It is easy to check that $N = \operatorname{Map}(r|\partial N)$.

Note that r as defined in the previous paragraph is not a piecewise linear map. But pseudo-radial projection (Rourke and Sanderson, 1972, pp. 20–21) can be used to replace $r|\partial N$ with a PL map that is linear on each simplex of $T'|\partial N$. If $\operatorname{mesh}(T) < \epsilon$, then each fiber has diameter less than ϵ because the fiber is contained in a simplex of T. Hence N is an ϵ-regular neighborhood. □

We are now ready for the fundamental definition.

Definition. A compact subset S of a PL manifold W is said to have *embedding dimension* $\leq k$, abbreviated $\operatorname{dem} X \leq k$, if for every $\epsilon > 0$ there exist a k-dimensional polyhedron $K \subset W$ and an ϵ-regular neighborhood N of K such that $X \subset \operatorname{Int} N \subset N \subset B(X;\epsilon)$. Say that X has *embedding dimension equal to k*, abbreviated as $\operatorname{dem} X = k$, if $\operatorname{dem} X \leq k$ and $\operatorname{dem} X \nleq k-1$.

The embedding dimension is also called the *dimension of embedding* or simply "demension." It is, as the name suggests, a property of the embedding rather than a topological property of the compactum itself. Two familiar Cantor sets in $\mathbb{R}^3$ highlight this dependence on the embedding.

Example 3.4.2. *If C is the standard (middle-thirds) Cantor set in $\mathbb{R}^3$ and A is Antoine's necklace, then* $\operatorname{dem} C = 0$ *while* $\operatorname{dem} A = 1$.

Proof. That $\operatorname{dem} C = 0$ is Exercise 3.4.1. The construction of Antoine's necklace assures A has arbitrarily close neighborhoods that are regular neighborhoods of a finite union of circles. Hence $\operatorname{dem} A \leq 1$. Since A is not flat, $\operatorname{dem} A \neq 0$ (Exercise 3.4.1), so $\operatorname{dem} A = 1$. □

Even though $\operatorname{dem} X$ does not necessarily equal $\dim X$, standard dimension theory facts show that the embedding dimension is bounded below by the topological dimension.

Theorem 3.4.3. *If X is a compact subset X of the PL manifold W, then* $\dim X \leq \operatorname{dem} X$.

Proof. This follows immediately from Alexandroff's Theorem (0.7.1) on approximation of compact metric spaces by polyhedra. □

Another effective way to interpret embedding dimension involves general position. If $\operatorname{dem} X \leq k$, then X has the general position properties of a k-dimensional polyhedron and conversely. That observation is recorded in the next theorem.

Theorem 3.4.4. *A compact subset X of a PL w-manifold W^w satisfies* $\operatorname{dem} X \leq k$ *if and only if for every subpolyhedron P of W^w with* $\dim P \leq w - k - 1$ *and for every $\epsilon > 0$ there exists a PL ϵ-push ψ of (W, X) such that $\psi(X) \cap P = \emptyset$.*

Corollary 3.4.5. *A compact subset X of a PL w-manifold W^w satisfies* $\operatorname{dem} X \leq k$ *if and only if, for every $\epsilon > 0$ and every $(w-k-1)$-dimensional polyhedron L in W^w, there exists an ϵ-push ψ of $(W^w, X \cap L)$ such that $\psi(L) \cap X = \emptyset$.*

Corollary 3.4.6. *If X is a compact subset of the PL manifold W^w and* $\operatorname{dem} X \leq w - 3$, *then X is LCC^1 in W^w.*

Corollary 3.4.7. *If X and Y are compact subsets of the PL manifold W^w and* $\operatorname{dem} X + \operatorname{dem} Y < w$, *then for every $\epsilon > 0$ there exists an ϵ-push ψ of $(W^w, X \cap Y)$ such that $\psi(X) \cap Y = \emptyset$.*

Proof. Exercise 3.4.2. □

The main result of the section is the following powerful converse to Corollary 3.4.6. Its LCC^1 hypothesis suggests—quite accurately, it turns out, but that is not a concern at the moment—that $\dim X \leq w - 3$.

Theorem 3.4.8. *If X is a compact subset of the PL w-manifold W, $w \geq 5$, and X is LCC^1 in W, then* $\operatorname{dem} X \leq \dim X$.

Proof. Assume $\dim X \leq k$. We will prove that $\operatorname{dem} X \leq k$. Let $\epsilon > 0$ be given.

Assume first that $k \geq 2$. Choose a PL ∂-manifold neighborhood M of X such that M is contained in the ϵ-neighborhood of X and let T be a triangulation of M that has mesh less than $\epsilon/9$. Define L to be the k-skeleton of T. Take a barycentric subdivision T' and define P to be the simplicial complement of L in T'. Then each simplex in T' is the join of a simplex in L and a simplex in P, so $\dim P = w-k-1$ and there is a natural join structure between P and L (see the remarks at the beginning of §3.1). Take N to be the simplicial neighborhood of L in the second barycentric subdivision T''; note that N is an $(\epsilon/9)$-regular neighborhood of L.[1]

By Bryant-Seebeck engulfing (Theorem 3.3.2) there is an $(\epsilon/9)$-push ψ of (W, X) such that $\psi(P) \cap X = \emptyset$ and the support of ψ is in $\operatorname{Int} M$. Every simplex in T' is the join of one in L' and one in P, so we can use the join structure to define a second push ϕ such that $\phi(\psi(N)) \supset X$. Now ϕ is a push across the image of the join structure under ψ, so ϕ will be an $(\epsilon/3)$-push. Originally N was an $(\epsilon/9)$-regular neighborhood, but it was stretched by the $(4\epsilon/9)$-push $\phi \circ \psi$, so now it is an ϵ-regular neighborhood of $K = \phi(\psi(L))$. Since K is k-dimensional and ϵ was arbitrary, this shows that $\operatorname{dem} X \leq k$.

The cases $k = 0$ and $k = 1$ are more difficult because in those cases we do not have $w - k - 1 \leq w - 3$ and therefore cannot engulf the dual $(w-k-1)$-skeleton. To prove the theorem in those cases we must use an additional *ad hoc* argument that is itself a miniature version of the proof of the engulfing theorem. We present an outline of the argument that omits some of the details regarding the size controls.

Consider $k = 1$ and $w \geq 6$. By Corollary 0.7.2 there is a compact 1-dimensional polyhedron $K \subset W$ and a map $f : X \to K$ such that $d(x, f(x)) < \epsilon$ for every $x \in X$. Since K is an ANR, the map f may be extended to a neighborhood U of X; we may also assume (by Theorem 0.6.3) that $f : U \to K$ is ϵ-homotopic to the inclusion $U \hookrightarrow W$. By the part of the theorem already proved we know that $\operatorname{dem} X \leq 2$, so for every $\delta > 0$ there exists a 2-dimensional polyhedron P and a δ-regular neighborhood N of P such that $X \subset \operatorname{Int} N \subset N \subset U$. The map $f|P : P \to K$ is ϵ-homotopic in W to the inclusion $P \hookrightarrow W$. Take a PL approximation to this homotopy and put it in general position. Then the singular set will have dimension 0 and its shadow will be 1-dimensional. Let K_+ be the 1-dimensional polyhedron consisting of K plus this shadow. Then the track of the homotopy collapses to K_+, so P can be engulfed with a regular neighborhood of K_+. A stretch across the mapping cylinder structure of N engulfs all of N, and therefore also X, into a regular neighborhood of K_+. The same kinds of controls that were placed on the motions in the proof of Bryant-Seebeck engulfing work here.

The case $k = 1$ and $w = 5$ is the hardest. We cannot proceed as in the last paragraph because this time the singular set will be 1-dimensional and its shadow 2-dimensional. Use Corollary 0.7.2 and Theorem 0.6.3 as above

[1] In order to avoid excessive wordiness, we are not distinguishing between a simplicial complex and its underlying polyhedron.

to find a 1-dimensional polyhedron K, a neighborhood U of X, and a map $f : U \to K$ such that $f : U \to K$ is ϵ-homotopic to the inclusion $U \hookrightarrow W$. Again use the fact that the theorem has already been proved for $k \geq 2$ to find a 2-dimensional polyhedron P and an ϵ-regular neighborhood N of P such that $X \subset \operatorname{Int} N \subset N \subset U$. Let $H : P \times [0,1] \to W$ be an ϵ-homotopy such that $H(P \times 1) \subset K$. Put H in general position, keeping $H|P \times 1$ fixed. Then $H(P^{(1)} \times [0,1))$ will be embedded in $W \smallsetminus K$, where $P^{(1)}$ is the 1-skeleton of P. Hence $K \cup H(P^{(1)} \times [0,1)) \searrow K$, so a regular neighborhood of K can be stretched out to cover $H(P^{(1)} \times [0,1))$. Now proceed as in the proof of an engulfing theorem to use the track of H to engulf the 2-simplices of P, one at a time. The only difference between this proof and the earlier one is that the 1-dimensional shadows that appear are not engulfed by induction, but are simply added to K to build a larger 1-dimensional polyhedron K_+. A regular neighborhood of K_+ can be stretched out to engulf P. A push across the mapping cylinder structure of N stretches the regular neighborhood of K_+ out to cover all of N and therefore also all of X.

Finally, suppose $k = 0$ and $w \geq 5$. First use the $k = 1$ case of the proof to see that $\operatorname{dem} X \leq 1$. There is a small map of a neighborhood U of X to a 0-dimensional polyhedron K (Corollary 0.7.2). Using the $k = 1$ case of the theorem we can find a 1-dimensional polyhedron P and a regular neighborhood N of P such that $X \subset \operatorname{Int} N \subset N \subset U$. This time the track of the homotopy of P can be embedded and so it is easy to use it to engulf P with a neighborhood of K. □

In codimensions one and two the 1-LCC hypothesis is not needed for showing $\operatorname{dem} X \leq \dim X$.

Theorem 3.4.9. *If X is a compact subset of the PL w-manifold W^w, $w \geq 4$, and* $\dim X = w - 2$, *then* $\operatorname{dem} X = \dim X$.

Proof. By Corollary 0.3.11, X is 0-LCC. Hence, Lemma 3.3.3 implies that 1-dimensional polyhedra can be homotoped off X. As long as $n \geq 4$, this allows 1-dimensional polyhedra to be engulfed by the complement of X, and so the proof of Theorem 3.4.8 shows that $\operatorname{dem} X \leq w - 2$. □

Theorem 3.4.10. *If X is a compact subset of the PL w-manifold W^w and* $\dim X = w - 1$, *then* $\operatorname{dem} X = \dim X$.

Proof. The fact that X contains no open subset of W means that points can be engulfed in the complement of X. Again the proof of Theorem 3.4.8 applies to show $\operatorname{dem} X \leq w - 1$. □

The following theorem summarizes the last few results and clarifies the relationship between dimension and demension.

Theorem 3.4.11 (Relationship between dimension and demension). *Let X be a compact subset of the PL w-manifold W^w.*

1. *If* $\dim X \geq w-2$ *and* $w \neq 3$, *then* $\operatorname{dem} X = \dim X$.
2. *If* $\dim X \leq w-3$ *and* $w \neq 4$, *then either*
   (a) $\operatorname{dem} X = \dim X$ *and X is LCC^1, or*
   (b) $\operatorname{dem} X = w-2$ *and X is not 1-LCC.*

A significant aspect of Theorem 3.4.8 is that its hypothesis is topological while its conclusion is piecewise linear. This observation suggests that the definition of embedding dimension can be formulated using either PL or topological regular neighborhoods. Combining this with Theorem 3.4.4 gives a total of four equivalent conditions, any one of which could be used as the definition of embedding dimension. Before stating the four conditions we spell out what is meant by a topological regular neighborhood.

Definition. Let M be a manifold and let X be a closed subset of M. A neighborhood N of X in M is a *mapping cylinder neighborhood* of X if N is a ∂-manifold and N is homeomorphic to $\operatorname{Map}(r)$ for some map $r : \partial N \to X$. It is required that the homeomorphism

$$H : \operatorname{Map}(r) = (\partial N \times [0,1] \sqcup X)/\{\langle z,1\rangle \sim r(z)\} \to N$$

be natural in the sense that $H(z,0) = z$ for each $z \in \partial N$ and that $H|X = \mathrm{Id}$. The neighborhood is called an *ϵ-mapping cylinder neighborhood* if for every $z \in \partial N$ we have $\operatorname{diam} H(\{z\} \times [0,1]) < \epsilon$.

Theorem 3.4.12 (Alternative definitions of embedding dimension). *If X is any compact subset of the PL n-manifold M^n, $n \geq 5$, and $k \leq n$, then the following conditions are equivalent.*

1. *For every $\epsilon > 0$ there exists an ϵ-regular neighborhood N of a k-dimensional polyhedron K such that* $X \subset \operatorname{Int} N \subset N \subset B(X;\epsilon)$.
2. *For every $\epsilon > 0$ and for every subpolyhedron P of M with* $\dim P \leq n-k-1$ *there exists a PL ϵ-push ψ of (M,X) with* $\psi(X) \cap P = \emptyset$.
3. *For every $\epsilon > 0$ and for every subpolyhedron P of M with* $\dim P \leq n-k-1$ *there exists a topological ϵ-push ψ of (M,X) such that* $\psi(X) \cap P = \emptyset$.

Moreover, if $k \leq n-3$, then these three conditions are equivalent to:

4. *For every $\epsilon > 0$ there exists an ϵ-mapping cylinder neighborhood N of a k-dimensional, 1-LCC compactum K such that* $X \subset \operatorname{Int} N \subset N \subset B(X;\epsilon)$.

Historical Notes. The concept of embedding dimension is implicit in work of Bryant, (1969) and (1971a), but the property was first formally identified and systematically studied by M. A. Štan′ko (1969). Edwards (1975b) provided an exposition in which he extended the theory of embedding dimension to σ-compact subsets of a manifold.

The restriction $w \neq 3$ is necessary in Part 1 of Theorem 3.4.11; McMillan-Row (1969) and H. Bothe (1964) have constructed an example of a compactum $X \subset \mathbb{R}^3$ such that $\dim X = 1$ but $\operatorname{dem} X = 2$. The example is a version of the Menger universal 1-dimensional continuum in $\mathbb{R}^3$ that is constructed by removing knotted tunnels instead of straight ones.

The mistake of thinking that the retraction r in the proof of Lemma 3.4.1 is PL is the "standard mistake" of (Rourke and Sanderson, 1972) and (Zeeman, 1963a). Further information on simplicial mapping cylinders may be found in (Hatcher, 2002) and (Bryant, 2002).

Exercises

3.4.1. A Cantor set $X \subset \mathbb{R}^n$ is flat if and only if $\operatorname{dem} X = 0$.

3.4.2. Prove Corollary 3.4.7.

3.4.3. If X is a compact subset of the PL manifold W^w, $w \geq 5$, then $\operatorname{dem} X = \operatorname{dem} h(X)$ for any $h \in \operatorname{Homeo}(W, W)$.

3.4.4. If X is a compact subset of the PL manifold W^w and $Y \subset X$ is compact, then $\operatorname{dem} Y \leq \operatorname{dem} X$.

3.4.5. Let (X, Y) be a compact pair such that $2 \dim X + 1 \leq n$ and $f : X \to \mathbb{R}^n$ a map such that $f|Y$ is an embedding with $\operatorname{dem} f(Y) = \dim Y$. Then f can be approximated by an embedding F such that $\operatorname{dem} F(X) = \dim X$ and $F|Y = f|Y$. [Hint: the usual embedding of X into $\mathbb{R}^n$, as defined in either (Hurewicz and Wallman, 1948) or (Munkres, 2000), has the correct embedding dimension.]

3.4.6. Let X be a polyhedron topologically embedded in the PL manifold W, $w \geq 5$. If X is locally tame, then $\operatorname{dem} X = \dim X$.

3.5. Embeddings of Menger continua

Here we aim to present a collection of examples of k-dimensional compacta in $\mathbb{R}^n$ that are ambiently universal for the class of all compacta in $\mathbb{R}^n$ of embedding dimension $\leq k$; in other words, for any compact $X \subset \mathbb{R}^n$ with $\operatorname{dem} X \leq k$, there exists a self-homeomorphism of $\mathbb{R}^n$ sending X into the universal space.

It pays to start off with some notation and terminology. As before, we let $I = [0,1]$ and we use $C^n = [0,1] \times \cdots \times [0,1] \subset \mathbb{R}^n$ to denote the *standard n-cube*. More generally, if $I_1, \ldots, I_n$ are closed intervals, we will call $Q = I_1 \times \cdots \times I_n$ an *n-cube* (even though the intervals may be of widely varying lengths). A *face* of Q is a subset of the form $f = f_1 \times \cdots \times f_n$ where each f_j is either an endpoint of I_j or equal to all of I_j. Since there are three choices for each f_j, Q has exactly 3^n (nonempty) faces. A face is k-dimensional if and only if precisely k of the f_j are equal to I_j. Thus there are exactly $\binom{n}{k}2^{n-k}$ faces of dimension k. The union of all the k-dimensional faces is called the *k-skeleton* of Q.

Construction of the Menger Continua. For each pair of integers n and k, $0 \leq k \leq n$, we construct a Menger continuum M_n^k of dimension k in $\mathbb{R}^n$. The construction of M_n^k is inductive. We begin by defining T_0 to be the trivial subdivision of I; i.e., $T_0 = \{\{0\}, \{1\}, [0,1]\}$. At the jth stage of the construction, T_j is a subdivision of I into precisely 3^j subintervals. We define T_{j+1} to be the subdivision of I that is obtained by subdividing each interval of T_j into three subintervals of equal length. Then T_{j+1} is a subdivision of I into 3^{j+1} subintervals and its n-fold self-product induces a subdivision T_{j+1}^n of I^n into $3^{(j+1)n}$ subcubes. Define $P_0 = I^n$ and define P_{j+1} to be the union of all subcubes of T_{j+1}^n that lie in P_j and intersect the k-skeleton of P_j. The *k-dimensional Menger continuum in $\mathbb{R}^n$* is defined by

$$M_n^k = \bigcap_{j=0}^{\infty} P_j.$$

Several examples of Menger continua of various dimensions are shown in Figures 3.11–3.13.

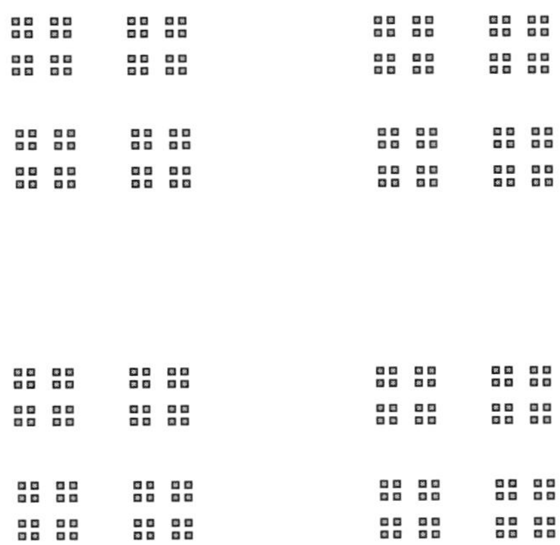

Figure 3.11. The standard Menger continuum M_2^0, which is a Cantor set

Modified Menger continua. The reason for defining the subdivisions T_j recursively in the preceding construction (rather than simply defining T_j to be the subdivision of I into 3^j subintervals of equal length) is to allow ourselves the freedom to modify the construction later. Whenever we use

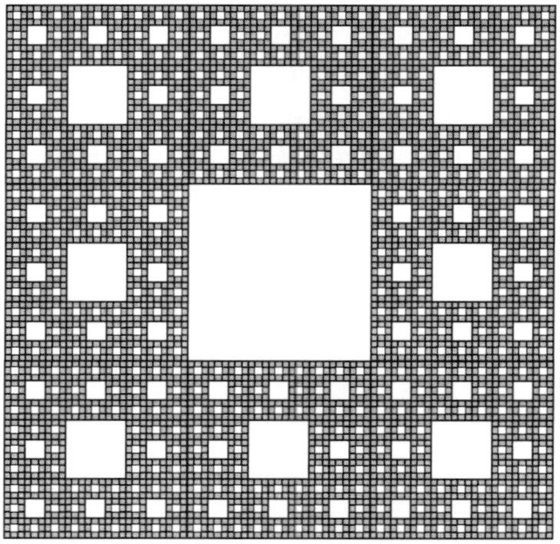

Figure 3.12. The standard Menger continuum M_2^1, commonly known as the "Sierpiński carpet"

Figure 3.13. The standard Menger continuum M_3^1, commonly called the "Menger sponge"

subintervals of equal length, as we did above, we refer to the continuum constructed as a *standard Menger continuum* and denote it by M_n^k.

In this section and the next we will want to modify the construction to allow for the possibility of intervals of different lengths. In this way we will produce Menger-like continua having additional useful properties. In the more general case, the only restriction on the subdivisions T_j is that the length of each interval in T_{j+1} must be less than half the length of the interval in T_j that contains it. When we use subintervals of varying lengths, we will refer to the continuum constructed as a *modified Menger continuum* and will denote it by $\widetilde{M}_n^k$.

In the forthcoming proof of Theorem 3.5.1, we will make the middle interval very short and the other two intervals correspondingly longer. This will result in a modified Menger continuum that is relatively fat and able to absorb other compact subsets of $\mathbb{R}^n$. In the next section we will do exactly the opposite: we will make the middle interval very long and the other two much shorter. This will result in a modified Menger continuum that is relatively thin so that its Hausdorff dimension is minimized.

Every modified Menger continuum $\widetilde{M}_n^k$ is equivalent to M_n^k in the strong sense of there being an isotopy $h : \mathbb{R}^n \times [0,1] \to \mathbb{R}^n$ such that h has compact support, $h_0 = \mathrm{Id}_M$ and $h_1(\widetilde{M}_n^k) = M_n^k$. Proposition 3.5.2 spells out the construction of a modified Menger continuum in more detail and establishes this equivalence.

Definition. Let $Q = I_1 \times \cdots \times I_n$ be an n-cube and let f be a k-dimensional face of Q. Then $f = f_1 \times \cdots \times f_n$ with $f_j = I_j$ for exactly k distinct values of j and f_j an endpoint of I_j otherwise. The *dual face associated with* f is $\hat{f} = \hat{f}_1 \times \cdots \times \hat{f}_n$ defined by $\hat{f}_j = I_j$ if $f_j \neq I_j$ and $\hat{f}_j$ is the midpoint of I_j otherwise. It is clear that $\hat{f}$ is an $(n-k)$-cube. The union of all the dual $(n-k)$-cubes is called the *dual* $(n-k)$*-skeleton of* Q.

Embedding dimension of Menger continua. At each stage of the construction of the Menger continuum we keep all the subcubes that intersect the k-skeleton of the previous stage and discard the others. Another way to describe this construction is to say that we remove the interior of the union of all the cubes that intersect the dual $(n-k-1)$-skeleton of the previous stage. As a result it is easy to push $(n-k-1)$-dimensional polyhedra off M_n^k and so we see that $\operatorname{dem} M_n^k \leq k$. Since the k-skeleton of I^n is a subset of M_n^k, we have $\dim M_n^k \geq k$. Hence $\dim M_n^k = \operatorname{dem} M_n^k = k$.

Universality of the Menger continua. The Menger continua were originally constructed to serve as examples of (compact) universal n-dimensional spaces. They are universal in the sense that every n-dimensional separable metric space can be embedded in M_{2n+1}^n. Our next theorem generalizes that result (at least for compacta) by showing that M_n^k is ambiently universal for compact subspaces of $\mathbb{R}^n$ having embedding dimension $\leq k$. An approximation theorem treated in Chapter 5 will give the stronger conclusion that M_n^k is universal for all k-dimensional compact subsets of $\mathbb{R}^n$.

Theorem 3.5.1. *Let X be a compact subset of $\mathbb{R}^n$ and let k be an integer with $k \leq n$. Then* $\operatorname{dem} X \leq k$ *if and only if there is an ambient isotopy of $\mathbb{R}^n$, with compact support, that moves X into M_n^k.*

The idea is to build a modified Menger continuum containing X and then to show that all modified Menger continuum are ambient isotopic to a standard one. We begin by proving the equivalence of modified and standard Menger continua.

Proposition 3.5.2. *If $\widetilde{M_n^k}$ is any modified Menger continuum in $\mathbb{R}^n$, then there exists an isotopy $h : \mathbb{R}^n \times [0,1] \to \mathbb{R}^n$ such that $h_0 = \mathrm{Id}$, h has compact support and $h_1(\widetilde{M_n^k}) = M_n^k$.*

Proof. We use the same notation as in the construction of M_n^k: T_j denotes the subdivision of I into 3^j intervals of equal length; T_j^n is the corresponding subdivision of I^n into 3^{jn} cubes; $P_0 = I^n$; and P_j is the union of all subcubes in T_j^n which lie in P_{j-1} and intersect the k-skeleton of P_{j-1}. The modified Menger continuum $\widetilde{M_n^k}$ is also defined by a sequence $\{\widetilde{T}_j\}$ of subdivisions of I. It begins with $\widetilde{T}_0 = T_0$ and each $\widetilde{T}_{j+1}$ is a subdivision of $\widetilde{T}_j$. In the inductive step of the construction, each interval in $\widetilde{T}_j$ is replaced by three subintervals in $\widetilde{T}_{j+1}$. The choice of subintervals is arbitrary, except for the restriction that the middle interval in $\widetilde{T}_{j+1}$ must contain the midpoint of the corresponding interval in $\widetilde{T}_j$. It follows that the length of each interval in $\widetilde{T}_j$ is less than $1/2^j$. As before, we use $\widetilde{T}_j^n$ to denote the subdivision of I^n into 3^{jn} cubes that is induced by $\widetilde{T}_j$. Define $\widetilde{P}_0 = I^n$ and define $\widetilde{P}_j$ to be the union of all the cubes in $\widetilde{T}_j^n$ which intersect the k-skeleton of $\widetilde{P}_{j-1}$. Finally, $\widetilde{M_n^k} = \cap_{j=0}^{\infty} \widetilde{P}_j$.

For each j, there is a linear isomorphism $\phi_j : \widetilde{T}_j \to T_j$. The isomorphism ϕ_j induces a PL homeomorphism $g_j : I \to I$. Since mesh $T_j < 3^{-j}$ and mesh $\widetilde{T}_j < 2^{-j}$, $g = \lim_{j\to\infty} g_j$ defines a homeomorphism of I to itself. The n-fold product of g defines a homeomorphism $f : I^n \to I^n$ such that $f(\widetilde{M_n^k}) = M_n^k$.

In order to complete the proof we must show that f can be realized by an ambient isotopy with compact support. First notice that g_j can be extended via the identity to all of $\mathbb{R}$. In addition, there is a $1/2^j$ isotopy from g_{j-1} to g_j. (Take g_0 to be the identity.) Thus there is an isotopy $F : \mathbb{R} \times [0,1] \to \mathbb{R}$ such that $F_0 = \mathrm{Id}$, F_t is the identity outside I for each t, and $F_1 = g$. We use F to define an isotopy G of $\mathbb{R}^n$. On $\mathbb{R} \times I^{n-1}$, $G_t = F_t \times \mathrm{Id}$. Let N be a collar neighborhood of I^{n-1} in $\mathbb{R}^{n-1}$. We extend G_t to $\mathbb{R} \times N$ in such a way that $G_t | \mathbb{R} \times \partial N = \mathrm{Id}$ by using less and less of the isotopy F_t as we move along the collar away from I^{n-1}. Extend G_t via the identity to all of $\mathbb{R}^n$. Notice that G is an ambient isotopy from $G_0 = \mathrm{Id}$ to some homeomorphism $G_1 : \mathbb{R}^n \to \mathbb{R}^n$ with the properties that G_1 changes only the first coordinates of points in $\mathbb{R}^n$ and equals g on the first coordinate of I^n.

Finally, to obtain the isotopy in the conclusion of the Proposition, we first do the isotopy G, then do an analogous isotopy that changes second coordinates, then third coordinates and so forth through the nth coordinate.

This succession of isotopies, strung together, forms the isotopy that moves $\widetilde{M}_n^k$ to M_n^k. □

Proof of Theorem 3.5.1. If such an isotopy h_t exists, then $\operatorname{dem} X = \operatorname{dem} h_1(X) \leq \operatorname{dem} M_n^k = k$. Conversely, suppose $\operatorname{dem} X \leq k$. The strategy of the proof is this: construct simultaneously a modified Menger continuum $\widetilde{M}_n^k$ and an ambient isotopy of X into $\widetilde{M}_n^k$. This will suffice because Proposition 3.5.2 then provides a further isotopy of $\mathbb{R}^n$ that pushes $\widetilde{M}_n^k$ to M_n^k.

Now consider a compactum $X \subset \mathbb{R}^n$ with $\operatorname{dem} X \leq k$. We may assume that $X \subset \operatorname{Int} I^n$. Let $\widetilde{T}_0$ denote the trivial subdivision of I. Since $\operatorname{dem} X \leq k$, there is a small PL homeomorphism $h_1 : \mathbb{R}^n \to \mathbb{R}^n$, fixed outside I^n, that shifts X off the dual $(n-k-1)$-skeleton of I^n (Theorem 3.4.4). Choose a subdivision $\widetilde{T}_1$ of I into three intervals with the middle interval very short and centered about the midpoint of I. Just how short the middle interval should be is determined by the distance from $h_1(X)$ to the dual $(n-k-1)$-skeleton of I^n. As before, let $\widetilde{T}_1^n$ denote the collection of n-dimensional cubes in I^n formed by taking products of intervals in $\widetilde{T}_1$. A cube $Q \in \widetilde{T}_1^n$ misses the k-skeleton of I^n if and only if it is a product of intervals that include at least $k+1$ short intervals. This means it is completely contained in a small neighborhood of a dual cell of dimension $\leq n-k-1$. We choose the middle interval in $\widetilde{T}_1$ to be so short that $h_1(X)$ misses the union of all such cubes. In this way we construct $\widetilde{T}_1$ so that $h_1(X)$ is completely contained in $\operatorname{Int} \widetilde{P}_1$, where $\widetilde{P}_1$ denotes the union of all parallelepipeds in $\widetilde{T}_1^n$ that intersect the k-skeleton of I^n.

Next use Theorem 3.4.4 again to find a PL homeomorphism h_2 fixed outside $\widetilde{P}_1$ and shifting $h_1(X)$ off the dual $(n-k-1)$-skeleton of every n-cube in $\widetilde{T}_1^n$. Then find a subdivision $\widetilde{T}_2$ of I that subdivides each interval of $\widetilde{T}_1$ into three subintervals and makes the middle one so short that no n-cube in $\widetilde{T}_2^n$ intersects both the dual $(n-k-1)$-skeleton of $\widetilde{P}_1$ and $h_2(h_1(X))$. Let $\widetilde{P}_2$ denote the union of all the n-cubes of $\widetilde{T}_2^n$ that lie in $\widetilde{P}_1$ and intersect the k-skeleton of $\widetilde{P}_1$. This procedure is continued inductively and results in a modified Menger compactum $\widetilde{M}_n^k$.

At the jth stage of the construction we must choose a homeomorphism h_j. We have the freedom to make that homeomorphism as close to the identity as we like, with the closeness possibly depending on the choice of homeomorphism at the previous stage. As a result we can choose the homeomorphisms h_j in such a way that the sequence $\{h_j \circ h_{j-1} \circ \cdots \circ h_1\}$ converges to a homeomorphism (Proposition 2.2.2). If

$$h = \lim_{j \to \infty} h_j \circ h_{j-1} \circ \cdots \circ h_1,$$

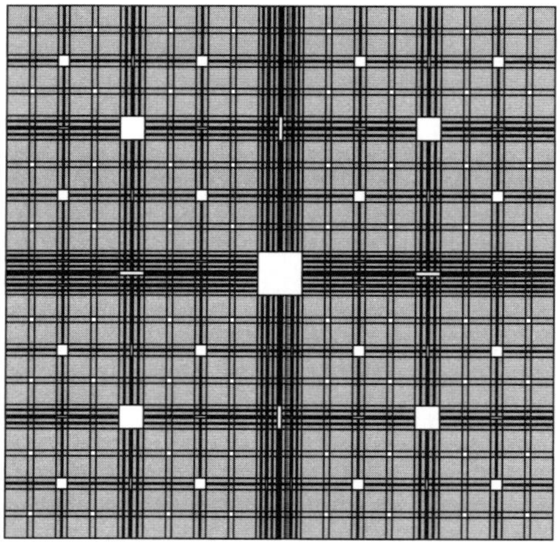

Figure 3.14. This modified Menger curve $\widetilde{M}_2^1$ is relatively fat so that it can easily absorb 1-dimensional compacta

then $h(X) \subset \widetilde{M}_n^k$. Since there is a short isotopy from $h_j \circ h_{j-1} \circ \cdots \circ h_1$ to $h_{j+1} \circ h_j \circ \cdots \circ h_1$, the homeomorphism h is isotopic to the identity. $\square$

Historical Notes. K. Menger (1926) originally constructed M_n^k as an example of a universal n-dimensional space. He described the space and conjectured that M_n^k should be universal for k-dimensional subsets of $\mathbb{R}^n$, but he did not supply a proof. The first complete proof of the universality of M_{2n+1}^n for n-dimensional separable metric spaces is due to S. Lefschetz (1931). Theorem 3.5.1 and the stronger version to be proved later are due to M. A. Štan′ko (1971b).

3.6. Embedding dimension and Hausdorff dimension

Anyone who reads popular accounts of the theory of fractals, such as (Mandelbrot, 1982), will notice a resemblance between the Menger continua in the previous section and fractals. The superficial resemblance hints at something deeper. There is a meaningful relationship between wildly embedded compacta and fractals, and there is also a close connection between the embedding dimension of a compactum and its Hausdorff dimension. This section offers a digression exploring those connections. The results will not be used in the sequel, so the entire section may be omitted without serious consequence.

In order to proceed, it is necessary to have some understanding of Hausdorff dimension. We cannot develop a complete exposition of the theory of Hausdorff dimension and fractals, but we attempt to include enough information so that the reader has some grounding in the relationships being discussed.

Hausdorff dimension is defined for any separable metric space X. It depends on the metric for X and is not a topological invariant. In most cases we will not specify an explicit metric but will consider only subsets $X \subset \mathbb{R}^n$ and will treat the standard Euclidean metric inherited from $\mathbb{R}^n$. We use $\mathcal{U}$ to denote a countable cover of X (not necessarily by open sets), $\operatorname{diam} U$ to denote the diameter of $U \in \mathcal{U}$ and $\operatorname{mesh} \mathcal{U}$ to denote $\sup\{\operatorname{diam} U \mid U \in \mathcal{U}\}$.

Definitions. Suppose X is a separable metric space. Let p be a real number in the range $0 \leq p < \infty$ and let $\epsilon > 0$. First define

$$m_{p,\epsilon}(X) = \inf\left\{ \sum_{U \in \mathcal{U}} (\operatorname{diam} U)^p \;\middle|\; \mathcal{U} \text{ a countable cover of } X, \operatorname{mesh} \mathcal{U} < \epsilon \right\}.$$

(Of course $m_{p,\epsilon}(X) = \infty$ is possible.) Next define the *Hausdorff p-measure* of X by

$$m_p(X) = \sup\{m_{p,\epsilon}(X) \mid \epsilon > 0\}.$$

For each p and n, m_p is a countably additive measure on the Borel sets in $\mathbb{R}^n$ (see (Edgar, 1990, pp. 135 and 147) or (Keesling, 1986, Theorem 2.2)). For a fixed X, $m_p(X)$ is a monotonically decreasing function of p. In fact, if $m_p(X) < \infty$ for some p, then $m_q(X) = 0$ for all $q > p$ (see (Hurewicz and Wallman, 1948, Theorem VII.1.B) or (Federer, 1969, p. 171)). This last fact makes the following definition reasonable: the *Hausdorff dimension* of X is defined by

$$\dim_{\mathrm{h}} X = \inf\{p \mid m_p(X) = 0\}.$$

A compactum $X \subset \mathbb{R}^n$ is said to be a *fractal* if its Hausdorff dimension is greater than its topological dimension, i.e., if $\dim_{\mathrm{h}} X > \dim X$.

Remark. The definition of fractal is analogous to the following fact about polyhedra, which will be proved over the course of the next two chapters: a codimension-three polyhedron $X \subset \mathbb{R}^n$ is wild if and only if its embedding dimension is greater than its topological dimension; i.e., X is a fractal if and only if $\operatorname{dem} X > \dim X$.

Examples. Let C denote the standard middle-thirds Cantor set in $\mathbb{R}^1$. It is easy to compute the Hausdorff dimension of C using standard techniques. (See (Edgar, 1990) or (Keesling, 1986), for example.) The result of the computation is $\dim_{\mathrm{h}} C = \log 2/ \log 3$. Thus $\dim_{\mathrm{h}} C > \dim C = 0$ and C is a fractal. If C' is the Cantor set in $\mathbb{R}^1$ constructed by removing "middle-halves" rather than middle-thirds, then $\dim_{\mathrm{h}} C' = 1/2$. Since C and C' are equivalently embedded, we see that Hausdorff dimension is not an invariant of the embedding class.

The Cantor set is a special case of a Menger continuum, namely, $C = M_1^0$. In general, the standard k-dimensional Menger continuum in $\mathbb{R}^n$ satisfies

$$\dim_h M_n^k = \frac{\log F(n,k)}{\log 3},$$

where $F(n,k)$ is the number of faces of dimension $\leq k$ in the n-dimensional cube. Since the n-cube has 3^n faces, it is clear that $3^k < F(n,k) < 3^n$ (assuming $k < n$). Thus $k < \dim_h M_n^k < n$ and every standard Menger continuum is a fractal. In Lemma 3.6.9 we will explain how to vary the construction of the Menger continuum in such a way as to produce a modified Menger continuum $\widetilde{M}_n^k$ which has $\dim_h \widetilde{M}_n^k = k$ and is therefore not a fractal. In order to accomplish this it is necessary to subdivide the cubes faster and faster as the construction proceeds. Since the modified Menger continuum $\widetilde{M}_n^k$ constructed in Lemma 3.6.9 is equivalent by ambient isotopy to the standard Menger continuum M_n^k (Proposition 3.5.2), the property of being a fractal is also not an invariant of the embedding class.

Antoine's necklace $A \subset \mathbb{R}^3$ satisfies $\dim_h A \geq 1$ and so it too is a fractal. The precise value of $\dim_h A$ depends on the number of solid tori used at each stage in the construction and on the exact thicknesses of the solid tori. (See (Rushing, 1992) for details.)

The main purpose of this section is to explore connections among Hausdorff measure, Hausdorff dimension, and embedding dimension. Theorem 3.6.1 exposes the basic relationship between Hausdorff measure and embedding dimension, while Theorem 3.6.2 exposes the relationship between Hausdorff dimension and embedding dimension. An immediate consequence is that any codimension-three compactum that is not 1-LCC is a fractal.

Theorem 3.6.1. *If X is a compact subset of $\mathbb{R}^n$ and k is a nonnegative integer such that $m_{k+1}(X) = 0$, then* $\operatorname{dem}(X) \leq k$.

Theorem 3.6.2. *If X is a compact subset of $\mathbb{R}^n$, then* $\operatorname{dem} X \leq \dim_h \phi(X)$ *for every homeomorphism $\phi : \mathbb{R}^n \to \mathbb{R}^n$. Furthermore, there exists some homeomorphism $\phi : \mathbb{R}^n \to \mathbb{R}^n$ such that* $\operatorname{dem} X = \dim_h \phi(X)$.

Corollary 3.6.3. *If X is a compactum in $\mathbb{R}^n$ such that* $\dim X \leq n - 3$ *and X is not 1-LCC, then X is a fractal.*

It follows from Theorems 3.6.2 and 3.4.11 that the Hausdorff dimension of any compactum that does not satisfy the 1-LCC condition must be in the interval $[n-2, n]$. For each real number r in that range, Rushing (1992) constructs an example of a wildly embedded Cantor set in $\mathbb{R}^n$ that has Hausdorff dimension equal to r.

It is not the case, however, that every codimension-three fractal fails to be 1-LCC. To the contrary, for every real number in the interval $[0, n]$, it is relatively easy to construct a tame Cantor set in $\mathbb{R}^n$ that has that real number as its Hausdorff dimension. (See Rushing (1992), Example 1.)

Before addressing Theorems 3.6.1 and 3.6.2 we state three properties of measure that will be needed. The first property is a quick consequence of the definitions, so we will include a proof. It indicates that measure 0 is a PL invariant.

Lemma 3.6.4. *Suppose P and Q are polyhedra, $X \subset P$, and $f : P \to Q$ is a PL map. If $m_p(X) = 0$ for some p, then $m_p(f(X)) = 0$.*

Proof. Suppose first that $P = \Delta$ is a simplex and $f : \Delta \to Q$ is a linear map. Then f is a Lipschitz map; i.e., there exists a real number $L > 0$ such that $d(f(x), f(y)) \leq L \cdot d(x, y)$ for all x and y in Δ. It is clear from the definition of measure that $m_p(f(X)) \leq L^p \cdot m_p(X)$ for every $X \subset \Delta$. Thus $m_p(X) = 0$ implies $m_p(f(X)) = 0$ in this special case. The general case follows from the special case along with the countable additivity of m_p. □

The next two properties are intuitively plausible, but supplying rigorous proofs would lead us too far afield. As a result we merely state them without proof and refer the reader to (Federer, 1969) for details. Lemma 3.6.5, a product theorem for Hausdorff measure, is a restatement of (Federer, 1969, Theorem 2.10.45). If X and Y are metric spaces, then we assume that $X \times Y$ is given the metric defined by

$$d^2(\langle x_1, y_1\rangle, \langle x_2, y_2\rangle) = d_X^2(x_1, x_2) + d_Y^2(y_1, y_2).$$

Lemma 3.6.5. *Suppose n is a nonnegative integer and A is a metric space. If $m_k(A) < \infty$ for some k, then there exists a number c such that*

$$m_{k+n}(A \times B) = c\, m_k(A)\, m_n(B)$$

for every $B \subset \mathbb{R}^n$.

Corollary 3.6.6. *If P is a p-dimensional polyhedron and $m_q(X) = 0$, then $m_{q+p}(X \times P) = 0$.*

Proof. Let Δ be a simplex of P. Since Δ is isometric to a compact subset of $\mathbb{R}^p$, Lemma 3.6.5 gives $m_{q+p}(X \times \Delta) = m_q(X)\, m_p(\Delta) = 0$. The corollary then follows from the countable additivity of m_{q+p}. □

The third property is a slice theorem for Hausdorff measure. A special case of (Federer, 1969, 2.10.27), it is a consequence of the fact that the measure of a set can be calculated by integrating the measures of the cross-sections.

Lemma 3.6.7. *Suppose n and k are nonnegative integers, P is a polyhedron, and D is a subset of $\mathbb{R}^n \times P$. For each $x \in \mathbb{R}^n$, define*

$$D_x = \{y \in P \mid \langle x, y\rangle \in D\}.$$

If $m_k(D) = 0$, then $m_{k-n}(D_x) = 0$ for almost all $x \in \mathbb{R}^n$. In particular, if $m_n(D) = 0$, then $D_x = \emptyset$ for almost all $x \in \mathbb{R}^n$.

The following lemma provides the main step in the proof of Theorem 3.6.1.

Lemma 3.6.8. *Suppose P is a compact p-dimensional polyhedron in B^n and $X \subset B^n$ is compact with $m_q(X) = 0$. If $p + q \le n$, then for every $\epsilon > 0$ there exists a PL homeomorphism $h : B^n \to B^n$ such that*

(1) *$h(x) = x$ for every $x \in \partial B^n$,*

(2) *$d(x, h(x)) < \epsilon$ for every $x \in B^n$, and*

(3) *$h(P) \cap X \cap \operatorname{Int} B^n = \emptyset$.*

Proof. For each fixed $z \in \operatorname{Int} B^n$, define $h_z : B^n \to B^n$ by the formula $h_z(y) = y + (1 - \|y\|)z$ (where $\|y\|$ denotes the norm of y). Note that h_z is the PL homeomorphism that is fixed on ∂B^n, moves 0 to z, and is extended conewise. It suffices to show that h_z satisfies conclusion (3) for almost all $z \in \operatorname{Int} B^n$.

Let $P' = P \cap \operatorname{Int} B^n$ and define $\psi : \operatorname{Int} B^n \times P' \to \operatorname{Int} B^n \times P'$ by $\psi(z, y) = \langle h_z(y), y\rangle$. Then it is not too difficult to check that ψ is a PL embedding. In fact, $\psi : \operatorname{Int} B^n \times P' \to \psi(\operatorname{Int} B^n \times P')$ has a two-sided inverse defined by the formula

$$\psi^{-1}(z, y) = \left\langle \frac{z - y}{1 - \|y\|}, y \right\rangle.$$

By Corollary 3.6.6, $m_{p+q}(X \times P') = m_q(X) m_p(P') = 0$. As a result, $m_n(X \times P') = 0$. Let $D = \psi^{-1}(X \times P')$. By Lemma 3.6.4, $m_n(D) = 0$. Hence Lemma 3.6.7 gives $D_z = \emptyset$ for almost all $z \in B^n$.

Suppose $X \cap h_z(P') \neq \emptyset$ for some $z \in \operatorname{Int} B^n$. Then there exists $x \in X$ and $y \in P'$ such that $x = h_z(y)$. Thus $\langle x, y\rangle = \psi(z, y)$ and so $\langle z, y\rangle \in D$; in other words, $D_z \neq \emptyset$. It follows that $X \cap h_z(P') = \emptyset$ for almost all z. □

Proof of Theorem 3.6.1. Suppose $X \subset \mathbb{R}^n$ is compact and $m_{k+1}(X) = 0$. Let $p = n - k - 1$. It suffices to prove the following: For every compact p-dimensional polyhedron P in $\mathbb{R}^n$ and for every $\epsilon > 0$ there exists a homeomorphism $h : \mathbb{R}^n \to \mathbb{R}^n$ such that $d(x, h(x)) < \epsilon$ for every $x \in \mathbb{R}^n$ and $h(P) \cap X = \emptyset$.

Suppose P is such a polyhedron. Specify a triangulation T of $\mathbb{R}^n$ so P underlies a subcomplex, and let $v_1, \dots, v_\ell$ denote the vertices of T in P. For each i, set $C_i = \text{star}(v_i, T)$. Fix a PL homeomorphism $\phi_i : C_i \to B^n$. By applying Lemma 3.6.8, we can find a PL homeomorphism $h_1 : B^n \to B^n$ arbitrarily close to the identity such that $h_1|\partial B^n = \text{Id}$ and $h_1(\phi_1(P \cap C_1)) \cap \phi_1(X \cap C_1) \cap \text{Int}\, B^n = \emptyset$. Choose h_1 so close to the identity that if $g_1 = \phi_1^{-1} h_1 \phi_1$, then $d(x, g_1(x)) < \epsilon/2$ for every $x \in C_1$. Extend g_1 via the identity to a PL homeomorphism of all of $\mathbb{R}^n$. Notice that $P \cap X \subset \cup_{i \geq 1} \text{Int}\, C_i$, but $g_1(P) \cap X \subset \cup_{i \geq 2} \text{Int}\, C_i$.

Now consider X and $g_1(P)$. Apply Lemma 3.6.8 again to produce a PL homeomorphism $g_2 : \mathbb{R}^n \to \mathbb{R}^n$ such that $d(x, g_2(x)) < \epsilon/4$, $g_2|(\mathbb{R}^n \smallsetminus \text{Int}\, C_2) = \text{Id}$, and $g_2(g_1(P)) \cap X \cap \text{Int}\, C_2 = \emptyset$. Since $g_2|(C_1 \smallsetminus \text{Int}\, C_2) = \text{Id}$, $g_2(g_1(P)) \cap X \subset \cup_{i \geq 3} \text{Int}\, C_i$. This process is continued inductively through C_ℓ. The result is a homeomorphism $h = g_\ell \circ \cdots \circ g_1$, which satisfies the conclusion of the theorem. □

Turning toward the proof of Theorem 3.6.2, we need a modified Menger continuum that is extremely thin in order to minimize its Hausdorff dimension.

Lemma 3.6.9. *For each $k \leq n$ there is a modified Menger continuum $\widetilde{M}_n^k$ in $\mathbb{R}^n$ such that $\dim_{\text{h}} \widetilde{M}_n^k = k$.*

Proof. We first describe the construction of a specific modified Menger continuum $\widetilde{M}_n^k$ and then prove that it has the correct Hausdorff dimension. Begin, as before, with T_0, the trivial subdivision of I. Define T_1 to be the subdivision of I into three subintervals, the first and last having length $1/4$ and the middle interval having length $1/2$. In general, T_j is constructed from T_{j-1} as follows: given an interval $J \in T_{j-1}$, subdivide J into three subintervals, the first and last having length $(1/2^2)(1/2^3)\cdots(1/2^{j+1})$ and the middle interval filling the rest of J. Let T_j^n be the collection of all subcubes of I^n obtained by taking products of intervals in T_j. As before, $P_0 = I^n$ and P_j is the union of all cubes in T_j^n that are contained in P_{j-1} and intersect the k-skeleton of P_{j-1}. Define $\widetilde{M}_n^k = \cap_{j=1}^{\infty} P_j$.

To complete the proof of Lemma 3.6.9 we must show that $\dim_{\text{h}} \widetilde{M}_n^k \leq k$. The proof is quite technical because it is derived directly from the definition of Hausdorff dimension. We will construct a sequence $\{\mathcal{U}_j\}$ of covers of $\widetilde{M}_n^k$ such that

$$\lim_{j \to \infty} \sum_{U \in \mathcal{U}_j} (\text{diam}\, U)^p = 0$$

for every $p > k$. A quick review of the definitions at the beginning of this section shows that this will suffice.

Let S_1 be the partition of I into 4 intervals of length $\frac{1}{4}$ and let S_2 be the partition of I into 32 intervals of length $\frac{1}{4} \cdot \frac{1}{8} = \frac{1}{32}$. In general, let S_j be the partition of I obtained by partitioning each interval of S_{j-1} into 2^{j+1} subintervals of equal length. Define S_j^n to be the collection of n-dimensional cubes obtained by taking products of intervals in S_j and define $\mathcal{U}_j$ to be the covering of P_j consisting of all the cubes in S_j^n that are contained in P_j. Since the lengths of the intervals in T_j are all multiples of the length of an interval in S_j, each cube in T_j^n is the union of cubes in S_j^n. Hence $\mathcal{U}_j$ exactly covers P_j and, in addition, $\mathcal{U}_j$ covers $\widetilde{M}_n^k$.

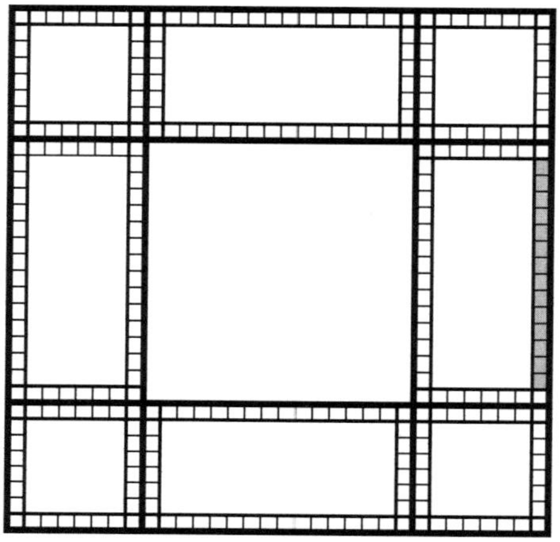

Figure 3.15. The cover $\mathcal{U}_2$ of $\widetilde{M}_2^1$. The small squares are the cubes in S_2^2; one "cube" of T_2^2 is shaded

We now find an upper bound for $\sum_{U \in \mathcal{U}_j} (\operatorname{diam} U)^p$. To accomplish this we compute the diameter of each cube in $\mathcal{U}_j$ and get an upper bound on the number of cubes in $\mathcal{U}_j$. If U is a cube in $\mathcal{U}_j$, then each edge of U has length

$$2^{-2}2^{-3} \cdots 2^{-(j+1)} = 2^{-j(j+3)/2}.$$

Hence the diameter of U is $\sqrt{n2^{-j(j+3)}} = n^{1/2}2^{-j(j+3)/2}$.

The number of cubes in $\mathcal{U}_1$ that intersect one of the k-dimensional faces of I^n is 4^k. Let r be the number of k-dimensional faces of an n-cube. Since a cube can intersect more than one k-dimensional face, the total number of cubes in $\mathcal{U}_1$ is strictly less than $4^k r$. Fix a cube Q in $\mathcal{U}_1$ and let Q' be a cube in $\mathcal{U}_2$ such that $Q' \subset Q$. Then Q' intersects the k-skeleton of P_1, so Q' must intersect one of the k-dimensional faces of Q. Thus the number of cubes in $\mathcal{U}_2$ that are contained in Q is less than $8^k r$. Continuing inductively, we see that the total number of cubes in $\mathcal{U}_j$ is less than

$$4^k r 8^k r \cdots 2^{(j+1)k} r = 2^{j(j+3)k/2} r^j.$$

Therefore

$$\sum_{U \in \mathcal{U}_j} (\operatorname{diam} U)^p < 2^{j(j+3)k/2} r^j \left(n^{1/2} 2^{-j(j+3)/2}\right)^p = n^{p/2} \left(r 2^{(k-p)(j+3)/2}\right)^j .$$

Since $p > k$, the last quantity approaches 0 as j approaches ∞. □

Proof of Theorem 3.6.2. Let X be a compact subset of $\mathbb{R}^n$. If $\dim_{\mathrm{h}} X = k$, then by definition $m_{k+1}(X) = 0$. Hence $\operatorname{dem} X \leq k = \dim_{\mathrm{h}} X$ by Theorem 3.6.1. But $\operatorname{dem} X$ is invariant under homeomorphisms of $\mathbb{R}^n$, so $\operatorname{dem} X = \operatorname{dem} \phi(X) \leq \dim_{\mathrm{h}} \phi(X)$ for every homeomorphism $\phi : \mathbb{R}^n \to \mathbb{R}^n$.

Now suppose $k = \operatorname{dem} X$. Let $\widetilde{M_n^k}$ be as in Lemma 3.6.9. By Theorem 3.5.1 and Proposition 3.5.2, there exists a homeomorphism $\phi : \mathbb{R}^n \to \mathbb{R}^n$ such that $\phi(X) \subset \widetilde{M_n^k}$. Thus $\dim_{\mathrm{h}} \phi(X) \leq \dim_{\mathrm{h}} \widetilde{M_n^k} = k = \operatorname{dem} X$. By the previous paragraph we have $\operatorname{dem} X \leq \dim_{\mathrm{h}} \phi(X)$, so $\dim_{\mathrm{h}} \phi(X) = \operatorname{dem} X$. □

Historical Notes. Hausdorff dimension was originally defined by Hausdorff (1919). It is also defined on page 107 of (Hurewicz and Wallman, 1948). More recent expositions of Hausdorff dimension may be found in (Edgar, 1990), (Federer, 1969), and (Keesling, 1986). The term "fractal" was coined by B. B. Mandelbrot (1982).

Theorem 3.6.1 is due to J. Luukkainen and J. Väisälä (1977); Theorem 3.6.2 is due to Väisälä (1979). Theorem 3.6.1 should be viewed as a geometric version of (Hurewicz and Wallman, 1948, Theorem VII.3). Lemma 3.6.7 is a generalization of (Hurewicz and Wallman, 1948, Theorem VII.2.A). The examples mentioned on page 137 are included in most standard works on the subject of Hausdorff dimension and even appear in Hausdorff's original work (1919).

Chapter 4

Trivial-range Embeddings

The *trivial range* relative to dimension n comprises all dimensions k for which $2k + 2 \leq n$. In that dimension range one can derive the strongest embedding and unknotting theorems.

This chapter presents two fundamental results regarding trivial range embeddings: a taming theorem for embeddings of polyhedra and an unknotting theorem for embeddings of compacta. The first asserts that 1-LCC embeddings of polyhedra are tame and the second that homotopic 1-LCC embeddings of compacta are equivalent by ambient isotopy, provided, in both cases, that the dimension of the embedded object is in the trivial range relative to the ambient dimension. The techniques developed to prove the first result set the stage for similar investigations in later chapters; for instance, in Chapter 5 we will see that there is a similar taming theorem for codimension-three embeddings, and in subsequent chapters that there are some appropriate generalizations to codimensions two and one. Examples of linking in Chapter 1 show that the unknotting theorem does not generalize to smaller codimensions, at least not without additional hypotheses.

The taming and unknotting theorems in this chapter demonstrate once more the power of the 1-LCC property. For that reason it includes, near the end, a section(§4.6) describing ways in which the 1-LCC property can be detected. The chapter concludes with new examples of wildness in dimensions greater than three that stem from the existence of wild Cantor sets there.

4.1. Unknotting PL embeddings of polyhedra

According to (Rourke and Sanderson, 1972, Corollary 5.9), homotopic PL embeddings of a compact k-manifold into an n-manifold in the trivial range are unknotted in a strong sense: they are PL ambient isotopic. The primary aim of this section is to derive a similar unknotting theorem for embeddings of polyhedra, not merely of manifolds, and to impose controls limiting the amount of motion. Theorem 4.1.1, the most general version of these unknotting results, provides adequate control for our needs. When $2k+2<n$ it is possible to develop even better control, and Proposition 4.1.3 does so in a form that is obviously optimal for embeddings in that slightly more restricted range.

Theorem 4.1.1 (Unknotting PL embeddings in the trivial range). *Let K be a finite simplicial k-complex, M a PL n-manifold with $2k+2 \le n$, λ_0 and λ_1 two PL embeddings of K in M, b a positive number, and $\mu : K \times I \to M$ a homotopy between λ_0 and λ_1 such that* $\operatorname{diam} \mu(y \times I) < b$ *for all $y \in K$. Then there exists a compactly supported PL isotopy $\Theta : M \times I \to M$ such that $\Theta_0 = \mathrm{Id}_M$ and $\Theta_1 \lambda_0 = \lambda_1$; moreover, Θ can be regulated so* $\operatorname{diam} \Theta(z \times I) < 2(k+1) \cdot b$, *for all $z \in M$.*

Familiar examples like linked versus unlinked k-sphere pairs in S^{2k+1}, mentioned is §1.1, illustrate sharpness of the dimension restrictions in Theorem 4.1.1. Its proof involves two key steps, one detailing the fundamental move (Lemma 4.1.2), and the other imposing motion controls on a concatenation of fundamental moves.

Two k-dimensional complexes K_1,K_2 in a PL-manifold M are said to differ by a *cellular move* if there exists a PL embedded $(k+1)$-cell D in M that meets K_1,K_2 in complementary faces of ∂D (i.e., $|K_i| \cap D = |K_i| \cap \partial D$ is a k-cell β_i, where $\beta_1 \cup \beta_2 = \partial D$ and $\beta_1 \cap \beta_2 = \partial\beta_1 = \partial\beta_2$) and $|K_1| \setminus D = |K_2| \setminus D$. Rourke and Sanderson demonstrate the usefulness of this concept for PL embedded manifolds; the next result establishes an analog of their Proposition 4.15 for polyhedra.

Lemma 4.1.2 (Cellular Move). *Suppose K_1 and K_2 are PL k-complexes in a PL n-manifold M, $k \le n-3$, that differ by a cellular move across a PL $(k+1)$-cell D, and suppose N is a subcomplex of M containing a neighborhood of $D \setminus \partial(|K_1| \cap \partial D)$. Then there exists a PL isotopy of M carrying $|K_1|$ to $|K_2|$ and fixing all points of $M \setminus |N|$.*

Proof. In this argument $N(A,T)$ denotes the simplicial neighborhood of A in the simplicial complex T.

Restrict N so $|N \cap K_i| = D \cap |K_i|$. Choose a triangulation T of M containing D, N, and $K_i (i=1,2)$ as full subcomplexes, and let T' denote its first barycentric subdivision. Set $D^{\#} = \text{Cl}(D \setminus N(\partial(D \cap |K_1|), T'))$. Check that $D^{\#}$ is another $(k+1)$-cell, as it is just the complement in D of a collar on $A \subset \partial D$, where $A \cong S^{k-1} \times I$ equals a regular neighborhood of $\partial(D \cap |K_1|)$ in ∂D. Use J to denote the subcomplex of T' carried by $D^{\#}$. One can easily confirm that J is full in T'.

It follows from the Simplicial Neighborhood Theorem (Rourke and Sanderson, 1972, 3.11) that $B = N(D^{\#}, T') \subset |N|$ is a regular neighborhood of $D^{\#}$ and, hence, is an n-ball. Clearly, $D \subset B$, $D \cap \partial B = \partial(D \cap |K_1|)$, and $B \cap |K_1 \cup K_2| = \partial D$. (See Figure 4.1.) This means that each $(B, B \cap |K_i|)$ is an (n,k)-ball pair $(i = 1, 2)$, which is unknotted by (Rourke and Sanderson, 1972, Theorem 7.1). As a result, there exists an isotopy of B to itself fixing ∂B and carrying $B \cap |K_1|$ onto $B \cap |K_2|$, which extends via the identity to the desired isotopy of M. □

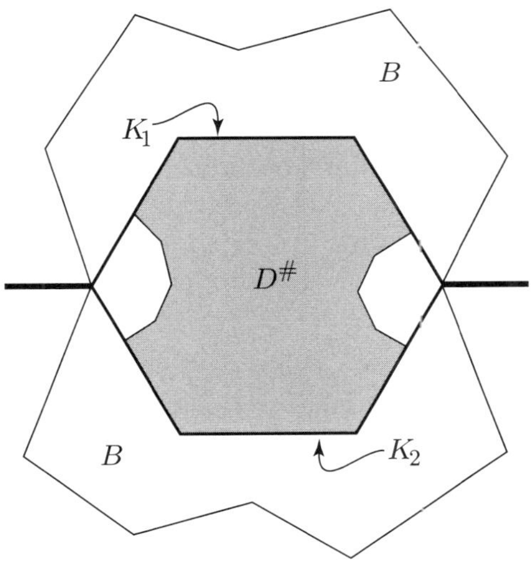

Figure 4.1. $(B, B \cap |K_1|)$ and $(B, B \cap |K_2|)$ are both unknotted ball pairs

Proposition 4.1.3. *Let K be a finite simplicial complex of dimension k, M a PL n-manifold with $2k + 2 < n$, λ_0 and λ_1 two PL embeddings of K in M, b a positive number, and $\mu : K \times I \to M$ a homotopy between λ_0 and λ_1 such that* $\text{diam}\, \mu(y \times I) < b$ *for all $y \in K$. Then there exists a compactly supported PL isotopy $\Phi : M \times I \to M$ such that $\Phi_0 = \text{Id}_M$, $\Phi_1 \lambda_0 = \lambda_1$, and* $\text{diam}\, \Phi(z \times I) < b$, *for all $z \in M$.*

Proof. Perform a general position adjustment of λ_1, if necessary, to make its image disjoint from that of λ_0. The special dimension restriction allows approximation of μ by a PL general position embedding $\Omega : K \times I \to M$ with $\Omega_0 = \lambda_0$ and $\Omega_1 = \lambda_1$. Subdivide the triangulation of K with such small mesh that $\operatorname{diam}\Omega(\sigma\times I)<b$ for each $\sigma\in K$. As $\Omega(\sigma\times I)$ is a PL cell, its regular neighborhoods are n-balls; for each k-simplex σ specify an open set $U_\sigma\cong\mathbb{R}^n$ with $\Omega(\sigma\times I)\subset U_\sigma$ and $\operatorname{diam}U_\sigma<b$. Recursively, for the various $\tau\in K$ choose open sets $U_\tau\supset\Omega(\tau\times I)$ with $U_\tau\cong\mathbb{R}^n$ and $U_\tau\supset U_\gamma$ whenever γ is a face of τ. This can be done so that $U_\tau\cap U_\gamma\neq\emptyset$ if and only if $\tau\cap\gamma\neq\emptyset$.

Claim. *There exists a PL isotopy $\Phi: M\times I\to M$ starting at the identity, carrying $\lambda_0(K)$ to $\lambda_1(K)$, and moving points so that for each $z\in M$ for which $\Phi(z\times I)\neq\{z\}$ there is a $\tau\in K$ such that $\Phi(z\times I)\subset U_\tau$.*

The claim is proved by induction on k, and the $k=0$ case is routine. We assume Ψ is an isotopy carrying the $(k-1)$-skeleton of $\lambda_0(K)$ to that of $\lambda_1(K)$ in such a way that the track of any point moved under Ψ is contained in some U_τ, $\tau\in K^{(k-1)}$. Due to nesting features of $\{U_\gamma\}$, $\Psi_1\lambda_0(\sigma)\cup\lambda_1(\sigma)\subset U_\sigma$ for each k-simplex $\sigma\in K$. Adjust $\Psi_1\lambda_0(\sigma)$ to make $\Psi_1\lambda_0(\sigma)\cup\lambda_1(\sigma)$ be a k-sphere, which then is unknotted in U_σ (Rourke and Sanderson, 1972, Theorem 7.1). General position ensures $\Psi_1\lambda_0(\sigma)$, $\lambda_1(\sigma)$ differ by a cellular move across a $(k+1)$-cell $D_\sigma\subset U_\sigma$ satisfying $\operatorname{Int}D_\sigma\cap(\Psi_1\lambda_0(K)\cup\lambda_1(K))=\emptyset$. Upon making another general position adjustment we can assume the D_σ's have pairwise disjoint interiors, after which we specify pairwise disjoint subcomplexes N_σ of M, where $D_\sigma\subset N_\sigma\subset U_\sigma$ and N_σ contains a neighborhood of $D_\sigma\smallsetminus\lambda_1(\partial\sigma)$. Multiple application of the Cellular Move Lemma (4.1.2) yields an isotopy Ψ' supported in $\cup N_\sigma$ and carrying $\Psi_1\lambda_0(K)$ to $\lambda_1(K)$. It follows that, if $z\in M$ is nonstationary under the isotopy $\Phi=\Psi'\Psi$, then there exists $\sigma\in K$ with $\Phi(z\times I)\subset U_\sigma$. This completes the proof of the Claim and of Proposition 4.1.3 as well. □

Proposition 4.1.4. *Let K be a finite simplicial complex of dimension k, M a PL n-manifold with $2k+2\le n$, b a positive number and $\Omega: K\times I\to M$ a PL embedding such that* $\operatorname{diam}\Omega(y\times I)<b$ *for all $y\in K$. Then there exists a compactly supported PL isotopy $\Theta: M\times I\to M$ such that $\Theta_0=\mathrm{Id}_M$, $\Theta_1\Omega_0|K=\Omega_1|K$, and* $\operatorname{diam}\Theta(z\times I)<(k+1)\cdot b$ *for all $z\in M$.*

Proof. Specify a triangulation T of K with $\operatorname{diam}\Omega(\sigma\times I)<b$ when $\sigma\in T$. For $i=0,1,\ldots,k$ let A_i denote the union of all $\Omega(\mathrm{St}(\beta_\tau,T'')\times I)$, where β_τ denotes the barycenter of an i-simplex $\tau\in T$, and $\mathrm{St}(\beta_\tau,T'')$ denotes its star in T'', a second derived subdivision of T. Again observe that regular neighborhoods of $\Omega(\mathrm{St}(\beta_\tau,T'')\times I)$, being collapsible, are n-balls. Construct an open set $W_i\supset A_i$, each component of which is PL equivalent to $\mathbb{R}^n$ and has diameter less than b.

We claim that there exists a PL isotopy $\Psi^0: M\times I\to M$ supported in W_0 such that $\Psi^0_1\Omega_0|A_0=\Omega_1|A_0$. Actually, Ψ^0 arises from a sequence of such isotopies, one to transport $\Omega_0(L\cap A_0)$ to $\Omega_1(L\cap A_0)$, where L

denotes the $(k-1)$-skeleton of T'', and the others indexed by the k-simplices σ in A_0 (regarded as a subcomplex of T''). Proposition 4.1.3 promises a PL isotopy $\Phi : M \times I \to M$ supported in W_0 such that $\Phi_0 = \mathrm{Id}_M$ and $\Phi_1\Omega_0|L \cap A_0 = \Omega_1|L \cap A_0$; the methodology allows for strict regulation ensuring that $\Phi_t\Omega(A_i \times I) \subset W_i$ for all i and all $t \in I$. Applying general position, we assume $\Phi_1(\Omega_0(K)) \cap \Omega_1(K) = \Omega_1(L)$. Consider a typical k-sphere $\Sigma = \Phi_1\Omega_0(\sigma) \cup \Omega_1(\sigma)$, $\sigma \in A_0$: as before, being unknotted in (a component of) W_0 (Rourke and Sanderson, 1972), Σ bounds a PL $(k+1)$-cell $D_\sigma \subset W_0$. Although we cannot make the various D_σ be pairwise disjoint, we have enough room to perform a general position alteration on $\mathrm{Int}\, D_\sigma$ yielding $D_\sigma \cap \Phi_1\Omega_0(K) \cup \Omega_1(K) = \Sigma = \partial D_\sigma$ and $\mathrm{Int}\, D_\sigma \cap \Phi_1\Omega(L \times I) = \emptyset$. Thicken D_σ to a simplicial complex N_σ with

$$N_\sigma \cap [\Phi_1\Omega_0(K) \cup \Omega_1(K) \cup \Phi_1\Omega(L \times I)] = \partial D_\sigma$$

and with $D_\sigma \setminus \partial(D_\sigma \cap \Omega_1(K)) \subset \mathrm{Int}\, N_\sigma \subset W_0$. Use Lemma 4.1.2 to move $\Phi_1\Omega_0(\sigma)$ to $\Omega_1(\sigma)$ via an ambient isotopy supported in N_σ, keeping other points of $\Phi_1(\Omega_0(K) \cup \Omega(L \times I))$ in place. After some portion P of $\Phi_1\Omega_0(A_0)$ has been moved to $\Omega_1(A_0)$, P can be left fixed under repetitions of the procedure, simply by general positioning subsequent $D_{\sigma'}$ off P and choosing support $N_{\sigma'}$ for the next portion of the isotopy with $P \cap \mathrm{Int}\, N_{\sigma'} = \emptyset$. The choice of W_0 ensures that the track of any point moved under Ψ^0 has diameter less than b.

In exactly the same way, there exist PL isotopies $\Psi^1, \ldots, \Psi^k$ of M supported in $W_1, \ldots, W_k$, respectively, such that

$$\Psi_1^i \Psi_1^{i-1} \cdots \Psi_1^0 \Omega_0 | A_i = \Omega_1 | A_i,$$

where Ψ_i is fixed on $\Psi_1^{i-1} \cdots \Psi_1^0 \Omega_0(A_j) = \Omega_1(A_j)$, $j < i$, and where

$$\Psi_i \Psi_{i-1} \cdots \Psi_0 \Omega((L \cap A_j) \times I) \subset W_j, \quad j > i.$$

Reparametrize these isotopies so Ψ^i acts only during the t interval $[i/(k+1), (i+1)/(k+1)]$ (i.e., Ψ_t^i equals the identity when $t < i/(k+1)$ and equals Ψ_1^i when $t > (i+1)/(k+1)$). The composite isotopy $\Theta = \Psi^k\Psi^{k-1} \ldots \Psi^0$ carries Ω_0 to Ω_1. Moreover, the track of any $z \in M$ under Θ has diameter less than $(k+1)(b)$, since

$$\Theta(\{z\} \times I) \subset \\ \Psi^0(\{z\} \times I) \cup \Psi^1(\Psi_1^0(z) \times I) \cup \cdots \cup \Psi^k(\Psi_1^{k-1}\Psi_1^{k-2} \cdots \Psi_1^0(z) \times I),$$

and each of the tracks in this (connected) union has diameter less than b. □

Let K be a compact k-dimensional polyhedron and M an n-dimensional PL manifold. In case $2k+2 \le n$, a general position map $\mu : K \times I \to M$ will not quite be an embedding. But the singular set of μ will be simple: it will consist of a finite number of double points, all of them in $K \times (0,1)$. The next two lemmas show how to simplify the singular set even more, so that μ is replaced by a map that embeds each of $K \times [0,1/2]$ and $K \times [1/2,1]$.

In the remainder of this section we use proj to denote the projection $K \times I \to K$.

Lemma 4.1.5. *Let σ be a k-simplex and let Ξ be a finite subset of* $\operatorname{Int}\sigma \times (0,1)$. *Then there exists $\psi \in \mathrm{Homeo}_{PL}(\sigma \times I, \sigma \times I)$ such that $\psi|\partial(\sigma \times I) = \mathrm{Id}$ and* $\mathrm{proj} \circ \psi \mid \Xi$ *is one-to-one.*

Proof. Consider $z = \langle x,t\rangle$, $z' = \langle x,t'\rangle \in \Xi$. Choose $z^* = \langle x^*,t\rangle \in \sigma \times I$, $x^* \neq x$, and define $\psi \in \mathrm{Homeo}_{\mathrm{PL}}(\sigma \times I, \sigma \times I)$ to be the linear extension of the function sending z to z^* and fixing $\partial(\sigma \times I)$ pointwise. When z^* is sufficiently close to z, the singular set of $\mathrm{proj} \circ \psi \mid \Xi$ is a subset of that of $\mathrm{proj} \mid \Xi$, and clearly $\mathrm{proj} \circ \psi(z) \neq \mathrm{proj} \circ \psi(z')$ (see Figure 4.2). This process is continued until all multiple points have been eliminated. □

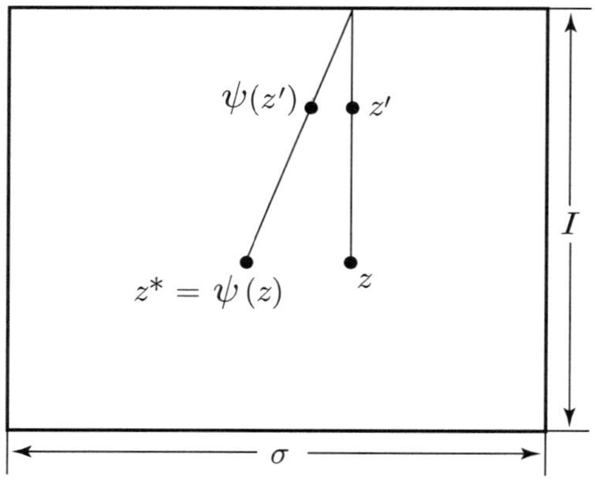

Figure 4.2. Proof of Lemma 4.1.5

Lemma 4.1.6. *Let K be a finite k-complex, M a PL n-manifold, $2k+2 = n$, and $\mu : K \times I \to M$ a PL general position map such that S_μ, the singular set of μ, consists only of double point singularities and that* $\mathrm{proj} : K \times I \to K$ *restricts to an injection on S_μ. Then there exists $\kappa \in \mathrm{Homeo}_{PL}(K \times I, K \times I)$ such that $\kappa|K \times \{0,1\} = \mathrm{Id}$, $\mathrm{proj} \circ \kappa = \mathrm{proj}$, and both $\mu\kappa|K \times [0,1/2]$ and $\mu\kappa|K \times [1/2,1]$ are PL embeddings.*

Proof. Impose a triangulation T on K with $S_\mu \cap (T^{(k-1)} \times I) = \emptyset$ and with small enough mesh that no $\sigma \times I$, $\sigma \in T$, contains more than one point of S_μ. Write S_μ as $\{p_1, q_1, \ldots, p_t, q_t\}$ where $\mu(p_i) = \mu(q_i)$, for $i = 1, \ldots, t$. By a sequence of vertical conewise adjustments, like the horizontal adjustments described in Lemma 4.1.5, specify $\kappa \in \mathrm{Homeo}_{\mathrm{PL}}(K \times I, K \times I)$ such that

$\kappa|(K \times \{0,1\}) \cup (T^{(k-1)} \times I) = \text{Id}$, $\text{proj} \circ \kappa = \text{proj}$, $\kappa(p_i) \in K \times \{1/4\}$ and $\kappa(q_i) \in K \times \{3/4\}$. □

Proof of Theorem 4.1.1. Approximate μ by a general position map having at worst double point singularities, with $S_\mu \subset K \times (0,1)$. Choose a triangulation T of K so that $\operatorname{diam} \mu(\sigma \times I) < b$ for all $\sigma \in T$ and $\text{proj}(S_\mu) \cap T^{(k-1)} = \emptyset$. Apply 4.1.5 to ensure $\text{proj} : K \times I \to K$ restricts to a 1-1 map on S_μ, and identify a PL homeomorphism $\kappa : K \times I \to K \times I$ fulfilling the conclusions of 4.1.6. Two applications of Proposition 4.1.4—guided by $\mu\kappa \mid K \times [0,1/2]$ and $\mu\kappa \mid K \times [1/2,1]$, respectively—provide PL isotopies $\Theta,\Theta' : M \times I \to M$ for which $\Theta_0 = \text{Id}_M$, $\Theta_1\lambda_0(z) = \mu\kappa(z \times \{\frac{1}{2}\})$, $\Theta' = \text{Id}_M$, $\Theta'_1(\mu\kappa(z \times \{\frac{1}{2}\})) = \mu\kappa(z \times \{1\}) = \lambda_1(z)$, $\operatorname{diam} \Theta(x \times I) < (k+1)(b)$, and $\operatorname{diam} \Theta'(x \times I) < (k+1)(b)$. As in 4.1.4 modify the t parameter to have Θ be constant on the interval $[1/2,1]$ and to have Θ' act as the identity on the interval $[0,1/2]$; now the composite isotopy $\Theta' \circ \Theta$ moves λ_0 to λ_1 with the prescribed control. □

Corollary 4.1.7. *Let M denote either a compact PL n-manifold or $\mathbb{R}^n$. For every $\epsilon > 0$ there exists $\delta > 0$ such that, given any two embeddings $\lambda, \lambda' \in \text{Emb}_{PL}(K,M)$ of a finite k-complex K, $2k+2 \le n$, for which $\rho(\lambda',\lambda) < \delta$, there is an ϵ-push ψ of $(M, \lambda(K))$ with $\psi\lambda = \lambda'$.*

Proof. Set $b = \epsilon/2(k+1)$. For any compact manifold M there exists $\delta > 0$ such that any two δ-close maps into M are b-homotopic; the same obviously holds for $\mathbb{R}^n$. □

Corollary 4.1.8. *Let $\lambda : K \to M$ be a PL embedding of a finite simplicial k-complex K in a PL n-manifold M, $2k+2 \le n$. For every $\epsilon > 0$ there exists $\delta > 0$ such that, given any $\lambda' \in \text{Emb}_{PL}(K,M)$ for which $\rho(\lambda',\lambda) < \delta$, there is an ϵ-push ψ of $(M, \lambda(K))$ with $\psi\lambda = \lambda'$.*

Historical Notes. Gugenheim (1953) established a global version of Theorem 4.1.1 without motion control. Bing and Kister exploited a technical device called *ratio-changing general position* to prove Proposition 4.1.4 (for $M = \mathbb{R}^n$) with stricter motion control than is developed here.

Exercise

4.1.1. If K is any k-complex ($k > 2$), $\text{Emb}_{PL}(K, \mathbb{R}^{2k})$ fails to be dense in $C(K, \mathbb{R}^{2k})$. [Hint: first do this for the disjoint union of two k-cells.]

4.1.2. Let K be a finite k-complex and $f : K \times I \to M$ a map to a PL $(2k+2)$-manifold. Then f can be approximated by a PL map $F : K \times I \to M$ such that F_t is an embedding for all $t \in I$.

4.1.3. Let X be a compact k-dimensional space and let $g : X \times I \to M$ be a map to a PL n-manifold, $n \ge 2k+2$. Then g can be approximated by a map $G : X \times I \to M$ such that G_t is a 1-LCC embedding for all $t \in I$. Moreover, if g_t is already a 1-LCC embedding for $t \in C$, where C is a closed subset of I, then G can be obtained with $G|X \times C = g|X \times C$.

4.2. Spaces of embeddings and taming of polyhedra

The proofs of both the taming theorem and the unknotting theorem require an ambient isotopy that moves one embedding to another. The two proofs exploit a remarkable common construction that achieves this goal in an indirect way. Rather than moving the first embedding directly to the second, the argument trades on a back-and-forth process that gradually moves the two embeddings closer and closer to each other; in the limit each of the original embeddings is pushed to a common limiting embedding. This section lays out the technical details of that construction. The theorems in this section will be applied in the next two sections to derive the advertised taming and unknotting theorems.

Throughout this section M will denote a PL n-manifold endowed with a bounded, complete metric d, and X will denote a compact subset of M. The object of study will be the embedding space $\mathrm{Emb}(X, M)$, under the sup-norm metric ρ. Our central concern will be to understand which elements of $\mathrm{Emb}(X, M)$ are equivalent under ambient homeomorphism.

Definition. A subset Λ of $\mathrm{Emb}(X, M)$ is *solvable* if to each $\epsilon > 0$ there corresponds $\delta > 0$ such that given any $\lambda,\ \lambda' \in \Lambda$ with $\rho(\lambda, \lambda') < \delta$, there is an ϵ-push ψ of $(M, \lambda(X))$ satisfying $\psi\lambda = \lambda'$.

Here the ordering of λ,λ' carries mild formal significance, for by Observation 3.3.1, ψ^{-1} carries λ' back to λ and ψ^{-1} is an ϵ-push of $(M, \lambda(X))$ but perhaps at best a 2ϵ-push of $(M, \lambda'(X))$.

The solvability of a space of embeddings assures a uniform local path connectedness. For finite k-complexes K, $2k + 2 \leq n$, a rephrasing of Corollary 4.1.7 promises that $\mathrm{Emb}_{\mathrm{PL}}(K, \mathbb{R}^n)$ is a (dense and) solvable subset of $\mathrm{Emb}(K, \mathbb{R}^n)$ and, similarly, for compact PL n-manifolds M, $\mathrm{Emb}_{\mathrm{PL}}(K, M)$ is a (dense and) solvable subset of $\mathrm{Emb}(K, M)$.

Remark. Although $\mathrm{Emb}(K, \mathbb{R}^n)$ is pathwise connected, it is definitely *not* the case that any two of its elements are equivalent via an ambient isotopy. The existence of wildly embedded k-cells makes the impossibility of ambient equivalence transparent.

Here is a statement of the principal result. Meaningful technical variations appear toward the end of the section.

Theorem 4.2.1. *Let M denote a manifold endowed with a complete metric d and let X be a compact subset of M. The union of any two dense, solvable subsets of* $\mathrm{Emb}(X, M)$ *is dense and solvable.*

Lemma 4.2.2. *Let φ be an ϵ-push of (M, X) and let ψ be an η-push of $(M, \varphi(X))$ realized by the isotopy Ψ_t. Then $\Psi_t\varphi$ is supported in the $(\epsilon+\eta)$-neighborhood of X and $\rho(\Psi_t\varphi, \varphi) < \eta$ for all $t \in [0, 1]$.*

Proof. Obviously $\varphi(X) \subset B(X; \epsilon)$, so $B(\varphi(X); \eta) \subset B(X; \epsilon+\eta)$, indicating that the support of $\Psi_t\varphi$ resides in the $(\epsilon + \eta)$-neighborhood of X. That $\rho(\Psi_t\varphi, \varphi) < \eta$ holds is elementary. □

Corollary 4.2.3. *For $i = 1, 2, \ldots, r$, let ϵ_i denote a positive number and let $\psi_i \in$* $\mathrm{Homeo}(M, M)$ *be an ϵ_i-push of $(M, \psi_{i-1} \cdot \cdots \cdot \psi_1(X))$, with Ψ_t^r the ϵ_r-isotopy realizing the push ψ_r. Then the isotopy $\Phi_t = \Psi_t^r \cdot \psi_{r-1} \cdot \cdots \cdot \psi_1$ is supported in the $(\epsilon_1+\epsilon_2+\cdots+\epsilon_r)$-neighborhood of X and $\rho(\Phi_t, \psi_{r-1} \cdot \cdots \cdot \psi_1) < \epsilon_r$ for all $t \in [0, 1]$.*

Corollary 4.2.4. *Assume*

(1) $\{\epsilon_1, \epsilon_2, \ldots\}$ is an infinite sequence of positive numbers such that the series $\Sigma_i \epsilon_i$ converges to S, and

(2) $\{\psi_1, \psi_2, \ldots\}$ is an infinite sequence of homeomorphisms of M such that ψ_1 is an ϵ_1-push of (M, X) and for $i > 1, \psi_i$ is an ϵ_i-push of $(M, \varphi_{i-1}(X))$, where $\varphi_{i-1} = \psi_{i-1} \cdot \cdots \cdot \psi_2 \cdot \psi_1$.

Then the sequence $\{\varphi_1, \varphi_2, \ldots\}$ converges in $C(M, M)$ to $g \in$ $\mathrm{Surj}(M, M)$ *with $\rho(g, \mathrm{Id}_M) < S$, and g acts as the identity on $M \setminus B(X; S)$.*

Proof. Repeated application of Corollary 4.2.3 yields

$$\rho(\varphi_i, \varphi_j) < \epsilon_{i+1} + \epsilon_{i+2} + \cdots + \epsilon_j$$

for $i < j$. Convergence of the infinite series $\Sigma_i \epsilon_i$ ensures that $\{\varphi_i\}$ is a Cauchy sequence in $\mathrm{Homeo}(M, M) \subset \mathrm{Surj}(M, M)$, so $\{\varphi_i\}$ converges to some $g \in \mathrm{Surj}(M, M)$ by completeness. □

Lemma 4.2.5. *Under the hypotheses of Corollary 4.2.4, there exists a map $\Phi : M \times [0, 1] \to M$ such that $\Phi_0 = \mathrm{Id}_M$, Φ_1 is the map $g \in$* $\mathrm{Surj}(M, M)$ *from the conclusion of Corollary 4.2.4, and for all $t \in [0, 1)$, Φ_t is an S-push of (M, X). Moreover, when g is one-to-one, g also is an S-push of (M, X) and Φ_t is the associated isotopy.*

Proof. Let $\{\psi_1, \psi_2, \ldots, \psi_r, \ldots\}$ be the infinite sequence of ϵ_r-pushes and let $\{\Psi_t^1, \Psi_t^2, \ldots, \Psi_t^r, \ldots\}$ be the associated sequence of isotopies realizing these pushes. Let

$$\Phi_t^r = \Psi_t^r \cdot \psi_{r-1} \cdot \cdots \cdot \psi^1 = \Psi_t^r \cdot \Psi_1^{r-1} \cdot \cdots \cdot \Psi_1^1.$$

Corollary 4.2.4 ensures that the sequence $\{\Phi_1^1, \Phi_1^2, \Phi_1^3, \ldots\}$ converges to a map $g \in \mathrm{Surj}(M, M)$. For $t \in [0, 1)$ choose $r \in \mathbb{Z}_+$ with $t \in [(r-1)/r, r/(r+1)]$, express t as $s \cdot (r-1)/r + (1-s)(r/(r+1))$, $s \in [0, 1]$, and define $\Phi : M \times [0, 1] \to M$ as

$$\Phi(z, t) = \Phi_s^r(z) \text{ for } t < 1 \text{ and}$$
$$\Phi(z, 1) = g(z).$$

Continuity of Φ stems from the uniform convergence of

$$\{\Phi_{t(r)}^r \mid t(r) \in [0, 1],\ r = 1, 2, \ldots\}.$$

The other features required of Φ follow from Corollary 4.2.4. □

The function Φ in Lemma 4.2.5 is usually called a pseudo-isotopy of M. To be precise, a *pseudo-isotopy of M* is a continuous function $\Phi : M \times [0, 1] \to M$ such that $\Phi_0 = \mathrm{Id}_M$, $\Phi_t \in \mathrm{Homeo}(M, M)$ for every $t < 1$, and $\Phi_1 \in \mathrm{Surj}(M, M)$.

For the most potent applications of Lemma 4.2.5, one should recall the conditions listed in Proposition 2.2.2 under which a sequence of homeomorphisms like $\{\varphi_1, \varphi_2, \ldots\}$ converges to a homeomorphism.

Proof of Theorem 4.2.1. Let Λ and Λ' be two dense, solvable subsets of $\mathrm{Emb}(X, M)$. Clearly $\Lambda \cup \Lambda'$ is dense in $\mathrm{Emb}(X, M)$, so the real issue is solvability. To each $\epsilon > 0$ there corresponds $\delta(\epsilon) \in (0, \epsilon)$ satisfying the condition defining solvability of Λ; symmetrically, to each $\epsilon' > 0$ there corresponds $\delta'(\epsilon') \in (0, \epsilon')$ from the condition defining solvability of Λ'. Fix $\epsilon > 0$. Choose $\delta(\epsilon/6)$ for Λ, choose $\delta'(\epsilon/6)$ for Λ', and set

$$\delta = \min\{\delta(\epsilon/6), \delta'(\epsilon/6)\}.$$

We will show that any two elements of $\Lambda \cup \Lambda'$ that are δ-close to each other are equivalent via an ϵ-push.

Consider $\lambda, \lambda' \in \Lambda \cup \Lambda'$ with $\rho(\lambda, \lambda') < \delta$. If $\lambda, \lambda' \in \Lambda$ or if $\lambda, \lambda' \in \Lambda'$, existence of the ϵ-push follows immediately from the preceding choice of δ, based on the solvability of Λ and Λ' themselves. Assume then that $\lambda \in \Lambda$ and $\lambda' \in \Lambda'$.

We will construct:

- a sequence $\{\lambda_0 = \lambda, \lambda_1, \lambda_2, \ldots\}$ of elements of Λ;
- a sequence $\{\lambda_0' = \lambda', \lambda_1', \lambda_2', \ldots\}$ of elements of Λ';
- two sequences $\{\psi_1, \psi_2, \psi_3, \ldots\}$ and $\{\psi_1', \psi_2', \psi_3', \ldots\}$ of elements of $\mathrm{Homeo}(M, M)$; and
- two sequences $\{\epsilon_0, \epsilon_1, \epsilon_2, \ldots\}$ and $\{\epsilon_0', \epsilon_1', \epsilon_2', \ldots\}$ of positive real numbers.

For notational convenience, associated with the sequences $\{\psi_1, \psi_2, \psi_3, \ldots\}$ and $\{\psi'_1, \psi'_2, \psi'_3, \ldots\}$ from Homeo(M, M) will be two additional sequences $\{\varphi_1, \varphi_2, \varphi_3, \ldots\}$ and $\{\varphi'_1, \varphi'_2, \varphi'_3, \ldots\}$ from Homeo(M, M) defined by:

$$\varphi_i = \psi_i \cdot \psi_{i-1} \cdot \cdots \cdot \psi_1 \quad \text{and} \quad \varphi'_i = \psi'_i \cdot \psi'_{i-1} \cdot \cdots \cdot \psi'_1.$$

The sequences from Λ and Λ' will be Cauchy sequences (with distances between successive elements limited by the sequences of positive numbers) that converge in Emb(X, M), and the sequences $\{\varphi_i\}$ and $\{\varphi'_i\}$ also will be Cauchy sequences that converge in Homeo(M, M) to $(\epsilon/3)$ pushes ξ and ξ' of $(M, \lambda(X))$ and $(M, \lambda'(X))$, respectively. The crucial additional feature is to interweave these arrangements to obtain $\xi \cdot \lambda = \xi' \cdot \lambda'$.

Specifically, these sequences will be constructed subject to the following conditions:

(1) $\rho(\lambda_i, \lambda'_{i-1}) < \delta'(\epsilon'_{i-1}) < \epsilon'_{i-1}$
(2) $\rho(\lambda_i, \lambda_{i-1}) < \delta(\epsilon_{i-1}) < \epsilon_{i-1}$
(3) $\psi_i \lambda_{i-1} = \lambda_i$
(4) ψ_i is an ϵ_{i-1}-push of $(M, \lambda_{i-1}(X))$
(5) $2\epsilon_i < \epsilon_{i-1}$ and $\epsilon_0 = \epsilon/6$
(6) $4\epsilon_i < d(\varphi_i(z), \varphi_i(z'))$ for all $z, z' \in M$ satisfying $d(z, z') \geq 1/i$

(1′) $\rho(\lambda'_i, \lambda_i) < \delta(\epsilon_i)$
(2′) $\rho(\lambda'_i, \lambda'_{i-1}) < \delta'(\epsilon'_{i-1})$
(3′) $\psi'_i \lambda'_{i-1} = \lambda'_i$
(4′) ψ'_i is an ϵ'_{i-1}-push of $(M, \lambda'_{i-1}(X))$
(5′) $2\epsilon'_i < \epsilon'_{i-1}$ and $\epsilon'_0 = \epsilon/6$
(6′) $4\epsilon'_i < d(\varphi'_i(z), \varphi'_i(z'))$ for all $z, z' \in M$ satisfying $d(z, z') \geq 1/i$.

To begin, set $\lambda_0 = \lambda$, $\lambda'_0 = \lambda'$, and $\epsilon_0 = \epsilon/6 = \epsilon'_0$. Condition (1′) holds for $i = 0$ because

$$\rho(\lambda_0, \lambda'_0) = \rho(\lambda, \lambda') < \delta \leq \delta(\epsilon/6) = \delta(\epsilon_0).$$

The second parts of conditions (5) and (5′) also hold, but no other conditions make sense yet at this preliminary stage.

Assume that for each of these sequences all terms subscripted by j, $0 \leq j < i$, have been found subject to conditions (1) through (6) and (1′) through (6′).

Step i. As Λ is dense in Emb(X, M), choose $\lambda_i \in \Lambda$ such that

$$\rho(\lambda_i, \lambda'_{i-1}) < \min\{\delta'(\epsilon'_{i-1}), \delta(\epsilon_{i-1}) - \rho(\lambda_{i-1}, \lambda'_{i-1})\}.$$

Then not only do we have $\rho(\lambda_i, \lambda'_{i-1}) < \delta'(\epsilon'_{i-1})$, but also

$$\begin{aligned} \rho(\lambda_i, \lambda_{i-1}) &\leq \rho(\lambda_i, \lambda'_{i-1}) + \rho(\lambda'_{i-1}, \lambda_{i-1}) \\ &< (\delta(\epsilon_{i-1}) - \rho(\lambda_{i-1}, \lambda'_{i-1})) + \rho(\lambda'_{i-1}, \lambda_{i-1}) \\ &= \delta(\epsilon_{i-1}), \end{aligned}$$

fulfilling conditions (1) and (2) for i. Since $\rho(\lambda_i, \lambda_{i-1}) < \delta(\epsilon_{i-1})$, the solvability of Λ provides an ϵ_{i-1}-push ψ_i of $(M, \lambda_{i-1}(X))$, where $\psi_i \cdot \lambda_{i-1} = \lambda_i$; this yields conditions (3) and (4) for this case. Choose ϵ_i satisfying conditions (5) and (6) to complete Step i.

Step i'**.** By condition (1) from Step i, we have $\rho(\lambda_i, \lambda'_{i-1}) < \delta'(\epsilon'_{i-1})$. As Λ' is dense in $\mathrm{Emb}(X, W)$, we can choose $\lambda'_i \in \Lambda'$ satisfying

$$\rho(\lambda'_i, \lambda_i) < \min\{\delta(\epsilon_i), \delta'(\epsilon'_{i-1}) - \rho(\lambda_i, \lambda'_{i-1})\}.$$

Then not only do we have $\rho(\lambda'_i, \lambda_i) < \delta(\epsilon_i)$, but also

$$\begin{aligned}\rho(\lambda'_i, \lambda'_{i-1}) &\le \rho(\lambda'_i, \lambda_i) + \rho(\lambda_i, \lambda'_{i-1}) \\ &< (\delta'(\epsilon'_{i-1}) - \rho(\lambda_i, \lambda'_{i-1})) + \rho(\lambda_i, \lambda'_{i-1}) \\ &= \delta'(\epsilon'_{i-1}),\end{aligned}$$

yielding conditions (1′) and (2′) for i. Since $\rho(\lambda'_i, \lambda'_{i-1}) < \delta'(\epsilon'_{i-1})$, the solvability of Λ' provides an ϵ'_{i-1}-push ψ'_i of $(M, \lambda'_{i-1}(X))$ with $\psi'_i \cdot \lambda'_{i-1} = \lambda'_i$, which takes care of conditions (3′) and (4′) for this case. Finally, choose ϵ'_i satisfying conditions (5′) and (6′) to complete Step i'.

Now conditions (1) through (6) and (1′) through (6′) are all fulfilled for this choice of i. The six sequences are constructed recursively in this manner.

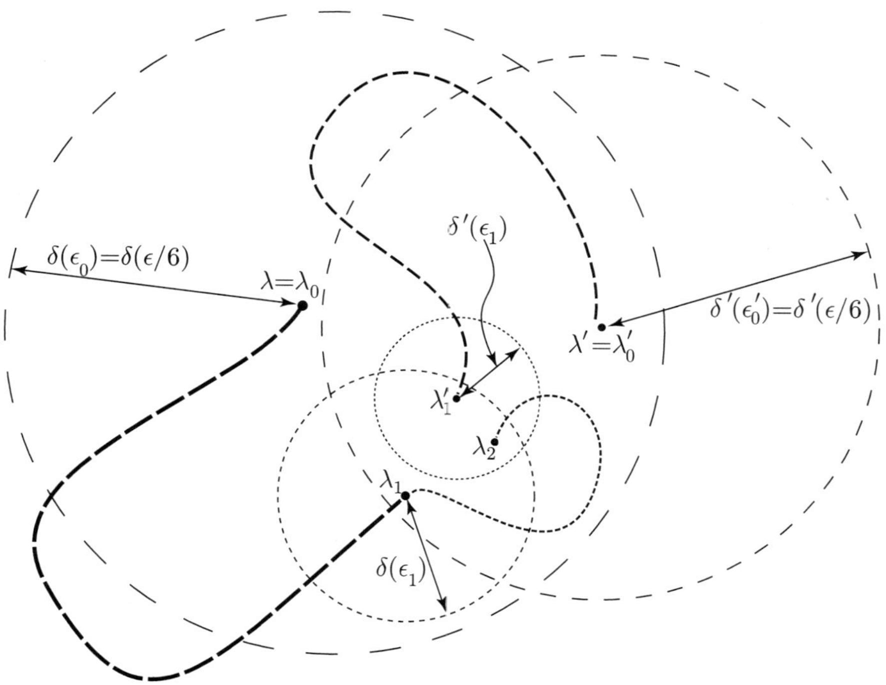

Figure 4.3. Solvability of $\Lambda \cup \Lambda'$ in the function space $\mathrm{Emb}(K, M)$

Conditions (5) and (5′) guarantee both $\Sigma_i \epsilon_i < \epsilon/3$ and $\Sigma_i \epsilon'_i < \epsilon/3$. Since ψ_i can be realized by an ϵ_{i-1}-push of $(M, \lambda_{i-1}(X))$, Proposition 2.2.2 and Lemma 4.2.5 imply that the sequence $\{\varphi_1, \varphi_2, \ldots\}$ converges uniformly to $\xi \in \text{Homeo}(M, M)$ which can be realized by an $(\epsilon/3)$-push of $(M, \lambda(X))$. Similarly, the sequence $\{\varphi'_1, \varphi'_2, \ldots\}$ converges uniformly to $\xi' \in \text{Homeo}(M, M)$ which can be realized by an $(\epsilon/3)$-push of $(M, \lambda'(X))$.

It follows from the definitions of φ_i, φ'_i and conditions (3) and (3′) that $\varphi_i\lambda = \psi_i\varphi_{i-1}\lambda = \lambda_i$ and $\varphi'_i\lambda' = \psi'_i\varphi'_{i-1}\lambda' = \lambda'_i$. Thus, by condition (1′), $\rho(\varphi_i\lambda, \varphi'_i\lambda') < \delta(\epsilon_{i-1})$. Since $\epsilon_{i-1} \to 0$, $\delta(\epsilon_{i-1}) \to 0$ as well, and therefore

$$\xi \cdot \lambda = \lim_{i\to\infty} \varphi_i\lambda = \lim_{i\to\infty} \varphi'_i\lambda' = \xi' \cdot \lambda'.$$

Since ξ' is an $(\epsilon/3)$-push of $(M, \lambda'(X))$, Observation 3.3.1 indicates that $(\xi')^{-1}$ can be realized as a $(2\epsilon/3)$-push of $(M, \xi'\lambda'(X)) = (M, \xi\lambda(X))$, and Lemma 4.2.2 promises that $(\xi')^{-1} \cdot \xi$ can be realized as an ϵ-push of $(M, \lambda(X))$. Since $(\xi')^{-1}\xi\lambda = \lambda'$, $\Lambda \cup \Lambda'$ is solvable. □

Our first application of Theorem 4.2.1 localizes tameness. The ϵ in the terminology "ϵ-tame", introduced below, does not stand for one positive number but for the phrase, " for every $\epsilon > 0$ there exists ...".

Definition. Let $\lambda : K \to M$ be an embedding of a complex in a PL manifold. Say that λ is *ϵ-tame* if for every $\epsilon > 0$ there exists an ϵ-push ψ of $(M, \lambda(K))$ with $\psi\lambda \in \text{Emb}_{PL}(K, M)$.

It is obvious that ϵ-tame implies tame; Theorem 4.2.1 yields a trivial range converse.

Corollary 4.2.6. *Let M be a compact PL n-manifold, K a finite k-complex where $2k + 2 \le n$, $\theta \in \text{Emb}_{PL}(K, M)$, and $h \in \text{Homeo}(M, M)$. Then $h\theta$ is ϵ-tame.*

Proof. Write $\Lambda = \text{Emb}_{PL}(K, M)$ and $\Lambda' = h(\text{Emb}_{PL}(K, M)) = \{h \cdot \lambda \mid \lambda \in \text{Emb}_{PL}(K, M)\}$. As Λ and Λ' are solvable, dense subsets of $\text{Emb}(K, M)$, Theorem 4.2.1 assures that $\Lambda \cup \Lambda'$ is solvable. Given $\epsilon > 0$, apply density of Λ to obtain $\lambda \in \Lambda$ close to $h\theta \in \Lambda'$ and then use solvability of $\Lambda \cup \Lambda'$ to obtain an ϵ-push ψ of $(M, h\theta(K))$ with $\psi h\theta = \lambda$. □

Corollary 4.2.7 (Tame Implies ϵ-Tame). *Let M be a compact PL n-manifold, K a finite k-complex where $2k + 2 \le n$, and $f \in \text{Emb}(K, M)$ a tame embedding. Then f is ϵ-tame.*

Proof. If f is tame, then there exists $g \in \text{Homeo}(M, M)$ such that $gf \in \text{Emb}_{PL}(K, M)$. Appeal to Corollary 4.2.6 with $\theta = gf$ and $h = g^{-1}$. □

The remainder of the section consists of some mild but extremely useful generalizations of Theorem 4.2.1. Among the applications will be the extension of Corollary 4.2.7 to non-compact manifolds M.

Definition. Say that a subset Λ of $\mathrm{Emb}(X, M)$ is *locally solvable* if for each $\lambda \in \Lambda$ and $\epsilon > 0$ there exists $\delta = \delta(\lambda, \epsilon) > 0$ such that given any $\lambda' \in \Lambda$ with $\rho(\lambda, \lambda') < \delta$, there is an ϵ-push ψ of $(M, \lambda(X))$ satisfying $\psi\lambda = \lambda'$.

For non-compact PL manifolds M and trivial range finite complexes K, $\mathrm{Emb}_{PL}(K, M)$ is a locally solvable subset of $\mathrm{Emb}(K, M)$.

Theorem 4.2.8. *Suppose Λ and Λ' are dense, locally solvable subsets of $\mathrm{Emb}(X, M)$. Then $\Lambda \cup \Lambda'$ is a dense and locally solvable subset of $\mathrm{Emb}(X, M)$.*

Proof. The argument supplied for Theorem 4.2.1 essentially gives this localized version. The minor adjustments needed mostly occur at the outset. Given some $f \in \Lambda \cup \Lambda'$ and $\epsilon > 0$, assume $f = \lambda = \lambda_0 \in \Lambda$. Of course, any $\lambda^* \in \Lambda$ within $\delta(\lambda, \epsilon/6)$ can be obtained from λ as the end of an ϵ-push, so examine instead any $\lambda' = \lambda'_0 \in \Lambda'$ for which $\rho(\lambda_0, \lambda'_0) < \delta(\lambda_0, \epsilon/6)$. Repeat the earlier argument, using the localized $\delta(\lambda_i, \epsilon_i)$ and $\delta'(\lambda'_i, \epsilon'_i)$, as bounds on the distance to the choices of later λ_{i+1} and λ'_{i+1}. □

The equivalence of tame and ϵ-tame for non-compact M now follows, as before.

Corollary 4.2.9 (Tame Implies ϵ-Tame). *Let M be a PL n-manifold, K a finite k-complex where $2k + 2 \leq n$, and $f \in \mathrm{Emb}(K, M)$ a tame embedding. Then f is ϵ-tame.*

Definition. A subset E of $\mathrm{Emb}(X, M)$ is *full* if E is invariant under the action of $\mathrm{Homeo}(M, M)$; i.e., $\psi(E) = E$ for all $\psi \in \mathrm{Homeo}(M, M)$.

Example 4.2.10. *For compact spaces X, the space of all 1-LCC embeddings of X in M is a full subset of $\mathrm{Emb}(X, M)$. However, if K is a k-complex, $k > 0$, $\mathrm{Emb}_{PL}(K, M)$ is not full in $\mathrm{Emb}(K, M)$.*

Theorem 4.2.11 (Local Solvability Criterion). *Suppose Λ is a full, dense subset of $\mathrm{Emb}(X, M)$ satisfying:*

> (*) *for each $\lambda \in \Lambda$ and $\epsilon > 0$ there exists $\delta = \delta(\lambda, \epsilon) > 0$ such that for all $\lambda' \in \Lambda$ with $\rho(\lambda', \lambda) < \delta$ and for all $\eta > 0$, there exists an ϵ-push φ of $(M, \lambda(X))$ with $\rho(\varphi\lambda, \lambda') < \eta$.*

Then Λ is locally solvable; moreover, if $\delta(\lambda, \epsilon)$ in (*) depends only on ϵ, then Λ is solvable.

Proof. Repeat the proof of Theorem 4.2.8. Start with $\lambda = \lambda_0 \in \Lambda$ and a $\lambda' = \lambda'_0 \in \Lambda$ nearby. Instead of choosing some $\lambda_1 \in \Lambda$ close to λ'_0, use (*)

to obtain $\lambda_1 \in \text{Emb}(X, M)$ close to λ'_0 to which λ_0 can be ϵ_0-pushed. Here λ_1 will belong to Λ, due to hypothesized fullness of Λ. Such replacement of a chosen element—say the λ_i near λ'_{i-1} or the λ'_i near λ_i—by one to which the earlier element (λ_{i-1} or λ'_{i-1}) can be pushed, with control, is the modification required. □

Historical Notes. The first to observe the usefulness of dense, solvable subsets were T. Homma (1962) and H. Gluck (1963). Homma deserves credit for the meshing of structures idea central to the proof of Theorem 4.2.1; Gluck introduced the solvability terminology and extended its applications. J. Dancis (1970) localized the concept. Ultimately Homma obtained a result comparable to Corollary 4.2.6; Gluck, as well as C. Greathouse (1963), showed that locally tame embeddings in the trivial range are ϵ-tame (Corollary 4.2.4).

Exercises

4.2.1. The existence of a dense, solvable subset implies $\text{Emb}(X, M)$ is uniformly locally pathwise connected.

4.2.2. Suppose $\Lambda \subset \text{Emb}(X, M)$ is solvable and $\Lambda' \subset \text{Emb}(X, M)$ satisfies: for each $\lambda' \in \Lambda'$ and each $\epsilon > 0$ there is an ϵ-push ψ of $(M, \lambda'(X))$ with $\psi\lambda' \in \Lambda$. Then $\Lambda \cup \Lambda'$ is solvable.

4.2.3. Prove Theorem 4.2.8.

4.2.4. Prove Theorem 4.2.11.

4.3. Taming 1-LCC embeddings of polyhedra

This section relies upon Bryant-Seebeck engulfing methods to establish the fundamental 1-LCC characterization of tameness for trivial range embeddings of polyhedra. The following central pushing lemma will also prove useful in the next section.

Lemma 4.3.1. *Let X be a k-dimensional compactum 1-LCC embedded in a PL n-manifold M, $n \geq 5$, $2k + 2 \leq n$. Then for every $\epsilon > 0$ there exists $\delta > 0$ such that, for any $f \in C(X, M)$ with $\rho(f, \text{incl}_X) < \delta$ and for any $\eta > 0$ there exists an ϵ-push ψ of (M, X) such that $\rho(\psi|X, f) < \eta$.*

Proof. The strategy is to extend f to a neighborhood of X, use the theory of embedding dimension to find a smaller neighborhood whose spine is a k-dimensional polyhedron, and then apply the unknotting theorem for polyhedra to that spine. While the strategy is simple, its implementation requires careful choices of neighborhoods.

Let $\epsilon > 0$ be given. First use Theorem 0.6.3 to choose $\delta > 0$ such that any map within δ of $\text{incl}_X : X \to M$ is $(\epsilon/4(k + 1))$-homotopic to

incl_X. Then specify $f \in C(X, M)$ with $\rho(f, \mathrm{incl}_X) < \delta$, and let η be another positive number, as in the statement of the lemma.

Since M is an ANR, f can be extended to a neighborhood of X. Fix both a compact PL ∂-manifold W such that $X \subset \mathrm{Int}\, W \subset W \subset M$ and an extension $F : W \to M$. Make W so close to X so that $\rho(F, \mathrm{incl}_W) < \delta$. Choose a number $\zeta \in (0, \epsilon/2)$ such that if $x, x' \in W$ and $d(x, x') < \zeta$, then $d(F(x), F(x')) < \eta/3$. Now the two conditions $2k + 2 \leq n$ and $n \geq 5$ imply $k \leq n - 3$, so Theorem 3.4.8 shows that $\mathrm{dem}\, X \leq k$. Hence there exist a k-dimensional polyhedron K and a ζ-regular neighborhood N of K such that $X \subset \mathrm{Int}\, N \subset N \subset \mathrm{Int}\, W$. Let $r : N \to K$ be the ζ-retraction of N to the spine K.

Let $F' : K \to M$ be a PL general position approximation to $F|K$. Take such a close approximation that $\rho(F', F|K) < \eta/3$ and we still have $\rho(F', \mathrm{incl}_K) < \delta$. By Theorem 4.1.1 and the choice of δ, there is an $(\epsilon/2)$-push ϕ of (M, X) such that $\phi|K = F'$. Choose $\gamma > 0$ such that if $x, x' \in W$ and $d(x, x') < \gamma$, then $d(\phi(x), \phi(x')) < \eta/3$. Let θ be an isotopy that squeezes the regular neighborhood N so close to K that $d(\theta(x), r(x)) < \gamma$ for each $x \in N$. We can extend θ over all of M to obtain θ as a ζ-push of (M, X).

Define $\psi = \phi \circ \theta$. Note that ψ is an $(\epsilon/2 + \zeta)$-push and $\zeta < \epsilon/2$, so ψ is an ϵ-push of (M, X). Pick $x \in X$. Then

$$\begin{aligned} d(\psi(x), f(x)) &= d(\phi\theta(x), f(x)) \\ &\leq d(\phi\theta(x), \phi r(x)) + d(\phi r(x), F r(x)) + d(F r(x), f(x)) \\ &= d(\phi\theta(x), \phi r(x)) + d(F' r(x), F r(x)) + d(F r(x), F(x)) \\ &< \eta/3 + \eta/3 + \eta/3 = \eta. \end{aligned}$$

Hence $\rho(\psi|X, f) < \eta$ and the proof is complete. □

The lemma promptly implies the pivotal tameness characterization.

Theorem 4.3.2 (1-LCC Taming). *If K is a finite k-complex, M is a PL n-manifold, $n \geq 5$, $2k+2 \leq n$, and $\lambda : K \to M$ is a 1-LCC embedding, then λ is ϵ-tame.*

Proof. Consider $\Lambda = \{\lambda : K \to M \mid \lambda \text{ is a 1-LCC embedding}\}$. By Theorem 4.2.11, Λ is locally solvable. Since $\mathrm{Emb}_{PL}(X, M)$ is a dense subset of Λ, each $\lambda \in \Lambda$ admits an ϵ-push to a PL embedding. □

Corollary 4.3.3. *Suppose M is a PL n-manifold, $n \geq 5$, and $\lambda : K \to M$ is an embedding of a finite k-complex K, $2k+2 \leq n$, such that* $\mathrm{dem}(\lambda(K)) < n - 2$. *Then $\lambda(K)$ is ϵ-tame.*

Corollary 4.3.4. *Suppose M is a PL n-manifold, $n \geq 5$, and $\lambda : K \to M$ is an embedding of a finite k-complex, $2k+2 \leq n$. Then $\lambda(K)$ is tame if and only if* $\operatorname{dem}(\lambda(K)) = k$.

Corollary 4.3.5. *Every locally tame embedding of a finite k-complex K in a PL n-manifold, $n \geq 5$ and $2k+2 \leq n$, is ϵ-tame.*

Corollary 4.3.6. *Suppose M is a PL n-manifold, $n \geq 5$, K is a finite k-complex with $2k+2 \leq n$, $\lambda_0, \lambda_1 \in \operatorname{Emb}(K, M)$ whose images are 1-LCC embedded in M, and $f_t : K \to M$ is a homotopy between λ_0, λ_1 satisfying $\rho(f_t, f_0 = \lambda_0) < b$ for some $b > 0$. Then there exists a $2(k+1)b$-push ψ of $(M, \lambda_0(K))$ such that $\psi\lambda_0 = \lambda_1$.*

Proof. Compactness yields $b' \in (0, b)$ for which f_t is a b'-homotopy. Set $\zeta = (b - b')/3$. Theorem 4.3.2 promises ζ-pushes φ_i of $(M, \lambda_i(K))$ with $\varphi_i\lambda_i$ PL $(i = 0, 1)$, and Theorem 4.1.1 yields a $2(k+1)(2\zeta + b')$-push Θ of $(M, \varphi_0\lambda_0(K))$ such that $\Theta\varphi_0\lambda_0 = \varphi_1\lambda_1$. Put $\psi = \varphi_1^{-1}\Theta\varphi_0$. □

Corollary 4.3.7. *For $n \geq \max\{5, 2k+2\}$, any two 1-LCC embeddings of a finite k-complex K in $\mathbb{R}^n$ or S^n are ambient isotopic.*

4.4. Unknotting 1-LCC embeddings of compacta

We now turn to unknotting theorems for compacta. They present further evidence to reinforce the contention that 1-LCC compacta behave very much like polyhedra. There are both a local and a global unknotting result; the proofs are short because all the groundwork already has been laid in the two preceding sections.

Theorem 4.4.1 (Local Solvability for 1-LCC Compacta). *Suppose X is a compact, k-dimensional space and M is a PL n-manifold, where $n \geq 5$ and $2k+2 \leq n$. Then the collection of all 1-LCC embeddings $\lambda : X \to M$ is a dense and locally solvable subset of $Emb(X, M)$.*

Proof. Density is treated in Exercise 3.4.5. Local solvability is assured by Lemma 4.3.1 and Theorem 4.2.11. □

Theorem 4.4.2. *Suppose X is a compact, k-dimensional space and M is a PL n-manifold, where $n \geq 5$ and $2k+2 \leq n$. If $\lambda_0, \lambda_1 : X \to M$ are two topological embeddings whose images are 1-LCC in M and $f_t : X \to M$ is a homotopy between λ_0 and λ_1, then there exists a compactly supported ambient isotopy Ψ of M such that $\Psi_0 = \mathrm{Id}_M$ and $\Psi_1\lambda_0 = \lambda_1$.*

Proof. By Theorem 3.4.8 we have $\operatorname{dem}\lambda_0(X) = \operatorname{dem}\lambda_1(X) = k$. Thus Theorem 3.4.7 allows us to adjust λ_1 so that $\lambda_0(X) \cap \lambda_1(X) = \emptyset$. Focus on the $2k+3 \leq n$ case. Since $f|X \times \partial I$ is an embedding and $2(k+1) < n$,

$f : X \times I \to M$ can be approximated by an embedding F that agrees with f on $X \times \partial I$ (see Exercise 3.4.5). The approximation can be done in such a way that $\mathrm{dem}(F(X \times I)) = k+1$ (see Exercise 3.4.5).

Now Theorem 4.4.1 yields that the set S of all $t \in [0,1]$ for which there exists a compactly supported ambient isotopy Φ of M, starting at Id_M, such that $\Phi_1\lambda_0 = F_t$ is both open and closed in [0,1], implying that $S = [0,1]$, and completing the proof of the $2k+2 < n$ case.

The extreme case $2k+2 = n$ can be handled in almost the same way. The only change is to use Exercise 4.1.3 for obtaining an approximation of $f : X \times I$ by a map $F : X \times I \to M$ such that $F|X \times \partial I = f|X \times \partial I$ and F_t is an embedding with 1-LCC image for all $t \in I$. $\square$

Corollary 4.4.3. *For $n \geq \max\{5, 2k+2\}$, any two 1-LCC embeddings of a compact, k-dimensional metric space X in $\mathbb{R}^n$ or S^n are ambient isotopic.*

Corollary 4.4.4. *Let X be a k-dimensional compactum topologically embedded in a PL n-manifold M, $n \geq \max\{5, 2k+2\}$. Then there exist a 1-LCC embedding $\lambda : K \to M$ and a pseudo-isotopy θ_t of M such that $\theta_1\lambda = \mathrm{incl}_K : K \to M$.*

The definition of pseudo-isotopy was given on page 154.

Historical Notes. J. L. Bryant and C. L. Seebeck, III, developed the techniques reproduced in §4.3 and §4.4. They proved Theorem 4.3.3 in (Bryant and Seebeck, 1969). At the same time they also established Theorem 4.4.2 for compact absolute retracts. Later Bryant (1969) (1971a) generalized this trivial range unknotting theorem to arbitrary compacta.

4.5. Chart-by-chart analysis of topological manifolds

Operating in manifolds with PL (or Differentiable) structure is a great convenience. The PL structure allows one to make simplifying general position adjustments globally and, after that, to exploit powerful tools such as engulfing. Many standard results about embeddings in PL manifolds also hold for embeddings in topological manifolds; the more general setting simply demands extra effort. Progress comes about by covering the target with charts admitting PL structures and then by improving the situation one chart at a time.

The following illustrates an elementary procedure.

Proposition 4.5.1. *If K is a finite k-complex and M^n is a topological n-manifold, $2k+1 \leq n$, then every map $f : K \to M$ can be approximated by an embedding.*

Proof. Relative general position methods readily give:

Lemma 4.5.2. *Under the hypotheses of 4.5.1, let U be an open subset of M^n with a PL structure, $f : K \to M^n$ a map and $\epsilon > 0$. Then there exists a map $g : K \to M^n$ satisfying (i) $g|K \setminus f^{-1}(U) = f|K \setminus f^{-1}(U)$, (ii) $gf^{-1}(U) \subset U$, (iii) $\rho(g, f) < \epsilon$ and (iv) $g|f^{-1}(U)$ is 1-1.*

Produce a sequence of n-balls $C_1, C_2, \ldots$ covering M^n, where each C_i is contained in a Euclidean chart $U_i \approx \mathbb{R}^n$. For $i, m = 1, 2, \ldots$ let

$$A_i^m = \{g \in C(K, M^n) \mid \operatorname{diam} g^{-1}(c) < 1/m \text{ for all } c \in C_i\}.$$

It is easy to check that A_i^m is an open subset of $C(K, M^n)$; by Lemma 4.5.2 it is also a dense subset. Thus $\bigcap_{i,m} A_i^m$ consists of embeddings, and they are dense in $C(K, M^n)$. □

Corollary 4.5.3. *Every map from a k-dimensional, compact metric space into a topological n-manifold, $2k + 1 \le n$, can be approximated by an embedding.*

4.6. Detecting 1-LCC embeddings

The tameness and unknotting results of §4.3 and §4.4 impart substantial inducements for seeking conditions under which embeddings are 1-LCC. This section presents a variety of such conditions. Surprisingly, perhaps, none imposes trivial range restrictions. This has more potential than one has a right to expect at this juncture. Although the immediate applications do occur with trivial range limitations, in the next chapter we shall prove that comparable controlled 1-LCC taming theorems and related 1-LCC local unknotting theorems hold up to codimension three.

Being LCC^1 is a hereditary property in codimension three.

Proposition 4.6.1. *If X is a k-dimensional space LCC^1 embedded as a closed subset of an n-manifold M^n, $k \le n - 3$, and Y is a closed subset of X, then Y is LCC^1 in M^n.*

Proof. Given a neighborhood U of $y \in Y$, the hypotheses about X being LCC^1 promises a neighborhood V of y such that loops in $V \setminus X$ are inessential in $U \setminus X$. Since any loop in $V \setminus Y$ is homotopic there to a loop in $V \setminus X$, the result follows. □

Theorem 4.6.2. *Suppose X is a k-dimensional space embedded in an n-manifold M^n as a closed subset, $k \le n - 3$, and Y is a closed subset of X such that Y is LCC^1 in M^n and X is LCC^1 in M^n at each point of $X \setminus Y$. Then X is LCC^1 in M^n.*

Proof. Given a neighborhood U of $x \in X$, find a smaller neighborhood V of x such that $\pi_1(V) \to \pi_1(U)$ is trivial. Take any loop $f : \partial I^2 \to V \setminus X$

and extend it to a map $F_1 : I^2 \to U$. Make successive applications of π_1-negligibility (Lemma 3.3.4), first for the 1-LCC subset $Y \cap U$ of U and then for the 1-LCC subset $X \cap (U \setminus Y)$ of $U \setminus Y$, to obtain maps $F_2 : I^2 \to U \setminus Y$ and $F_3 : I^2 \to U \setminus X$, both agreeing with f on ∂I^2. □

Let A be a closed subset of the polyhedron (respectively, manifold) P topologically embedded in a PL manifold M. We say that P is *locally tame* (respectively, *locally flat*) *modulo* A provided P is locally tame (respectively, locally flat) at all points of $P \setminus A$.

Corollary 4.6.3. *If P is a p-dimensional polyhedron topologically embedded in the PL n-manifold M^n as a closed subset, $p \leq n-3$, and C is a closed, countable subset of P such that P is locally tame in M^n modulo C, then P is 1-LCC in M^n. Hence, P is ϵ-tame provided $2p+2 \leq n$ and $n \geq 5$.*

Corollary 4.6.4. *Suppose P is a p-dimensional polyhedron topologically embedded in the PL n-manifold M^n as a closed subset, $p \leq n-3$, and K is a k-dimensional polyhedron tamely embedded in M^n, $k \leq n-3$, such that P is locally tame in M^n modulo $P \cap K$. Then P is 1-LCC in M^n.*

Proof. Since it is tame, K is 1-LCC in M^n, by Proposition 1.3.3; Proposition 4.6.1 assures that $P \cap K$ is 1-LCC as well. Another application of 1.3.3 gives that $P \setminus (P \cap K)$ is 1-LCC. □

Corollary 4.6.5. *Suppose B_1 and B_2 are k-cells in $\mathbb{R}^n$, $2k+2 \leq n$, such that $B_1 \cap B_2 = \partial B_1 = \partial B_2$, where B_1 is flat and B_2 is locally flat modulo ∂B_2. Then $B_1 \cup B_2$ is a flat k-sphere.*

Corollary 4.6.6. *If B is a k-cell in $\mathbb{R}^n$, $2k+2 \leq n$, that can be expressed as a union of k-cells B_1 and B_2, where B_1 is flat and B_2 is locally flat modulo $B_1 \cap B_2$, then B is flat.*

Lemma 4.6.7. *Let X be a closed, $(n-2)$-dimensional subset of a PL n-manifold M, $n \geq 4$. Then X contains a 0-dimensional F_σ-set F such that all compact subsets of $X \setminus F$ are 1-LCC.*

Proof. Consider the 2-skeleton K of some triangulation of M, and let L_i denote the 1-skeleton of a subdivision $K^{(i)}$ of K such that $\text{mesh}(K^{(i)}) < 1/i$. Fix $\epsilon > 0$. Since X is $(n-2)$-dimensional, it is 0-LCC, so Lemma 3.3.3 yields a small homotopy that pushes L_1 off X; Theorem 4.1.1 transforms that homotopy into an $(\epsilon/2)$-push of M. Next push L_2 off X with an $(\epsilon/4)$-homeomorphism that is fixed on the image of L_1. With repeated application of this process, adjust the various L_i via a sequence of homeomorphisms of M so that they miss X. Imposing the controls of Proposition 2.2.2 forces the sequence to converge to an ϵ-homeomorphism $h : M \to M$ with $h(\cup L_i) \cap X = \emptyset$. Then $h(K) \cap X$ is 0-dimensional, since $h(K) \cap X \subset h(K \setminus \cup L_i)$.

By naming a sequence K_j of 2-skeleta of triangulations T_j of M with mesh $T_j < 1/j$, and then adjusting each K_j as above via a homeomorphism of M moving points less than $1/j$ to make its intersection F_j with X be 0-dimensional, we produce an F_σ set $F = \cup F_j \subset X$.

Consider any compact subset C of $X \smallsetminus F$. A small loop in the complement of C is null-homotopic in a small subset of M. The track of that homotopy can be pushed into the 2-skeleton of a fine triangulation. The homeomorphisms constructed earlier in the proof move the 2-skeleton so that its intersection with X is contained in F; in particular, the adjusted 2-skeleton misses C. Hence the loop is null-homotopic in a small subset of $M \smallsetminus C$ and thus C is 1-LCC. □

Theorem 4.6.8. *Each 2-cell D in a PL n-manifold M^n, $n \geq 5$, contains tame arcs. Specifically, for each embedding $\lambda : I \to D$ with $\lambda(I) \cap \partial D \subset \lambda(\partial I)$ and $\epsilon > 0$, there exists $\theta \in \mathrm{Homeo}(D, D)$ such that $\rho(\theta, \mathrm{Id}_D) < \epsilon$ and $\theta\lambda(I)$ is tame in M^n.*

Proof. To do this, one produces $\theta \in \mathrm{Homeo}(D, D)$ so that $\theta\lambda(I)$ misses the F_σ-subset F of D provided by Lemma 4.6.7. Tameness is ensured by Theorem 4.3.2. □

Proposition 4.6.9. *Each $(n-3)$-dimensional closed subset C of an $(n-1)$-hyperplane in $\mathbb{R}^n$ is 1-LCC in $\mathbb{R}^n$.*

Proof. Assume $C \subset \mathbb{R}^{n-1}$, and express $\mathbb{R}^n \smallsetminus C$ as $U_+ \cup U_-$, where

$$U_+ = [\mathbb{R}^{n-1} \times (0, \infty)] \cup [(\mathbb{R}^{n-1} \smallsetminus C) \times (-\infty, \infty)],$$
$$U_- = [\mathbb{R}^{n-1} \times (-\infty, 0)] \cup [(\mathbb{R}^{n-1} \smallsetminus C) \times (-\infty, \infty)].$$

Since U_+, U_- both are contractible and $U_+ \cap U_- = (\mathbb{R}^{n-1} \smallsetminus C) \times (-\infty, \infty)$ is connected, the Seifert-van Kampen Theorem indicates that $\mathbb{R}^n \smallsetminus C$ is 1-connected. Localization of this argument is routine. □

Corollary 4.6.10. *Each finite k-complex embedded in an $(n-1)$-hyperplane in $\mathbb{R}^n$, $2k+2 \leq n$ and $n \geq 5$, is tame in $\mathbb{R}^n$.*

Corollary 4.6.11. *Any two embeddings $\lambda_1, \lambda_2 : X \to \mathbb{R}^{n-1} \subset \mathbb{R}^n$ of a k-dimensional compactum X, $2k+2 \leq n$ and $n \geq 5$, are equivalent as embeddings into $\mathbb{R}^n$.*

The next result generalizes Theorem 4.6.9 considerably.

Theorem 4.6.12. *Suppose X is a k-dimensional closed subset of $\mathbb{R}^{n+1} = \mathbb{R}^n \times \mathbb{R}^1$, $k \leq n-2$, and $\mathcal{E}$ is a dense subset of $\mathbb{R}$ such that, for each $t \in \mathcal{E}$, $X \cap (\mathbb{R}^n \times t)$ is 1-LCC in $\mathbb{R}^n \times t$. Then X is 1-LCC in $\mathbb{R}^{n+1}$.*

The proof of Theorem 4.6.12 is based on two lemmas.

Lemma 4.6.13. *Under the hypotheses of Theorem 4.6.12 each map* $g : I^2 \to \mathbb{R}^n \times \{t\}$ *can be approximated, arbitrarily closely, by a map* $g' : I^2 \to \mathbb{R}^{n+1} \setminus X$*; moreover, if* $g(\partial I^2) \cap X = \emptyset$*, then approximations* g' *can be obtained such that* $g' \mid \partial I^2 = g \mid \partial I^2$.

Proof. When $t \in \mathcal{E}$ see Lemma 3.3.3; in general, translate via a homotopy to put the image of g in such a special level. □

Lemma 4.6.14. *Under the hypotheses of Theorem 4.6.12, suppose* $f : D = I \times [r, r'] \to \mathbb{R}^n \times \mathbb{R}^1$ *is a map such that* $f(\partial D) \cap X = \emptyset$*,* f *sends each level* $I \times s \subset I \times [r, r']$ *to a level* $\mathbb{R}^n \times t(s)$ *of* $\mathbb{R}^{n+1}$*, and* f *is "vertical" on* $\partial I \times [r, r']$ *in the sense that to* $x \in \partial I$ *corresponds* $z \in \mathbb{R}^n$ *with* $f(\langle x, s\rangle) = \langle z, t(s)\rangle$ *for all* $s \in [r, r']$*. Then for each neighborhood* U *of* $f(D)$ *there exists a map* $g : D \to \mathbb{R}^{n+1}$ *such that* $g|\partial D = f|\partial D$ *and* $g(D) \subset U \setminus X$.

Proof. Choose $\epsilon > 0$ with $B(f(D); 2\epsilon) \subset U$ and $B(f(\partial D); \epsilon) \subset U \setminus X$. For notational simplicity, let $f_s : I \to \mathbb{R}^n$ be the composite map

$$I \longrightarrow I \times s \xrightarrow{f|I\times s} \mathbb{R}^{n+1} = \mathbb{R}^n \times \mathbb{R}^1 \xrightarrow{\text{proj}} \mathbb{R}^n \quad (s \in [r, r']).$$

Since $\dim X \leq n-2$, there exists a map $\alpha_s : I \to \mathbb{R}^n$ such that $\alpha_s|\partial I = f_s|\partial I$, $\rho(\alpha_s, f_s) < \epsilon$, and $X \cap (\alpha_s(I) \times t(s)) = \emptyset$; at the endpoints, take $\alpha_r = f_r$ and $\alpha_{r'} = f_{r'}$. There exists $\delta(s) > 0$ such that $X \cap B(\alpha_s(I) \times t(s); 2\delta(s)) = \emptyset$ and $\rho(f_z, \alpha_s) < \epsilon$ for all $z \in [t(s)\delta(s), t(s)+\delta(s)] \cap [r, r']$. Let $\lambda > 0$ be a Lebesgue number for the associated cover $\{(t(s)-\delta(s), t(s)+\delta(s)) \mid s \in [r, r']\}$ of $[r, r']$. Specify numbers

$$r(0) = r < r(1) < \cdots < r(m) = r'$$

with $r(i) - r(i-1) < \lambda(i = 1, \ldots, m)$ and $r(i) \in \varphi(i = 1, \ldots, m-1)$. This yields maps $\alpha_i = \alpha_{r(i)} : I \to \mathbb{R}^n$ satisfying $\alpha_i|\partial I = f_{r(i)}|\partial I$, $\alpha_0 = f_r$, $\alpha_m = f_{r'}$ and

$$\alpha_i(I) \times [r(i), r(i+1)] \subset B(f(D); \epsilon) \setminus X \subset U \setminus X.$$

For $i = 1, 2, \ldots, m$, $\alpha_{i-1}(I) \cup \alpha_i(I) \subset B(f_{r(i)}(I); \epsilon)$. The straight line homotopy between α_{i-1} and α_i gives a singular disk in $B(f_{r(i)}(I); \epsilon)$, which includes naturally as a singular disk D_i' in $U \cap (\mathbb{R}^n \times t(r(i)))$. Regard D_i' as the image of a map $\beta_i : I^2 \to \mathbb{R}^n \times t(r(i))$, and $(1 \leq i < m)$ use Lemma 4.6.13 to approximate β_i by $\beta_i' : I^2 \to U \setminus X$ with $\beta_i'|\partial I^2 = \beta_i|\partial I^2$. Set D_i'' equal to the image of β_i' $(0 < i < m)$ and $D_m'' = D_m'$. Now it should be clear that $f|\partial D$ is null-homotopic in

$$(\cup_{i=0}^{m-1} \alpha_i(I) \times [r(i), r(i+1)]) \cup (\cup_{i=1}^{m} D_i'') \subset U \setminus X,$$

as required. □

Proof of Theorem 4.6.12. The proof rests on approximating small loops in $\mathbb{R}^{n+1} \smallsetminus X$ by polygonal curves comprised of segments parallel to either the $\mathbb{R}^n$ or the $\mathbb{R}^1$ coordinate factor and on representing the result as a finite product of loops, each bounding singular disks satisfying the hypothesis of either Lemma 4.6.13 or Lemma 4.6.14, where the union of all these singular disks is small. The details are left as an exercise. □

Theorem 4.6.15. *If C is a closed, $(n-2)$-dimensional subset of $\mathbb{R}^n$, $n > 2$, then each $(k-1)$-dimensional compactum $X \subset C \times \mathbb{R}^k \subset \mathbb{R}^n \times \mathbb{R}^k$ is 1-LCC in $\mathbb{R}^{n+k} = \mathbb{R}^n \times \mathbb{R}^k$.*

Proof. Starting with an arbitrary neighborhood U of $x \in X \subset C \times \mathbb{R}^k$, restrict U, if necessary, to a basic neighborhood of the form $U = U' \times U'' \subset \mathbb{R}^n \times \mathbb{R}^k$, where $U'' \subset \mathbb{R}^k$ is contractible. Express x as $\langle z', z''\rangle \in \mathbb{R}^n \times \mathbb{R}^k$. Since $\dim(X \cap (z' \times U'')) \leq k-1 < \dim U''$, there exists $y'' \in U''$ with $\langle z', y''\rangle \notin X$. Now choose a contractible neighborhood $W' \subset U' \subset \mathbb{R}^n$ of z' such that $(W' \times y'') \cap X = \emptyset$.

To see that $V = W' \times U''$ is an appropriate smaller neighborhood of x fulfilling the definition of 1-LCC, consider a map $f : \partial I^2 \to V \smallsetminus X$, and express f as a product $f_1 \times f_2$ of coordinate functions. Compute the distance η between $f(\partial I^2)$ and X. Since $\dim C \leq n-2$, there is an η-homotopy $F_t : \partial I^2 \to W'$ between f_1 and a map $g_1 : \partial I^2 \to W'$ with $g_1(\partial I^2) \subset W' \smallsetminus C$; then the product $F_t \times f_2$ serves as a homotopy in $V \smallsetminus X$ between $f_1 \times f_2$ and $g_1 \times f_2$. Next, f_2 is homotopic in U'' to the constant mapping g_2 to y''; the product of g_1 with the latter homotopy has image in $(W' \smallsetminus C) \times U'' \subset V \smallsetminus X$. Finally, the image of $g_1 \times g_2$ lives in $W' \times y''$, so $g_1 \times g_2$ is null-homotopic in $W' \times y'' \subset V \smallsetminus X$. The net effect is a null-homotopy of f in $V \smallsetminus X \subset U \smallsetminus X$. □

Theorem 4.6.16. *Suppose C is a closed, $(n-3)$-dimensional subset of $\mathbb{R}^n$, and $K \subset C \times \mathbb{R}^k \subset \mathbb{R}^n \times \mathbb{R}^k$ is homeomorphic to a finite k-complex. Then K is 1-LCC in $\mathbb{R}^{n+k} = \mathbb{R}^n \times \mathbb{R}^k$.*

The proof, which is similar to that of Theorem 4.6.15, is left as an exercise.

Theorem 4.6.17. *If K is a finite k-complex and M is an n-manifold, $k \leq n-3$, then most embeddings of K in M are 1-LCC; i.e., the 1-LCC embeddings form a dense, G_δ-subset of* $\mathrm{Emb}(K, M)$.

Proof. Specify a countable collection of maps $f_1, f_2, \ldots$ that form a dense subset of $C(I^2, M)$. For $j = 1, 2, \ldots$, define O_j to be the collection of all $\lambda \in Emb(K, M)$ such that there exists $f'_j \in C(I^2, M)$ with $\rho(f'_j, f_j) < 1/j$ and $f'_j(I^2) \cap \lambda(K) = \emptyset$. It is easy to see that O_j is a dense, open subset of $Emb(K, M)$ and that every $\lambda \in \cap_j O_j$ is a 1-LCC embedding. □

Corollary 4.6.18. *If K is a finite k-complex and M is a PL n-manifold, $2k+2 \leq n$, then most embeddings of K in M are tame.*

We conclude the section by relating the 1-LCC property of embeddings to a property of maps.

Definition. A compact set $C \subset Y$ has *Property 1-UV* if for each open set $U \supset C$ there exists an open set V such that $C \subset V \subset U$, and each map $\partial I^2 \to V$ extends to a map $I^2 \to U$. As usual, $f \in C(Y,S)$ is a *1-UV map* if $f^{-1}(s)$ has Property 1-UV for each $s \in S$.

In some contexts we will emphasize the ambient space Y and say "C has Property 1-UV in Y," but usually Y will not be mentioned. This is justified by Exercise 4.6.5, which shows that Property 1-UV is a topological property and does not depend on the embedding.

The following theorem demonstrates the fundamental connection between Property 1-UV for maps and the 1-LCC property for embeddings whose images have mapping cylinder neighborhoods. Examples developed in §7.11 will reveal that wildly embedded objects can possess manifold mapping cylinder neighborhoods — such objects, of course, are not 1-LCC and the maps that determine the mapping cylinders fail to be 1-UV.

Theorem 4.6.19. *Suppose M is a compact manifold and $f : M \to X$ is a map such that $2 + \dim X \leq \dim M$ and $\mathrm{Map}(f)$ is a ∂-manifold. Then X is LCC^1 in $\mathrm{Map}(f)$ if and only if f is 1-UV.*

Proof. Assume f is a 1-UV map. As $\dim(\mathrm{Map}(f)) \geq \dim X + 3$, X must be both (-1)-LCC and 0-LCC in $\mathrm{Map}(f)$. To check that it is also 1-LCC, let W be a neighborhood in $\mathrm{Map}(f)$ of $x \in X$. Let $q : M \times [0,1] \cup X \to \mathrm{Map}(f)$ denote the mapping cylinder quotient map. Find a neighborhood U in M of $f^{-1}(x)$ and $t \in (0,1)$ such that $q(U \times (t,1)) \subset W$. Since $f^{-1}(x)$ has Property 1-UV, it has a neighborhood $V \subset U$ such that loops in V are null-homotopic in U. The reader should confirm that $q((V \times (t,1)) \cup (X \smallsetminus f(M \smallsetminus V)))$ is a neighborhood W' of x in $\mathrm{Map}(f)$. Clearly each map $\mu : \partial I^2 \to W' \smallsetminus X = q(V' \times (t,1))$ is null-homotopic in $q(U \times (t,1)) \subset W \smallsetminus X$.

The proof of the other implication is left as an exercise. □

Historical Notes. Corollaries 4.6.5 and 4.6.6, as well as their generalizations beyond the trivial range, were proved by A. V. Černavskiĭ (1965) via an elegant straightening technique. Corollary 4.6.6 represents an example of what Rushing called a "β-statement," a set of results about unions of cells comprehensively analyzed in (Rushing, 1972, Section 5.2). In this book the β-statements, as well as related γ-statements, receive less prominent treatment, as consequences of the 1-LCC characterization of flatness/tameness.

Theorem 4.6.8 was proved by Seebeck (1971); earlier, Bing (1962b) had shown that all disks embedded in 3-manifolds contain many tame arcs, and then Sher (1971) did the same for disks embedded in 4-manifolds. In contrast, examples of wildly embedded cells containing no tame 2-cells whatsoever were constructed by Daverman (1975). Corollary 4.6.10 about polyhedra embedded in hyperplanes was originally proved by Bing and Kister (1964); Gillman found counterexamples to unknotting of arbitrary polyhedra in hyperplanes (1967). Theorem 4.6.12 was proved by Bryant (1971b). Theorems 4.6.15 and 4.6.16 were proved by Daverman (1973a). Theorem 4.6.19 is part of the folklore of the subject.

Exercises

4.6.1. If Q is a compact $(n-2)$-manifold in an $(n-1)$-dimensional hyperplane in $\mathbb{R}^n$, then Q is 1-alg in $\mathbb{R}^n$.

4.6.2. Fill in the details in the proof of Theorem 4.6.12.

4.6.3. Suppose X is a k-dimensional closed subset of $\mathbb{R}^{n+1} = \mathbb{R}^n \times \mathbb{R}^1$, $k \leq n-2$, such that for each $t \in \mathbb{R}^1$ and each $\epsilon > 0$ there exists an ϵ-push ψ of $(\mathbb{R}^{n+1}, X)$ with $\psi(X) \cap (\mathbb{R}^n \times t)$ LCC^1 in $\mathbb{R}^n \times t$. Then X is LCC^1 in $\mathbb{R}^{n+1}$.

4.6.4. Prove Theorem 4.6.16.

4.6.5. (Topological invariance of Property 1-UV) Suppose X is a compact subset of an ANR Y such that X has Property 1-UV in Y and e is an embedding of X into another ANR Y'. Then $e(X)$ has Property 1-UV in Y'.

4.6.6. If $f : M \to X$ is a map from a compact manifold M onto a compactum X such that $\text{Map}(f)$ is a ∂-manifold with $\partial\,\text{Map}(f) = M$ and X is LCC^1 in $\text{Map}(f)$, then f is a 1-UV map.

4.6.7. If $f : M^n \to N^n$, $n \geq 5$, is a map between compact PL n-manifolds such that $\text{Map}(f)$ is an $(n+1)$-dimensional ∂-manifold and $\partial\,\text{Map}(f)$ is the disjoint union of M^n and N^n, then f is cellular.

4.7. More wild embeddings

Example 4.7.1. *There exists a wild Cantor set in $\mathbb{R}^n$, $n \geq 3$, with non-simply connected complement.*

We will produce a wild Cantor set in $\mathbb{R}^4$; examples in higher dimensions can be constructed by a straightforward generalization of the process.

That process builds on the construction of Antoine's Necklace, the wild Cantor set with non-simply connected complement in $\mathbb{R}^3$ given as Example 2.1.1. The starting point for Antoine's Necklace is a solid torus $A_0 \cong$

$S^1 \times B^2$. Treat A_0 as a subset of Int $\mathbb{R}^3_+$, so it can be spun to produce $B_0 = \text{Spin}^1(A_0) \subset \text{Spin}^1(\mathbb{R}^3_+) \approx \mathbb{R}^4$. Inside B_0 we will spin the Antoine construction twice to produce a collection of small copies of B_0 with useful π_1 features.

Name a homeomorphism $g : A_0 \times S^1 \to B_0$. Insert multiple solid tori $T_1, T_2, \ldots, T_{k(1)}$ in IntA_0, arranged in linked fashion just as in Figure 2.1, with each T_i small enough that diam $g(T_i \times s) < 1/4$ for all $s \in S^1$, and set $B_1 = g(\cup_i T_i \times S^1)$.

Equate each $T_i \times S^1$ with $[S^1 \times B^2 \times S^1]_i$. Interchanging the roles of the two S^1 factors, choose solid tori $T'_1, T'_2, \ldots, T'_{k(2)} \subset B^2 \times S^1$, again arranged in linked fashion just as in Figure 2.1, only now with each T'_j chosen small enough that diam $g([S^1 \times T'_j]_i) < 1/2$ for all i, j. Set $B_2 = \cup_i g(\cup_j [S^1 \times T'_j]_i)$.

Iterate the two-step process, producing a nested, decreasing sequence of compact ∂-manifolds $B_0, B_1, B_2, \ldots, B_{2k-1}, B_{2k}, \ldots$, where for each component C of each B_i there exist a homeomorphism $g_C : S^1 \times B^2 \times S^1 \to C$ and a collection of solid tori $\tau_1, \ldots, \tau_{k(i)}$ situated in $S^1 \times B^2$ according to the Antoine linking pattern of Figure 2.1, so $C \cap B_{i+1} = g_C(\cup_j \tau_j \times S^1)$. Moreover, repeating the methodology of decreasing sizes first in one S^1 direction, then the other, carry out these constructions in such a way that all components of B_{2k} have diameter at most $1/2^k$. The Cantor set of interest is $A = \cap_i B_i$.

Some additional terminology will streamline the analysis of $\pi_1(\mathbb{R}^4 \setminus A)$.

Definition. Let W be a ∂-manifold and $H \subset \mathbb{R}^2$ a compact, 2-dimensional ∂-manifold. A faithful map $f : H \dashrightarrow W$ is said to be *interior essential*, or, more simply, *I-essential* if there exists no map $F : H \to \partial W$ such that $F|\partial H = f|\partial H$; otherwise, f is *interior inessential*. Moreover, W is said to have the *interior inessential property* if every such faithful map $f : H \to W$ is interior inessential.

Remark. If a connected ∂-manifold W has the interior inessential property, clearly then ∂W also must be connected and $\pi_1(\partial W) \to \pi_1(W)$ must be one-to-one.

Lemma 4.7.2. *Suppose the n-manifold W contains ∂-manifolds W_0 and W_1, each of which is closed in W, such that W_0 is connected, $W = W_0 \cup W_1$, $W_0 \cap W_1 = \partial W_0 = \partial W_1$ and W_1 has the interior inessential property. Then $\pi_1(W \setminus W_1) \to \pi_1(W)$ is one-to-one.*

An appropriate, relevant model to have in mind would be the manifolds $W = \mathbb{R}^3 \setminus A_1, W_0 = \mathbb{R}^3 \setminus \text{Int } A_0$ and $W_1 = A_0 \setminus A_1$ from Figure 2.1.

Proof. Consider a map $f : I^2 \to W$ with $f(\partial I^2) \subset W \setminus W_1$. First focus on the case where W_1 is a PL subset of W, and assume f is in general position with respect to W_1. Then $f^{-1}(W_1)$ is a compact, 2-dimensional ∂-manifold in $\operatorname{Int} I^2$, and the interior inessential property promises a map $F : I^2 \to W_0$ such that the restrictions of F and f agree on $I^2 \smallsetminus f^{-1}(\operatorname{Int} W_1)$. Use a collar on $\partial W_1 = \partial W_0$ in W_0 to adjust F slightly by sliding the image off W_1.

The same process can be used in the general case, outside the PL situation, by covering $f^{-1}(W_1)$ with a compact ∂-manifold $H \subset \operatorname{Int} I^2$ that maps under f into a small collared neighborhood of W_1, and then by exploiting the crucial property to send all of such H into the collar on ∂W_1 without any change on $I^2 \smallsetminus \operatorname{Int} H$. □

Repeated application of the same argument yields:

Lemma 4.7.3. *Suppose $W_0, W_1, W_2, \ldots$ are ∂-manifolds and closed subsets of the n-manifold W such that W_0 is connected, $\cup_i W_i = W$, $W_i \cap W_j \neq \emptyset$ if and only if $|i-j| \leq 1$, $W_i \cap W_{i+1} \subset \partial W_i \cap \partial W_{i+1}$, $\partial W_0 \subset \partial W_1$, $\partial W_i = \partial W_i \cap (\partial W_{i-1} \cup \partial W_{i+1})$ $(i > 0)$, and $W_i \smallsetminus W_{i+1}$ has the interior inessential property. Then $\pi_1(W_0 \smallsetminus W_1) \to \pi_1(W)$ is one-to-one.*

Returning to the analysis of $\pi_1(\mathbb{R}^4 \smallsetminus A)$, we apply Lemma 4.7.3 with $W = \mathbb{R}^4 \smallsetminus A, W_0 = \mathbb{R}^4 \smallsetminus \operatorname{Int} B_0$ and $W_i = B_{i-1} \smallsetminus \operatorname{Int} B_i$ for $i > 0$. It follows from Exercise 2.1.2 that $A_0 \smallsetminus A_1$ has the interior inessential property. By a straightforward argument $(A_0 \smallsetminus A_1) \times S^1$ possesses the same property. In other words, $B_0 \smallsetminus B_1$ has it and, by construction, so does $B_i \smallsetminus B_{i+1}$. Note for the record that $\pi_1(\mathbb{R}^4 \smallsetminus B_0) \neq 1$, since

$$H_1(\mathbb{R}^4 \smallsetminus B_0) \cong H^3_c(\mathbb{R}^4, B_0) \cong H^2(B_0) \cong \mathbb{Z}.$$

By Lemma 4.7.3, $\pi_1(W_0 \smallsetminus W_1 = \mathbb{R}^4 \smallsetminus B_0)$ injects in $\pi_1(W = \mathbb{R}^4 \smallsetminus A)$, so the latter is non-trivial.

Remark. If X is the wild Cantor set in $\mathbb{R}^n$ of Example 4.7.1, then for every $\epsilon > 0$ there exists an ϵ-push ϕ of $(\mathbb{R}^n, X)$ such that $\phi(X) \cap X = \emptyset$. Cover X by a compact PL n-manifold S with small components, each homeomorphic to $S^1 \times \cdots \times S^1 \times B^2$, and push each component close to but just outside itself.

Question. Does every wild Cantor set in $\mathbb{R}^n$ admit an ϵ-push off itself?

Example 4.7.4. *There exists a wild k-cell γ^k in $\mathbb{R}^n$, $2 \leq k \leq n$ and $n \geq 4$, with non-simply connected complement that is locally flat modulo a Cantor set A flatly embedded in $\partial\gamma^k$.*

The construction of the Antoine 3-cell in $\mathbb{R}^3$ (see Example 2.1.7) serves as a model for producing a wild n-cell γ^n containing the wild Cantor set A of Example 4.7.1 as a flat subset of its boundary. Insist that γ^n avoids some loop J that is homotopically essential in $\mathbb{R}^n \setminus A$ to insure the non-simple connectedness of $\mathbb{R}^n \setminus \gamma^n$. The other γ^k arise as standardly embedded subsets of $\partial \gamma^n$.

Example 4.7.5. *There exists a wild k-sphere Σ^k in $\mathbb{R}^n$, $1 \leq k < n$ and $n \geq 4$, with non-simply connected complement that is locally flat modulo a Cantor set A flatly embedded in Σ^k.*

Historical Notes. The wild Cantor set of Example 4.7.1 was developed by W. A. Blankinship (1951).

For $n > 3$, McMillan (1978) produced a wild arc α in a PL n-manifold having no neighborhood that embeds in $\mathbb{R}^n$. D. G. Wright (1977) used α to prove that some arc in $\mathbb{R}^n$ cannot be pushed off itself. Wright's example contrasts with the unsolved question preceding Example 4.7.4 about pushing Cantor sets off themselves.

Exercises

4.7.1. If S is a manifold and W is a ∂-manifold with the interior inessential property, then $S \times W$ has the interior inessential property.

4.8. Even more wild embeddings

The crux of this section is a construction procedure for ramifying the wildness of certain Cantor sets, like the Antoine and Blankinship examples. The procedure turns out a Cantor set's worth of Cantor sets, each one wildly embedded just like its prototype; it gives rise to examples of embedded k-cells K in $\mathbb{R}^n$ $(2 < k < n)$ containing 2-cells D such that all disks D' in K sufficiently close to D are wildly embedded in $\mathbb{R}^n$. In contrast, Theorem 4.6.8 assures that 2-cells in $\mathbb{R}^n$ contain many tame arcs, but soon we will see that 3-cells need not contain many tame disks. This construction procedure will be revisited in §7.10 to delineate wild phenomena in codimension one.

Let X denote a Cantor set. A sequence $\mathscr{S} = \{\mathscr{Y}_i \mid i = 1, 2, \ldots\}$ is called an *abstract defining sequence for* X if (1) each $\mathscr{Y}_i$ is a finite set consisting of pairwise disjoint, nonvoid, compact subsets of X, (2) the union of the elements of each $\mathscr{Y}_i$ equals X, (3) $\mathscr{Y}_{i+1}$ refines $\mathscr{Y}_i$ for each i and (4) $d_i \to 0$ as $i \to \infty$, where d_i denotes the diameter of the largest element of $\mathscr{Y}_i$. Then we go on to call $\{\mathscr{Y}_i\}$ a *special abstract defining sequence for* X if (1) $\{\mathscr{Y}_i\}$ is a defining sequence for X, (2) $\mathscr{Y}_1$ has cardinality $k(0) > 1$ and every element of $\mathscr{Y}_1$ contains exactly $k(0)$ elements of $\mathscr{Y}_2$ and (3) for each positive integer m there is an integer $k(m) > 1$ such that every element of $\mathscr{Y}_{2m}$ contains exactly $k(m)$ elements of $\mathscr{Y}_{2m+1}$ and, likewise, every element of $\mathscr{Y}_{2m+1}$ contains exactly $k(m)$ elements of $\mathscr{Y}_{2m+2}$.

Given a special abstract defining sequence $\mathscr{S} = \{\mathscr{Y}_i\}$ for X, we say that $A \subset X$ is *admissible with respect to* $\mathscr{S}$ if A is nonvoid and compact and, whenever $j > 0$ is an odd integer, $Y \in \mathscr{Y}_j$ meets A and $Y' \in \mathscr{Y}_{j+1}$ satisfies $Y' \subset Y$, then $Y' \cap A \neq \emptyset$. In such an X the rich collection of admissible subsets submits to a notable mixing feature.

Lemma 4.8.1 (Mixing). *Let X be a Cantor set equipped with a special defining sequence $\mathscr{S} = \{\mathscr{Y}_i\}$. Then there exists a mixing homeomorphism $\tau : X \to X$ such that, for any two admissible subsets A, A' of X with respect to $\mathscr{S}$, $\tau(A) \cap A' \neq \emptyset$.*

Proof. Label the elements of $\mathscr{Y}_1$ as $Y_1, Y_2, \ldots, Y_{k(0)}$, and label those of $\mathscr{Y}_2$ as $Y_{j(1),j(2)}$, where $1 \leq j(i) \leq k(0)$, and where each $Y_{j(1),j(2)}$ is a subset of $Y_{j(1)}$. Generally, label the elements of $\mathscr{Y}_i$ as $Y_{j(1),j(2),\ldots,j(i)}$, where $1 \leq j(i) \leq k(m)$ when $i \in \{2m+1, 2m\}$, with the understanding that $Y_{j(1),j(2),\ldots,j(i)} \subset Y_{j(1),j(2),\ldots,j(i-1)}$. Then each $x \in X$ can be uniquely labelled as $x = \langle j(1), j(2), \ldots, j(i), \ldots \rangle$ by the rule

$$x = \langle j(1), j(2), \ldots, j(i), \ldots \rangle \text{ if and only if } \{x\} = \cap_{i=1}^{\infty} Y_{j(1),j(2),\ldots,j(i)}.$$

Using this representation, define the mixing homeomorphism $\tau : X \to X$ as the infinite transpose sending $x = \langle j(1), j(2), \ldots, j(2m-1), j(2m), \ldots \rangle$ to $\langle j(2), j(1), \ldots, j(2m), j(2m-1), \ldots \rangle$. A salient point is that τ simply permutes the elements of $\mathscr{Y}_{2m}$ for each $m \geq 1$, from which it follows routinely that the function τ is a homeomorphism.

Now consider admissible subsets A, A' of X. We must show that $\tau(A)$ meets A'. It suffices to prove that for each $m \geq 1$ some $Y \in \mathscr{Y}_{2m}$ meets both $\tau(A)$ and A'. This is done by induction. To get started, note there exist $r(1), s(1) \in \{1, \ldots, k(0)\}$ such that $A \cap Y_{r(1)} \neq \emptyset \neq A' \cap Y_{s(1)}$. By definition of admissibility,

$$A \cap Y_{r(1),s(1)} \neq \emptyset \neq A' \cap Y_{s(1),r(1)},$$

and $\tau(Y_{r(1),s(1)}) = Y_{s(1),r(1)}$ by the definition of τ. Essentially the same argument disposes of the inductive step. $\square$

Turning to Cantor sets embedded in manifolds, let $\mathscr{S} = \{\mathscr{M}_i\}$ be a sequence where each $\mathscr{M}_i$ consists of finitely many compact connected, n-dimensional ∂-manifolds PL embedded in S^n, no two of which intersect, and let $|\mathscr{M}_i| = \cup\{M \mid M \in \mathscr{M}_i\}$. Such a sequence $\mathscr{S}$ is called a *geometric defining sequence for a Cantor set* $X \subset S^n$ if $|\mathscr{M}_{i+1}| \subset \text{Int}\,|\mathscr{M}_i|$, $\cap_i |\mathscr{M}_i| = X$, and each element of each $\mathscr{M}_i$ contains a point of X. Associated with any geometric defining sequence $\{\mathscr{M}_i\}$ for X is an abstract defining sequence $\{\mathscr{Y}_i\}$, where

$$\mathscr{Y}_i = \{M \cap X \mid M \in \mathscr{M}_i\}.$$

Furthermore, we call such a sequence $\{\mathscr{M}_i\}$ a *special (geometric) defining sequence* if (1) for $i \geq 1$ each $M \in \mathscr{M}_i$ is the product of some $(n-2)$-manifold with B^2 and (2) the abstract defining sequence $\{\mathscr{Y}_i\}$ for X is special. We will say that a special geometric defining sequence $\mathscr{S} = \{\mathscr{M}_i\}$ for a Cantor set has the *interior inessential property* if $|\mathscr{M}_i| \smallsetminus |\mathscr{M}_{i+1}|$ has that property for each integer $i > 0$ and that $\mathscr{S}$ has the *strong interior inessential property* if, in addition, for each $m \geq 0$, each component $M \in \mathscr{M}_{2m+1}$ and each component $M' \in \mathscr{M}_{2m+2}$ with $M' \subset M$, $M \smallsetminus M'$ has the interior inessential property.

Lemma 4.8.2. *Let W', W'' be ∂-manifolds such that $W' \cap W'' = \partial W''$ is a union of components of $\partial W'$, and suppose $W' \smallsetminus \partial W''$ has the interior inessential property. Then each I-essential map $f : H \to W' \cup W''$ defined on a disk with holes H satisfies not only that $f(H) \cap W'' \neq \emptyset$ but also, if $f^{-1}(W'')$ is a ∂-manifold, then for some component H'' of $f^{-1}(W'')$, $f|H'' : H'' \to W''$ is I-essential.*

The argument is straightforward. The lemma immediately applies to each component C of each B_i in the defining sequence for the Blankinship Cantor set to assure that for each I-essential map $f : H \to C$, $f(H) \cap B_{i+1} \neq \emptyset$.

Corollary 4.8.3. *If $\mathscr{S} = \{\mathscr{M}_i\}$ is a special defining sequence for a Cantor set X and $\mathscr{S}$ has the interior inessential property, then for each pair of positive integers $\{i, k\}$, $|\mathscr{M}_i| \smallsetminus |\mathscr{M}_{i+k}|$ has the interior inessential property.*

In Cantor sets determined by a class of special geometric defining sequences, admissible subsets arise for geometric reasons: all singular disks bounded by certain loops contain admissible subsets.

Proposition 4.8.4. *Suppose $\mathscr{S} = \{\mathscr{M}_i\}$ is a special geometric defining sequence for a Cantor set $X \subset S^n$, where $\mathscr{S}$ has the strong interior inessential property, and suppose $f : \partial I^2 \to S^n \smallsetminus |\mathscr{M}_1|$ is a homotopically essential map. Then for each map $F : I^2 \to S^n$ extending f, $F(I^2)$ contains an admissible subset of X.*

Proof. Given an extension F of f, adjust F slightly, without introducing any new preimages of points of X, to put $F(I^2)$ in general position with respect to the boundary of each $|\mathscr{M}_i|$, thereby assuring that all components of $F^{-1}(|\mathscr{M}_i|)$ are disks with holes. Then for $i = 1, 2, \ldots$ let A_i denote the union of all components H of $F^{-1}(|\mathscr{M}_i|)$ for which $F|H$ is I-essential. It is a direct consequence of Lemma 4.8.2 and the relevant definitions that $F(\cap_i A_i)$ is an admissible subset of X. □

Proposition 4.8.5. *For $n \geq 3$ there exists a Cantor set X_n in S^n having a special geometric defining sequence $\mathscr{S} = \{\mathscr{M}_i\}$ with the strong interior inessential property.*

Proof. Recall the construction of Antoine's necklace in S^3 and of Blankinship's Cantor set in S^n, $n > 3$. Each is determined by means of a special geometric defining sequence $\{\mathscr{T}_i \mid i = 0, 1, 2, \ldots\}$ with the interior inessential property; moreover, for all $i \geq 0$ there exists an integer $k(i) > 1$ such that each $T \in \mathscr{T}_i$ contains exactly $k(i)$ elements of $\mathscr{T}_{i+1}$ and all such T are the product of the $(n-2)$-dimensional torus $S = S^1 \times \cdots \times S^1$ with B^2. Assume that the initial stage $\mathscr{T}_0$ consists of a single copy of $S \times B^2$.

First, define $\mathscr{M}_1 = \mathscr{T}_1$. For each $T \in \mathscr{T}_1 = \mathscr{M}_1$ specify a PL homeomorphism $\Phi_T : S \times B^2 \to T$, choose $k(0)$ pairwise disjoint disks $B_1, \ldots, B_{k(0)}$ in $\operatorname{Int} B^2$, and let

$$\mathscr{M}_2 = \{\Phi_T(S \times B_i) \mid T \in \mathscr{M}_1 \text{ and } 1 \leq i \leq k(0)\}.$$

Thus, each $M \in \mathscr{M}_2$ is a slightly shrunken version of some $T \in \mathscr{T}_1 = \mathscr{M}_1$; furthermore, $T \smallsetminus M$ retracts to ∂T, so $T \smallsetminus M$ has the interior inessential property. Name a PL homeomorphism $\phi_i : B^2 \to B_i$ for each i. Choose an index $t(1) > 1$ such that for each $T \in \mathscr{T}_1$ and each $T' \in \mathscr{T}_{t(1)}$ contained in T, $\operatorname{diam} \phi_T(\operatorname{Id} \times \phi_i)(T') < 1/2$. Set

$$\mathscr{M}_3 = \{\Phi_T(\operatorname{Id} \times \phi_i)(T') \mid T \in \mathscr{T}_1, T' \in \mathscr{T}_{t(1)}, T' \subset T \text{ and } 1 \leq i \leq k(0)\}.$$

Let $k(1)$ be the number of elements from $\mathscr{M}_3$ contained in any $M \in \mathscr{M}_2$. Again, for each $T \in \mathscr{M}_3$ specify a PL homeomorphism $\Phi_T : S \times B^2 \to T$, choose $k(1)$ pairwise disjoint disks $B_1, \ldots, B_{k(1)}$ in $\operatorname{Int} B^2$ and set

$$\mathscr{M}_4 = \{\Phi_T(S \times B_i) \mid T \in \mathscr{M}_3 \text{ and } 1 \leq i \leq k(1)\}.$$

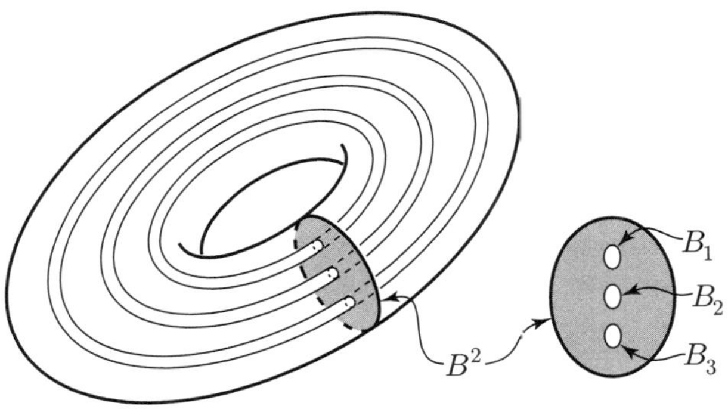

Figure 4.4. The ramification step inside an odd-numbered stage

Continue in this manner, determining a sequence of integers $1 = t(0) < t(1) < t(2) < \cdots$ such that for each $M \in \mathscr{M}_{2m}$, $m \geq 1$, the pair $(M, M \cap |\mathscr{M}_{2m+1}|)$ is homeomorphic to $(T, T \cap |\mathscr{T}_{t(m)}|)$ for some (all) $T \in \mathscr{T}_{t(m-1)}$. This causes $|\mathscr{M}_{2m}| \smallsetminus |\mathscr{M}_{2m+1}|$ to have the interior inessential property, by Corollary 4.8.3. Choose the $t(m)$ to assure, in addition, that diameters of elements of $\mathscr{M}_{2m+1}$ tend to 0 as $m \to \infty$. Each $M \cong S \times B^2 \in \mathscr{M}_{2m+1}$ then should contain a fixed number $k(m)$ of slightly shrunken parallel copies equivalent to $S \times B_i$ in $\mathscr{M}_{2m+2}$, where $k(m)$ also equals the number of components of $T \cap |\mathscr{T}_{t(m)}|$ for (all) $T \in \mathscr{T}_{t(m-1)}$. This guarantees, given $m \geq 0$, $M \in \mathscr{M}_{2m+1}$ and $M' \in \mathscr{M}_{2m+2}$ with $M' \subset M$, that $M \smallsetminus M'$ has the interior inessential property, since it retracts to ∂M. Of course, this means that $|\mathscr{M}_{2m+1}| \smallsetminus |\mathscr{M}_{2m+2}|$ has the interior inessential property for all $m \geq 0$. Moreover, $\mathscr{S} = \{\mathscr{M}_i\}$ is a special defining sequence for a Cantor set $X_n = \cap_m \mathscr{M}_m \subset S^n$, and $\mathscr{S}$ has the strong interior inessential property. □

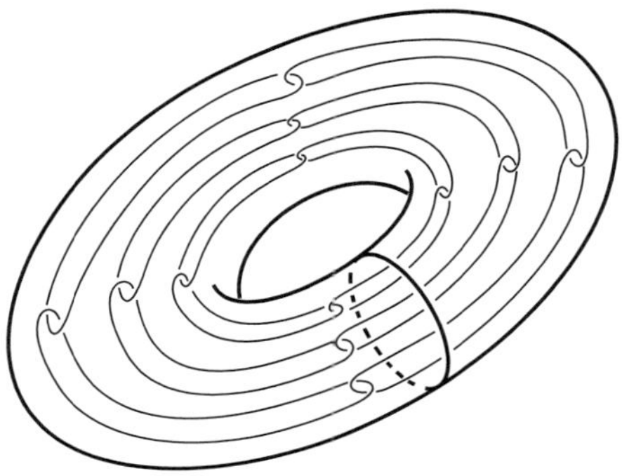

Figure 4.5. One stage beyond the ramification step

Subsequently we refer to this replication at even stages of multiple shrunken copies of elements from some original stage as the *ramification step.*

Corollary 4.8.6. *Let $X_n \subset S^n$ denote the Cantor set of Proposition 4.8.5. There exists a loop $f : \partial I^2 \to S^n \smallsetminus |\mathscr{M}_1|$ for which the image of every extension $F : I^2 \to S^n$ contains a subset of X_n admissible with respect to $\mathscr{S}$.*

Proof. Here f can be any homotopically nontrivial loop in $S^n \smallsetminus |\mathscr{M}_1|$; loops that homologically link $|\mathscr{M}_1|$ do exist (by Alexander duality, as in the proof of Lemma 4.7.3). Proposition 4.8.4 attests that the image of any singular disk bounded by f contains an admissible subset of X_n. □

Definition. Two Cantor sets X and X' equipped with special abstract or geometric defining sequences $\{\mathscr{M}_i\}$ and $\{\mathscr{M}'_i\}$, respectively, are *compatible* if for each $m \geq 0$ the number $k(m)$ of elements from $\mathscr{M}_{j+1}$ contained in any $M \in \mathscr{M}_j$ equals the number $k'(m)$ of those from $\mathscr{M}'_{j+1}$ in any $M' \in \mathscr{M}'_j$, for $j = 2m, 2m+1$. In this setting one should regard both $\mathscr{M}_0$ and $\mathscr{M}'_0$ as consisting of a single ∂-manifold.

Remark. If the Cantor sets X and X' are equipped with compatible special abstract or geometric defining sequences, Mixing Lemma 4.8.1 clearly applies to provide a homeomorphism $\tau : X \to X'$ mixing their admissible subsets.

Lemma 4.8.7. *For $n \geq 4$ there exist Cantor sets $X_{n-1} \subset S^{n-1}$ and $X_n \subset S^n$ equipped with compatible special geometric defining sequences, each with the strong interior inessential property.*

Proof. Given a Cantor set $Z \subset S^k$ having a special geometric defining sequence $\{\mathscr{M}_i\}$, we describe how to add exactly one element of $\mathscr{M}_{i+1}$ inside a preassigned element M' of $\mathscr{M}_i$ without changing $\mathscr{M}_j$, $j \leq i$. Choose $M \in \mathscr{M}_{i+1}$ with $M \subset M'$. By hypothesis M is topologically $S \times B^2$, where S is a compact $(k-2)$-manifold. Select disjoint disks B_1 and B_2 in $\operatorname{Int} B^2$ and define

$$\mathscr{M}^*_{i+1} = \{M^* \in \mathscr{M}_{i+1} \mid M^* \neq M\} \cup \{S \times B_1\} \cup \{S \times B_2\}.$$

Name a first-coordinate preserving homeomorphism $h_e : S \times B^2 \to S \times B_e$ $(e = 1, 2)$. Then, for $j > i + 1$, define

$$\mathscr{M}^*_j = \{M^* \in \mathscr{M}_j \mid M^* \cap M = \emptyset\} \cup \{h_e(M^*) \mid M^* \in \mathscr{M}_j,\, M^* \subset M, \text{ and } e = 1, 2\},$$

and for $j \leq i$ define $\mathscr{M}^*_j = \mathscr{M}_j$. Then $\{\mathscr{M}^*_i\}$ is a special geometric defining sequence for another Cantor set $Z^* \subset S^k$. We will refer to this procedure as the *supplementation step.* Although qualitatively similar to the replication step, it is directed toward a different purpose.

Note that diameters of elements of $\mathscr{M}^*_{i+1}$ are bounded by those of $\mathscr{M}_{i+1}$. Note also that if $|\mathscr{M}_j| \smallsetminus |\mathscr{M}_{j+1}|$ has the interior inessential property for each $j \geq 1$, then the same holds for the modified defining sequence $\{\mathscr{M}^*_j\}$; all cases except for $j = i$ are completely obvious, and there

$$|\mathscr{M}^*_i| \smallsetminus |\mathscr{M}^*_{i+1}| = |\mathscr{M}_i| \smallsetminus |\mathscr{M}^*_{i+1}|$$

is boundary-preserving equivalent to a subset of $|\mathscr{M}_i| \smallsetminus |\mathscr{M}_{i+1}|$, from which the property readily follows.

The construction itself repeats that of Lemma 4.8.5, with the addition of multiple applications of the supplementation step to make the elements at any given even stage (including the initial stage 0) contain the same number of elements at the next stage. Start with special geometric defining sequences $\{\mathscr{M}_i\}$ and $\{\mathscr{N}_i\}$ for Cantor sets in S^{n-1} and S^n, respectively, with the interior essential property. Assume that the diameter of each element of $\mathscr{M}_1 \cup \mathscr{N}_1$ is less than 1, and let $k(0)$ be the maximum cardinality of $\mathscr{M}_1, \mathscr{N}_1$. Add elements to the smaller of $\mathscr{M}_1, \mathscr{N}_1$, using the supplementation procedure, so both have cardinality $k(0)$, and let $\mathscr{R}_1$ and $\mathscr{S}_1$ denote the resulting first stages. Now do a ramification step, determining $\mathscr{R}_2$ and $\mathscr{S}_2$, so that each $R \in \mathscr{R}_1$ contains exactly $k(0)$ elements of $\mathscr{R}_2$, each a slightly shrunken copy of R, and the same for $\mathscr{S}_1, \mathscr{S}_2$. Ramification further modifies $\{\mathscr{M}_i^*\}$ and $\{\mathscr{N}_i^*\}$, of course. To assure the eventual outputs are Cantor sets, choose $t(1) > 2$ such that each element of the further modified $\mathscr{M}_{t(1)}^* \cup \mathscr{N}_{t(1)}^*$ has diameter less than $1/2$. Find $k(1) > 2$ such that elements of $\mathscr{R}_2$, $\mathscr{S}_2$ contain at most $k(1)$ elements of $\mathscr{M}_{t(1)}^*$, $\mathscr{N}_{t(1)}^*$, respectively. Add elements as needed via the supplementation step to determine $\mathscr{R}_3, \mathscr{S}_3$ such that each $R \in \mathscr{R}_2$ contains exactly $k(1)$ elements of $\mathscr{R}_3$, and similarly for $S \in \mathscr{S}_2$. Ramify to produce $\mathscr{R}_4$ and $\mathscr{S}_4$. Generally, at odd stages, go deep into the modified defining sequences to pick out a stage where the components have small diameter, and supplement to obtain uniformity in the number of elements at that stage of the two sequences in any element of the preceding stage; at even stages, ramify. The resulting $\{\mathscr{R}_i\}$ and $\{\mathscr{S}_i\}$ are compatible, special defining sequences for Cantor sets in S^{n-1} and S^n, each satisfying the strong interior inessential property. □

Example 4.8.8. *For $3 \leq k < n$ there exist a k-cell $K \subset \mathbb{R}^n$ and a 2-cell $D \subset K$ such that, for any homeomorphism $\theta : K \to K$ sufficiently close to $\mathrm{Id} : K \to K$, $\theta(D)$ is wildly embedded in $\mathbb{R}^n$.*

Using Lemma 4.8.7 we find Cantor sets X and X' in $\operatorname{Int} B^k$ and S^n, respectively, with compatible special defining sequences, each of which possesses the strong interior inessential property. Applying Mixing Lemma 4.8.1 we obtain a homeomorphism $\tau : X \to X'$ mixing their admissible subsets. Let γ denote a loop outside the first stage, $\mathscr{T}_1$, of the defining sequence for X', with γ homotopically essential in $S^n \setminus |\mathscr{T}_1|$. There exists an embedding $e : B^k \to \mathbb{R}^n \setminus \gamma$ such that $e|X = \tau$ and $e(B^k)$ is locally flat modulo $e(X) = X'$ (this possibly could have been established, using much greater care with the methods for Exercise 2.1.5, but we use other methods to flesh out this claim in the conclusion of this section, culminating in Corollary 4.8.12). It is possible to position X in K so that the boundary of some 2-cell $D \subset K$ misses the first stage, $\mathscr{S}_1$, of the special defining sequence for X and is homotopically essential in $K \setminus |\mathscr{S}_1|$.

Under any sufficiently small adjustment $\theta : K \to K$, $e\theta(D)$ will be wild; it cannot have simply connected complement. With control on θ, $\theta(\partial D)$ will still be homotopically essential in $K \setminus |\mathscr{T}_1|$, so $\theta(D)$ will contain an admissible subset A of X, by Proposition 4.8.4. Similarly, the image of any contraction of the loop γ in S^n must contain an admissible subset A' of X'. The mixing properties of $e|X = \tau : X \to X'$ assure that $e(A) \cap A' \neq \emptyset$; as a result, $\pi_1(S^n \setminus e\theta(D)) \neq \{1\}$, so $e\theta(D)$ is wild. This completes the description of Example 4.8.8.

Lemma 4.8.9. *Let C be a compact 0-dimensional subset of the unit interval I and let $\lambda : I^2 \to B^n$, $n \geq 5$, be an embedding such that $\lambda(I^2) \cap \partial B^n = \lambda(I \times \{0\})$ and $\lambda(I^2)$ is 1-LCC in $\operatorname{Int} B^n$. Then the decomposition G of B^n into the arcs $\{\lambda(c \times I) \mid c \in C\}$ and the singletons from $B^n \setminus \lambda(C \times I)$ is shrinkable.*

Proof. Consider any neighborhood W of $\lambda(C \times I) \subset B^n$ and any $\epsilon > 0$. Determine an open set $U \supset \lambda(C \times \{0\})$ having components of diameter less than ϵ. Regard W as the preimage of a small open neighborhood of the image of $\lambda(C \times I)$ in B^n/G. Hence, to confirm shrinkability of G, it suffices to obtain an engulfing homeomorphism $\Phi : B^n \to B^n$, fixed on $\partial B^n \cup (B^n \setminus W)$, such that $U \supset \Phi\lambda(C \times I)$. Cover C by a compact 1-manifold with boundary K in I such that $\lambda(K \times I) \subset W$ and $\lambda(K \times \{0\}) \subset U$. As $\lambda(K \times (0, 1])$ is 1-LCC in $\operatorname{Int} B^n$, it is locally flat there. Consequently, one can push $\lambda(K \times I)$ from the $\lambda(K \times \{1\})$ end toward the $\lambda(K \times \{0\})$ end and into $W \cap U$, while keeping points of $\partial B^n \cup (B^n \setminus W)$ fixed throughout. □

Lemma 4.8.10. *Let X denote a Cantor set in an m-cell K, $1 \leq m$. Then for $n > m$ there exists an embedding $e : K \to B^n$ such that $e(K) \cap \partial B^n = e(X)$ is a flat Cantor set in ∂B^n and $e(K)$ is locally flat in $\operatorname{Int} B^n$ at each point of $e(K \setminus X)$.*

Proof. When $n \geq m + 2$, start with a standard (flat) embedding $e : K \to \partial B^n$ and adjust, pushing the image of $K \setminus X$ into $\operatorname{Int} B^n$, keeping $e(X)$ fixed. Then $e(X)$ is flat in ∂B^n by the Klee trick (Corollary 2.5.3).

The case $n = m + 1$ requires more care, since the Klee trick does not apply. For simplicity we treat only the cases $n \geq 5$; the other low-dimensional cases require extra effort. Let B' be another round n-cell in $\operatorname{Int} B^n$, centered at the origin, and let $\theta : K \to \partial B'$ be an embedding. Find an embedding $\lambda : I^2 \to B^n \setminus \operatorname{Int} B'$ such that $\lambda(I^2) \cap \partial B^n = \lambda(I \times \{0\})$, $\lambda(I^2) \cap \partial B' = \lambda(I \times \{1\})$, $\lambda(I \times \{0\})$ is flat in ∂B^n and $\lambda(C \times \{1\}) = \theta(X)$, where $C \subset I$ denotes the standard Cantor set in I. By taking a general position approximation rel $\lambda(I \times \{1\})$, if necessary, we may assume that λ is 1-LCC in B^n at points of $I \times [0, 1)$. Then λ is 1-LCC at all points of I^2, by Corollary 4.6.4. According to Lemma 4.8.9 the decomposition G of B^n into points and the arcs $\lambda(c \times I)$, $c \in C$, is shrinkable, so there is a map $\mu : B^n \to B^n$ realizing G. Note that $\mu : \partial B^n \to \partial B^n$ is a homeomorphism, $\mu|\theta(K)$ is 1-1 and $\mu\theta(K)$ is locally flat mod $\mu\lambda(C \times \{1\})$. The embedding $e = \mu\theta : K \to B^n$ has the required properties, as $e(X) = \mu\lambda(C \times \{1\}) = \mu\lambda(C \times \{0\})$ is flat in ∂B^n. □

Proposition 4.8.11. *Let X be a Cantor set tamely embedded in the boundary of an n-cell B, X' a Cantor set in a connected n-manifold M, and $h : X \to X'$ a homeomorphism. Then there exists an embedding $e : B \to M$*

such that $e|X = h$*; moreover, given any* $e' : B \to M$ *with* $\epsilon'(X) = X'$*, there exists such an embedding* e *with* $e(B) = e'(B)$ *and* $e|X = h$*.*

Proof. We consider only $n \geq 6$; the result is true in the other dimensions.

Methods from Chapter 2 (Exercise 2.1.5) promise an embedding $e' : B \to M$ such that $e'(X) = X'$. Then $h^{-1}e'|X : X \to X$ is a homeomorphism. Corollary 4.4.3 gives an ambient isotopy $\Psi_t : \partial B \to \partial B$ such that $\Psi_0 = \mathrm{Id}_{\partial B}$ and $\Psi_1|X = h^{-1}e'|X$. Determine a collar $\Lambda : \partial B \times [0,1] \to B$ on ∂B in B, parameterized so that $\Lambda(\partial B \times [0,1)) \subset \operatorname{Int} B$ and $\Lambda_1 = \mathrm{Id}_{\partial B}$. Define $\psi : B \to B$ as the identity outside the collar and essentially as Ψ_t on $\Lambda(\partial B \times t)$. Then $e = e'\psi^{-1}$ has the desired properties. □

Corollary 4.8.12. *Let* X *be a Cantor set in an* m*-cell* K*,* X' *a Cantor set in a connected* n*-manifold* M*, where* $n > m$*, and* $h : X \to X'$ *a homeomorphism. Then there exists an embedding* $e : K \to M$ *such that* $e|X = h$ *and* $e(K)$ *is locally flat mod* $e(X)$*.*

Proof. Use Lemma 4.8.10 to embed K nicely in B^n, with the image of X tame in ∂B^n, and apply Proposition 4.8.11. □

Historical Notes. The ramification process was introduced in (Daverman, 1973c) in order to describe wildly embedded cells containing few tame disks. A later construction (Daverman, 1975) gave wildly embedded cells containing no tame disks whatsoever.

Exercises

4.8.1. Let X denote a Cantor set. For $n \geq 3$ there exists an embedding $\lambda : X \times X \to \mathbb{R}^n$ such that $\lambda(X \times \{x\})$ is wildly embedded for each $x \in X$; moreover, given $x, x' \in X$ there exists a homeomorphism $\theta_{x,x'} : \mathbb{R}^n \to \mathbb{R}^n$ such that $\theta_{x,x'}\lambda(X \times \{x\}) = \lambda(X \times \{x'\})$.

Chapter 5

Codimension-three Embeddings

Next, moving beyond the trivial range, we launch an investigation of embeddings for which the codimension is relatively small. There are two ways in which the trivial range is special: first, continuous maps can be approximated by embeddings and, second, tame embeddings unknot—specifically, if X is a k-dimensional compactum and M is an n-dimensional PL manifold, then any map of X into M is homotopic to an embedding provided $n \geq 2k+1$ and any two homotopic 1-LCC embeddings of X in M are ambient isotopic provided $n \geq 2k+2$. The chapter contains examples which show that the dimension restrictions in these results are sharp and that neither of them generalizes completely to lower codimensions.

The basic trivial-range theorems regarding embeddings of polyhedra do generalize to codimension three provided suitable hypotheses are added. In most cases it is necessary to assume some degree of connectivity before a trivial-range theorem can be promoted to codimension three. The general pattern of the proofs is that a straightforward attempt to generalize the trivial-range proof runs into an obstruction in the form of a singular set for a certain map. The connectivity hypotheses are used to replace the singular set with a different one. This new singular set has lower dimension than the original provided the codimension of the embedding is at least three; an inductive argument eliminates the singular set altogether.

The chapter also includes proofs that it is possible to tame all 1-LCC codimension-three embeddings of polyhedra and that it is possible to approximate any topological embedding of a codimension-three compactum with 1-LCC embeddings. These last two results are completely general and require no special hypotheses.

5.1. Constructing PL embeddings of polyhedra

The first question considered is that of existence: under what conditions is a map homotopic to an embedding? This section contains some answers for maps of polyhedra and the next section takes up the problem for maps of manifolds. We begin with an example which shows that it is not possible to improve the dimension restriction $n \geq 2k + 1$ in the classical embedding theorem.

Example 5.1.1. *There is no topological embedding of K^k, the k-skeleton of the $(2k + 2)$-simplex, into $\mathbb{R}^{2k}$.*

The statement of this example is widely known, but the details of the proof are not easily found in the literature. We will include, therefore, a complete exposition of the proof. It requires some new terminology and a few preliminary lemmas.

Definition. A compact metric space X is an *absolute self-intersector in* $\mathbb{R}^m$ provided there exists no continuous function f from cX, the cone over X, to $\mathbb{R}^m$ with $f|X$ one-to-one and $f^{-1}f(X) = X$.

If a compact metric space X embeds in $\mathbb{R}^m$, then clearly cX embeds in $\mathbb{R}^{m+1}$; in other words, no absolute self-intersector in $\mathbb{R}^{m+1}$ can be embedded in $\mathbb{R}^m$. We will show that the K of Example 5.1.1 cannot be embedded in $\mathbb{R}^{2k}$ by showing that K is an absolute self-intersector in $\mathbb{R}^{2k+1}$. In fact we will establish the following stronger result: *For each map $f : cK \to \mathbb{R}^{2k+1}$ there are disjoint simplices $\sigma, \tau \subset K$ and points $a \in \sigma$, $b \in c\tau$ such that $f(a) = b$.*

Define $L(K) \subset cK \times cK$ to be the collection of all ordered pairs $\langle a, b\rangle$ such that at least one of a and b lies in K and there exist disjoint simplices σ and τ in K such that $a \in c\sigma$ and $b \in c\tau$. Two points in $L(K)$ that differ only in the order of their coordinates are said to *symmetrically situated in $L(K)$*. Given $f : cK \to \mathbb{R}^{2k+1}$, define an associated map $\Psi_f : L(K) \to \mathbb{R}^{2k+1}$ by $\Psi_f(a, b) = f(a) - f(b)$. The following lemma is obvious.

Lemma 5.1.2. *If the origin $\mathbf{0}$ lies in $\Psi_f(L(K))$ for every $f \in C(cK, \mathbb{R}^{2k+1})$, then K is an absolute self-intersector in $\mathbb{R}^{2k+1}$.*

Before we can take the next step in the proof of Example 5.1.1, we need to specify K more precisely. Let T^{2k+2} be a $(2k+2)$-simplex that is the convex hull of vertices $\{v_0, v_1, \ldots, v_{2k+2}\}$ in $\mathbb{R}^{2k+2}$. The vertices $\{v_0, v_1, \ldots, v_{2k+2}\}$ can be thought of as vectors and chosen to have the following properties.

(1) $\sum_{i=0}^{2k+2} v_i = \mathbf{0}$ (so the origin $\mathbf{0}$ is the centroid of T), and

(2) no proper subset of the vertices of T^{2k+2} is linearly dependent.

Observe that conditions (1) and (2) together imply that $\sum_{i=0}^{2k+2} \lambda_i v_i = \mathbf{0}$ if and only if $\lambda_i =$ constant.

The complex K is defined to be the k-dimensional skeleton of T^{2k+2}; i.e., K is the union of all faces of T^{2k+2} that have dimension $\leq k$. The *antipodal map* is the map $\alpha : \mathbb{R}^{2k+2} \to \mathbb{R}^{2k+2}$ defined by $\alpha(x) = -x$. Two points $x, y \in \mathbb{R}^{2k+2}$ are said to be *antipodal points* if $y = \alpha(x)$.

Define $U(K)$ to be the complex consisting of all simplices of the form $\sigma * \alpha(\tau)$, where σ and τ are disjoint simplices in K, $\alpha(\tau)$ is the image of τ under the antipodal map, and "$*$" denotes join. Since the vertices of $\sigma \cup \tau$ are linearly independent, the vertices of $\sigma \cup \alpha(\tau)$ are also linearly independent. Consequently each simplex $\sigma * \alpha(\tau)$ of $U(K)$ is realized geometrically as the convex hull of $\sigma \cup \alpha(\tau)$ and $\mathbf{0} \notin \sigma * \alpha(\tau)$. Furthermore, each point of x in $\sigma * \alpha(\tau)$ satisfies the following condition.

($*$) x can be written as a linear combination $x = \sum_{i=0}^{2k+2} \lambda_i v_i$ in which at most $k+1$ of the coefficients $\{\lambda_i\}$ are positive, at most $k+1$ of the coefficients $\{\lambda_i\}$ are negative, and $\sum_{i=0}^{2k+2} |\lambda_i| = 1$.

Lemma 5.1.3. *Each point $x \in \mathbb{R}^{2k+2}$ can have at most one representation of the form ($*$).*

Proof. Suppose $\sum_{i=0}^{2k+2} \lambda_i v_i = \sum_{i=0}^{2k+2} \mu_i v_i$ are two representations satisfying condition ($*$). Reorder the vertices so that $\lambda_0 \geq \lambda_1 \geq \cdots \geq \lambda_{2k+2}$. As observed above, the fact that $\sum_{i=0}^{2k+2} (\lambda_i - \mu_i) v_i = \mathbf{0}$ implies that there is a constant c such that $\lambda_i - \mu_i = c$. Thus the $\{\mu_i\}$ also form a decreasing sequence. Combining this fact with condition ($*$) yields $\mu_{k+1} = \lambda_{k+1} = 0$. Since $\{v_0, v_1, \ldots, v_{2k+2}\} \setminus \{v_{k+1}\}$ is a linearly independent set, we can conclude that $\mu_i = \lambda_i$ for every i. □

Lemma 5.1.3 implies that the entire complex $U(K)$ is realized geometrically as the set of points in $\mathbb{R}^{2k+2}$ that satisfy ($*$). Note that $\alpha(\sigma * \alpha(\tau)) = \alpha(\sigma) * \tau \subset U(K)$, so α restricts to an involution of $U(K)$.

Lemma 5.1.4. *$U(K)$ is homeomorphic to S^{2k+1} via a map $\theta : U(K) \to S^{2k+1}$ that preserves antipodal points.*

Proof. Since $\mathbf{0} \notin U(K)$, radial projection from the origin in $\mathbb{R}^{2k+2}$ provides a continuous map from $U(K)$ to S^{2k+1} that preserves antipodal points. We will prove that this map is a homeomorphism by showing that each ray from the origin intersects $U(K)$ in exactly one point.

Let u be a unit vector in $\mathbb{R}^{2k+2}$. The vectors $\{v_i\}$ span $\mathbb{R}^{2k+2}$, so $u = \sum_{i=0}^{2k+2} \lambda_i v_i$ for some $\{\lambda_i\}$. Reorder the vertices so that $\lambda_0 \geq \lambda_1 \geq \cdots \geq \lambda_{2k+2}$. Since $\sum_{i=0}^{2k+2} v_i = \mathbf{0}$, we can write

$$u = \sum_{i=0}^{2k+2} \lambda_i v_i - \lambda_{k+1} \sum_{i=0}^{2k+2} v_i = \sum_{i=0}^{2k+2} (\lambda_i - \lambda_{k+1}) v_i.$$

Define $\nu = \sum_{j=0}^{2k+2} |\lambda_j - \lambda_{k+1}|$ and define $\mu_i = (\lambda_i - \lambda_{k+1})/\nu$ for each i. Then $(1/\nu)u = \sum_{i=0}^{2k+2} \mu_i v_i$ satisfies condition $(*)$, so the ray from the origin through u intersects $U(K)$ at $(1/\nu)u$.

The proof that the ray from the origin through u intersects $U(K)$ in at most one point is essentially the same as the proof of Lemma 5.1.3. $\square$

Lemma 5.1.5. *$U(K)$ is homeomorphic to $L(K)$ via a PL map $\phi : L(K) \to U(K)$ that sends symmetrically situated points to antipodal points.*

Proof. Any point in $L(K)$ can be uniquely written in the form

$$(\lambda a + (1 - \lambda)c, \mu b + (1 - \mu)c),$$

where a and b lie in disjoint simplices of K, $0 \leq \lambda, \mu \leq 1$, and at least one of λ and μ is 1. Define ϕ by

$$\phi(\lambda a + (1 - \lambda)c, \mu b + (1 - \mu)c) = \begin{cases} (1 - \frac{1}{2}\mu)a + \frac{1}{2}\mu\alpha(b) & \text{if } \lambda = 1 \\ \frac{1}{2}\lambda a + (1 - \frac{1}{2}\lambda)\alpha(b) & \text{if } \mu = 1. \end{cases}$$

It is easy to check that ϕ has the desired properties. $\square$

Proof of Example 5.1.1. As noted above, it suffices to show that K is an absolute selfintersector for $\mathbb{R}^{2k+1}$. Let $f : cK \to \mathbb{R}^{2k+1}$ be a continuous map. Then $\Psi_f \circ \phi^{-1} \circ \theta^{-1}$ defines a map from S^{2k+1} to $\mathbb{R}^{2k+1}$. The Borsuk-Ulam Theorem (Munkres, 1984, Theorem 68.6) implies that there exist antipodal points in S^{2k+1} that have the same image in $\mathbb{R}^{2k+1}$. By the properties of ϕ and θ specified in Lemmas 5.1.4 and 5.1.5, this means that there are symmetrically situated points $\langle a, b\rangle$ and $\langle b, a\rangle$ in $L(K)$ such that $\Psi_f(a, b) = \Psi_f(b, a)$. But $\Psi_f(a, b) = -\Psi_f(b, a)$ for every $\langle a, b\rangle \in C(K)$, so $\Psi_f(a, b) = \mathbf{0}$ and thus K is an absolute selfintersector by Lemma 5.1.2. $\square$

Even though a map of a polyhedron need not be homotopic to an embedding, it is often possible to improve a map within its homotopy class. Consider, for example, a compact, connected 1-dimensional polyhedron K. In general it will not be possible to embed K in $\mathbb{R}^2$. But if we form a quotient polyhedron K' by shrinking out a maximal tree in K, then the new polyhedron K' is just a finite wedge of circles and can be embedded in $\mathbb{R}^2$. The next theorem generalizes that simple idea. It asserts that, given certain homotopy conditions, a map of a polyhedron is homotopic to an embedding "up to simple-homotopy type." Before stating the theorem we must define the terms used.

Definition. Let K and L be simplicial complexes. A *simple-homotopy equivalence* from K to L is a finite sequence of collapses and expansions that begins at K and ends at L. More specifically, say that K *is simple-homotopy equivalent to* L if there exists a sequence

$$K = K_0 \nearrow K_1 \searrow K_2 \nearrow \cdots \searrow K_m = L$$

in which each arrow represents a simplicial expansion or collapse. (An expansion is the inverse of a collapse.)

It is clear that simple-homotopy equivalence implies homotopy equivalence; the converse is true for some fundamental groups but not for others.

Definition. A map $f : K \to L$ is said to be *r-connected* if f induces isomorphisms $\pi_i(K) \to \pi_i(L)$ for $i < r$ and a surjection $\pi_r(K) \to \pi_r(L)$.

We can now state the basic embedding theorem.

Theorem 5.1.6 (Embedding up to simple-homotopy type). *Let $f : K^k \to M^n$ be a map of a compact polyhedron into a PL manifold such that*

(1) *$k \leq n - 3$ and*
(2) *f is $(2k - n + 1)$-connected.*

Then there exist a subpolyhedron $L \subset M$ and a simple-homotopy equivalence $g : K \to L$ such that g and f are homotopic in M.

Corollary 5.1.7. *Any compact, r-connected, k-dimensional polyhedron is simple-homotopy equivalent to a subpolyhedron of $\mathbb{R}^{2k-r}$, $k - r \geq 3$.*

Remark. Corollary 5.1.7 is true when the restriction $k - r \geq 3$ is reduced to $k \geq r$. Special *ad hoc* arguments are needed for the additional cases.

The proof of Theorem 5.1.6 requires some technical facts about mapping cylinders. Let $f : K \to L$ be a PL map. Subdivide so that f is simplicial and triangulate Map(f) as a simplicial mapping cylinder. We claim that Map$(f) \searrow L$. This follows from the inductive construction of the simplicial mapping cylinder on page 124, since for each simplex $\sigma \subset K$, we know by induction that Map$(f|\partial\sigma) \searrow L$ and the cone on a collapsible set collapses to its base (Rourke and Sanderson, 1972, Example (3), page 40).

For any subpolyhedron K_0 of K we define the *reduced mapping cylinder*, Map(f, K_0), to be the quotient space formed from Map(f) by shrinking each of the fibers $\{x\} \times [0, 1]$, $x \in K_0$, to a point. As usual, we identify K with $K \times \{0\}$ and consider K to be a subset of Map(f, K_0). Just as in the unreduced case, a simplicial structure on $\mathrm{Map}(f, K_0)$ may be defined inductively and $\mathrm{Map}(f, K_0) \searrow L$.

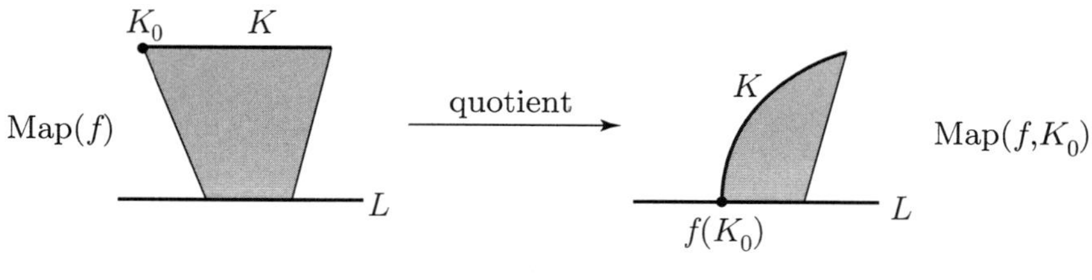

Figure 5.1. The mapping cylinder and the reduced mapping cylinder

The retraction $\pi_f : \mathrm{Map}(f, K_0) \to L$ that maps each $\{x\} \times [0,1]$ to $f(x)$ for $x \in K$ and is the identity on L is called the *mapping cylinder retraction.* Note that π_f is a quotient map. In case P is a polyhedron such that $\mathrm{Map}(f, K_0) \subset P$, the quotient map may be extended over the domain P. We use P/f to denote the quotient space whose points are

$$\{\pi_f^{-1}(x) \mid x \in L\} \cup \{\{x\} \mid x \in P \smallsetminus \mathrm{Map}(f, K_0)\}.$$

Lemma 5.1.8. *If K, L, and P are compact polyhedra, $f : K \to L$ is a PL map, and $\mathrm{Map}(f) \subset P$, then the quotient map $q : P \to P/f$ is a simple-homotopy equivalence.*

Sketch of proof. Let $E = P/f$. It is clear that $\mathrm{Map}(q) \searrow E$; the proof is completed by showing that $\mathrm{Map}(q) \searrow P$ as well. Note that $\mathrm{Map}(\pi_f) \subset \mathrm{Map}(q)$. Since q is one-to-one on $P \smallsetminus \mathrm{Map}(f)$, we have $\mathrm{Map}(q) \searrow P \cup \mathrm{Map}(\pi_f)$. Furthermore, the fact that $\mathrm{Map}(f) \searrow L$ means that $\mathrm{Map}(\pi_f) \searrow L \times [0,1] \cup \mathrm{Map}(f)$. Hence $\mathrm{Map}(q) \searrow P \cup \mathrm{Map}(\pi_f) \searrow P \cup L \times [0,1] \searrow P$. $\square$

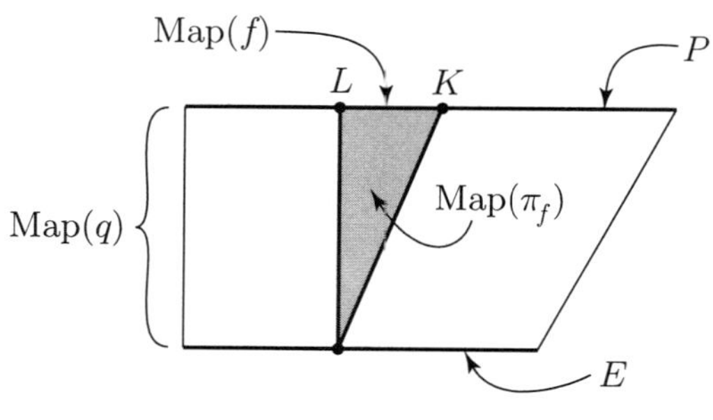

Figure 5.2. Proof of Lemma 5.1.8

Lemma 5.1.9. *If $f : K \to L$ is r-connected, A is an r-dimensional polyhedron, and B is a subpolyhedron of A, then any map $(A, B) \to (\mathrm{Map}(f), K)$ is homotopic rel B to a map into K.*

Proof. The lemma follows from Exercise 5.1.1 and Theorem 0.5.2. □

We use $S(f)$ to denote the singular set of the map f (see page 100).

Lemma 5.1.10. *Suppose $f : K \to M$ is a PL map of a compact polyhedron into a PL manifold, $s = \dim S(f)$, and f is $(s+1)$-connected. Then there exist compact polyhedra $P \supset K$ and $L \supset f(K)$, and PL maps $\phi_1 : P \to K$, $\phi_2 : P \to L$, and $g : L \to M$ such that*

(1) *ϕ_1 and ϕ_2 are simple-homotopy equivalences,*
(2) *g extends the inclusion $f(K) \hookrightarrow M$,*
(3) $\dim(L \setminus f(K)) \leq s+2$, *and*
(4) *the diagram*

$$\begin{array}{ccc} P & \xrightarrow{\phi_2} & L \\ {\scriptstyle \phi_1}\downarrow & & \downarrow{\scriptstyle g} \\ K & \xrightarrow[f]{} & M \end{array}$$

commutes up to homotopy.

Proof. Let $f : K \to M$ be as in the lemma.

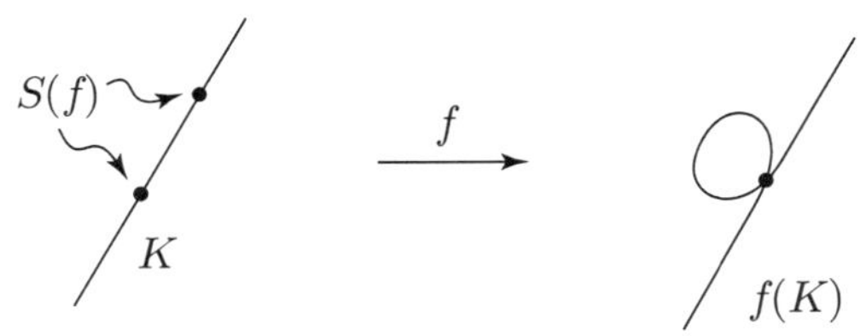

Figure 5.3. The singular set in the domain and range

Define $\bar{f} = f|S(f) : S(f) \to f(S(f))$. Note that $\mathrm{Map}(\bar{f}) \subset \mathrm{Map}(f)$ and $\dim(\mathrm{Map}(\bar{f})) = s+1$. By Lemma 5.1.9 there exists a PL homotopy $h_t : \mathrm{Map}(\bar{f}) \to \mathrm{Map}(f)$ such that h_0 is the inclusion, $h_t|S(f) \times \{0\}$ is the inclusion for every t, and $h_1(\mathrm{Map}(\bar{f})) \subset K \times \{0\}$. Define $P = \mathrm{Map}(\bar{h}, S(f))$, where $\bar{h} : \mathrm{Map}(\bar{f}) \to K$ is defined by $h_1(x) = \langle \bar{h}(x), 0 \rangle$. There is a natural map $\psi : P \to \mathrm{Map}(f)$ defined by $\psi(x,t) = h_t(x)$ for $\langle x, t \rangle \in \mathrm{Map}(\bar{f}) \times [0,1]$ and $\psi(x) = \langle x, 0 \rangle$ for $x \in K$.

The map ϕ_1 is simply the mapping cylinder retraction $\mathrm{Map}(\bar{h}, S(f)) \to K$. Define L to be the quotient polyhedron of P formed by collapsing out the fibers of $\mathrm{Map}(\bar{f})$. The map ϕ_2 is the projection map from P to the quotient space L. (See Figure 5.4.) Since there is a natural PL homeomorphism from $\phi_2(K)$ to $f(K)$, we can identify the two and consider $f(K)$ to be a subset of L. Finally, g is defined by $g(x) = x$ for $x \in \phi_2(K \cup \mathrm{Map}(\bar{f}))$

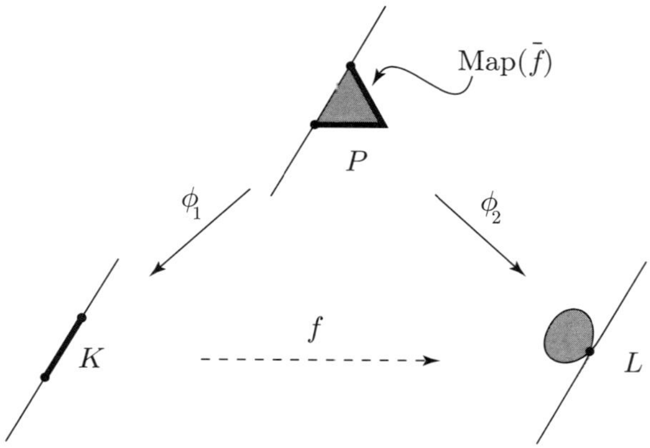

Figure 5.4. The reduced mapping cylinder of $\bar{h}$ is the intermediate space in the simple-homotopy equivalence

and $g(x) = \pi_f \psi \phi_2^{-1}(x)$ for $x \in L \setminus \phi_2(K \cup \mathrm{Map}(\bar{f}))$. It is clear that ϕ_1 is a simple-homotopy equivalence; ϕ_2 is a simple-homotopy equivalence by Lemma 5.1.8. □

Proof of Theorem 5.1.6. Define $K_1 = K$ and $f_1 = f$. First shift f_1 into general position to obtain $\dim S(f_1) \leq 2k - n$. By Lemma 5.1.10 there exist a polyhedron K_2 and a map $f_2 : K_2 \to M$ such that K_2 is simple-homotopy equivalent to K_1, $K_2 \supset f_1(K_1)$, f_2 extends the inclusion $f_1(K_1) \hookrightarrow M$, and $\dim(K_2 \setminus f_1(K_1)) \leq 2k - n + 2$. ($K_2$ is the L in the conclusion of the lemma.) Shift f_2 into general position, keeping $f_1(K_1)$ fixed, so we have $\dim S(f_2) \leq k + (2k - n + 2) - n \leq 2k - n - 1$ (since $k \leq n - 3$). Thus one application of the lemma has lowered the dimension of the singular set by 1.

Next apply Lemma 5.1.10 to $f_2 : K_2 \to M$. This gives a polyhedron K_3 and a map $f_3 : K_3 \to M$ such that K_3 is simple-homotopy equivalent to K_2, $K_3 \supset f_2(K_2)$, f_3 extends the inclusion $f_2(K_2) \hookrightarrow M$, and $\dim(K_3 \setminus f_2(K_2)) \leq 2k - n + 1$. Again shift f_3 into general position, keeping $f_2(K_2)$ fixed, so $\dim S(f_3) \leq k + (2k - n + 1) - n \leq 2k - n - 2$. Hence, two applications of the lemma have reduced the dimension of the singular set by 2. The process is continued until the singular set is empty. □

Historical Notes. Example 5.1.1 is due to E. R. van Kampen (1932) and A. Flores (1934). The proof presented in the section is based on those in (Flores, 1934) and (Grünbaum, 1970). Other expositions of the proof may be found in (Matoušek, 2003) and (Grünbaum, 2003). A. Shapiro (1957) developed an obstruction theory for embeddings of polyhedra in Euclidean space.

Theorem 5.1.6 and its corollary are contained in unpublished work of Stallings (1965a). An alternate proof of Theorem 5.1.6 is included in (Hudson, 1969a, Chapter XII).

The book by M. Cohen (1973) is a rich source of further information regarding simple-homotopy theory. In particular, additional detail on the proof of Lemma 5.1.8 may be found in (Cohen, 1973, §5).

Bing's house with two rooms (Exercise 5.1.2) was described by Bing (1964) as part of a construction of a nonshellable triangulation of B^3. The name "dunce hat" (Exercise 5.1.3) was introduced by Zeeman (1964).

Exercises

5.1.1. Let $f : K \to L$ be a map and identify K with the subset $K \times \{0\}$ of $\mathrm{Map}(f)$. Prove that the map f is r-connected if and only if the pair $(\mathrm{Map}(f), K)$ is r-connected.

5.1.2. *Bing's house with two rooms* is the 2-dimensional polyhedron constructed as follows. Start with the surface of a cube. Add a horizontal floor that separates an upper room from a lower room. Make a passageway through the upper room into the lower room by cutting a hole in the floor and a hole in the ceiling of the upper room and connecting the two holes with a cylinder. Construct a similar passageway through the lower room into the upper room. For each passageway, add a panel that connects the passageway to an adjacent wall. The finished house is shown in Figure 5.5. Prove that Bing's house with two rooms is simple-homotopy equivalent to a point but is not collapsible.

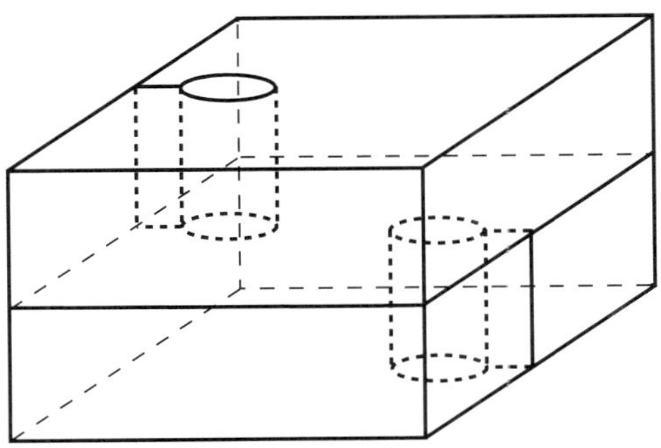

Figure 5.5. Bing's house with two rooms

5.1.3. The *dunce hat* is the space formed by identifying the three edges of a triangle as indicated in Figure 5.6. Prove that the dunce hat is simple-homotopy equivalent to a point but is not collapsible.

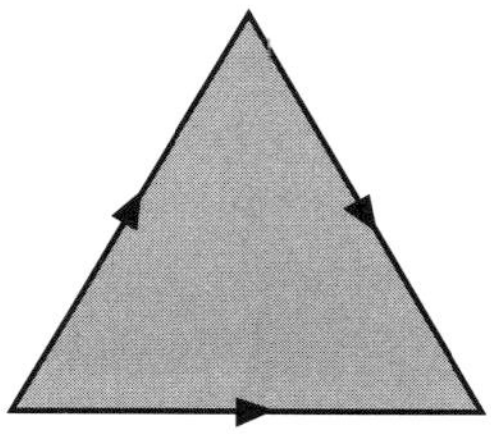

Figure 5.6. The topological dunce hat

5.2. Constructing PL embeddings of manifolds

An essential ingredient in the proof of Theorem 5.1.6 was the mapping cylinder construction of Lemma 5.1.10. Since it was necessary to shrink out some of the fibers of the mapping cylinder, the topological type of the polyhedron K was changed during the construction. The result was an embedding of K only up to simple-homotopy type, rather than an actual PL embedding of K. In case the domain of the continuous map is a manifold, however, this shrinking can be done without changing the topological type of the domain. Thus the techniques of the previous section can be retooled to establish PL embedding theorems in the manifold setting. Of course the same connectivity hypotheses are still needed, but the conclusions are much stronger.

Theorem 5.2.1 (Existence of PL Embeddings). *Let $f : Q^k \to M^n$ be a map from a compact PL manifold Q into a PL manifold M. If*

(1) *$k \leq n-3$, and*

(2) *f is $(2k-n+1)$-connected,*

then f is homotopic to a PL embedding.

Corollary 5.2.2. *Let $f : Q^k \to M^n$ be a map from a compact PL manifold Q into a PL manifold M. If*

(1) *$k \leq n-3$,*

(2) *Q is $(2k-n)$-connected, and*

(3) *M is $(2k-n+1)$-connected,*

then f is homotopic to a PL embedding.

Corollary 5.2.3. *Any closed, r-connected, k-dimensional PL manifold can be PL embedded in $\mathbb{R}^{2k-r}$, $k-r \geq 3$.*

In particular, any connected manifold of dimension k can be embedded in $\mathbb{R}^{2k}$. By contrast, Example 5.1.1 shows that a connected polyhedron of dimension k might not be embeddable in $\mathbb{R}^{2k}$.

Corollary 5.2.4. *If M^n is an r-connected PL n-manifold, then any element of $\pi_k(M)$ can be represented by a PL embedded k-sphere provided that $k \leq \frac{1}{2}(n+r-1)$.*

Before taking up the theorem, let us look at some examples. The next two examples show that the connectivity conditions in Theorem 5.2.1 are sharp and cannot be improved. In the first example f fails to induce a monomorphism on $\pi_{2k-n} = \pi_1$. In the second example f fails to induce an epimorphism on $\pi_{2k-n+1} = \pi_1$.

Example 5.2.5. *If k is a power of 2, then there is no topological embedding of $\mathbb{R}\mathrm{P}^k$ in $\mathbb{R}^{2k-1}$. Moreover, there is no subpolyhedron of $\mathbb{R}^{2k-1}$ that has the homotopy type of $\mathbb{R}\mathrm{P}^k$.*

Sketch of proof. The fact that $\mathbb{R}\mathrm{P}^k$ cannot be embedded in $\mathbb{R}^{2k-1}$ is proved by computing Stiefel-Whitney classes—see (Milnor and Stasheff, 1974, pages 120 and 50). If there were a subpolyhedron of $\mathbb{R}^{2k-1}$ with the homotopy type of $\mathbb{R}\mathrm{P}^k$, then Theorem 5.2.1 would give an embedding of $\mathbb{R}\mathrm{P}^k$ in a regular neighborhood. □

Example 5.2.6. *Let $f : S^k \to \mathbb{R}^{2k}$ be a PL immersion with exactly one transverse double point and let M be a regular neighborhood of $f(S^k)$ in $\mathbb{R}^{2k}$. Then f is not homotopic to a topological embedding in M. Moreover, there is no subpolyhedron L of M such that f is homotopic in M to a homotopy equivalence from S^k onto L.*

Proof. We begin by constructing the map f. Write S^k as $S^k = B_0^k \cup (S^{k-1} \times I) \cup B_1^k$, where B_0^k and B_1^k are k-cells and the three pieces have disjoint interiors. Map B_0^k and B_1^k onto two transverse linear k-cells that span B^{2k}. Then extend to a map f of all of S^k that embeds $S^{k-1} \times I$ in the closure of $\mathbb{R}^{2k} \setminus B^{2k}$. One way to construct the extension is to view the sphere ∂B^{2k} as the join $\partial B_0^k * \partial B_1^k$ and embed $S^{k-1} \times I$ in ∂B^{2k} as the diagonal $\bigcup\{\{x\} * \{h(x)\} \mid x \in \partial B_0^k\}$, where $h : \partial B_0^k \to \partial B_1^k$ is a PL homeomorphism. (The existence of the extension also follows from Theorem 5.2.8, below.) Note that the preimage under f of the origin consists of two points and that f is otherwise one-to-one.

Now let α denote an arc in S^k connecting the two preimages of the origin in $\mathbb{R}^{2k}$. Then $f(\alpha)$ is a simple closed curve and the regular neighborhood M of $f(S^k)$ in $\mathbb{R}^{2k}$ consists of a tubular neighborhood of $f(\alpha)$ together with a regular neighborhood of a k-dimensional disk that is attached to the neighborhood of $f(\alpha)$ in a homotopically inessential way. Hence the universal cover $\widetilde{M}$ consists of $\mathbb{R}^1 \times B^{2k-1}$ with a sequence of thickened k-disks attached. There is a sequence of lifts $f_i : S^k \to \widetilde{M}$ of f such that each f_i is an

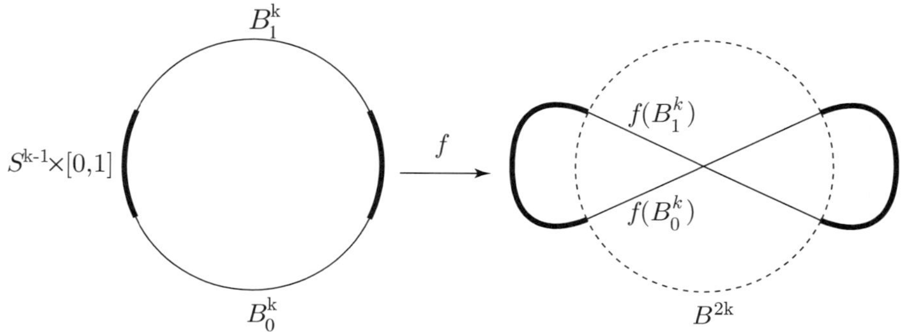

Figure 5.7. An immersion of S^k in $\mathbb{R}^{2k}$

embedding and $f_i(S^k)$ intersects $f_{i+1}(S^k)$ in a single point. The intersection number of $f_i(S^k)$ and $f_{i+1}(S^k)$ is ± 1.

Suppose f is homotopic to a topological embedding $e : S^k \to M$. Then there exists a sequence of lifts $e_i : S^k \to \widetilde{M}$ with e_i homotopic to f_i. Since e is an embedding, each e_i is an embedding and $e_i(S^k) \cap e_j(S^k) = \emptyset$ for $i \neq j$. Thus the intersection number of $e_i(S^k)$ and $e_{i+1}(S^k)$ is zero. But this contradicts the fact that the intersection number of spheres in the middle dimension is a homotopy invariant—see (Freedman and Quinn, 1990, §1.7), (Wall, 1999, page 46), or (Rourke and Sanderson, 1972, page 68). □

The remainder of this section will be devoted to the proof of Theorem 5.2.1. The proof given here is only valid in the range $k \leq (2/3)n - 1$ (the "metastable range"). The full codimension-three proof is so technically complicated that it is not appropriate for inclusion in this text. Since our proof of Theorem 5.2.1 is not valid in codimension three, we will give a separate proof covering all cases of Corollary 5.2.2.

We need a lemma that will allow us to shrink out the fibers of a mapping cylinder embedded in a PL manifold. Suppose $f : K \to L$ is a PL map, $K_0 \subset K$ is a subpolyhedron, $\pi_f : \mathrm{Map}(f, K_0) \to L$ is the mapping cylinder retraction, M is a PL manifold, and $h : \mathrm{Map}(f, K_0) \to M$ is a PL embedding. We use M/h to denote the quotient space of M whose points are

$$\{h(\pi_f^{-1}(x)) \mid x \in L\} \cup \{\{x\} \mid x \in M \smallsetminus h(\mathrm{Map}(f, K_0))\}.$$

By Lemma 5.1.8, we know that the quotient map $M \to M/h$ is a simple-homotopy equivalence. In case M is a manifold, we can reach a much stronger conclusion. The proof of the following lemma will be postponed until the next section where we will address several related technical issues.

Lemma 5.2.7. *Let $f : K \to L$ be a PL map of compact k-dimensional polyhedra, let K_0 be a subpolyhedron of K, let M be a PL n-manifold, and* *let $h : \mathrm{Map}(f, K_0) \to M$ be a PL embedding. If $2k + 3 \leq n$, or else $K_0 = \emptyset$ and either $n \leq 3$ or $k \leq n - 4$, then $M/h \cong M$.*

Proof of Theorem 5.2.1 for the case $k \leq (2/3)n - 1$. Let $f : Q^k \to M^n$ be a map from a compact PL manifold into a PL manifold. Shift f into general position and define $\bar{f} = f|S(f) : S(f) \to f(S(f))$. Then $\dim(S(f)) \leq 2k - n$, so $\dim(\mathrm{Map}(\bar{f})) \leq 2k - n + 1$. As in the proof of Lemma 5.1.10, the connectivity of f implies that the inclusion $S(f) \hookrightarrow Q$ extends to a map $h : \mathrm{Map}(\bar{f}) \to Q$. If we put h in general position, then $\dim(S(h)) \leq 2(2k - n + 1) - k = 3k - 2n + 2$. The hypothesis $k \leq (2/3)n - 1$ implies that $S(h) = \emptyset$, so h is an embedding.

We use notation similar to that in the proof of Lemma 5.1.10. In particular, let $P = \mathrm{Map}(h, S(f))$, let L be the quotient of P obtained by shrinking out the fibers of $\mathrm{Map}(\bar{f})$ and let $\phi : P \to L$ be the quotient map. There is a natural PL homeomorphism from $\phi(Q)$ to $f(Q)$, so we identify $\phi(Q) \subset L$ with $f(Q) \subset M$. As in the proof of Lemma 5.1.10, there is a map $g : L \to M$ that extends the inclusion $f(Q) \hookrightarrow M$. Shift g into general position; then $\dim S(g) \leq k + (2k - n + 2) - n = 3k - 2n + 2 \leq -1$, so g is an embedding. Define $P' = \overline{P \setminus Q}$ and $L' = \overline{L \setminus f(Q)}$. Since h is an embedding, P' is a product. Thus we can reverse the $[0, 1]$ coordinates and view P' as $\mathrm{Map}(h^{-1}, S(f))$. Similarly, we can view L' as $\mathrm{Map}(f|h(\mathrm{Map}(\bar{f})), S(f))$. Let $g' = g|L'$.

The two conditions $k \leq n - 3$ and $k \leq (2/3)n - 1$ combine to show that either $k \leq 3$ or $\dim(\mathrm{Map}(\bar{f})) \leq k - 3$. Thus Lemma 5.2.7 applies and we can conclude that $Q/h \cong Q$. The same two conditions combine to show that $2\dim(\mathrm{Map}(\bar{f})) + 3 \leq n$, so the lemma applies again to give $M/g' \cong M$. The proof of the theorem is completed by observing that f induces an embedding of Q/h into M/g'. Since $Q/h \cong Q$ and $M/g' \cong M$, this gives an embedding of Q in M. □

As noted earlier, our proof of Theorem 5.2.1 is valid only in the range $k \leq (2/3)n - 1$, which is known as the *metastable range*. The complicated technical difficulties encountered in the proof of the codimension-three case of Theorem 5.2.1 do not arise in the proof of Corollary 5.2.2. For completeness we include a separate proof of the general case of that Corollary. It is essentially the same as the proof of (Rourke and Sanderson, 1972, Theorem 7.12).

Proof of Corollary 5.2.2. Let $S = S(f)$, the singular set of f. The objective of the proof is to construct collapsible polyhedra $C \subset Q$ and $D \subset \mathrm{Int}\, M$ such that $S \subset \mathrm{Int}\, C$ and $f^{-1}(D) = C$. Once we have C and D, the proof is easily completed since regular neighborhoods of C and D are PL balls. The embedding in the conclusion of the theorem is defined to agree with f on the complement of a regular neighborhood N of C and is defined to map N conewise into a regular neighborhood of D.

Assume f is in general position, so $\dim S \leq 2k - n$. Engulfing Theorem 3.1.3 implies that there exists a PL ball B_1 that contains S in its interior. Since B_1 is collapsible, we can use Shadow Building Lemma 3.1.5 to find a collapsible polyhedron $C_1 \subset B_1$ such that $S \subset C_1$ and $\dim C_1 \leq 2k - n + 1$. Applying the engulfing theorem again gives a PL ball B_1' such that $f(C_1) \subset \operatorname{Int} B_1'$ and applying the shadow building lemma again yields a collapsible polyhedron D_1 in B_1' such that $f(C_1) \subset D_1$ and $\dim D_1 \leq 2k - n + 2$. Now C_1 and D_1 are collapsible, but they are not the collapsible polyhedra needed because $f^{-1}(D_1)$ might not equal C_1; the best we can say is that $C_1 \subset f^{-1}(D_1)$. Write $f^{-1}(D_1) = C_1 \cup E_1$ with $E_1 = \overline{f^{-1}(D_1) \smallsetminus C_1}$; note that $\dim E_1 \leq (2k - n + 2) + k - n \leq 2k - n - 1$.

Apply the engulfing theorem once more to engulf $C_1 \cup E_1$ with B_1. This gives a PL ball $B_2 \subset Q$ such that $C_1 \cup E_1 \subset \operatorname{Int} B_2$. By Shadow Building Lemma 3.1.5 there is a collapsible polyhedron C_2 such that $B_2 \searrow C_2$, $C_1 \cup E_1 \subset C_2$, and $\dim(C_2 \smallsetminus C_1) \leq 2k - n$. The engulfing theorem and the shadow building lemma combine to give another collapsible polyhedron $D_2 \supset D_1$ such that $f(C_2) \subset D_2$ and $\dim(D_2 \smallsetminus D_1) \leq 2k - n + 1$. If we define E_2 by $f^{-1}(D_2) = C_2 \cup E_2$, we find that $\dim E_2 \leq 2k - n - 2$.

This construction is continued inductively until the dimension of E_m is negative. Then $E_m = \emptyset$ and $f^{-1}(D_m) = C_m$ so the proof is complete. $\square$

The proof above makes it particularly clear why codimension three is so special for the engulfing arguments used in this chapter. Several of the proofs yet to come involve just this sort of inductive argument. A singular set is identified at each stage of the proof; this set is thought of as an "error term" (the E_i in the proof above). A shadow is constructed in the domain and a shadow of the shadow is constructed in the range. The double shadow building increases dimension by two, but then the double shadow is intersected with a codimension-three polyhedron to reduce the dimension by three. The net effect is that the dimension of the error term is reduced by one and progress is made toward elimination of the error.

In applications of the embedding theorem, it is often necessary to have a relative version of the theorem. The relative version of Theorem 5.2.1 does not follow from the absolute one, but the proof can be modified to fit the relative case. We leave the required modifications as an exercise.

Definition. An embedding $\lambda : Q \to M$ of ∂-manifolds is called *faithful* if $\lambda^{-1}(\partial M) = \partial Q$.

Theorem 5.2.8. *Let $f : Q^k \to M^n$ be a map of a compact PL ∂-manifold Q into a PL ∂-manifold M such that $f|\partial Q$ is a PL embedding of ∂Q into ∂M. If*

(1) $k \leq n - 3$,

(2) *Q is $(2k - n)$-connected, and*

(3) *M is $(2k - n + 1)$-connected,*

then f is homotopic, keeping ∂Q fixed, to a faithful PL embedding.

Historical Notes. The first embedding theorem beyond the trivial range is due to H. Whitney (1944a). Whitney used his famous "Whitney Trick" to prove that any smooth n-manifold can be smoothly embedded in $\mathbb{R}^{2n}$. Whitney's technique suffices to prove Theorem 5.2.1 in case $n = 2k$; see Rourke and Sanderson (1972), Theorems 5.5 and 5.12. The next step is due to Penrose, Whitehead, and Zeeman (Penrose et al., 1961), who essentially proved Corollary 5.2.3. (They did not at that time have the Poincaré Conjecture available to them, so they did not state the theorem in its full generality.) Corollary 5.2.2 is due to M. C. Irwin (1965); proofs can also be found in (Rourke and Sanderson, 1972) and (Hudson, 1969a).

J. F. P. Hudson, (1966b) and (1967), proved Theorem 5.2.1 with the additional hypothesis that Q is $(3k - 2n + 2)$-connected. The general case follows from (Stallings, 1965a, Theorem 4.1) (= our Theorem 5.1.6) along with (Wall, 1970, Corollary 11.3.4). The simply connected case of (Wall, 1970, Corollary 11.3.4) is usually attributed to Browder, Casson, and Sullivan, although they did not publish proofs.

It is tempting to think that the metastable-range proof of Theorem 5.2.1 could be pushed through to codimension three by exploiting the kind of inductive argument that was used in the proof of Theorem 5.1.6. A straightforward attempt at generalization runs into difficulties, however. The basic reason for the complication is the fact that a singular image of the mapping cylinder does not have a natural collapse associated with it. Despite the problems encountered, the proof can be made to work. H. W. Berkowitz and J. Dancis (1970a) have a fairly elementary argument that pushes the proof to the (3/4)-range. The idea is that, in the (3/4)-range, the singularities of the retraction of the mapping cylinder to Q are simple enough that the collapsible sets can be amalgamated into slightly more complicated collapsible sets. Eventually Bryant (1990) worked out all the technical machinery needed to make the argument work in codimension three.

Exercise

5.2.1. Prove Theorem 5.2.8.

5.3. Unknotting PL embeddings of manifolds

We next turn our attention to the Unknotting Problem: under what conditions are two homotopic maps ambiently isotopic? The case of manifolds is considered in this section; polyhedra are considered in the next. Rourke and Sanderson (1972) prove unknotting for PL ball pairs and PL sphere pairs in codimension three. In this section we extend those results to PL embeddings of more general PL manifolds. Later in the chapter we will generalize further to the case of 1-LCC topological embeddings of manifolds. While the main theorems stated in this section apply only to manifolds, much of the technical machinery developed here will be useful in the following section when we take up the problem of unknotting polyhedra.

Let us begin with a statement of the unknotting theorem.

Theorem 5.3.1 (Unknotting PL Embeddings). *Let $\lambda_0, \lambda_1 : Q^k \to M^n$ be homotopic PL embeddings of a compact PL manifold Q into a PL manifold M. If*

(1) $k \leq n - 3$,

(2) *Q is $(2k - n + 1)$-connected, and*

(3) *M is $(2k - n + 2)$-connected,*

then there exists compactly supported PL isotopy $\Phi_t : M \to M$ such that $\Phi_0 = \mathrm{Id}_M$ and $\lambda_1 = \Phi_1 \circ \lambda_0$.

The following examples show that the connectivity conditions in this unknotting theorem are best possible. The first of them shows that the hypothesis on the domain is sharp, while the second shows that the hypothesis on the range is sharp.

Example 5.3.2. *$S^0 \times S^k$ knots in S^{2k+1}.*

Proof. Consider one embedding whose image consists of two k-spheres that bound disjoint $(k+1)$-cells and a second embedding whose image is the union of the two factors in the join structure $S^{2k+1} = S^k * S^k$. The two embeddings are indicated in Figure 5.8. The embeddings are not equivalent because each k-sphere of the first embedding is null-homologous in the complement of the other while in the second embedding that is not the case. □

Example 5.3.3. *S^k knots in $S^1 \times S^{2k}$.*

Proof. Start with two simply linked k-spheres in $[0, 1] \times B^{2k} \subset S^1 \times S^{2k}$. Construct an embedding $e : S^k \to S^1 \times S^{2k}$ by connecting the two k-spheres with a tube that runs around the S^1 factor of $S^1 \times S^{2k}$ as shown in Figure 5.9.

Figure 5.8. Two inequivalent embeddings of $S^0 \times S^k$ in S^{2k+1}

Then e is not equivalent to an embedding that extends to a map of B^{k+1}; the proof is left as Exercise 5.3.1. □

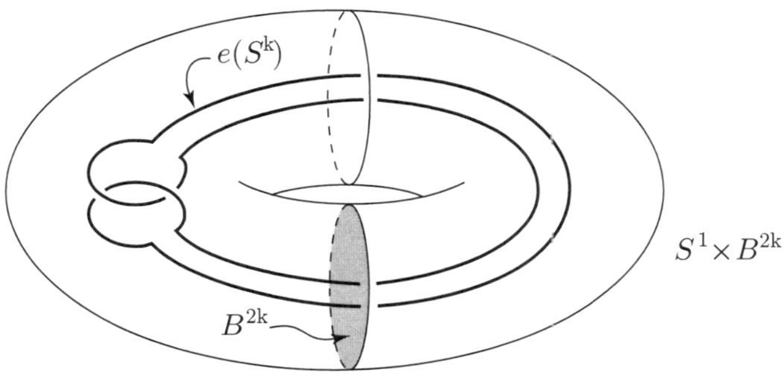

Figure 5.9. A knotted embedding $e : S^k \to S^1 \times B^{2k} \subset S^1 \times S^{2k}$

There is also a connected version of Example 5.3.2. It necessarily fails to be simply connected.

Example 5.3.4. *$S^1 \times S^k$ knots in S^{2k+2}.*

Proof. Begin with a pair of linked k-spheres in the equator S^{2k+1} of S^{2k+2}. Use the fact that the equator is bicollared to extend to an embedding of $S^0 \times S^k \times [-1,1]$ into $S^{2k+1} \times [-1,1] \subset S^{2k+2}$ that preserves the last coordinate. Then add an annulus $[0,1] \times S^k$ in each of the levels $S^{2k+1} \times \{-1\}$ and $S^{2k+1} \times \{+1\}$ to complete the embedding of $S^1 \times S^k$. (The annuli are constructed in the same way as in Example 5.2.6.) The resulting embedding $f : S^1 \times S^k \to S^{2k+2}$ is illustrated in Figure 5.10.

In order to prove that f is knotted we introduce a special *ad hoc* invariant. For each point $x \in S^1$, let $S_x = f(\{x\} \times S^k)$. Choose three distinct points $a, b, c \in S^1$ such that S_a, S_b, and S_c bound PL embedded $(k+1)$-cells A, B, and C in S^{2k+2} that intersect $f(S^1 \times S^k)$ only along their boundaries. Shift A, B, and C into general position; then any two of them will intersect in a finite number of points. Let AB denote the number of points in $A \cap B$, counted modulo two, and define BC and AC in a similar way. The invariant $k(f, a, b, c)$ is defined by $k(f, a, b, c) = AB + BC + AC \pmod 2$.

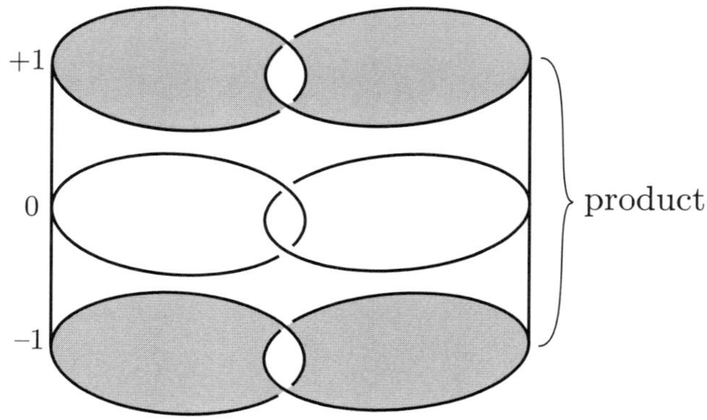

Figure 5.10. A knotted $S^1 \times S^k$ in S^{2k+2}

We claim that $k(f,a,b,c)$ depends only on f, a, b, and c and is independent of the particular $(k+1)$-cells A, B, and C. Let $[bc]$ denote the closed interval from b to c in S^1 that does not contain a. Note that $AB + AC$ is equal to the homological linking number of S_a with the $(k+1)$-sphere $B \cup f([bc] \times S^k) \cup C$. It follows that $AB + AC$ is independent of the particular spanning disk A that is used and thus $k(f,a,b,c)$ does not depend on A. The invariant $k(f,a,b,c)$ is defined for any PL embedding $f : S^1 \times S^k \to S^{2k+2}$ and any $a, b, c \in S^1$ as long as A, B, and C with the required properties exist. (It can be shown that they always exist—see Exercise 5.3.2.)

For the embedding f of the example, choose a, b, and c so that S_a is one of the k-spheres in the equator, S_b is at the $(+1)$-level, and S_c is at the (-1)-level, as indicated in Figure 5.11. Let B be the $(k+1)$-cell obtained by coning S_b from a point above the $(+1)$-level and let C be a similar cone below. There is an annulus in the 0-level between S_a and the sphere S'_a shown in Figure 5.11; the annulus is disjoint from $f(S^1 \times S^k)$ except where it intersects along S_a. (Use one-half of the annulus that forms part of the sphere in the (± 1)-levels.) Let A consist of this annulus together with a vertical annulus from S'_a straight up to the $(1+\epsilon)$-level and a $(k+1)$-cell in the $(1+\epsilon)$-level. Since S_a simply links $B \cup f([bc] \times S^k) \cup C$, we see that $k(f,a,b,c) = 1$.

On the other hand, if $e : S^1 \times S^k \to S^{2k+2}$ is any embedding that extends to $S^1 \times B^{k+1}$, then $k(e,a,b,c) = 0$ for every $a, b, c \in S^1$. (See Figure 5.11.) It follows that f and e are not equivalent embeddings. Thus $S^1 \times S^k$ knots in S^{2k+2}. □

The proof of Theorem 5.3.1 involves two results about concordances that are of independent interest. In the remainder of this section we use I to denote the closed unit interval $[0,1]$.

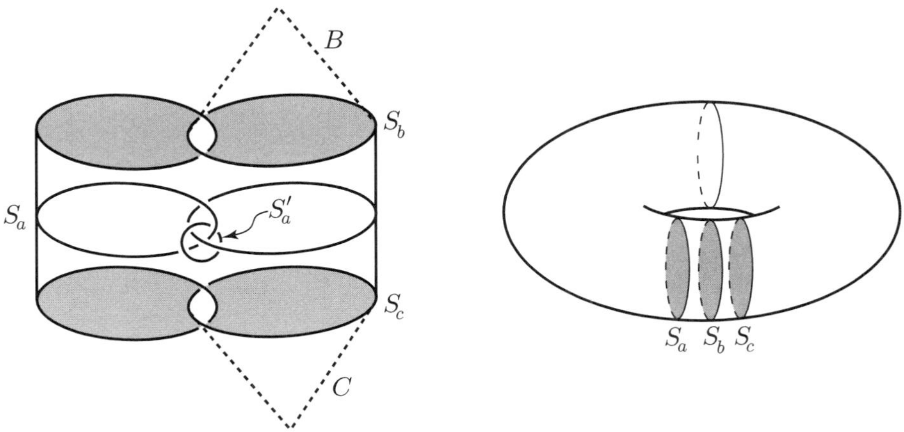

Figure 5.11. Calculation of $k(f, a, b, c)$ and $k(e, a, b, c)$

Definition. Let $\lambda_0, \lambda_1 : K^k \to M^n$ be two PL embeddings of a compact polyhedron K into a PL manifold M. A *concordance* from λ_0 to λ_1 is an embedding $\Phi : K \times I \to M \times I$ such that

(1) $\Phi(x, 0) = \langle \lambda_0(x), 0 \rangle$ for every $x \in K$,
(2) $\Phi(x, 1) = \langle \lambda_1(x), 1 \rangle$ for every $x \in K$, and
(3) $\Phi(K \times (0, 1)) \subset \mathrm{Int}(M \times I)$.

Two embeddings λ_0 and λ_1 are *concordant* if there exists a concordance Φ from λ_0 to λ_1. The embeddings are *PL concordant* if Φ is a PL embedding.

Theorem 5.3.5 (Existence of Concordances). *Let $\lambda_0, \lambda_1 : Q^k \to M^n$ be homotopic PL embeddings of a compact PL manifold Q into a PL manifold M. If*

(1) $k \leq n - 3$,
(2) *Q is $(2k - n + 1)$-connected, and*
(3) *M is $(2k - n + 2)$-connected,*

then λ_0 and λ_1 are PL concordant.

Proof. The fact that λ_0 and λ_1 are homotopic implies that there exists a continuous map $\phi : Q \times I \to M \times I$ such that $\phi(x, 0) = \langle \lambda_0(x), 0 \rangle$ and $\phi(x, 1) = \langle \lambda_1(x), 1 \rangle$ for every $x \in Q$. An application of Theorem 5.2.8 to the map ϕ gives the required PL embedding $\Phi : Q \times I \to M \times I$. □

Theorem 5.3.6 (Concordance Implies Unknotting). *Suppose $\lambda_0, \lambda_1 : K^k \to M^n$ are PL concordant embeddings of a compact polyhedron K into a PL n-manifold M. If*

(1) $k \leq n-3$, *and*

(2) M *is simply connected,*

then there exists a PL homeomorphism $h : M \to M$ *such that* $\lambda_1 = h \circ \lambda_0$.

The proof of Theorem 5.3.6 requires a technical result on neighborhoods of embedded products which assures that the concordance extends to an embedding of a product neighborhood. The proof of the Product Neighborhood Theorem is quite complicated and is linked to the proofs of several other technical results that will be needed in the next section. Before addressing those issues we first complete the proofs of Theorems 5.3.6 and 5.3.1.

Proof of Theorem 5.3.6. We may assume that $n \geq 5$. Let $\Phi : K \times I \to M \times I$ be a PL concordance from λ_0 to λ_1. By Product Neighborhood Theorem 5.3.9 (below), there is a neighborhood N of $\Phi(K \times I)$ in M, a PL ∂-manifold $P \supset K$, and a PL homeomorphism $f : P \times I \to N$ such that

(1) $f(P \times \{0\})$ is a regular neighborhood of $\Phi(K \times \{0\})$ in $M \times \{0\}$,

(2) $f(P \times \{1\})$ is a regular neighborhood of $\Phi(K \times \{1\})$ in $M \times \{1\}$,

(3) $f(P \times (0,1)) \subset M \times (0,1)$, and

(4) $f(x,t) = \Phi(x,t)$ for every $\langle x,t \rangle \in K \times I$.

Define $W = \overline{M \times [0,1] \setminus N}$, $N_0 = h(P \times \{0\})$ and $N_1 = h(P \times \{1\})$. Then W is a cobordism-with-boundary from $M_0 = M \times \{0\} \setminus \operatorname{Int} N_0$ to $M_1 = M \times \{1\} \setminus \operatorname{Int} N_1$. Since $k \leq n-3$, each of W, M_0, and M_1 is simply connected. Further, Duality Theorem 0.3.1 gives $H_i(W, M_0) \cong H^{n+1-i}(N, N_0) = 0$ for every i. Hence we may apply the Relative h-Cobordism Theorem (Rourke and Sanderson, 1972, Theorem 6.18) to conclude that the product structure on $\partial N \setminus (\operatorname{Int} N_0 \cup \operatorname{Int} N_1)$ from ∂N_0 to ∂N_1 can be extended to a product structure on W. This product structure defines the PL homeomorphism h in the conclusion of the theorem. □

Proof of Theorem 5.3.1. Theorem 5.3.1 follows immediately from Theorems 5.3.5 and 5.3.6. □

The next three theorems will all be proved together. Note that the existence of non-locally flat codimension-two PL embeddings means that the codimension restriction in each of the three theorems is necessary. In §5.4 we will need relative versions of the three theorems that include more technical hypotheses and conclusions. For now we present only the absolute versions because that is all we need to complete the proof of Unknotting Theorem 5.3.1 and because the structure of the proofs can be seen more clearly in this relatively simple setting.

Definition. A map $f : (X, A) \to (Y, B)$ of pairs is said to be *faithful* if $f^{-1}(B) = A$.

Let K be a polyhedron and let cK be the cone on K. We identify K with the base of the cone. We also identify B^n with the cone on S^{n-1}. This allows us to consider cK to be a subset of B^n whenever $K \subset S^{n-1}$.

Theorem 5.3.7 (Cone Unknotting). *If K is a compact, k-dimensional subpolyhedron of S^{n-1}, $k \leq n-4$, and $e : (cK, K) \to (B^n, S^{n-1})$ is a faithful PL embedding such that $e|K = \text{incl}_K$, then there exists a PL isotopy $\psi_t : B^n \to B^n$ such that $\psi_0 = \text{Id}$, $\psi_t|\partial B^n =$ inclusion for every t, and $\psi_1 \circ e = \text{incl}_K$.*

Definition. Let K be a polyhedron and let M be a PL ∂-manifold. A PL embedding $g : K \times I \to M$ is *locally flat* at $\langle x, t\rangle \in K \times I$ if there exists a neighborhood U of $g(x,t)$ in M such that $U \cong V \times [a, b]$ via a homeomorphism that extends the product structure on $g(K \times [a, b])$.

Theorem 5.3.8 (Local Flatness). *If K is a compact k-dimensional polyhedron, M is an n-dimensional PL ∂-manifold, $k \leq n-4$, and $g : (K \times I, K \times \partial I) \to (M, \partial M)$ is a faithful PL embedding, then g is locally flat at $\langle x, t\rangle$ for every $\langle x, t\rangle \in K \times I$.*

Definition. Assume K is a compact polyhedron, M is a PL ∂-manifold, and $g : (K \times I, K \times \partial I) \to (M, \partial M)$ is a faithful PL embedding. A regular neighborhood N of $g(K \times I)$ in M is said to *meet the boundary regularly* if $N \cap \partial M = N_0 \cup N_1$, where $N_0 \cap N_1 = \emptyset$ and N_i is a regular neighborhood of $g(K \times \{i\})$ in ∂M for $i = 0, 1$.

If T is a triangulation of M that contains a triangulation of $g(K \times [0,1])$ as a full subcomplex, then the simplicial neighborhood of $g(K \times [0,1])$ in a derived subdivision T' is a regular neighborhood that meets the boundary regularly. A regular neighborhood that meets the boundary regularly is just a special case of the regular neighborhoods in a pair that are described in (Rourke and Sanderson, 1972, pages 52–54).

Theorem 5.3.9 (Product Neighborhood). *Let M be an n-dimensional ∂-manifold, let K be a compact k-dimensional polyhedron, $k \leq n-4$, let $g : (K \times I, K \times \partial I) \to (M, \partial M)$ be a faithful PL embedding, and let N be a regular neighborhood of $g(K \times I)$ in M that meets ∂M regularly. If $N \cap \partial M$ is the disjoint union of N_0 and N_1 where N_0 is a regular neighborhood of $g(K \times \{0\})$, then there exists a PL homeomorphism $\lambda : N_0 \times I \to N$ such that $\lambda(x, 0) = \langle x, 0\rangle$ for every $x \in N_0$ and $\lambda(g(x, 0), t) = g(x, t)$ for every $\langle x, t\rangle \in K \times I$.*

As mentioned above, the three theorems just stated are linked in one large inductive proof. In order to describe the structure of the proof succinctly, we make the following abbreviations.

CU(n)	=	Cone Unknotting Theorem 5.3.7 in dimension n
LF(n)	=	Local Flatness Theorem 5.3.8 in dimension n
PN(n)	=	Product Neighborhood Theorem 5.3.9 in dimension n

We will prove that CU($n-1$) $\Rightarrow$ LF(n) $\Rightarrow$ PN(n) $\Rightarrow$ CU(n). Since the theorems are all easily seen to be true for $n \leq 4$, this suffices to prove all three theorems in all dimensions. The one catch is that the PN(n) $\Rightarrow$ CU(n) step uses the h-Cobordism Theorem and is therefore valid only for $n \geq 6$. We will plug that gap in the inductive proof by giving a separate *ad hoc* proof that PN(5) $\Rightarrow$ CU(5).

PN(n) $\Rightarrow$ CU(n), $n \geq 6$. Let $e : (cK, K) \to (B^n, S^{n-1})$ be a faithful PL embedding such that $e|K$ is the inclusion. We may assume that e maps the cone point of cK to the origin of B^n. Choose a small convex PL n-cell neighborhood C of the origin in B^n such that $D = e^{-1}(C)$ is a subcone of cK and $e|D : D \to C$ is a cone map.

The radial structure of B^n determines a product structure on $B^n \setminus \operatorname{Int} C \cong S^{n-1} \times I$ in which S^{n-1} is identified with $S^{n-1} \times \{0\}$. We can also identify $\overline{cK \setminus D}$ with $K \times I$. Thus e determines a faithful PL embedding $g : (K \times I, K \times \partial I) \to (S^{n-1} \times I, S^{n-1} \times \partial I)$ such that $g(x, 0) = \langle x, 0 \rangle$ for each $x \in K$ and $g(K \times \{1\}) \subset S^{n-1} \times \{1\}$. We will prove the existence of a homeomorphism $\phi : S^{n-1} \times I \to S^{n-1} \times I$ such that $\phi|S^{n-1} \times \{0\} =$ inclusion and $\phi(g(x,t)) = \langle x, t \rangle$ for each $\langle x, t \rangle \in K \times I$. Once such a homeomorphism has been found, it can be extended conewise to C to produce a homeomorphism $\Phi : B^n \to B^n$ such that $\Phi \circ e =$ inclusion. The isotopy ψ_t connecting the identity to Φ is the Alexander isotopy (Rourke and Sanderson, 1972, Proposition 3.22).

It remains to construct ϕ. By PN(n), $g(K \times I)$ has a product neighborhood N. The relative h-Cobordism Theorem can be applied exactly as in the proof of Theorem 5.3.6 to extend the product structure on N to a product structure on all of $S^{n-1} \times I$. The homeomorphism ϕ is the one that takes this new product structure to the standard product structure. □

PN(5) $\Rightarrow$ CU(5). Let K be a compact 1-dimensional subpolyhedron of S^4. As in the proof of the $n \geq 6$ case, it suffices to prove the following: if $g : (K \times I, K \times \partial I) \to (S^4 \times I, S^4 \times \partial I)$ is faithful PL embedding such that $g(x, 0) = \langle x, 0 \rangle$ for every $x \in K$ and $g(K \times \{1\}) \subset S^4 \times \{1\}$, then there exists a homeomorphism $\phi : S^4 \times I \to S^4 \times I$ such that $\phi|S^4 \times \{0\} =$ inclusion and $\phi(g(x,t)) = \langle x, t \rangle$ for each $\langle x, t \rangle \in K \times I$.

We claim that there is a PL homeomorphism $h : S^4 \times I \to S^4 \times I$ such that $h|S^4 \times \partial I = \mathrm{Id}$ and $S^4 \times I \searrow S^4 \times \{0\} \cup hg(K \times I)$. Let us assume the claim for now and complete the proof. Let C be a short collar on $S^4 \times \{0\}$ in $S^4 \times I$. By PN(5), $hg(K \times I)$ has a product neighborhood N in $S^4 \times I$ that meets the boundary regularly. We can arrange that the product structures on C and N are compatible where they overlap (use Rourke and Sanderson (1972), Addendum 4.21, if necessary) and, after trimming N slightly, we can construct a homeomorphism of pairs $\phi_1 : (S^4 \times I, K \times I) \to (C \cup N, hg(K \times I))$ such that $\phi_1(x,t) = hg(x,t)$ for each $\langle x,t\rangle \in K \times I$. By the claim, $S^4 \times I \searrow S^4 \times \{0\} \cup hg(K \times I)$ and so the Uniqueness of Regular Neighborhoods Theorem supplies a second homeomorphism $\phi_2 : C \cup N \to S^4 \times I$ that is the identity on $S^4 \times \{0\} \cup hg(K \times I)$; the composition $\phi = \phi_1^{-1} \circ \phi_2^{-1} \circ h$ is the homeomorphism needed to complete the proof.

In order to prove the claim, we begin by shifting g into general position and taking the shadow of its image:

$$\Sigma = \mathrm{Sh}(g(K \times I)) = \{\langle x,s\rangle \in S^4 \times I \mid \langle x,t\rangle \in g(K \times I) \text{ for some } t \geq s\}.$$

Let $\pi : S^4 \times I \to S^4 \times \{0\}$ be the projection map. By general position, we know that the singular set of $\pi \circ g$ consists of a finite number of double points each of which is in the interior of $\sigma \times I$, where σ is a 1-simplex in K. A small adjustment will ensure that no two of the double points lie on the same vertical fiber in $K \times I$. For each of these double points, we wish to push the lower point of intersection off the shadow of the upper point. More specifically, let us suppose that $\langle y,a\rangle, \langle z,b\rangle$ are points in $K \times I$ such that $\pi g(y,a) = \pi g(z,b)$. Subdivide so that y and z are in the interiors of distinct 1-simplices of K. There exists $x \in S^4$ such that $g(y,a) = \langle x,t\rangle$ and $g(z,b) = \langle x,s\rangle$ with, say, $s < t$. Let α be a path in Σ that is below $g(\{x\} \times [a,1])$, above $\pi g(\{x\} \times [a,1])$, and joins $g(z,b)$ to a point below $g(y,1)$. Push $g(\{z\} \times I)$ along this path and off the end of Σ as illustrated in Figure 5.12. This move is accomplished by a homeomorphism h_z of $S^4 \times I$ such that $h_z g$ differs from g only on the interior of the 1-simplex containing z.

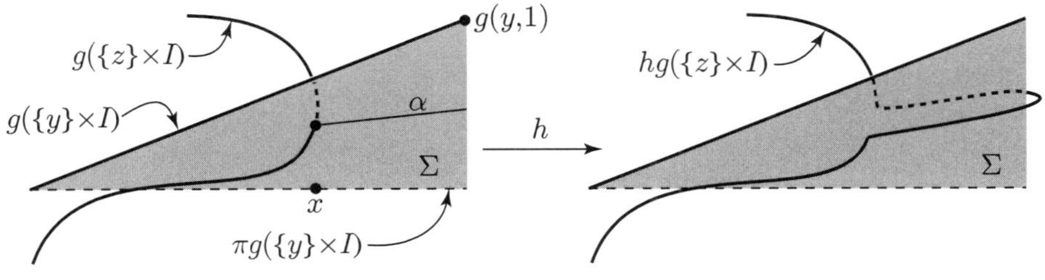

Figure 5.12. Push $g(\{z\} \times I)$ off the end of Σ

Let h be the homeomorphism of $S^4 \times I$ that accomplishes all the pushes described in the previous paragraph. Observe that $\pi|hg(K \times I)$ is one-to-one. The shadow of each i-simplex in $hg(K \times I)$ is a convex $(i+1)$-cell in $S^4 \times I$ and these cells have disjoint interiors. We can collapse $S^4 \times I$ to $S^4 \times \{0\} \cup hg(K \times I)$ in two stages. First, collapse straight down to achieve $S^4 \times I \searrow S^4 \times \{0\} \cup \mathrm{Sh}(hg(K \times I))$. Then collapse $S^4 \times \{0\} \cup \mathrm{Sh}(hg(K \times I))$ to $S^4 \times \{0\} \cup hg(K \times I)$ by collapsing out one of the convex cells at a time. The collapse $K \times I \searrow K \times \{0\}$ serves as a guide for the collapse of $\mathrm{Sh}(hg(K \times I))$: when a simplex σ in $K \times I$ is collapsed across a face τ, we collapse the convex cell below σ across the face below τ. $\square$

CU$(n-1)$ $\Rightarrow$ LF(n). Let $g : K^k \times [0,1] \to M^n$ be as in the statement of LF(n). Pick a point $\langle x_0, t_0 \rangle \in K \times [0,1]$. Assume, first, that $0 < t_0 < 1$. Choose triangulations T of $K \times [0,1]$ and S of M relative to which g is simplicial and $g(K \times [0,1])$ is full. We may further assume that both $\langle x_0, t_0 \rangle$ and $g(x_0, t_0)$ are vertices and that $D = \mathrm{St}(\langle x_0, t_0 \rangle, T')$, the star of $\langle x_0, t_0 \rangle$ in a first derived subdivision T' of T, can be written as $D = D_a \times [a,b]$, where D_a is a cone neighborhood of $\langle x_0, a \rangle \in K \times \{a\}$ and $0 \le a < t_0 < b \le 1$. Let $C = \mathrm{St}(g(x_0, t_0), S')$. Since M is a PL manifold, C is an n-cell. Observe that D has a natural cone structure as a cone from $\langle x_0, t_0 \rangle$ and that C is a cone from $g(x_0, t_0)$; furthermore, $g|D : D \to C$ is a cone map.

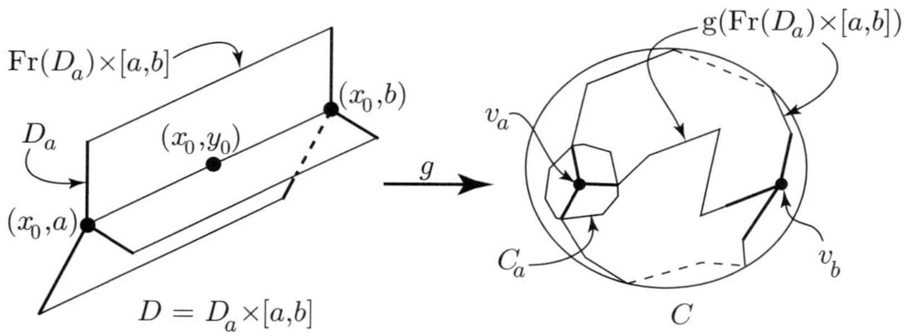

Figure 5.13. The neighborhoods C and D in the proof of CU$(n-1) \Rightarrow$ LF(n)

Let $v_a = g(x_0, a)$ and $v_b = g(x_0, b)$. The star of v_a in ∂C is a PL $(n-1)$-cell neighborhood C_a of v_a in ∂C such that C_a is a cone from v_a. Furthermore, $g^{-1}(C_a) = D_a \times \{a\}$ and $g|D_a \times \{a\} : D_a \times \{a\} \to C_a$ is a cone map. By CU$(n-1)$, $C'_a = \overline{\partial C \setminus C_a}$ has a cone structure $C'_a \cong v_b * \partial C_a$ such that the embedding

$$g|(\mathrm{Fr}\, D_a \times [a,b]) \cup (D_a \times \{b\}) : (\mathrm{Fr}\, D_a \times [a,b]) \cup (D_a \times \{b\}) \to C'_a$$

is a cone on $g|\,\mathrm{Fr}\, D_a \times \{a\}$ in this structure. Thus ∂C has a double cone structure extending the double cone structure on $\mathrm{Fr}\, D = (D_a \times \{a,b\}) \cup$

$(\operatorname{Fr} D_a \times [a,b])$. Hence there is a PL homeomorphism $\phi : \partial C \to \partial(C_a \times [a,b])$ such that $\phi(y) = \langle y, a\rangle$ for each $y \in C_a$ and $\phi(g(x,t)) = \langle g(x,a), t\rangle$ for each $\langle x,t\rangle \in \operatorname{Fr} D = (D_a \times \{a,b\}) \cup (\operatorname{Fr} D_a \times [a,b])$. Extend ϕ conewise to a homeomorphism $\phi : C \to C_a \times [a,b]$. Then $\phi(g(x,t)) = \langle g(x,a), t\rangle$ for every $\langle x,t\rangle \in D$, so the proof is complete for the case in which $0 < t_0 < 1$.

In case $t_0 = 0$, the only change that must be made in the proof is that $a = t_0 = 0$. In case $t_0 = 1$, we simply take $b = t_0 = 1$. □

The proof of LF$(n) \Rightarrow$ PN(n) requires a lemma that allows an isotopy of an embedded product to be extended to an ambient homeomorphism. After confirming PN(n), we will use it to establish a much more general covering isotopy theorem.

Lemma 5.3.10. *Assume LF(n). If K is a compact k-dimensional polyhedron, M is an n-dimensional PL ∂-manifold, $k \le n-4$, $g : (K\times I, K\times\partial I) \to (M, \partial M)$ is a faithful PL embedding, and $0 < a < b < 1$, then there exists a PL homeomorphism $h : I \to I$ such that $h|\partial I = \mathrm{incl}_{\partial I}$ and $h(a) = b$ and there exists a PL homeomorphism $\phi : M \to M$ such that $\phi|\partial M = \mathrm{incl}_{\partial M}$ and $\phi(g(x,t)) = g(x, h(t))$ for every $\langle x,t\rangle \in K \times I$.*

Proof. Use LF(n) to cover $g(K \times [0,1])$ with product neighborhoods. Find a triangulation T of K and a partition $0 = t_0 < t_1 < \cdots < t_m = 1$ such that $g(\sigma \times [t_i, t_{i+1}])$ is contained in the interior of one of these product neighborhoods for every $\sigma \in T$ and for every i. We may assume that there exist i and j such that $a = t_i$ and $b = t_j$. It suffices to prove that the t_i-level can be moved to the t_{i+1}-level. In order to simplify notation let us assume that $a = t_i$ and $b = t_{i+1}$.

Pick a vertex $v \in K$ and a product neighborhood U of $g(\{v\} \times [a,b])$. Define f_1 to be a homeomorphism of $g(\{v\} \times I)$ that moves $g(v,a)$ to $g(v,b)$ and is the identity outside a compact subset of $U \cap g(\{v\} \times I)$. Use the cone structure on a neighborhood of v in K to extend f_1 to $g(K \times I)$ and then use the product structure on U to extend f_1 to U in such a way that f_1 reduces to the identity outside a close neighborhood of $g(\{v\} \times I)$ and preserves fibers of $g(K \times I)$. Do this for each vertex of K, working in disjoint neighborhoods, and extend f_1 via the identity to a homeomorphism of all of M.

Next consider a 1-simplex $\sigma \in T$. Let f_2 be a homeomorphism of $g(\sigma \times I)$ that moves $f_1 g(\sigma \times \{a\})$ to $g(\sigma \times \{b\})$. Extend f_2 to $g(K \times I)$ by using the join structure on star(σ, T), then to a product neighborhood of $g(\sigma \times [a,b])$, and then to all of M via the identity. Figure 5.14 illustrates the combined action of f_1 and f_2.

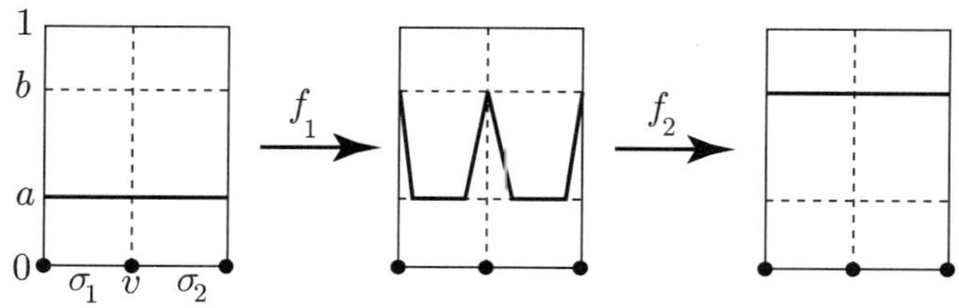

Figure 5.14. In this diagram K consists of two 1-simplices σ_1 and σ_2; the a-level is moved to the b-level in two steps

This process is continued inductively, moving up through the skeleta of increasing dimension. The final homeomorphism ϕ is the composition of the various f_i. Care should be taken in the construction of the f_i to ensure that there is one PL homeomorphism $h : I \to I$ such that $\phi(g(x,t)) = g(x, h(t))$ for every $\langle x, t\rangle \in K \times I$. $\square$

LF$(n) \Rightarrow$ **PN**(n). Let $g : (K^k \times I, K \times \partial I) \to (M^n, \partial M^n)$, $k \leq n-4$, be a faithful PL embedding into a PL ∂-manifold and let N be a regular neighborhood of the image that meets ∂M regularly. Define M^+ to be the ∂-manifold obtained from M by adding an exterior collar. Use the product structure on the collar to extend g to a faithful PL embedding $g : (K \times [-2,2], K \times \{-2,2\}) \to (M^+, \partial M^+)$ as illustrated in Figure 5.15.

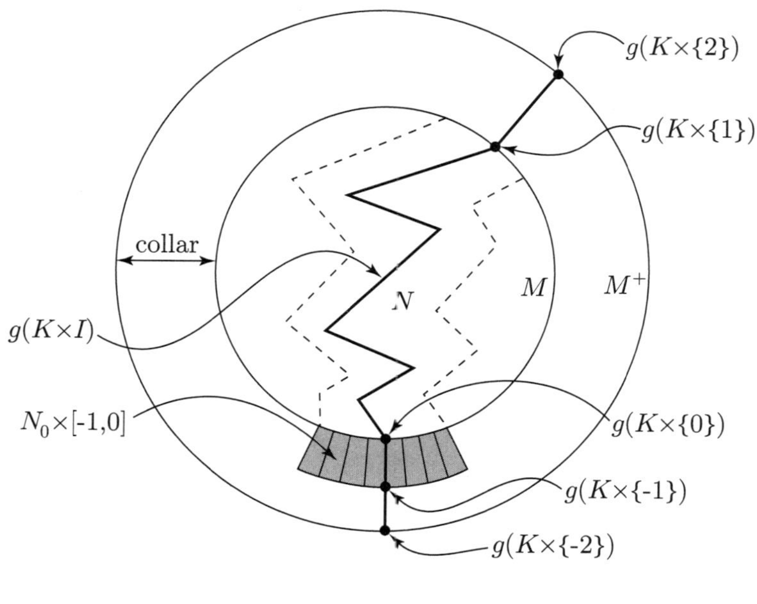

Figure 5.15. Add a collar to form M^+ and extend to $g : K \times [-2,2] \to M^+$

Since N meets ∂M regularly, $N \cap \partial M$ is a disjoint union $N_0 \cup N_1$, where N_i is a regular neighborhood of $g(K \times \{i\})$ in ∂M. Let $N_0 \times [-1, 0]$ be a product neighborhood of $g(K \times [-1, 0])$ in the inner half of the collar. (See Figure 5.15.) Two applications of Lemma 5.3.10 yield a PL homeomorphism $\phi : M^+ \to M^+$ that slides $g(K \times [-1, 0])$ to $g(K \times I)$. The ∂-manifold $P = \phi(N_0 \times [-1, 0])$ is almost the neighborhood we seek: P has the required product structure, but it is not necessarily a neighborhood of $g(K \times I)$ in M because $\phi(N_0 \times \partial[-1, 0])$ may not match up correctly with ∂M. We will provide an argument involving collars to correct this deficiency.[1]

Observe that there is a collar $c : N_0 \times [0, 1] \to N$ and a number $\epsilon > 0$ such that $c(g(x, 0), t) = g(x, t)$ for every $\langle x, t\rangle \in K \times [0, \epsilon]$. This observation is proved using the fact that $\mathrm{LF}(n)$ gives local collars and these local collars can be amalgamated into a global collar using the same argument as in Theorem 2.4.10. The details are worked out in (Rourke and Sanderson, 1972, Theorem 4.5 and Addendum 4.21).

Let T be a triangulation of M^+ that contains triangulations of N and P as subcomplexes and contains a triangulation of $g(K \times I)$ as a full subcomplex. Subdivide T so that $N(N_0, T) \cap c(N_0 \times [0, 1]) = c(N_0 \times [0, \delta])$ for some positive number δ. Subdivide further so that, in addition, the projection $N_0 \times [0, \delta] \to N_0$ and the restricted collar $c : N_0 \times [0, \delta] \to M$ are simplicial relative to T. Let T' be a subdivision of T derived near $g(K \times I)$; i.e., T' is a derived subdivision obtained from T by adding a new vertex in the interior of each simplex that intersects $g(K \times I)$ but is not contained in $g(K \times I)$. The vertices in $T' \smallsetminus T$ can be chosen in such a way that $P' = N(g(K \times I), T') \cap P$ is a subproduct of P; specifically, if $B_0 = N(g(K \times \{0\}), T') \cap \partial P$, then there is a homeomorphism $\lambda' : B_0 \times I \to P'$ such that $\lambda'(g(x, 0), t) = g(x, t)$ for every $\langle x, t\rangle \in K \times I$. Define $N' = N(g(K \times I), T') \cap N$.

The proof is completed by removing a small collar from P' and replacing it with a collar from N'. The result is a new ∂-manifold that is a product and is a regular neighborhood of $g(K \times I)$ in M. Define $A_0 = N(g(K \times \{0\}), T') \cap \partial N'$, $A_1 = N(g(K \times \{0\}), T') \cap N'$, and $B_1 = N(g(K \times \{0\}), T') \cap P'$. By construction there is a product structure $A_1 \cong A_0 \times [0, \delta]$ that extends the product structure on $g(K \times [0, \delta])$ and there is also a product structure $B_1 \cong B_0 \times [0, \delta]$ that extends the product structure on $g(K \times [0, \delta])$. Furthermore, $\overline{P' \smallsetminus B_1}$ has a product structure $\overline{P' \smallsetminus B_1} \cong B_0 \times [\delta, 1]$ that extends the product structure on $g(K \times [\delta, 1])$.

[1]The proof of the theorem could be completed quickly if we were to make use of the theory of relative regular neighborhoods (Cohen, 1969). In particular, $\mathrm{LF}(n)$ can be used to verify that each of N and P is a relative regular neighborhood in M^+ of $g(K \times I)$ mod $g(K \times \partial I)$ and so the uniqueness theorem for relative regular neighborhoods (Cohen, 1969, Theorem 3.1) implies that there is a PL homeomorphism of P to N that fixes $g(K \times I)$. We supply the alternative proof utilizing collars because we are using (Rourke and Sanderson, 1972) as our reference for PL topology and Rourke and Sanderson do not consider relative regular neighborhoods.

Consider $P'' = \overline{P' \setminus B_1} \cup A_1$. Each piece of P'' is a product, but we cannot quite claim that the union is a product because $A_0 \times \{\delta\}$ does not exactly match $\lambda'(B_0 \times \{\delta\})$. The product structures define projections $\overline{P' \setminus B_1} \to B_0 \times \{\delta\}$ and $A_1 \to A_0 \times \{\delta\}$. Choose a regular neighborhood C_0 of $g(K \times \{\delta\})$ in $\partial A_1 \cap \partial B_1$ (which we are identifying with $B_0 \times \{\delta\} \cap A_0 \times \{\delta\}$). Define P''' to be the union of the preimages of C_0 under these two projection maps. Then P''' has a product structure that extends the product structure on $g(K \times I)$ and the 0-level of this product structure is a regular neighborhood of $g(K \times \{0\})$ in ∂M.

Make a similar modification near $g(K \times \{1\})$. The result is a regular neighborhood P^* of $g(K \times I)$ in M that meets ∂M regularly and has a product structure that extends the product structure on $g(K \times I)$. It follows from the uniqueness of regular neighborhoods of pairs (Rourke and Sanderson, 1972, Theorem 4.11) that there is a PL homeomorphism $P^* \cong N$ that is the identity on $g(K \times I)$, so N also has the required product structure. □

Remark. The proof of LF(n) $\Rightarrow$ PN(n) does not require that $g(K \times I)$ have codimension three, only that it be locally flat. As a result, the product neighborhood theorem is valid in any codimension provided PL local flatness is assumed as a hypothesis. This observation will be used in the proof of Lemma 5.2.7, below.

We conclude this section with two applications of Product Neighborhood Theorem 5.3.9: first a proof of Shrinking Lemma 5.2.7 and then a general Covering Isotopy Theorem. These two applications serve to illustrate the power of the tools we have developed.

Lemma 5.3.11. *If $f : K \to L$ is a PL map of compact polyhedra and K_0 is a subpolyhedron of K, then there exists a quotient map $q : \mathrm{Map}(f) \to \mathrm{Map}(f)$ that preserves fibers in the sense that $q(\{x\} \times I) = \{x\} \times I$ for each $x \in K$ and whose only nontrivial point preimages are sets of the form $\{x\} \times [1/2, 1]$ for $x \in K_0$. Furthermore, $q|K \cup L = \mathrm{Id}$ and if U is any neighborhood of K_0 in K, then q can be constructed so that $q|(K \setminus U) \times I$ is the identity.*

Proof. Let $\mu : K \to [1/2, 1]$ be a continuous PL map such that $\mu(K \setminus U) = 1/2$, $\mu(K_0) = 1$, and $\mu(x) < 1$ for every $x \in K \setminus K_0$. For each $x \in K$, define $\lambda_x : [0,1] \to [0,1]$ to be the map such that $\lambda_x(0) = 0$, $\lambda_x(1/2) = \mu(x)$, $\lambda_x(1) = 1$, and λ_x is linear on each of $[0, 1/2]$ and $[1/2, 1]$. Then define $q : \mathrm{Map}(f) \to \mathrm{Map}(f)$ by $q(x,t) = \langle x, \lambda_x(t)\rangle$ for $\langle x, t\rangle \in K \times [0,1]$ and $q(y) = y$ for $y \in L$. It is easy to check that q has the required properties. □

Proof of Lemma 5.2.7. Let $f : K \to L$ be a PL map of a compact k-dimensional polyhedron K into a polyhedron L, let M be a PL n-manifold, and let $h : \mathrm{Map}(f, K_0) \to M$ be a PL embedding. Assume, first, that $K_0 = \emptyset$

and $k \leq n-4$. Let T be a triangulation of M that contains a triangulation of $h(\mathrm{Map}(f))$ as a full subcomplex and take the barycentric subdivision T' of T. Define N_1 to be the simplicial neighborhood of $h(L)$ in T'. Note that N_1 is a mapping cylinder neighborhood of $h(L)$.

The triangulation T can be chosen so that $N_1 \cap h(\mathrm{Map}(f))$ is the image of $K \times [1-\epsilon, 1] \sqcup L$ for some small positive number ϵ and the two mapping cylinder retractions agree on the overlap. Now h embeds $K \times [0, 1-\epsilon]$ in $M \smallsetminus \mathrm{Int}\, N_1$. By taking a simplicial neighborhood of $h(K \times [\delta, 1-\epsilon])$ in $T'|(M \smallsetminus \mathrm{Int}\, N_1)$ for some $\delta > 0$, we can construct a PL ∂-manifold N_2 such that $h|K \times [0, 1-\epsilon] : (K \times [0, 1-\epsilon], K \times \partial[0, 1-\epsilon]) \to (N_2, \partial N_2)$ is a faithful embedding. By PN(n) we may assume that N_2 is a product neighborhood of $h(K \times [0, 1-\epsilon])$.

Let $N = N_1 \cup N_2$. Note that N has the structure of a mapping cylinder associated with a map $\partial N \to h(L)$ and that it contains $h(\mathrm{Map}(f))$ as a sub-mapping cylinder. Add an exterior collar to N to form N^+. Then N^+ is also a mapping cylinder neighborhood of $h(L)$ and it can be reparametrized so that $h(\mathrm{Map}(f))$ corresponds to $K \times [1/2, 1]$. An application of Lemma 5.3.11 gives a homeomorphism $N^+ \to N^+/h$ that is the identity on ∂N^+. This homeomorphism extends via the identity to a homeomorphism $M \to M/h$.

In case $n \leq 3$ and k is unrestricted, LF(n) does not apply. But in low ambient dimensions the local flatness of PL embeddings is automatic, so the proof above still works.

Now suppose $K_0 \neq \emptyset$. Observe that $\mathrm{Map}(f, K_0)$ is naturally homeomorphic to a subset of $\mathrm{Map}(f)$. The hypothesis $2\dim(\mathrm{Map}(f))+1 \leq n$ allows us to extend h to an embedding of all of $\mathrm{Map}(f)$. The proof then proceeds as above, except that Lemma 5.3.11 must be replaced by a lemma that shrinks out only those fibers of $\mathrm{Map}(f)$ still present in $\mathrm{Map}(f, K_0)$. $\square$

Definition. Let K be a polyhedron and let $\lambda : K \times I \to M$ be an embedding of $K \times I$ into M. An isotopy $\phi_t : \lambda(K \times I) \to \lambda(K \times I)$ is said to *preserve fibers* if $\phi_t(\lambda(\{x\} \times I)) = \lambda(\{x\} \times I)$ for every $x \in K$ and every $t \in I$.

Definition. Assume $X \subset Y$ and $\phi : X \times I \to X$ is an isotopy. An isotopy $\Phi : Y \times I \to Y$ is said to *cover* ϕ if $\Phi|X \times I = \phi$. If $\phi_0 = \mathrm{Id}_X$, then it is required that $\Phi_0 = \mathrm{Id}_Y$ as well.

There is a minor distinction between "isotopy extension" and "covering isotopy." We speak of *extending* an isotopy $\phi : X \times I \to Y$ of X in Y to an isotopy $\tilde{\phi} : Z \times I \to Y$, where $Z \supset X$. We use the terminology *covering* an isotopy when the given isotopy ϕ is an isotopy of X in itself rather than an isotopy of X in Y. In that case we "cover" ϕ with an ambient isotopy.

As usual, we are using ϕ_t to denote the homeomorphism of X defined by $\phi_t(x) = \phi(x, t)$.

Theorem 5.3.12 (Covering Isotopy). *Let M be an n-dimensional PL manifold, K a compact k-dimensional polyhedron, $k \leq n-4$, and $\lambda : K \times I \to M$ a PL embedding. If ϕ_t is any fiber preserving PL isotopy of $\lambda(K \times I)$ such that $\phi_t|\lambda(K \times \partial I) =$ inclusion, then there exists a PL isotopy Φ_t of M that covers ϕ_t. Furthermore, if $\epsilon > 0$ is specified, then it is possible to construct Φ_t in such a way that for each $y \in M$ either $\Phi_t(y) = y$ for every t, or there exists one $x \in K$ such that the path $\Phi_t(y)$ is contained in the ϵ-neighborhood of $\lambda(\{x\} \times I)$.*

Proof. The idea is to find a product neighborhood N of $\lambda(K \times I)$ and then to use the mapping cylinder and product structures of N for extending ϕ_t over N. The extension can be tapered off along the mapping cylinder structure to make it the identity on the boundary of N, and then it can be extended via the identity to all of M. The plan is simple, but executing it requires careful specification of several functions.

Begin by constructing an n-dimensional PL ∂-manifold $N \subset M$ such that λ is a faithful embedding of $(K \times I, K \times \partial I)$ in $(N, \partial N)$. By PN(n), we may assume that there is a regular neighborhood N_0 of $\lambda(K \times \{0\})$ in ∂N and a PL homeomorphism $h : N_0 \times I \to N$ such that $h(x, 0) = x$ for every $x \in N_0$ and $h(\lambda(x, 0), s) = \lambda(x, s)$ for each $x \in K$ and $s \in I$. Since N_0 is a regular neighborhood of $\lambda(K \times \{0\})$, it can be identified with Map(r) for some PL map $r : \partial N_0 \to \lambda(K \times \{0\})$. Thus each point in $N_0 \smallsetminus \lambda(K \times \{0\})$ can be specified as $\langle x, u\rangle$, where $x \in \partial N_0$ and $u \in [0, 1)$, and each point in $N \smallsetminus \lambda(K \times I)$ has the form $h(\langle x, u\rangle, s)$ with $s \in I$. Define a function $\mu : K \times I \times I \to [-1, 1]$ by $\phi_t(\lambda(x, s)) = \lambda(x, s + \mu(x, s, t))$. (The function μ simply measures how far along the fiber $\lambda(\{x\} \times I)$ the point $\lambda(x, s)$ is moved by ϕ_t.) We then can define $\Phi_t : N \to N$ by

$$\Phi_t(h(\langle x, u\rangle, s)) = h(\langle x, u\rangle, s + \mu(r(x), s, ut))$$

for any point $h(\langle x, u\rangle, s) \in N \smallsetminus \lambda(K \times I)$ and $\Phi_t|\lambda(K \times I) = \phi_t$. It is easy to check that $\Phi_0 = \text{Id}$, $\Phi_t|\partial N_0 \times I = \text{Id}$, and that Φ_t is a homeomorphism for each t. The fact that $\phi_t|\lambda(K \times \partial I) = \text{Id}$ implies that $\Phi_t|N_0 \times \partial I = \text{Id}$, so $\Phi_t|\partial N = \text{Id}$. Therefore Φ_t can be extended via the identity to all of M.

The uniform continuity of h implies the existence of a number $\delta > 0$ such that if $d(x_1, x_2) < \delta$, then $d(h(x_1, s), h(x_2, s)) < \epsilon$ for every $x_1, x_2 \in N_0$ and for every $s \in I$. The last condition in the theorem is achieved by trimming N_0 so that $d(x, r(x)) < \delta$ for every $x \in N_0$. □

Historical Notes. Theorem 5.3.1 is due to Zeeman, (1960) and (1963a), and is known as Zeeman's Unknotting Theorem. Hudson (1966b) proved a different version of the unknotting theorem. The Concordance Implies Unknotting Theorem (Theorem 5.3.6) is due to Hudson, (1966a) and (1970), who proves the much stronger statement: concordance implies ambient isotopy. Hudson's theorem is valid without the simply connected hypothesis. Examples 5.3.2 and 5.3.4 come from (Zeeman, 1963a); Zeeman attributes Example 5.3.2 to Irwin and Example 5.3.4 to Hudson. Examples 5.3.2 and 5.3.3 are obviously generalizations of the familiar Hopf and Whitehead links in $\mathbb{R}^3$.

The Cone Unknotting Theorem (Theorem 5.3.7) is due to W. B. R. Lickorish (1965), while the Product Neighborhood Theorem (Theorem 5.3.9) is due to Lickorish and Siebenmann (1969). The fact that CU($n-1$) is needed to prove PN(n) was pointed out by Edwards (1975a), who also outlined the inductive proof CU($n-1$) $\Rightarrow$ LF(n) $\Rightarrow$ PN(n) $\Rightarrow$ CU(n). Both Lickorish (1965) and Edwards (1975a) base their proofs of the Cone Unknotting Theorem on "sunny collapsing," which is a technique introduced by Zeeman (1963b) to prove the unknotting of ball pairs theorem and later reinterpreted by Hudson (1969b) as a way to factor collapses.

Exercises

5.3.1. Prove that the embedding e in Example 5.3.3 is knotted in the sense that $e(S^k)$ does not bound a $(k+1)$-cell in $S^1 \times S^{2k}$.

5.3.2. Let $f : S^1 \times S^k \to S^{2k+2}$ be a PL embedding.
  (a) For every $a \in S^1$ there exists a PL $(k+1)$-cell A such that $\partial A = f(\{a\} \times S^k)$ and $\operatorname{Int} A \cap f(S^1 \times S^k) = \emptyset$. [Hint: Use the Whitney Lemma (Rourke and Sanderson, 1972, Theorem 5.12). You may need to introduce additional points of intersection first.]
  (b) The invariant $k(f, a, b, c)$ of Example 5.3.4 is defined for every $a, b, c \in S^1$ and depends only on f.

5.3.3. Use the Product Neighborhood Theorem to give a new proof of Trivial Range Unknotting Theorem 4.1.1.

5.4. Unknotting PL embeddings of polyhedra

Polyhedra do not unknot as readily as manifolds. In particular, the connectivity conditions of the previous section do not suffice to assure unknottedness of embedded polyhedra, as the following example illustrates.

Example 5.4.1. *If K is the polyhedron obtained by connecting two disjoint copies of S^k with an arc, then K can be knotted in S^{2k+1}.*

Proof. The example is obtained from the knotted embedding of $S^k \times S^0$ in Example 5.3.2 by connecting the two components with an arc. If the embedding of the polyhedron could be unknotted, then the embedding of $S^k \times S^0$ could be unknotted. See Figure 5.16. $\square$

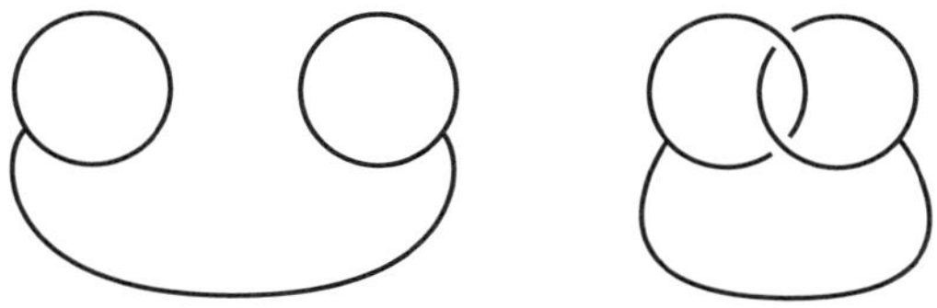

Figure 5.16. An unknotted pair of eyeglasses and a knotted pair

The example shows that we cannot expect to derive a global unknotting theorem for embedded polyhedra comparable in strength to the theorems in the previous section for embedded manifolds. Instead we will add the hypothesis that the PL embeddings to be unknotted are both close to some specified topological embedding. The topological embedding serves as a template that rectifies the potential global knotting of the embeddings considered. When the existence of this template embedding is assumed, no connectivity conditions are required.

Theorem 5.4.2 (Unknotting Close Polyhedra). *Let $e : K^k \to M^n$ be a topological embedding of a compact polyhedron K into a PL manifold M with $k \leq n-3$. For every $\epsilon > 0$ there exists a $\delta > 0$ such that if $\lambda_0, \lambda_1 : K \to M$ are two PL embeddings each pointwise within δ of e, then there is a PL ϵ-push ψ of $(M, e(K))$ such that $\lambda_0 = \psi \circ \lambda_1$. Moreover, if K_0 is a subpolyhedron of K for which $\lambda_0|K_0 = \lambda_1|K_0$, then the supporting isotopy Ψ_t can be chosen so that $\Psi_t|\lambda_1(K_0) =$ inclusion for every t.*

Example 5.4.3. *Theorem 5.4.2 fails without the topological embedding e.*

Proof. For every $\epsilon > 0$, it is possible to construct the nonequivalent embeddings of Example 5.4.1 so that the distance between them is less than ϵ. □

Example 5.4.4. *Theorem 5.4.2 fails if the domain is a codimension-three compactum rather than a polyhedron.*

Proof. For example, let X be the compactum constructed by stringing together a null-sequence of k-spheres along an arc. The obvious embedding of X into S^{2k+1} can be arbitrarily closely approximated by embeddings of the sort illustrated in Figure 5.17. □

The proof of Theorem 5.4.2 is based on two somewhat more technical results: an engulfing theorem and an unknotting theorem for embeddings in a regular neighborhood. Those two results are stated next.

Theorem 5.4.5. *Let K be a compact k-dimensional polyhedron and let $e : K \to M$ be a topological embedding of K into an n-dimensional PL manifold, $n \geq k+3$. For every $\epsilon > 0$ there exists a $\delta > 0$ such that if $\lambda_0, \lambda_1 : K \to M$ are two PL embeddings within δ of e and N_0 is any regular*

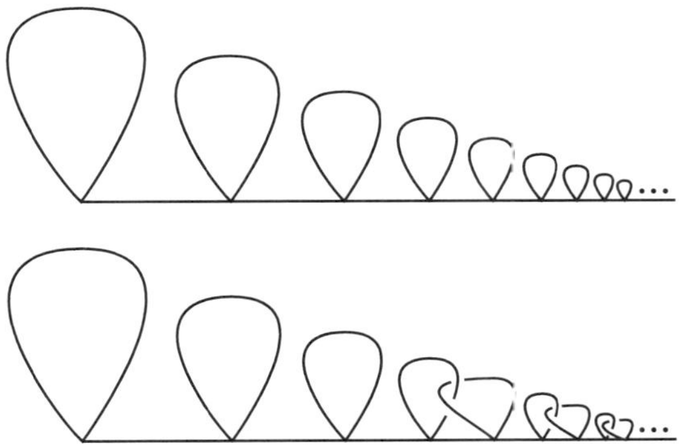

Figure 5.17. A compactum and a nearby knotted version

neighborhood of $\lambda_0(K)$ *in* M*, then there exists a PL* ϵ*-push* ψ *of* $(M, e(K))$ *such that* $\psi(\lambda_1(K)) \subset N_0$*. Moreover, if* K_0 *is a subpolyhedron of* K *and* $\lambda_0|K_0 = \lambda_1|K_0$*, then* ψ_t *can be chosen so that* $\psi_t|\lambda_1(K_0) =$ *inclusion for every* t*.*

Proof. This is a standard engulfing argument. The existence of the necessary short homotopies follows from Corollary 0.6.5. The existence of the ϵ-push is confirmed by an application of Generalized Controlled Engulfing Theorem 3.3.7. □

Remark. It is in the proof of Theorem 5.4.5 that the template embedding e is needed. Because e is known at the outset, we can use Corollary 0.6.5 to specify an arbitrarily long but finite sequence of neighborhoods of $e(K)$ such that for each neighborhood there is a short homotopy through the previous neighborhood into $\lambda_0(K)$. The key fact is that the neighborhoods themselves can be specified first, before the particular PL embeddings λ_0 and λ_1 are known. If the embedding e were determined at the same time as λ_0 and λ_1, we still would be able to find the first of the required homotopies, but there would be no way to obtain the others.

Definition. Let K be a compact polyhedron. An *abstract regular neighborhood of* K is any compact PL ∂-manifold N such that $K \subset \operatorname{Int} N$ and $N \searrow K$. The retraction $r : N \to K$ induced by the mapping cylinder structure of N is called a *contractible retraction* because the point preimages under r are contractible subsets of N.

Theorem 5.4.6 (Unknotting in a Regular Neighborhood)**.** *Let* K *be a compact* k*-dimensional polyhedron and let* n *be an integer with* $n \geq k + 3$*. For every* $\epsilon > 0$ *there exists a* $\delta > 0$ *such that if* N *is any* n*-dimensional abstract regular neighborhood of* K *with contractible retraction* $r : N \to K$ *and*

$\lambda : K \to \operatorname{Int} N$ is a PL embedding such that $\rho(r \circ \lambda, \mathrm{Id}_K) < \delta$, then there exists a PL isotopy $\psi_t : N \to N$ such that

(1) $\psi_0 = \mathrm{Id}_N$,

(2) $\psi_t | \partial N = \mathrm{Id}_{\partial N}$ *for every* t,

(3) $\rho(r \circ \psi_t, r) < \epsilon$ *for every* t, *and*

(4) $\psi_1 \circ \lambda = \mathrm{Id}_K$.

Moreover, if K_0 is any subpolyhedron of K such that $\lambda | K_0 =$ inclusion, then ψ_t can be constructed so that $\psi_t | K_0 =$ inclusion for every t.

Remark. It is critically important to note that the number δ in Theorem 5.4.6 depends only on K, n, and ϵ and is independent of the particular abstract regular neighborhood N.

The proof of Theorem 5.4.6 will occupy most of the remainder of this section. For now let us see how it implies Theorem 5.4.2.

Proof of Theorem 5.4.2. Let $\epsilon > 0$ and $e : K \to M$ be given. Choose $\delta_1 > 0$ to satisfy the conclusion of Theorem 5.4.6 with input $\epsilon/2$. (Use the metric on K determined by considering $e(K)$ to be a subspace of M.) Then choose $\delta > 0$ to satisfy the conclusion of Theorem 5.4.5 with input $\delta_1/4$.

Now consider PL embeddings $\lambda_0, \lambda_1 : K \to M$ as in the theorem. Take a close regular neighborhood N_0 of $\lambda_0(K)$ with collapsible retraction $r : N_0 \to \lambda_0(K)$ such that the distance (in M) from x to $r(x)$ is less than δ. By the choice of δ there is a $\delta_1/8$-push ψ_1 such that $\psi_1(\lambda_1(K)) \subset N_0$. It is easy to check that the distance from λ_0 to $r \circ \psi_1 \circ \lambda_1$ is less than δ_1 (as measured in $e(K)$). The choice of δ_1 guarantees the existence of an $(\epsilon/2)$-push from $\psi_1 \circ \lambda_1$ to λ_0. The concatenation of the two pushes is the push we need.

If K_0 is a subpolyhedron of K for which $\lambda_0 | K_0 = \lambda_1 | K_0$, then both Theorem 5.4.6 and Theorem 5.4.5 allow $\lambda_1(K_0)$ to be kept fixed. Hence the final push will keep $\lambda_1(K_0)$ fixed as well. □

We turn now to the proof of Unknotting in a Regular Neighborhood Theorem 5.4.6. The proof consists of four parts.

Step 1: The polyhedron K is subdivided into cones that fit together nicely and for which there is a corresponding handle decomposition of the abstract regular neighborhood N.

Step 2: A slicing theorem allows the embedded polyhedron to be adjusted so that the frontiers of the cones are mapped setwise into the boundaries of the corresponding handles of N.

Step 3: The $(n-1)$-dimensional case of the Unknotting in a Regular Neighborhood Theorem is applied in the boundaries of the handles to make the frontiers of the cones pointwise match the corresponding subsets of K.

Step 4: The Cone Unknotting Theorem is applied to straighten out the embedding on the interiors of the cones.

Step 3 is really a double induction: the frontiers of the cones are straightened out inductively (first the frontiers of the cones in the 0-handles, then those in the 1-handles, etc.) and each frontier is straightened by applying a lower-dimensional case of the theorem itself. When the frontiers of the cones in the i-handles are straightened, the frontiers of those in the $(i-1)$-handles have already been straightened and it is necessary to keep them fixed. For this reason all the results in the remainder of this section must be stated and proved for pairs of polyhedra. This adds considerable complexity to the statements of the various lemmas, but it is an unavoidable consequence of the overall structure of the proof of the unknotting theorem. We proceed by stating relative versions of the technical tools from the previous section.

Theorem 5.4.7 (Relative Cone Unknotting). *Let K be a compact k-dimensional polyhedron and let L be a subpolyhedron with* $\dim L \leq k-1$. *If the pair (K, L) is faithfully contained in $(B^n, \partial B^n)$, $n \geq k+3$, and $e : (cK, K) \to (cB^n, B^n)$ is a faithful PL embedding such that $e^{-1}(c\partial B^n, \partial B^n) = (cL, L)$ and* $e|K \cup cL = \text{incl}$, *then there exists a PL homeomorphism $\phi : cB^n \to cB^n$ such that* $\phi|B^n \cup c\partial B^n = \text{Id}$ *and* $\phi \circ e = \text{incl}_{cK}$.

Definition. Suppose (K, L) is a pair of compact polyhedra, (Q, W) is a pair of PL ∂-manifolds such that W is a codimension-zero submanifold of ∂Q, the pair $(K \times I, L \times I)$ is faithfully contained in (Q, W), and

$$(K \times I) \cap (\partial Q \setminus \operatorname{Int} W) = K \times \partial I.$$

A regular neighborhood pair (N, M) of $(K \times I, L \times I)$ in (Q, W) is said to *meet the boundary regularly* if it is possible to write $N \cap (\partial Q \setminus \operatorname{Int} W) = N_0 \sqcup N_1$ and $M \cap (\partial Q \setminus \operatorname{Int} W) = M_0 \sqcup M_1$ in such a way that (N_i, M_i) is a regular neighborhood pair for $(K \times \{i\}, L \times \{i\})$ in $(\partial Q \setminus \operatorname{Int} W, \partial W)$ and $(M, M_0 \cup M_1)$ is a regular neighborhood of $(L \times I, L \times \partial I)$ in $(W, \partial W)$. (See Figure 5.18.)

Theorem 5.4.8 (Relative Product Neighborhood). *Let K, L, Q, W, M, N, M_0, M_1, N_0, and N_1 be as in the definition above, where* $\dim L \leq \dim K - 1$ *and* $\dim K \leq \dim Q - 4$. *Then there exists a PL homeomorphism $\phi : N_0 \times I \to N$ such that $\phi(M_0 \times I) = M$, $\phi(N_0 \times \{1\}) = N_1$, $\phi(M_0 \times \{1\}) = M_1$ and $\phi|(N_0 \times \{0\}) \cup (K \times I) = \text{Id}$.*

Theorems 5.4.7 and 5.4.8 are proved in the same way as are their absolute counterparts in §5.3. In particular, the Relative Cone Unknotting

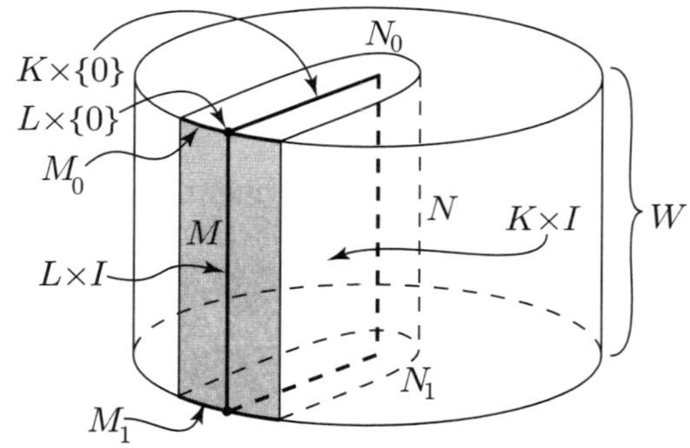

Figure 5.18. In this figure Q is the solid cylinder, W consists of the vertical walls of the cylinder, K is a segment, L is an endpoint of K, and M is the shaded rectangle

Theorem follows from the Relative Product Neighborhood Theorem by essentially the same argument as in the absolute case. The Relative Cone Unknotting Theorem in ambient dimension $n-1$ implies a Relative Local Flatness Theorem in ambient dimension n which, in turn, implies the Relative Product Neighborhood Theorem by the same argument as in §5.3. The following Relative Covering Isotopy Theorem is a consequence of the Relative Product Neighborhood Theorem.

Theorem 5.4.9 (Relative Covering Isotopy). *Let K be a compact k-dimensional polyhedron, let L be a subpolyhedron with $\dim L \le k-1$, and let M be an n-dimensional PL ∂-manifold, $k \le n-4$. If $\lambda : (K\times\mathbb{R}, L\times\mathbb{R}) \to (M, \partial M)$ is a faithful PL embedding and $\phi_t : \lambda(K\times\mathbb{R}) \to \lambda(K\times\mathbb{R})$ is a fiber preserving isotopy with compact support, then there exists an ambient isotopy Φ_t of M that covers ϕ_t and also has compact support. If $\phi_t|\lambda(L\times\mathbb{R}) = \mathrm{Id}$, then $\Phi_t|\partial M = \mathrm{Id}$. Furthermore, if $\epsilon > 0$ is specified, then it is possible to construct Φ_t in such a way that for each $y \in M$ either $\Phi_t(y) = y$ for every t or there exists one $x \in K$ such that the path $\Phi_t(y)$ is contained in the ϵ-neighborhood of $\lambda(\{x\}\times\mathbb{R})$.*

Proof. There is a compact interval J such that the support of every ϕ_t is contained in $\lambda(K\times J)$. Find a PL ∂-manifold $M_1 \subset M$ such that $M_1 \cap \lambda(K\times\mathbb{R}) = \lambda(K\times J)$ and $\lambda|K\times J : (K\times J, K\times\partial J) \to (M_1, \partial M_1)$ is faithful. The proof of Covering Isotopy Theorem 5.3.12, with the absolute Product Neighborhood Theorem replaced by the Relative Product Neighborhood Theorem, gives an isotopy of M_1 that covers $\phi_t|\lambda(K\times J)$. Extend via the identity to construct the required isotopy of M. □

We are now ready for the Slicing Lemma. In the statement of the Slicing Lemma there are several projection maps. We use π to denote the projections $\pi : M\times\mathbb{R} \to \mathbb{R}$ and $\pi : K\times[-1,1] \to [-1,1] \subset \mathbb{R}$, while we use π_M

to denote the projection $\pi_M : M \times \mathbb{R} \to M$. We make the following *ad hoc* definition in order to simplify the statement of the theorem.

Definition. A map $f : K \times [-1,1] \to M \times \mathbb{R}$ is *sliced* if $f(K \times [-1,0)) \subset M \times (-\infty, 0)$, $f(K \times \{0\}) \subset M \times \{0\}$, and $f(K \times (0,1]) \subset M \times (0,\infty)$.

Theorem 5.4.10 (Slicing Lemma). *Let (K, L) be a pair of compact polyhedra with $\dim K = k$ and $\dim L \leq k-1$ and let M be an n-dimensional PL ∂-manifold, $n \geq k+3$. For every $\epsilon > 0$ there exists a $\delta > 0$ such that if $\lambda : (K \times [-1,1], L \times [-1,1]) \to (M \times \mathbb{R}, \partial M \times \mathbb{R})$ is a faithful PL embedding such that $f|L \times [-1,1] : L \times [-1,1] \to \partial M \times \mathbb{R}$ is sliced and $d(\pi(\lambda(w)), \pi(w)) < \delta$ for every $w \in K \times [-1,1]$, then there exists a PL isotopy ψ_t of $M \times \mathbb{R}$ such that $\psi_0 = \mathrm{Id}$, $\psi_t|\partial M \times \mathbb{R} = \mathrm{Id}$, and $\psi_1\lambda : K \times [-1,1] \to M \times \mathbb{R}$ is sliced.*

For every $\eta > 0$, ψ_t can be constructed to have the following additional property: For each $z \in M \times \mathbb{R}$, either the path $\psi_t(z)$ is constant or there exists one $x \in K$ such that the entire path $\psi_t(z)$ lies in $N_\eta(\pi_M(\lambda(\{x\} \times [-\epsilon, \epsilon]))) \times [-\epsilon, \epsilon]$, where N_η denotes the η-neighborhood in M. Moreover, if K_0 is a subpolyhedron of K such that $\lambda(x,t) = \langle \pi_M \lambda(x,0), t\rangle$ for every $\langle x, t\rangle \in K_0 \times [-1,1]$, then ψ_t can be constructed so that $\psi_t|\lambda(K_0 \times [-1,1]) =$ inclusion.

The Slicing Lemma has an interesting corollary regarding stability of embeddings. The proof of the corollary is left as an exercise.

Corollary 5.4.11 (Stability of Embedding). *Let K be a compact k-dimensional polyhedron and let M be an n-dimensional compact PL manifold, $n \geq k+3$. Then K can be PL embedded in M if and only if $K \times \mathbb{R}$ can be properly PL embedded in $M \times \mathbb{R}$.*

The proof of Slicing Theorem 5.4.10 is based on shadow building techniques. Let us clarify what is meant by a shadow in this context. Assume K is a polyhedron and M is a ∂-manifold. For $X \subset K \times [-1,1]$, define the *shadow* of X by

$$\mathrm{Sh}(X) = \{\langle x, s\rangle \in K \times [-1,1] \mid \langle x, t\rangle \in X \text{ for some } t \geq s\}.$$

For $Y \subset M \times \mathbb{R}$ the *shadow* of Y, $\mathrm{Sh}(Y)$, is defined similarly. Note that the same notation is used for shadows in either setting.

If $C \subset K \times [-1,1]$ is a subpolyhedron such that $C \supset K \times [-1,0]$ and $\mathrm{Sh}(C) = C$, then C has a regular neighborhood N in $K \times [-1,1]$ whose frontier intersects each vertical fiber $\{x\} \times [-1,1]$ in exactly one point. Such a PL neighborhood will be called *vertically simple*. Given any neighborhood U of C in $K \times [-1,1]$, there is a vertically simple regular neighborhood N of C such that $C \subset N \subset U$. Vertically simple neighborhoods of subsets of

$M \times \mathbb{R}$ are defined similarly. A subset D of $M \times \mathbb{R}$ such that $D \supset M \times (-\infty, 0]$ and $\mathrm{Sh}(D) = D$ has close vertically simple regular neighborhoods in $M \times \mathbb{R}$ (see Figure 5.19).

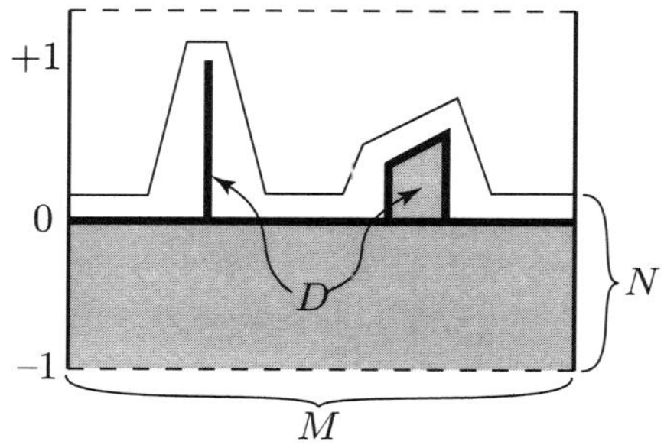

Figure 5.19. N is a vertically simple neighborhood of D

Proof of Theorem 5.4.10. Let us first consider the case in which $L, \partial M$, and K_0 are all empty. The idea of the proof is to construct subpolyhedra C of $K \times [-1, 1]$ and D of $M \times (-\infty, 1]$ satisfying the following conditions.

(1) $C = \mathrm{Sh}(C)$ and $C \supset K \times [-1, 0]$,

(2) $D = \mathrm{Sh}(D)$ and $D \supset M \times (-\infty, 0]$,

(3) $C = \lambda^{-1}(D)$,

(4) $C \setminus K \times [-1, 0] \subset K \times (0, \delta_1)$ for some small number $\delta_1 > 0$, and

(5) $D \setminus M \times (-\infty, 0] \subset N_\epsilon(\lambda(K \times [-1, 1]) \cap M \times \{0\})$.

Once the polyhedra C and D have been constructed, it is quite easy to obtain the isotopy ψ_t needed to complete the proof. Triangulate $K \times [-1, 1]$ and $M \times \mathbb{R}$ compatibly and then take a derived subdivision of each so that the simplicial neighborhood N_1 of C in $K \times [-1, 1]$ is a vertically simple regular neighborhood and so that $N = \lambda(N_1)$ is the simplicial neighborhood of D in $M \times \mathbb{R}$. Choose a nearby vertically simple neighborhood N' of D. There is a fiber preserving isotopy of $K \times [-1, 1[$ that pushes $K \times \{0\}$ to ∂N_1. By Theorem 5.4.9 this isotopy is covered by an isotopy of $M \times \mathbb{R}$ that pushes $\lambda(K \times \{0\})$ into the boundary of N. There is a second isotopy of $M \times \mathbb{R}$ that pushes N to N' and keeps D fixed and there is a vertical isotopy of $M \times \mathbb{R}$ that takes the boundary of N' to $M \times \{0\}$. The concatenation of the three isotopies is the isotopy ψ_t of the conclusion of the theorem. The motion in Theorem 5.4.9 can be kept arbitrarily near to fibers of λ and the other isotopies are both small in $M \times \mathbb{R}$, so the final isotopy satisfies the size constraints specified in the theorem.

In order to complete the proof of the special case of the theorem it remains only to construct C and D. The construction involves the back and

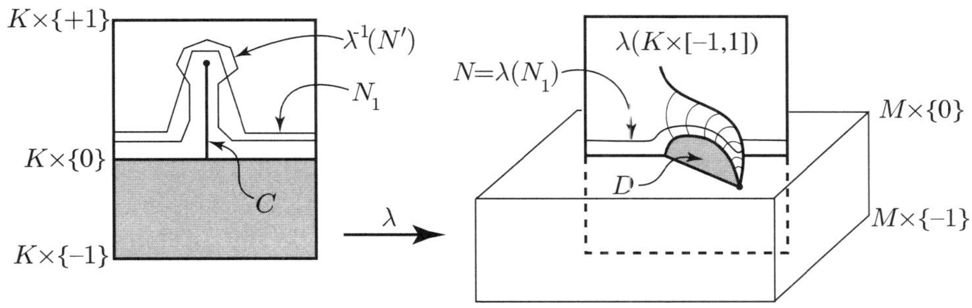

Figure 5.20. In this figure, $\lambda(K \times (0,1]) \cap M \times (-\infty, 0]$ consists of a single point.

forth inductive shadow building that is common to most codimension-three engulfing proofs. To begin the induction, shift λ into general position and define $C_0 = \mathrm{Sh}(\lambda^{-1}(M \times (-\infty, 0])) \cup K \times [-1, 0]$ and $D_0 = \mathrm{Sh}(\lambda(C_0)) \cup M \times (-\infty, 0]$. In general, we must expect that $\dim C_0 = k + 1$.

The next step is to shift λ into general position again, this time keeping C_0 fixed, and then define $C_1 = C_0 \cup \mathrm{Sh}(\lambda^{-1}(D_0))$ and $D_1 = D_0 \cup \mathrm{Sh}(\lambda(C_1))$. Note that $\dim(C_1 \setminus C_0) \leq k$. Now shift λ into general position once more, keeping C_1 fixed, and define $C_2 = C_1 \cup \mathrm{Sh}(\lambda^{-1}(D_1))$ and $D_2 = D_1 \cup \mathrm{Sh}(\lambda(C_2))$. Then $\dim(C_2 \setminus C_1) \leq k - 1$, so the induction is under way. It results, after k steps, in the polyhedra C and D that we need.

In case L and ∂M are nonempty, we must take care to ensure that $C \cap (L \times [-1, 1]) = L \times [-1, 0]$ and $D \cap (\partial M \times \mathbb{R}) = \partial M \times (-\infty, 0]$. If that is done, then the isotopy ψ_t will be the identity on $\partial M \times \mathbb{R}$. Finally, if $K_0 \neq \emptyset$, the same proof still works. We merely need to add the requirement that $\lambda(K_0 \times [-1, 1])$ be kept fixed when the embedding is shifted into general position in $\mathrm{Int}\, M \times \mathbb{R}$. Then, for each i, $C_i \cap (K_0 \times (0, 1]) = \emptyset$ and $D_i \cap (\lambda(K_0 \times (0, 1]) = \emptyset$, so the isotopy constructed will leave $\lambda(K_0 \times [-1, 1])$ fixed. □

Next we generalize the definition of abstract regular neighborhood to the case of pairs. Using the expanded terminology we can state the relative version of the Unknotting in a Regular Neighborhood Theorem that will form the basis for the inductive proof.

Definition. Let (K, L) be a pair of compact polyhedra. An n-dimensional *abstract regular neighborhood of* (K, L) is a PL ∂-manifold pair (N^n, M^{n-1}) such that $M \subset \partial N$, $L = K \cap \partial N$, $(K, L) \subset (\mathrm{Int}\, N \cup \mathrm{Int}\, M, \mathrm{Int}\, M)$, and there exists a contractible retraction $r : N \to K$ for which $r|M$ is a contractible retraction of M to L.

Figure 5.21 shows an example of an abstract regular neighborhood of a pair. Observe that $M = r^{-1}(L) \cap \partial N$ is *not* required.

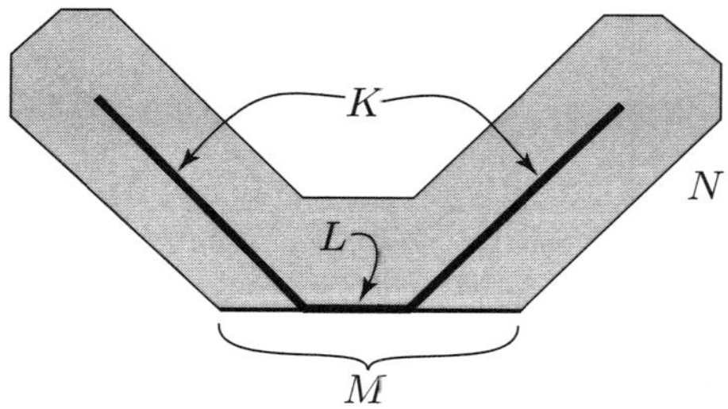

Figure 5.21. An abstract regular neighborhood of the pair (K, L)

Theorem 5.4.12 (Unknotting in a Regular Neighborhood of a Pair). *Let (K, L) be a compact polyhedral pair with $\dim K = k \leq n - 3$ and $\dim L \leq k - 1$. For every $\epsilon > 0$ there exists a $\delta > 0$ such that if (N, M) is any n-dimensional abstract regular neighborhood of (K, L) with contractible retraction $r : N \to K$ and $\lambda : (K, L) \to (\operatorname{Int} N \cup \operatorname{Int} M, \operatorname{Int} M)$ is a faithful PL embedding such that $\lambda | L = \operatorname{Id}_L$ and $\rho(r \circ \lambda, \operatorname{Id}_K) < \delta$, then there exists a PL isotopy $\psi_t : (N, M) \to (N, M)$ such that*

(1) $\psi_0 = \operatorname{Id}_N$,

(2) $\psi_t | \partial N = \operatorname{Id}_{\partial N}$ *for every* t,

(3) $\rho(r \circ \psi_t, r) < \epsilon$ *for every* t, *and*

(4) $\psi_1 \circ \lambda = \operatorname{Id}_K$.

Moreover, if P is any subpolyhedron of K such that $\lambda | P = \operatorname{incl}_P$, then ψ_t can be constructed so that $\psi_t | P = \operatorname{incl}_P$ for every t.

It is clear that Theorem 5.4.6 is a special case of Theorem 5.4.12, so it suffices to prove the latter.

The handle-like decomposition. As indicated earlier, the proof of Theorem 5.4.12 uses a decomposition of K into cones. We now construct that decomposition. The process is based on the standard construction of a handle decomposition of a PL manifold from a second derived subdivision and the cones constructed in the polyhedron K share many of the properties of the handles in a handle decomposition of a manifold. For this reason we will refer to the decomposition as a *handle-like decomposition* of K. The background for the construction is the material on pages 81–83 of (Rourke and Sanderson, 1972).

Let $\epsilon > 0$ and the compact polyhedral pair (K, L) be given. Fix a triangulation T of K. Subdivide if necessary so that T restricts to a triangulation of L. For each $\sigma \in T$, let $\hat{\sigma}$ denote the barycenter of σ. Let T'' denote the second barycentric subdivision of T. Define H_σ to be the underlying polyhedron of the star of $\hat{\sigma}$ in T'' and define G_σ to be the intersection with L; i.e., $H_\sigma = \mathrm{St}(\hat{\sigma}, T'')$ and, in case $\sigma \subset L$, $G_\sigma = \mathrm{St}(\hat{\sigma}, T''|L)$. We may assume that the mesh of T is small enough that $\operatorname{diam} H_\sigma < \epsilon/2$ for every $\sigma \in T$.

The decomposition $\{H_\sigma\}$ of the polyhedron K is like a handle decomposition in that the H_σ cover K, each H_σ is a cone, and $H_\sigma \cap H_\tau = \emptyset$ whenever $\dim \sigma = \dim \tau$ and $\sigma \neq \tau$. There is a natural PL homeomorphism $\phi : H_\sigma \to \sigma \times (\hat{\sigma} * Lk(\sigma, T))$. In case $\sigma \subset L$, ϕ is a homeomorphism of pairs

$$\phi : (H_\sigma, G_\sigma) \cong (\sigma \times (\hat{\sigma} * Lk(\sigma, T)), \sigma \times (\hat{\sigma} * Lk(\sigma, T|L))).$$

We use ϕ to identify the following subsets of H_σ (see Figure 5.22).

$$A_\sigma = \phi^{-1}(\sigma \times (\mathrm{Lk}(\sigma, T))),$$
$$B_\sigma = \phi^{-1}(\partial\sigma \times (\hat{\sigma} * \mathrm{Lk}(\sigma, T))),$$
$$C_\sigma = \phi^{-1}(\partial\sigma \times (\mathrm{Lk}(\sigma, T))) = A_\sigma \cap B_\sigma, \text{ and}$$
$$D_\sigma = \phi^{-1}(\sigma \times (\mathrm{Lk}(\sigma, T|L))) \text{ (in case } \sigma \subset L).$$

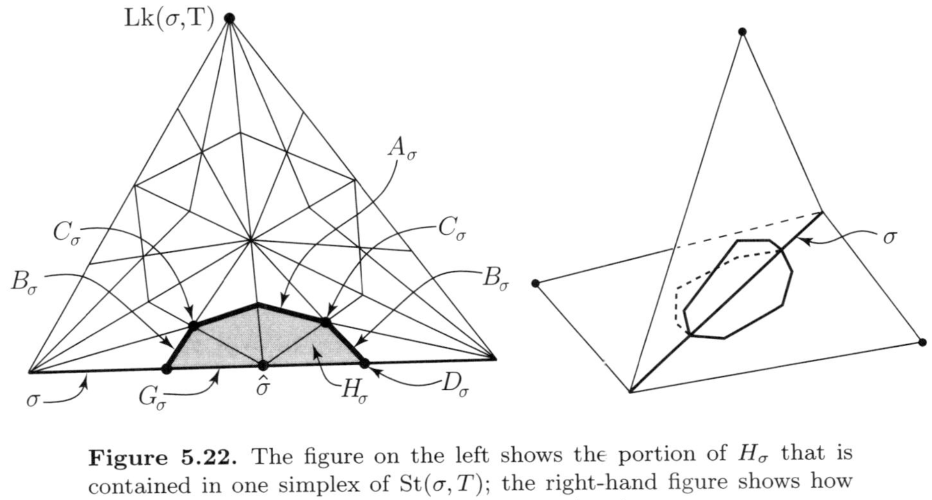

Figure 5.22. The figure on the left shows the portion of H_σ that is contained in one simplex of $\mathrm{St}(\sigma, T)$; the right-hand figure shows how several such pieces combine to form $H_\sigma \subset \mathrm{St}(\sigma, T)$.

Let us define

$$H_i = \bigcup\{H_\sigma \mid \dim \sigma = i\} \text{ and } G_i = \bigcup\{G_\sigma \mid \dim \sigma = i\}.$$

Define A_i, B_i, and C_i similarly. Then define $K_i = \cup\{H_j \mid j \leq i\}$ and $L_i = \cup\{G_j \mid j \leq i\}$. Observe that $K_0 \subset K_1 \subset \cdots \subset K_k = K$ and $L_0 \subset L_1 \subset \cdots \subset L_{k-1} = L$. The essential properties of the handle-like decompositions $\{H_\sigma\}$, $\{G_\sigma\}$ and the filtrations $\{K_i\}$, $\{L_i\}$ are as follows.

(1) Each H_σ and each G_σ is a cone; in particular, $H_\sigma = \hat{\sigma} * \mathrm{Lk}(\hat{\sigma}, T'')$ and $G_\sigma = \hat{\sigma} * \mathrm{Lk}(\hat{\sigma}, T''|L)$.

(2) $\mathrm{Fr}\, H_\sigma = \mathrm{Lk}(\hat{\sigma}, T'') = A_\sigma \cup B_\sigma$. (Here Fr denotes the frontier in K.)

(3) $\mathrm{Fr}\, K_{i-1} = (\mathrm{Fr}\, K_i \setminus A_i) \cup B_i$.

(4) $(K_i, \mathrm{Fr}\, K_i, L_i) = (K_{i-1}, \overline{\mathrm{Fr}\, K_{i-1} \setminus B_i}, L_{i-1}) \cup (H_i, A_i, G_i)$.

(5) $(B_i, C_i, D_i) = (K_{i-1}, \overline{\mathrm{Fr}\, K_{i-1} \setminus B_i}, L_{i-1}) \cap (H_i, A_i, G_i)$.

(6) The pair $(B_i, C_i \cup D_i)$ is collared in $(K_{i-1}, \overline{\mathrm{Fr}\, K_{i-1} \setminus B_i} \cup L_{i-1})$ and in $(H_i, A_i \cup G_i)$ and is, therefore, bicollared in $(K_i, \mathrm{Fr}\, K_i \cup L_i)$.

Next we state a lemma that formalizes the inductive step in the proof of Theorem 5.4.12. The lemma says, roughly, that if K can be split into two pieces in such a way that the intersection of the two pieces is collared in each, then there exist a corresponding decomposition of N and an isotopy that pushes λ restricted to the intersection of the two pieces back to the identity and pushes λ restricted to each piece into the appropriate part of N.

Lemma 5.4.13. *Assume Theorem 5.4.12 in ambient dimension $n-1$ and let (K, L) be a compact polyhedral pair with $\dim L \leq \dim K - 1 \leq n-4$. Suppose (K, L) can be written as $(K, L) = (K_1, L_1) \cup (K_2, L_2)$, where K_1 and K_2 are subpolyhedra of K and $L_i = L \cap K_i$, in such a way that the intersection $(K_0, L_0) = (K_1 \cap K_2, L_1 \cap L_2)$ is bicollared in (K, L). Then for every $\epsilon > 0$ there exists a $\delta > 0$ such that, given*

(1) *an n-dimensional abstract regular neighborhood (N, M) of (K, L) with contractible retraction $r : N \to K$, and*

(2) *a faithful PL embedding $\lambda : (K, L) \to (N, M)$ such that $\lambda|L = \mathrm{incl}_L$ and $\rho(r \circ \lambda, \mathrm{Id}_K) < \delta$,*

there exist

(3) *a regular neighborhood (N', M') of (K, L) in (N, M) and a contractible retraction $r' : N' \to K$ arbitrarily close to $r|N'$ such that*
   (a) *$\lambda(K) \subset M'$,*
   (b) *$(N_0, M_0) = ((r')^{-1}(K_0), (r'|M)^{-1}(L_0))$ is an abstract regular neighborhood of (K_0, L_0), and*
   (c) *$(N_i, M_i) = ((r')^{-1}(K_i), N_0 \cup (r'|M)^{-1}(L_i))$, $i = 1, 2$, is an abstract regular neighborhood of $(K_i, K_0 \cup L_i)$, with contractible retractions induced by r', and*

(4) *a PL isotopy ψ_t of N such that*
   (a) *$\psi_0 = \mathrm{Id}_N$,*
   (b) *$\psi_t|\partial N = \mathrm{Id}_{\partial N}$,*
   (c) *$\rho(r\psi_t, r) < \epsilon$,*
   (d) *$\psi_1 \lambda | K_0 = \mathrm{incl}_{K_0}$, and*

(e) $\psi_1\lambda|K_i$ *is a faithful embedding of* $(K_i, K_0 \cup L_0)$ *into* $(N_i, \partial N_i)$ *for* $i = 1, 2$.

Furthermore, if P *is a subpolyhedron of* K *such that* $P_0 = P \cap K_0$ *is bicollared in* K *compatibly with* (K_0, L_0) *in* (K, L) *and if* $\lambda|P = \mathrm{Id}_K$, *then* ψ_t *may be chosen in such a way that* $\psi_t|P = \mathrm{incl}_P$.

Proof of Theorem 5.4.12. Let $\lambda : (K, L) \to (N, M)$ be as in the statement of the theorem. Lemma 5.4.13 will be used to successively adjust the following polyhedra.

$$(K_k, L_k) = (K_{k-1}, L_{k-1}) \cup (H_k, G_k)$$

$$(K_{k-1}, \mathrm{Fr}\, K_{k-1} \cup L_{k-1}) = (K_{k-2}, \overline{\mathrm{Fr}\, K_{k-2} \setminus B_{k-1}} \cup L_{k-2}) \cup (H_{k-1}, A_{k-1} \cup G_{k-1})$$

$$\vdots$$

$$(K_1, \mathrm{Fr}\, K_1 \cup L_1) = (H_0, \overline{\mathrm{Fr}\, K_0 \setminus B_1} \cup G_0) \cup (H_1, A_1 \cup G_1).$$

Even though Lemma 5.4.13 will be used to adjust K_k first, then $K_{k-1}, \ldots,$ and finally K_1, the lemma must actually be applied in reverse order so that all of the ϵ's and δ's are chosen before the first adjustment is made. Let ϵ be the positive number given in the statement of Theorem 5.4.12 and use $\epsilon/4$ as the input for the application of Lemma 5.4.13 to the last pair of polyhedra listed. The δ that results from this application of the lemma becomes the ϵ for the application of the lemma to the preceding polyhedra, and so on. The δ of the conclusion of Theorem 5.4.12 is the δ that results from the final application of Lemma 5.4.13 to (K_k, L_k). [Note that $L_k = L_{k-1}$ and that $G_k = \emptyset$.] The relationship between ϵ and δ in Lemma 5.4.13 depends only on the polyhedra involved and does not depend on the particular abstract regular neighborhoods or embeddings. Since we have identified all the K_i and L_i before the beginning of this proof, it is possible to choose the ϵ's and δ's in this way.

Now suppose (N, M), $r : N \to K$, and $\lambda : K \to N$ are as in Theorem 5.4.12 with δ chosen as in the preceding paragraph. Apply Lemma 5.4.13 to $(K, L) = (K_k, L_k) = (K_{k-1}, L_{k-1}) \cup (H_k, G_k)$ to obtain a new regular neighborhood N', a new contractible retraction $r' : N' \to K$, and the PL isotopy ψ_t. Define $(N_{k-1}, M_{k-1}) = ((r')^{-1}(K_{k-1}), (r')^{-1}(\mathrm{Fr}\, K_{k-1} \cup L_{k-1}))$. Observe that $\overline{N' \setminus N_{k-1}}$ consists of a finite number of n-cells N_σ, one for each k-simplex $\sigma \subset K$, and that $\psi_1 \circ \lambda|H_\sigma : H_\sigma \to N_\sigma$ is a faithful PL embedding that is the identity on $\mathrm{Fr}\, H_\sigma$. Next apply Lemma 5.4.13 to (K_{k-1}, L_{k-1}), decomposed as above, along with the regular neighborhood (N_{k-1}, M_{k-1}) and the PL embedding $\psi_1 \circ \lambda|K_{k-1}$. The result is a smaller regular neighborhood and new PL isotopy. Restrict to K_{k-2} and apply the lemma again. Continue this process, successively applying the lemma to the polyhedra listed at the beginning of this proof.

After all the applications of the lemma have been completed, λ has been isotoped to a new PL embedding of K. In addition, for each simplex $\sigma \subset K$ we have identified an abstract regular neighborhood (N_σ, M_σ) of (H_σ, G_σ) such that the various N_σ have disjoint interiors. If we let λ_σ denote the restriction of the new embedding to H_σ, then $\lambda_\sigma : (H_\sigma, \mathrm{Fr}\, H_\sigma \cup G_\sigma) \to (N_\sigma, M_\sigma)$ is a faithful PL embedding with $\lambda_\sigma | \mathrm{Fr}\, H_\sigma \cup G_\sigma = \mathrm{Id}$. We can therefore apply Relative Cone Unknotting Theorem 5.4.7 to each ball-cone pair (N_σ, H_σ) to unknot λ_σ by an ambient isotopy that is the identity on $\partial N_\sigma \cup G_\sigma$. The size of this last isotopy is limited by the size of the H_σ, so it is small when measured in K. Since the N_σ have disjoint interiors we can unknot all the H_σ at once to isotope λ back to the identity. $\square$

This leaves only the proof of Lemma 5.4.13. The conclusion of the lemma asserts the existence of a new abstract regular neighborhood N' and the existence of the PL isotopy ψ_t. The next proof shows that we really only need to construct N'; then the existence of ψ_t will follow from the Slicing Lemma and the inductive hypothesis.

Proof of Lemma 5.4.13. Given $K = K_1 \cup K_2$ as in the statement of Lemma 5.4.13, apply Lemma 5.4.14, below, to replace N with an abstract regular neighborhood N' that is a product $N_0 \times \mathbb{R}$ near $K_0 = K_1 \cap K_2$. Then use Slicing Theorem 5.4.10 to adjust λ so that K_0 is mapped setwise into N_0. Finally, apply the inductive hypothesis in N_0 to isotope $\lambda | K_0$ to the identity. The isotopy can be extended to N' using the product structure.

Given $\epsilon > 0$, the inductive hypothesis provides a $\delta_0 > 0$ so that if $\rho(\lambda | K_0, \mathrm{incl}) < \delta_0$, then the isotopy of N_0 is limited in size by ϵ. Use δ_0 as the ϵ in the Slicing Lemma to obtain a second δ. This δ can serve as the δ of the conclusion of Lemma 5.4.13 because the adjustment in Lemma 5.4.14 can be made arbitrarily small, independent of any other size considerations. $\square$

Lemma 5.4.14. *Let (K, L), (N, M), $r : N \to K$, and $\lambda : K \to N$ be as in the statement of Lemma 5.4.13. Assume that the bicollar on (K_0, L_0) is realized by the inclusion $(K_0 \times [-2, 2], L_0 \times [-2, 2]) \subset (K, L)$. Then for every $\eta > 0$ there exist*

(1) *an abstract regular neighborhood (N', M') of (K, L) and a contractible retraction $r' : N' \to K$ such that*
  (a) $(\lambda(K), \lambda(L)) \subset (N', M')$,
  (b) $\rho(r', r | N') < \eta$, *and*

(c) $(N_0, M_0) = ((r')^{-1}(K_0), (r'|M')^{-1}(L_0))$ *is an abstract regular neighborhood of* (K_0, L_0) *with contractible retraction* $r_0 = r'|N_0$, *and*

(2) *a PL homeomorphism*

$$\phi : ((r')^{-1}(K_0 \times [-1,1]), (r'|M')^{-1}(L_0 \times [-1,1])) \to (N_0 \times [-1,1], M_0 \times [-1,1])$$

such that

(a) $\phi | N_0 \cup K_0 \times [-1,1] = \mathrm{Id}$, *and*

(b) $r'|(r')^{-1}(K_0 \times [-1,1]) = (r_0 \times \mathrm{Id}) \circ \phi$.

Proof. We begin by constructing a regular neighborhood N' that contains the required bicollar. We will then modify the initial regular neighborhood so that the other conclusions of the lemma are satisfied as well.

Let T be a triangulation of N that contains K as a full subcomplex. Subdivide so that, for some small positive number γ, $K_0 \times [-1-2\gamma, 1+2\gamma]$ is a subcomplex and there are no vertices in either $K_0 \times (-1-2\gamma, -1)$ or $K_0 \times (1, 1+2\gamma)$. Define (N', M') to be $(N(K, T'), N(L, T'|M))$, the simplicial neighborhoods of K and L in a first derived subdivision of T, and let $q : N' \to K$ be an associated contractible retraction. We may assume that q is simplicial relative to T' and, by appropriate choice of the derived subdivision T', we can arrange that q is arbitrarily close to the identity.

Define subpolyhedra Q and P of N' and M' by

$$(Q, P) = (q^{-1}(K_0 \times [-1-\gamma, 1+\gamma]), (q|M)^{-1}(L_0 \times [-1-\gamma, 1+\gamma])).$$

Observe that $Q \cap K = K_0 \times [-1-\gamma, 1+\gamma]$, Q is a PL ∂-manifold that collapses to $K_0 \times [-1-\gamma, 1+\gamma]$, and $(q|Q)^{-1}(K_0 \times \{-1-\gamma\}) \subset \partial Q$.[2] Define

$$(Q_0, P_0) = ((q|Q)^{-1}(K_0 \times \{-1-\gamma\}), (q|P)^{-1}(L_0 \times \{-1-\gamma\})).$$

It follows from the Simplicial Neighborhood Theorem (Rourke and Sanderson, 1972, Theorem 3.11) that (Q_0, P_0) is a regular neighborhood pair in $(\partial Q, \partial P)$. By Relative Product Neighborhood Theorem 5.4.8, there is a homeomorphism

$$\xi : (Q_0 \times [-1-\gamma, 1+\gamma], P_0 \times [-1-\gamma, 1+\gamma]) \to (Q, P)$$

such that $\xi(x, -1-\gamma) = x$ for each $x \in Q_0$ and $\xi | K_0 \times [-1-\gamma, 1+\gamma] =$ inclusion.

Let $\theta : [1, 1+\gamma] \to [-1-\gamma, 1+\gamma]$ be the linear homeomorphism with $\theta(1+\gamma) = 1+\gamma$ and define $\bar{\theta} : K_0 \times [-1-\gamma, 1+\gamma] \to K_0 \times [1, 1+\gamma]$ by

[2] The reason Q is a ∂-manifold is that we have arranged for there to be no vertices in the preimage of either $-1-\gamma$ or $1+\gamma$. As a result, the frontier of Q is collared. Values such as $-1-\gamma$ and $1+\gamma$ are called *regular values* of the projection onto the interval.

$\bar{\theta}(x,t) = \langle x, \theta^{-1}(t)\rangle$. Then define $r' : N' \to K$ by

$$r'(z) = \begin{cases} q(z) & \text{if } z \in \overline{N' \smallsetminus Q} \\ \langle q(x), t\rangle & \text{if } z = \xi(x,t) \text{ and } -1-\gamma \le t \le 1 \\ \bar{\theta} q \xi(x, \theta(t)) & \text{if } z = \xi(x,t) \text{ and } 1 \le t \le 1+\gamma \end{cases}$$

Define $(N_0, M_0) = ((r')^{-1}(K_0 \times \{0\}), (r'|M')^{-1}(L_0 \times \{0\}))$ and define $\phi : N_0 \times [-1,1] \to \xi(Q_0 \times [-1,1])$ by $\phi(y,t) = \xi(x,t)$, where $x \in N_0$ and $y = \xi(x,0)$. Then ϕ and N_0 satisfy all parts of Conclusion 2 of the lemma. In addition, Conclusion 1(c) is satisfied. To achieve Conclusion 1(b), simply restrict to a tighter neighborhood of $K_0 \times [-2,2]$.

To achieve Conclusion 1(a) we use the mapping cylinder structure of N to stretch N' out to cover $\lambda(K)$. Note that $N = \mathrm{Map}(r|\partial N) = \partial N \times [0,1) \cup K$ and that r simply collapses out the mapping cylinder fibers. Choose numbers a and b, $0 < a < b < 1$, so that $\lambda(K) \subset \partial N \times (a,1] \cup K$ and $\partial N \times [b,1) \cup K \subset \operatorname{Int} N'$. Let $\zeta : N \to N$ be the homeomorphism that slides the a-level of the mapping cylinder down the mapping cylinder fibers to the b-level and keeps $\partial N \cup K$ fixed. Replace N' with $N'' = \zeta^{-1}(N')$ and replace r' with $r'' = r'\zeta|N''$. Since $r\zeta = r$ we have

$$\rho(r'', r|N'') = \rho(r'\zeta|N'', r|N'') = \rho(r'\zeta|N'', r\zeta|N'') = \rho(r', r|N') < \eta.$$

This completes the proof. □

Historical Notes. Theorem 5.4.2 is due to Edwards (1975b). The first unknotting theorem beyond the trivial range was proved by Černavskiĭ (1965); his result is similar to Theorem 5.4.2 except that the template embedding e is a PL embedding and the dimension k of the embedded polyhedron is in the metastable range relative to the ambient dimension n; i.e., $k \le (2/3)n - 1$. Later Miller (1970) derived a similar theorem for $k \le n-3$ and e a PL embedding of a PL manifold while Connelly (1970a) (1970b) developed a codimension-three theorem in which e is a PL embedding of a compact polyhedron. Bryant and Seebeck (1970) proved the theorem in codimension three with e a topological embedding of a PL manifold.

Engulfing Theorem 5.4.5 is due to Bryant and Seebeck (1968a), (1968b), and (1970).

Exercise

5.4.1. Prove Corollary 5.4.11 (Stability of Embedding).

5.4.2. Let $e : K^k \to M^n$ be a topological embedding of a compact polyhedron K into a PL manifold M with $k \le n-3$. Prove that there is a PL embedding $\lambda : K \to M$ and a pseudo-isotopy $\Phi_t : M \to M$ such that $\Phi_1 \circ \lambda = e$.

5.5. 1-LCC approximation of embeddings of compacta

Approximation is a recurring theme in the theory of topological embeddings. The ability to approximate a given embedding by one that is nicer in some way is often useful, and sometimes necessary. Over the course of the next four sections we will prove three codimension-three approximation theorems. The first, proved in what immediately follows, applies to arbitrary compacta. The second, tackled in §5.6, gives PL approximations to topological embeddings of manifolds. The third, building on the second and laid out in §5.8, provides PL approximations to topological embeddings of polyhedra.

We begin with the problem of approximating embeddings of compacta. In that case we must first decide what we mean by a "nice" embedding. Results about embedding dimension presented in Chapter 3 justify defining 1-LCC embeddings of codimension-three compacta to be *tame*. We prove that every topological embedding of a codimension-three compactum can be approximated by a tame embedding.

Theorem 5.5.1 (1-LCC Approximation). *Let X be a compact subset of an n-manifold M^n with $\dim X \leq n-3$. Then for each $\epsilon > 0$ there exists a 1-LCC embedding $\lambda \in \mathrm{Emb}(X, M)$ ϵ-close to the inclusion $X \to M^n$.*

Throughout §5.5 we use $I^1 = [-1,1] \subset \mathbb{R}$ and $I^n = [-1,1]^n \subset \mathbb{R}^n$. For $\delta > 0$ we use δI^n to represent $[-\delta, \delta]^n$ and $\hat{I}^n$ to represent $2I^n$, an often-used special case. We also employ $\mathbf{0}$ in conjunction with sets $A \subset \mathbb{R}^k$ to denote the $(n-k)$-tuple of zeroes, so $A \times \mathbf{0} \subset \mathbb{R}^n$.

In addressing 5.5.1 we assume $n > 5$, since the result obviously holds for $n \leq 5$ (see Exercise 3.4.5).

Consider a countable dense subset $\{\kappa_i\}$ of $C(I^2, M)$; without loss of generality, each κ_i is a locally flat embedding. Note that any closed subset of $M \setminus \cup_i \kappa_i(I^2)$ is LCC^1 (see Exercise 3.3.3). This suggests a strategy, which we will pursue, striving to reembed X in $M \setminus \cup_i \kappa_i(I^2)$. To that end, the central issue is the following:

Lemma 5.5.2 (Fundamental lemma). *Suppose X is a closed subset of $\mathbb{R}^n$, $\dim X \leq n-3$, such that $X \cap (\partial I^2 \times I^{n-2}) = \emptyset$ and $\epsilon > 0$. Then there exists $g \in \mathrm{Emb}(X, \mathbb{R}^n)$ such that $\rho(g, \mathrm{incl}_X) < \epsilon$, $g(x) = x$ for $x \in X \setminus I^n$, $g(X \cap I^n) \subset I^n$, and $g(X) \cap (I^2 \times \mathbf{0}) = \emptyset$.*

The argument depends upon an elementary, direct manipulation, strikingly unusual in that the first of infinitely many stages to the process removes the image from $I^2 \times \mathbf{0}$ at the expense of introducing singularities. Later stages, trading on essentially the same manipulation, shrink diameters of these singularities (measured in the domain) and ultimately, on passage to the limit, rectify the problem. The construction requires the following *ad hoc* definition. Call a map $g : X \to \mathbb{R}^n$ a *nice immersion* if there exists a finite collection $D_1, D_2, \dots, D_r$ of pairwise disjoint n-cells in $\mathbb{R}^n$, each factoring as $D_j = D_j^2 \times D_j^{n-2}$, and if g can be represented as a union $g = g_1 \cup g_2$ of two embeddings $g_i : U_i \to \mathbb{R}^n$ $(i = 1, 2)$, where $X = U_1 \cup U_2$ with $U_1 = X \smallsetminus g_2^{-1}(\cup D_j)$ and $U_2 = X \smallsetminus g_1^{-1}(\cup D_j)$ representing open subsets of X, and

$$g_1(U_1) \cap (\cup_j \partial D_j^2 \times D_j^{n-2}) = \emptyset = g_2(U_2) \cap (\cup_j D_j^2 \times \partial D_j^{n-2}).$$

In particular, the 2-to-1 map g is an immersion, and the image of its singular set lives in $\cup D_j$; these cells $D_1, \dots, D_r$ are called the *singularity cells* of g. If, in addition, both $\operatorname{diam} D_j < \epsilon$ and $\operatorname{diam} g^{-1}(D_j) < \epsilon$ for all j, g is called a *nice* ϵ*-immersion.*

Fundamental Construction. In the setting of Fundamental Lemma 5.5.2, $n > 5$, there exists $g \in C(X, \mathbb{R}^n)$ such that $\rho(g, \operatorname{incl}_X) < \epsilon$, $g(x) = x$ for $x \in X \smallsetminus I^n$, $g(X \cap I^n) \subset I^n$, $g(X) \cap (I^2 \times \mathbf{0}) = \emptyset$ and g is a nice ϵ-immersion of X into $\mathbb{R}^n \smallsetminus (I^2 \times \mathbf{0})$ whose singularity n-cells lie in $\operatorname{Int}(I^n) \smallsetminus (I^2 \times \mathbf{0})$.

Proof. First we reduce to the case $\epsilon = \infty$. Partition $I^2 \times \mathbf{0}$ by a fine, 1-dimensional grid $L \subset I^2 \times \mathbf{0}$ into finitely many squares $\{I_j^2 \times \mathbf{0}\}$ for which both $\operatorname{diam} I_j^2 \times \mathbf{0} < \epsilon$ and $\operatorname{diam} e^{-1}(I_j^2 \times \mathbf{0}) < \epsilon$. By Theorem 3.4.7 we can assume $e(X)$ misses $L = \cup \partial I_j^2 \times \mathbf{0}$. Choose $\delta > 0$ so small that, for each j, $\operatorname{diam}(I_j^2 \times \delta I^{n-2}) < \epsilon$ and $X \cap (\partial I_j^2 \times \delta I^{n-2}) = \emptyset$. Now apply the $\epsilon = \infty$ case to each n-cell $I_j^2 \times \delta I^{n-2}$, compile the outcomes, and note that the result solves the ϵ-controlled problem.

Turn now to the $\epsilon = \infty$ case. A crucial beginning step is to produce a connected, orientable surface (= 2-dimensional ∂-manifold) $M^2 \subset I^n \smallsetminus X$ such that $M^2 \cap \partial I^n = \partial M^2 = \partial I^2 \times \mathbf{0}$. This depends upon the consequence of Alexander Duality that $\partial I^2 \times \mathbf{0}$ is null-homologous in $I^n \smallsetminus X$ and on:

Lemma 5.5.3. *If $e : \partial \Delta^2 \to Y$ is an embedding of a simple closed curve $\partial \Delta^2$ such that $e_* : H_1(\partial \Delta^2; \mathbb{Z}) \to H_1(Y; \mathbb{Z})$ is trivial, then $e(\partial \Delta^2)$ bounds a singular, orientable surface in Y; that is, there exists a map $\beta : S \to Y$, defined on a compact, orientable surface S such that $\beta | \partial S : \partial S \to e(\partial \Delta^2) \subset Y$ is injective.*

Proof. Here is a sketch, which fleshes out and slightly improves upon the Geometric Interpretation of Homology in (Rourke and Sanderson, 1972)[pp. 98-99] for the 2-dimensional case. Use Δ^2 to represent the standard source 2-simplex for 2-dimensional singular homology; regard Δ^2 as the convex hull of vertices v_0, v_1, v_2, and use $\xi_i : [0, 1] \to \Delta^2$ to represent the obvious linear embedding onto the edge of Δ^2 spanned by the two vertices other than v_i.

By hypothesis there exists a singular 2-chain $c = \epsilon_1 f_1 + \cdots \epsilon_N f_N$, where $\epsilon_i \in \{-1, 1\}$ and $f_i : \Delta^2 \to Y$ is continuous, and where ∂c is the 1-cycle $e_{\#}(\xi_0 - \xi_1 + \xi_2)$. Let Σ represent the disjoint union of copies $\sigma_1, \ldots, \sigma_N$ of Δ^2. Treat f_i as defined on σ_i, thereby equipping Σ with a map $f : \Sigma \to Y$ determined componentwise. Identify the edges of Σ in pairs, insofar as possible, gluing $\xi_i([0,1]) \subset \sigma_k$ to (at most one) $\xi_j([0,1]) \subset \sigma_m$ if $(-1)^i \epsilon_i = -(-1)^j \epsilon_j$ and $f_i \xi_i = f_j \xi_j$. The resulting identification space is a pinched surface PS—a space that is a 2-manifold with boundary except possibly at the images of the vertices of Σ. The compatibility condition along edges assures that f determines a map $\beta' : PS \to Y$. Each exceptional point w of PS has a small closed neighborhood N_w equivalent to the cone over L_w, a finite union of arcs and closed curves. Cut out each N_w, and along each component C of L_w abstractly attach a cone over C. The result is a surface S, and the obvious cone-preserving map $\nu : S \to PS$ leads to a map $\beta = \beta' \nu : S \to Y$. A fixed orientation on Δ^2 produces global coherent orientations on Σ, PS and, most importantly, S. By the pairing requirement, each point of $e(\partial I^2)$ must be the image under β of exactly one point from ∂S. □

In the setting at hand we adjust the singular surface to an embedded 2-cell-with-handles $M^2 \subset I^n \setminus X$. Since $n \geq 6$, M^2 is ambient isotopic in $\mathbb{R}^n$ to a copy standardly embedded in $I^3 \times \mathbf{0}$. Hence, we regard M^2 as being obtained from I^2 by attaching a finite number of unknotted, pairwise unlinked 1-handles

$$\{H_j^3 = B_j^1 \times B_j^2 \mid 1 \leq j \leq r\}$$

lying in I^3; that is, $M^2 = [(I^2 \times \{0\}) \cup (\cup_j (B_j^1 \times \partial B_j^2))] \setminus \cup_j (\partial B_j^1 \times \operatorname{Int} B_j^2)$. Here we presume $I^2 \cap H_j^3 = \partial B_j^1 \times B_j^2$.

At this stage our goal is to define, for each elongated handle $H_j^{3,+}$, described below, a reembedding $\psi_j : H_j^{3,+} \to I^2 \times [-1, 0)$ beneath $I^2 \times \{0\}$, as suggested by Figure 5.25, and then to extend this to an embedding $\Psi_j : H_j^{3,+} \times \mu I^{n-3} \to I^n \setminus I^2 \times \mathbf{0}$, as suggested in Figure 5.25.

We begin by spelling out precise structure for the handles of M^2. Much of this data is best understood by viewing Figure 5.23. For each j, let C_j^2 be a 2-cell in $(-1,1) \times \{0\} \times (-1,1) \subset \mathbb{R}^3$ such that $C_j^2 = A_j^2 \cup D_j^2 \cup F_j^2 \cup G_j^2 \cup H_j^2$ is the union of five 2-cells as shown in Figure 5.23(a), where

$$C_j^2 \cap (-1,1) \times \{0\} \times \{0\} = (D_j^2 \cup H_j^2) \cap (A_j^2 \cup F_j^2 \cup G_j^2).$$

Suppose that $(F_j^2 \cup G_j^2 \cup H_j^2, H_j^2) \approx (\hat{I}^1 \times I^1, I^2)$ in such a manner that $(F_j^2 \cup G_j^2 \cup H_j^2) \cap (I^1 \times \{0\} \times \{0\})$ corresponds to $\partial B_j^1 \times I^1$, $H_j^2 \cap D_j^2$ to $B_j^1 \times \{-1\}$, and $(F_j^2 \cup G_j^2) \cap A_j^2$ to $(\hat{I}^1 \setminus \operatorname{Int}(I_j^1)) \times \{-1\}$. Elongate H_j^2 by defining $H_j^{2,+} =$

$H_j^2 \cup F_j^2 \cup G_j^2$ and let $\xi H_j^{2,+} \subset H_j^2$ be the subset corresponding to $\partial \hat{I}^1 \times I$. We can suppose that $H_j^3 = H_j^2 \times [-\xi, \xi]$ for some $\xi > 0$ ($[-\xi,\xi]$ is really the second factor in $I \times [-\xi,\xi] \times I \subset I^3$), with $B_j^2 \subset H_j^3 = B_j^1 \times B_j^2$ corresponding to $[-\xi,\xi] \times I$ and $X \cap (C_j^2 \times [-\xi,\xi]) \subset E_j^2 \times (-\xi,\xi) \cup \mathrm{Int}(D_j^2 \times [-\xi,\xi])$, where E_j^2 is the open core of $H_j^{2,+}$ corresponding to $B_j^{1,+} \times (-1,1)$. Define $H_j^{3,+} = H_j^{2,+} \times [-\xi,\xi]$. Choose $\mu > 0$ so small that

$$\begin{aligned} &X \cap (C_j^2 \times [-\xi,\xi] \times \mu I^{n-3}) \\ &\quad \subset [(E_j^2 \times (-\xi,\xi)) \cup (\mathrm{Int}(D_j^2) \times [-\xi,\xi))] \times \mu I^{n-3}. \end{aligned}$$

Define $D_j^n = D_j^2 \times [-\xi,\xi] \times \mu I^{n-3}$ and let D_j^{n-2} correspond to the factor $[-\xi,\xi] \times \mu I^{n-3}$ in D_j^n. We assume that the cells $\{C_j^2 \times [-\xi,\xi] \times \mu I^{n-3}\}$ are pairwise disjoint. The immersion g is obtained simply by pushing the handles H_j^3 underneath $I^2 \times \{0\}$ in I^3 and by tapering the push in the μI^{n-3} direction. Specifically, let $\Phi_{j,s}$ be an isotopy of $H_j^{2,+}$ in C_j^2, fixed on $\partial C_j^2 \cap H_j^{2,+}$, such that $\Phi_{j,0} =$ identity and

$$\Phi_{j,1}(H_j^{2,+}) \subset A_j^2 \cup F_j^2 \cup G_j^2 \setminus (I \times \{0\} \times \{0\}).$$

Let $\psi_{j,s} = \Phi_{j,s} \times \mathrm{Id}_{[-\xi,\xi]} : H_j^{3,+} \to C_j^2 \times [-\xi,\xi]$ and let $\varphi_s = \cup_j \psi_{j,s}$ be the disjoint union of these isotopies. Let the embedding

$$\Xi : \cup_j H_j^{3,+} \times \mu I^{n-3} \to I^n$$

be the natural extension of φ_1 given by

$$\Xi | \cup_j H_j^{3,+} \times \{w\} = \varphi_{1-\|w\|/\mu} : \cup_j H_j^{3,+} \times \{w\} \to I^3 \times \{w\}$$

for each $w \in \mu I^{n-3}$, where $\| w \|$ is the maximum-of-coordinates norm.

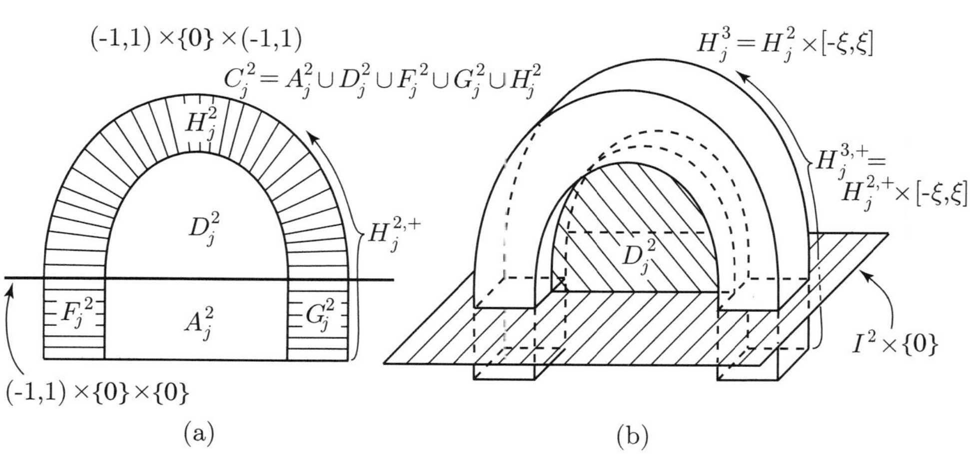

Figure 5.23. The handle structures

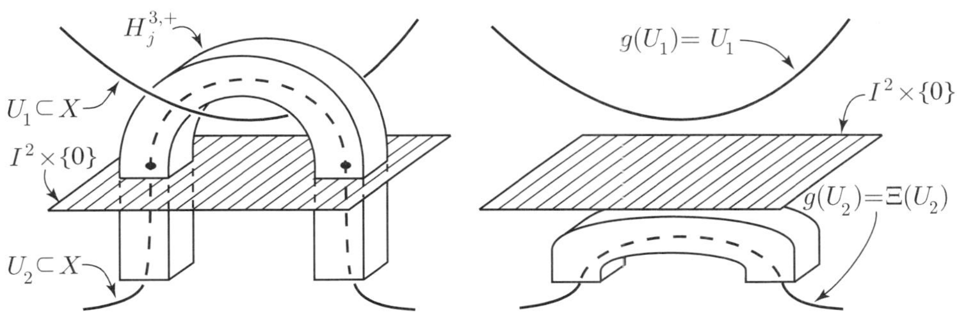

Before: the inclusion of X After: the immersion g

Figure 5.24. The immersion g

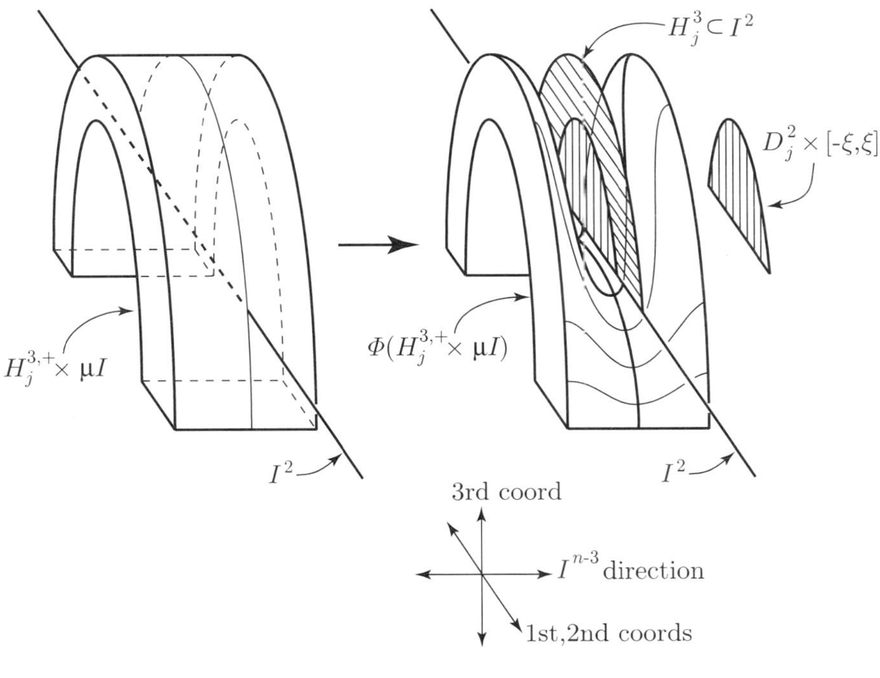

Figure 5.25. A thickened handle and its re-embedding

Let $W = X \cap (\cup_j H_j^3 \times \mu I^{n-3})$ and define $g : X \to \mathbb{R}^n$ as Ξ on W and as the inclusion elsewhere. Then g is the desired nice immersion, being the union of the embeddings $g|U_1$ and $g|U_2$, where $U_1 = X \setminus W$ and $U_2 = (X \setminus \cup_j D_j^n) \cup W$. The singularity cells $\{D_j^n\}$ can be compressed slightly to miss $I^2 \times \mathbf{0}$. $\square$

Proof of Fundamental Lemma 5.5.2. This involves repeated application of the Fundamental Construction. Choose a sequence of positive numbers $\epsilon_0, \epsilon_1, \ldots$ with $\Sigma\epsilon_i < \epsilon$. The idea is to construct a sequence $g_0 = \mathrm{incl}_X, g_1, g_2, \ldots$ of maps $X \to \mathbb{R}^n$, where each $g_i (i > 0)$ is a nice ϵ_i-immersion whose singularity cells lie in $\mathbb{R}^n \smallsetminus I^2 \times \mathbf{0}$, $g_i(x) = x$ for $x \notin I^n$, $g_i(X \cap I^n) \subset I^n$, and $\rho(g_i, g_{i-1}) < \epsilon_i$. The immersion g_{i+1} will be obtained from g_i using the Fundamental Construction to remove the singularities of g_i, at the expense of introducing new singularities for g_{i+1}, but paying off by significantly reducing their size. The map $g = \lim_{i\to\infty} g_i$ will be the desired embedding.

Figure 5.26 suggests all the 2-cells-with-handles $\{M^2\}$ constructed amidst the countable number of applications of the Fundamental Construction employed in this argument. The g_i's reembed the solid handles of this collection of M^2's, pulling them down along the spanning membranes. As we shall see later, one could regard the entire collection of M^2's as being constructed first and then the reembeddings g_i performed all at once, but the inductive proof given here offers the advantage of brevity.

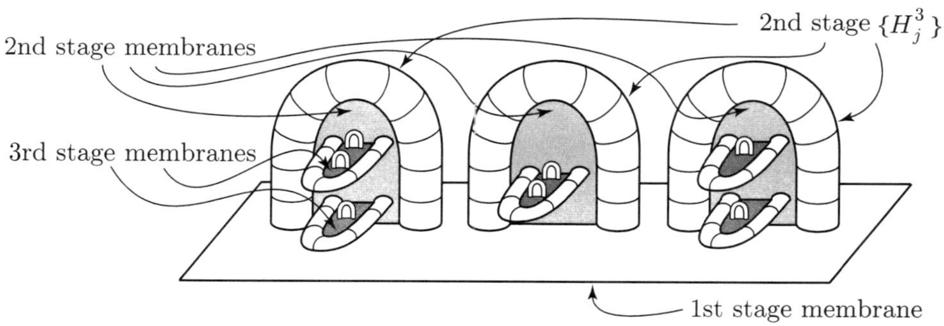

Figure 5.26. The first three membrane stages

To start, apply the Fundamental Construction to $g_0 = \mathrm{incl}_X$ to obtain a nice ϵ_1-immersion g_1 within ϵ_0 of g_0. (Reminder: the image of g_0 misses $I^2 \times \mathbf{0}$; the purpose of the succeeding g_i is to convert g_0 from an immersion to an embedding.) Generally, given g_i, let $\mathcal{D}_i = \{D_j = D^2_j \times D^{n-2}_j\}$ denote its collection of singularity n-cells. Then obtain g_{i+1} by altering g_i on $g_i^{-1}(\cup D_j)$ as follows. Suppose $g_{i,1} : U_1 \to \mathbb{R}^n$ and $g_{i,2} : U_2 \to \mathbb{R}^n$ are the promised embeddings with $g_i = g_{i,1} \cup g_{i,2}$. Apply the Fundamental Construction separately (with $\epsilon = \epsilon_{i+1}$) to each of the embeddings $g_{i,1}|g_{i,1}^{-1}(D_j), j \in J_i$, to move them off $\cup\{D^2_j \times \mathbf{0}) : j \in J_i\}$ and amalgamate the results as a nice ϵ_{i+1}-immersion $h_{i,1} : U_1 \to \mathbb{R}^n$. Thus, $h_{i,1}(U_1) \cap (D^2_j \times \mathbf{0}) = \emptyset$ for each $j \in J_i$, and the singularity n-cells in $\mathcal{D}_{i+1}$ (which arise in the construction of $h_{i,1}$) lie in $\cup(\mathrm{Int}(D^n_j) \smallsetminus D^2_j \times \mathbf{0})$. Let $h_{i,2} : U_2 \to \mathbb{R}^n$ be an embedding resulting from $g_{i,2}$ by compressing each image set $g_{i,2}(U_2) \cap D_j$ toward $D_j^2 \times \mathbf{0}$ by sliding along the D_j^{n-2} factor, keeping ∂D_j fixed, to make

$$h_{i,2}(U_2 \smallsetminus U_1) \cap (h_{i,1}(U_1) \cup \cup\mathcal{D}_{i+1}) = \emptyset.$$

Then $g_{i+1} = h_{i,1} \cup h_{i,2}$ is the next nice immersion in the sequence. Its associated singularity cells, which form $\mathcal{D}_{i+1}$, are subsets of the interiors of the members of $\mathcal{D}_i$. For $A \subset X$ let $D(A, i)$ denote

$$\cup\{D_j : D_j \text{ is a singularity } n\text{-cell of } g_i \text{ such that } A \cap g_i^{-1}(D_j) \neq \emptyset\}.$$

Then $g(A) \subset g_i(A \cup D(A, i))$. To show g is an embedding, suppose B is closed in X and $x \in X \smallsetminus B$. Choose r so large that $2\epsilon_r < d_X(x, B)$, which by diameter restrictions on the singularity cells yields

$$(g_r(B) \cup D(B, r)) \cap (g_r(x) \cup D(\{x\}, r)) = \emptyset,$$

implying $g(x) \notin g(B)$. □

Proof of Theorem 5.5.1. Let $\{D_i^2\}$ be a collection of PL embedded 2-cells in $\mathbb{R}^n$ representing a dense subset of $C(I^2, \mathbb{R}^n)$, and let $\varphi_j : \mathbb{R}^n \to M^n$ be a collection of coordinate neighborhoods that covers M^n. Reindex the collection $\{\varphi_j(D_i^2)\}_{i,j}$ as $\{T_k : k = 1, 2, \ldots\}$, and define

$$A_k = \{g \in \operatorname{Emb}(X, M^n) \mid g(X) \subset M^n \smallsetminus T_k\}.$$

By Fundamental Lemma 5.5.2, each A_k is (open and) dense in $\operatorname{Emb}(X, M^n)$; hence, $\operatorname{incl}_X : X \to M^n$ can be ϵ-approximated by $\lambda \in \cap_k A_k$. Since $\lambda(X) \cap \varphi_j(D_i^2) = \emptyset$ for all i, j, it follows that λ is a 1-LCC embedding. □

The 1-LCC re-embedding theorem allows us to complete the proof of the universality of the Menger continuum M_n^k for k-dimensional compact subsets of $\mathbb{R}^n$.

Corollary 5.5.4. *Every compact, k-dimensional subset X of $\mathbb{R}^n$ admits an embedding in M_n^k.*

Proof. When $k \leq n - 3$, which is the central issue, Theorem 5.5.1 gives a 1-LCC embedding $h : X \to \mathbb{R}^n$. The image has embedding dimension k. By Theorem 3.5.1 $h(X)$ can be pushed into M_n^k. When $k \geq n - 2$, $\dim X = \operatorname{dem} X$ is automatic, so Theorem 3.5.1 applies directly to X itself. □

At this point we are about to undertake the usual step of presenting an alternate proof of the Fundamental Lemma. This is not intended to devalue the one just completed, which is ingenious, direct and appealing; moreover, it has the benefit of providing an introduction to a collection of objects about to be defined in a fairly abstract manner. The alternate proof has its own benefits, presenting techniques that will be reapplied in §7.7 toward proving a Locally Flat Approximation Theorem concerning codimension one manifolds. Since that argument will place considerable demands on the reader, it seems advantageous to present some of its key ideas in a less complicated setting.

Štan′ko Move. The basic move—kin to the nice immersion of 5.5.2—is described in terms of a specific homeomorphism of the n-cube $\hat{I}^n = [-2,2]^n$. Certain subsets of $\hat{I}^2$ are prominently featured (see Figure 5.27):

$$\begin{aligned} A &= [-2,2] \times [-2,2] \setminus (-\tfrac{1}{2},\tfrac{1}{2}) \times (-\tfrac{1}{2},\tfrac{1}{2}) \subset \hat{I}^2;\\ B &= [-\tfrac{1}{2},\tfrac{1}{2}] \times [-\tfrac{1}{2},\tfrac{1}{2}] \subset \hat{I}^2;\\ C &= [\tfrac{5}{4},\tfrac{7}{4}] \times [-2,2] \subset \hat{I}^2;\\ D &= [-1,1] \times [-1,1] \subset \hat{I}^2;\\ e &= [1,2] \times \{0\} \subset \hat{I}^2. \end{aligned}$$

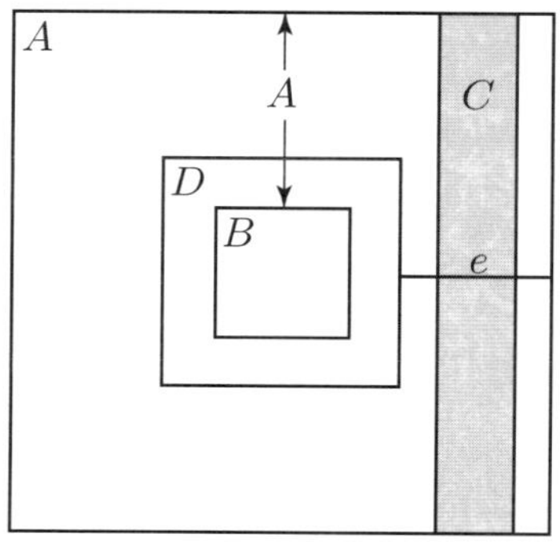

Figure 5.27. The standard template $(A, B, C, D, e) \subset \hat{I}^2$

Now let $n \geq 3$ and consider the sets $\mathscr{A} = A \times \hat{I}^{n-2}$, $\mathscr{B} = B \times \hat{I}^{n-2}$, $\mathscr{C} = C \times [-1,1] \times \hat{I}^{n-3}$, $\mathscr{D} = D \times \mathbf{0} \subset \hat{I}^n$ and $e \times \hat{I} = e \times \hat{I} \times \mathbf{0} \subset \hat{I}^n$. Here $\mathscr{A} \cup \mathscr{B} = \hat{I}^n$ and $\mathscr{B} \cap \mathscr{C} = \emptyset$. In the $n = 3$ case $\mathscr{B}$ and $\mathscr{C}$ may be viewed as disjoint 1-handles cutting through $\hat{I}^3$ in orthogonal directions. The basic Štan′ko move restricted to $\hat{I}^3$ simply will pull $C \times [-1,1] = \mathscr{C} \cap \hat{I}^3$ through $B \times [-2,2] = \mathscr{B} \cap \hat{I}^3$.

The specific homeomorphism $\Phi_n : \hat{I}^n \to \hat{I}^n$ mentioned above is fixed on $\partial \hat{I}^n$ and shifts the set $\mathscr{C}$ relative to $\mathscr{B}$ in the following way: define an isotopy $\phi : \hat{I} \times [0,1] \to \hat{I}$ of $\hat{I}$ which, for $t \in [0,1]$, fixes -2 and 2, shifts the segment $[5/4, 7/4]$ $3t$ units to the left and is linear on the segments $[-2, 5/4]$ and $[7/4, 2]$. Then, for an arbitrary point $\langle s, t\rangle \in \hat{I} \times \hat{I}^{n-1} = \hat{I}^n$, set

$$\Phi_n(s,t) = \begin{cases} \langle \phi(s,1), t\rangle & \text{if } \| t \| \in [0,1], \text{ and}\\ \langle \phi(s, 2 - \| t \|), t\rangle & \text{if } \| t \| \in [1,2], \end{cases}$$

where $\| t \|$ is the maximum-of-coordinates norm. The case $n = 2$ is pictured in Figure 5.28. We note the following:

(i) $\Phi_n(C \times [-1,1] \times [-1,1]^{n-3}) \subset \Phi_2(C) \times [-1,1] \times \hat{I}^{n-3} \subset \hat{I}^n \setminus \mathscr{B}$,

(ii) $\Phi_n(\mathscr{C}) \cap \mathscr{B} \neq \emptyset$ if $n > 3$.

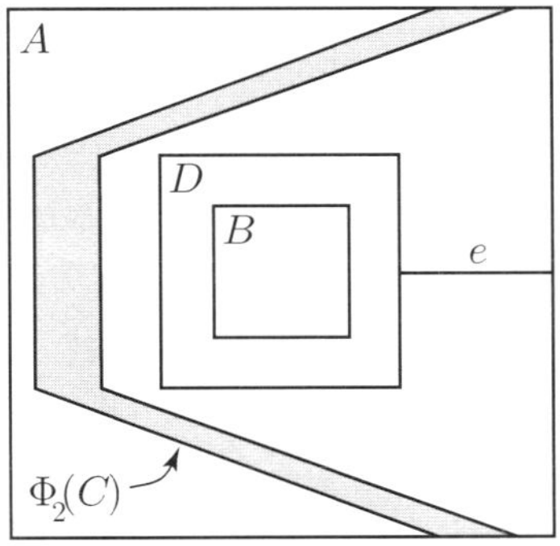

Figure 5.28. The action of Φ_2 on C

The basic move is determined using Φ_n as follows. Suppose X is a closed subset of an n-manifold M and $\alpha : \hat{I}^n \to M$ is an embedding such that $\alpha^{-1}(X) \subset \mathscr{B} \cup \mathscr{C}$. Then the map $\nu : X \to M$ defined as

$$\nu(x) = \begin{cases} \alpha \circ \Phi_n(x) \circ \alpha^{-1}(x) & \text{if } x \in \alpha(\mathscr{C}) \\ x & \text{otherwise} \end{cases}$$

is a *basic Štan′ko move.* When $n \leq 3$ ν is a re-embedding of X because of (i) above; however, for $n > 3$ the map ν will be an immersion, not injective, in general, because of (ii). Nevertheless, in both cases the basic Štan′ko move may undo some linking in X.

We will summarize the above information as a symbol (A, B, C, D, e) called a *template.* We will use the (+) symbol to denote finite disjoint topological union; accordingly, a disk (+) is a space having finitely many components, each of which is a disk. This allows us to describe constructions easily, component by component, and to economize with notation. Eventually we will expand all template data by the (+) convention to admit a host of templates and associated homeomorphisms $\Phi_n : \mathscr{A} \cup \mathscr{B} \to \mathscr{A} \cup \mathscr{B}$. This means of course that a replica of Φ_n acts upon each component of $\mathscr{A} \cup \mathscr{B}$.

We turn next to the structures that will guide the embeddings of templates for basic moves.

Semi-capped surfaces and Štan′ko complexes. Let D denote an oriented PL disk; let $L_1, \ldots, L_k$ and $L'_1, \ldots L'_k$ denote compatibly oriented, pairwise disjoint subdisks of $\operatorname{Int} D$ irreducibly joined in pairs by mutually exclusive PL arcs $\gamma_1, \ldots, \gamma_k$ in Int D, where $\partial\gamma_j \subset L_j \cup L'_j$ for each j; let

$$\theta : L_1 \cup \cdots \cup L_k \to L'_1 \cup \cdots L'_k$$

denote an orientation-reversing homeomorphism sending the point $\gamma_j \cap L_j$ to $\gamma_j \cap L'_j$ for $j = 1, \ldots, k$. Let D^* denote the identification space D/θ, called a *semi-capped surface*. Set $E = L_1 \cup \cdots \cup L_k \cup L'_1 \cup \cdots \cup L'_k$ and $\Gamma = \gamma_1 \cup \cdots \cup \gamma_k$. We will refer to such data as a *Delta structure* $\Delta = (D, E, \Gamma)$, with the identification map $(^*) : D \to D^*$ implicitly understood. The image disks $E^* = (^*)(E)$ and simple closed curves $\Gamma^* = (^*)(\Gamma)$ will play essential roles.

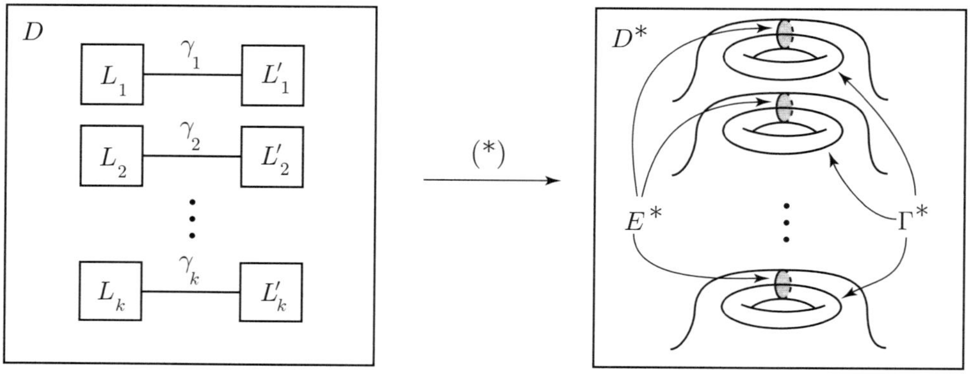

Figure 5.29. A Delta structure Δ

A *branching system* is a system $(\Delta, q) : \Delta_0 \to \Delta_1 \to \Delta_2 \to \cdots$ where each Δ_i is a Delta structure $(+)\Delta_i = (D_i, E_i, \Gamma_i)$ and for each $i \geq 0$ there is a PL homeomorphism $q(i) : \Gamma_i^* \to \partial D_{i+1}^*$. The identification space

$$C(\Delta, q) = D_0^* \sqcup_{q(0)} \sqcup D_1^* \sqcup_{q(1)} \sqcup D_2^* \sqcup_{q(2)} \cdots$$

is called a *Štan′ko complex*. Generally, we identify each D_i^* with its image in $C(\Delta, q)$. Then we take $\Gamma_i^* = \partial D_{i+1}^*$, suppress the homeomorphisms $q(i)$, shorten (Δ, q) to Δ and, similarly, shorten $C(\Delta, q)$ to $C(\Delta)$. We combine the various identifications $(^*) : D_i \to D_i^*$ into a single map

$$(^*) : D_0 \sqcup D_1 \sqcup \cdots \quad \text{(disjoint union)} \to C(\Delta) = D_0^* \cup D_1^* \cup \cdots .$$

The semi-capped surfaces D_i^* will serve as precursors of what were called membranes in the first proof of the Fundamental Lemma. Analogs to membranes will arise by transmuting D_i^* via surgery on E_i^*—thickening those 2-cells to a disjoint union of 3-cells $\psi(E_i^* \times I)$, where $(D_i^* \setminus \text{Int } E_i^*) \cap \psi(E_i^* \times I) = \psi(\partial E_i^* \times I)$, removing the thickened 2-cells and replacing with $\psi(E_i^* \times \partial I)$. The thickened 3-cells $\phi(E_i^* \times I)$ will act like solid handles (the H_j^3) attached to that membrane.

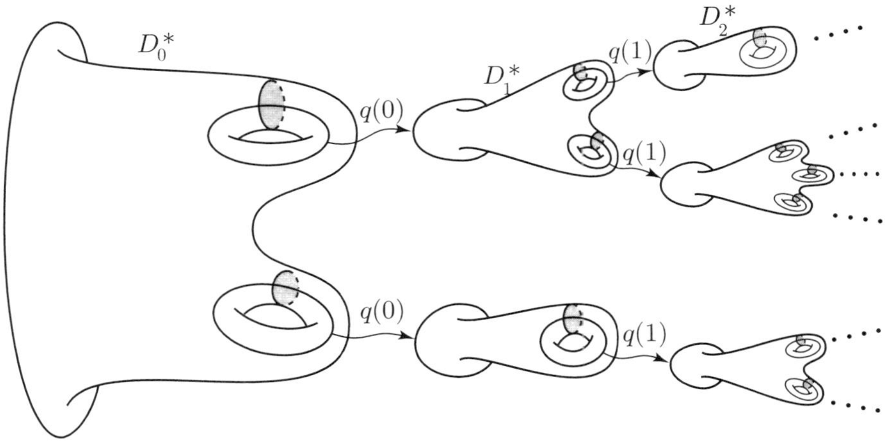

Figure 5.30. A branching system (Δ, q)

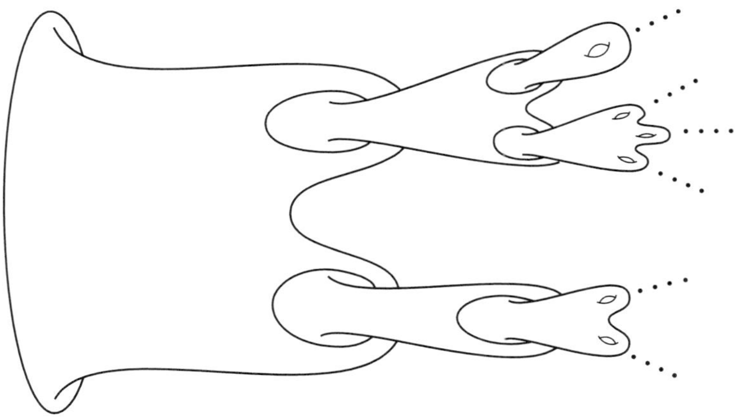

Figure 5.31. A Štan′ko complex $C(\Delta)$

Theorem 5.5.5 (Štan′ko Complex Embedding Theorem). *Suppose X is a closed, $(n-3)$-dimensional subset of $\mathbb{R}^n$ such that $X \cap (I^2 \times \mathbf{0})$ is a 0-dimensional subset of $\operatorname{Int}(I^2 \times \mathbf{0})$, and suppose $\epsilon > 0$. Let F denote the obvious embedding $D_0 = I^2 \to I^2 \times \mathbf{0} \subset \mathbb{R}^n$. Then there exists a branching system $\Delta : \Delta_0 \to \Delta_1 \to \Delta_2 \to \cdots$, with D_0 of $\Delta_0 = (D_0, E_0, \Gamma_0)$ equal to I^2, and there exists a PL embedding $h : C(\Delta) \to \mathbb{R}^n$ satisfying:*

(1) $h(C(\Delta) \setminus \operatorname{Int} E^*) \subset \mathbb{R}^n \setminus X$,

(2) $h(D_i^* \cup D_{i+1}^* \cup D_{i+2}^* \cup \cdots) \subset B(X; \epsilon/2^{i-1})$ *for* $i > 0$,

(3) $\rho(h \circ (^*)|D_0, F) < \epsilon$,

(4) $h \circ (^*)|\partial D_0 = F|(\partial D_0 = \partial I^2)$,

(5) $\operatorname{diam} h(P_i) < \epsilon/2^i$ *for* $i > 0$ *and each component* P_i *of* $D_i^* \cup E_{i-1}^*$.

The Embedding Theorem is a relatively straightforward consequence of iterated applications of the following lemma.

Lemma 5.5.6. *Suppose* X *as in Theorem 5.5.5 and* $G : (D, \partial D) \to (\mathbb{R}^n, \mathbb{R}^n \smallsetminus X)$ *is a PL embedding of pairs, where* D *is an oriented disk, and* $\delta > 0$. *Then there exist a Delta structure* $\Delta = (D, E, \Gamma)$ *and a PL embedding* $h : D^* \to \mathbb{R}^n$ *satisfying:*

(1′) $h(D^* \smallsetminus \operatorname{Int} E^*) \subset \mathbb{R}^n \smallsetminus X$,

(2′) $h(E^*) \subset B(X; \delta)$,

(3′) $\rho(h \circ (^*)|D, G) < \delta$,

(4′) $h \circ (^*)|\partial D = G|\partial D$, *and*

(5′) $\operatorname{diam} h(K) < \delta$ *for each component* K *of* E^* *or* Γ^*.

Proof. Impose a triangulation T on D with mesh so small that $\operatorname{diam} G(\sigma) < \delta/2$ for all $\sigma \in T$. Adjust G, fixing $G|\partial D$, so the image of the 1-skeleton misses X (Theorem 3.4.7).

For each 2-simplex $\sigma \in T$ and neighborhood U_σ of $G(\sigma)$, $G(\partial\sigma)$ is null homologous in $U_\sigma \smallsetminus X$. Lemma 5.5.3 assures the existence of an orientable disk-with-handles Q_σ bounded by $\partial\sigma$ as well as a continuous extension $h_\sigma : Q_\sigma \to \mathbb{R}^n \smallsetminus X$ of $G|\partial\sigma$ that sends Q_σ to a set of diameter less than $\delta/2$. When $G(\sigma) \cap X = \emptyset$, insist that $Q_\sigma = \sigma$ and $h_\sigma = G|\sigma$.

In the interior of each Q_σ identify complete sets $\Gamma_\sigma, \Gamma'_\sigma$ of handle curves: i.e., Γ_σ (resp. Γ'_σ) is a finite union of mutually exclusive simple closed curves, each meeting Γ'_σ (resp. Γ_σ) transversely in a single point, with Γ_σ and Γ'_σ maximal collections with respect to these properties. Of course, should $G(\sigma)$ miss X, then $\Gamma_\sigma = \emptyset = \Gamma'_\sigma$.

For each $\sigma \in T^{(2)}$ there exist a finite disjoint union E_σ of disks having boundary Γ'_σ (with $E_\sigma = \emptyset$ if and only if $\Gamma'_\sigma = \emptyset$) and a continuous extension $h : E_\sigma \to \mathbb{R}^n$ of $G|\partial\sigma$ taking each component of E_σ to a set of diameter less than $\delta/2$. Here $h(E_\sigma) \cap X \neq \emptyset$ is permitted, of necessity. Define

$$D^* = |T^{(1)}| \cup [\cup_\sigma (Q_\sigma \cup E_\sigma)], \quad E^* = \cup_\sigma E_\sigma, \quad \Gamma^* = \cup_\sigma \Gamma_\sigma.$$

It should be obvious that D^* is the semi-capped surface of a Delta structure $\Delta = (D, E, \Gamma)$, where $(^*)|\,|T^{(1)}| = G|\,|T^{(1)}|$ and $(^*)$:$\sigma \to Q_\sigma \cup E_\sigma$. The map $h : D^* \to \mathbb{R}^n$ can be approximated, rel $|T^{(1)}|$, by a PL embedding satisfying Conditions (1′) – (5′). □

Proof of Embedding Theorem 5.5.5. Lemma 5.5.6 provides both a Delta structure $\Delta_0 = (D_0, E_0, \Gamma_0)$, where $D_0 = I^2$, and a PL embedding $h : D_0^* \to$

$\mathbb{R}^n$ satisfying, for $f_0 = \mathit{inclusion} : (D_0 = I^2) \to \mathbb{R}^n$, the $j = 0$ case of the following conditions:

(1_j) $h(D_j^* \smallsetminus \operatorname{Int} E_j^*) \subset \mathbb{R}^n \smallsetminus X$,

(2_j) $h(E_j^*) \subset B(X; \epsilon/2^{j+2})$,

(3_j) $\rho(h \circ (^*)|D_j, f_j) < \epsilon/2^{j+2}$,

(4_j) $h \circ (^*)|\partial D_j = f_j|\partial D_j$, and

(5_j) $\operatorname{diam} h(K) < \epsilon/2^{j+3}$ for each component K of E_j^* or Γ_j^*.

Assume inductively that $\Delta_0 \to \cdots \to \Delta_{i-1}$, $f_j : D_j \to \mathbb{R}^n$ and a PL embedding $h : D_0^* \cup \cdots \cup D_{i-1}^* \to \mathbb{R}^n$ have been obtained satisfying (1_j)–(5_j) for each $j \in \{0, \dots, i-1\}$ as well as

(6_j) $\operatorname{diam} f_j(K') < \epsilon/2^{j+2}$ for $j > 0$ and each component K' of D_j, and

(7_j) for $0 \le k < j$, $h(D_k^*)$ has a neighborhood U_k such that $\operatorname{diam} h(\gamma) < d(h(\gamma), \overline{U}_k)$ for all components γ of Γ_j^*.

(Condition (7_0) is vacuous.) For each component γ of Γ_{i-1}^*, let $D(\gamma)$ denote an abstract disk with boundary γ. By (5_{i-1}), $\operatorname{diam} h(\gamma) < \epsilon/2^{i+2}$, so we have a continuous extension $f_i(\gamma) : D(\gamma) \to \mathbb{R}^n$ of $h|\gamma$, the image of which has diameter less than $\epsilon/2^{i+2}$, yielding (6_i); moreover, we require that $f_i(\gamma)(D_\gamma)$ lie very near the convex hull of $h(\gamma)$, so $f_i(\gamma)(D_\gamma) \cap \overline{U}_k = \emptyset$ for $0 \le k \le i-2$, by (7_{i-1}). Set $D_i = \cup_\gamma D(\gamma)$ and define f_i on D_i as $f_i = \cup_\gamma f_i(\gamma)$; obviously this gives ($6_i$). Assume f_i is a PL embedding, in general position with respect to $h(D_0^* \cup \cdots \cup D_{i-1}^*)$. Let D_i^- denote a slightly smaller finite union of disks in $\operatorname{Int} D_i$, where $\operatorname{Int} D_i^- \supset f_i^{-1}(X)$. Choose $\delta > 0$ less than the distance from $f_i(D_i^-)$ to $U_0 \cup \cdots \cup U_{i-2} \cup h(D_{i-1}^*)$. Set $U_{i-1} = B(h(D_{i-1}^*); \delta/3)$. Again by Lemma 5.5.6, applied with positive number $\delta' = \{\delta/3, \epsilon/2^{i+3}\}$, there exist a Delta structure (+) $\Delta_i = (D_i, E_i, \Gamma_i)$ and a PL embedding $h|D_i^* : D_i^* \to \mathbb{R}^n$ satisfying (1_i) – (5_i). A standard general position adjustment of $h|D_i^*$ assures that h is 1-1 on $D_0^* \cup \cdots \cup D_i^*$. Checking that ($7_i$) holds as well is routine. This completes the inductive construction of

$$\Delta : \Delta_0 \to \Delta_1 \to \Delta_2 \to \cdots \quad \text{and} \quad h : C(\Delta) \to \mathbb{R}^n.$$

Conditions (1), (3) and (4) of the Embedding Theorem obviously hold here. For each component $P \cup Q$ of $D_i^* \cup E_{i-1}^*$ ($i \ge 1$), $P \subset D_i^*$, $Q \subset E_{i-1}^*$, we have

$$\operatorname{diam} h(P) \le 2\rho(h \circ (^*)|P, f_i|P) + \operatorname{diam} f_i(P) < 3\epsilon/2^{i+2}$$

by (3_i) and (6_i), and $\operatorname{diam} h(Q) < \epsilon/2^{i+2}$ by (5_{i-1}); thus (5) is satisfied. Also, for $i \ge 1$, $d(h(P), X) < \epsilon/2^{i+1}$ by (2_{i-1}), since $h(P) \cap h(E_{i-1}^*) \ne \emptyset$. But $\operatorname{diam} h(P) < \epsilon/2^i$, so $h(P) \subset B(X; \epsilon/2^{i-1})$, and (2) is satisfied. This yields, in conjunction with (1_k), that no sequence from $h(C(\Delta) \smallsetminus D_k^*)$

can have a limit point in $h(\text{Int } D_k^* \setminus \text{Int } E_k^*)$; moreover, no sequence from $h(C(\Delta) \setminus \text{Int } E_k^*)$ can have a limit point in $h(\text{Int } E_k^*)$ by (7_j), $j \geq k+1$. Hence, h is an embedding. □

Lemma 5.5.7. *Let $C(\Delta)$ denote a Štan′ko complex PL embedded in a connected PL n-manifold M, $n \geq 5$, and let Z denote a compact subset of $C(\Delta)$. Then Z is contained in a collapsible finite 2-dimensional subpolyhedron P^+ of M, where P^+ PL embeds in I^3.*

Proof. Find an integer $N \geq 0$ such that $Z \subset D_0^* \cup \cdots \cup D_N^*$, and set

$$f = (^*)|\partial D_{N+1} : \partial D_{N+1} \to \partial D_{N+1}^* = \Gamma_{N+1}.$$

Each component of the identification space $P = (D_0^* \cup \cdots \cup D_N^*) \cup_f D_{N+1}$ is collapsible and embeds in I^3. Also, P admits an obvious map to M. Join the components of P by a wedge of arcs to obtain a single collapsible complex P^+. Since M is connected and $n \geq 5$, P^+ admits an embedding in M for which the image contains $(D_0^* \cup \cdots \cup D_N^*) \supset Z$. □

Lemma 5.5.8 (Unknotting). *Suppose $C(\Delta)$ is a Štan′ko complex PL embedded in a connected PL n-manifold M $(n \geq 6)$ and Z is a compact subset of $C(\Delta)$. Then there exist a PL 3-cell Y_Z^3 and a PL embedding $\psi : Y_Z^3 \times \hat{I}^{n-3} \to M$ such that $Z \subset \psi(Y_Z^3 \times \mathbf{0})$.*

Proof. By Lemma 5.5.7, M contains a PL embedded, collapsible finite 2-complex $P^+ \supset Z$. Name a regular neighborhood B of P^+. Being collapsible, B is an n-ball, and $P^+ \subset B \approx I^n$ can be regarded as $P^+ \subset I^3 \times \mathbf{0}$, since P^+ does embed in I^3 and any two PL embeddings of P^+ in $\text{Int } I^n$ are ambient isotopic. □

Second Proof of Fundamental Lemma 5.5.2. We take Δ and h from the conclusion of the Štan′ko Complex Embedding Theorem and then identify $C(\Delta)$ with $h(C(\Delta))$ via the homeomorphism h. Recall the combined identification map

$$(^*) : D_0 \sqcup D_1 \sqcup \cdots \to C(\Delta) = D_0^* \cup D_1^* \cup \cdots = h(C(\Delta)) \subset \mathbb{R}^n.$$

For each $i > 0$ we identify that D_i associated with the Delta structure $(+)\Delta_i = (\Delta_i, E_i, \Gamma_i)$ with the D_i from the template $(+)(A_i, B_i, C_i, D_i, e_i)$ in such a manner that $E_i \cup \Gamma_i \subset \text{Int } B_i$ and $(D_i \cap e_i)^* = D_i^* \cap E_{i-1}^* \subset C(\Delta) \subset \mathbb{R}^n$. That $X \cap C(\Delta) \subset \cup_i E_i^* \subset \cup_i B_i^*$ deserves heavy emphasis.

By Unknotting Lemma 5.5.8 there exist a regular neighborhood N_i of $D_i^* \cup E_{i-1}^*$ in $C(\Delta)$, a PL 3-cell Y_i and a PL product structure $Y_i \times \hat{I}^{n-3} \subset \mathbb{R}^n$ such that $N_i \subset Y_i = Y_i \times \{\mathbf{0}\} \subset Y_i \times \hat{I}^{n-3} \subset \mathbb{R}^n$. For each $i > 0$ we will use the sets $D_i^* \cup E_{i-1}^* \subset N_i$ and the product structure $\hat{I}^3 \times \hat{I}^{n-3}$ to construct an embedding $\alpha_i : \hat{I}^n = \mathscr{A}_i \cup \mathscr{B}_i \to \mathbb{R}^n$ suitable for use in a basic Štan′ko move. The embedding α_i will be constructed in three steps.

Step 1: Constructing α_i on $(A_i \times \{0\}) \cup (e_i \times \hat{I}) \subset \hat{I}^3$. Define α_i on $(A_i \cap D_i) \times \{0\}$ as $(^*)|(A_i \cap D_i) \times \{0\}$. Since $(D_i \cap e_i)^* = D_i^* \cap E_{i-1}^*$, we may extend α_i to carry $e_i \times \hat{I}$ onto E_{i-1}^*, with

$$X \cap E_{i-1}^* \subset \alpha_i[\left(\frac{5}{4}, \frac{7}{4}\right) \times \{0\} \times (-1,1)] \subset \alpha_i(e_i \times \hat{I}).$$

This embedding, in turn, can be extended to the remainder of $A_i \times \{0\} \subset \hat{I}^3$ so as to take $(A_i \setminus (D_i \cup e_i)) \times \{0\}$ into the intersection of $\mathbb{R}^n \setminus C(\Delta)$ with the $(\epsilon/2^i)$-neighborhood (component by component) of $D_i^* \cup E_{i-1}^*$, so that

$$X \cap \alpha_i(A_i \times \{0\} \cup e_i \times \hat{I}) \subset \alpha_i[\left(\frac{5}{4}, \frac{7}{4}\right) \times \{0\} \times (-1,1)] \subset \alpha_i(C_i \times (-1,1)).$$

We require the sets $\alpha_i[(A_i \times \{0\}) \cup (e_i \times \hat{I})]$ to be pairwise disjoint.

Remark. In Steps 2 and 3 we extend the definition of α_i to $\mathscr{A}_i$ and to $\mathscr{B}_i$. In doing so, certain basic precautions should be taken. They are:

(1) each component of $\operatorname{Im}(\mathscr{A}_i \cup \mathscr{B}_i)$ has diameter less than $\epsilon/2^i$;

(2) of the sets $[\operatorname{Im} D_0 \cap A_0]$, $[\operatorname{Im} B_0]$, $[\operatorname{Im} A_1, \operatorname{Im} \mathscr{A}_1]$, $[\operatorname{Im} B_1, \operatorname{Im} \mathscr{B}_1]$, $[\operatorname{Im} A_2, \operatorname{Im} \mathscr{A}_2]$, $[\operatorname{Im} B_2, \operatorname{Im} \mathscr{B}_2], \ldots,$ only the ones listed in the same or adjacent square brackets can intersect.

Step 2. Constructing $\alpha_i|\mathscr{A}_i$. Recall that $\mathscr{A}_i$ equals $A_i \times \hat{I} \times \hat{I}^{n-3}$. Since $\alpha_i(A_i \times \{0\})$ is bicollared in $Y_i \times \mathbf{0}$, it is clearly possible to extend α_i over $A_i \times \hat{I}$, taking each fiber $\{a\} \times \hat{I}$ to a bicollar fiber $\alpha_i(a) \times \hat{I}$ in $Y_i \times \mathbf{0}$. Then one can extend α_i to all of $A_i \times \hat{I} \times \hat{I}^{n-3}$ by sending $\langle a, t\rangle \times \hat{I}^{n-3}$ to $\alpha_i(a,t) \times \hat{I}^{n-3}$ in the natural way. By shortening the bicollar fibers and the $\hat{I}^{n-3}$ fibers of $Y_i \times \hat{I}^{n-3}$, if necessary, we can protect conditions (1) and (2) of the preceding Remark and insist upon the following:

(3) $X \cap \alpha_i(\mathscr{A}_i) \subset \alpha_i(\mathscr{C}_i)$;

(4) $\alpha_i(\mathscr{A}_i \cap \Phi_n(\mathscr{C}_i)) \subset \mathbb{R}^n \setminus C(\Delta)$.

(Condition (4) can be obtained because α_i can be extended over $\mathscr{A}_i$ so $\alpha_i(\mathscr{A}_i) \cap C(\Delta) \subset \alpha_i(A_i \times [-2,2] \times \mathbf{0}) \cap C(\Delta) \subset N_i \subset Y_i \times \mathbf{0}$ and

$$\alpha_i([A_i \setminus (D_i \cup e_i)] \times [-2,2] \times \mathbf{0}) \cap C(\Delta) = \emptyset;$$

then the image of the portion of $\mathscr{A}_i \cap \Phi_n(\mathscr{C}_i)$ in $A_i \times [-2,2] \times \mathbf{0}$ misses $C(\Delta)$, and clearly the image of the rest of $\mathscr{A}_i \cap \Phi_n(\mathscr{C}_i)$ does the same, as it lies in

$$\alpha_i(A_i \times [-2,2] \times (\hat{I}^{n-3} \setminus \mathbf{0})) \subset Y_i \times (\hat{I}^{n-3} \setminus \mathbf{0}) \subset \mathbb{R}^n \setminus N_i.)$$

Step 3. Constructing $\alpha_i|\mathscr{B}_i$. The embedding $\alpha_i|\mathscr{A}_i \cap \mathscr{B}_i$ has already been defined. So has $\alpha_{i+1}|A_{i+1}$, and the latter is of crucial importance for (5) and (6) below. The process which extends α_i over the rest of $\mathscr{B}_i$ runs as follows. Cut B_i^* apart via surgery along E_i^* in $Y_i \times \mathbf{0}$ to obtain an embedding of B_i in $Y_i \times \mathbf{0}$. Insist that the images of all modifications lie very close to $\alpha_{i+1}(\mathscr{C}_{i+1})$. Use a bicollar on the embedded B_i in Y_i and the product structure $Y_i \times \hat{I}^{n-3}$ to extend the embedding, fiber by fiber, to $\mathscr{B}_i$. One must keep the embedded B_i very close to B_i^* and must use very short bicollar fibers and $\hat{I}^{n-3}$ fibers (except near $\alpha_i(A_i \cap B_i)$, where such fibers have already been chosen) not only to ensure conditions (1) and (2) of the Remark preceding Step 2, but also to obtain the following:

(5) $X \cap \alpha_i(\mathscr{B}_i) \subset \alpha_{i+1}(\mathscr{C}_{i+1})$;

(6) $\alpha_i(\mathscr{B}_i) \cap \alpha_{i+1}(\mathscr{A}_{i+1} \cap \Phi_n(\mathscr{C}_{i+1})) = \emptyset$;

(7) $\alpha_i(\mathscr{B}_i \cap \Phi_n(\mathscr{C}_i)) \subset \mathbb{R}^n \smallsetminus C(\Delta)$.

Condition (5) can be obtained since $\alpha_{i+1}(\mathscr{C}_{i+1})$ is a neighborhood of Int $E_i^* \supset X \cap B_i^*$ in $\mathbb{R}^n$. Condition (6) can be obtained because $B_{i+1}^* \subset C(\Delta)$ while $\alpha_{i+1}(\mathscr{A}_{i+1} \cap \Phi_n(\mathscr{C}_{i+1})) \subset \mathbb{R}^n \smallsetminus C(\Delta)$ by (4). Condition (7) is automatic for small fibers by the definition of Φ_n.

Finally, to define the infinite Štan′ko move on X, which provides the re-embedding of X and establishes the Fundamental Lemma, set

$$g'(x) = \begin{cases} \alpha_i \circ \Phi_n(x) \circ \alpha_i^{-1}(x) & \text{if } x \in X \cap \alpha_i(\mathscr{C}_i) \\ x & \text{otherwise.} \end{cases}$$

The continuity of g' is easy to verify, since the various $\alpha_i(\mathscr{C})$ are pairwise disjoint and their component diameters tend to 0 as $i \to \infty$. Note that the set $g'(X \cap \alpha_i(\mathscr{C}_i))$ lies in $\alpha_i \Phi_n(\mathscr{C}_i)$. The sets $\alpha_i \Phi_n(\mathscr{C}_i)$ and $\alpha_j \Phi_n(\mathscr{C}_j)$ miss one another if $j \neq i-1, i, i+1$. Moreover, $\alpha_i \Phi_n(\mathscr{C}_i) \cap \alpha_{i+1} \Phi_n(\mathscr{C}_{i+1})$ lies in $\alpha_i(\mathscr{B}_i \cap \Phi_n(\mathscr{C}_i)) \cap \alpha_{i+1}(\mathscr{A}_{i+1} \cap \Phi_n(\mathscr{C}_{i+1}))$ by (2), and the latter intersection is empty by (6). Consequently, g' is injective. Since it has compact support, g' is an embedding; moreover, $\rho(g', \mathrm{incl}_X) < \epsilon$ by (1).

Among the $\mathscr{A}_i \cup \mathscr{B}_i$, only the image of $\mathscr{A}_1 \cup \mathscr{B}_1$ meets D_0^*, and there $D_0^* \cap \alpha_1(\hat{I}^n) \subset B_0^*$, by initial prearrangement. Moreover, $X \cap D_0^* \subset E_0^* \subset \alpha_1(\mathscr{C}_1)$, and

$$D_0^* \cap g'(X) = B_0^* \cap \lambda(X \cap \alpha_1(\mathscr{C}_1)) \subset B_0^* \cap \alpha_1(\mathscr{A}_1 \cap \Phi_n(\mathscr{C}_1)) = \emptyset,$$

by (6) and the requirement that $B_0^* \cap \alpha_1(\mathscr{B}_1) = \emptyset$. To conclude, one can approximate $(*) : D_0 \to \mathbb{R}^n$, rel ∂D_0, by a PL embedding ψ. (Here ψ might be $\alpha_0 | D_0$, had α_0 been defined.) Then one can produce a controlled ambient isotopy transforming ψ to the inclusion $D_0 \to \mathbb{R}^n$. The end of that ambient isotopy provides a further adjustment g to g', where $g(X) \cap (I^2 \times \mathbf{0}) = \emptyset$. □

The real secret to the success of this argument is the insistence that all the $\alpha_j | \mathscr{A}_j$ be defined before doing so for any of the $\alpha_i | \mathscr{B}_i$; this means, quite specifically, that α_{i+1} is prescribed on $\mathscr{C}_{i+1}$ before any consideration is given to the definition of α_i on $\mathscr{B}_i$. In effect, $\alpha_{i+1}(\mathscr{C}_{i+1})$ then determines a neighborhood of $\operatorname{Int} E_i^*$ from which X will be cleared away and it frees up desirable space for $\alpha_i(\mathscr{B}_i)$.

Historical Notes. Štan′ko (1971a) gave the first proof of Theorem 5.5.1 and developed the key ideas. He used it to confirm that the Menger k-dimensional compactum in $\mathbb{R}^n$ contains a copy of each k-dimensional compactum that does embed in $\mathbb{R}^n$ (1971b). Edwards (1975b) reformulated Štan′ko's argument in the more familiar language of immersions, imitated here, and extended it to embeddings of σ-compacta. The second proof of the Fundamental Lemma is modelled on a related codimension-one argument in (Ancel and Cannon, 1979).

Exercises

5.5.1. If X is a cell-like subset of a PL n-manifold M^n, $\dim X \leq n-3$, then X can be embedded in $\mathbb{R}^n$.

5.5.2. If $C(\Delta)$ is a branching system and $k \geq 0$ is an integer, then ∂D_0 is null-homotopic in $[(D_0^* \cup \cdots \cup D_k^*) \setminus E^*] \cup D_{k+1}^*$.

5.6. PL approximation of embeddings of manifolds

Having twice demonstrated that embeddings of codimension-three compacta can be approximated by 1-LCC embeddings, we now return to the PL category, where the new aim is to establish that codimension-three topological embeddings of PL manifolds can be approximated by PL embeddings. The issue will be addressed in the noncompact setting, which is particularly meaningful because a PL approximation theorem for noncompact manifolds will be an absolutely necessary ingredient later in the chapter when deriving a PL approximation theorem for embeddings of compact polyhedra.

Theorem 5.6.1. *Let $h : Q \to M$ be a topological embedding of the k-dimensional PL ∂-manifold Q into an n-dimensional PL manifold M, $k \leq n-3$. For every continuous function $\epsilon : Q \to (0,\infty)$ there exists a PL embedding $h' : Q \to M$ such that $d(h(x), h'(x)) < \epsilon(x)$ for every $x \in Q$.*

Most of the section will be devoted to producing PL approximations to embeddings of cells in $\mathbb{R}^n$.

Proposition 5.6.2. *Let $h : D^k \to \mathbb{R}^n$ be a topological embedding of the k-cell D^k into $\mathbb{R}^n$, $k \leq n-3$. For every positive number ϵ there exists a PL embedding $h' : D^k \to \mathbb{R}^n$ such that $d(h(x), h'(x)) < \epsilon$ for every $x \in D^k$.*

The proof involves a complicated construction that will be described below. With the tools at hand the proof that the special case of k-cells in $\mathbb{R}^n$ implies the general case is comparatively simple, however, so we dispose of that matter at the outset. The idea is to decompose Q into handles, to thicken the handles slightly so that they overlap, to approximate h on each of the thickened handles using Proposition 5.6.2, and then to push these approximations together where they overlap using Unknotting Theorem 5.4.2.

Proof of Theorem 5.6.1. Let $\mathscr{H}$ be a handle decomposition of Q. We may assume the handles are small enough that for each handle $H \in \mathscr{H}$, $h(H)$ is contained in a chart in M which is PL homeomorphic to $\mathbb{R}^n$. Define $Q_i = \cup\{H \in \mathscr{H} \mid \text{index}(H) \leq i\}$. We will prove by induction on i that $h|Q_i$ can be approximated arbitrarily closely by PL embeddings.

First consider the case $i = 0$. For each 0-handle $H_j^0 \in \mathscr{H}$, define $\epsilon_j = \min\{\epsilon(x) \mid x \in H_j^0\}$. Make ϵ_j smaller, if necessary, so the ϵ_j-neighborhood of $h(H_j^0)$ is disjoint from the ϵ_ℓ-neighborhood of $h(H_\ell^0)$ whenever $H_j^0 \neq H_\ell^0$. Then apply Proposition 5.6.2 to construct a PL embedding $h_j : H_j^0 \to M$ within ϵ_j of $h|H_j^0$. The union of the PL embeddings h_j defines a PL embedding $h' : Q_0 \to M$ that is an $\epsilon(x)$-approximation to $h|Q_0$.

Now consider the general case. To simplify notation, let $A = Q_{i-1}$, let B denote a regular neighborhood in Q_i of the union of the i-handles, and let $C = A \cap B$. Observe that the i-handles are pairwise disjoint, so each component of C is compact. Let $\epsilon' : A \cup B \to (0, \infty)$ be a continuous map such that

(1) $\epsilon'(x) < (1/2)\epsilon(x)$ for every x,

(2) $x \in A \smallsetminus B$ implies $\epsilon'(x) < (1/2)d(x, B \smallsetminus A)$, and

(3) $x \in B \smallsetminus A$ implies $\epsilon'(x) < (1/2)d(x, A \smallsetminus B)$.

For each component C_j of C, define $\epsilon'_j = \min\{\epsilon'(x) \mid x \in C_j\}$. We may assume that the ϵ'_j-neighborhood of C_j is disjoint from the ϵ'_ℓ-neighborhood of C_ℓ for $C_j \neq C_\ell$. Apply Unknotting Theorem 5.4.2 to each component C_j of C to produce $\delta_j > 0$ such that if λ_0 and λ_1 are two PL embeddings of C_j that are within δ_j of $h|C_j$, then there is a PL $(\epsilon'_j/2)$-push ψ_t of $(M, h(C_j))$ such that $\psi_1\lambda_0 = \lambda_1$. Let $\delta : A \cup B \to (0, \infty)$ be a continuous function such that $\delta(x) < \delta_j$ whenever $x \in C_j$. Reduce δ, if necessary, to achieve $\delta(x) < \epsilon(x)/2$ for each $x \in A \cup B$.

Induction provides a PL embedding $\lambda_0 : A \to M$ with $d(\lambda_0(x), h(x)) < \delta(x)$ for each $x \in A$. The $i = 0$ case (proved above) gives a PL embedding $\lambda_1 : B \to M$ such that $d(\lambda_1(x), h(x)) < \delta(x)$ for each $x \in B$. By the choice of δ (and hence by Theorem 5.4.2), there exists a PL $(\epsilon'/2)$-push ψ_t of $(M, h(C))$ such that $\psi_1\lambda_0|C = \lambda_1|C$. Define a PL embedding $h' : A \cup B \to M$

by

$$h'(x) = \begin{cases} \psi_1(\lambda_0(x)) & \text{if } x \in A \\ \lambda_1(x) & \text{if } x \in B. \end{cases}$$

It is easy to check that h' is a PL embedding and that $d(h(x), h'(x)) < \epsilon(x)$ for every $x \in Q_i = A \cup B$. □

Throughout this section we use D^k to denote the k-cell $[0,1]^k$. Assume for the remainder of the section that $h : D^k \to \mathbb{R}^n$ is a fixed topological embedding of D^k into $\mathbb{R}^n$ and that $k \leq n-3$. We will prove that h can be approximated by PL embeddings. The strategy is to start with a thin k-cell that approximates h on a proper face of D^k and then to use the methods of controlled engulfing to stretch this thin k-cell out until it approximates all of $h(D^k)$. Since the face is proper, we need only codimension-four engulfing techniques. The process employed in this section is, in a sense, the opposite of the standard engulfing argument: instead of using the inverse of a collapse to expand an open set out to cover something, we will use the collapse itself to stretch a subpolyhedron in a controlled way. It will be clear to the reader that the techniques used here are based on the methods of controlled engulfing, but they are sufficiently different from the usual techniques that they must be developed from first principles and are not consequences of the controlled engulfing theorem stated earlier.

We will describe the basic construction in the simplest possible setting initially. After the details of the construction have been mastered we will make several observations about how the same construction could also be accomplished in other settings.

Three homotopies. Three special homotopies will be useful in the proof that follows: the first is a strong deformation retraction of a neighborhood of $h(D^k)$ to $h(D^k)$, the second is a homotopy of D^k to a face, and the third combines these two motions.

Fix an integer j, $1 \leq j \leq k$. As usual, we identify D^j with the subset $D^j = D^j \times \mathbf{0} \subset D^k$. Thus the given embedding $h : D^k \to \mathbb{R}^n$ induces an embedding $h|D^j : D^j \to \mathbb{R}^n$ for each $j \leq k$. Since $h(D^j)$ is an ANR, there exist a neighborhood U_j of $h(D^j)$ and a retraction $r^j : U_j \to h(D^j)$. We may assume that $d(x, r^j(x)) \leq 1/2$ for every $x \in U_j$. As the ambient manifold is $\mathbb{R}^n$, we have the straight line homotopy connecting r^j to the identity; we want to parametrize this homotopy in a particular way. For a fixed x, define $\lambda_j(x,t)$ to be the linear function of $\mathbb{R}$ such that $\lambda_j(x,0) = 0$ and $\lambda_j(x, d(x, r^j(x))) = 1$. Define a deformation retraction $r^j_t : U_j \to M$ by

$$r^j_t(x) = \begin{cases} (1-\lambda_j(x,t))x + \lambda_j(x,t)r^j(x) & \text{if } 0 \leq t \leq d(x, r^j(x)), \\ r^j(x) & \text{if } d(x, r^j(x)) \leq t \leq 1. \end{cases}$$

Observe that r_t^j moves x to $r^j(x)$ at unit speed during the time interval $[0, d(x, r^j(x))]$ and then keeps x stationary.

Let $\theta_t^j : D^j \to D^j$ be the homotopy defined by

$$\theta_t^j(x_1, \ldots, x_j, 0, \ldots, 0) = \begin{cases} \langle x_1, \ldots, x_j, 0, \ldots, 0\rangle & \text{if } 0 \leq t \leq x_j, \\ \langle x_1, \ldots, x_{j-1}, t, 0, \ldots, 0\rangle & \text{if } x_j \leq t \leq 1. \end{cases}$$

The homotopy θ_t^j deforms D^j in the jth direction to the face $x_j = 1$. Its action is rather like raising a window shade: a point $\langle x_1, \ldots, x_j\rangle$ does not move as long as $t \leq x_j$ and then it is raised linearly to the $x_j = 1$ level. (Even though the window shade analogy helps to gain an intuitive understanding of what the homotopy θ_t^j does, be warned that it will be convenient to draw the diagrams in this section with the 0-level at the top and the 1-level at the bottom—so the window is upside down.)

For a fixed $x \in U_j$, let $\mu(j, x, t)$ be the linear function of t such that $\mu(j, x, d(x, r^j(x))) = 0$ and $\mu(j, x, 1) = 1$. Define $\psi_t^j : U_j \to M$ by

$$\psi_t^j(x) = \begin{cases} r_t^j(x) & \text{if } 0 \leq t \leq d(x, r^j(x)), \\ h \circ \theta_{\mu(j,x,t)}^j \circ h^{-1} & \text{if } d(x, r^j(x)) \leq t \leq 1. \end{cases}$$

The homotopy ψ_t^j simultaneously squeezes U_j to $h(D^j)$ and deforms $h(D^j)$ to the image of the face $x_j = 1$. Each point x uses the time interval $[0, r^j(x)]$ to move into $h_j(D^j)$ and then uses the remaining time to move through $h(D^j)$ to the image of the face $x_j = 1$.

The basic construction. The basic construction starts with a polyhedron and ends with a second, larger polyhedron containing the first and collapsing toward a face of $h(D^k)$. Throughout the construction, $j \leq k$ is fixed.

Specify an integer p in the range $1 \leq p < j$ and a neighborhood $V_p \subset U_j$ of $h(D^j)$. Choose a sequence of neighborhoods $V_0 \subset V_1 \subset \cdots \subset V_{p-1} \subset V_p$ of $h(D^j)$ such that $\psi_t^j(V_i) \subset V_{i+1}$. Let $P_1 \subset V_0$ be a compact polyhedron of dimension $\leq p$. Define $f_1 : P_1 \times [0,1] \to V_1$ by $f_1(x,t) = \psi_t^j(x)$. Put f_1 in general position and let $P_2 = f_1(\mathrm{Sh}(S(f_1)))$. (Here $S(f_1)$ denotes the singular set of f_1 and $\mathrm{Sh}(S(f_1))$ is the shadow of $S(f_1)$ under the vertical collapse $P_1 \times [0,1] \searrow P_1 \times \{1\}$; refer to page 100 for the definitions of singular set and shadow.) Define $C_1 = f_1(P_1 \times [0,1])$ and $E_1 = f_1(P_1 \times \{1\})$. Note that $P_1 \subset C_1$, $\dim C_1 \leq p + 1 \leq j$, $\dim(P_2) \leq 2(p+1) - n + 1 \leq p - 1$ and $C_1 \searrow P_2 \cup E_1$ (see Figure 5.32).

Identify $P_2 \subset C_1$ with $P_2 \times \{0\} \subset P_2 \times [0,1]$ and consider the identification space $C_1 \cup (P_2 \times [0,1])$. Observe that the collapse $C_1 \searrow P_2 \cup E_1$ extends to a collapse

$$(*) \qquad C_1 \cup (P_2 \times [0,1]) \searrow E_1 \cup (P_2' \times [0,1]) \cup (P_2 \times \{1\}),$$

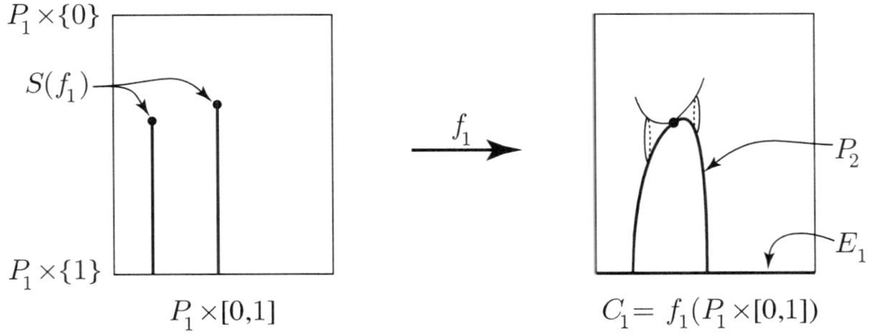

Figure 5.32. In this figure $S(f_1)$ consists of one pair of double points

where $P_2' = P_2 \cap E_1$ (see Figure 5.33). Define $f_2 : C_1 \cup (P_2 \times [0,1]) \to V_2$ by

$$f_2(y) = \begin{cases} y & \text{if } y \in C_1 \\ \psi_t^j(x) & \text{if } y = \langle x, t\rangle \in P_2 \times [0,1]. \end{cases}$$

Shift f_2 into general position and define $P_3 = f_2(\mathrm{Sh}(S(f_2)))$, where $\mathrm{Sh}(S(f_2))$ denotes the shadow of $S(f_2)$ under the collapse $(*)$. By general position, $\dim(P_3) \le (p+1) + p - n + 1 \le p - 2$. Define $C_2 = f_2(C_1 \cup (P_2 \times [0,1]))$ and $E_2 = f_2(E_1 \cup (P_2' \times [0,1]) \cup (P_2 \times \{1\}))$; then $C_2 \searrow P_3 \cup E_2$.

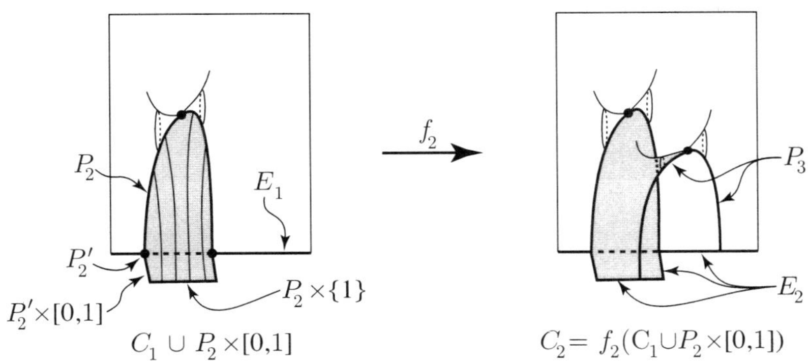

Figure 5.33. In this figure f_2 is an embedding except for one point where $f_2(C_1)$ meets $f_2(P_2 \times [0,1])$

This process is continued inductively. At the ith step we identify $P_i \subset C_{i-1}$ with $P_i \times \{0\} \subset P_i \times [0,1]$ and define $f_i : C_{i-1} \cup (P_i \times [0,1]) \to V_i$ to be the inclusion on C_{i-1} and ψ_t^j on $P_i \times [0,1]$. Shift f_i into general position and let $P_{i+1} = f_i(\mathrm{Sh}(S(f_i)))$, where this time $\mathrm{Sh}(S(f_i))$ is the shadow under the collapse $C_{i-1} \cup (P_i \times [0,1]) \searrow E_{i-1} \cup (P_i' \times [0,1]) \cup (P_i \times \{1\})$ and $P_i' = P_i \cap E_{i-1}$. Define

$$C_i = f_i(C_{i-1} \cup (P_i \times [0,1]) \text{ and } E_i = f_i(E_{i-1} \cup (P_i' \times [0,1]) \cup (P_i \times \{1\})).$$

Note that $\dim P_{i+1} \le p - i$ and $C_i \searrow P_{i+1} \cup E_i$.

The construction terminates after p steps. Specifically, $\dim S(f_{p-1}) \le 0$ and $\dim P_p \le 1$, so $f_p : C_{p-1} \cup (P_p \times [0,1]) \to V_p$ is an embedding. The end result of the construction is a pair of polyhedra

$$C = C_p = f_p(C_{p-1} \cup (P_p \times [0,1])) \text{ and}$$
$$E = E_p = f_p(E_{p-1} \cup (P_p' \times [0,1]) \cup (P_p \times \{1\}))$$

such that $P_1 \subset C$ and $C \searrow E$.

The following lemma summarizes the essential properties of C and E. As in §3.3, we use $B(A;\epsilon)$ to denote the ϵ-neighborhood of a set A.

Lemma 5.6.3. *For every $\epsilon > 0$ there exists $\delta > 0$ such that if P is any compact polyhedron in $B(h(D^j);\delta)$ with $\dim P < j$, then there exist a compact polyhedron $C \supset P$ with $\dim C = \dim P + 1$ and a compact subpolyhedron $E \subset C$ such that*

(1) $C \subset B(h(D^j);\epsilon)$,

(2) $E \subset B(h(D^{j-1} \times \{1\});\epsilon)$, *and*

(3) $C \searrow E$.

Furthermore, if N is a regular neighborhood E, then there exists a PL isotopy Φ_t of $\mathbb{R}^n$ such that

(4) $\Phi_0 = \mathrm{Id}$,

(5) $\Phi_1(C) \subset N$, *and*

(6) $\Phi_t | E \cup (\mathbb{R}^n \smallsetminus B(h(D^j);\epsilon)) =$ *inclusion.*

Proof. Given $\epsilon > 0$, begin the construction above by making sure that $V_p \subset B(h(D^j);\epsilon)$. Then choose δ so that $B(h(D^j);\delta) \subset V_0$ and so that $E \subset B(h(D^{j-1} \times \{1\});\epsilon)$. The first part of the lemma simply lays out the properties of the sets constructed above. The existence of Φ_t follows from regular neighborhood theory (Rourke and Sanderson, 1972, Chapter 3). □

Next we will examine the basic construction more carefully and will see how to impose controls on C and E that make the ambient isotopy Φ_t approximate the homotopy ψ_t^j. The control is imposed in two steps. First we will show how to make the collapse, and hence the ambient isotopy, follow the fibers of $h(D^j)$. We picture these fibers as vertical, so making the isotopy closely follow the fibers means that it is tightly controlled in the horizontal directions. Later, in the second step, we will impose controls in the vertical direction as well.

Horizontal control. A *fiber* of $h(D^j)$ is a set of the form

$$F = F^j(x_1, \ldots, x_{j-1}) = h(\{\langle x_1, \ldots, x_{j-1}\rangle\} \times [0,1])$$

for some fixed $\langle x_1, \ldots, x_{j-1}\rangle \in D^{j-1}$. The following elementary lemma asserts that if two fibers ever come close, then they are always close.

Lemma 5.6.4. *For every $\epsilon > 0$ there exists $\delta > 0$ such that if F_1 and F_2 are two fibers of $h(D^j)$ and $B(F_1; \delta) \cap B(F_2; \delta) \neq \emptyset$, then $F_2 \subset B(F_1; \epsilon)$.*

Proof. Let $\epsilon > 0$ be given. Use the uniform continuity of h to choose $\eta > 0$ such that if $x, x' \in D^j$ and $d(x, x') < \eta$, then $d(h(x), h(x')) < \epsilon$. Then use the uniform continuity of h^{-1} to choose $\delta > 0$ such that if $y, y' \in h(D^j)$ and $d(y, y') < 2\delta$, then $d(h^{-1}(y), h^{-1}(y')) < \eta$. It is routine to check that this δ works. □

Definition. An isotopy Φ_t of $\mathbb{R}^n$ is said to be *ϵ-parallel to fibers of $h(D^j)$* if for each $y \in \mathbb{R}^n$, either $\Phi_t(y) = y$ for every $t \in [0,1]$ or there exists one fiber F of $h(D^j)$ such that $\{\Phi_t(y) \mid 0 \leq t \leq 1\} \subset B(F; \epsilon)$.

Lemma 5.6.5. *The choice of $\delta > 0$ in Lemma 5.6.3 may be refined to allow the isotopy Φ_t to be constructed to have the following additional property:*

(7) *Φ_t is ϵ-parallel to fibers of $h(D^j)$.*

Proof. Given $\epsilon > 0$ and $p = \dim P$, apply Lemma 5.6.4 repeatedly to choose $\delta > 0$ having the property that if Ω is a path such that

$$\Omega \subset B(F_1; \delta) \cup B(F_2; \delta) \cup \cdots \cup B(F_p; \delta)$$

for some fibers $F_1, F_2, \ldots, F_p$ of $h(D^j)$, then $\Omega \subset B(F_1; \epsilon)$. We will assume that the δ of Lemma 5.6.3 satisfies this additional condition and prove that Φ_t can be constructed to be ϵ-parallel to fibers of $h(D^j)$.

The collapse $C \searrow E$ of Lemma 5.6.3 is composed of the collapses

$$f_1(P_1 \times [0,1]) \searrow E_1 \cup P_2 \subset E_1 \cup f_2(P_2 \times [0,1]) \\ \searrow E_2 \cup P_3 \subset E_2 \cup f_3(P_3 \times [0,1]) \searrow E_3 \cup P_4 \subset \cdots .$$

Each of these collapses, in turn, is the image under f_i of a vertical collapse in $P_i \times [0,1]$. Just as in the proof of the Controlled Engulfing Theorem, it is convenient to think of the inverse isotopy $\xi_t = \Phi_t^{-1}$, which expands a regular neighborhood of E to a regular neighborhood of C. The isotopy ξ_t is determined by the inverses of the collapses listed above.

Pick an integer i and consider (the inverses of) the collapses that take place in $f_i(P_i \times [0,1])$. First a neighborhood of E is pushed out to cover $f_i(\{v\} \times [0,1])$ for each vertex $v \in P_i$, then it is pushed farther to cover $f_i(\sigma \times [0,1])$ for each 1-simplex $\sigma \subset P_i$, and so forth. We can choose the neighborhood V_p at the beginning of the construction in such a way that for each $x \in P_i$ there exists a fiber F of $h(D^j)$ such that $f_i(\{x\} \times [0,1]) \subset B(F;\delta/2)$. We can then triangulate P_i so that for each simplex $\sigma \subset P_i$ there is a fiber F of $h(D^j)$ such that $f_i(\sigma \times [0,1]) \subset B(F;\delta)$. Identify disjoint neighborhoods of the various $f_i(\{v\} \times [0,1])$, where v is a vertex of P_i, and confine the motion involved in pushing out to cover $f_i(\{v\} \times [0,1])$ to these neighborhoods. In this way we can expand out to cover all the vertices with a motion that is δ-parallel to fibers of $h(D^j)$. Next expand out to cover $f_i(\sigma \times [0,1])$ for each 1-simplex $\sigma \subset P_i$. Since the boundaries of the 1-simplices have already been covered, we can use regular neighborhood theory to expand to cover the remaining parts of the various $f_i(\sigma \times [0,1])$ by motions that are confined to disjoint sets, one for each 1-simplex σ (Rourke and Sanderson, 1972, Lemma 3.25 and Theorem 3.26). Because the supporting sets are disjoint, the entire motion needed to cover the 1-simplices is δ-parallel to fibers. The process is continued, working up the skeleta.

It would appear that the combined motion determined by all the collapses in $f_i(P_i \times [0,1])$ could move each point through the neighborhoods of p different fibers. But in fact we can exercise enough care so that the motion involved in expanding out to cover $f_i(P_i \times [0,1])$ is δ-parallel to fibers. In order to accomplish this level of control, the open sets associated with $f_i(\sigma \times [0,1])$, for σ a simplex in P_i, must be chosen in such a way that if the open sets associated with two simplices intersect, then one of the simplices is a face of the other. If the open sets are chosen that way, then the combined motion associated with all the collapses in $f_i(P_i \times [0,1])$ will be δ-parallel to fibers of $h(D^j)$. (Note that this is exactly the kind of control achieved in the proof of the codimension-four case of Controlled Engulfing Theorem 3.3.7.)

In order to isotope a neighborhood of E to a neighborhood of C, we must push across polyhedra of the form $f_i(P_i \times [0,1])$ for $i = p, p-1, \ldots, 1$. By the previous paragraph, each of the associated isotopies moves δ-parallel to fibers of $h(D^j)$. The combined motion will have the property that each point that is moved at all is moved in the union of p sets of the form $B(F;\delta)$. The choice of δ guarantees that the combined motion is ϵ-parallel to fibers.

The result of this construction is an isotopy ξ_t that is ϵ-parallel to fibers of $h(D^j)$ and moves a regular neighborhood N of E to a regular neighborhood of C. The isotopy Φ_t of the conclusion is ξ_{1-t}, the reverse isotopy. □

Vertical control. We now turn our attention to control in the vertical direction. We will modify the isotopy Φ_t so that its motion in the vertical direction approximates that of the homotopy θ_t^j. In order to accomplish this we will split each of the elementary collapses that make up $C \searrow E$ into a sequence of small collapses, then will rearrange their order and, finally, will adjust the timing of those small collapses.

Lemma 5.6.6. *The choice of $\delta > 0$ in Lemmas 5.6.3 and 5.6.5 may be further refined so the isotopy Φ_t to be constructed has the following additional properties for each t:*

(8) $\Phi_t(C) \subset B(h(D^{j-1} \times [t, 1]); \epsilon)$, *and*

(9) $\Phi_t|(\mathbb{R}^n \smallsetminus B(h(D^{j-1} \times [0, t]); \epsilon)) =$ *inclusion.*

Proof. Begin by choosing a number $\gamma > 0$ such that if $|s - t| < (j - 1)\gamma$, then $h(D^{j-1} \times \{s\}) \subset B(h(D^{j-1} \times \{t\}); \epsilon)$. Choose $\eta_1 > 0$ such that

$$B(h(D^{j-1} \times [0, t]); \eta_1) \cap B(h(D^{j-1} \times [t + \gamma, 1]); \eta_1) = \emptyset$$

for every $t \in [0, 1 - \gamma]$. Then choose η_2, $0 < \eta_2 \leq \eta_1/2$, such that if $\psi^j_s(x) \in B(h(D^{j-1} \times [0, t]); 2\eta_2)$ for some s, t, x, then $\psi^j_{s'}(x) \in B(h(D^{j-1} \times [0, t]); \eta_1)$ for every $s' \leq s$. Finally, choose $\delta > 0$ so that $\psi^j_t(x) \in B(h(D^{j-1} \times [t, 1]); \eta_2)$ for every $x \in B(h(D^j); \delta)$ and for every $t \in [0, 1]$. We will assume that the δ in the proofs of Lemmas 5.6.3 and 5.6.5 satisfies this additional restriction and prove that we can obtain conclusions (8) and (9) while maintaining the earlier conclusions.

In the construction thus far, it has not been necessary to specify the time parameter in Φ_t precisely. The isotopy Φ_t is the concatenation of a sequence of isotopies, each of which corresponds to one of the elementary expansions that make up the expansion from E to C. It is assumed that the interval $0 \leq t \leq 1$ is subdivided into disjoint subintervals and that the isotopies corresponding to the various elementary collapses are performed, one at a time, on these disjoint subintervals. The order in which the isotopies are run corresponds to the order of the expansions, which is the reverse of the order of the collapses. We can think of the collapses as happening at discrete times during the interval $0 \leq t \leq 1$ and of the isotopies as taking place in the intervening subintervals. We have some freedom to rearrange the order of these collapses; we must just make certain that any simplex that could obstruct the collapse across a given face is collapsed out before the collapse across that face. We intend to take advantage of this flexibility to adjust the timing of the collapses and alter the corresponding isotopy Φ_t.

Fix an integer i and consider the collapse $f_i(P_i \times [0, 1]) \searrow P_{i+1} \cup E_i$. This collapse is made up of a sequence of collapses of the form

$$(**) \qquad f_i(\sigma \times [0, 1]) \searrow f_i(\partial\sigma \times [0, 1] \cup \sigma \times \{1\})$$

(or of the form $f_i(\sigma \times [0, 1]) \searrow f_i(\partial\sigma \times [0, 1] \cup (\mathrm{Sh}(S(f_i)) \cap (\sigma \times [0, 1])))$ in case $f_i(\mathrm{Int}\,\sigma \times [0, 1] \cap S(f_i) \neq \emptyset)$. Choose a triangulation for P_i and a positive integer m such that $\mathrm{diam}\, f_i(\sigma \times [\ell/m, (\ell + 1)/m]) < \eta_2$ for every simplex σ

in P_i and for every ℓ. Then replace the collapse (**) by the sequence

$$\begin{aligned} f_i(\sigma \times [0,1]) &\searrow f_i(\partial\sigma \times [0,1] \cup \sigma \times [1/m, 1]) \\ &\searrow f_i(\partial\sigma \times [0,1] \cup \sigma \times [2/m, 1]) \\ &\ \vdots \\ &\searrow f_i(\partial\sigma \times [0,1] \cup \sigma \times \{1\}). \end{aligned}$$

This subdivides the single collapse (**) into a sequence of m collapses, each of which is supported on a set of diameter $< \eta_2$. The total isotopy Φ_t is unchanged by this modification, but it is broken up into the concatenation of a large number of smaller pushes.

Subdivide every collapse in $f_i(P_i \times [0,1])$ as described in the previous paragraph and consider the portion of the isotopy Φ_t that corresponds to the collapses in $f_i(P_i \times [0,1])$. We wish to reorder the timing of the collapses so that the collapse $f_i(\sigma \times [\ell/m, (\ell+1)/m]) \searrow f_i(\partial\sigma \times [\ell/m, (\ell+1)/m]) \cup f_i(\sigma \times \{(\ell+1)/m\})$ is done just as soon as possible after the first time t at which the condition

$$(\ddagger) \qquad f_i(\sigma \times [\ell/m, (\ell+1)/m]) \cap B(h(D^{j-1} \times [t,1]); \eta_2) = \emptyset$$

is satisfied. Fix one such collapse of $f_i(\sigma \times [\ell/m, (\ell+1)/m])$ and let t_0 be the smallest t for which the condition ($\ddagger$) is satisfied. If possible, perform the isotopy associated with this collapse immediately after time t_0. The isotopy should be performed during the time interval in which the cell $f_i(\sigma \times [\ell/m, (\ell+1)/m])$ is contained in $B(h(D^{j-1} \times [t_0, 1]); 2\eta_2)$. If this prescription results in two or more isotopies being performed at the same time, then subdivide the time interval more finely and perform the isotopies on disjoint subintervals. Any collapses for which ($\ddagger$) is never satisfied should be performed at the very end.

There are two reasons why the construction just described might not be possible. One is that a higher-dimensional simplex in the same level $f_i(P_i \times [\ell/m, (\ell+1)/m])$ does not yet satisfy ($\ddagger$). Another is that simplices at higher levels may not yet have been collapsed out. If σ is the face of a simplex τ and $f_i(\tau \times [\ell/m, (\ell+1)/m])$ has not already been collapsed because it does not yet satisfy condition ($\ddagger$), wait until it does satisfy the condition. Since $\operatorname{diam} f_i(\tau \times [\ell/m, (\ell+1)/m]) < \eta_2$, performing the isotopy of $f_i(\sigma \times [\ell/m, (\ell+1)/m])$ at that later time will be soon enough. If simplices above $f_i(\sigma \times [\ell/m, (\ell+1)/m])$ (i.e., with smaller second coordinate) have not yet been collapsed out, collapse them first. The choice of η_2 ensures that these collapses will take place in $B(h(D^{j-1} \times [0,t]); \eta_1)$.

When the collapses are reordered as prescribed in the last two paragraphs, they determine a PL isotopy ϕ_t^i having the following properties:

(8_i) $\phi_t^i(f_i(P_i \times [0,1])) \subset P_{i+1} \cup B(h(D^{j-1} \times [t,1]); \eta_1)$, and

(9_i) $\phi_t^i|(\mathbb{R}^n \setminus B(h(D^{j-1} \times [0,t]); \eta_1)) =$ inclusion.

Condition (8_i) is clearly satisfied for $t = 0$. A point in the slice $f_i(P_i \times [j/m, (j+1)/m])$ moves as soon as it is outside $B(h(D^{j-1} \times [t,1]); \eta_2)$ and it keeps moving until its image is either in P_{i+1} or in $B(h(D^{j-1} \times [t,1]); \eta_1)$, so condition ($8_i$) is satisfied at later times as well. In order to verify condition (9_i), observe that a point x is not moved by ϕ_t^i unless x is outside $B(h(D^{j-1} \times [t,1]); \eta_2)$. Since the entire construction takes place within $B(h(D^j); \eta_1)$, it must be the case that any movement of x before time t takes place in $B(h(D^{j-1} \times [0,t]); \eta_1)$.

Now we would like to define Φ_t as the isotopy obtained by running all of the isotopies ϕ_t^i simultaneously. This will not quite work because the faces we want to collapse across may not yet be free when we get to them. In order to remedy this, we first start running ϕ_t^1, then start ϕ_t^2 after a lag of $t = \gamma$, then start ϕ_t^3 with a time lag of $t = 2\gamma$, etc. This works because the choice of η_1 ensures that any simplex in $f_i(P_i \times [0,1])$ is collapsed out before a simplex in $f_{i+1}(P_{i+1} \times [0,1])$ that is attached to it must be collapsed. By subdividing the time interval and running the individual pieces of ϕ_t^i more quickly, if necessary, we may assume that only one isotopy is running at any particular time. Define Φ_t to be the isotopy obtained by combining the ϕ_t^i with the delays. The choice of γ guarantees that Φ_t will satisfy conclusions (8) and (9). None of the conclusions of Lemma 5.6.3 is affected by the modifications made above. The path $\Phi_t(y)$ remains unchanged, even though it is parametrized differently, so the new Φ_t still moves ϵ-parallel to fibers. Thus all of conclusions (1) through (9) are now satisfied by Φ_t. □

Tightness control. We must place one more form of control on the isotopy Φ_t; we will confine its action to tighter and tighter neighborhoods of C. More specifically, we momentarily stop the action of Φ_t after some of the collapses have been performed and the isotopy associated with those collapses has been determined and then restrict subsequent motion to a closer neighborhood of C. The main point of the next lemma is that the polyhedra C and E can be specified first, then the collapses and their timing can all be completely determined, and finally the isotopy Φ_t can be specified.

Setting. Let C, E, and Φ_t be as in the proof of Lemma 5.6.6 and assume the collapse $C \searrow E$ has been subdivided so as to coincide with the action of Φ_t as described in that proof. Pick a finite number of intermediate polyhedra $C = C_0 \supset C_1 \supset \cdots \supset C_m = E$ so that the collapse $C \searrow E$ is subdivided as $C_0 \searrow C_1 \searrow \cdots \searrow C_m$. Define t_ℓ to be the instant at which all the subisotopies associated with the collapses in $C \searrow C_\ell$ are completed.

Lemma 5.6.7. *Assume the setting above. Let* $W_0 \supset W_1 \supset \cdots \supset W_m$ *be a sequence of regular neighborhoods of* C. *The isotopy* Φ_t *can be constructed so that, for each* $\ell \geq 0$, *it has the following additional properties:*

(10) $\Phi_t | C_\ell = \text{incl}$ *for* $t \leq t_\ell$, *and*

(11) *the support of* Φ_t, $t \geq t_\ell$, *is contained in* $\Phi_{t_\ell}(W_\ell)$.

Moreover, the regular neighborhood W_ℓ *can be chosen after the initial isotopy* $\{\Phi_t \mid 0 \leq t \leq t_\ell\}$ *has been determined.*

Proof. The isotopy Φ_t is constructed as the concatenation of a sequence of isotopies, each associated with one of the elementary collapses that make up $C \searrow E$. An elementary collapse has the form $\sigma \times [a, b] \searrow \partial\sigma \times [a, b] \cup \sigma \times \{b\}$ and the associated isotopy is obtained by applying the Regular Neighborhood Theorem for Pairs (Rourke and Sanderson, 1972, Theorem 4.11). The advantage of applying the theorem for pairs is that it supplies an isotopy that preserves $\sigma \times [a, b]$ setwise: the part of the isotopy Φ_t associated with this elementary collapse squeezes $\sigma \times [a, b]$ into a region like the shaded region D shown in Figure 5.34 and keeps $\partial\sigma \times [a, b] \cup \sigma \times \{b\}$ fixed. Hence the composite isotopy can be chosen to satisfy conclusion (10). All subsequent motion is controlled by a subset of the image of C under the isotopy so far. Thus we are free to choose a tighter regular neighborhood of C at this stage of the construction of Φ_t and to confine all further movement to the image of this neighborhood. □

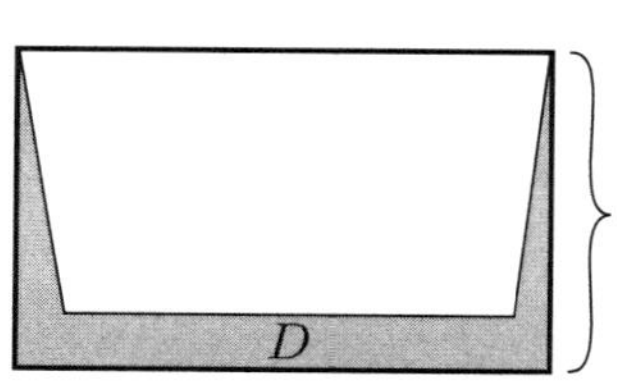

Figure 5.34. The collapse $\sigma \times [a, b] \searrow \partial\sigma \times [a, b] \cup \sigma \times \{b\}$ determines an isotopy that moves $\sigma \times [a, b]$ to D

This completes our description of the basic construction. It will be convenient to have a name for the polyhedra we have constructed.

Definition. A (j, ϵ)-*collapse* is a pair of compact polyhedra (C, E) for which there is a PL isotopy Φ_t such that C, E, Φ_t satisfy all the conclusions of Lemmas 5.6.3, 5.6.5, 5.6.6, and 5.6.7.

The entire construction is summarized in the following proposition.

Proposition 5.6.8. *Fix $j \leq k$. For every $\epsilon > 0$ there exists a $\delta > 0$ such that if P is a compact polyhedron in $B(h(D^j);\delta)$ and* $\dim P < j$, *then there exists a (j,ϵ)-collapse (C,E) such that $P \subset C$ and* $\dim C \leq \dim P + 1$.

We state two sample corollaries. One illustrates how to find maps into the boundary of a regular neighborhood of a collapse and the other illustrates how to enlarge a collapse to include a specified polyhedron.

Corollary 5.6.9. *For every $j \leq k$ and for every $\epsilon > 0$ there exist a (j,ϵ)-collapse (C,E), an ϵ-regular neighborhood W of C, and a PL map $g : D^{j-1} \to \partial W$ such that $\rho(g, h|D^{j-1}) < \epsilon$.*

Proof. Let $\gamma > 0$ be a number such that $\operatorname{diam}(h(\{x\} \times [0,\gamma])) < \epsilon/5$ for every $x \in D^{j-1}$. Choose $\eta > 0$ so that for every $x \in D^{j-1}$,

$$B(h(\{x\} \times [0,1]);\eta) \cap B(h(D^{j-1} \times [0,\gamma]);\eta) \subset B(h(\{x\} \times [0,\gamma]);\epsilon/5),$$

and so that

$$B(h(D^{j-1} \times \{0\});\eta) \cap B(h(D^{j-1} \times [\gamma,1]);\eta) = \emptyset.$$

Let δ be the δ given by Proposition 5.6.8 with input η. We may assume that $\eta < \epsilon/5$ and $\delta < \epsilon/5$.

Take $P = f(D^{j-1})$, where $f : D^{j-1} \to \mathbb{R}^n$ is a PL map such that $\rho(f, h|D^{j-1}) < \delta$. By the choice of δ, there exists a (j,η)-collapse (C,E) with $P \subset C$. Let W and $W_0 \subset \operatorname{Int} W$ be a pair of δ-regular neighborhoods of C. Then $W \setminus \operatorname{Int} W_0$ has a δ-product structure $W \setminus \operatorname{Int} W_0 \cong \partial W \times [0,1]$. Lemma 5.6.7 allows us to construct the isotopy Φ_t so that $\Phi_t|\partial W$ is the inclusion and $\Phi_\gamma(W_0) \subset B(h(D^{j-1} \times [\gamma,1]);\eta)$. The choice of γ and η ensure that $d(\Phi_\gamma(x),x) < 3\epsilon/5$ for every x and $P \cap \Phi_\gamma(W_0) = \emptyset$. The δ-product structure on $W \setminus \operatorname{Int} W_0$ has been stretched out to a $4\epsilon/5$-product structure on $W \setminus \Phi_\gamma(W_0)$. This product structure can be used to define a projection $\pi : W \setminus \Phi_\gamma(W_0) \to \partial W$. Then $g = \tau \circ f : D^{j-1} \to \partial W$ satisfies $\rho(g, h|D^{j-1}) < \epsilon$ and the proof is complete. □

Corollary 5.6.10. *For every $j \leq k$ and for every $\epsilon > 0$ there exists a $\delta > 0$ such that if (C,E) is a (j,δ)-collapse and $L \subset B(h(D^j);\delta)$ is a compact polyhedron with* $\dim L < j$, *then there exists a (j,ϵ)-collapse (C^*,E^*) such that $C \cup L \subset C^*$. Moreover, if $0 < \gamma < 1$ and $L \subset B(h(D^{j-1} \times [\gamma,1]);\delta)$, then $C \setminus B(h(D^{j-1} \times [\gamma,1]);\epsilon) = C^* \setminus B(h(D^{j-1} \times [\gamma,1]);\epsilon)$.*

Proof. Given $\epsilon > 0$, choose $\delta > 0$ exactly as in the proof of the proposition. Let (C,E) and L be as in the statement of the corollary. Triangulate $\mathbb{R}^n$ so that C, E, and L correspond to subcomplexes. Then any simplex of L whose interior intersects C is already contained in C, so it may be omitted from L. With those simplices omitted, $\dim(L \cap C) < \dim L$. Define $L^* =$

$L \cup \mathrm{Sh}(L \cap C)$, where $\mathrm{Sh}(L \cap C)$ is the shadow of $L \cap C$ under the collapse $C \searrow E$. Observe that $\dim L^* = \dim L$ and that $C \cup L^* \searrow E \cup L^*$. Thus the entire construction can be started over again by attaching a homotopy of L^* to C, shifting the union into general position, and then considering a new singular set. The end result of the construction is a (j, ϵ)-collapse (C^*, E^*) having the required properties. □

We are now ready to begin exploiting the construction to prove Proposition 5.6.2. The next proposition demonstrates how the existence of a (k, ϵ)-collapse can be utilized to construct a PL embedding that approximates h. It is worth observing that the hypotheses of Proposition 5.6.11 are just a little stronger than the conclusions of Corollary 5.6.9.

Proposition 5.6.11. *For every $j \le k$ and for every $\epsilon > 0$ there exists a $\delta > 0$ such that if*

(1) *(C, E) is a (j, δ)-collapse,*

(2) *W is a δ-regular neighborhood of C, and*

(3) *$g : D^{j-1} \to \partial W$ is a PL embedding with $\rho(g, h|D^{j-1}) < \delta$,*

then there exists a PL embedding $h' : D^j \to W$ such that $\rho(h', h|D^j) < \epsilon$.

Proof. First choose $\gamma > 0$ so that $\mathrm{diam}(h(\{x\} \times [t, t+\gamma])) < \epsilon/3$ for every $x \in D^{j-1}$ and for every $t \in [0, 1-\gamma]$. We may assume that $\gamma = 1/m$ for some integer m. Next choose $\eta > 0$ so that

$$B(h(\{x\} \times [0,1]); \eta) \cap B(h(D^{j-1} \times [t, t+\gamma]); \eta) \subset B(h(\{x\} \times [t, t+\gamma]); \epsilon/3)$$

for every $x \in D^{j-1}$ and for every $t \in [0, 1-\gamma]$. Let δ be the δ given by Proposition 5.6.8 with inputs j and $\epsilon = \eta$.

Assume C, E, W, and g are as in the statement of the proposition. Adjust the timing of the collapses in $C \searrow E$ and the isotopy Φ_t so that no collapse takes place at the instant $\ell\gamma = \ell/m$ for any integer ℓ and that none of the associated isotopies is in progress at any of those times. Pick intermediate polyhedra $C = C_0 \supset C_1 \supset \cdots \supset C_m = E$ so that the collapse $C \searrow E$ is subdivided as $C_0 \searrow C_1 \searrow \ldots \searrow C_m$ and so that for each ℓ the collapse $C_\ell \searrow C_{\ell+1}$ includes all the collapses that take place during the time subinterval $[\ell\gamma, (\ell+1)\gamma]$.

Let $W_0 = W$ and choose a sequence $W_0 \supset W_1 \supset \cdots \supset W_m$ of tighter regular neighborhoods of C such that $W_\ell \smallsetminus \mathrm{Int}\, W_{\ell+1}$ has a short collar structure for each ℓ. Use this collar structure to extend g to a PL embedding $g' : D^j \to W_0 \smallsetminus \mathrm{Int}\, W_\ell$ so that $g'(x, 0) = g(x)$ for each $x \in D^{j-1}$, $g'(D^{j-1} \times [\ell/m, (\ell+1)/m]) \subset W_\ell \smallsetminus \mathrm{Int}\, W_{\ell+1}$, and $\mathrm{diam}\, g'(\{x\} \times [0,1]) < \delta$ for each $x \in D^{j-1}$.

Let Φ_t, $0 \le t \le \gamma$, be a PL isotopy constructed in accordance with Proposition 5.6.8 so $\Phi_t|\partial W_0$ is the inclusion. Lemma 5.6.7 allows us to replace W_1 with a tighter regular neighborhood so that $\Phi_\gamma(W_1) \subset B(h(D^{j-1} \times [\gamma, 1]); \eta)$ and to construct the rest of Φ_t with support in $\Phi_\gamma(W_1)$. Now construct Φ_t for $\gamma \le t \le 2\gamma$ and use Lemma 5.6.7 to replace W_2 with a tighter neighborhood satisfying $\Phi_{2\gamma}(W_2) \subset B(h(D^{j-1} \times [2\gamma, 1]); \eta)$. Continue this process until the entire isotopy Φ_t has been defined.

A key observation is that

$$\Phi_1(W_\ell \smallsetminus \operatorname{Int} W_{\ell+1}) \subset B(h(D^{j-1} \times [\ell\gamma, (\ell+1)\gamma]); \eta)$$

for each ℓ. Thus the choice of η guarantees that the PL embedding $h' = \Phi_1 \circ g' : D^j \to W$ satisfies $\rho(h', h|D^j) < \epsilon$. □

There is a gap between Corollary 5.6.9 and Proposition 5.6.11. Notice, however, that the two results combine to prove the codimension-three approximation theorem in dimension $n = 2k$. In that case, general position allows the map g given by Corollary 5.6.9 to be approximated by a PL embedding and the approximation serves as the embedding required for the hypothesis of Proposition 5.6.11. Hence the construction described thus far allows us to advance the approximation theorem one step beyond the trivial range. This idea will be utilized to give an inductive proof of Proposition 5.6.2 in codimension three. Before we are in a position to complete this inductive argument we must make several observations about how the basic construction can be made to work in a more general setting.

The first step is to observe that the construction in the section can be done in a subset of $\mathbb{R}^n$. Assume that S is an s-dimensional PL submanifold of $\mathbb{R}^n$, $j \le s - 3$, and P is a subpolyhedron of S with $\dim P < j$. A *(j, ϵ)-collapse in S* is a pair $(C, E) \subset S$ satisfying all the conditions in the original definition of (j, ϵ)-collapse except that the isotopy Φ_t operates on S rather than on all of $\mathbb{R}^n$. It is also understood that any neighborhoods appearing in the construction are to be neighborhoods in S rather than in $\mathbb{R}^n$. A crucial observation is that the same construction can be used to construct a (j, ϵ)-collapse in S provided S contains the necessary homotopies. The next definition specifies what is required of the homotopies.

Definition. Let S be a PL submanifold of $\mathbb{R}^n$ and let P be a subpolyhedron of S. A *(j, ϵ)-homotopy of P in S* is a homotopy $f_t : P \to S$ such that $f_0 = \text{incl}$ and $d(f_t(x), \psi_t^j(x)) < \epsilon$ for every $x \in P$ and for every $t \in [0, 1]$.

The statement of Proposition 5.6.8 could be changed to say that for every $\epsilon > 0$ there exists a $\delta > 0$ such that a (j, ϵ)-collapse can be constructed in S provided S contains a (j, δ)-homotopy of P for every polyhedron $P \subset S$ with $\dim P < j$. Unfortunately we will not be able to construct submanifolds S

that satisfy such a strong hypothesis. Instead we will identify a polyhedron $P \subset S$ for which a (j, δ)-homotopy is required, and then we will replace S with a different submanifold S^* of $\mathbb{R}^n$ such that S^* contains the necessary homotopy.

Definition. Fix $j < k$ and $s \geq j + 3$. Assume that for each $\epsilon > 0$ there is a collection $\mathscr{S}_\epsilon$ of s-dimensional PL submanifolds of $\mathbb{R}^n$ such that $S \subset B(h(D^{j+1}); \epsilon)$ for each $S \in \mathscr{S}_\epsilon$. We say that $\{\mathscr{S}_\epsilon\}_{\epsilon > 0}$ has the (j, ϵ, δ)-*homotopy property* if for each $\epsilon > 0$ there exists a $\delta > 0$ such that if $S \in \mathscr{S}_\delta$ and L is a polyhedron in $S \cap B(h(D^j); \delta)$ with $\dim L < j$, then there exists $S_L \in \mathscr{S}_\epsilon$ such that $S_L \cap B(h(D^j); \delta) = S \cap B(h(D^j); \delta)$ and S_L contains the track of a (j, ϵ)-homotopy of L.

Lemma 5.6.12. *Fix $j < k$ and $s \geq j + 3$. Assume $\{\mathscr{S}_\epsilon\}_{\epsilon > 0}$ is a collection of s-dimensional PL submanifolds of $\mathbb{R}^n$ such that $\{\mathscr{S}_\epsilon\}_{\epsilon > 0}$ has the (j, ϵ, δ) homotopy property. For every $\epsilon > 0$ there exists a $\delta > 0$ such that if $S \in \mathscr{S}_\delta$, (C, E) is a (j, δ)-collapse in S, and $L \subset S$ is a compact polyhedron with $\dim L < j$, then there exists $S^* \in \mathscr{S}_\epsilon$ and a (j, ϵ)-collapse $(C^*, E^*) \subset S^*$ such that $C \cup L \subset C^*$.*

Proof. This lemma, a more elaborate version of Corollary 5.6.10, can be proved in essentially the same fashion. The basic construction described earlier in the section can be carried out in this setting because it really only required the existence of (j, δ)-homotopies for its success. Each time a polyhedron is identified and a homotopy of that polyhedron into a neighborhood of $h(D^{j-1} \times \{1\})$ is required, use the hypothesis of the lemma to replace the manifold S with a new manifold S^* that contains the track of the required homotopy. Since it is possible to anticipate how often this must be done, it is possible to choose the $\delta > 0$ required. Otherwise the proof is the same as the earlier one. □

Definition. A *tower of ϵ-collapses of height m* consists of sequences

$$(C_k, E_k), (C_{k-1}, E_{k-1}), \ldots, (C_{k-m}, E_{k-m}) \text{ and } W_k, W_{k-1}, \ldots, W_{k-m}$$

such that

(1) (C_k, E_k) is a (j, ϵ)-collapse in $\mathbb{R}^n$,

(2) W_k is an ϵ-regular neighborhood of C_k in $\mathbb{R}^n$,

(3) (C_j, E_j) is a (j, ϵ)-collapse in ∂W_{j+1} for $j < k$,

(4) W_j is an ϵ-regular neighborhood of C_j in ∂W_{j+1} for $j < k$, and

(5) there is a PL map $g : D^{k-m-1} \to \partial W_{k-m}$ with $\rho(g, h|D^{k-m-1}) < \epsilon$.

Proposition 5.6.13. *For every $\epsilon > 0$ and for every m there is tower of ϵ-collapses of height m.*

Proof. The proof is by induction on m. Corollary 5.6.9 is the $m = 0$ case. Assume that towers of ϵ-collapses of height $m-1$ exist for every $\epsilon > 0$. Start with a tower of height $m-1$ and raise its height to height m by adding the top level $C_{k-m} = E_{k-m} = W_{k-m} = \emptyset$. Conditions (1) through (4) of the definition of tower of ϵ-collapses are satisfied in this simple way. To complete the proof we must modify the entire tower so that it admits a map $g_m : D^{k-m-1} \to \partial W_{k-m}$ as required in part (5) of the definition. It suffices to construct the tower so that there is a map $g'_m : D^{k-m-1} \to C_{k-m}$ because the technique of the second paragraph of the proof of Corollary 5.6.9 can be used to push the image of g'_m out to the boundary of a regular neighborhood of C_{k-m}.

Induction provides a map $g_{m-1} : D^{k-m} \to \partial W_{k-m+1}$. The proof is completed by enlarging C_{k-m} so that it contains $g_{m-1}(D^{k-m-1} \times \{0\})$. Lemma 5.6.12 allows this to be done, provided the necessary homotopies exist. The homotopies are provided by Lemma 5.6.14, which will be proved next. □

Lemma 5.6.14. *Fix $m \leq k$. For every $\epsilon > 0$ there exists a $\delta > 0$ such that if $(C_k, E_k), \ldots, (C_{k-m+1}, E_{k-m+1})$; $W_k, \ldots, W_{k-m+1}$ is a tower of δ-collapses of height $m-1$ and $L \subset \partial W_{k-m+1} \cap B(h(D^{k-m}); \delta)$ is a compact polyhedron with $\dim L < k-m$, then there exists $(C^*_k, E^*_k), \ldots, (C^*_{k-m+1}, E^*_{k-m+1})$; $W^*_k, \ldots, W^*_{k-m+1}$, a tower of ϵ-collapses of height $m-1$, such that*

(1) *$C_j \subset C^*_j$ for every j,*

(2) *$C^*_{k-m+1} \cap B(h(D^{k-m}); \delta) = C_{k-m+1} \cap B(h(D^{k-m}); \delta)$, and*

(3) *C^*_{k-m+1} contains the image of a $(k-m, \epsilon)$-homotopy of L.*

Proof. The proof is by induction on m. Consider, first, the case $m = 1$. A tower of height 0 is just a (k, δ)-collapse (C, E) in $\mathbb{R}^n$ together with a regular neighborhood W of C and a map $g : D^{k-1} \to \partial W$. Let L be a compact polyhedron with $\dim L < k-1$. Consider the homotopy $f : L \times [0,1] \to \mathbb{R}^n$ that consists of the concatenation of the following homotopies: first the retraction of W to C, then the homotopy induced by the first part of the collapse that moves L near to $h(D^{k-1} \times \{\gamma\})$ for some $\gamma > 0$, then a retraction of a neighborhood of $h(D^{k-1} \times \{\gamma\})$ to $h(D^{k-1} \times \{\gamma\})$, and finally the homotopy $\theta^{k-1} \times \{\gamma\}$. This composite homotopy pushes L into $h(D^{k-2} \times \{1\} \times \{\gamma\})$. Choose γ small enough so that the homotopy is a $(k-1, \epsilon)$-homotopy. The first two parts of the homotopy are contained in W but the last two parts will likely stray out of W. By Corollary 5.6.10, we can replace C with a collapse C^* containing the track of the homotopy. The technique of the second paragraph of the proof of Corollary 5.6.9 can be used to push the track of the homotopy out to the boundary of a regular neighborhood of C^*.

The proof of the inductive step is exactly the same except that the inductive hypothesis is applied in place of Corollary 5.6.10. □

Proof of Proposition 5.6.2. If $2k < n$, then the proposition follows from general position. Otherwise choose $m = 2k - n$ and let ϵ_0 be the given positive number ϵ. Then apply Proposition 5.6.11 a total of $m+1$ times. The first time Proposition 5.6.11 is applied, take the j and ϵ of the hypothesis to be k and ϵ_0; define ϵ_1 to be the δ of the conclusion. The second time Proposition 5.6.11 is applied, take the j and ϵ of the hypothesis to be $k-1$ and ϵ_1; define ϵ_2 to be the δ of the conclusion. This pattern is continued in the obvious way. Use Proposition 5.6.13 to build a tower of ϵ_{m+1}-collapses of height m. The map $g : D^{k-m-1} \to \partial W_{k-m}$ associated with the top layer of the tower may be approximated by a PL embedding h_{k-m} using general position and the choice of m. The choice of ϵ_{k-j} allows us to find PL embeddings $h_j : D^j \to \partial W_{j+1}$ for progressively larger j. The PL embedding $h_k : D^k \to \mathbb{R}^n$ is the approximation we seek. □

Historical Notes. T. Homma (1966), (1968) was the first to prove a PL approximation theorem above the trivial range. Homma's theorem has a complicated statement and at first it was not clear in what generality his technique might apply. Berkowitz (1971) found a counterexample to part of the technique in codimension three and it was agreed that Homma's proof of the approximation theorem is only valid for embeddings of manifolds in the (2/3)-range (i.e., $k \leq (2/3)n$, also called the metastable range). Later Berkowitz and Dancis (1970a) extended Homma's technique to approximate embeddings of manifolds in the (3/4)-range.

Miller (1972) introduced a completely new technique that proved the PL approximation theorem for manifolds in codimension three. The proofs of Proposition 5.6.2 and Theorem 5.6.1 in the text are based on the proofs in (Miller, 1972). Bryant (1972) used Miller's approximation theorem for embeddings of manifolds to prove a PL approximation theorem for codimension-three embeddings of polyhedra; Bryant's proof will be presented in §5.8. Eventually Bryant (1990) developed the necessary technical machinery that allowed a modification of Homma's technique to succeed in codimension three. Theorem 5.6.1 was also announced by Černavskiĭ (1969a), (1969b), (1970).

The construction that was used to PL approximate topological embeddings of codimension-three cells can be made to work to PL approximate topological embeddings of codimension-two cells (Venema, 1987). Examples described in the next chapter demonstrate that it is not possible to approximate PL embeddings of arbitrary manifolds in codimension two.

5.7. Taming 1-LCC embeddings of polyhedra

§5.5 established that the 1-LCC embeddings of a codimension-three compactum X are dense in $\text{Emb}(X, M)$. We now investigate other ways in which 1-LCC embeddings are nice. The most prominent and useful result asserts that a 1-LCC embedding of a codimension-three polyhedron K is tame. Establishing this fact is accomplished in two steps. In this section a proof is provided under the additional hypothesis that the PL embeddings, $\text{Emb}_{\text{PL}}(K, M)$, are dense in $\text{Emb}(K, M)$. This intermediate result has immediate consequences for 1-LCC embeddings of PL manifolds, since Theorem 5.6.1 shows that the PL embeddings of manifolds are dense. In the next section this intermediate step is exploited in an inductive argument confirming that the extra assumption always holds, thereby yielding both the full strength codimension-three taming theorem and the PL approximation theorem at the same time.

Theorem 5.7.1 (Codimension-three 1-LCC taming). *Suppose K is a compact k-dimensional polyhedron and M^n is a PL n-manifold, where $k \leq n-3$ and $n \geq 5$, such that $\text{Emb}_{PL}(K, M^n)$ is a dense subset of $\text{Emb}(K, M^n)$. Then each 1-LCC topological embedding $e : K^k \to M^n$ is ϵ-tame.*

Corollary 5.7.2. *Every 1-LCC topological embedding $I^k \to M^n$ of a k-cell in a PL n-manifold M^n, $k \leq n-3$ and $n \geq 5$, is ϵ-tame.*

Corollary 5.7.3. *Every 1-LCC topological embedding $Q^k \to M^n$ of a compact, PL k-manifold in a PL n-manifold M^n, $k \leq n-3$ and $n \geq 5$, is ϵ-tame and thus locally flat.*

Corollary 5.7.4. *Let $f, g : Q^k \to M^n$ be homotopic, 1-LCC topological embeddings of a compact PL q-manifold into a PL n-manifold such that*

(1) $k \leq n-3$

(2) *Q^k is $(2k-n+1)$-connected, and*

(3) *M^n is $(2k-n+2)$-connected.*

Then f and g are ambient isotopic.

Corollary 5.7.5. *Any embedding of the k-sphere or k-cell into a PL n-manifold, $k \leq n-3$, that is locally flat modulo a flat cell or sphere is locally flat.*

Proof. The image is 1-LCC by Theorem 4.6.2. □

The proof of Theorem 5.7.1 is based on two lemmas.

Lemma 5.7.6 (Engulfing lemma). *Under the hypotheses of Theorem 5.7.1, suppose K is 1-LCC embedded in M and $\epsilon > 0$. Then there exists $\delta > 0$*

such that for each embedding $\lambda : K \to M$ within δ of the inclusion and for each open set $U \supset \lambda(K)$, there exists an ϵ-push φ of (M, K) with $\varphi(K) \subset U$.

Proof. The proof is almost exactly the same as that of Theorem 5.4.5. Since K is 1-LCC, $\operatorname{dem} K = k$ (Theorem 3.4.8). Thus, for each $\eta > 0$, K has a neighborhood N such that N is an η-regular neighborhood of a k-dimensional polyhedron K'. The idea is to engulf K' with U and then to use the product structure on $N \setminus K'$ to cover $N \supset K$. The required homotopies are supplied by Corollary 0.6.5. An application of Generalized Controlled Engulfing Theorem 3.3.7 completes the proof. □

Lemma 5.7.7. *Under the hypothesis of Theorem 5.7.1, suppose K is 1-LCC embedded in M and $\epsilon > 0$. Then there exists $\delta > 0$ such that for each embedding $\lambda : K \to M$ within δ of the inclusion and for any $\eta > 0$ there exists an ϵ-push ψ of (M, K) with $\rho(\psi\ K, \lambda) < \eta$.*

Proof. For an arbitrary positive number γ let $E(\gamma)$, the engulfing constraint, and $P(\gamma)$, the pushing constraint, be the positive numbers corresponding to K and γ given by Lemma 5.7.6 and Theorem 5.4.2, respectively.

Given $\epsilon > 0$, let $\delta = P(\epsilon/2)$. Consider an embedding $\lambda : K \to M$ within δ of $\text{incl} : K \hookrightarrow M$. Fix $\eta > 0$. Apply density of $\text{Emb}_{\text{PL}}(K, M)$ to produce a PL embedding $F : K \to M$ very close to λ. It suffices to determine an ϵ-push ψ of (M, K) with $\rho(\psi|K, F) < \eta$, for then $\psi|K$ will be close to λ as well. Compute $\eta' > 0$ such that if x and x' are points in K with $d(x, x') < 2\eta'$, then $d(F(x), F(x')) < \eta$. Set

$$\delta' = \min\{\delta, \eta', E(\eta'), E(\epsilon/2)\}.$$

Specify a PL embedding $G : K \to M$ not just within δ' of incl but so close that $d(F(x), F(x')) < \eta$ whenever $d(G(x), G(x')) < 2\eta'$.

Since F and G are $P(\epsilon/2)$-approximations of incl, there exists a PL $(\epsilon/2)$-push φ of (M, K) such that $\varphi G = F$. In addition, $G(K)$ has a neighborhood U such that $z, z' \in U$ and $d(z, z') < 2\eta'$ imply $d(\varphi(z), \varphi(z')) < \eta$.

In light of the choice of δ', there is a PL $\min\{\eta', \epsilon/2\}$-push φ' of (M, K) such that $\varphi'(K) \subset U$. Define ψ as $\varphi\varphi'$. For $x \in K$ clearly $d(G(x), \varphi'(x)) < 2\eta'$, and hence

$$d(\lambda(x), \psi(x)) \approx d(F(x), \psi(x)) = d(\varphi G(x), \varphi\varphi'(x)) < \eta$$

since $G(x), \varphi'(x) \in U$. This shows ψ to be the desired ϵ-push of (M, K). □

Remark. Observe that the hypothesis regarding denseness of the PL embeddings was employed in the proof of Lemma 5.7.7 but not in that of Lemma 5.7.6.

Proof of Theorem 5.7.1. Let $\Lambda = \mathrm{Emb}_{\mathrm{PL}}(K, M)$ and

$$\Lambda' = \{\lambda : K \to M \mid \lambda \text{ is a 1-LCC embedding}\}.$$

By hypothesis Λ is dense in $\mathrm{Emb}(K, M)$, and by Theorem 5.4.2 it is locally solvable. Lemma 5.7.7 and Theorem 4.2.11 ensure that $\Lambda' \supset \Lambda$ is locally solvable. Hence, each $\lambda' \in \Lambda'$ is ϵ-tame. □

Historical Notes. Bryant and Seebeck (1970) developed the techniques reproduced in this section and proved a result even stronger than Theorem 5.7.1.

5.8. PL approximation of embeddings of polyhedra

We are now ready to tackle the approximation theorem for topological embeddings of codimension-three polyhedra.

Theorem 5.8.1 (Approximating codimension-three polyhedra). *Let $e : K^k \to M^n$ be a topological embedding of a (not necessarily compact) k-dimensional polyhedron K into a PL n-manifold M, $k \leq n-3$, and let $\epsilon : K \to (0, \infty)$ be continuous. Then there exists a PL embedding $e' : K \to M$ such that $d(e(x), e'(x)) < \epsilon(x)$ for every $x \in K$. Moreover, if e is already PL on some subpolyhedron K_0 of K, then e' can be chosen so that $e'|K_0 = e|K_0$.*

The last part of the theorem follows easily from the first part: simply take a close PL approximation e' with no regard to fixing K_0 and then use Unknotting Theorem 5.4.2 to push $e'|K_0$ back to $e|K_0$. As a result, we need not mention the "moreover" statement again.

Once Approximation Theorem 5.8.1 has been established, the hypothesis that $\mathrm{Emb}_{\mathrm{PL}}(K, M^n)$ is a dense subset of $\mathrm{Emb}(K, M^n)$ can be removed from all the theorems in the preceding section. In particular, the taming theorem can be restated as follows.

Corollary 5.8.2 (Codimension-three taming). *If K is a compact k-dimensional polyhedron and M^n is a PL n-manifold, $k \leq n-3$ and $n \geq 5$, then each 1-LCC topological embedding $e : K^k \to M^n$ is ϵ-tame.*

Corollary 5.8.3. *For K and M^n as in Corollary 5.8.2, most embeddings of K in M^n are tame.*

Proof. See Theorem 4.6.17. □

As usual, the taming theorem can be used to prove flattening theorems for cells and spheres. In each of the next two corollaries, the 1-LCC hypothesis is verified by an application of Theorem 4.6.2.

Corollary 5.8.4. *If Σ is a k-sphere in S^n, $k \leq n-3$, $n \geq 5$, and Σ is locally flat modulo a cell or sphere that is flat in S^n, then Σ is flat.*

Corollary 5.8.5. *If B is a k-cell in $\mathbb{R}^n$, $k \leq n-3$, $n \geq 5$, that can be expressed as a union of k-cells B_1 and B_2, where B_1 is flat and B_2 is locally flat modulo $B_1 \cap B_2$, then B is flat.*

Combining Corollary 5.8.2 with Unknotting Theorem 5.4.2 yields a new unknotting theorem.

Corollary 5.8.6 (Unknotting 1-LCC embeddings of polyhedra). *Let $e : K^k \to M^n$ be a topological embedding of a compact k-dimensional polyhedron K into an n-dimensional PL manifold M with $k \leq n-3$ and $n \geq 5$. For every $\epsilon > 0$ there exists a $\delta > 0$ such that if $\lambda_0, \lambda_1 : K \to M$ are two 1-LCC topological embeddings, each pointwise within δ of e, then there is an ϵ-push ψ of M such that $\lambda_1 = \psi \circ \lambda_0$.*

Another interesting corollary assures that any two PL embeddings that are topologically equivalent are in fact PL equivalent. The next result is a taming theorem in the sense that it asserts that the isotopy between two PL embeddings can be "tamed."

Corollary 5.8.7 (Topological equivalence implies PL equivalence). *Suppose K is a compact k-dimensional polyhedron, M^n is a PL n-manifold, $k \leq n-3$, and $\lambda_0, \lambda_1 : K \to M$ are two PL embeddings. If λ_0 and λ_1 are equivalent via a topological isotopy of M, then λ_0 and λ_1 are equivalent via a PL isotopy of M.*

Proof. Let $\psi_t : M \to M$ be a topological isotopy such that $\psi_0 = \mathrm{Id}$ and $\psi_1\lambda_0 = \lambda_1$. Define S to be the set of all $t \in [0,1]$ such that for every $\eta > 0$ there exists a PL isotopy $\phi_s : M \to M$ with $\phi_0 = \mathrm{Id}_M$ and $\rho(\phi_1\lambda_0, \psi_t\lambda_0) < \eta$.

Obviously $0 \in S$. Let $t_0 \in S$. Apply Theorem 5.4.2 to $\psi_{t_0}\lambda_0$ with $\epsilon = 1$ to produce a number $\delta > 0$. There exists $\gamma > 0$ such that $\rho(\psi_t\lambda_0, \psi_{t_0}\lambda_0) < \delta$ whenever $|t_0 - t| < \gamma$. Every t with $|t_0 - t| < \gamma$ is in S because Theorem 5.8.1 allows us to approximate $\psi_t\lambda_0$ with a PL embedding; then the fact that $t_0 \in S$ allows us to find a PL isotopy that pushes λ_0 to a PL approximation of $\psi_{t_0}\lambda_0$ and Theorem 5.4.2 allows us to push it the rest of the way to the PL approximation to $\psi_t\lambda_0$. Thus S is open. The proof that S is closed is similar. If $t_0 \in \overline{S}$, then there is a $t \in S$ that is near to t_0. Hence there is a PL isotopy that pushes λ_0 very close to $\psi_t\lambda_0$ and then another small push by Theorem 5.4.2 taking the adjusted λ_0 to a PL approximation to $\psi_{t_0}\lambda_0$.

Since S is open, closed, and nonempty we have $S = [0,1]$ and can conclude that $1 \in S$. One final application of Theorem 5.4.2 yields the desired conclusion. □

The proof of Approximation Theorem 5.8.1 is linked in an inductive argument with the proof of a taming theorem. The taming theorem involved is closely related to the taming theorems of §5.7, but it incorporates more specialized hypotheses and conclusions. We need some new terminology in order to state it.

Recall that a subset L of a polyhedron K is a subpolyhedron of K if there is a triangulation T of K such that L is the underlying polyhedron of some subcomplex of T; in particular, L is a closed subset of K. We say that a map $f : K \to M$ is *PL on the open subset* $K \smallsetminus L$ if there is a locally finite subdivision of $T|K \smallsetminus L$ relative to which $f|(K \smallsetminus L)$ is simplicial. A map $f : K \to M$ is called *almost PL* if there exists a subpolyhedron $L \subset K$ such that f is PL on L and f is PL on $K \smallsetminus L$.

Theorem 5.8.8 (Taming almost PL embeddings). *Suppose K is a compact k-dimensional polyhedron, L is a subpolyhedron of K, and M^n is a PL n-manifold, where $k \leq n-3$ and $n \geq 5$. If $e : K \to M$ is a topological embedding such that $e|L$ and $e|K \smallsetminus L$ are both PL, then for every $\epsilon > 0$ there exists an ϵ-push φ of $(M, e(K))$ with supporting isotopy $\Phi : M \times [0,1] \to M$ such that*

(1) *$\varphi e : K \to M$ is PL,*

(2) *$\Phi_t e|L = e|L$ for every t, and*

(3) *$\Phi|(M \smallsetminus e(L)) \times [0,1]$ is PL.*

To describe the logical structure of the proofs of Theorems 5.8.1 and 5.8.8 succinctly, we exploit the following abbreviations.

Approx(n) = Approximation Theorem 5.8.1 in ambient dimension n
Taming(n) = Taming Theorem 5.8.8 in ambient dimension n

We will prove Approx(n) $\Rightarrow$ Taming(n) and Taming($n-1$) $\Rightarrow$ Approx(n).

The proof that Approx(n) $\Rightarrow$ Taming(n) follows the same outline as the proof of Theorem 5.7.1, but care must be taken to nail down the additional conclusions of Theorem 5.8.8. The next lemma replaces Lemma 5.7.6 in the argument.

Lemma 5.8.9. *Suppose K is a compact k-dimensional polyhedron, L is a subpolyhedron of K, M^n is a PL n-manifold, $k \leq n-3$ and $n \geq 5$, and $e : K \to M$ is an embedding that is PL on both L and on $K \smallsetminus L$. For every $\epsilon > 0$ there exists a $\delta > 0$ such that for each embedding $\lambda : K \to M$ with $\rho(e, \lambda) < \delta$ and $\lambda|L = e|L$, and for each open set $U \supset \lambda(K)$, there exists an ϵ-push ψ of (M, K) with $\psi e(K) \subset U$ and $\psi|L = \text{incl}_L$.*

Proof. The proof is just like those of Lemma 5.7.6 and Theorem 5.4.5. The objective is to engulf $e(K)$ with U. Corollary 0.6.5 supplies the required homotopies. Even though $e(K)$ is an infinite polyhedron, the part of $e(K)$ that is not already contained in U is covered by a finite polyhedron, so Generalized Controlled Engulfing Theorem 3.3.7 applies. $\square$

Proof that Approx(n) $\Rightarrow$ Taming(n). Assume Theorem 5.8.1 in ambient dimension n. First observe that this assumption, when combined with the proof of Lemma 5.7.7, yields a stronger version of Lemma 5.8.9. Specifically, for each $\eta > 0$, the ϵ-push ψ in the conclusion of 5.8.9 can be chosen so that the condition $\psi e(K) \subset U$ is replaced by the stronger condition $\rho(\psi e, \lambda) < \eta$. The proof of Taming Theorem 5.8.8 is then completed by means of essentially the same argument as that in the proof of Theorem 5.7.1. That proof was an application of Theorem 4.2.8; however, Theorem 4.2.8 as stated does not include provisions for conclusions (2) and (3) of the present theorem. Thus we will outline the proof of Theorem 4.2.8 for this special case and indicate how conclusions (2) and (3) are obtained.

Let $N_1, N_2, \ldots$ be a sequence of PL neighborhoods of L in K such that $N_{i+1} \subset \operatorname{Int} N_i$ and $\cap_{i=1}^{\infty} N_i = L$. Define $P_i = L \cup (K \smallsetminus \operatorname{Int} N_i)$. Start with a PL approximation $\lambda : K \to M$ that agrees with e on P_1. Construct sequences of small pushes $\{\psi_i\}_{i=1}^{\infty}$ and $\{\psi_i'\}_{i=1}^{\infty}$ of $(M, e(K))$ that satisfy

- $\psi = \lim_{i\to\infty} \psi_i \cdots \psi_1$ and $\psi' = \lim_{i\to\infty} \psi_i' \cdots \psi_1'$ are $(\epsilon/3)$-pushes of $(M, e(K))$,
- $\psi\lambda = \psi' e$,
- $\psi_i \cdots \psi_1 \lambda | P_i = \psi_i' \cdots \psi_1' e | P_i$,
- the supporting isotopies for ψ_{i+1} and ψ_{i+1}' are fixed on $\psi_i \cdots \psi_1 \lambda | P_i$, and
- each point of $M \smallsetminus e(L)$ has a neighborhood that is moved by at most a finite number of the supporting isotopies.

The push ψ_i is obtained by an application of Unknotting Theorem 5.4.2 and the push ψ_i' is obtained by an application of Lemma 5.7.7, strengthened as described at the beginning of this proof. Those two theorems provide the controls specified above. The homeomorphism $\varphi = \psi^{-1}\psi'$ is the ϵ-push required for the conclusion of the theorem. In particular, the fact that each point of $M \smallsetminus e(L)$ has a neighborhood that is moved by at most a finite number of the pushes implies that the supporting isotopy Φ is PL on $(M \smallsetminus e(L)) \times [0, 1]$. $\square$

The proof that Taming($n-1$) $\Rightarrow$ Approx(n) requires a number of technical lemmas. Several of them are counterparts to familiar results or to results already proved earlier in the text, but the statements and proofs must be modified somewhat to fit a noncompact setting.

Lemma 5.8.10. *Let M^n be a compact PL ∂-manifold and let $f : B^k \to \operatorname{Int} M$, $k \leq n-3$, be a topological embedding such that $f|\operatorname{Int} B^k$ is PL. Triangulate $M \smallsetminus f(\partial B^k)$ so that for each $\epsilon > 0$ there are only a finite number of simplices of diameter greater than ϵ and define V to be the simplicial neighborhood of $f(\operatorname{Int} B^k)$ in the second derived subdivision. Then $\overline{V} = V \cup f(\partial B^k)$ and there exists an extension of f to a homeomorphism $h : B^n \to \overline{V}$ such that $h|B^n \smallsetminus \partial B^k$ is PL.*

Proof. Exercise 5.8.1. □

Notation. The following notation will be assumed for the remainder of the section. Let K be a (not necessarily compact) k-dimensional polyhedron. Fix a triangulation T of K and let $\sigma \in T$ be a simplex with $\dim \sigma \leq k-1$. Define

$$\begin{aligned} S &= |\operatorname{St}(\sigma, T)|, \text{ the underlying polyhedron of the star of } \sigma \text{ in } T, \\ L &= |\operatorname{Lk}(\sigma, T)|, \text{ the underlying polyhedron of the link of } \sigma \text{ in } T, \\ \operatorname{Fr} S &= \text{ the frontier of } S \text{ in } K, \\ \operatorname{Int} S &= S \smallsetminus \operatorname{Fr} S, \text{ and} \\ \hat{\sigma} &= \text{ barycenter of } \sigma. \end{aligned}$$

Observe that $\operatorname{Fr} S$ naturally decomposes as $\operatorname{Fr} S = \partial\sigma * L$, the join of $\partial\sigma$ to L, while S can be decomposed as a join in two different ways. On the one hand, S has a cone structure $S = \hat{\sigma} * \operatorname{Fr} S$. On the other hand, $S = \sigma * L$. This latter join structure allows us to write $\operatorname{Int} S \smallsetminus \operatorname{Int}\sigma = \operatorname{Int}\sigma \times L \times (0,1)$ and to define a natural projection map $\pi : \operatorname{Int} S \to \operatorname{Int}\sigma$. If $P \subset \operatorname{Int} S$ is a subpolyhedron, we use $\operatorname{Sh}(P)$ to denote the shadow of P under the projection π.

Let $P \subset \operatorname{Int} S$ be a polyhedron such that $P \searrow \operatorname{Int}\sigma$. In order to simplify the statement of the next lemma we make the following *ad hoc* definition: a regular neighborhood U of P in $\operatorname{Int} S$ is a *nice neighborhood* if $\overline{U} = U \cup \partial\sigma$ and each of the open intervals in the product structure $\operatorname{Int} S \smallsetminus \operatorname{Int}\sigma = \operatorname{Int}\sigma \times L \times (0,1)$ intersects $\operatorname{Fr} U$ in exactly one point.

Lemma 5.8.11. *There exists a neighborhood N of $\operatorname{Int}\sigma$ in $\operatorname{Int} S$ such that for any neighborhood N' of $\operatorname{Int}\sigma$ with $\overline{N'} \smallsetminus \partial\sigma \subset N$, any polyhedron P in N', any nice neighborhood $U \subset N'$ of $\operatorname{Int}\sigma$, and any nice neighborhood $U' \subset N'$ of $\operatorname{Sh}(P) \cup \operatorname{Int}\sigma$, there is a PL homeomorphism $F : \operatorname{Int} S \to \operatorname{Int} S$ such that $F(U) = U'$, $F|\operatorname{Int} S \smallsetminus N = \operatorname{Id}$, and F extends via the identity to a homeomorphism $F : S \to S$.*

Proof. Simply push away from σ using the join structure described above. Choose N so that the distance a point is allowed to move approaches 0 as the point approaches $\partial\sigma$. (See Figure 5.35.) □

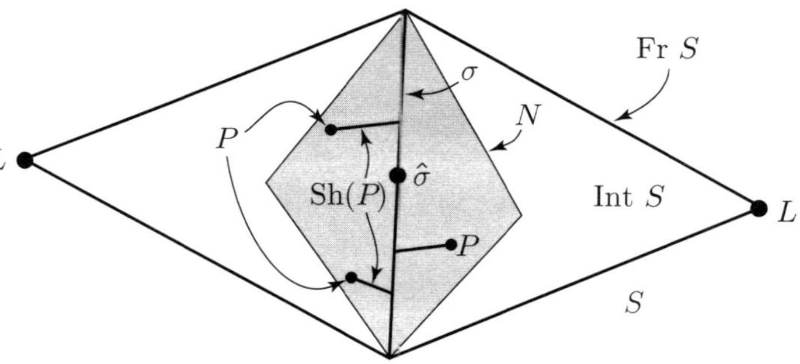

Figure 5.35. Use the join structure $S = \sigma * L$ to push a neighborhood U of σ out to cover $\mathrm{Sh}(P)$ while keeping $S \setminus N$ fixed

Lemma 5.8.12. *Suppose $f : S \to M$ is an embedding into a PL n-manifold M such that $f|\operatorname{Int} S \setminus \operatorname{Int}\sigma$ is PL. For every $\delta : \operatorname{Int} S \to (0,\infty)$ and for every neighborhood N of $\operatorname{Int}\sigma$ in $\operatorname{Int} S$ there exists a PL general position map $g : \operatorname{Int} S \to M$ such that*

(1) $g|\operatorname{Int} S \setminus N = f|\operatorname{Int} S \setminus N$,

(2) $g|\operatorname{Int}\sigma$ *is a PL embedding,*

(3) $d(f(x), g(x)) < \delta(x)$ *for every* $x \in \operatorname{Int} S$, *and*

(4) $S(g) \subset N$.

Proof. Use Estimated Homotopy Extension Theorem 0.6.4 to obtain $\eta : \operatorname{Int}\sigma \to (0,\infty)$ so that if $g : \operatorname{Int}\sigma \to M$ is any map with $d(f(x), g(x)) < \eta(x)$, then g is $\delta(x)$-homotopic to $f|\operatorname{Int}\sigma$. Since $\operatorname{Int}\sigma$ is a manifold, we may apply Theorem 5.6.1 to find a PL embedding $g : \operatorname{Int}\sigma \to M$ that is an $\eta(x)$-approximation to $f|\operatorname{Int}\sigma$. The choice of η allows g to be extended to a neighborhood of $\operatorname{Int}\sigma$ in such a way that the extension agrees with f on the boundary of the neighborhood. Then g may be further extended, via f, to all of $\operatorname{Int} S$. □

Lemma 5.8.13. *Suppose $f : S \to M$ is an embedding into a PL n-manifold M, $k \leq n-3$, such that $f|\operatorname{Int} S \setminus \operatorname{Int}\sigma$ is PL. For every $\epsilon : M \to (0,\infty)$ and for every neighborhood N of $\operatorname{Int}\sigma$ in $\operatorname{Int} S$ there exists $\delta : \operatorname{Int} S \to (0,\infty)$ such that if $g : \operatorname{Int} S \to M$ is a PL general position map satisfying the conclusions of Lemma 5.8.12, then there exist polyhedra $C \subset \operatorname{Int} S$ and $D \subset M$ such that*

(1) $\operatorname{Int}\sigma \cup S(g) \subset C$,

(2) $C = \mathrm{Sh}(C) \subset N$,

(3) $g(C) = D \cap g(\operatorname{Int} S)$, *and*

(4) $D \searrow g(\operatorname{Int}\sigma)$.

Moreover, if W is a regular neighborhood of D and V and V' are simplicial neighborhoods of $g(\operatorname{Int}\sigma)$ and D, respectively, in the second barycentric subdivision of W, then there exists a PL homeomorphism $F : M \smallsetminus g(\partial\sigma) \to M \smallsetminus g(\partial\sigma)$ such that

(5) $F(V) = V'$,

(6) $d(x, F(x)) < \epsilon(x)$ *for every* x,

(7) $F|M \smallsetminus W = \text{incl}$, *and*

(8) F *extends via the identity to a (topological) homeomorphism of* M.

Proof. This proof utilizes the familiar codimension-three shadow-building construction that has been employed several times before in this chapter. Except for the compactness issue, the construction is the same as the basic construction in §5.6 and it is also quite similar to the construction in the proof of Slicing Theorem 5.4.10. Since the proof is so much like those earlier proofs, we will merely outline the argument and trust that the reader can fill in the details.

Given ϵ, apply Corollary 0.6.5 to find a sequence of neighborhoods of $f(\operatorname{Int}\sigma)$ that ϵ-deform to $g(\operatorname{Int}\sigma)$. Choose δ small enough so that any g satisfying the conclusions of Lemma 5.8.12 will have its image in the smallest of these neighborhoods. Also require that $\delta(x) \to 0$ as $x \to f(\partial\sigma)$. Shrink N, if necessary, so that it satisfies the conclusions of Lemma 5.8.11. Let g be as in Lemma 5.8.12 and throw $g|\operatorname{Int} N$ into general position keeping $g|\operatorname{Int} S \smallsetminus \operatorname{Int} N$ fixed.

To start the construction, let $C_1 = \operatorname{Int}\sigma \cup \operatorname{Sh}(S(g))$. By the choice of δ there is a small homotopy $\theta_t^1 : g(C_1) \to M$ such that $\theta_0^1 = \text{incl}$, $\theta_t^1|g(\operatorname{Int}\sigma) = \text{incl}$ for every t, and $\theta_1^1(C_1) \subset g(\operatorname{Int}\sigma)$. Let B_1 be the quotient space of $g(C_1) \times [0,1]$ obtained by shrinking all arcs of the form $\{x\} \times [0,1]$, $x \in g(\operatorname{Int}\sigma)$, to points and also identifying each point $\langle x, 1\rangle$ with $\langle\theta_1^1(x), 1\rangle$ for $x \in g(C_1) \smallsetminus g(\operatorname{Int}\sigma)$. Another way to understand B_1 is to think of it as the reduced mapping cylinder $\operatorname{Map}(\theta_1^1, g(\operatorname{Int}\sigma))$; this view requires that we consider $g(\operatorname{Int}\sigma)$ to be the range of θ_1^1. Observe that $B_1 \searrow g(\operatorname{Int}\sigma)$. The homotopy θ_t^1 determines a map $\Theta_1 : B_1 \to M$. Put Θ_1 in general position, keeping $g(C_1) \cup g(\operatorname{Int}\sigma)$ fixed. Define $D_1 = \Theta_1(B_1)$. Conditions (1) and (2) are satisfied by C_1 and D_1, but conditions (3) and (4) probably are not. The set $E_1 = (D_1 \cap g(\operatorname{Int} S)) \cup S(\Theta_1)$ represents the error: if E_1 were empty, then conditions (3) and (4) would be satisfied. The dimension of E_1 is at most $2k - n + 2 \le n - 4$. As usual in a proof of this sort, the objective is to reduce the dimension of the error term and eventually eliminate it.

Define $C_2 = C_1 \cup \operatorname{Sh}(g^{-1}(D_1 \cap g(\operatorname{Int} S)))$. By the choice of δ there is a homotopy $\theta_t^2 : g(C_2) \cup \operatorname{Sh}(S(\Theta_1)) \to M$ that pushes $g(C_2) \cup \operatorname{Sh}(S(\Theta_1))$ into $g(\operatorname{Int}\sigma)$. (Here $\operatorname{Sh}(S(\Theta_1))$ denotes the shadow of the singular set of

Θ_1 under the collapse $B_1 \searrow g(\operatorname{Int}\sigma)$.) Define $B_2 = B_1 \cup \operatorname{Map}(\theta_1^2, g(\operatorname{Int}\sigma))$ and let $\Theta_2 : B_2 \to M$ be the map induced by θ_t^2. Put Θ_2 in general position, keeping B_1 fixed, and define $D_2 = \Theta_2(B_2)$. The new error term is $E_2 = (D_2 \cap g(\operatorname{Int} S)) \cup S(\Theta_2)$, which has dimension at most $n-5$. The construction is continued inductively until the error term is empty.

The output of this construction is a pair of polyhedra, C and D, that satisfy conclusions (1) through (4). The existence of the homeomorphism F follows from regular neighborhood theory. It is the composition of a sequence of homeomorphisms, one for each of the elementary collapses in $D \searrow g(\operatorname{Int}\sigma)$. Since there are infinitely many such collapses, we must be careful to make sure that each point of $M \smallsetminus f(\partial\sigma)$ is moved only a finite number of times. The "horizontal control" of §5.6 provides exactly what is needed to accomplish this. Finally, the fact that $\delta(x) \to 0$ as $x \to f(\partial\sigma)$ allows us to reach conclusion (8). □

The last lemma of this section will supply the inductive step in the proof of Taming$(n-1) \Rightarrow$ Approx(n).

Lemma 5.8.14. *Assume* Taming$(n-1)$. *Suppose $f : S \to M$ is an embedding into a PL n-manifold M, $k \le n-3$, such that $f|\operatorname{Int} S \smallsetminus \operatorname{Int}\sigma$ is PL. For every $\epsilon : \operatorname{Int} S \to (0,\infty)$ and for every neighborhood N of $\operatorname{Int}\sigma$ in $\operatorname{Int} S$ there exists an embedding $f' : S \to M$ such that*

(1) $f'|L \cup (S \smallsetminus N) = f|L \cup (S \smallsetminus N)$,

(2) $f'|\operatorname{Int} S$ *is PL, and*

(3) $d(f(x), f'(x)) < \epsilon(x)$ *for every* $x \in \operatorname{Int} S$.

Proof. Given ϵ, choose δ according to Lemma 5.8.13. Let $g : \operatorname{Int} S \to M$ be as in the conclusion of Lemma 5.8.12. We may assume that g extends via f to all of S. We will also call this extension g. Let C and D be the associated polyhedra given by Lemma 5.8.13. Choose nice regular neighborhoods U of $\operatorname{Int}\sigma$ and U' of C. Choose U so that $\overline{U}$ is a compact polyhedron that is a cone from $\hat{\sigma}$. By Lemma 5.8.11 there is a small PL homeomorphism $F_1 : \operatorname{Int} S \to \operatorname{Int} S$ such that $F_1(U) = U'$ and F_1 extends via the identity to all of S.

Let V' be a close regular neighborhood of D in $M \smallsetminus g(\partial\sigma)$. We may assume that $g(U') = V' \cap g(\operatorname{Int} S)$ by first finding triangulations in which g is simplicial and then defining U' and V' to be second derived neighborhoods relative to those triangulations. (The derived subdivision can be chosen so that U' is still a nice neighborhood of C.) Take V to be the simplicial neighborhood of $g(\operatorname{Int}\sigma)$ in a second derived subdivision of $M \smallsetminus g(\partial\sigma)$.

Let $j = \dim\sigma$ and fix a PL homeomorphism $\phi : \sigma \to B^j$. By Lemma 5.8.10 there is a homeomorphism $h : B^n \to \overline{V}$ such that $h|B^j = g\phi^{-1}$

and $h|B^n \setminus \partial B^j$ is PL. The choice of δ yields a PL homeomorphism $F_2 : M \setminus f(\partial\sigma) \to M \setminus f(\partial\sigma)$ such that $F_2(V) = V'$. We may assume that F_2 extends via the identity to a homeomorphism of all of M.

Consider the map

$$H = h^{-1}F_2^{-1}gF_1|\overline{U} : \overline{U} \to B^n.$$

Observe that $H(\operatorname{Fr} U) \subset \partial B^n$ and that $H|\operatorname{Fr} U$ is one-to-one. We wish to replace H with an embedding $\bar{\phi} : \overline{U} \to B$ that agrees with H on $\operatorname{Fr} U$, and is PL on $\operatorname{Int} U$. Both $H|\partial\sigma = h^{-1}g|\partial\sigma = \phi|\partial\sigma$ and $H|\operatorname{Fr} U \setminus \partial\sigma$ are PL, so Taming($n-1$) applies to the embedding $H|\operatorname{Fr} U : \operatorname{Fr} U \to \partial B^n$. Use the resulting isotopy to extend $H|\operatorname{Fr} U$ to an embedding $\bar{\phi}$ of a collar on $\operatorname{Fr} U$ in $\overline{U}$ onto a collar of $\partial B^n \setminus \partial B^j$ in B^n. The collar on $\operatorname{Fr} U$ is pinched along $\partial\sigma$ and should be chosen so that the remainder of $\overline{U}$ is a cone from $\hat{\sigma}$ on the inside boundary component. The collar on $\partial B^n \setminus \partial B^j$ should be chosen so that the inside boundary is a PL $(n-1)$-sphere that intersects ∂B^n precisely in ∂B^j and whose interior is convex. Since $\bar{\phi}$ is PL on the inside boundary of the collar, it can be extended to the remainder of $\overline{U}$ by defining $\bar{\phi}(\hat{\sigma}) = 0$ and extending conewise.

Define $f' : S \to M$ by

$$f'(x) = \begin{cases} F_2^{-1}gF_1(x) & \text{if } x \in S \setminus U \\ h\bar{\phi}(x) & \text{if } x \in \overline{U}. \end{cases}$$

This embedding has the required properties. □

Proof that Taming($n-1$) ⇒ Approx(n). Let $e : K^k \to M^n$ be a topological embedding of a k-dimensional polyhedron K into a PL n-manifold M, $k \leq n-3$, and let $\epsilon : K \to (0,\infty)$ be continuous. Fix a triangulation T of K such that $\operatorname{diam} e(\tau) < (1/2)\epsilon(x)$ for every $\tau \in T$ and for every $x \in \tau$. We will construct a PL embedding e' so that $e'(\tau) \subset B(e(\tau); \operatorname{diam} e(\tau))$ for each $\tau \in T$; then the choice of T will yield that $d(e(x), e'(x)) < \epsilon(x)$ for every $x \in K$.

For each $i \leq k$, let $K_i = |T^{(i)}|$ denote the underlying polyhedron of the i-dimensional skeleton of T. We prove by downward induction on i that e can be approximated by topological embeddings f such that $f|K \setminus K_i$ is PL. The base case $i = k-1$ follows from an application of Approximation Theorem 5.6.1 to $e|\operatorname{Int}\tau$ for each k-simplex τ. More specifically, apply Theorem 5.6.1 using a function $\epsilon' : \operatorname{Int}\tau \to (0,\infty)$ with the property that $\epsilon'(x)$ approaches 0 very rapidly as x approaches $\partial\tau$. In that case, the function defined to be the PL approximation on the interior of each k-simplex and e on K_{k-1} will be a topological embedding.

If $f : K \to M$ is a topological embedding that is PL on $K \setminus K_i$, Lemma 5.8.14 can be used to modify f near each i-simplex to produce an approximation f' that is PL on $K \setminus K_{i-1}$. This completes the inductive proof. □

Historical Notes. Theorem 5.8.1 and its proof are due to Bryant (1972). Theorem 5.8.8 was announced by J. Cobb (1968); the proof here is based on that in an appendix to (Bryant, 1972). The 4-dimensional case of Theorem 5.8.8 is due to Cantrell (1964); Bryant (1966a), (1966b) and Dancis (1966) proved it for embeddings in the trivial range; Rushing (1969) then did the same for the metastable range.

Exercises

5.8.1. Prove Lemma 5.8.10. [Hint: Start in the middle of $f(B^k)$ and use Product Neighborhood Theorem 5.3.9 to define the homeomorphism on concentric rings.]

5.8.2. For $k \geq 4$ there exists a finite, collapsible k-complex that admits no embedding in a $(2k-1)$-manifold.

5.8.3. Suppose K is a finite k-complex topologically embedded in a PL n-manifold M, $k \leq n-3$. Then there exist a PL embedding $\lambda : K \to M$ and a pseudo-isotopy $\Phi_t : M \to M$ such that $\Phi_0 = \mathrm{Id}_M$ and $\Phi_1 \lambda = \mathrm{incl}_K$.

Chapter 6

Codimension-two Embeddings

Topological embeddings behave quite differently in codimension two than they do in any other codimension. Very few of the codimension-three results of Chapter 5 translate directly to codimension two. In particular, there is no general unknotting theorem in codimension two and there are no generally applicable theorems on existence of PL embeddings or on approximation of topological embeddings. As a result, this chapter is largely devoted to the construction of examples that illustrate these phenomena.

A major theme of the chapter is that, in contrast with the situation in all other codimensions, the fundamental group is not a powerful enough invariant to detect either knotting or wildness in codimension two. We will illustrate this with examples of locally flat piecewise linear embeddings of spheres that are knotted even though their complements have good fundamental groups and with examples of topological embeddings of manifolds that are wild even though their complements have good local fundamental groups. Positive theorems require additional hypotheses on the higher homotopy groups. For example, we prove that a locally flat embedding of the $(n-2)$-sphere in S^n is flat provided all the homotopy groups of the complement are standard.

The invariants used to detect codimension-two knotting and wildness are based on the homology of an infinite cyclic cover of the complement. These same invariants detect the fact that certain highly connected maps are not homotopic to embeddings and that certain topological embeddings cannot be approximated by PL embeddings. The chapter is self-contained in the sense that it includes a complete description of the invariants and a thorough verification of the necessary algebraic properties.

6.1. Piecewise linear knotting and algebraic unknotting

We begin the study of codimension two by revisiting the global knotting of PL spheres. We construct locally flat PL spheres that are knotted even though their complements have infinite cyclic fundamental groups; these examples are the first indication that codimension-two knotting is too subtle to be detected by π_1 alone. In the light of those examples we reexamine the codimension-two unknotting theorem for locally flat PL sphere pairs to see how the algebraic hypotheses can be sharpened. The section concludes with an example which reveals that topological equivalence and PL equivalence are not the same for PL embeddings in codimension two.

Example 6.1.1. *For each integer k, $1 \leq k < n/2$, there is an embedding $\phi : S^{n-2} \to S^n$ such that*

(1) *ϕ is locally flat and PL,*
(2) *$\pi_i(S^n - \phi(S^{n-2})) \cong \pi_i(S^1)$ for $i < k$, but*
(3) *$\pi_k(S^n - \phi(S^{n-2})) \ncong \pi_k(S^1)$.*

The previous chapter establishes that if two codimension-three embeddings are both close to a model embedding, then they are equivalent via a short ambient isotopy (Theorem 5.4.2). There is no such result in codimension two. In fact, once the embeddings of Example 6.1.1 have been constructed, it is quite easy to see that they can be made arbitrarily close to the inclusion.

Addendum. *Given $\epsilon > 0$, the embedding ϕ in Example 6.1.1 can be constructed so that $d(x, \phi(x)) < \epsilon$ for every $x \in S^{n-2}$.*

The invariant used to detect knottedness in this section is the homology of the infinite cyclic cover of the knot complement, sometimes called the *Alexander invariant.* Before beginning the construction of the example we briefly describe the homology groups of the cover and an associated long exact sequence. Later in the chapter we will study these algebraic objects in more detail and will develop additional invariants from them.

The homology groups of a cover. In what follows it is important to distinguish the ring of integers from the infinite cyclic multiplicative group; we will use $\mathbb{Z}$ to denote the former and $\mathbb{J}$ to denote the latter. The distinction becomes most effective when the spaces considered have infinite cyclic fundamental groups and simultaneously we must deal with the integers as coefficients in homology groups. We will consistently use the notation $\mathbb{J}$ when referring to the fundamental group. We write $\mathbb{J}$ multiplicatively and use t to denote a generator of $\mathbb{J}$; thus

$$\mathbb{J} = \{ t^j \mid j = 0, \pm 1, \pm 2, \dots \}.$$

Let R be a ring and let $R[\mathbb{J}]$ denote the group ring. The elements of $R[\mathbb{J}]$ are Laurent polynomials in t with coefficients in R; thus

$$R[\mathbb{J}] = \left\{ \sum_{i=-m}^{m} q_i t^i \;\middle|\; m \in \mathbb{N}, q_i \in R \right\}.$$

In this section we will use $R = \mathbb{Z}$ while in later sections of the chapter we will use $R = \mathbb{Q}$.

Let X be a connected CW complex with $\pi_1(X) \cong \mathbb{J}$, let $\widetilde{X}$ denote the universal cover of X, and let $p : \widetilde{X} \to X$ denote the covering projection. Fix a cell structure on X and lift the cells of X via p to determine a corresponding cell structure on $\widetilde{X}$. Note that $\mathbb{J}$ acts on $\widetilde{X}$ as the group of deck transformations, so it is natural to think of $C_k(\widetilde{X}; \mathbb{Z})$, the group of cellular k-chains on $\widetilde{X}$, as a module over $\mathbb{Z}[\mathbb{J}]$. It is a free $\mathbb{Z}[\mathbb{J}]$-module with one generator for each k-cell in X. For this reason we use $C_k(X; \mathbb{Z}[\mathbb{J}])$ to denote $C_k(\widetilde{X}; \mathbb{Z})$. The module structure on $C_*(X; \mathbb{Z}[\mathbb{J}])$ induces a $\mathbb{Z}[\mathbb{J}]$-module structure on $H_*(\widetilde{X}; \mathbb{Z})$. When we wish to emphasize this module structure we denote $H_*(\widetilde{X}; \mathbb{Z})$ by $H_*(X; \mathbb{Z}[\mathbb{J}])$. Using this notation we can state the Hurewicz Theorem as follows: if $\pi_1(X) \cong \mathbb{J}$ and $\widetilde{H}_i(X; \mathbb{Z}[\mathbb{J}]) = 0$ for $i \leq k-1$, then $\pi_i(X) = 0$ for $2 \leq i \leq k-1$ and $\pi_k(X) \cong H_k(X; \mathbb{Z}[\mathbb{J}])$.

Remark. The groups $H_*(X; \mathbb{Z}[\mathbb{J}])$ defined in the preceding paragraph are known as homology groups of X with *local coefficients*. While we will not make use of cohomology with local coefficients, it is worth noting that it is possible to extend the definitions above to that context as well. Use $C^k(X; \mathbb{Z}[\mathbb{J}])$ to denote the group $\mathrm{Hom}_{\mathbb{Z}[\mathbb{J}]}(C_k(X; \mathbb{Z}[\mathbb{J}]), \mathbb{Z}[\mathbb{J}])$ and assume that X is a finite complex. Then $C^k(X; \mathbb{Z}[\mathbb{J}])$ is finitely generated over $\mathbb{Z}[\mathbb{J}]$, so a cochain in $C^k(X; \mathbb{Z}[\mathbb{J}])$ corresponds to a cochain in $C^k(\widetilde{X}; \mathbb{Z})$ that is nonzero on at most a finite number of cells. Hence $H^*(X; \mathbb{Z}[\mathbb{J}]) \cong H^*_c(\widetilde{X}; \mathbb{Z})$, the cohomology group with compact supports (Hatcher, 2002, Proposition 3H.5). In this notation, Poincaré-Lefschetz duality simply states that $H_k(X; \mathbb{Z}[\mathbb{J}]) \cong H^{n-k}(X, \partial X; \mathbb{Z}[\mathbb{J}])$ whenever X is a compact orientable n-dimensional ∂-manifold with $\pi_1(X) \cong \mathbb{J}$.

The Milnor sequence. There is a very useful long exact sequence associated with the universal cover of a space whose fundamental group is infinite cyclic. To construct it, one must examine the sequence

$$(*) \qquad 0 \to C_*(X; \mathbb{Z}[\mathbb{J}]) \xrightarrow{t-1} C_*(X; \mathbb{Z}[\mathbb{J}]) \xrightarrow{p_*} C_*(X; \mathbb{Z}) \to 0.$$

Theorem 6.1.2. *The sequence* $(*)$ *is a short exact sequence of chain complexes.*

Proof. It is clear that $t-1$ is a monomorphism and that p_* is an epimorphism. In addition, $p_*((t-1)(c)) = p_*(c) - p_*(c) = 0$, so $p_* \circ (t-1) = 0$. An element of $c \in C_*(X;\mathbb{Z}[\mathbb{J}])$ has the form $c = \sum_i q_i(t)\sigma_i$, where $q_i(t) \in \mathbb{Z}[\mathbb{J}]$ and σ_i is a cell in X. Note that $p_*(\sum_i q_i(t)\sigma_i) = \sum_i q_i(1)\sigma_i$. Thus $p_*(\sum_i q_i(t)\sigma_i) = 0$ implies that $q_i(1) = 0$ and consequently $t-1$ divides $q_i(t)$. Hence $p_*(\sum_i q_i(t)\sigma_i) = 0$ implies that $\sum_i q_i(t)\sigma_i$ is in the image of $t-1$. □

The sequence $(*)$ gives rise to the long exact sequence

$$(**) \qquad \cdots \to H_{k+1}(X;\mathbb{Z}) \xrightarrow{p_*} H_k(X;\mathbb{Z}[\mathbb{J}]) \xrightarrow{t-1} H_k(X;\mathbb{Z}[\mathbb{J}]) \xrightarrow{p_*} H_k(X;\mathbb{Z}) \to \cdots$$

of homology groups. The exactness of $(**)$ is a standard consequence of the Zig-zag Lemma ((Munkres, 1984, Lemma 24.1) or (Hilton and Stammbach, 1971, Theorem IV.2.1)).

We will refer to the sequence $(**)$ as a *Milnor sequence.* The name is chosen because of the important role the sequence plays in the fundamental paper (Milnor, 1968). The sequence $(**)$ is also known in other contexts as the Wang Sequence. (See, for example, (Spanier, 1966, p. 456) or (Hu, 1959, p. 282).)

The construction of Example 6.1.1, as well as that of other examples in the chapter, is described in terms of handle theory. The handle-theoretic terminology employed is consistent with that of (Rourke and Sanderson, 1972, Chapter 6).

Construction of Example 6.1.1. The case $k = 1$ was covered in Chapter 1, with its examples of knotted 1-spheres in S^3 whose complements have nonabelian fundamental groups. Higher-dimensional examples (with $n > 3$ and $k = 1$) are constructed by spinning the 3-dimensional ones.

Assume $k > 1$. Let (B^{n+1}, B^{n-1}) denote the standard unknotted ball pair and let $B^* = B^{n+1} \setminus B^{n-1}$. Then $\pi_1(\partial B^*) = \mathbb{J}$ and $H_j(\partial B^*;\mathbb{Z}[\mathbb{J}]) = 0$ for $j > 1$. Let M_1 be the manifold obtained by attaching a trivial k-handle to B^{n+1} along ∂B^*; that is, M_1 is the identification space $M_1 = B^{n+1} \cup h^{(k)}$, where $h^{(k)} = B^k \times B^{n-k+1}$ and $(\partial B^k) \times B^{n-k+1}$ is identified with a thin tubular neighborhood of an unknotted $(k-1)$-sphere in ∂B^*. In case $k = 2$, this attaching $(k-1)$-sphere should be inessential in ∂B^*. Define $M_1^* = M_1 \setminus B^{n-1}$.

We now compute $H_j(\partial M_1^*;\mathbb{Z}[\mathbb{J}])$ for $1 \le j \le k$. For $j < k$, general position shows that any singular j-sphere in ∂M_1^* can be homotoped off the belt sphere $\{0\} \times \partial B^{n-k+1}$ in ∂M_1^* and thence off the handle $h^{(k)}$.

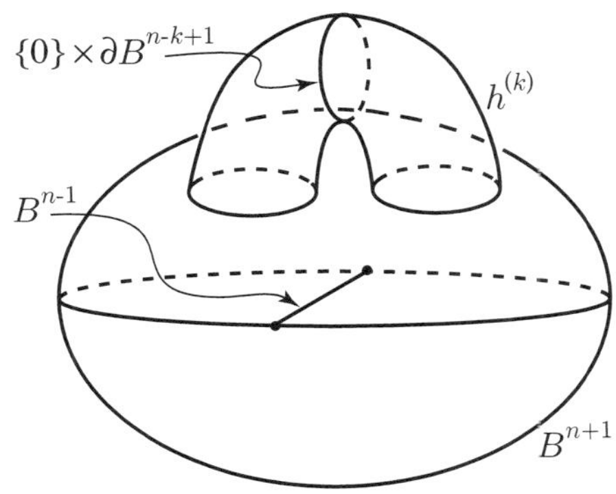

Figure 6.1. A trivial k-handle attached to an $(n+1, n-1)$-ball pair

General position, together with the requirement that $k < n/2$, shows that $\pi_j(\partial B^* \smallsetminus h^{(k)}) = 0$ for $1 < j < k$. So we have that $\pi_1(\partial M_1^*) = \mathbb{J}$ and $\pi_j(\partial M_1^*) = H_j(\partial M_1^*; \mathbb{Z}[\mathbb{J}]) = 0$ for $2 \leq j < k$. Furthermore, we have added a generator to $H_k(\partial M_1^*; \mathbb{Z}[\mathbb{J}])$, so $H_k(\partial M_1^*; \mathbb{Z}[\mathbb{J}])$ is naturally identified with $\mathbb{Z}[\mathbb{J}]$.

Let α denote a $\mathbb{Z}[\mathbb{J}]$-generator of $H_k(\partial M_1^*; \mathbb{Z}[\mathbb{J}])$ and let Σ denote a locally flat PL k-sphere in ∂M_1^* that represents the element $(2-t)\alpha$ in $H_k(\partial M_1^*; \mathbb{Z}[\mathbb{J}])$. The k-sphere Σ is pictured schematically in Figure 6.2. Now define M_2 to be the manifold obtained by attaching a $(k+1)$-handle to M_1 along Σ. Thus M_2 is the identification space $M_1 \cup h^{(k+1)}$ where $h^{(k+1)} = B^{k+1} \times B^{n-k}$ and $\partial B^{k+1} \times B^{n-k}$ is identified with a small regular neighborhood of Σ in ∂M_1^*. As before, use M_2^* to denote $M_2 \smallsetminus B^{n-1}$.

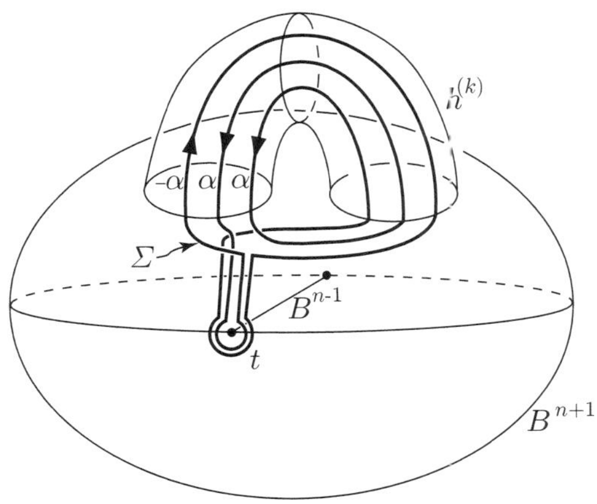

Figure 6.2. Σ is the attaching sphere for the $(k+1)$-handle

We next observe that, while M_2^* is a complicated space, M_2 is simply an $(n+1)$-ball. The reason for this is the fact that

$$M_2 = B^{n+1} \cup \bar{h}^{(k)} \cup h^{(k+1)}$$

and the two handles algebraically and geometrically cancel each other. Thus there is a homeomorphism $g : M_2 \to B^{n+1}$. Identify S^{n-2} with $\partial B^{n-1} \subset \partial M_2$. The embedding ϕ we seek is simply

$$\phi = g \mid S^{n-2} : S^{n-2} \to \partial B^{n+1} = S^n.$$

In order to complete the construction, we must check that $S^n \smallsetminus \phi(S^{n-2}) \cong \partial M_2^*$ has the correct homotopy groups. First note that $H_j(\partial M_2^*; \mathbb{Z}[\mathbb{J}]) = H_j(\partial M_1^*; \mathbb{Z}[\mathbb{J}])$ for $j < k$ because adding the $(k+1)$-handle $h^{(k+1)}$ does not change homology in dimensions below k. Then note that the addition of the $(k+1)$-handle adds a relation to H_k; the relation is represented by the sphere Σ. Thus $H_k(\partial M_2^*; \mathbb{Z}[\mathbb{J}]) \cong \mathbb{Z}[\mathbb{J}]/(2-t)$ and an application of the Hurewicz Theorem shows that $\pi_k(S^n \smallsetminus \phi(S^{n-2})) \cong \mathbb{Z}[\mathbb{J}]/(2-t)$. Since $\mathbb{Z}[\mathbb{J}]/(2-t) \neq 0$, this completes the proof. □

Remark. A *slice knot* is a PL $(n-2)$-sphere in S^n that bounds a locally flat PL $(n-1)$-ball in B^{n+1}. It is clear from the construction that the knots in Example 6.1.1 are slice knots whenever $k \geq 2$.

Proof of the Addendum to Example 6.1.1. Let D be a small PL ball in S^n that intersects $\phi(S^{n-2})$ in a standard $(n-2)$-cell and define $C = S^n \smallsetminus \operatorname{Int} D$. Choose a second small PL ball B such that $B \cap S^{n-2}$ is a standard $(n-2)$-cell. Cut out the pair $(B, B \cap S^{n-2})$ and replace it with $(C, C \cap \phi(S^{n-2}))$ to construct a PL embedding of S^{n-2} into S^n that is very close to the identity, but has complement homeomorphic to that of $\phi(S^{n-2})$. □

In Example 6.1.1, the first nonstandard homotopy group appears in dimension k for some $k < n/2$. The next two theorems are positive results which show that this must always hold: if all homotopy groups below the middle dimension are standard, then the sphere must be unknotted. We consider only the PL case in this section; the topological case will be investigated in subsequent sections of the chapter.

Theorem 6.1.3 (PL Unknotting). *Let $h : S^{n-2} \to S^n$ be a locally flat PL embedding, $n \geq 6$, such that $\pi_i(S^n \smallsetminus h(S^{n-2})) \cong \pi_i(S^1)$ for every $i < n/2$. Then there is a PL homeomorphism $g : S^n \to S^n$ such that $g(h(S^{n-2})) = S^{n-2}$.*

Theorem 6.1.3, as stated, asserts only a weak form of unknotting in which $g(h(S^{n-2})) = S^{n-2}$ setwise. However it follows routinely from (Rourke and

Sanderson, 1972, Proposition 4.18) that the conclusion can be strengthened to assure that $g \circ h : S^{n-2} \to S^n$ is pointwise equal to the inclusion.

Theorem 6.1.3 is a high-dimensional result—the full PL Unknotting Theorem is not known to hold in dimension four. It is known that a locally flat PL 2-sphere in S^4 is topologically flat provided its complement has the homotopy type of S^1 (Freedman and Quinn, 1990), but it is unknown whether such a sphere is PL unknotted. As a result, we cannot prove that a PL embedding $S^{n-2} \to S^n$, $n \geq 5$, that is topologically locally flat is locally flat in the PL sense. We will not address that subtle issue in this chapter; whenever the terms "PL" and "locally flat" are used together we will assume that the embedding is locally flat in the PL sense.

Theorem 6.1.3 is the same as the codimension-two unknotting theorem in (Rourke and Sanderson, 1972), except that the hypotheses are stated in an apparently weaker form. The next proposition promises that any knot Σ satisfying the hypotheses of Theorem 6.1.3 is *algebraically unknotted*; i.e., $\pi_i(S^n \smallsetminus \Sigma) \cong \pi_i(S^n \smallsetminus S^{n-2})$ for every i. This follows from Poincaré duality in case n is odd, but a delicate argument is needed to show that the homology vanishes in the middle dimension when n is even. We restrict attention to $n \geq 6$ so we can use handle theory to simplify the approach.

Proposition 6.1.4 (Algebraic Unknotting). *Let $h : S^{n-2} \to S^n$ be a locally flat PL embedding, $n \geq 6$. If $\pi_i(S^n \smallsetminus h(S^{n-2})) \cong \pi_i(S^1)$ for every $i < n/2$, then $S^n \smallsetminus h(S^{n-2})$ has the homotopy type of S^1.*

Proof. Define $k = \lfloor \frac{n-1}{2} \rfloor$, the greatest integer less than or equal to $(n-1)/2$, and $\Sigma = h(S^{n-2})$. Let C_1 be a small n-cell such that $(C_1, C_1 \cap \Sigma) \cong (B^n, B^{n-2})$ and let C_2 be a regular neighborhood of $\Sigma \smallsetminus \operatorname{Int} C_1$ in $S^n \smallsetminus \operatorname{Int} C_1$. Then $(C_2, C_2 \cap \Sigma)$ is an unknotted ball pair by (Rourke and Sanderson, 1972, Corollary 4.14)[1]. Define $N = C_1 \cup C_2$ and $X = S^n \smallsetminus \operatorname{Int} N$. Note that $C_1 \cap C_2 = \partial C_1 \cap \partial C_2 \cong B^2 \times S^{n-3}$. Furthermore,

$$\overline{\partial C_i \smallsetminus \partial C_j} \cong S^1 \times B^{n-2}$$

for $(i, j) = (1, 2)$ and $(2, 1)$. Hence ∂N can be expressed as the union of two copies of $S^1 \times B^{n-2}$ sewn together along their boundaries. Any singular sphere in ∂N of dimension less than $n - 2$ can be homotoped off one copy of $S^1 \times B^{n-2}$ and into a regular neighborhood of the other. Thus $\pi_i(\partial N)$ is trivial for $2 \leq i \leq n - 3$. Furthermore, the S^1 factor of ∂N homologically links Σ, so $\pi_1(\partial N) = \mathbb{J}$. Therefore $\pi_i(\partial N) \cong \pi_i(S^1)$ for $1 \leq i \leq k$.

Now X is a strong deformation retract of $S^n \smallsetminus h(S^{n-2})$, so $\pi_1(X) = \mathbb{J}$ and $\pi_i(X) = 0$ for $2 \leq i \leq k$. It follows that $(X, \partial X)$ is a k-connected

[1] Rourke and Sanderson state the local flatness hypothesis on the bottom of page 51 and it is assumed throughout the remainder of Chapter 4.

pair. If we treat X as a cobordism based on ∂X and consider a handle decomposition of that cobordism, then we can cancel all handles of index $\leq k$ by standard handle cancellation techniques. (See Rourke and Sanderson (1972), Theorems 6.13, 6.15, and 6.16.) Thus the dual handle decomposition has no handles of index $n-k$ or greater, which means that X has the homotopy type of a finite CW complex of dimension $n-k-1$. Hence we have $H_i(X;\mathbb{Z}[\mathbb{J}]) = 0$ for $1 \leq i \leq k$ and for $i \geq n-k$.

In case n is odd, we have $H_i(X;\mathbb{Z}[\mathbb{J}]) = 0$ for every $i \geq 1$, so the universal cover $\widetilde{X}$ is contractible and the proof is complete in that case.

Suppose, then, that n is even. We already know that $\widetilde{H}_i(X;\mathbb{Z}[\mathbb{J}]) = 0$ for $i \neq n/2 = k+1$. To complete the proof we must show that $H_{k+1}(X;\mathbb{Z}[\mathbb{J}]) = 0$ as well. Since X has the homotopy type of a CW complex of dimension $n-k-1 = k+1$, the cellular chain complex of $\widetilde{X}$ has the form

$$0 \to C_{k+1}(X;\mathbb{Z}[\mathbb{J}]) \to C_k(X;\mathbb{Z}[\mathbb{J}]) \to \cdots .$$

It follows that $H_{k+1}(X;\mathbb{Z}[\mathbb{J}])$ is a submodule of the finitely generated free $\mathbb{Z}[\mathbb{J}]$-module $C_{k+1}(X;\mathbb{Z}[\mathbb{J}])$. The Milnor Sequence

$$\cdots H_{k+2}(X;\mathbb{Z}) \to H_{k+1}(X;\mathbb{Z}[\mathbb{J}]) \xrightarrow{t-1} H_{k+1}(X;\mathbb{Z}[\mathbb{J}]) \xrightarrow{p_*} H_{k+1}(X;\mathbb{Z}) \cong 0$$

shows that the homomorphism $(t-1) : H_{k+1}(X;\mathbb{Z}[\mathbb{J}]) \to H_{k+1}(X;\mathbb{Z}[\mathbb{J}])$ is onto. Hence every element of $H_{k+1}(X;\mathbb{Z}[\mathbb{J}])$ is divisible by arbitrarily high powers of $(t-1)$. The only element of a free $\mathbb{Z}[\mathbb{J}]$-module with this property is 0, so $H_{k+1}(X;\mathbb{Z}[\mathbb{J}]) = \{0\}$. Hence $S^n \smallsetminus h(S^{n-2})$ is aspherical, and an application of the Whitehead Theorem completes the proof. □

Proof of Theorem 6.1.3. Let $h : S^{n-2} \to S^n$ be an embedding satisfying the hypotheses of the theorem. Then $S^n \smallsetminus h(S^{n-2})$ has the homotopy type of S^1 by Proposition 6.1.4. Hence the theorem follows from (Rourke and Sanderson, 1972, Theorem 7.6). □

Remark. It should be noted that (Rourke and Sanderson, 1972) does not contain a complete proof of the codimension-two unknotting theorem. Rourke and Sanderson's argument is based on two results they do not prove. The first is the s-cobordism Theorem and the second is the fact that the Whitehead group of the integers, $\mathrm{Wh}(\mathbb{Z})$, is trivial. In the next section we will give a complete proof of a topological version of the unknotting theorem that is independent of those results.

The final example in the section illustrates another way in which codimension-two embeddings differ from those in codimension three: two PL embeddings can be topologically equivalent but PL inequivalent. This contrasts with Corollary 5.8.7.

Example 6.1.5. *For $n \geq 5$ there exist PL embeddings $h_1, h_2 : S^{n-2} \to S^n$ such that h_1 and h_2 are topologically equivalent but not PL equivalent. More precisely, there is a topological homeomorphism $g : S^n \to S^n$ such that $gh_1 = h_2$ but there is no PL homeomorphism $g' : S^n \to S^n$ with $g'h_1(S^{n-2}) = h_2(S^{n-2})$. Furthermore, the topological homeomorphism g can be made PL on $h_1(S^{n-2})$ and on $S^n \setminus \{p\}$, where $p \in h_1(S^{n-2})$.*

Construction of Example 6.1.5. The construction of Example 6.1.5 requires ingredients that are beyond the scope of this book. In particular, it requires the existence of a group π with two properties: the Whitehead group of π must be nontrivial and π must be the fundamental group of the complement of a locally flat PL $(n-3)$-sphere in S^{n-1}. The group $\pi = \mathbb{J} \times G$, where G is the binary icosahedral group of order 120, is such a group. The fact that $\mathrm{Wh}(\pi) \neq 0$ follows from (Siebenmann and Sondow, 1966/1967, Lemma 2.4). The five-twist spin of the trefoil knot is an example of a locally flat PL 2-sphere in S^4 whose complement has fundamental group π (Zeeman, 1965). Higher-dimensional examples are obtained by (untwisted) spinning of the 4-dimensional example (see Exercise 1.4.4). We will assume those two results and sketch the remainder of the construction.

Fix $n \geq 6$. Let $K \subset S^{n-1}$ be a locally flat PL $(n-3)$-sphere such that $\pi_1(S^{n-1} \setminus K) = \pi$ and let τ be a nonzero element of $\mathrm{Wh}(\pi)$. Define $(W_0, M_0) = (S^{n-1} \times [0,1], K \times [0,1])$. Then construct a second cobordism (W_1, M_1) by attaching handles to (W_0, M_0) using τ. To be more specific, find an $m \times m$ matrix over $\mathbb{Z}[\pi]$ that represents τ; attach m trivial 2-handles to W_0 along $(S^{n-1} \setminus K) \times \{1\}$ and then attach an equal number of 3-handles in such a way that τ represents the intersection matrix between the 2- and 3-handles; define W_1 to be W_0 plus these handles and define M_1 to be the image of M_0 under the inclusion map $W_0 \hookrightarrow W_1$. Because W_0 is simply connected and the Whitehead group of the trivial group is trivial, we see that $W_1 \cong W_0$. On the other hand, $W_1 \setminus M_1$ is a nontrivial cobordism based on $S^{n-1} \setminus K$ whose Whitehead torsion is τ. As a result there is no PL homeomorphism $(W_0, M_0) \cong (W_1, M_1)$.

For $i = 0, 1$, define (S^n, Σ_i) to be the PL sphere pair obtained by capping the two ends of (W_i, M_i) with cones. Note that each Σ_i is a PL $(n-2)$-sphere in S^n with exactly two non-locally flat points and that Σ_0 is just the suspension of K. We will show that Σ_0 and Σ_1 have the properties specified in the example.

Suppose there did exist a PL homeomorphism $g' : (S^n, \Sigma_0) \to (S^n, \Sigma_1)$. Then g' would necessarily map the two non-locally flat points of Σ_0 to the non-locally flat points of Σ_1. Excising a cone about each of the non-locally flat points would result in a PL homeomorphism $(W_0, M_0) \cong (W_1, M_1)$. Hence we can conclude that no PL homeomorphism g' exists.

Now we show that there does exist a topological homeomorphism $g : (S^n, \Sigma_0) \to (S^n, \Sigma_1)$. Let $(\partial_+ W_1, \partial_+ M_1)$ and $(\partial_- W_1, \partial_- M_1)$ denote the standard and nonstandard boundary components of (W_1, M_1), respectively. Attach a half-open collar $(\partial_- W_1 \times [0, 1), \partial_- M_1 \times [0, 1))$ to the cobordism (W_1, M_1) to form (W_1^+, M_1^+). By Collaring Theorem 2.4.10,

$$(W_1^+, M_1^+) \cong (W_1 \setminus \partial_- W_1, M_1 \setminus \partial_- M_1)$$

and by Weak h-Cobordism Theorem 3.1.8,

$$(W_1 \setminus \partial_- W_1, M_1 \setminus \partial_- M_1) \cong (\partial_+ W_1 \times [0, 1), \partial_+ M_1 \times [0, 1)).$$

One-point compactification then shows that the space obtained by attaching a cone to (W_1, M_1) along $(\partial_- W_1, \partial_- M_1)$ is homeomorphic to the cone on $(\partial_+ W_1, \partial_+ M_1)$. It follows that the topological homeomorphism g exists. Since the homeomorphisms provided by the Collaring Theorem and the Weak h-Cobordism Theorem are both PL, g is PL in the complement of a point. A little extra care makes $g|\Sigma_0$ PL as well.

This completes the construction of the example in case $n \geq 6$. The 5-dimensional case requires some modification in the construction of W_1; details may be found in (Siebenmann and Sondow, 1966/1967). □

Historical Notes. A special case of Example 6.1.1 appears as Theorem V in (Kervaire, 1965a), where the special case is attributed to Stallings. The generalized example is described by D. W. Sumners in (1970); see also (Sumners, 1966). Knots with fundamental group $\mathbb{J}$ can also be constructed by plumbing techniques—see (Rolfsen, 1990, §7F). M. Kervaire (1965a), (1965b) gives conditions which characterize the groups that can be fundamental groups of locally flat PL $(n-2)$-spheres in S^n, $n \geq 5$. The conditions in Exercise 6.1.1 are sufficient, but not necessary.

Theorem 6.1.3 is known as Levine's Unknotting Theorem. J. Levine (1965) does not state the theorem in this form, but the statement given here is what he actually proves in (Levine, 1965). Our statement can be found on page 74 of (Kervaire and Weber, 1978) where it is attributed to Levine. The 5-dimensional case was first announced in (Wall, 1965b) and a complete proof appears in (Wall, 1970, §16). The 5-dimensional theorem was also proved independently by J. L. Shaneson (1968). The 3-dimensional case is the classical Dehn's Theorem, which is a consequence of Dehn's Lemma (Papakyriakopoulos, 1957). It is not known whether Theorem 6.1.3 is valid in dimension 4, although a weak version of the 4-dimensional theorem, with a topological conclusion, is known (Freedman and Quinn, 1990, Theorem 11.7A).

Algebraic Unknotting (Proposition 6.1.4) is implicit in (Levine, 1965). Another proof, based on (Wall, 1966), may be found in (Hirschhorn and Ratcliffe, 1980). The 4-dimensional case is attributed to A. Kawauchi (1974a).

Other proofs of the 4-dimensional case appear in (Swarup, 1975, Remark 3), (Freedman and Quinn, 1990, Proposition 11.6C(1)), and Exercise 6.5.1.

Example 6.1.5 is taken from (Siebenmann and Sondow, 1966/1967). A similar construction may be found in (Stallings, 1965a, §4). Siebenmann and Sondow construct an infinite family of embeddings that are all topologically equivalent but pairwise PL distinct. The last part of the argument in the construction of Example 6.1.5 can also be completed using the inverse of τ in the Whitehead group and the Mazur swindle (§2.9) in place of the Weak h-Cobordism Theorem.

Exercises

6.1.1. Let π be a finitely presented group of deficiency one and weight one.[2] Prove that for each $n \geq 5$ there is a locally flat embedding $h : S^{n-2} \to S^n$ such that $\pi_1(S^n \setminus h(S^{n-2})) \cong \pi$. [Hint: Start with B^{n+1}. Attach one 1-handle to B^{n+1} for each generator of π and attach a 2-handle corresponding to each relation. Attach a final 2-handle along a weight element. Prove that the boundary of the resulting manifold is homeomorphic to S^n and that the belt sphere of the final handle is the desired $(n-2)$-sphere.]

6.1.2. Let $p(t) \in \mathbb{Z}[\mathbb{J}]$ be a polynomial with $p(1) = \pm 1$. Prove that for each $n \geq 5$ and for each k, $2 \leq k < n/2$, there exists a locally flat PL embedding $h : S^{n-2} \to S^n$ such that $\pi_i(S^n \setminus h(S^{n-2})) \cong \pi_i(S^1)$ for $i < k$ and $\pi_k(S^n \setminus h(S^{n-2})) \cong \mathbb{Z}[\mathbb{J}]/(p(t))$.

6.1.3. Prove the $n = 3$ and $n = 5$ cases of Theorem 6.1.4.

6.1.4. The *Hauptvermutung*[3] is the conjecture that two topologically homeomorphic polyhedra are PL homeomorphic. Use a construction like that in Example 6.1.5 to produce compact counterexamples to the *Hauptvermutung*.

6.2. Topological flattening and algebraic knotting

Next we consider topological embeddings in codimension two. We give a complete proof of a topological version of the codimension-two unknotting theorem that is based on Stallings engulfing. As part of the argument we prove several other results of independent interest. Foremost among them are the monotone union theorem and an unknotting theorem for embeddings of $\mathbb{R}^{n-2}$ in $\mathbb{R}^n$. The monotone union theorem states that any manifold that

[2] A group has *deficiency one* if it has a presentation in which the number of generators exceeds the number of relations by one. A group π has *weight one* if there is one element $z \in \pi$ whose normal closure is π. Such an element z is called a *weight element*.

[3] At one time the *Hauptvermutung* was the "main conjecture" of geometric topology.

can be written as a monotone union of n-cells is homeomorphic to $\mathbb{R}^n$; we prove a relative version of the theorem.

Theorem 6.2.1 (Topological flattening). *Let $h : S^{n-2} \to S^n$ be a locally flat topological embedding, $n \geq 5$, such that $S^n \setminus h(S^{n-2})$ has the homotopy type of S^1. Then there is a topological homeomorphism $g : S^n \to S^n$ such that $g(h(S^{n-2})) = S^{n-2}$.*

The hypothesis about $S^n \setminus h(S^{n-2})$ having the homotopy type of S^1 could be replaced by the weaker algebraic hypotheses of Theorem 6.1.3, but extension of the algebraic unknotting principle to the locally flat topological category requires rather delicate algebraic arguments that we prefer to omit. While the algebraic unknotting principle does extend to locally flat topological embeddings, the principle definitely does not extend to general topological embeddings. Later in the section we will present a collection of examples exhibiting a new kind of wildness; they disclose that the first non-standard homotopy group of the complement of a topologically embedded sphere can appear in any dimension up to $n-2$.

We begin the proof of Theorem 6.2.1 by removing a point from $h(S^{n-2})$ to form an embedded copy of $\mathbb{R}^{n-2}$ in $\mathbb{R}^n$. This leads us to consider the problem of unknotting embeddings of $\mathbb{R}^{n-2}$ in $\mathbb{R}^n$.

Definition. Suppose Y is a closed subset of $\mathbb{R}^n$ such that Y is homeomorphic to $\mathbb{R}^{n-2}$. Say that Y is *unraveled at infinity* if for every compact subset K of $\mathbb{R}^n$ there exists a second compact subset D with $K \subset D \subset \mathbb{R}^n$ such that the pair $(\mathbb{R}^n \setminus Y, \mathbb{R}^n \setminus (Y \cup D))$ is 2-connected.

It is easy to see that Y is unraveled at infinity if $\pi_2(\mathbb{R}^n \setminus Y) = 0$ and for every compact subset K of $\mathbb{R}^n$ there exists a second compact subset D with $K \subset D \subset \mathbb{R}^n$ such that the usual inclusion induced homomorphism $\pi_1(\mathbb{R}^n \setminus (Y \cup D)) \to \pi_1(\mathbb{R}^n \setminus Y)$ is an isomorphism.

Proposition 6.2.2. *If $h : \mathbb{R}^{n-2} \to \mathbb{R}^n$ is a locally flat, proper, topological embedding such that $\mathbb{R}^n \setminus h(\mathbb{R}^{n-2})$ has the homotopy type of S^1 and $h(\mathbb{R}^{n-2})$ is unraveled at infinity, then $h(\mathbb{R}^{n-2})$ is flat.*

Proof of Theorem 6.2.1. Pick $x_0 \in h(S^{n-2})$. Set $\Sigma = h(S^{n-2})$ and $\Sigma^- = \Sigma \setminus \{x_0\}$. Stereographic projection from x_0 gives a homeomorphism $H : S^n \setminus \{x_0\} \to \mathbb{R}^n$ and $H|(\Sigma^-)$ is a proper embedding of $\Sigma^- \cong \mathbb{R}^{n-2}$ into $\mathbb{R}^n$. Since $\mathbb{R}^n \setminus H(\Sigma^-) \cong S^n \setminus \Sigma$, we see that $\mathbb{R}^n \setminus H(\Sigma^-)$ has the homotopy type of S^1. Furthermore, local flatness of Σ at x_0 means that there exist arbitrarily small neighborhoods U of x_0 such that $(U, U \cap \Sigma) \cong (\mathbb{R}^n, \mathbb{R}^{n-2})$. Note that $\pi_1(U \setminus \Sigma) \to \pi_1(S^n \setminus \Sigma)$ is an isomorphism, so Exercise 0.5.3 implies that $(S^n \setminus \Sigma, U \setminus \Sigma)$ is 2-connected. Taking complements, one obtains arbitrarily large compact subsets $K = H(S^n \setminus U)$ of $\mathbb{R}^n$ such that

$(\mathbb{R}^n \smallsetminus H(\Sigma^-), \mathbb{R}^n \smallsetminus (K \cup H(\Sigma^-))$ is 2-connected and $H(\Sigma^-)$ is unraveled at infinity. Proposition 6.2.2 shows that there is a homeomorphism $g : \mathbb{R}^n \to \mathbb{R}^n$ such that $g(H(\Sigma^-)) = \mathbb{R}^{n-2}$. The map $H^{-1} \circ g \circ H : S^n \smallsetminus \{x_0\} \to S^n \smallsetminus \{x_0\}$ extends via the identity to a homeomorphism of S^n that moves Σ to a flat $(n-2)$-sphere. □

We now turn our attention to the proof of Proposition 6.2.2. The proof is based on engulfing and a relative version of the "Monotone Union Theorem." Before stating that relative result we must extend some standard notation. Suppose $k < n$. As usual we identify $\mathbb{R}^k$ with the subspace $\mathbb{R}^k \times \mathbf{0}$ of $\mathbb{R}^n$ and B^k with the subset $B^k \times \mathbf{0}$ of B^n. Suppose $(B_1, A_1) \subset (B_2, A_2)$ are unknotted (n, k)-ball pairs. This means that for each $i = 1, 2$ there is a homeomorphism $h_i : (B_i, A_i) \to (B^n, B^k)$. We use $\operatorname{Int} B_i$ to denote the interior of B_i and $\operatorname{Int} A_i = A_i \cap \operatorname{Int} B_i$. We say that *$C_1$ is a collar of (B_1, A_1) in (B_2, A_2)* if $C_1 \subset \operatorname{Int} B_2$, $C_1 \cap B_1 = \partial B_1$, and there exists a homeomorphism $\phi : \partial B_1 \times [0, 1) \to C_1$ such that $\phi(x, 0) = x$ for each $x \in \partial B_1$ and $\phi(\partial A_1 \times [0, 1)) \subset A_2$.

The basic move in the proof of the Monotone Union Theorem is the push described in the following lemma.

Lemma 6.2.3. *Assume $k < n$, $(B_1, A_1) \subset (B_2, A_2)$ are unknotted (n, k)-ball pairs, and C_1 is a collar of (B_1, A_1) in (B_2, A_2). For every compact set $K \subset \operatorname{Int} B_2$ there exists a homeomorphism $g : B_2 \to B_2$ such that*

(1) *$g|B_1$ is the identity,*
(2) *$g(B_1 \cup C_1) \supset K$, and*
(3) *$g(C_1 \cap A_2) \subset A_2$.*

Proof. Pick $x_0 \in \operatorname{Int} A_1$ and a small neighborhood $U \subset \operatorname{Int} B_1$ of x_0 that is round relative to the structure on B_2. There exists a radially compressing homeomorphism $\rho : B_2 \to B_2$ such that $\rho(B_1) \subset U$ and the support of ρ is contained in a compact subset of $B_1 \cup C_1$; this compression is radial with respect to the radial structure of $B_1 \cup C_1 \cong \mathbb{R}^n$. There is also a stretching homeomorphism $\sigma : B_2 \to B_2$ such that $\sigma|U$ is the identity and $\sigma(B_1 \cup C_1) \supset K \cup B_1 \cup C_1$; this stretch is radial with respect to the structure on $\operatorname{Int} B_2 \cong \mathbb{R}^n$. Finally, define $g = \rho^{-1} \circ \sigma \circ \rho$. It is easy to check that g has the desired properties. □

Theorem 6.2.4 (Relative Monotone Union). *Fix $k < n$. If $\{(B_i, A_i)\}_{i=1}^{\infty}$ is a sequence of unknotted (n, k)-ball pairs such that for each i, $(B_i, A_i) \subset \operatorname{Int}(B_{i+1}, A_{i+1})$, then $\cup_{i=1}^{\infty}(B_i, A_i) \cong (\mathbb{R}^n, \mathbb{R}^k)$.*

Proof. Let $(M, Y) = \cup_{i=1}^{\infty}(B_i, A_i)$. Begin by choosing a collared (n, k)-ball pair $(B_0, A_0) \subset \operatorname{Int}(B_1, A_1)$ and a collar C_0 of (B_0, A_0) in (B_1, A_1). There

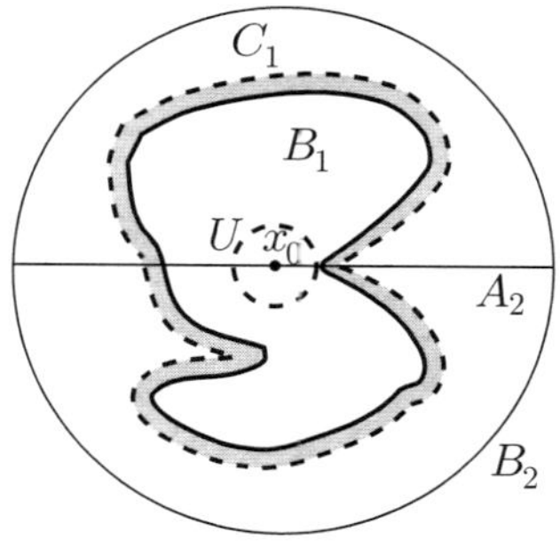

Figure 6.3. Proof of Lemma 6.2.3

exists a homeomorphism $\phi : \partial B_0 \times [0, 1) \to C_0$ such that $\phi(x, 0) = x$ for each $x \in \partial B_0$ and $\phi(\partial A_0 \times [0, 1)) \subset A_-$. Split C_0 into a sequence $C_1, C_2, \ldots$ of collars by defining $C_i = \phi(\partial B_0 \times [(i-1)/i, i/(i+1)))$. By Lemma 6.2.3 there is a homeomorphism $g_1 : M \to M$ such that $g_1|B_0$ is the identity, $g_1(B_0 \cup C_1) \supset B_1$, the support of g_1 is contained in B_2, and $g_1(A_2) \subset A_2$. Inductively applying Lemma 6.2.3 gives a sequence of homeomorphisms $g_i : M \to M$ such that g_i is the identity on $g_{i-1} \circ \cdots \circ g_1(B_0 \cup C_1 \cup \cdots \cup C_{i-1})$, $g_i \circ \cdots \circ g_1(B_0 \cup C_1 \cup C_2 \cup \cdots \cup C_i) \supset B_i$, the support of g_i is contained in B_{i+1}, and $g_i(A_{i+1}) \subset A_{i+1}$. Define $G = \lim_{i\to\infty} g_i \circ \cdots \circ g_1|B_0 \cup C_0$. Then G defines a homeomorphism from $(B_0 \cup C_0, (B_0 \cup C_0) \cap Y) \cong (\mathbb{R}^n, \mathbb{R}^k)$ onto (M, Y), so the proof is complete. □

In order to use the Monotone Union Theorem to prove Proposition 6.2.2, we must produce n-cells that cover large compact subsets of $\mathbb{R}^n$ and intersect $h(\mathbb{R}^{n-2})$ standardly. To do so we will take a small n-cell that intersects $h(\mathbb{R}^{n-2})$ standardly and stretch it out to cover a specified compact set K. This stretch is accomplished in two stages: first the n-cell is stretched out along $h(\mathbb{R}^{n-2})$ to cover $K \cap h(\mathbb{R}^{n-2})$ and then the image cell is stretched via engulfing to cover the rest of K. The next two lemmas use a topological version of shelling (Rourke and Sanderson, 1972, Theorem 3.26) to stretch along $h(\mathbb{R}^{n-2})$.

Lemma 6.2.5. *Suppose M is an n-manifold, Y is a closed subset of M, and U is an open subset of M such that $(U, U \cap Y) \cong (\mathbb{R}^n, \mathbb{R}^{n-2})$. If $g : Y \to Y$ is a homeomorphism such that the support of g is contained in a compact subset of $U \cap Y$, then there exists a homeomorphism $G : M \to M$ such that $G|Y = g$ and $G|(M \setminus U)$ is the identity.*

Proof. Fix a homeomorphism $H : (U, U \cap Y) \to (\mathbb{R}^n, \mathbb{R}^{n-2})$. Define $g' : \mathbb{R}^{n-2} \to \mathbb{R}^{n-2}$ by $g'(x) = H(g(H^{-1}(x)))$. Then g' is a homeomorphism with compact support so there exists an $(n-2)$-cell C in $\mathbb{R}^{n-2}$ such that the support of g' is contained in the interior of C. Define

$$S = \mathbf{0} \times S^1 \subset \mathbb{R}^{n-2} \times \mathbb{R}^2 = \mathbb{R}^n,$$

and let $C * S \subset \mathbb{R}^n$ denote the join of C and S. By (Rourke and Sanderson, 1972, Proposition 2.23), $C * S$ is an n-cell. Extend g' over S via the identity and then over $C * S$ to a homeomorphism $G' : C * S \to C * S$ using the join operator. Note that $G'|\partial(C * S)$ is the identity. Hence $G = H^{-1} \circ G' \circ H$ extends via the identity to a homeomorphism of M. This is the homeomorphism G we seek. □

Lemma 6.2.6. *Suppose M is an n-manifold, $h : \mathbb{R}^{n-2} \to M$ is a proper, locally flat topological embedding and $Y = h(\mathbb{R}^{n-2})$. For every compact subset K of Y there exists an open subset V of M such that $K \subset V$ and $(V, V \cap Y) \cong (\mathbb{R}^n, \mathbb{R}^{n-2})$.*

Proof. Choose a rectilinear cube Q in $\mathbb{R}^{n-2}$ large enough that $h(Q) \supset K$. Use local flatness of h to cover Y with open sets $\{U_\alpha\}$ for which $(U_\alpha, U_\alpha \cap Y) \cong (\mathbb{R}^n, \mathbb{R}^{n-2})$, and let δ be a Lebesgue number for the open cover $\{h^{-1}(U_\alpha)\}$ of Q. Write $Q = Q_1 \cup Q_2 \cup \cdots \cup Q_m$, where each Q_i is a cube of diameter less than δ, $Q_1 \cup \cdots \cup Q_i$ is an $(n-2)$-cell for each i, and $(Q_1 \cup \cdots \cup Q_i) \cap Q_{i+1}$ is an $(n-3)$-cell for each i. Then for each i there is a homeomorphism $g_i : \mathbb{R}^{n-2} \to \mathbb{R}^{n-2}$ such that $g_i(Q_1 \cup \cdots \cup Q_i) = Q_1 \cup \cdots \cup Q_{i+1}$ and the support of g_i is contained in a small neighborhood of Q_{i+1}. We may assume that the support of g_i has diameter less than δ. As a result there is a U_α such that the support of g_i is contained in $h^{-1}(U_\alpha)$. By Lemma 6.2.5, each $h \circ g_i \circ h^{-1}$ extends to a homeomorphism $G_i : \mathbb{R}^n \to \mathbb{R}^n$. Select a U_β containing $h(Q_1)$. Then $G_k \circ \cdots \circ G_1(U_\beta)$ has the desired properties. □

Lemma 6.2.7. *Let $h : \mathbb{R}^{n-2} \to \mathbb{R}^n$ be a proper, locally flat, topological embedding and let $Y = h(\mathbb{R}^{n-2})$. If $\mathbb{R}^n \smallsetminus Y$ has the homotopy type of S^1 and Y is unraveled at infinity, then for every compact subset K of $\mathbb{R}^n$ there exists an open subset U of $\mathbb{R}^n$ such that $K \subset U$ and $(U, U \cap Y) \cong (\mathbb{R}^n, \mathbb{R}^{n-2})$.*

Proof. Since Y is unraveled at infinity, there is a compact subset D of $\mathbb{R}^n$ such that $K \subset D$ and $(\mathbb{R}^n \smallsetminus Y, \mathbb{R}^n \smallsetminus (Y \cup D))$ is 2-connected. By Lemma 6.2.6, there is an open subset V of $\mathbb{R}^n$ such that $D \cap Y \subset V \cap Y$ and $(V, V \cap Y) \cong (\mathbb{R}^n, \mathbb{R}^{n-2})$. The idea of the proof is to use engulfing to stretch V out to cover all of D (and hence all of K); more specifically, we will engulf a 2-skeleton with a neighborhood of infinity and engulf the dual $(n-3)$-skeleton with V. Special care is required because the polyhedra involved are not compact.

Choose a homeomorphism $\lambda : (\mathbb{R}^n, \mathbb{R}^{n-2}) \to (V, V \cap Y)$. Since $\lambda^{-1}(D \cap Y)$ is compact, there exists a round n-ball B_1 in $\mathbb{R}^n$ of sufficiently large radius that $\lambda^{-1}(D \cap Y) \subset \operatorname{Int} B_1$. Define $C_1 = \lambda(B_1)$ and $C_2 = \lambda(B_2)$, where B_2 is a concentric ball that contains B_1 in its interior. The relationship between these sets is illustrated schematically in Figure 6.4.

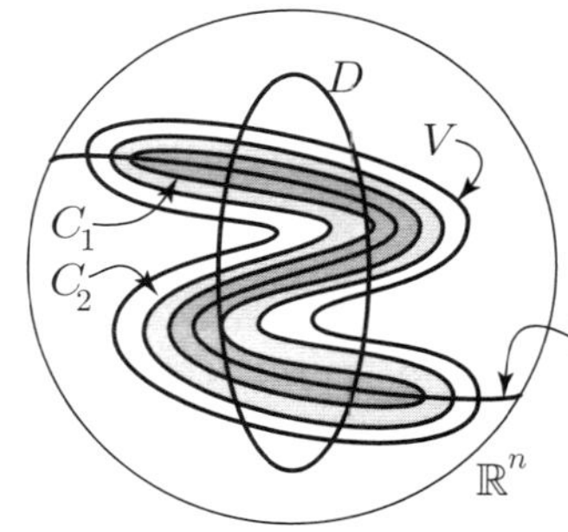

Figure 6.4. Proof of Lemma 6.2.7

Fix a triangulation T of $\mathbb{R}^n \smallsetminus Y$. We may assume that T is locally finite and that for each $\epsilon > 0$ there are only finitely many simplices in T of diameter greater than ϵ. Refine T so that, for any simplex $\sigma \in T$, $\sigma \cap C_1 \neq \emptyset$ implies $\sigma \subset \operatorname{Int} C_2$ and $\sigma \cap C_2 \neq \emptyset$ implies $\sigma \subset V$.

Define P to be the union of all simplices $\sigma \in T$ such that $\dim \sigma \leq 2$ and $\sigma \cap (\mathbb{R}^n \smallsetminus \operatorname{Int} C_1) \neq \emptyset$. Note that only a finite number of the simplices in P will intersect D since T is locally finite and $D \smallsetminus \operatorname{Int} C_1$ is compact. By Stallings engulfing (Theorem 3.1.3), there is a homeomorphism $\phi_1 : \mathbb{R}^n \smallsetminus Y \to \mathbb{R}^n \smallsetminus Y$ such that $P \subset \phi_1(\mathbb{R}^n \smallsetminus (Y \cup D))$ and $\phi_1 | (\mathbb{R}^n \smallsetminus (Y \cup E))$ is the identity for some compact subset E of $\mathbb{R}^n \smallsetminus Y$. We may assume E to be large enough that $D \subset C_2 \cup E$.

Let P_+ be the union of P and all simplices of T (regardless of dimension) that do not intersect $E \cup D$. Note that $P_+ \subset \phi_1(\mathbb{R}^n \smallsetminus (Y \cup D))$ since all the simplices of $P_+ \smallsetminus P$ just added are already outside of D and are not moved by ϕ_1. Define L to be the dual polyhedron of P_+; i.e., L is the simplicial complement of P_+ in the first barycentric subdivision T' of T.

Every simplex $\tau \in T'$ such that $\tau \subset L$ and $\tau \cap (\mathbb{R}^n \smallsetminus C_2) \neq \emptyset$ is contained in a simplex $\sigma \in T$ that is disjoint from C_1. Since the 2-skeleton of σ is contained in P_+, $\dim \tau \leq n-3$. Hence $L \smallsetminus C_2$ is $(n-3)$-dimensional. Furthermore, P_+ contains all simplices of T that do not intersect $D \cup E$, so every simplex in L is contained in a simplex of T that intersects either D or E. Any simplex that intersects D but not E must intersect C_2 and is therefore completely contained in V. Hence $L \smallsetminus V$ is covered by the finite complex consisting of all simplices in T (and their faces) that intersect E.

Since $\pi_1(V \smallsetminus (Y \cup C_2)) \cong \mathbb{J} \cong \pi_1(\mathbb{R}^n \smallsetminus (Y \cup C_2))$ and both groups are generated by a circle that is the homology dual to Y, the inclusion induced homomorphism $\pi_1(V \setminus (Y \cup C_2)) \to \pi_1(\mathbb{R}^n \setminus (Y \cup C_2))$ is an isomorphism. All higher homotopy groups are zero, so $(\mathbb{R}^n \setminus (Y \cup C_2), V \setminus (Y \cup C_2))$ is $(n-3)$-connected. Thus we can apply Stallings engulfing to obtain a homeomorphism $\phi_2 : \mathbb{R}^n \setminus (Y \cup C_2) \to \mathbb{R}^n \setminus (Y \cup C_2)$ with compact support such that $\phi_2(V \setminus (Y \cup C_2)) \supset L \setminus C_2$. Because it has compact support, ϕ_2 extends via the identity to $\mathbb{R}^n \setminus Y$. The extended homeomorphism still has compact support and also satisfies $\phi_2(V \setminus Y) \supset L$.

Because P_+ and L are dual polyhedra, there exists a push across the join structure, $\theta : \mathbb{R}^n \setminus Y \to \mathbb{R}^n \setminus Y$ such that $\mathbb{R}^n \setminus Y = \phi_1(\mathbb{R}^n \setminus (Y \cup D)) \cup \theta(\phi_2(V \setminus Y))$. Since ϕ_1, ϕ_2 have compact support, they can be extended via the identity to all of $\mathbb{R}^n$. In addition, the fact that the simplices of T near Y have small diameter means that θ can also be extended via the identity to $\mathbb{R}^n$. The extended homeomorphisms satisfy

$$\mathbb{R}^n = \phi_1(\mathbb{R}^n \setminus D) \cup \theta(\phi_2(V)),$$

so $U = \phi_1^{-1}(\theta(\phi_2(V)))$ fulfills the requirements of the conclusion of the lemma. □

Proof of Proposition 6.2.2. Proposition 6.2.2 follows almost immediately from Theorem 6.2.4 and Lemma 6.2.7. Write $\mathbb{R}^n$ as the ascending union of a countable collection of compact sets $K_1 \subset K_2 \subset \cdots$ with $\cup_{i=1}^{\infty} K_i = \mathbb{R}^n$. By Lemma 6.2.7, there exists an open set U_1 such that $U_1 \supset K_1$ and $(U_1, U_1 \cap Y) \cong (\mathbb{R}^n, \mathbb{R}^{n-2})$. Choose a compact ball $B_1 \subset U_1$ such that $(B_1, B_1 \cap Y) \cong (B^n, B^{n-2})$ and B_1 is large enough to contain K_1. Apply Lemma 6.2.7 again to find U_2 such that $U_2 \supset B_1 \cup K_2$ and $(U_2, U_2 \cap Y) \cong (\mathbb{R}^n, \mathbb{R}^{n-2})$. Cut U_2 back to a compact ball B_2 such that $\operatorname{Int} B_2 \supset B_1 \cup K_2$ and $(B_2, B_2 \cap Y) \cong (B^n, B^{n-2})$. This process is continued inductively to produce an infinite sequence $B_1 \subset B_2 \subset \cdots$ such that $\{(B_i, B_i \cap Y)\}$ satisfies the hypotheses of Theorem 6.2.4. □

In Proposition 6.2.2 the only hypothesis at infinity is about being unraveled there, which is merely a π_1 condition. This means that it is not necessary to assume that the topological embedding in Theorem 6.2.1 is locally flat at every point; it suffices to assume that the embedding is locally flat at every point but one and that it has good local π_1 at the exceptional point. The following statement makes that observation precise; its proof is essentially the same as that of Theorem 6.2.1, except for using Modified Stallings Engulfing to engulf 2-complexes.

Proposition 6.2.8. *Let $h : S^{n-2} \to S^n$, $n \geq 5$, be a topological embedding that is locally flat except possibly at one point $x_0 \in h(S^{n-2})$. If $S^n \setminus h(S^{n-2})$ has the homotopy type of S^1 and $h(S^{n-2})$ is 1-alg at x_0, then there is a topological homeomorphism $g : S^n \to S^n$ such that $g(h(S^{n-2})) = S^{n-2}$.*

Proof. Exercise 6.2.1. □

We conclude the section with an example demonstrating that the algebraic unknotting principle (Proposition 6.1.4) does not extend to topological embeddings: in the topological category the first nonstandard homotopy group of the complement of a knot can occur in any dimension up to $n-2$.

Example 6.2.9. *For each k, $1 < k \leq n-2$, there is a (wild) topological embedding $h : S^{n-2} \to S^n$ such that*

(1) *h is locally flat and PL except at one point,*
(2) *$\pi_i(S^n \smallsetminus h(S^{n-2})) \cong \pi_i(S^1)$ for $i < k$, but*
(3) *$\pi_k(S^n \smallsetminus h(S^{n-2})) \neq \{0\}$.*

Construction of Example 6.2.9. The construction consists of two parts. First, we construct an open subset W of S^n that has the unusual homotopy properties necessary to be the complement of the knot we seek. Second, we explain how to reembed W in S^n so that $S^n \smallsetminus W$ is a topological $(n-2)$-sphere.

Pursuing that strategy, we begin with W. The reader may find it helpful to note the parallel between the construction of W and the construction of the Whitehead manifold in Chapter 3. Since $k \geq 2$, $\pi_1(W)$ must be infinite cyclic. In addition, W must have $\pi_i(W) = 0$ for $1 < i < k$ and $H_k(W;\mathbb{Z}) = 0$ but $\pi_k(W) \neq 0$. Since we are working in the topological category we can use an infinite construction to accomplish this. We will construct W as an ascending union $W_1 \subset W_2 \subset W_3 \subset \cdots$ of compact PL ∂-manifolds in S^n. Each W_m will be a regular neighborhood of a copy of $S^k \vee S^1$. (Recall that $S^k \vee S^1$ denotes the wedge, or one point union, of S^k and S^1.)

Choose a standard unknotted copy $A_2 \vee B_2$ of $S^k \vee S^1$ in S^n and let W_2 be a regular neighborhood of $A_2 \vee B_2$. Notice that $\pi_1(W_2) \cong \mathbb{J}$. The groups $\pi_i(W_2)$ are all trivial for $1 < i < k$ and $\pi_k(W_2)$ is naturally isomorphic to $\mathbb{Z}[\mathbb{J}]$. Inside W_2 we embed a second copy, A_1, of S^k. Choose the embedding in such a way that A_1 consists of two disjoint parallel copies of A_2 connected by a tube that goes around B_2. (Here we need $k < n-1$.) The connection should be made in such a way that A_1 represents the element $t-1$ in $\pi_k(W_2) \cong \mathbb{Z}[\mathbb{J}]$. We also make sure that A_1 is unknotted in S^n. Let B_1 be a circle in W_2 such that B_1 is parallel to B_2 and $A_1 \vee B_1$ is unknotted in S^n; in fact, let us just say that $B_1 = B_2$. Define W_1 to be a thin regular neighborhood of $A_1 \vee B_1$ in $\operatorname{Int} W_2$.

Having defined $W_1 \subset \operatorname{Int} W_2$, we proceed to define W. Both $A_1 \vee B_1$ and $A_2 \vee B_2$ are unknotted in S^n, so there is an isotopy $h_t : S^n \to S^n, 0 \leq t \leq 1$, with $h_0 = \mathrm{Id}$ and $h_1(W_1) = W_2$. We recursively define $W_m, m \geq 3$, by $W_m = h_1(W_{m-1})$ and define $W = \cup_{m=1}^{\infty} W_m$. It is obvious that $\pi_1(W) = \mathbb{J}$

and $\pi_i(W) = 0$ for $1 < i < k$. In addition, the inclusion $W_1 \hookrightarrow W_2$ induces the trivial homomorphism on H_k, so $H_k(W;\mathbb{Z}) = 0$. We claim that the generator of $\pi_k(W_1)$ represents a nonzero element of $\pi_k(W)$. If not, this generator would be null-homotopic in W_m for some m. But the generator of $\pi_k(W_1)$ goes to $(t-1)^{m-1}$ times the generator of $\pi_k(W_m) \cong \mathbb{Z}[\mathbb{J}]$ and thus does not represent the zero element in that group. Thus we have constructed an open subset of S^n with the properties needed for our example; it remains only to show how to make arrangements so that W is the complement of a topological knot.

Define $V_m = S^n \smallsetminus \operatorname{Int} W_m$. Then $S^n \smallsetminus W = \cap_{m=1}^{\infty} V_m$ and $V_1 \supset V_2 \supset V_3 \supset \cdots$. Now W_m collapses to $A_m \vee B_m$ and $A_m \vee B_m$ is an unknotted copy of $S^k \vee S^1$, so V_m is a regular neighborhood of a polyhedron $C_m \vee D_m$ with $C_m \cong S^{n-k-1}$ and $D_m \cong S^{n-2}$. Figure 6.5 shows the relationships between A_2, B_2, C_2, and D_2.

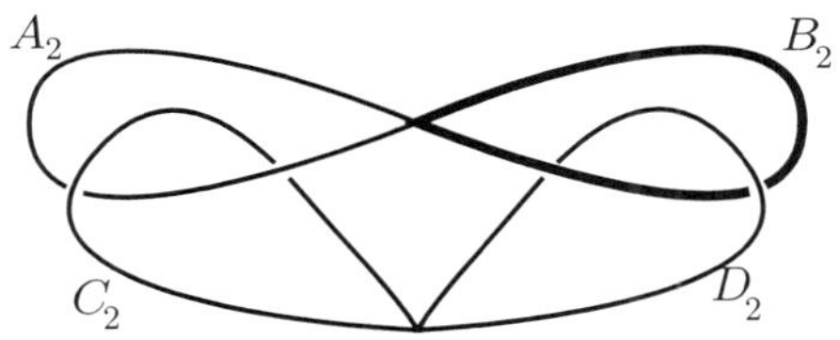

Figure 6.5. The embedding of $(A_2 \vee B_2) \cup (C_2 \vee D_2)$

Since $B_m = B_{m+1}$, we may assume that $D_m = D_{m+1}$. We need to understand how $C_{m+1} \vee D_{m+1}$ is embedded in V_m. We concentrate on the case $m = 1$. Figure 6.6 shows the way in which C_2 and D_2 are situated in V_1 (which is the complement of a regular neighborhood of $A_1 \vee B_1$).

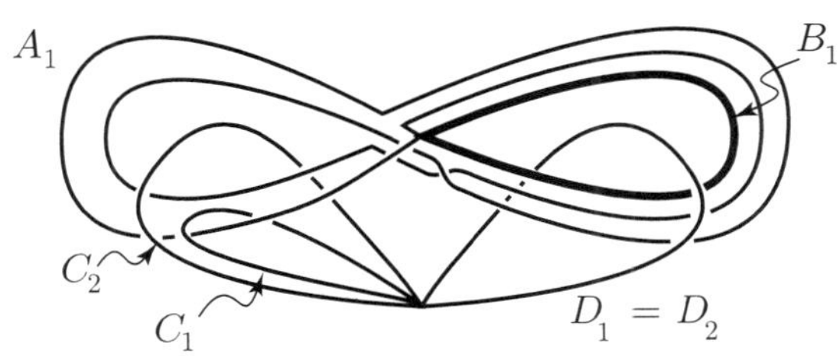

Figure 6.6. The embedding of $A_1 \vee B_1$ in the complement of $C_2 \vee D_2$

In Figure 6.7 we see what $C_2 \vee D_2$ looks like after we have performed an isotopy that straightens out $A_1 \vee B_1$. As one would expect, $D_2 = D_1$ while C_2 consists of two copies of C_1 joined by a tube that winds around D_2. Since the tube can be made very small, we can arrange things so that C_2 lies in a close regular neighborhood of C_1.

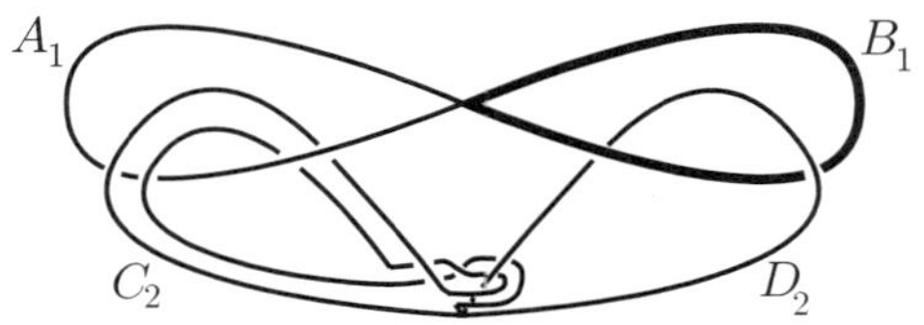

Figure 6.7. The embedding of $C_2 \vee D_2$ in the complement of $A_1 \vee B_1$

Perform an isotopy in each of the regular neighborhoods V_m that makes C_{m+1} lie in a close regular neighborhood of C_m. Then $S^n \setminus W = \cap_{m=1}^{\infty} V_m = C \vee D$, where $D = D_1$ is a locally flat PL $(n-2)$-sphere, $C = \cap_{m=1}^{\infty} N(C_m)$, and each $N(C_m)$ is a regular neighborhood of C_m. Furthermore, each C_{m+1} is a codimension-3 subpolyhedron of S^n and C_{m+1} is null homotopic in $N(C_m)$. It follows that C is a cellular set (by Theorem 3.2.3 in case $n \neq 4$ or (Freedman and Quinn, 1990) in case $n = 4$) so we can shrink it to a point; i.e., there is a map $f : S^n \to S^n$ whose only nondegenerate point inverse is C. The topological knot we are looking for is just $f(D) = f(C \cup D)$. $\square$

So far we have only examined the global homotopy properties of the sphere in Example 6.2.9; it has unusual local homotopy properties as well and they will be investigated in the next section.

Historical Notes. Theorem 6.2.1 is due to Stallings (1963). The theorem is valid in dimensions 3 and 4 as well. In dimension 3, Theorem 6.2.1 follows from Theorem 6.1.3 because every locally flat simple closed curve is equivalent to a PL one (Bing, 1954). In dimension 4 the result follows from (Freedman and Quinn, 1990, Theorem 11.7A).

The (nonrelative) monotone union theorem is due to M. Brown (1961). Brown proves the theorem for a monotone union of open n-cells. Since we only need the version for a monotone union of closed n-cells, that is what we prove; this allows us to omit one step in the proof. The relative version of the monotone union theorem was observed by Stallings (1963).

Proposition 6.2.8 can also be proved in dimensions $n \leq 4$. To do this in dimension $n = 3$, one would first apply (Boyd and Wright, 1973) or (Cannon, 1973a) to conclude that the simple closed curve is locally flat and then apply (Bing, 1954) to conclude that it is equivalent to a PL simple curve. After that one can apply the 3-dimensional case of Theorem 6.1.3. The 4-dimensional case follows from the work of Freedman and Quinn (1990).

A special case of Example 6.2.9 first appeared in (Liem and Venema, 1993) and the general case appeared in (Venema, 1995). Extension of algebraic unknotting to the locally flat topological category is based on a generalization of Milnor Duality (Theorem 6.5.13) to noncompact manifolds—see (Venema, 1995).

Exercise

6.2.1. Prove Theorem 6.2.8.

6.3. Local flatness and local homotopy properties

As we have seen in earlier chapters, even PL embeddings can fail to be locally flat in codimension two. This section contains a number of examples that localize the phenomena described in the last two sections and illustrate various ways in which codimension-two embeddings can be locally knotted or wild. All the examples of codimension-two non-local flatness described in previous chapters have the property that the local knotting or wildness is detected by the failure of the 1-alg condition; the examples in this section reveal that the 1-alg condition alone is not sufficient to guarantee local flatness. This means that codimension two is the only codimension in which a local π_1 condition does not imply local flatness. On the positive side, the section contains the statement of a theorem asserting that a codimension-two embedding that is locally homotopically unknotted is locally flat. The section explores the range of possibilities between the extremes of assuming local homotopy information only in dimension one and assuming it in all dimensions. It ends with a demonstration that most codimension-two embeddings fail to be locally flat.

Example 6.3.1. Take a locally flat PL embedding of the $(n-3)$-sphere into S^{n-1}, constructed as in Example 6.1.1 with $k > 1$, and suspend it; the result is a PL embedding of the $(n-2)$-sphere in S^n. If $\Sigma^{n-2} \subset S^n$ is such an embedded $(n-2)$-sphere, then Σ is locally flat except at the two suspension points and at each of those points Σ is 1-alg and i-LCC for $1 < i < k$. However, Σ fails to be k-LCC at the suspension points and therefore is not locally flat. It is possible to construct a PL example of this kind for each k in the range $1 < k < (n-1)/2$. □

The following consequence of the global theorems in §6.1 indicates that the bound on k in Example 6.3.1 is sharp.

Theorem 6.3.2. *Let $f : Q^{n-2} \to M^n$ be a PL embedding of a PL $(n-2)$-manifold Q into the PL n-manifold M. If $f(Q)$ is 1-alg and k-LCC for $2 \leq k < (n-1)/2$ at every point of $f(Q)$, then f is (PL) locally flat.*

Proof. Exercise 6.3.1. □

In the topological category the first nonstandard local homotopy group can occur in a slightly higher dimension. We offer two different kinds of examples of this phenomenon.

Example 6.3.3. Fix k, $1 < k < n/2$, and select a codimension-two sphere $\Sigma_0 \subset S^n$ that satisfies the conclusions of Example 6.1.1. Take an infinite sequence of disjoint copies $\Sigma_1, \Sigma_2, \Sigma_3, \ldots$ of Σ_0 converging to some point $p \in S^n$ such that $\lim_{i\to\infty} \operatorname{diam} \Sigma_i = 0$. For each i, remove a small $(n-2)$-cell from each of Σ_i and Σ_{i+1} and connect the two punctured spheres with a PL tube (a copy of $S^{n-3} \times [0,1]$). Form a single $(n-2)$-sphere Σ from the union of all the punctured spheres, all the tubes, and $\{p\}$. Notice that Σ is not PL, but that it can be made locally PL and locally flat at every point other than p. At the point p, Σ is 1-alg and i-LCC for $1 < i < k$. Again Σ is not k-LCC at p and thus is not locally flat at p. (Proofs of these assertions are outlined in the exercises at the end of the section.) It follows that Σ is wildly embedded despite being 1-alg at every point. It is possible to construct an example of this kind for each k in the range $1 < k < n/2$. □

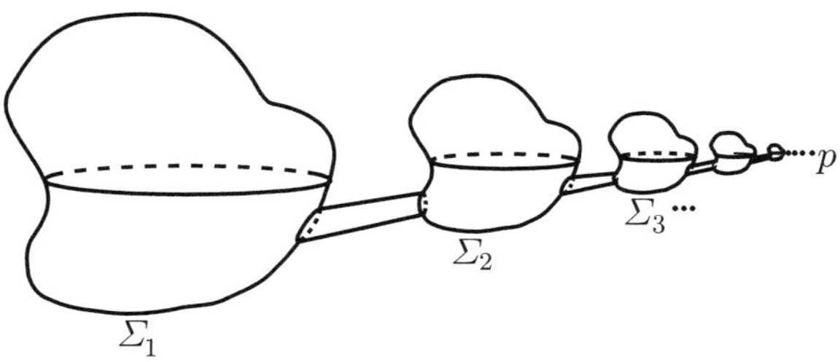

Figure 6.8. The sphere of Example 6.3.3

Example 6.3.4. Fix k, $1 < k \leq n-2$, and let $\Sigma_0 \subset S^n$ be the corresponding codimension-two sphere of Example 6.2.9. By construction, Σ_0 is PL and locally flat at every point but one. At the exceptional point Σ_0 fails to be k-LCC and fails to be $(n-k-1)$-LCC, but is 1-alg and i-LCC for $1 < i < \min\{k, n-k-1\}$. Those claims will be proved below. Observe that, regardless of the value of k, the embedding fails to be i-LCC for some i in the range $1 < i < n/2$.

Assume all the notation in the construction of Example 6.2.9. To see that Σ_0 fails to be k-LCC at the exceptional point, observe that any k-sphere that simply links C_m is homotopic to A_m. The proof of the example shows that A_m is essential in W. Since there are such spheres in any neighborhood of the point $f(C)$, $f(D)$ is not k-LCC at $f(C)$.

To see that Σ_0 is not $(n-k-1)$-LCC at the exceptional point, observe that an $(n-k-1)$-sphere parallel to C_2 is essential in the complement of $A_1 \cup D_1$. This becomes clear when one considers the universal cover of W, since a parallel copy of C_2 lifts to a sphere $\widehat{C}_2$ that is homologically linked with each of two consecutive lifts $\widehat{A}_1$ and $t(\widehat{A}_1)$ of A_1 (see Figure 6.9). Thus it is impossible to shrink C_2 to a point in $W \smallsetminus (A_1 \cup D_1)$. A similar proof shows that for every m an $(n-k-1)$-sphere parallel to C_m is essential in the complement of $A_1 \cup D_1$ and is therefore essential in a deleted neighborhood of $f(D)$. Since there are such spheres in any neighborhood of the point $f(C)$, $f(D)$ is not $(n-k-1)$-LCC at $f(C)$.

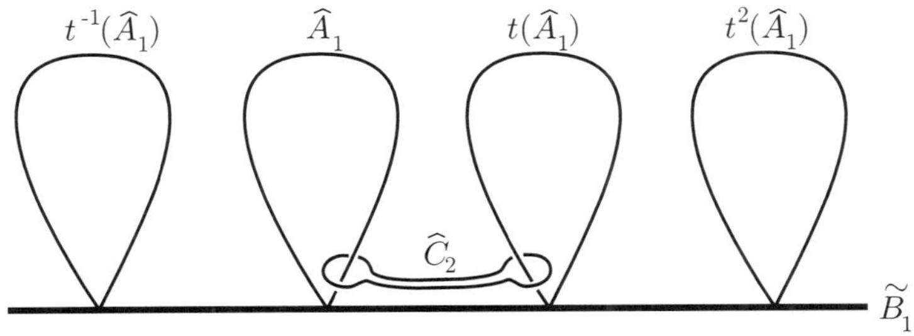

Figure 6.9. In the universal cover of W

Fix a positive integer m and decompose V_m into $U_m \cup H_m$, where U_m is a regular neighborhood of C_m and H_m is an $(n-2)$-handle attached to U_m. Consider a map $g : S^i \to U \smallsetminus \Sigma_0$, $1 < i < \min\{k, n-k-1\}$. Since $\pi_i(W) = \{0\}$, g is null homotopic in W. General position allows the track of the homotopy to be pushed off $A_m \vee B_m$ and into $V_m \smallsetminus D$. But $\operatorname{dem} C = k$, so the track of the homotopy can be pushed off C as well; thus g is null homotopic in $V_m \smallsetminus C \cup D$. By general position we can push the track of the homotopy off the cocore of H_m and conclude that g is null-homotopic in $U_m \smallsetminus \Sigma_0$. Since the exceptional point of $f(D)$ has arbitrarily small neighborhoods of the form $f(U_m)$, we can conclude that $f(D)$ is i-LCC at $f(C)$. A similar proof shows that $f(D)$ is 1-alg at $f(C)$. □

The next two theorems are positive results that give local analogs of Algebraic Unknotting Theorem 6.1.4 and Topological Flattening Theorem 6.2.1. The proofs of both theorems require techniques that go beyond those expounded in this book, so the proofs are omitted. The statements are included to complete the picture of local flatness in codimension two.

Theorem 6.3.5. *Let $f : Q^{n-2} \to M^n$ be a topological embedding of an $(n-2)$-manifold Q into the PL manifold M. If $f(Q)$ is 1-alg and k-LCC for $2 \leq k < n/2$ at every point of $f(Q)$, then $f(Q)$ is locally homotopically unknotted at every point of $f(Q)$.*

Theorem 6.3.6. *Let $f : Q^{n-2} \to M^n$ be a topological embedding of an $(n-2)$-manifold Q into an n-manifold M. If $f(Q)$ is locally homotopically unknotted at each of its points, then f is a locally flat embedding.*

The two theorems combined show that if a codimension-two embedding is not locally flat, then there must be points at which it either fails to be 1-alg or fails to be i-LCC for some i in the range $1 < i < n/2$. Note, however, that Theorem 6.3.5 does *not* assert that if an embedding is 1-alg at a point p and i-LCC for i in the range $1 < i < n/2$ at the same point p, then the embedding is locally homotopically unknotted at p. In order to conclude that the embedding is locally homotopically unknotted at a particular point, it is necessary for the local homotopy conditions to be satisfied at every point of the manifold (or at least at every point in a neighborhood of the original point). The last example in the section illustrates this subtle issue by showing that it is possible for the first nonstandard local homotopy group at a single point to occur in any dimension up to $n-2$. That cannot happen at an isolated wild point, however; there must be other nearby wild points at which the local homotopy groups go bad in lower dimensions.

Example 6.3.7. Let $\Sigma_0 \subset S^n$ be the codimension-two sphere of Examples 6.2.9 and 6.3.4. Apply the same infinite connected sum construction to Σ_0 as was used in Example 6.3.3. The resulting $(n-2)$-sphere has a distinguished point p at which it is 1-alg and i-LCC for $1 < i < k$ but not k-LCC. The proofs of these assertions are essentially the same as those of the corresponding assertions regarding Example 6.3.3 since those proofs use only the global homotopy properties of the complement. It is possible to construct an example of this kind for each k in the range $1 < k \leq n-2$. Observe that Σ has a sequence of non-locally flat points converging to p and that at each of those points the embedding fails to be both k-LCC and $(n-k-1)$-LCC. One of k and $n-k-1$ is less than $n/2$ regardless of the value of k. □

Unlike what occurs in other codimensions (see Theorem 4.6.17 and Exercise 7.9.3), local flatness is not the dominant characteristic of embeddings in codimension two, not even for embeddings of spheres in spheres.

Proposition 6.3.8. *It is not true that most embeddings of S^{n-2} in S^n are locally flat; that is,* $\mathrm{Emb}(S^{n-2}, S^n)$ *contains no dense G_δ-subset of locally flat embeddings.*

Proof. Suppose to the contrary that $V_1 \supset V_2 \supset \cdots$ are dense, open subsets of $\mathrm{Emb}(S^{n-2}, S^n)$ and each $\lambda \in \Lambda = \bigcap_{j=1}^{\infty} V_j$ is locally flat. Note that Λ is a Baire space.

Given $\lambda \in \mathrm{Emb}(S^{n-2}, S^n)$, we use $\operatorname{rank} \lambda(S^{n-2})$ to denote the minimum number of generators required for $\pi_1(S^n \setminus \lambda(S^{n-2}))$. Work of Fox (1950), coupled with the spin construction of §1.4, assures that S^n $(n \geq 3)$ contains locally flatly embedded codimension-two spheres of arbitrarily large (finite) rank. For $r = 1, 2, \ldots$, let W_r consist of all $\lambda \in \operatorname{Emb}(S^{n-2}, S^n)$ for which rank $\lambda(S^{n-2}) \geq r$.

Each $W_r \cap \Lambda$ is an open subset of Λ. Every locally flat embedding λ has a neighborhood N_λ homeomorphic to $S^{n-2} \times B^2$ such that $\pi_1(S^n \setminus \operatorname{Int} N_\lambda) \cong \pi_1(S^n \setminus \lambda(S^{n-2}))$. (Look ahead to Theorem 6.8.1, which promises that $\lambda(S^{n-2})$ has a disk bundle neighborhood; even when $n = 4$ the only possible disk bundle neighborhood in S^4 is the product bundle.) Consider $\lambda' \in W_r$ for which $\lambda' : S^{n-2} \to N_\lambda$ is a homotopy equivalence. Then the inclusion induced $\pi_1(\partial N_\lambda) \to \pi_1(N_\lambda \setminus \lambda'(S^{n-2}))$ has a left inverse since $\pi_1(\partial N_\lambda) \cong \mathbb{Z} \cong H_1(\partial N_\lambda; \mathbb{Z})$ and $H_1(\partial N_\lambda; \mathbb{Z}) \to H_1(N_\lambda \setminus \lambda'(S^{n-2}); \mathbb{Z})$ is an isomorphism, by duality. It is a straightforward consequence of the Seifert-van Kampen Theorem that $\pi_1(S^n \setminus \lambda'(S^{n-2}))$ retracts to $\pi_1(S^n \setminus \operatorname{Int} N_\lambda) \cong \pi_1(S^n \setminus \lambda(S^{n-2}))$. Thus, rank $\lambda'(S^{n-2}) \geq$ rank $\lambda(S^{n-2}) \geq r$.

In addition, each $W_r \cap \Lambda$ is dense in Λ. Any locally flat embedding can be approximated by one with large rank, simply by replacing a small, standard ball pair with a knotted ball pair of large rank (here the boundary of the knotted pair must be standard, to match up properly). Again the Seifert-van Kampen Theorem yields that the fundamental group of the newly constructed knot complement retracts to that of the replacing ball pair, which assures large rank. Density of Λ in $\operatorname{Emb}(S^{n-2}, S^n)$ implies $W_r \cap \Lambda$ is dense in Λ.

The Baire Category Theorem promises that $\Lambda' = \bigcap_{r=1}^{\infty} W_r \neq \emptyset$. For each $\lambda \in \Lambda'$ we have rank $\lambda(S^{n-2}) \geq r$ for all integers r, so λ must be wildly embedded, which is impossible. □

There are two possible explanations for Proposition 6.3.8. It might be that in a given dimension n there are wild codimension-two spheres that cannot be approximated by locally flat embeddings; indeed, §6.6 provides examples of embeddings $Q^{n-2} \to S^n$, where Q^{n-2} is non-simply connected and compact and $n = 4$ or $n \geq 6$, that cannot approximated by locally flat ones. However, if all embeddings $S^{n-2} \to S^n$ can be approximated by locally flat ones, for some choice of n, then in fact most of them are wild, by an argument similar to the above.

Historical Notes. Theorem 6.3.6 is due to Chapman (1979). Theorem 6.3.5 appears in (Venema, 1997). The argument for Proposition 6.3.8 is essentially due to Milnor (1964), who proved that most knots in S^3 are wild.

Exercises

6.3.1. Prove Theorem 6.3.2.

6.3.2. If the $(n-1)$-manifold Q^{n-1} is a closed subset of the n-manifold M^n, then $Q^{n-1} \times \{0\}$ is locally homotopically unknotted in $M^n \times \mathbb{R}$.

6.3.3. Show that most embeddings of S^1 in S^3 are wild.

In the following exercises, $\Sigma \subset S^n$ denotes the $(n-2)$-sphere of Example 6.3.3 and all the notation of that example is assumed. Let $B_1 \supset B_2 \supset \ldots$ be a nested sequence of concentric round n-balls centered at p such that $\cap_j B_j = \{p\}$. We will assume that ∂B_j cuts across the tube connecting Σ_j to Σ_{j+1} so that, for every j, $\Sigma_j \subset B_j \smallsetminus B_{j+1}$ and $(\partial B_j, \partial B_j \cap \Sigma)$ is an unknotted PL $(n-1, n-3)$-sphere pair.

6.3.4. Prove that $\pi_1(B_j \smallsetminus \Sigma) \cong \mathbb{J}$ for every j and that the inclusion induced homomorphism $\pi_1(B_{j+1} \smallsetminus \Sigma) \to \pi_1(B_j \smallsetminus \Sigma)$ is an isomorphism for every j. Conclude that Σ is 1-alg at p.

By Exercise 6.3.4, the homology of $B_j \smallsetminus \Sigma$ with coefficients in $\mathbb{Z}[\mathbb{J}]$ is defined and the universal cover of $B_{j+1} \smallsetminus \Sigma$ is naturally identified with a subset of the universal cover of $B_j \smallsetminus \Sigma$.

6.3.5. Prove that $H_i(B_j \smallsetminus \Sigma; \mathbb{Z}[\mathbb{J}]) = 0$ for $1 < i < k$ and for every j. Conclude that Σ is i-LCC at p for $1 < i < k$.

6.3.6. Prove that $H_k(B_j \smallsetminus \Sigma; \mathbb{Z}[\mathbb{J}]) \neq 0$ and that the inclusion induced homomorphism $H_k(B_{j+1} \smallsetminus \Sigma; \mathbb{Z}[\mathbb{J}]) \to H_k(B_j \smallsetminus \Sigma; \mathbb{Z}[\mathbb{J}])$ is one-to-one for every j. Conclude that Σ is not k-LCC at p.

6.4. The homology of an infinite cyclic cover

This section is devoted to the study of the homology of an infinite cyclic cover. The Alexander polynomial, which is the invariant to be used in the remainder of the chapter, is defined and methods for its computation are developed. The emphasis in the section is on geometry; the more technical, algebraic properties of the Alexander polynomial will be examined in the next section. The example calculations that are included will be needed later in the chapter.

Let us begin by generalizing the definitions of §6.1. Assume X is a connected (possibly infinite) CW complex and $\gamma : \pi_1(X) \to \mathbb{J}$ is an epimorphism. The *infinite cyclic cover of* X *determined by* γ is the connected (regular) covering space $p : \widetilde{X} \to X$ such that $p_*(\pi_1(\widetilde{X})) = \ker \gamma \subset \pi_1(X)$. The existence of a unique connected regular covering space with this property is a standard fact from elementary topology—see (Munkres, 2000, Theorems 82.1 and 79.2), for example. A different construction of the infinite cyclic cover is indicated in Exercise 6.4.1.

The group of deck transformations (or covering transformations) of $\widetilde{X}$ is isomorphic to $\pi_1(X)/p_*(\pi_1(\widetilde{X})) \cong \mathbb{J}$. As before, we use t to denote a generator of $\mathbb{J}$. We also use the same symbol t to denote both the associated deck transformation and the induced homomorphisms on the homotopy and homology groups of $\widetilde{X}$. It is usually clear from the context which of these objects t denotes, so the fact that the symbol t serves multiple purposes should not cause undue confusion.

Throughout this section and the next three, we use coefficients in $\mathbb{Q}$, the field of rational numbers, for most homology and cohomology groups. We use Λ to denote the group ring $\mathbb{Q}[\mathbb{J}]$; thus

$$\Lambda = \left\{ \sum_{i=-m}^{m} q_i t^i \;\middle|\; m \in \mathbb{N}, q_i \in \mathbb{Q} \right\},$$

the ring of Laurent polynomials in t with coefficients in $\mathbb{Q}$. It is easy to see that the units in Λ are exactly the monomials. We use the symbol $\doteq$ to indicate that two elements of Λ are equal up to multiplication by a unit. Since $\mathbb{Q}$ is a field, Λ is a principal ideal domain.[4] An important consequence is the following fact: if A is a finitely generated module over Λ and B is a submodule of A, then B is finitely generated over Λ (Hungerford, 1974, Corollary IV.6.2). Any module over Λ is also a vector space over $\mathbb{Q}$.

Just as before, the cellular chain complex $C_*(\widetilde{X};\mathbb{Q})$ and the homology groups $H_*(\widetilde{X};\mathbb{Q})$ have natural Λ-module structures; when we wish to emphasize this structure we denote them by $C_*(X;\Lambda)$ and $H_*(X;\Lambda)$, respectively.

Example 6.4.1. If $X = S^1 \vee S^2$ and $\gamma : \pi_1(X) \to \mathbb{J}$ is an isomorphism, then $\widetilde{X}$ is a line with an infinite number of 2-spheres attached. (See Figure 6.10, in which $\widehat{S^2}$ is a particular lift of S^2.) Thus $H_2(X;\mathbb{Q}) \cong \mathbb{Q}$ and $H_2(X;\Lambda) \cong \Lambda$. Note that $H_2(X;\Lambda) = H_2(\widetilde{X};\mathbb{Q})$ is infinite-dimensional as a vector space over $\mathbb{Q}$.

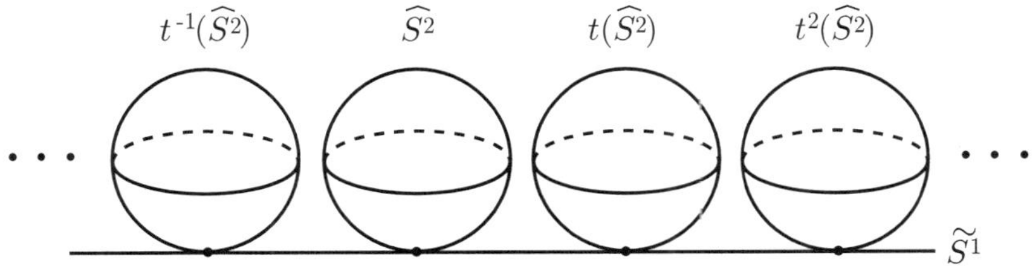

Figure 6.10. An infinite cyclic cover of $S^1 \vee S^2$

The Milnor sequence. As with $\mathbb{Z}$ coefficients, there is a *Milnor sequence*

$$\cdots \to H_k(X;\Lambda) \xrightarrow{t-1} H_k(X;\Lambda) \xrightarrow{p_*} H_k(X;\mathbb{Q}) \to H_{k-1}(X;\Lambda) \to \cdots,$$

which relates the homology of the covering space to that of the base space. This sequence is exact; the proof of exactness is precisely the same as before.

[4]This and other algebraic properties of Λ will be discussed in the next section.

The order of a module. Any finitely generated Λ-module B can be written as a sum of cyclic submodules:

$$B \cong \frac{\Lambda}{(p_1(t))} \oplus \cdots \oplus \frac{\Lambda}{(p_n(t))}$$

(Hungerford, 1974, Theorem IV.6.12). By analogy with finite abelian groups, the product $p_1(t) \cdots p_n(t)$ is called the *order* of B and the ideal in Λ generated by $p_1(t) \cdots p_n(t)$ is called the *order ideal* of B. The order ideal is denoted by $\mathcal{O}(B)$. Note that $\mathcal{O}(B)$ is a subset of Λ.

While the definition of order is intuitively simple, it is not always easy to apply directly because the modules we encounter may not be presented as sums of cyclic modules. To facilitate computations we give an alternative approach to the definition of order.

Assume B is a finitely generated Λ-module. Name a finite set $\{x_1, \ldots, x_n\}$ of generators for B and then form the free Λ-module F_n generated by $\{x_1, \ldots, x_n\}$. Let K be the kernel of the natural Λ-homomorphism $\phi : F_n \to B$ defined by $\phi(x_n) = x_n$. Since Λ is a principal ideal domain, K will also be a finitely generated Λ-module; let $\{y_1, \ldots, y_m\}$ be a generating set for K. Define an $m \times n$ matrix $P = (b_{ij})$ by $y_i = \sum_{j=1}^{n} b_{ij} x_j$. The matrix P is called a *presentation matrix* for B.

Definition. Let B be a finitely generated Λ-module and let P be a presentation matrix for B. The *order ideal* of B is defined to be the ideal $\mathcal{O}(B)$ in Λ generated by all the $n \times n$ minors of P provided $n \leq m$. Define $\mathcal{O}(B) = \{0\}$ in case $n > m$. A generator of $\mathcal{O}(B)$ is called the *order* of B.

In most cases of interest, the presentation matrix P is square and $\mathcal{O}(B)$ is the principal ideal generated by $\det P$. If B is written as

$$B \cong \frac{\Lambda}{(p_1(t))} \oplus \cdots \oplus \frac{\Lambda}{(p_n(t))},$$

then the diagonal matrix with the polynomials $p_i(t)$ on the diagonal is a presentation matrix. Thus the following proposition shows both that the order is well defined and that the second definition of order ideal agrees with the first.

Proposition 6.4.2. *The order ideal $\mathcal{O}(B)$ is well defined; i.e., $\mathcal{O}(B)$ depends only on B and does not depend on the particular presentation matrix that is used to describe B. The order of B is well defined up to multiplication by a unit.*

Sketch of proof. This is a fairly standard algebraic argument. One approach is to write down a set of operations that may be performed on the presentation matrix and then to prove that any two presentation matrices differ at most by a finite sequence of these operations and that the operations do not change the order ideal. The details of this argument may be found in (Zassenhaus, 1958, Chapter III) or (Crowell and Fox, 1977, Theorem VII.4.2); see also (Rolfsen, 1990, Theorem 8B1). Another approach is to use the first definition of order along with the Jordan-Hölder Theorem for modules (Hungerford, 1974, Theorem VIII.1.10)—see (Milnor, 1968, §1). □

Corollary 6.4.3. *If B_1, B_2, and B_3 are finitely generated Λ-modules such that $B_1 = B_2 \oplus B_2$, then the order of B_1 is the product of the orders of B_2 and B_3.*

Recall that any finitely generated Λ-module B naturally decomposes as $B = T \oplus F$, where T is a torsion module and F is a free module over Λ. This decomposition is unique (Hungerford, 1974, Theorem IV.6.6). The following lemma records some useful relationships between this decomposition and the order ideal.

Lemma 6.4.4. *Let B be a finitely generated Λ-module decomposed into torsion and free parts as $B = T \oplus F$.*

(1) $\mathcal{O}(B) = \Lambda$ *if and only if* $B = \{0\}$.
(2) $\mathcal{O}(B) = \{0\}$ *if and only if* $F \neq \{0\}$.
(3) $\mathcal{O}(B) \neq \{0\}$ *if and only if* $B = T$.

Proof. Exercise 6.4.3. □

The Alexander polynomial. We come now to the most important definition in the section—that of the Alexander polynomial. It is the invariant that will be used in §6.6 and §6.7 to verify that the codimension-three approximation and existence theorems do not extend to codimension two.

Assume $\gamma : \pi_1(X) \to \mathbb{J}$ is an epimorphism and that $H_1(\widetilde{X}; \mathbb{Q})$ is finitely generated over Λ. Decompose $H_1(\widetilde{X}; \mathbb{Q})$ into its torsion and free parts:

$$H_1(\widetilde{X}; \mathbb{Q}) = T_1(\widetilde{X}; \mathbb{Q}) \oplus F_1(\widetilde{X}; \mathbb{Q}).$$

The module $T_1(\widetilde{X}; \mathbb{Q})$ is called the *torsion submodule* of $H_1(\widetilde{X}; \mathbb{Q})$. The advantage of working with the torsion submodule is that its order is always nonzero.

Definition. Suppose $\gamma : \pi_1(X) \to \mathbb{J}$ is an epimorphism and that $H_1(\widetilde{X}; \mathbb{Q})$ is finitely generated over Λ. The *Alexander polynomial* $A(X, \gamma; t)$ of the pair (X, γ) is defined to be a generator of the order ideal $\mathcal{O}(T_1(\widetilde{X}; \mathbb{Q}))$.

By Proposition 6.4.2, $A(X, \gamma; t)$ is well-defined up to a product with units. Since we have used only the torsion submodule of $H_1(\widetilde{X}; \mathbb{Q})$ in the definition, $A(X, \gamma; t)$ is never zero (Lemma 6.4.4, Part 3). We also have $A(X, \gamma; t) \doteq 1$ if and only if $T_1(\widetilde{X}; \mathbb{Q}) = \{0\}$ (Lemma 6.4.4, Part 1).

Several useful properties of the Alexander polynomial will be developed in the next section. In the remainder of this section we explain how to calculate $A(X, \gamma; t)$ for a special class of spaces X and epimorphisms γ. The spaces in which we are interested are 3-dimensional ∂-manifolds associated with a link in S^3; their definitions come next.

Construction of the spaces $E(L)$ and $M(L)$. Suppose $L = \{\ell_1, \ldots, \ell_k\}$ is a k-component oriented PL link in S^3 having the property that the homological linking number $\operatorname{lk}(\ell_i, \ell_j)$ is 0 for $i \neq j$. For each component ℓ_i, choose a meridian μ_i and a longitude λ_i. The meridian is simply a small unknotted circle that links ℓ_i once and the longitude is a curve that is parallel to ℓ_i and satisfies $\operatorname{lk}(\ell_i, \lambda_i) = 0$. The meridian should be oriented so that $\operatorname{lk}(\mu_i, \ell_i) = +1$. Both μ_i and λ_i lie on the boundary of a regular neighborhood N_i of ℓ_i and intersect in a single point. (See Figure 1.4.) We will assume that the neighborhoods $\{N_i\}$ are pairwise disjoint. Let $E(L)$ denote the exterior of L and let $M(L)$ denote the manifold obtained from S^3 by surgery on L. Specifically,

$$E(L) = \overline{S^3 \smallsetminus \cup_{i=1}^k N_i},$$

and $M(L)$ is the manifold constructed by removing the neighborhoods N_i from S^3 and then sewing them back in with the meridians and longitudes interchanged. There is a natural inclusion $E(L) \hookrightarrow M(L)$.

Definition of the epimorphism γ_m. Observe that $H_1(E(L); \mathbb{Z})$ is the free abelian group with $\{\mu_i\}$ as generators. Because $\operatorname{lk}(\ell_i, \ell_j) = 0$ for $i \neq j$, each longitude λ_i is null homologous in $E(L)$. Hence the inclusion induced homomorphism $H_1(E(L); \mathbb{Z}) \to H_1(M(L); \mathbb{Z})$ is an isomorphism. For each m, $1 \leq m \leq k$, define an epimorphism $\gamma_m : \pi_1(M(L)) \to \mathbb{J}$ by $\gamma_m(\mu_i) = t$ for $i \leq m$ and $\gamma_m(\mu_i) = 1$ for $i > m$. The epimorphism γ_m is uniquely determined by the orientations on $\{\mu_i\}$. We use $\widetilde{M}_m(L)$ to denote the infinite cyclic cover of $M(L)$ determined by γ_m and $\widetilde{E}_m(L)$ to denote the infinite cyclic cover of $E(L)$ determined by $\gamma_m|E(L)$. (If $\gamma : \pi_1(X) \to \mathbb{J}$ is a homomorphism and $Y \subset X$, then $\gamma|Y$ denotes the composite homomorphism $\pi_1(Y) \to \pi_1(X) \xrightarrow{\gamma} \mathbb{J}$.) Again there is a natural inclusion $\widetilde{E}_m(L) \hookrightarrow \widetilde{M}_m(L)$.

Construction of the covering spaces $\widetilde{E}_m(L)$ and $\widetilde{M}_m(L)$. We give a geometric description of the covering spaces $\widetilde{E}_m(L)$ and $\widetilde{M}_m(L)$. For simplicity we begin with the case $m = k$. Find a connected, bicollared PL surface $\Sigma \subset E(L)$ such that $\partial\Sigma = \lambda_1 \cup \cdots \cup \lambda_k$. Such a surface is called a *Seifert surface* for the link. A good way to construct one is to find a PL map $f : E(L) \to S^1$ that induces γ_k on $\pi_1(E(L))$ and define Σ to be the preimage under f of a regular point. Alternatively, the Seifert Algorithm (Rolfsen, 1990, Theorem 5A4) can be used to construct Σ. The surface Σ is homeomorphic to a disk with strips attached and can be pictured as in Figure 6.12.

Thicken Σ along a bicollar in $E(L)$ to form a handlebody C. Choose C in such a way that, for each $i \leq k$, $C \cap N_i$ is a bicollar of λ_i in $\partial N_i \subset \partial E(L)$. Split $E(L)$ open along Σ to form $E'(L)$. Specifically, $E'(L) = \overline{E(L) \smallsetminus C}$. Then $\partial C \cap \partial E'(L)$ consists of two copies, Σ^+ and Σ^-, of Σ and there is a PL homeomorphism $H : \Sigma \times [-1,1] \to C$ such that $H(\Sigma \times \{-1\}) = \Sigma^-$, $H(\Sigma \times \{0\}) = \Sigma$, and $H(\Sigma \times \{1\}) = \Sigma^+$. Define $\widetilde{E}_k(L)$ to be the space obtained from the disjoint union of $E'(L) \times \mathbb{J}$ and $C \times \mathbb{J}$ by making the following identifications: for each $x \in \Sigma^-$, identify the point $\langle x, t^n \rangle$ in $C \times \mathbb{J}$ with the corresponding point in $E'(L) \times \mathbb{J}$ and for each $x \in \Sigma^+$, identify the point $\langle x, t^n \rangle$ in $C \times \mathbb{J}$ with $\langle x, t^{n+1} \rangle$ in $E'(L) \times \mathbb{J}$. We will use the notation $t^n x$ for the pair $\langle x, t^n \rangle$. See Figure 6.11 for a schematic diagram of $\widetilde{E}_k(L)$.

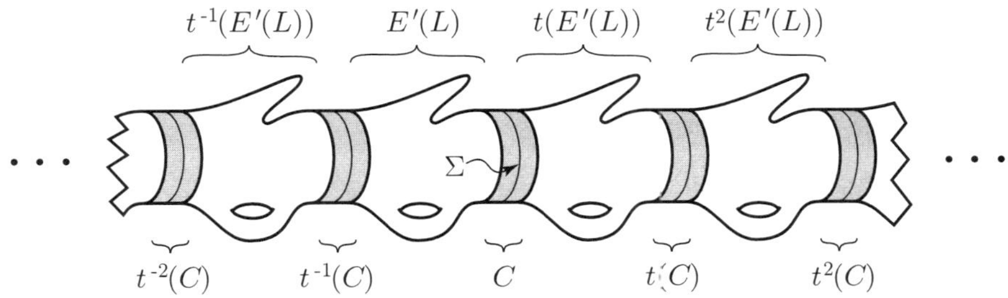

Figure 6.11. The cyclic cover $\widetilde{E}_k(L)$

The space $\widetilde{E}_k(L)$ could have been constructed from just $E'(L) \times \mathbb{J}$ by making identifications of the form $t^n x^- = t^{n+1} x^+$ for each $x \in \Sigma$. The reason a copy of C is inserted between consecutive copies of $E'(L)$ is to ensure that there is an obvious covering map $p : \widetilde{E}_k(L) \to E(L)$ and to facilitate the use of a Mayer-Vietoris sequence in the computation of the homology.

Observe that a loop α in $E(L)$ is in $\ker \gamma_k$ if and only if the homological intersection number $\alpha \cdot \Sigma$ is zero. Since loops α with $\alpha \cdot \Sigma = 0$ are precisely the loops that lift to closed loops in $\widetilde{E}_k(L)$, we see that $\widetilde{E}_k(L)$ is the covering space corresponding to $\ker \gamma_k$.

To construct $\widetilde{E}_m(L)$ for $m < k$, begin by finding a Seifert surface Σ for the partial link $\{\ell_1, \ldots, \ell_m\}$. Because $\operatorname{lk}(\ell_i, \ell_j) = 0$ for $i \neq j$, it is possible to add handles to Σ so that $\Sigma \cap N_i = \emptyset$ for $i > m$. Construct the cover corresponding to Σ exactly as above and then remove the interiors of all the lifts of N_i, $i > m$. The resulting space is $\widetilde{E}_m(L)$.

Now ∂N_i lifts to a copy of $S^1 \times \mathbb{R}^1$ in $\partial \widetilde{E}_m(L)$ in case $i \leq m$ and it lifts to $\partial N_i \times \mathbb{J}$ in case $i > m$. To construct $\widetilde{M}_m(L)$ from $\widetilde{E}_m(L)$, attach a copy of $B^2 \times \mathbb{R}^1$ to the preimage of each ∂N_i for which $i \leq m$ and attach $N_i \times \mathbb{J}$ to the preimage of each ∂N_i for which $i > m$. The resulting space is the infinite cyclic cover $\widetilde{M}_m(L)$.

Computation of $H_1(\widetilde{E}_k(L); \mathbb{Q})$. Assume $m = k$. The Seifert surface Σ consists of a disk with s strips attached. For each $i = 1, \ldots, s$, let a_i denote the curve on Σ that runs from a base point to the strip, follows the centerline of the strip and then returns to the base point (see Figure 6.12). The curves $\{a_1, \ldots, a_s\}$ form a basis for $H_1(\Sigma; \mathbb{Q})$. Let a_i^+ and a_i^- denote the copies of a_i on Σ^+ and Σ^-, respectively. For each i, also choose a curve α_i in $E'(L)$ that simply links the ith strip as indicated in Figure 6.12.

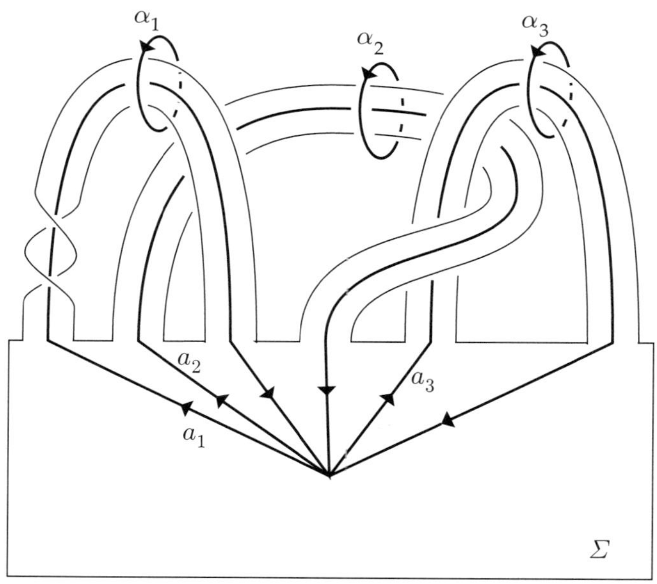

Figure 6.12. The Seifert surface Σ

By duality, $\{\alpha_1, \ldots, \alpha_s\}$ forms a basis for $H_1(E'(L); \mathbb{Q})$. Each α_i can be (uniquely) lifted to a curve $\widetilde{\alpha}_i$ in $E'(L) \times \{1\}$ and the set $\{\widetilde{\alpha}_1, \ldots, \widetilde{\alpha}_s\}$ freely generates $H_1(E'(L) \times \mathbb{J}; \mathbb{Q})$ as a module over Λ. Every loop in $\widetilde{E}_k(L)$ is homologous to one that misses the preimage of Σ, so the lifts $\{\widetilde{\alpha}_1, \ldots, \widetilde{\alpha}_s\}$ also generate $H_1(\widetilde{E}_k(L); \mathbb{Q})$ as a module over Λ. But they do not freely generate $H_1(\widetilde{E}_k(L); \mathbb{Q})$ since there are relations that are introduced by the identifications. Specifically, for each i, $1 \leq i \leq s$, there is a relation of the form $a_i^+ = ta_i^-$. The fact that the relations have this form is intuitively clear since inserting a copy of C between two adjacent copies of $E'(L)$ has the effect of identifying a_i^+ with a_i^-. The proof that $H_1(\widetilde{E}_k(L); \mathbb{Q})$ has a presentation with generators $\{\widetilde{\alpha}_1, \ldots, \widetilde{\alpha}_s\}$ and relations of the form $a_i^+ = ta_i^-$ can be made precise using a Mayer-Vietoris argument (Rolfsen, 1990, §6B and §8C) or an abelianized version of the Seifert-van Kampen Theorem (Burde and Zieschang, 1985, Theorem 8.8).

The Seifert and Alexander matrices. In case $m = k$, the Alexander polynomial can be computed as the determinant of a matrix. Let $x \in H_1(E'(L); \mathbb{Q})$. We wish to express x in terms of the basis $\{\alpha_1, \ldots, \alpha_s\}$. Since $\operatorname{lk}(a_i, \alpha_j) = \delta_{ij}$, x is homologous to the sum $\sum_{j=1}^{s} \operatorname{lk}(x, a_j)\alpha_j$. If we set $x = a_i$, the relation $a_i^- = ta_i^+$ becomes

$$\sum_{j=1}^{s} \operatorname{lk}(a_i^-, a_j)\widetilde{\alpha}_j = t \cdot \sum_{j=1}^{s} \operatorname{lk}(a_i^+, a_j)\widetilde{\alpha}_j.$$

Let $V = (v_{ij})$ be the matrix defined by $v_{ij} = \operatorname{lk}(a_i^+, a_j)$; the matrix V is called a *Seifert matrix*. Now $\operatorname{lk}(a_i^-, a_j) = \operatorname{lk}(a_i, a_j^+) = \operatorname{lk}(a_j^+, a_i) = v_{ji}$, so the relation can be rewritten as

$$\sum_{j=1}^{s} v_{ji}\widetilde{\alpha}_j = t \cdot \sum_{j=1}^{s} v_{ij}\widetilde{\alpha}_j.$$

In other words, the matrix $P = V^T - tV$ is a presentation matrix for $H_1(\widetilde{E}_k(L); \mathbb{Q})$. The matrix P is called an *Alexander matrix*. Since P is square, $\det P$ is a generator for $\mathcal{O}(H_1(\widetilde{E}_k(L); \mathbb{Q}))$. Hence $A(E(L), \gamma_k; t) \doteq \det(V^T - tV)$ whenever $\det P \neq 0$. This will be the case for all the examples we consider in the remainder of the section.

The basis used in the previous paragraph is not always the most convenient. Suppose $\{a_i'\}$ is another basis for $H_1(\Sigma; \mathbb{Q})$. Let $C = (c_{ij})$ be the change of basis matrix defined by $a_i' = \sum_{j=1}^{s} c_{ij}a_j$. We will assume that the c_{ij} are all integers so that each a_i' is represented by a loop on Σ. Let $V' = (v_{ij}')$ be the Seifert matrix $v_{ij}' = \operatorname{lk}(a_i'^+, a_j')$. Then $V' = CVC^T$, so $\det(V'^T - tV') \doteq \det(V^T - tV)$ (because $\det C$ is a constant and therefore a unit in Λ). As a result we may use the Seifert matrix associated with any basis for $H_1(\Sigma; \mathbb{Q})$ when computing $A(E(L), \gamma_k; t)$, provided each element of the basis is represented by a loop on Σ.

There is a simple relationship between $A(E(L), \gamma_k; t)$ and $A(M(L), \gamma_k; t)$, which is exposed in the next lemma.

Lemma 6.4.5. *$A(E(L),\gamma_k;t)$ is divisible by $(t-1)^{k-1}$. If $H_1(\widetilde{E}_k(L);\mathbb{Q}) = T_1(\widetilde{E}_k(L);\mathbb{Q})$, then $A(M(L),\gamma_k;t) \doteq A(E(L),\gamma_k;t)/(t-1)^{k-1}$.*

Proof. Here we will supply a simple geometric proof of the first part of the lemma. We will postpone the proof of the second part, which requires more algebraic preliminaries, until the next section.

By the classification theorem for compact, orientable surfaces, the Seifert surface Σ is homeomorphic to the surface obtained from S^2 by cutting out k holes and attaching a finite number of handles. Such a surface is represented in Figure 6.13 as a disk with strips attached. There are exactly $(k-1)$ strips like those shown on the left while the number of pairs of additional strips depends on the particular Seifert surface used. The diagram shows only an abstract picture of the surface itself and does not indicate the way in which the surface is embedded in S^3. In reality the handles will be twisted, knotted, and entangled with each other. Despite that limitation, the diagram does enable us to make the following observation: for $i \le k-1$ and for j arbitrary, $\mathrm{lk}(a_i^+, a_j) = \mathrm{lk}(a_i^-, a_j)$. Hence $v_{ij} = v_{ji}$ and $p_{ji} = (1-t)v_{ji}$, where p_{ji} is an entry in the ith column of the Alexander matrix P, $i \le k-1$. Thus $(1-t)$ divides each of the first $(k-1)$ columns of P and the first part of the conclusion of the lemma follows. □

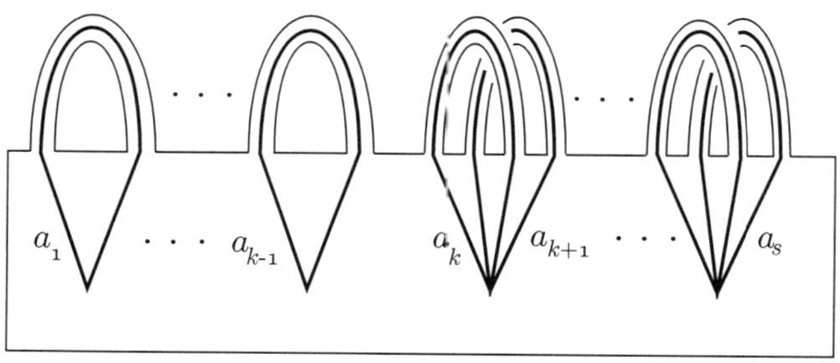

Figure 6.13. An abstract Seifert surface

We are now ready for some calculations. In the following examples we compute the Alexander polynomials that will be needed in the remainder of the chapter. To foreshadow what lies ahead, Examples 6.4.6 and 6.4.7 as well as Exercise 6.4.5 are required in the construction of a topological embedding that cannot be approximated by PL embeddings (§6.6) while Example 6.4.8 is needed in the construction of a homotopy equivalence that is not homotopic to an embedding (§6.7).

Example 6.4.6. If L is the link pictured in Figure 6.14, then

$$A(M(L),\gamma_3;t) \doteq (t-1)^2(2-3t+2t^2).$$

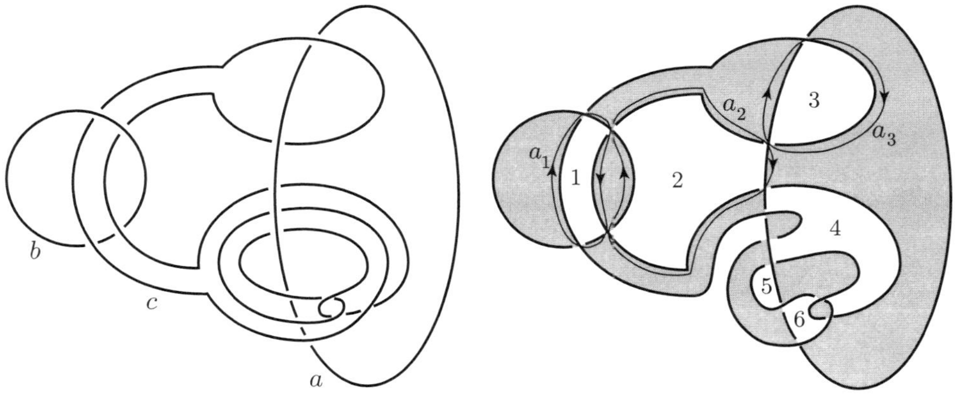

Figure 6.14. The Matsumoto link L and a Seifert surface for L

To compute $A(E(L), \gamma_3; t)$ we use the Seifert surface Σ shown in Figure 6.14. As a basis for $H_1(\Sigma; \mathbb{Q})$ we choose the curves $\{a_1, \ldots, a_6\}$, where a_i is the loop that goes clockwise around the region labeled i. Just a_1, a_2, and a_3 are shown in order to avoid cluttering up the picture. Despite the way they are drawn, the a_i intersect each other at the crossing points of the link (where there is a twist in the Seifert surface). The Seifert matrix corresponding to this basis is

$$V = \begin{pmatrix} 0 & 1 & 0 & 0 & 0 & 0 \\ 1 & -1 & -1 & 1 & 0 & 0 \\ 0 & 0 & 1 & 0 & 0 & 0 \\ 0 & 0 & 0 & 0 & 0 & -1 \\ 0 & 0 & 0 & -1 & 1 & 0 \\ 0 & 0 & 0 & -1 & -1 & 1 \end{pmatrix}.$$

A routine computation[5] shows that $A(E(L), \gamma_3; t) \doteq (t-1)^4(2 - 3t + 2t^2)$. By Lemma 6.4.5 we have $A(M(L), \gamma_3; t) \doteq (t-1)^2(2-3t+2t^2)$. □

Example 6.4.7. If L is the link of Example 6.4.6 and ℓ_1 is the middle component, then

$$A(M(L), \gamma_1; t) \doteq (t-1)^2.$$

Label the middle component c and label the other two components a and b. The link can be redrawn as in Figure 6.15. Using the shaded disk as Σ, one can view the covering space $\widetilde{E}_1(L)$ as in Figure 6.16; it consists of $\mathbb{R}^1 \times B^2$ with regular neighborhoods of each of the curves $t^n a$ and $t^n b$ removed. Each such neighborhood is a solid torus. Observe that $H_1(\widetilde{E}_1(L); \mathbb{Q}) \cong \Lambda \oplus \Lambda$. The generators are α and β, where α is a meridian of a and β is a meridian of b.

[5] A computer algebra system such as *Mathematica* is helpful.

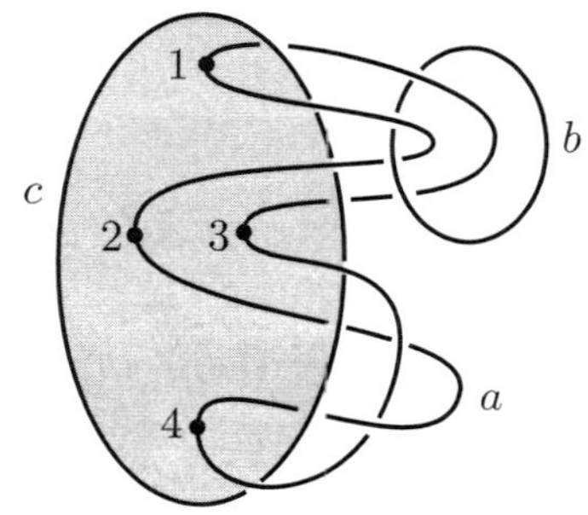

Figure 6.15. A different picture of the link L

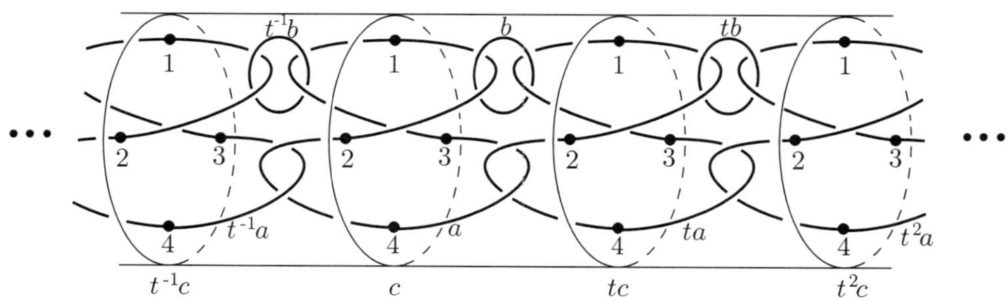

Figure 6.16. The infinite cyclic cover of the complement of L

To construct $\widetilde{M}_1(L)$ we must attach a copy of $B^2 \times \mathbb{R}^1$ to the outside of $\widetilde{E}_1(L)$ and then sew the solid tori corresponding to a and b back in with longitudes and meridians interchanged. Sewing in the b torus introduces the relation $\alpha - t\alpha = 0$ and sewing in the a torus introduces the relation $\beta - t\beta + t^{-1}\alpha + t\alpha = 0$. Thus $H_1(\widetilde{M}_1(L); \mathbb{Q})$ has presentation matrix

$$P = \begin{pmatrix} 1-t & 0 \\ t^{-1}+t & 1-t \end{pmatrix}$$

and $A(M(L), \gamma_1; t) = (t-1)^2$. □

Example 6.4.8. If $L_1, L_2, \ldots$ is the sequence of two-component links shown in Figure 6.17, then

$$A(M(L_n), \gamma_2; t) \doteq (t-1)^{2n}.$$

Using the surface and basis $\{a_1, a_2, a_3\}$ shown on the left in Figure 6.18, we obtain the Seifert matrix

$$V_1 = \begin{pmatrix} -1 & 1 & 0 \\ 0 & 1 & -1 \\ 0 & -1 & 0 \end{pmatrix}$$

for L_1. Hence $A(E(L_1), \gamma_2; t) \doteq \det(V_1^T - tV_1) \doteq (t-1)^3$.

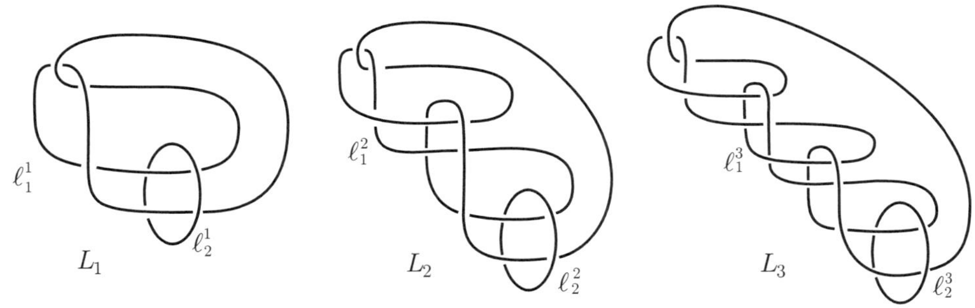

Figure 6.17. The links L_1, L_2, and L_3

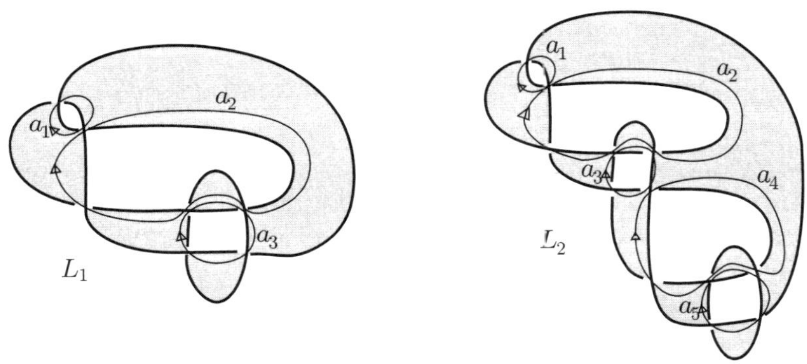

Figure 6.18. Seifert surfaces for L_1 and L_2

The surface shown on the right in Figure 6.18 and the basis $\{a_1, \ldots, a_5\}$ indicated in the figure yield the following Seifert matrix for L_2.

$$V_2 = \begin{pmatrix} -1 & 1 & 0 & 0 & 0 \\ 0 & 0 & -1 & 0 & 0 \\ 0 & -1 & 0 & 1 & 0 \\ 0 & 0 & 0 & 1 & -1 \\ 0 & 0 & 0 & -1 & 0 \end{pmatrix}.$$

Thus $A(E(L_2), \gamma_2; t) \doteq \det(V_2^T - tV_2) \doteq (t-1)^5$.

Observe that L_{i+1} has one more loop at the bottom than does L_i and there is one crossing change from L_i to L_{i+1}. Based on that observation we obtain the recursive relationship

$$V_{n+1} = \left(\begin{array}{c|cc} & 0 & 0 \\ V_n' & \vdots & \vdots \\ & 0 & 0 \\ & 1 & 0 \\ \hline 0 & 1 & -1 \\ & -1 & 0 \end{array}\right),$$

where V_n' is the matrix obtained from V_n by replacing the 1 in position $(2n, 2n)$ by a 0. Induction gives

$$A(E(L_n), \gamma_2; t) \doteq \det(V_n^T - tV_n) \doteq (t-1)^{2n+1}.$$

It follows from Lemma 6.4.5 that $A(M(L_n), \gamma_2; t) \doteq (t-1)^{2n}$. □

Historical Notes. Many of the ideas in this section and the next have their roots in the beautiful paper of Milnor (1968) and in the earlier paper by R. C. Blanchfield (1957).

Our approach to the definition of the Alexander polynomial is based on work of Kawauchi (1975), (1977), (1980). In particular, we have followed Kawauchi in defining the Alexander polynomial to be the order of the torsion part of $H_1(\widetilde{X}; \mathbb{Q})$. Many authors define $A(X, \gamma; t)$ using all of $H_1(\widetilde{X}; \mathbb{Q})$, not just the torsion submodule, so that $A(X, \gamma; t) = 0$ is possible. For example, that is the approach taken by Rolfsen (1990). Additional information about computing Alexander polynomials can be found in (Kawauchi, 1996), (Rolfsen, 1990), (Burde and Zieschang, 1985), and (Crowell and Fox, 1977).

The first part of Lemma 6.4.5 is due to G. Torres (1953), (1954) and F. Hosokawa (1958). A proof of the complete lemma may be found in (Kawauchi, 1977, Theorem 3.13).

Exercises

6.4.1. Let X be a connected CW complex and let $\gamma : \pi_1(X) \to \mathbb{J}$ be an epimorphism. Suppose $f : X \to S^1$ is a continuous map that induces γ in the sense that $f_* = \gamma : \pi_1(X) \to \pi_1(S^1) = \mathbb{J}$. Let $e : \mathbb{R}^1 \to S^1$ be the universal covering map $e(r) = \exp(2\pi i r)$. Prove that $\widetilde{X}$, the infinite cyclic cover of X determined by γ, is the pull-back of the following diagram.

$$\begin{array}{ccc} & & \mathbb{R}^1 \\ & & \downarrow e \\ X & \xrightarrow{f} & S^1 \end{array}$$

Specifically, prove that $\widetilde{X} = \{\langle x, r\rangle \in X \times \mathbb{R}^1 \mid f(x) = e(r)\}$ together with $p : \widetilde{X} \to X$ defined by $p(x, r) = x$ is a covering space and that $p_*(\pi_1(\widetilde{X})) = \ker \gamma \subset \pi_1(X)$. ($\widetilde{X}$ is given the subspace topology it inherits from $X \times \mathbb{R}^1$.) Show, in addition, that $\mathbb{J}$ acts on $\widetilde{X}$ according to the rule $t \cdot \langle x, r\rangle = \langle x, r + 1\rangle$.

6.4.2. Let $X = S^1 \vee S^1$ and let a_1, a_2 be the two generators of $\pi_1(X)$ corresponding to the two copies of S^1. Define $\gamma_1 : \pi_1(X) \to \mathbb{J}$ to be the epimorphism that takes a_1 to t and a_2 to 1; define $\gamma_2 : \pi_1(X) \to \mathbb{J}$ to be the epimorphism that takes both of the generators to t. Find (i.e., draw a picture of) the infinite cyclic covers of X determined by γ_1 and γ_2.

6.4.3. Prove Lemma 6.4.4.

6.4.4. Show that $A(E(L), \gamma_k; t)$ is symmetric in t; i.e., $A(E(L), \gamma_k; t) \doteq A(E(L), \gamma_k; t^{-1})$.

6.4.5. Let L be the Borromean rings, the link shown in Figure 6.19. (Note that the link is symmetric, so it makes no difference which component is labeled ℓ_1.) Compute $A(E(L), \gamma_i; t)$ and $A(M(L), \gamma_i; t)$ for $i = 1, 2$, and 3. [Answer: $A(M(L), \gamma_i; t) \doteq (t - 1)^2$ for every i.]

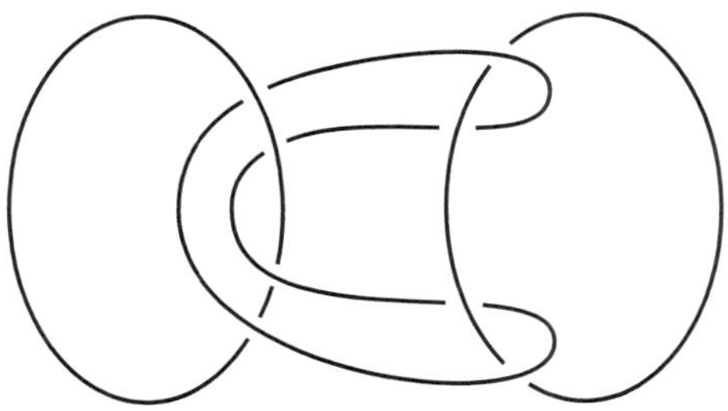

Figure 6.19. The Borromean rings

6.5. Properties of the Alexander polynomial

The proofs in the remainder of the chapter are based on arguments involving the Alexander polynomial. In order to use this tool effectively, we must first master some of its technical aspects. The various algebraic properties of the Alexander polynomial that will be required are collected in this section.

First we need a deeper understanding of the algebraic structure of Λ itself. The next several results are elementary, but we include proofs because the results are not part of the standard algebra literature. Note that $\Lambda = \mathbb{Q}[t, t^{-1}]$ contains $\mathbb{Q}[t]$, the ring of polynomials over $\mathbb{Q}$, as a subring. We will refer to polynomials in $\mathbb{Q}[t]$ as *ordinary polynomials* in order to distinguish them from the Laurent polynomials. Any nonzero Laurent polynomial $p(t)$ can be written as $p(t) = t^k \tilde{p}(t)$, where $k \in \mathbb{Z}$, $\tilde{p}(t)$ is an ordinary polynomial, and $\tilde{p}(0) \neq 0$. The *degree* of $p(t)$ is defined to be the ordinary degree of $\tilde{p}(t)$.

Lemma 6.5.1. *Λ is a principal ideal domain.*

Proof. Degree defines an integer-valued norm on Λ that gives Λ the structure of a Euclidean domain. Hence Λ is a principal ideal domain (Hungerford, 1974, Theorem III.3.9). □

Many familiar results regarding ordinary polynomials over a field are also true of Laurent polynomials. The next proposition is such a result.

Proposition 6.5.2. *For any nonzero Laurent polynomial $p(t)$,*

$$\dim_{\mathbb{Q}}(\Lambda/(p(t))) = \deg p(t).$$

Proof. Since t^k is a unit in Λ, we may assume that $p(t)$ is an ordinary polynomial with $p(0) \neq 0$. Let $s = \deg p(t)$. If $s = 0$, then $p(t)$ is a unit and $\Lambda/(p(t)) = \{0\}$. Thus we may assume $s > 0$.

We claim that $\{1, t, t^2, \ldots, t^{s-1}\}$ is a basis for $\Lambda/(p(t))$. It is clear that $\{1, t, t^2, \ldots, t^{s-1}\}$ is linearly independent over $\mathbb{Q}$ since no polynomial of degree less than s can be written as a multiple of $p(t)$. In order to show that $\{1, t, t^2, \ldots, t^{s-1}\}$ spans $\Lambda/(p(t))$ we will show that any $f(t) \in \Lambda$ can be expressed as $f(t) = q(t)p(t) + r(t)$, where $r(t)$ is an ordinary polynomial and $\deg r(t) < s$. Then $f(t) = r(t)$ in the factor ring $\Lambda/(p(t))$, so the proof will be complete.

First write $f(t) = f_1(t) + f_2(t^{-1})$, where each of $f_1(t)$ and $f_2(t)$ is an ordinary polynomial. The division algorithm for polynomials over a field shows that $f_1(t)$ can be expressed in the required form (Hungerford, 1974, Theorem V.1.6(iv)). Apply the division algorithm to $t^{-(s-1)} f_2(t^{-1})$ and $t^{-s}p(t)$, both of which are ordinary polynomials in t^{-1}. The result is

$$t^{-(s-1)} f_2(t^{-1}) = q(t^{-1}) t^{-s} p(t) + r_1(t^{-1}),$$

where q and r_1 are ordinary polynomials and $\deg r_1 \leq s - 1$. Multiply by t^{s-1} to get

$$f_2(t^{-1}) = t^{-1} q(t^{-1}) p(t) + r_2(t),$$

where $r_2(t) = t^{s-1} r_1(t^{-1})$ is an ordinary polynomial of degree $\leq s - 1$. The last displayed equation shows that $f_2(t^{-1})$ is equal to $r_2(t)$ in the factor ring $\Lambda/(p(t))$. This completes the proof of the proposition. □

The proposition indicates, in particular, that $\dim_{\mathbb{Q}}(\Lambda/(p(t)))$ is finite when $p(t) \neq 0$. Combining that information with Lemma 6.4.4, Part 3, yields the following corollary.

Corollary 6.5.3. *If B is a finitely generated Λ-module, then $\mathcal{O}(B) \neq \{0\}$ if and only if* $\dim_{\mathbb{Q}}(B) < \infty$.

The next corollary helps explain the relevance of Proposition 6.5.2 to the study of Alexander polynomials.

Corollary 6.5.4. *If $\gamma : \pi_1(X) \to \mathbb{J}$ is an epimorphism and $H_1(\widetilde{X}; \mathbb{Q})$ is finitely generated over Λ, then* $\deg(A(X, \gamma; t)) = \dim_{\mathbb{Q}}(T_1(\widetilde{X}; \mathbb{Q}))$.

The following proposition is a generalization of Corollary 6.4.3.

Proposition 6.5.5. *If B_1, B_2, and B_3 are finitely generated Λ-modules such that $B_1 = B_2/B_3$ and $q_i(t)$ is the order of B_i, then $q_2(t) \doteq q_1(t)q_3(t)$.*

Proof. If $q_1(t) = 0$, then the Λ-free part of B_1 is nonzero by Lemma 6.4.4, Part 2. Hence the free part of B_2 is nonzero and $q_2(t) = 0$. If the free part of B_3 is nonzero, then the free part of B_2 must also be nonzero. Thus $q_3(t) = 0$ implies $q_2(t) = 0$. If the free part of B_2 is nonzero, then the free part of either B_3 or B_1 must be nonzero. Hence $q_2(t) = 0$ implies that either $q_1(t) = 0$ or $q_3(t) = 0$. These observations show that the result holds if $q_i(t) = 0$ for any i. We may therefore assume that $q_i(t) \neq 0$ for every i. It follows that each B_i is a torsion module.

Let $\{x_1, \ldots, x_n\}$ be a generating set for B_3. Since B_3 is a torsion module, we may assume that this generating set corresponds to a square presentation matrix P_3. In the same way, let $\{\tilde{y}_1, \ldots, \tilde{y}_m\}$ be a generating set for B_1 that corresponds to a square presentation matrix P_1. For each j, $1 \leq j \leq m$, choose $y_j \in B_2$ such that $\tilde{y}_j = y_j + B_3$. Observe that $\{x_1, \ldots, x_n, y_1, \ldots, y_m\}$ is a generating set for B_2. Each row of P_1 corresponds to a combination of $\{\tilde{y}_j\}$ that is zero in B_1. Hence the corresponding combination of $\{y_j\}$ is in B_3 and can be written as a combination of $\{x_i\}$. Thus there is a matrix P_2 of the form

$$P_2 = \begin{pmatrix} P_3 & 0 \\ X & P_1 \end{pmatrix}$$

in which each row represents a combination of $\{x_1, \ldots, x_n, y_1, \ldots, y_m\}$ that is zero in B_2. It is not difficult to see that P_2 is a presentation matrix for B_2. Since $q_i(t) \doteq \det(P_i)$, this completes the proof. □

Now we are ready to begin applying the algebraic results just obtained to the study of Alexander polynomials. The proofs in the next two sections depend crucially on the fact that homology cobordisms preserve certain essential features of the Alexander polynomial. We should clarify what is meant by cobordism in this context.

Definition. Suppose M_1 and M_2 are n-manifolds and that $\gamma_1 : \pi_1(M_1) \to \mathbb{J}$ and $\gamma_2 : \pi_1(M_2) \to \mathbb{J}$ are epimorphisms. We say that (M_1, γ_1) is *homology* *cobordant* to (M_2, γ_2) if there exist an $(n+1)$-dimensional ∂-manifold W whose boundary is the disjoint union of M_1 and M_2 and an epimorphism $\gamma : \pi_1(W) \to \mathbb{J}$ such that $\gamma_i = \gamma|M_i$ and each of the inclusions $M_i \hookrightarrow W$ induces an isomorphism on homology (with $\mathbb{Z}$ coefficients).

One of the features that is preserved is the number of factors of $t-1$. It is convenient to give that number a name.

Definition. Suppose X is a connected CW complex, that $\gamma : \pi_1(X) \to \mathbb{J}$ is an epimorphism, and that $H_1(\widetilde{X}; \mathbb{Q})$ is finitely generated over Λ. Write $A(X, \gamma; t) = (t-1)^k B(t)$, where $B(1) \neq 0$ and $k \geq 0$. The integer $k = k(X, \gamma)$ is the *Kawauchi invariant* of the pair (X, γ).

The Kawauchi invariant can be defined this way since $A(X, \gamma; t) \neq 0$. The following theorem, one of two major results in the section, attests that the Kawauchi invariant is preserved by homology cobordism. It is a key ingredient in the construction of an extraordinary homotopy equivalence later in the chapter.

Theorem 6.5.6. *Suppose M_1 and M_2 are compact PL n-manifolds and $\gamma_1 : \pi_1(M_1) \to \mathbb{J}$ and $\gamma_2 : \pi_1(M_2) \to \mathbb{J}$ are epimorphisms. If (M_1, γ_1) is homology cobordant to (M_2, γ_2) via a compact PL ∂-manifold W, then $k(M_1, \gamma_1) = k(M_2, \gamma_2)$.*

The statement of Theorem 6.5.6 has the advantage of being brief and simple to understand, but in later applications we will not always have at hand the full force of a homology cobordism. Lemma 6.5.11, below, spells out precisely the technical hypotheses required for reaching the desired conclusion; in our main application we will appeal directly to that lemma rather than to Theorem 6.5.6. In that connection it is worth observing that the compactness hypotheses in the preceding theorem imply that each of $H_*(\widetilde{M_1}; \mathbb{Q})$, $H_*(\widetilde{M_2}; \mathbb{Q})$, $H_*(\widetilde{W}, \widetilde{M_1}; \mathbb{Q})$, and $H_*(\widetilde{W}, \widetilde{M_2}; \mathbb{Q})$ is finitely generated as a module over Λ.

As a first step toward the proof of Theorem 6.5.6 we will verify that $k(X, \gamma)$ is equal to the $\mathbb{Q}$-dimension of a certain submodule of $H_1(\widetilde{X}; \mathbb{Q})$.

Definition. For any Λ-module C, the $(t-1)$*-primary submodule of* C is defined to be

$$C_{(t-1)} = \{\, x \in C \mid (t-1)^m x = 0 \text{ for some } m \geq 0 \,\}.$$

Assume C is finitely generated over Λ. Then it has a decomposition of the form

$$C \cong \frac{\Lambda}{({p_1}^{s_1})} \oplus \cdots \oplus \frac{\Lambda}{({p_n}^{s_n})},$$

in which each p_i is a prime polynomial in Λ and s_i is a positive integer (Hungerford, 1974, Theorem IV.6.12). Such a decomposition is called the *primary decomposition* of C. The $(t-1)$-primary submodule $C_{(t-1)}$ is the direct sum of all the factors for which $p_i \doteq t-1$. It therefore follows from Proposition 6.5.2 that $C_{(t-1)}$ is finite-dimensional as a vector space over $\mathbb{Q}$ whenever C is finitely generated as a module over Λ.

Lemma 6.5.7. *If $\widetilde{X}$ is an infinite cyclic cover of X corresponding to γ and $H_1(\widetilde{X};\mathbb{Q})$ is finitely generated over Λ, then*

$$k(X,\gamma) = \dim_{\mathbb{Q}}\left(H_1(\widetilde{X};\mathbb{Q})_{(t-1)}\right).$$

Proof. Observe that $H_1(\widetilde{X};\mathbb{Q})_{(t-1)} \subset T_1(\widetilde{X};\mathbb{Q})$ and that $H_1(\widetilde{X};\mathbb{Q})_{(t-1)} = T_1(\widetilde{X};\mathbb{Q})_{(t-1)}$. Let

$$T_1(\widetilde{X};\mathbb{Q}) \cong \frac{\Lambda}{({p_1}^{s_1})} \oplus \cdots \oplus \frac{\Lambda}{({p_n}^{s_n})}$$

be a primary decomposition of $T_1(\widetilde{X};\mathbb{Q})$. Then $\dim_{\mathbb{Q}}(H_1(\widetilde{X};\mathbb{Q})_{(t-1)})$ equals the sum of the s_i for which $p_i \doteq t-1$. Since that sum is also equal to $k(X,\gamma)$, the proof is complete. □

Let C be a Λ-module. Multiplication by $t-1$ defines a Λ-homomorphism from C to C, *viz.* $t-1 : C \to C$ defined by $(t-1)(x) = tx - x$. The following proposition sets forth another algebraic property of Λ-modules that will be useful in the remainder of the chapter.

Proposition 6.5.8. *If C is a finitely generated Λ-module and $t-1 : C \to C$ is onto, then $C_{(t-1)} = \{0\}$ and C is finite-dimensional as a vector space over $\mathbb{Q}$.*

Proof. It is clear that $x \in C_{(t-1)}$ if and only if $(t-1)x \in C_{(t-1)}$; hence $(t-1)(C_{(t-1)}) \subset C_{(t-1)}$. The fact that $t-1 : C \to C$ is onto implies that $(t-1)|C_{(t-1)} : C_{(t-1)} \to C_{(t-1)}$ is also onto. Thus $(t-1)^m|C_{(t-1)} : C_{(t-1)} \to C_{(t-1)}$ is onto for every m. But $C_{(t-1)}$ is finitely generated over Λ, so there exists an integer m such that $(t-1)^m y = 0$ for every $y \in C_{(t-1)}$. It follows that $C_{(t-1)} = \{0\}$.

Now C can be decomposed into free and torsion submodules. Since $t-1$ is onto, each element of C can be divided by arbitrarily high powers of $t-1$. But the only element of a free module possessing that property is 0. Therefore the free part of C is $\{0\}$ and C is finite-dimensional by Corollary 6.5.3. □

Corollary 6.5.9. *Let $p : \widetilde{Y} \to Y$ be an infinite cyclic cover of the polyhedron Y and let $A \subset Y$ be a subpolyhedron. If $H_q(Y,A;\mathbb{Q}) = \{0\}$ and $H_q(\widetilde{Y},p^{-1}(A);\mathbb{Q})$ is finitely generated over Λ, then*

(1) $t-1 : H_q(\widetilde{Y}, p^{-1}(A); \mathbb{Q}) \to H_q(\widetilde{Y}, p^{-1}(A); \mathbb{Q})$ *is onto,*

(2) $H_q(\widetilde{Y}, p^{-1}(A); \mathbb{Q})_{(t-1)} = \{0\}$, *and*

(3) $H_q(\widetilde{Y}, p^{-1}(A); \mathbb{Q})$ *is finite-dimensional over* $\mathbb{Q}$.

Proof. It follows from the Milnor sequence

$$H_q(\widetilde{Y}, p^{-1}(A); \mathbb{Q}) \xrightarrow{t-1} H_q(\widetilde{Y}, p^{-1}(A); \mathbb{Q}) \xrightarrow{p_*} H_q(Y, A; \mathbb{Q}) = 0$$

that $t-1 : H_q(\widetilde{Y}, p^{-1}(A); \mathbb{Q}) \to H_q(\widetilde{Y}, p^{-1}(A); \mathbb{Q})$ is onto, so Proposition 6.5.8 applies. □

Lemma 6.5.10. *Suppose* $A \xrightarrow{\alpha} B \xrightarrow{\beta} C \xrightarrow{\gamma} D$ *is an exact sequence of* Λ*-modules and* Λ*-homomorphisms. If* $t-1 : A \to A$ *is onto,* B *is finitely generated over* Λ*, and* $D_{(t-1)} = \{0\}$*, then* $B_{(t-1)} \cong C_{(t-1)}$.

Proof. It is clear that $\beta(B_{(t-1)}) \subset C_{(t-1)}$, so we will show that $\beta|B_{(t-1)} : B_{(t-1)} \to C_{(t-1)}$ is an isomorphism.

Since B is finitely generated over Λ, there exists one number n such that $(t-1)^n w = 0$ for every $w \in B_{(t-1)}$. Suppose $x \in B_{(t-1)}$ and $\beta(x) = 0$. By exactness there exists $y \in A$ such that $\alpha(y) = x$. Since $t-1 : A \to A$ is onto, there exists $z \in A$ such that $y = (t-1)^n z$. Thus $x = \alpha(y) = (t-1)^n \alpha(z)$. Since $x \in B_{(t-1)}$, this last equation implies that $\alpha(z) \in B_{(t-1)}$. Hence $(t-1)^n \alpha(z) = 0$. It follows that $x = 0$ and so $\beta|B_{(t-1)}$ is one-to-one.

Fix $y \in C_{(t-1)}$. Then $\gamma(y) \in D_{(t-1)} = \{0\}$. Hence there exists $x \in B$ such that $\beta(x) = y$. Since $y \in C_{(t-1)}$, there exists an m such that $(t-1)^m y = 0$. Hence $\beta((t-1)^m x) = (t-1)^m \beta(x) = (t-1)^m y = 0$. By exactness there is $z \in A$ such that $\alpha(z) = (t-1)^m x$. The fact that $t-1 : A \to A$ is onto means that there exists $w \in A$ such that $z = (t-1)^m w$. It follows that $(t-1)^m x = \alpha(z) = (t-1)^m \alpha(w)$ and thus $(t-1)^m (x - \alpha(w)) = 0$. Therefore $(x - \alpha(w)) \in B_{(t-1)}$. Furthermore $\beta(x - \alpha(w)) = \beta(x) - \beta(\alpha(w)) = \beta(x) = y$. Hence $\beta|B_{(t-1)}$ is onto. □

Lemma 6.5.11. *Let* $X \subset Y$ *be polyhedra and let* $\gamma : \pi_1(Y) \to \mathbb{J}$ *be a homomorphism such that* $\gamma|X$ *is onto. If both* $H_1(\widetilde{X}; \mathbb{Q})$ *and* $H_1(\widetilde{Y}, \widetilde{X}; \mathbb{Q})$ *are finitely generated over* Λ *and* $H_1(Y, X; \mathbb{Q}) = \{0\} = H_2(Y, X; \mathbb{Q})$*, then* $k(Y, \gamma) = k(X, \gamma|X)$.

Proof. The Milnor sequence shows that $t-1 : H_i(\widetilde{Y}, \widetilde{X}; \mathbb{Q}) \to H_i(\widetilde{Y}, \widetilde{X}; \mathbb{Q})$ is onto for $i = 1, 2$. Corollary 6.5.9 gives $H_1(\widetilde{Y}, \widetilde{X}; \mathbb{Q})_{(t-1)} = \{0\}$. An application of Lemma 6.5.10 to the exact sequence

$$H_2(\widetilde{Y}, \widetilde{X}; \mathbb{Q}) \to H_1(\widetilde{X}; \mathbb{Q}) \to H_1(\widetilde{Y}; \mathbb{Q}) \to H_1(\widetilde{Y}, \widetilde{X}; \mathbb{Q})$$

establishes that $H_1(\widetilde{X}; \mathbb{Q})_{(t-1)} \cong H_1(\widetilde{Y}; \mathbb{Q})_{(t-1)}$, and Lemma 6.5.7 yields the desired conclusion. □

Proof of Theorem 6.5.6. The theorem follows from two applications of Lemma 6.5.11. □

The next result furnishes an elementary but useful condition under which the Kawauchi invariant dies.

Proposition 6.5.12. *Assume X is connected and $\gamma : \pi_1(X) \to \mathbb{J}$ is an epimorphism such that $H_1(\widetilde{X};\mathbb{Q})$ is finitely generated over Λ. If $H_1(X;\mathbb{Q}) \cong \mathbb{Q}$, then $k(X,\gamma) = 0$.*

Proof. Consider the Milnor Sequence

$$H_1(\widetilde{X};\mathbb{Q}) \xrightarrow{t-1} H_1(\widetilde{X};\mathbb{Q}) \xrightarrow{\alpha} H_1(X;\mathbb{Q}) \xrightarrow{\beta} H_0(\widetilde{X};\mathbb{Q}) \xrightarrow{t-1} H_0(\widetilde{X};\mathbb{Q}).$$

Each of the last three terms in the sequence is isomorphic to $\mathbb{Q}$, and the final homomorphism is zero. This means that β is onto. But every epimorphism of $\mathbb{Q}$ is an isomorphism, so $\alpha = 0$. Hence $t-1 : H_1(\widetilde{X};\mathbb{Q}) \to H_1(\widetilde{X};\mathbb{Q})$ is onto, so Corollary 6.5.9 and Lemma 6.5.7 combine to give the desired conclusion. □

The proofs in the remainder of the section depend on a special form of duality that applies to infinite cyclic covers. This specialized duality is known as *Milnor duality*.

Theorem 6.5.13 (Milnor Duality). *Let W be a compact, connected, n-dimensional PL ∂-manifold and let $p : \widetilde{W} \to W$ be an orientable infinite cyclic cover. If $H_*(\widetilde{W},\partial\widetilde{W};\mathbb{Q})$ is finitely generated as a vector space over $\mathbb{Q}$, then for each $q \geq 0$ the cup product induces a dual pairing*

$$\smile : H^{q-1}(\widetilde{W},\partial\widetilde{W};\mathbb{Q}) \times H^{n-q}(\widetilde{W};\mathbb{Q}) \to H^{n-1}(\widetilde{W},\partial\widetilde{W};\mathbb{Q}) \cong \mathbb{Q}.$$

Moreover, if $x \in H^{q-1}(\widetilde{W},\partial\widetilde{W};\mathbb{Q})$ and $y \in H^{n-q}(\widetilde{W};\mathbb{Q})$, then

$$(tx) \smile (ty) = t(x \smile y) = x \smile y.$$

Let A, B, and C be finite-dimensional vector spaces over $\mathbb{Q}$ with $C \cong \mathbb{Q}$. Recall (Munkres, 1984, page 400) that a bilinear function $f : A \times B \to C$ is called a *dual pairing* if $\dim A = \dim B$ and there exist bases $\{a_1,\ldots,a_m\}$ for A and $\{b_1,\ldots,b_m\}$ for B such that $f(a_i,b_j) = \delta_{ij}\gamma$, where γ is the multiplicative identity in C. Obviously the existence of a dual pairing means that A and B are isomorphic. The fact that the cup product induces a pairing implies the existence of additional structure that will be exploited in forthcoming applications.

One form of the Poincaré-Lefschetz Duality Theorem states that if W is a compact orientable n-dimensional ∂-manifold, then for each $q \geq 0$ the cup product induces a dual pairing[6]

$$H^q(W,\partial W;\mathbb{Q}) \times H^{n-q}(W;\mathbb{Q}) \xrightarrow{\smile} H^n(W,\partial W;\mathbb{Q}) \cong \mathbb{Q}.$$

Hence Duality Theorem 6.5.13 can be paraphrased as saying that an infinite cyclic cover of a compact n-dimensional manifold has the duality properties of a compact $(n-1)$-dimensional manifold (provided the finiteness hypotheses are satisfied). It is helpful to think of the infinite cyclic cover as being homologically like a compact $(n-1)$-dimensional manifold crossed with $\mathbb{R}^1$. In order to make the relationship with ordinary Poincaré duality more transparent, we restate the result:

Corollary 6.5.14. *Let W and $\widetilde{W}$ be as in the statement of Theorem 6.5.13. Then for each $q \geq 0$,*

$$H_q(\widetilde{W},\partial\widetilde{W};\mathbb{Q}) \cong H^{n-q-1}(\widetilde{W};\mathbb{Q}) \text{ and } H^q(\widetilde{W},\partial\widetilde{W};\mathbb{Q}) \cong H_{n-q-1}(\widetilde{W};\mathbb{Q}).$$

This restatement is a corollary of the theorem due to the use of field coefficients. Even though we have only assumed that $H_q(\widetilde{W},\partial\widetilde{W};\mathbb{Q})$ is finitely generated, it follows from the existence of the pairing in Theorem 6.5.13 that $H_*(\widetilde{W},\mathbb{Q})$ is also finitely generated over $\mathbb{Q}$. Thus $H_i(\widetilde{W};\mathbb{Q}) \cong H^i(\widetilde{W};\mathbb{Q})$ and $H_i(\widetilde{W},\partial\widetilde{W};\mathbb{Q}) \cong H^i(\widetilde{W},\partial\widetilde{W};\mathbb{Q})$ for every i (Munkres, 1984, Corollary 53.6). We illustrate the utility of Corollary 6.5.14 by using it to complete the proof of Lemma 6.4.5.

Proof of Lemma 6.4.5, Part 2. Let L, $E(L)$, $M(L)$, γ_k, $\widetilde{E}_k(L)$, $\widetilde{M}_k(L)$, and $\overline{M(L) \setminus E(L)} = N_1 \cup \cdots \cup N_k$ be as in §6.4.

By hypothesis $H_1(\widetilde{E}_k(L);\mathbb{Q})$ is finite-dimensional over $\mathbb{Q}$. In addition, $H_q(\widetilde{E}_k(L);\mathbb{Q})$, $q \geq 2$, is finite-dimensional over $\mathbb{Q}$ by the Milnor sequence and Lemma 6.5.8. Each ∂N_i lifts to a copy $\partial\widetilde{N}_i$ of $S^1 \times \mathbb{R}^1$ in $\widetilde{E}_k(L)$, so

$$H_1(\partial\widetilde{E}_k(L)) \cong \oplus_{i=1}^k H_1(\partial\widetilde{N}_i) \cong \mathbb{Q}^k \cong [\Lambda/(t-1)]^k$$

and $H_q(\partial\widetilde{E}_k(L);\mathbb{Q}) = 0$ for $q \geq 2$. Thus $H_*(\partial\widetilde{E}_k(L);\mathbb{Q})$ is finite-dimensional over $\mathbb{Q}$. Since both $H_*(\widetilde{E}_k(L);\mathbb{Q})$ and $H_*(\partial\widetilde{E}_k(L);\mathbb{Q})$ are finite-dimensional, we can conclude from the exact sequence of the pair $(\widetilde{E}_k(L),\partial\widetilde{E}_k(L))$ that $H_*(\widetilde{E}_k(L),\partial\widetilde{E}_k(L);\mathbb{Q})$ is finite-dimensional over $\mathbb{Q}$. Hence Corollary 6.5.14 applies to give

$$H_2(\widetilde{E}_k(L);\mathbb{Q}) \cong H^0(\widetilde{E}_k(L),\partial\widetilde{E}_k(L);\mathbb{Q}) \cong \{0\}$$

and

$$H_2(\widetilde{E}_k(L),\partial\widetilde{E}_k(L);\mathbb{Q}) \cong H^0(\widetilde{E}_k(L);\mathbb{Q}) \cong \mathbb{Q} \cong \Lambda/(t-1).$$

---

[6] The formulation of Poincaré duality in terms of cup products and field coefficients is explained in (Hatcher, 2002, pp. 249–251). If $\mathbb{Z}$ coefficients are used, it is necessary to factor out the torsion (Hatcher, 2002, Proposition 3.38) (Munkres, 1984, Theorem 68.1).

The following represents a portion of the exact sequence of the pair $(\widetilde{E}_k(L), \partial\widetilde{E}_k(L))$.

$$H_2(\widetilde{E}_k(L)) \to H_2(\widetilde{E}_k(L), \partial\widetilde{E}_k(L)) \to H_1(\partial\widetilde{E}_k(L)) \xrightarrow{\alpha} H_1(\widetilde{E}_k(L))$$

(All homology groups are assumed to have coefficients in $\mathbb{Q}$.) Exactness implies that $\operatorname{Im}(\alpha) \cong [\Lambda/(t-1)]^k/[\Lambda/(t-1)]$, so Proposition 6.5.5 indicates that the order of $\operatorname{Im}(\alpha)$ is $(t-1)^{k-1}$.

For each i, the generator of $\pi_1(N_i)$ is taken by γ_k to $t \in \mathbb{J}$, so the preimage of N_i in $\widetilde{M}_k(L)$ is $\widetilde{N}_i \cong B^2 \times \mathbb{R}^1$. Hence $H_1(\cup_{i=1}^k \widetilde{N}_i; \mathbb{Q}) \cong \oplus_{i=1}^k H_1(\widetilde{N}_i; \mathbb{Q}) = \{0\}$. By exactness of the following Mayer-Vietoris sequence

$$H_1(\partial\widetilde{E}_k(L);\mathbb{Q}) \xrightarrow{\alpha\oplus 0} H_1(\widetilde{E}_k(L);\mathbb{Q})\oplus\Big(H_1(\cup_{i=1}^k \widetilde{N}_i;\mathbb{Q})\Big) \to H_1(\widetilde{M}_k(L);\mathbb{Q}) \to 0$$

we have $H_1(\widetilde{M}_k(L);\mathbb{Q}) \cong H_1(\widetilde{E}_k(L);\mathbb{Q})/\operatorname{Im}(\alpha)$. The conclusion of the lemma follows from the previous paragraphs and Proposition 6.5.5. □

Milnor duality has many other applications; e.g., it can be used to give a new proof of algebraic unknotting that is valid in all dimensions, including dimension four—see Exercise 6.5.1.

The following notation will be used throughout the proof of the duality theorem. Let W be a compact, connected, n-dimensional PL ∂-manifold and let $p : \widetilde{W} \to W$ be an infinite cyclic cover. Choose K to be a compact, connected PL ∂-manifold contained in $\widetilde{W}$ such that $p(K) = W$. Then

$$\widetilde{W} = \bigcup_{j\in\mathbb{Z}} t^j(K).$$

For each integer r, define

$$N_r = \bigcup_{j\geq r} t^j(K) \text{ and } N'_r = \bigcup_{j\leq -r} t^j(K).$$

Note that

$$H^q_c(\widetilde{W}, \partial\widetilde{W};\mathbb{Q}) = \lim_{r\to\infty} H^q(\widetilde{W}, \partial\widetilde{W}\cup N_r \cup N'_r;\mathbb{Q}).$$

Lemma 6.5.15. *If $H_j(\widetilde{W}, \partial\widetilde{W};\mathbb{Q})$ is finitely generated over $\mathbb{Q}$ for $q-2 \leq j \leq q$, then the coboundary operator induces a natural isomorphism*

$$\phi : H^{q-1}(\widetilde{W}, \partial\widetilde{W};\mathbb{Q}) \xrightarrow{\cong} H^q_c(\widetilde{W}, \partial\widetilde{W};\mathbb{Q}).$$

Proof. First observe that $H_j(\partial\widetilde{W} \cup N_0, \partial\widetilde{W}; \mathbb{Q})$ is finitely generated over $\mathbb{Q}$ for $q-2 \le j \le q$. This follows from the Mayer-Vietoris sequence

$$\begin{aligned} &H_j(\partial\widetilde{W} \cup (N_0 \cap N_0'), \partial\widetilde{W}; \mathbb{Q}) \\ &\qquad \to H_j(\partial\widetilde{W} \cup N_0, \partial\widetilde{W}; \mathbb{Q}) \oplus H_j(\partial\widetilde{W} \cup N_0', \partial\widetilde{W}; \mathbb{Q}) \\ &\qquad\qquad \to H_j(\widetilde{W}, \partial\widetilde{W}; \mathbb{Q}) \end{aligned}$$

since $N_0 \cap N_0'$ is a compact polyhedron. Next consider the exact sequence of the triple $(\widetilde{W}, \partial\widetilde{W} \cup N_0, \partial\widetilde{W})$:

$$H_j(\widetilde{W}, \partial\widetilde{W}; \mathbb{Q}) \to H_j(\widetilde{W}, \partial\widetilde{W} \cup N_0; \mathbb{Q}) \to H_{j-1}(\partial\widetilde{W} \cup N_0, \partial\widetilde{W}; \mathbb{Q}).$$

The first term is finitely generated by hypothesis and the last one is finitely generated by the observation above. Hence $H_j(\widetilde{W}, \partial\widetilde{W} \cup N_0; \mathbb{Q})$ is finitely generated for $q-1 \le j \le q$.

Since $H_j(\widetilde{W}, \partial\widetilde{W} \cup N_0; \mathbb{Q})$ is finitely generated, there exists $s > 0$ such that the inclusion induced homomorphism

$$H_j(\partial\widetilde{W} \cup N_{-s}, \partial\widetilde{W} \cup N_0; \mathbb{Q}) \to H_j(\widetilde{W}, \partial\widetilde{W} \cup N_0; \mathbb{Q})$$

is onto. It follows from the exact sequence of the triple $(\widetilde{W}, N_{-s} \cup \partial\widetilde{W}, N_0 \cup \partial\widetilde{W})$ that the inclusion induced homomorphism

$$H_j(\widetilde{W}, \partial\widetilde{W} \cup N_0; \mathbb{Q}) \to H_j(\widetilde{W}, \partial\widetilde{W} \cup N_{-s}; \mathbb{Q})$$

is the zero homomorphism. Translating by t^{r+s} gives

$$H_j(\widetilde{W}, \partial\widetilde{W} \cup N_{s+r}; \mathbb{Q}) \xrightarrow{0} H_j(\widetilde{W}, \partial\widetilde{W} \cup N_r; \mathbb{Q}).$$

Hence the dual cohomology homomorphisms have the following property: For $j = q-1$ and $j = q$ there exists an $s > 0$ such that the inclusion induced homomorphism

$$H^j(\widetilde{W}, \partial\widetilde{W} \cup N_r; \mathbb{Q}) \to H^j(\widetilde{W}, \partial\widetilde{W} \cup N_{s+r}; \mathbb{Q})$$

is zero for every r. A similar proof shows

$$H^j(\widetilde{W}, \partial\widetilde{W} \cup N_r'; \mathbb{Q}) \xrightarrow{0} H^j(\widetilde{W}, \partial\widetilde{W} \cup N_{s+r}'; \mathbb{Q}).$$

The conclusion of the lemma now follows by taking a limit with respect to r in the Mayer-Vietoris sequence

$$\begin{aligned} &H^{q-1}(\widetilde{W}, \partial\widetilde{W} \cup N_r; \mathbb{Q}) \oplus H^{q-1}(\widetilde{W}, \partial\widetilde{W} \cup N_r'; \mathbb{Q}) \\ &\qquad \to H^{q-1}(\widetilde{W}, \partial\widetilde{W}; \mathbb{Q}) \xrightarrow{\delta} H^q(\widetilde{W}, \partial\widetilde{W} \cup N_r \cup N_r'; \mathbb{Q}) \\ &\qquad\qquad \to H^q(\widetilde{W}, \partial\widetilde{W} \cup N_r; \mathbb{Q}) \oplus H^q(\widetilde{W}, \partial\widetilde{W} \cup N_r'; \mathbb{Q}). \end{aligned}$$

□

Proof of Theorem 6.5.13. By Poincaré-Lefschetz Duality (Corollary 0.3.2) and Lemma 6.5.15 we have

$$H_{n-q}(\widetilde{W};\mathbb{Q}) \cong H_c^q(\widetilde{W},\partial\widetilde{W};\mathbb{Q}) \cong H^{q-1}(\widetilde{W},\partial\widetilde{W};\mathbb{Q}),$$

so $H_{n-q}(\widetilde{W};\mathbb{Q})$ is finitely generated over $\mathbb{Q}$. Thus $H^{n-q}(\widetilde{W};\mathbb{Q})$ is dual to $H_{n-q}(\widetilde{W};\mathbb{Q})$ and the cup product induces a dual pairing

$$H_c^q(\widetilde{W},\partial\widetilde{W};\mathbb{Q}) \times H^{n-q}(\widetilde{W};\mathbb{Q}) \xrightarrow{\smile} H_c^n(\widetilde{W},\partial\widetilde{W};\mathbb{Q}).$$

Lemma 6.5.15 allows us to replace the first term by $H^{q-1}(\widetilde{W},\partial\widetilde{W};\mathbb{Q})$ and the third one by $H^{n-1}(\widetilde{W},\partial\widetilde{W};\mathbb{Q})$. Explicitly, we have the following commutative diagram:

$$\begin{array}{ccc} H^{q-1}(\widetilde{W},\partial\widetilde{W};\mathbb{Q}) \times H^{n-q}(\widetilde{W};\mathbb{Q}) & \xrightarrow{\smile} & H^{n-1}(\widetilde{W},\partial\widetilde{W};\mathbb{Q}) \\ \downarrow{\scriptstyle \phi\times\mathrm{Id}} & & \downarrow{\scriptstyle\phi} \\ H_c^q(\widetilde{W},\partial\widetilde{W};\mathbb{Q}) \times H^{n-q}(\widetilde{W};\mathbb{Q}) & \xrightarrow{\smile} & H_c^n(\widetilde{W},\partial\widetilde{W};\mathbb{Q}). \end{array}$$

The bottom arrow is a dual pairing by ordinary Poincaré-Lefschetz duality and the vertical arrows are isomorphisms by Lemma 6.5.15, so the top arrow is also a dual pairing.

Let $x \in H^{q-1}(\widetilde{W},\partial\widetilde{W};\mathbb{Q})$, $y \in H^{n-q}(\widetilde{W};\mathbb{Q})$, and $z \in H_0(\widetilde{W};\mathbb{Q})$. Since $\widetilde{W}$ is connected, $H_0(\widetilde{W};\mathbb{Q}) \cong \mathbb{Q}$ and $tz = z$; thus

$$\phi(t(x \smile y))(z) = t(\phi(x \smile y))(z) = \phi(x \smile y)(tz) = \phi(x \smile y)(z).$$

But ϕ is an isomorphism, so we can conclude that $t(x \smile y) = x \smile y$. □

In our main application of Milnor duality, W will be a 4-dimensional ∂-manifold and we will want to apply the duality result to both W itself and to ∂W even though ∂W is not connected. The following addendum spells out the relationship between the two pairings.

Addendum to Theorem 6.5.13. *Assume $n = 2m$, $p : \widetilde{W} \to W$ is induced by the epimorphism $\gamma : \pi_1(W) \to \mathbb{J}$, and ∂W has two components M_1 and M_2 such that the restricted homomorphism $\gamma|M_j$ is an epimorphism for $j = 1, 2$. Then the cup product induces a dual pairing*

$$H^{m-1}(\partial\widetilde{W};\mathbb{Q}) \times H^{m-1}(\partial\widetilde{W};\mathbb{Q}) \xrightarrow{\smile} \mathbb{Q},$$

which is compatible with the pairing

$$H^m(\widetilde{W},\partial\widetilde{W};\mathbb{Q}) \times H^{m-1}(\widetilde{W};\mathbb{Q}) \xrightarrow{\smile} H^{2m-1}(\widetilde{W},\partial\widetilde{W};\mathbb{Q}) \cong \mathbb{Q}$$

of the theorem in the following sense: If

$$H^{m-1}(\widetilde{W};\mathbb{Q}) \xrightarrow{i^*} H^{m-1}(\partial\widetilde{W};\mathbb{Q}) \xrightarrow{\delta} H^m(\widetilde{W},\partial\widetilde{W};\mathbb{Q})$$

is part of the exact sequence of the pair $(\widetilde{W}, \partial\widetilde{W})$, $x \in H^{m-1}(\widetilde{W};\mathbb{Q})$, $y \in H^{m-1}(\partial\widetilde{W};\mathbb{Q})$, *and* $z \in H_{2m-1}(\widetilde{W}, \partial\widetilde{W};\mathbb{Q})$, *then*

$$(\delta(y) \smile x)(z) = (y \smile i^*(x))(\partial z).$$

Proof of the Addendum. Let $\widetilde{M}_j = p^{-1}(M_j)$. Since $H_*(\widetilde{W}, \partial\widetilde{W};\mathbb{Q})$ is finitely generated over $\mathbb{Q}$, the proof of Theorem 6.5.13 shows that $H_*(\widetilde{W};\mathbb{Q})$ is finitely generated as well. It follows from the exact sequence of the pair $(W, \partial W)$ that $H_*(\partial\widetilde{W};\mathbb{Q})$ is finitely generated. As $H_*(\partial\widetilde{W};\mathbb{Q})$ is naturally isomorphic to $H_*(\widetilde{M}_1;\mathbb{Q}) \oplus H_*(\widetilde{M}_2;\mathbb{Q})$, each $H_*(M_j;\mathbb{Q})$ must be finitely generated. Thus Theorem 6.5.13 applies to the restricted cover $\widetilde{M}_j \to M_j$ and the cup product induces a pairing

$$H^{m-1}(\widetilde{M}_j;\mathbb{Q}) \times H^{m-1}(\widetilde{M}_j;\mathbb{Q}) \xrightarrow{\smile} H^{2m-2}(\widetilde{M}_j;\mathbb{Q}) \cong \mathbb{Q}.$$

Now $H^{m-1}(\partial\widetilde{W};\mathbb{Q}) \cong H^{m-1}(\widetilde{M}_1;\mathbb{Q}) \oplus H^{m-1}(\widetilde{M}_2;\mathbb{Q})$ and the two subspaces are orthogonal relative to the cup product, so the two separate pairings combine to produce a dual pairing

$$H^{m-1}(\partial\widetilde{W};\mathbb{Q}) \times H^{m-1}(\partial\widetilde{W};\mathbb{Q}) \xrightarrow{\smile} \mathbb{Q}.$$

Naturality of cup products implies that the diagram

$$\begin{array}{ccccc} H^{m-1}(\widetilde{W};\mathbb{Q}) & \xrightarrow{i^*} & H^{m-1}(\partial\widetilde{W};\mathbb{Q}) & \xrightarrow{\delta} & H^{m}(\widetilde{W},\partial\widetilde{W};\mathbb{Q}) \\ \downarrow{\smile x} & & \downarrow{\smile i^*(x)} & & \downarrow{\smile x} \\ H^{2m-2}(\widetilde{W};\mathbb{Q}) & \longrightarrow & H^{2m-2}(\partial\widetilde{W};\mathbb{Q}) & \xrightarrow{\delta} & H^{2m-1}(\widetilde{W},\partial\widetilde{W};\mathbb{Q}) \end{array}$$

is commutative (Munkres, 1984, Exercise 48.2), so

$$(\delta(y) \smile x)(z) = \delta(y \smile i^*(x))(z) = (y \smile i^*(x))(\partial z)$$

for each $y \in H^{m-1}(\partial\widetilde{W};\mathbb{Q})$, and $z \in H_{2m-1}(\widetilde{W}, \partial\widetilde{W};\mathbb{Q})$. □

The next theorem, which is the second major result of the section, specifies another way in which the structure of the Alexander polynomial is preserved by homology cobordism. It is a generalization of the familiar fact that the Alexander polynomial of a slice knot in S^3 has the form $f(t)f(t^{-1})$ (Rolfsen, 1990, Theorem 8.20).

Theorem 6.5.16. *Suppose* M_1 *and* M_2 *are compact connected 3-manifolds and that* $\gamma_1 : \pi_1(M_1) \to \mathbb{J}$ *and* $\gamma_2 : \pi_1(M_2) \to \mathbb{J}$ *are epimorphisms such that* $H_1(\widetilde{M}_1;\mathbb{Q})$ *and* $H_1(\widetilde{M}_2;\mathbb{Q})$ *are finitely generated as vector spaces over* $\mathbb{Q}$. *If* (M_1, γ_1) *is homology cobordant to* (M_2, γ_2) *via a compact orientable* ∂-*manifold, then there exists a polynomial* $f(t) \in \Lambda$ *such that*

$$A(M_1, \gamma_1; t)A(M_2, \gamma_2; t) \doteq f(t)f(t^{-1}).$$

Proof. Let W be a compact, oriented, 4-dimensional ∂-manifold that exhibits the homology cobordism from M_1 to M_2 and let $\gamma : \pi_1(W) \to \mathbb{J}$ be the associated epimorphism extending γ_1 and γ_2. Let $p : \widetilde{W} \to W$ be the infinite cyclic cover determined by γ and define $\widetilde{M}_1 = p^{-1}(M_1)$ and $\widetilde{M}_2 = p^{-1}(M_2)$.

Step 1: *$H_*(\partial\widetilde{W};\mathbb{Q})$ is finitely generated over $\mathbb{Q}$.* First see that $H_*(\widetilde{M}_1;\mathbb{Q})$ is finitely generated over $\mathbb{Q}$. This is so because $H_0(\widetilde{M}_1;\mathbb{Q}) \cong \mathbb{Q}$ and $H_1(\widetilde{M}_1;\mathbb{Q})$ is finitely generated by hypothesis. Ordinary Poincaré duality shows that $H_q(\widetilde{M}_1;\mathbb{Q}) \cong H_c^{3-q}(\widetilde{M}_1;\mathbb{Q})$. For $q \geq 2$ Lemma 6.5.15 yields $H_c^{3-q}(\widetilde{M}_1;\mathbb{Q}) \cong H^{2-q}(\widetilde{M}_1;\mathbb{Q})$ and it follows that

$$H_q(\widetilde{M}_1;\mathbb{Q}) \cong H^{2-q}(\widetilde{M}_1;\mathbb{Q}) \cong H_{2-q}(\widetilde{M}_1;\mathbb{Q})$$

is finitely generated over $\mathbb{Q}$. In the same way, $H_*(\widetilde{M}_2;\mathbb{Q})$ is finitely generated over $\mathbb{Q}$. Consequently $H_*(\partial\widetilde{W};\mathbb{Q}) \cong H_*(\widetilde{M}_1;\mathbb{Q}) \oplus H_*(\widetilde{M}_2;\mathbb{Q})$ is finitely generated over $\mathbb{Q}$.

Step 2: *$H_*(\widetilde{W};\mathbb{Q})$ and $H_*(\widetilde{W},\partial\widetilde{W};\mathbb{Q})$ are both finitely generated over $\mathbb{Q}$.* Consider the exact sequence

$$H_q(\partial\widetilde{W};\mathbb{Q}) \to H_q(\widetilde{W};\mathbb{Q}) \to H_q(\widetilde{W},\partial\widetilde{W};\mathbb{Q}) \to H_{q-1}(\partial\widetilde{W};\mathbb{Q}).$$

The first and last terms are finitely generated by the previous paragraph. The inclusion-induced homomorphism $H_q(\widetilde{W};\mathbb{Q}) \to H_q(\widetilde{W},\partial\widetilde{W};\mathbb{Q})$ can be factored through $H_q(\widetilde{W},\widetilde{M}_1;\mathbb{Q})$, which is finitely generated by Corollary 6.5.9. Hence the image of $H_q(\widetilde{W};\mathbb{Q}) \to H_q(\widetilde{W},\partial\widetilde{W};\mathbb{Q})$ is finitely generated. It follows that $H_q(\widetilde{W},\partial\widetilde{W};\mathbb{Q})$ is finitely generated over $\mathbb{Q}$. But then $H_q(\widetilde{W};\mathbb{Q})$ is trapped between two finitely generated terms in the exact sequence and is itself finitely generated over $\mathbb{Q}$.

Step 3: *Apply Duality Theorem 6.5.13 and its Addendum.* Steps 1 and 2 allow the application of Duality Theorem 6.5.13 and its Addendum to $\widetilde{W}$ and $\partial\widetilde{W}$. Hence the cup product operation induces dual pairings

$$H^1(\partial\widetilde{W};\mathbb{Q}) \times H^1(\partial\widetilde{W};\mathbb{Q}) \xrightarrow{\smile} \mathbb{Q}$$

and

$$H^2(\widetilde{W},\partial\widetilde{W};\mathbb{Q}) \times H^1(\widetilde{W};\mathbb{Q}) \xrightarrow{\smile} H^3(\widetilde{W},\partial\widetilde{W};\mathbb{Q}) \cong \mathbb{Q}.$$

The exact sequence

$$H^1(\widetilde{W};\mathbb{Q}) \xrightarrow{i^*} H^1(\partial\widetilde{W};\mathbb{Q}) \xrightarrow{\delta} H^2(\widetilde{W},\partial\widetilde{W};\mathbb{Q})$$

relates these pairings as explained in the addendum.

Before taking the next step in the proof, we need a definition. The *orthogonal complement* of a subspace A of $H^1(\partial W;\mathbb{Q})$ is the set

$$A^\perp = \{y \in H^1(\partial W;\mathbb{Q}) \mid y \smile a = 0 \text{ for every } a \in A\}.$$

Observe that $\dim_{\mathbb{Q}} A + \dim_{\mathbb{Q}} A^\perp = \dim_{\mathbb{Q}}(H^1(\partial\widetilde{W};\mathbb{Q}))$.

Step 4: *The subspace* $\operatorname{Im} i^*$ *is self-orthogonal.* Suppose $(y \smile i^*(x))(\partial z) = 0$ for every $x \in H^1(\widetilde{W};\mathbb{Q})$ and for every $z \in H_3(\widetilde{W},\partial\widetilde{W};\mathbb{Q})$. Then, by the addendum, $(\delta(y) \smile x)(z) = 0$ for every $x \in H^1(\widetilde{W};\mathbb{Q})$ and for every $z \in H_3(\widetilde{W},\partial\widetilde{W};\mathbb{Q})$. Because the pairing is nonsingular, $\delta(y) = 0$, which, in turn, means that $y \in \operatorname{Im} i^*$. Hence $(\operatorname{Im} i^*)^\perp \subset \operatorname{Im} i^*$. Elementary linear algebra then shows that $(\operatorname{Im} i^*)^\perp = \operatorname{Im} i^*$ and

$$\begin{aligned}\dim_{\mathbb{Q}}(\operatorname{Im} i^*) &= \frac{1}{2}\dim_{\mathbb{Q}}(H^1(\partial\widetilde{W};\mathbb{Q}))\\ &= \dim_{\mathbb{Q}}(H^1(\partial\widetilde{W};\mathbb{Q})/\operatorname{Im} i^*)\\ &= \dim_{\mathbb{Q}}(\operatorname{Im}\delta).\end{aligned}$$

In particular, $\operatorname{Im} i^*$ is isomorphic (as a vector space) to $H^1(\partial\widetilde{W};\mathbb{Q})/\operatorname{Im} i^*$. Furthermore, the proof above shows that the cup product induces a nonsingular pairing

$$\operatorname{Im} i^* \times H^1(\partial\widetilde{W};\mathbb{Q})/\operatorname{Im} i^* \xrightarrow{\smile} \mathbb{Q}.$$

Step 5: *Relate the action of t to the cup product.* Let $x, y \in H^1(\partial\widetilde{W};\mathbb{Q})$. By the "moreover" part of Duality Theorem 6.5.13,

$$tx \smile y = tx \smile t \cdot t^{-1}y = t(x \smile t^{-1}y) = x \smile t^{-1}y.$$

Combining that fact with the $\mathbb{Q}$-linearity of the cup product yields $f(t)x \smile y = x \smile f(t^{-1})y$ for every $f(t) \in \Lambda$.

Step 6: *Show that if $f(t)$ is the order of* $\operatorname{Im} i^*$, *then $f(t^{-1})$ is the order of $H^1(\partial\widetilde{W};\mathbb{Q})/\operatorname{Im} i^*$.* Let $f(t)$ be the order of $\operatorname{Im} i^*$ and $g(t)$ the order of $H^1(\partial\widetilde{W};\mathbb{Q})/\operatorname{Im} i^*$. Fix $y \in H^1(\partial\widetilde{W};\mathbb{Q})/\operatorname{Im} i^*$. Then for every $x \in \operatorname{Im} i^*$,

$$x \smile (f(t^{-1})y) = (f(t)x) \smile y = 0,$$

and the nonsingularity of the pairing implies $f(t^{-1})y = 0$. Thus $g(t)|f(t^{-1})$. But Proposition 6.5.2 gives

$$\deg(g(t)) = \dim_{\mathbb{Q}}(H^1(\partial\widetilde{W};\mathbb{Q})/\operatorname{Im} i^*) = \dim_{\mathbb{Q}}(\operatorname{Im} i^*) = \deg(f(t)),$$

so $g(t) \doteq f(t^{-1})$.

Step 7: Since $H^1(\partial\widetilde{W};\mathbb{Q}) \cong H^1(\widetilde{M}_1;\mathbb{Q}) \oplus H^1(\widetilde{M}_2;\mathbb{Q})$, application of Corollary 6.4.3 and Proposition 6.5.5 completes the proof of the theorem. □

Historical Notes. Theorems 6.5.6 and 6.5.16 are both due to Kawauchi (1978), (1980). The invariant used in Theorem 6.5.6 is named for Kawauchi because of the prominent role this particular invariant played in (Kawauchi, 1980). Duality Theorem 6.5.13 is due to Milnor (1968). Kawauchi (1974b), (1977) has generalized this result; some of Kawauchi's work is based on earlier work of Levine, (1966) and (1969).

Exercises

6.5.1. Use Milnor duality to give a proof of the algebraic unknotting principle (Proposition 6.1.4) that is valid in all even dimensions, including dimension four.

6.5.2. Prove that the hypothesis $H_1(\widetilde{E}_k(L); \mathbb{Q}) = T_1(\widetilde{E}_k(L); \mathbb{Q})$ in the second part of Lemma 6.4.5 is always satisfied in case L is a knot (i.e., if $k = 1$).

6.6. A topological embedding that cannot be approximated by PL embeddings

Both PL approximation and locally flat approximation fail in codimension two. In this section we construct an example of a topological embedding of the 2-torus $T^2 = S^1 \times S^1$ into S^4 that cannot be approximated by PL embeddings. Later in the chapter we will see that this topological embedding is so wild that it also cannot be approximated by locally flat embeddings.

Example 6.6.1. *There exist a compact PL ∂-manifold $N \subset S^4$ and a topological embedding $g : T^2 \to N$ of the 2-torus T^2 in N such that g is a homotopy equivalence but g is not homotopic in N to a PL embedding. Hence g cannot be approximated by PL embeddings.*

The particular example also shows that the codimension-three theorem about highly connected maps between manifolds being homotopic to PL embeddings (Theorem 5.2.1) fails in codimension two. In the next section we will see even more dramatic examples of this failure. There the manifolds are simply connected and have arbitrary dimension (although the target is non-compact, unlike in 6.6.1) while the map between them is a homotopy equivalence that is homotopic to no embedding whatsoever.

In this section and the next we will provide counterexamples to the basic approximation and existence of embedding results in codimension two, but we will not do a systematic study of conditions under which embeddings exist. It should be pointed out, however, that such a study can be carried out, at least if one restricts attention to the compact case. In particular, the following problem has been studied extensively by many authors: if $f : Q^{n-2} \to M^n$ is a homotopy equivalence from a compact PL $(n-2)$-manifold Q to a compact n-dimensional PL ∂-manifold M, then is f homotopic to a locally flat PL embedding? Cappell and Shaneson (1974), (1976), and (1977), Kato (1970), Kato and Matsumoto (1972), and Matsumoto (1973), (1975a), (1975b), and (1979b) have developed codimension-two ambient surgery theories that answer this question in case $n \geq 6$. There is an obstruction that lives in a surgery group associated with a certain cyclic extension of $\pi_1(M)$. The obstruction vanishes if and only if f is homotopic to a locally flat PL embedding. It is interesting to note that, in case M is simply connected, the obstruction group is the same as the knot cobordism group and so a (not necessarily locally flat) PL embedding always exists (Kato and Matsumoto, 1972, Theorem A) or (Cappell and Shaneson, 1976, Theorem 6.1). In case n is odd, the obstruction group is the same as the ordinary surgery obstruction group (Kato and Matsumoto, 1972, Theorem C) or (Matsumoto, 1973, Theorem 5.11 (i)) and again a (possibly nonlocally flat) embedding always exists provided both Q and M are orientable (Cappell and Shaneson, 1976, Theorem 6.1).

One consequence of the results just quoted is that it will be difficult to find a counterexample to PL approximation in odd dimensions. If $f : Q^{n-2} \to M^n$, n odd, is a topological embedding and $f(Q)$ is contained in a compact PL ∂-manifold neighborhood P such that $f(Q) \hookrightarrow P$ is a homotopy equivalence, then f is at least homotopic to a PL embedding in P. It is not known whether topological embeddings of odd-dimensional PL manifolds can be PL approximated in general; any counterexample would either have to be so wild that its image cannot be homotopically captured in a PL neighborhood or would require an invariant sensitive enough to register that the embedding cannot be closely approximated by PL embeddings even though it is homotopic to one.

The even-dimensional situation is clearer: high-dimensional embeddings that cannot be approximated by PL embeddings arise upon taking Cartesian products of our 4-dimensional example with $\mathbb{C}\mathrm{P}^2$.

Addendum (to Example 6.6.1). *Let Q denote a finite product of copies of $\mathbb{C}\mathrm{P}^2$. The topological embedding $g \times \mathrm{Id}_Q : T^2 \times Q \to N \times Q$ cannot be approximated by PL embeddings.*

We do not prove this addendum, as it requires much deeper techniques than those developed here. In order to address it, one would start by using the obstruction theory of Matsumoto *et al.*, mentioned in the last section, to prove that the topological embedding g in Example 6.6.1 cannot be approximated by PL embeddings. That entails showing that the embedding g corresponds to a nonzero element of the 4-dimensional obstruction group (which is done in (Matsumoto, 1975a)). The addendum is then a consequence of the periodicity of the obstruction groups (Matsumoto, 1973, Theorem 5.12).

The following definition plays a central role in the geometric constructions of this section.

Definition. For any map $f : X \to X$, the *mapping torus of f* is the identification space $T(f) = X \times [0,1]/\sim$, where the identification $\sim$ is defined by $\langle x, 0\rangle \sim \langle f(x), 1\rangle$ for every $x \in X$.

Construction of the ∂-manifold N. Start with the solid torus $S^1 \times B^2$. Let Γ_0 be the centerline $S^1 \times \{0\}$ and let Γ_1 be the curve in $S^1 \times B^2$ that is drawn in Figure 6.20.

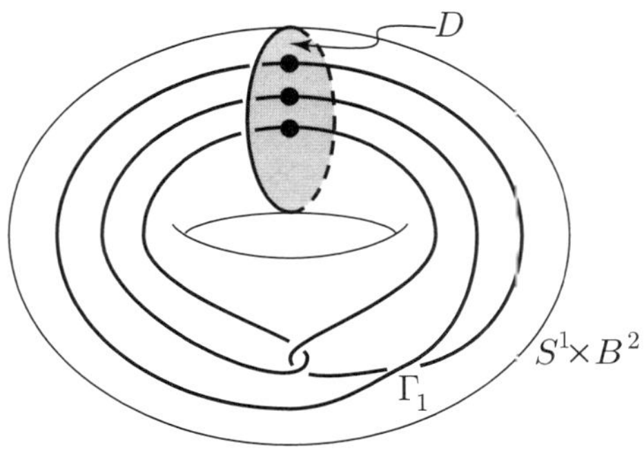

Figure 6.20. The curve Γ_1

Choose a regular neighborhood R_1 of Γ_1 in $S^1 \times B^2$ and an orientation-preserving homeomorphism $f : S^1 \times B^2 \to R_1$. The homeomorphism f should be chosen in such a way that f maps Γ_0 to Γ_1 and f maps each of the boundary circles $S^1 \times \{*\}$ to a longitude of R_1 having homological linking number 0 with Γ_1. Then N is defined to be the mapping torus of f.

It is clear that N is a compact, 4-dimensional PL ∂-manifold. Since Γ_1 is homotopic to Γ_0 in $S^1 \times B^2$, N has the homotopy type of T^2. Note that it is possible to construct N as a PL subset of $\mathbb{R}^4$. The basic reason is that Γ_1 is the unknot when considered as a subset of $\mathbb{R}^3$. We can use the unknotting isotopy to find a level-preserving embedding of $S^1 \times B^2 \times [1,2]$ into $\mathbb{R}^3 \times [1,2] \subset \mathbb{R}^4$ whose image intersects $\mathbb{R}^3 \times \{2\}$ in a round solid torus and intersects $\mathbb{R}^3 \times \{1\}$ in $R_1 \times \{1\}$. The union of this embedded copy of $S^1 \times B^2 \times [1,2]$ with $S^1 \times B^2 \times [0,1]$ is the heart of N. To complete the embedding of N, swing the bottom solid torus $S^1 \times B^2 \times \{0\}$ through a circle of large radius in $\mathbb{R}^4$ until it matches $S^1 \times B^2 \times \{2\}$. This completes a PL embedding of N into $\mathbb{R}^4$ (and hence into S^4). □

The important properties of the example are spelled out in the next two theorems.

Theorem 6.6.2. *There exists a topological embedding $g : T^2 \to N$ such that g is a homotopy equivalence.*

Theorem 6.6.3. *There exists no PL embedding $h : T^2 \to N$ such that h is a homotopy equivalence.*

Since any close approximation to the homotopy equivalence g must also be a homotopy equivalence, it will follow that g cannot be approximated by PL embeddings. We will prove Theorem 6.6.2 directly by constructing the topological embedding. Of course the proof of the nonexistence of a PL embedding that is asserted in Theorem 6.6.3 requires the use of an algebraic invariant. As indicated earlier, the algebraic invariant will be the Alexander polynomial.

Proof of Theorem 6.6.2. It suffices to construct an annulus $A \subset S^1 \times B^2 \times [0,1]$ such that $A \cap \partial(S^1 \times B^2 \times [0,1]) = \partial A$ and the two components of ∂A are $\partial^+ A = \Gamma_1 \times \{1\}$ and $\partial^- A = \Gamma_0 \times \{0\}$. In the next paragraph we describe a "shift-spin" construction used to build A.

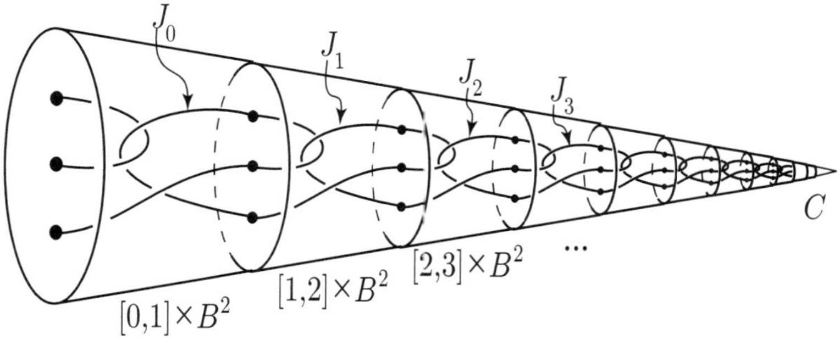

Figure 6.21. The 3-cell C contains a wild Fox-Artin arc

Cut the solid torus $S^1 \times B^2$ open along the disk D shown in Figure 6.20 to form $[0,1] \times B^2$. The disk D intersects Γ_1 in three points and Γ_1 is cut into three arcs in $[0,1] \times B^2$. We use J_0 to denote their union. Let C be the one-point compactification of $\cup_{i=0}^{\infty} \left([i, i+1] \times B^2\right) = [0, \infty) \times B^2$. Note that $C = ([0,\infty) \times B^2) \cup \{\infty\}$ is the 3-cell pictured in Figure 6.21.

Let J_i denote the translate of J_0 in $[i, i+1]$ and let $J = (\cup_{i=0}^{\infty} J_i) \cup \{\infty\} \subset C$. Then J is the union of two arcs, one of which is a Fox-Artin arc. Let $\tau : C \to C$ be the map that translates each point one unit to the right; i.e., $\tau(r, x) = \langle r+1, x\rangle$ and $\tau(\infty) = \infty$. Now let $T = T(\tau)$ be the mapping torus of τ. Note that τ is a very simple map of C to C and T is homeomorphic to $S^1 \times B^2 \times [0,1]$. Furthermore, $\tau(J) \subset J$ and $A = T(\tau|J) \subset T(\tau) = T$ is an annulus. As shown in the schematic diagram Figure 6.22, one boundary component of A is a copy of Γ_1 and the other boundary component of A is a copy of Γ_0.

Note that A is locally flat except along the boundary component $\partial^- A$. The embedding g is constructed by identifying the two boundary components of A and hence $g(T^2)$ has a circle of wild points. $\square$

The proof of Theorem 6.6.3 is based on the calculation of certain Alexander polynomials. In order to facilitate those calculations we give an alternate description of N, this time using a handle decomposition rather than the mapping torus structure. There is a useful geometric notation that is used

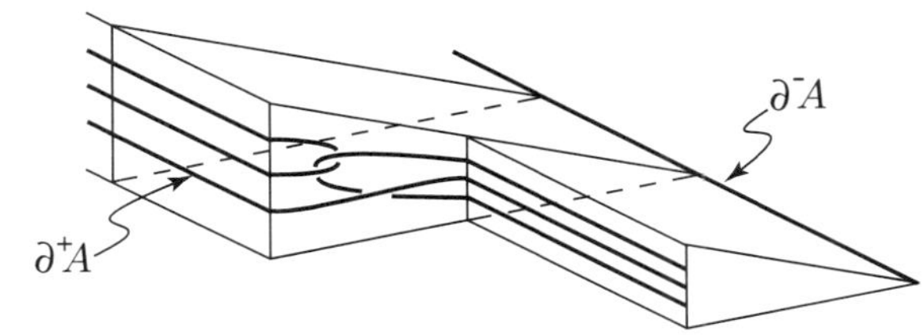

Figure 6.22. A schematic diagram of part of T

to describe ∂-manifolds formed by attaching handles to B^4. The handle decomposition itself is described by a link diagram in S^3 while homeomorphisms of the associated ∂-manifold are described via a set of operations on the diagrams. This method of studying 4-dimensional ∂-manifolds is known as the "link calculus" of Kirby, or simply the "Kirby calculus." Before describing the Kirby calculus itself, let us review handle decompositions of 4-dimensional ∂-manifolds. Handle decompositions in this dimension require special care because there is an issue of framing.

Manifolds constructed by attaching handles to B^4. Suppose ℓ is a PL loop in S^3. We say that *the ∂-manifold W is obtained from B^4 by attaching a 2-handle along ℓ* if

$$W = B^4 \cup_f (B^2 \times B^2),$$

where f is a homeomorphism from $(\partial B^2) \times B^2$ onto a regular neighborhood of ℓ in $S^3 = \partial B^4$ and $f(\partial B^2 \times \{0\}) = \ell$. The *core* of the 2-handle is $B^2 \times \{0\}$ and the *cocore* is $\{0\} \times B^2$. The *framing* of the handle is $\operatorname{lk}(f(\partial B^2 \times \{p\}), \ell)$ for some $p \neq 0$. We will be interested only in the case in which the framing is zero. In that case any one of the curves $f(\partial B^2 \times \{p\})$, $p \in \partial B^2$, is called a *longitude* for ℓ, while any one of the curves $f(\{p\} \times \partial B^2)$, $p \in \partial B^2$, is called a *meridian* of ℓ. If L is a link in S^3, we can similarly define the ∂-manifold obtained from B^4 by attaching one 2-handle to B^4 along each component of the link (with zero framing). In case ℓ is the unknot, we can also use ℓ to remove a 2-handle from B^4. This operation is performed as follows: Start with a PL 2-disk $D^2 \subset S^3$ that has ℓ as its boundary. Push the interior of D^2 into $\operatorname{Int} B^4$ and then remove a relative regular neighborhood of the new disk from B^4. Note that the ∂-manifold obtained by removing a 2-handle from B^4 is homeomorphic to the ∂-manifold obtained by adding a 1-handle to B^4.

The Kirby calculus. Consider a handle decomposition of a compact, connected, 4-dimensional ∂-manifold that has just one 0-handle and no handles of dimension greater than two. Such a handle decomposition can be described via a link in S^3. Each 1-handle is represented by an unknotted circle with a dot on it. This dotted circle is the boundary of a disk that is removed from B^4; as observed above, removing a 2-handle from B^4 is equivalent to adding a 1-handle to B^4. Each 2-handle is represented by a loop that is labeled with an integer. The integer indicates the framing of the 2-handle that is to be attached along the curve. Such link diagrams can then be manipulated by isotopy of the link, handle slides, and the birth or death of (1,2)-handle pairs. The various operations on the diagrams correspond to homeomorphisms of the ∂-manifold. This method of describing 4-dimensional ∂-manifolds and their homeomorphisms is called the *Kirby calculus.*

Second construction of the ∂-manifold N. Let us describe N using a Kirby diagram. The construction of N again starts with $S^1 \times B^2 \times [0,1]$, which is now understood as the union of a 0-handle and a 1-handle. Then the curve Γ_0 in the 0-level is identified with the curve Γ_1 in the 1-level. This can be accomplished in two stages: First add a 1-handle to identify one point of Γ_0 with a point of Γ_1 and then attach a 2-handle to complete the identification. Figure 6.23 shows the Kirby calculus diagram for N (cf. Figure 6.14). The two circles on the left and right represent 1-handles and the curve in the middle is the attaching curve for the 2-handle. Note that the attaching curve for the 2-handle is a band sum of Γ_0 and Γ_1 and that it is unknotted when considered as a subset of S^3. □

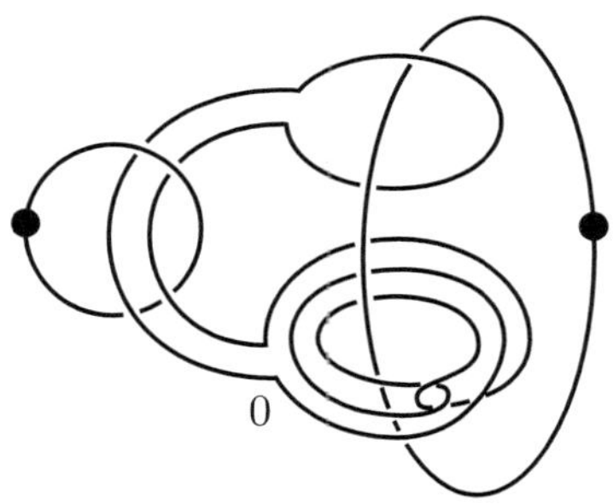

Figure 6.23. A Kirby diagram of N

Proof of a special case of Theorem 6.6.3. We begin the proof of Theorem 6.6.3 by showing that there can be no locally flat PL embedding that is a homotopy equivalence. The added complications associated with non-locally flat points will be faced later.

Suppose there exists a locally flat PL embedding $h : T^2 \to N$ such that h is a homotopy equivalence. We may assume that $h(T^2) \subset \operatorname{Int} N$. Let P be a regular neighborhood of $h(T^2)$ in $\operatorname{Int} N$. Since h is a homotopy equivalence, $\overline{N \smallsetminus P}$ is a homology cobordism from ∂N to ∂P. We will calculate

Alexander polynomials of ∂N and ∂P and then apply Theorem 6.5.16 to obtain a contradiction.

Let L denote the link in Figure 6.23. The boundary of the manifold obtained by adding a 2-handle along an unknotted curve on ∂B^4 is the same as the boundary of the manifold obtained by removing a 2-handle from the interior of B^4. Therefore $\partial N \cong M(L)$, where $M(L)$ is the 3-manifold associated with L that was described in §6.4. By Example 6.4.6, $A(\partial N, \gamma_3; t) \doteq (t-1)^2(2-3t+2t^2)$.

Now $h(T^2)$ has a handle decomposition consisting of one 0-handle, two 1-handles, and a 2-handle. Because h is locally flat, this handle decomposition induces a handle decomposition of P (Rourke and Sanderson, 1972, Corollary 4.14). A Kirby diagram of P is shown in Figure 6.24. Hence $A(\partial P, \gamma_3; t) \doteq (1-t)^2$ by Exercise 6.4.5.

Since $\overline{N \setminus P}$ is a homology product and both $\gamma_3|\pi_1(\partial M)$ and $\gamma_3|\pi_1(\partial P)$ send all three homology generators to t, γ_3 may be extended to an epimorphism $\pi_1(\overline{N \setminus P}) \to \mathbb{J}$. Hence Theorem 6.5.16 implies that there exists a polynomial $f(t) \in \Lambda$ such that $A(\partial N, \gamma_3; t) \cdot A(\partial P, \gamma_3; t) \doteq f(t)f(t^{-1})$. But this is impossible since the number of irreducible factors in $f(t)f(t^{-1})$ is even while the number of irreducible factors in $A(\partial N, \gamma_3; t) \cdot A(\partial P, \gamma_3; t) \doteq (t-1)^4(2-3t+2t^2)$ is five. □

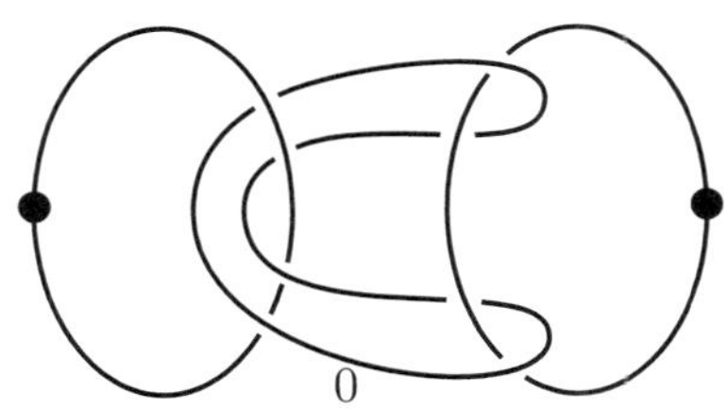

Figure 6.24. A Kirby diagram of P

In order to complete the proof of the general case of Theorem 6.6.3, we must analyze the regular neighborhood of a PL torus with nonlocally flat points. We will amalgamate all the nonlocally flat points into one and then find a decomposition of a regular neighborhood of $h(T^2)$ that is just like the handle decomposition in the preceding proof except that the 2-handle is replaced by a more complicated object.

Lemma 6.6.4. *If $h : T^2 \to S^4$ is a PL embedding, then for every neighborhood U of $h(T^2)$ there exists a PL embedding $h' : T^2 \to U$ such that h' is homotopic to h in U and h' is locally flat except possibly at one point.*

Proof. Suppose $h : T^2 \to S^4$ is a PL embedding. Since h has codimension two, there may be points at which h is not locally flat. It is easy to see from the PL structure that $h(T^2)$ will be locally flat at every point that is not a vertex; hence there are at most a finite number of non-locally flat points. Choose a PL arc $\alpha \subset T^2$ such that α runs through all the vertices at which h is not locally flat. Shrink α to a point in the domain and shrink $h(\alpha)$ to a point in the range; this results in a new PL embedding $h' : T^2 \to S^4$ such that h' has only one point at which it is not locally flat. Furthermore, $h'(T^2)$ can be realized in any arbitrarily small neighborhood of $h(T^2)$ and h' is homotopic to h in this neighborhood. □

Structure of a neighborhood of $h(T^2)$. Let $h : T^2 \to S^4$ be a PL embedding and let P be a regular neighborhood of $h(T^2)$. By Lemma 6.6.4, we may assume that there is a vertex v of T^2 such that h is locally flat modulo v. Let D be a small 2-cell neighborhood of $h(v)$ in $h(T^2)$. Define C to be the preimage of D under the regular neighborhood collapse $P \searrow h(T^2)$. Then C is a 4-cell, $A = \overline{P \setminus C}$ collapses to $S^1 \vee S^1$, and $A \cap C = \partial A \cap \partial C \cong S^1 \times B^2$. So P can be constructed as follows: start with a 4-ball; add two 1-handles (to form A); then attach a 4-cell C by identifying a solid torus in the boundary of A with a solid torus in the boundary of C. The attaching solid torus will always be embedded in ∂A in the same standard way: it will be a regular neighborhood of a commutator of the two 1-handles. The defining curves for the two 1-handles together with the core of the attaching solid torus will form the Borromean rings (Figure 6.24). In ∂C, which is homeomorphic to S^3, the core of the solid torus will represent some knot κ. Since this knot κ determines the PL homeomorphism type of P, let us say that P *has knot type* κ.

Structure of the boundary of a neighborhood of $h(T^2)$. Suppose P has knot type κ. Let L_{BR} denote the Borromean rings. The boundary of P can be constructed as follows: start with S^3, remove disjoint tubular neighborhoods of the three components of L_{BR}, sew two of the neighborhoods back in with longitudes and meridians interchanged, and then sew in the exterior of κ in place of the third (again with meridian and longitude interchanged). Another way to construct ∂P is to start with $M(L_{\mathrm{BR}})$ and then cut out a solid torus in $M(L_{\mathrm{BR}})$ and replace it with $E(\kappa)$, the exterior of κ. Let us use μ to denote the meridian of the third component of the Borromean rings. It is one of the generators of $H_1(\partial P; \mathbb{Q}) \cong \mathbb{Q}^3$.

Proposition 6.6.5. *Suppose $h : T^2 \to S^4$ is a PL embedding and P is a regular neighborhood of $h(T^2)$. Then there exists a polynomial $\Delta(t) \in \Lambda$ such that $A(\partial P, \gamma; t) = (t-1)^2\Delta(t)$ for every $\gamma : \pi_1(\partial P) \to \mathbb{J}$ satisfying $\gamma(\mu) = t$.*

Proof. Suppose P has knot type κ. Regarding κ as a link of one component, we have the epimorphism $\gamma_1 : \pi_1(S^3 - \kappa) \to \mathbb{J}$ that was described in §6.4. Define $\Delta(t) = A(E(\kappa), \gamma_1; t)$.

Let $\gamma : \pi_1(\partial P) \to \mathbb{J}$ be an epimorphism with $\gamma(\mu) = t$ and let L_{BR} denote the Borromean rings. Now ∂P is formed by removing from $M(L_{\mathrm{BR}})$ a solid torus that has μ as a meridian and replacing it with $E(\kappa)$. As a result, the covering space $\widetilde{\partial P}$ is constructed from $\widetilde{M}(L_{\mathrm{BR}})$ by removing a copy of $B^2 \times \mathbb{R}^1$ and replacing it with $\widetilde{E}(\kappa)$. Let us use $\widetilde{B}$ to denote the complement of $\operatorname{Int} B^2 \times \mathbb{R}^1$ in $\widetilde{\partial P}$. We have a Mayer-Vietoris sequence

$$H_1(S^1 \times \mathbb{R}^1) \xrightarrow{\alpha} H_1(\widetilde{B}) \oplus H_1(\widetilde{E}(\kappa)) \to H_1(\widetilde{\partial P}) \to \widetilde{H}_0(S^1 \times \mathbb{R}^1) \cong 0$$

in which α is the zero homomorphism. ($\mathbb{Q}$ coefficients are assumed.) Thus

$$H_1(\widetilde{\partial P}; \mathbb{Q}) \cong H_1(\widetilde{B}; \mathbb{Q}) \oplus H_1(\widetilde{E}(\kappa); \mathbb{Q})$$

and the conclusion of the proposition follows from Exercise 6.4.5. □

Proof of the general case of Theorem 6.6.3. Suppose there exists a PL embedding $h : T^2 \to N$ with h a homotopy equivalence. We may assume that $h(T^2) \subset \operatorname{Int} N$. Let P be a regular neighborhood of $h(T^2)$ in $\operatorname{Int} N$. As before, $\overline{N \setminus P}$ is a homology cobordism from ∂N to ∂P. From Examples 6.4.6 and 6.4.7 we know that there exist epimorphisms $\gamma_1, \gamma_3 : \pi_1(\partial N) \to \mathbb{J}$ such that $A(\partial N, \gamma_1; t) \doteq (t-1)^2$ and $A(\partial N, \gamma_3; t) \doteq (t-1)^2(2 - 3t + 2t^2)$. Since $\overline{N \setminus P}$ is a homology product, both γ_1 and γ_3 extend to $\overline{N \setminus P}$. By Theorem 6.5.16, there exists $f_1(t), f_3(t) \in \Lambda$ such that

$$A(\partial N, \gamma_1|\partial N; t) A(\partial P, \gamma_1|\partial P; t) \doteq f_1(t) f_1(t^{-1})$$

and

$$A(\partial N, \gamma_3|\partial N; t) A(\partial P, \gamma_3|\partial P; t) \doteq f_3(t) f_3(t^{-1}).$$

Both γ_1 and γ_3 map μ to t, so Proposition 6.6.5 implies that there exists $\Delta(t) \in \Lambda$ such that $A(\partial P, \gamma_1|\partial P; t) \doteq (t-1)^2 \Delta(t) \doteq A(\partial P, \gamma_3|\partial P; t)$. Combining all this information gives

$$(2 - 3t + 2t^2) f_1(t) f_1(t^{-1}) \doteq f_3(t) f_3(t^{-1}).$$

But this is impossible because $(2 - 3t + 2t^2)$ is irreducible over $\mathbb{Q}$. □

Historical Notes. The spineless 4-dimensional ∂-manifold in Example 6.6.1 first appeared in (Matsumoto, 1975a) and is usually called the *Matsumoto manifold.* The curve $\Gamma_1 \subset S^1 \times B^2$ is called the *Mazur curve* because it had been used earlier by Mazur (1961a) to construct an interesting contractible, 4-dimensional ∂-manifold. The proof that the Matsumoto manifold has no PL spine is based on a similar argument of Kawauchi (1980), (1978). The curve Γ_2 in Exercise 6.6.1 is called the "false Mazur curve" (Matsumoto, 1975a).

The shift-spin construction is due to C. Giffen (1977). The diagrams of the Giffen annulus used in the section come from (Eaton, 1977) and (Daverman, 1986). The Kirby calculus of links is described in (Kirby, 1978) and (Kirby, 1989). Another good reference is (Gompf and Stipsicz, 1999).

Exercises

6.6.1. Suppose the 4-dimensional ∂-manifold N of Example 6.6.1 is constructed using the curve Γ_2 shown in Figure 6.25 in place of Γ_1. Construct a locally flat PL embedding of T^2 into this new manifold that is a homotopy equivalence. [This shows that the Alexander polynomial detects rather subtle features of Γ_1 since Γ_1 and Γ_2 appear at first glance to have the same properties.]

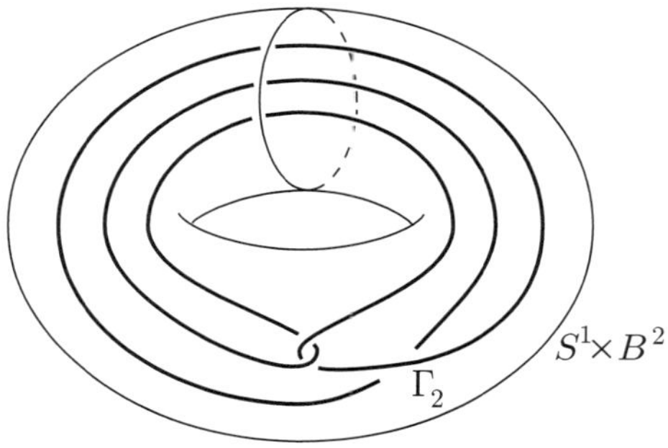

Figure 6.25. The curve Γ_2

6.6.2. Prove that the topological embedding constructed in the proof of Theorem 6.6.2 fails to be 1-alg at each point of the exceptional circle.

6.6.3. Prove that any PL embedding of a connected surface into a PL 4-manifold can be pointwise approximated by a PL embedding that has at most one nonlocally flat point.

6.7. A homotopy equivalence that is not homotopic to an embedding

This section contains examples which show that a highly connected map of manifolds need not be homotopic to any embedding whatsoever in codimension two. Thus the codimension-three results on existence of embeddings do not carry over to codimension two. Quite obviously the first dimension in which such examples can occur is ambient dimension four; hence our basic example is 4-dimensional. We begin with a statement of its properties.

Example 6.7.1. *There exist an open subset W of S^4 and a map $f : S^2 \to W$ such that*

(1) *f is a homotopy equivalence, but*

(2) *f is not homotopic to a topological embedding.*

The example shows that the codimension-three existence of embeddings theorem (Theorem 5.2.1) fails in codimension two. In fact, it fails in the strongest possible way: there is no topological embedding, not even a wild one. The proof of the PL case also shows that there is no homotopy equivalence to a compact subpolyhedron of W, so the codimension-three embedding up to simple-homotopy type result (Theorem 5.1.6) also fails in codimension two.

High-dimensional examples are constructed by taking Cartesian products with spheres.

Addendum. *The map $f \times \mathrm{Id} : S^2 \times S^m \to W \times S^m \subset S^{4+m}$ is not homotopic to a topological embedding if $m \geq 2$.*

The manifolds $W(L)$ and $W_0(L)$. Let $L = \{\ell_1, \ell_2\}$ be a two-component link in S^3, with ℓ_2 the unknot, such that $\mathrm{lk}(\ell_1, \ell_2) = 0$. We will associate two different 4-dimensional ∂-manifolds with L. The first, denoted $W(L)$, is the manifold obtained by attaching two 2-handles to B^4 along L using zero framings. The second, denoted $W_0(L)$, is obtained by attaching a 2-handle to B^4 along ℓ_1 (with zero framing) and then using the unknot ℓ_2 to remove a 2-handle from B^4. Figure 6.26 shows Kirby diagrams for $W(L)$ and $W_0(L)$ in case L is the Whitehead link.

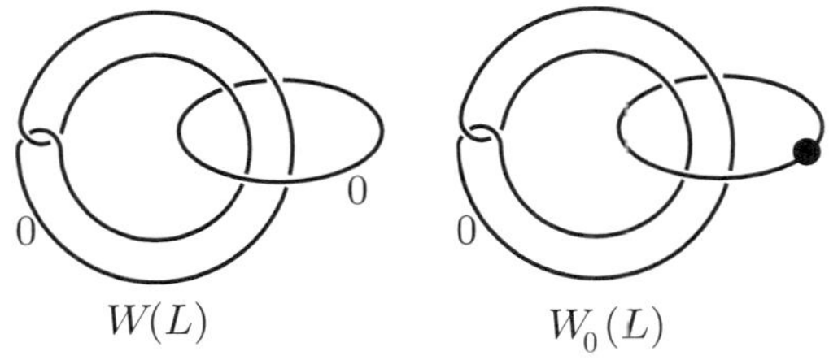

Figure 6.26. Kirby diagrams of $W(L)$ and $W_0(L)$

While $W(L)$ and $W_0(L)$ are very different, they share the same boundary. In fact the common boundary is the manifold $M(L)$ studied in §6.4.

Lemma 6.7.2. $\partial W(L) \cong \partial W_0(L) \cong M(L)$.

Proof. Both $\partial W(L)$ and $\partial W_0(L)$ can be constructed from S^3 as follows: First remove a regular neighborhood $N(L)$ of L from S^3. Note that $N(L)$ is the disjoint union of two solid tori. Second, sew the two components of $N(L)$ back in, but with the longitudinal and meridional curves interchanged. □

There is another simple relationship between $W(L)$ and $W_0(L)$.

Lemma 6.7.3. *There is a natural inclusion of $W_0(L) \subset W(L)$ and*

$$\overline{W(L) \setminus W_0(L)} \cong S^2 \times B^2.$$

Proof. The natural inclusion comes from the fact that $W_0(L)$ is formed by removing a 2-handle from B^4 while $W(L)$ is formed by attaching a 2-handle to B^4 along the very same curve. The two 2-handles fit together to form a copy of $S^2 \times B^2$. The union of the core of the 2-handle and the disk removed from B^4 is a locally flat PL 2-sphere Σ. This is illustrated in Figure 6.27. □

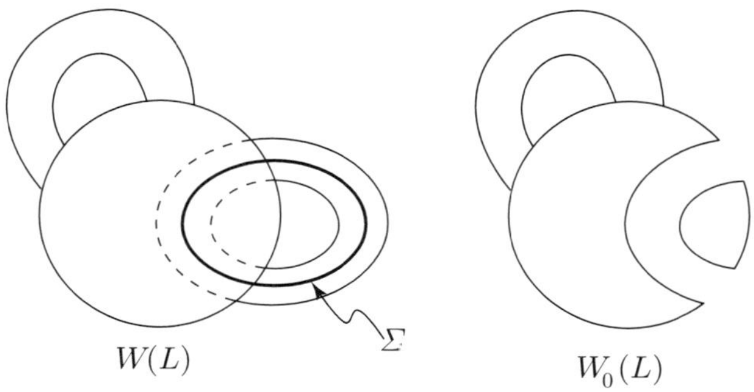

Figure 6.27. $W_0(L)$ is a subset of $W(L)$

Since $W(L)$ is built by attaching two 2-handles to B^4, it is easy to see that $H_1(W(L);\mathbb{Z}) = 0$, and $H_2(W(L);\mathbb{Z}) \cong \mathbb{Z} \oplus \mathbb{Z}$. Because $\operatorname{lk}(\ell_1, \ell_2) = 0$ and ℓ_2 is unknotted, ℓ_1 is homotopically inessential in $S^3 \setminus \ell_2$. Hence $H_1(W_0(L);\mathbb{Z}) \cong \mathbb{Z}$, and $H_2(W_0(L);\mathbb{Z}) \cong \mathbb{Z}$.

The noncompact manifold W of Example 6.7.1 will be constructed as a union of the compact ∂-manifolds $W_0(L_n)$, where the links $\{L_n\}$ are the two component links of Example 6.4.8. We label the components of L_n so that $L_n = \ell_1^n \cup \ell_2^n$ and ℓ_2^n is the round component at the bottom in Figure 6.17. The next proposition explains how to correctly nest the ∂-manifolds $W_0(L_n)$ in S^4.

Proposition 6.7.4. *The manifolds $W_0(L_n)$ can be constructed ambiently in S^4 so that*

(1) $W_0(L_n) \subset \operatorname{Int} W_0(L_{n+1}) \subset W_0(L_{n+1}) \subset S^4$,

(2) *the inclusion induced homomorphism $\pi_1(W_0(L_n)) \to \pi_1(W_0(L_{n+1}))$ is trivial, and*

(3) *the inclusion induced homomorphism* $H_2(W_0(L_n)) \to H_2(W_0(L_{n+1}))$ *is an isomorphism.*

Proof. We proceed inductively to show that $W_0(L_n)$ can be constructed as in the statement of the proposition. Let us begin with $W_0(L_1)$. Notice that each component of L_1 is an unknotted curve in S^3. Thus we can take $B^4 \subset S^4$, remove a neighborhood of a disk bounded by ℓ_2^1 to form the 1-handle, and then realize the 2-handle by adding a neighborhood of a disk in $S^4 \setminus \operatorname{Int} B^4$ that has ℓ_1^1 as its boundary. This construction gives an embedding of $W_0(L_1)$ in S^4.

Next we explain how to attach a 1-handle and a 2-handle to $W_0(L_1)$ in S^4 to form $W_0(L_2)$. First thicken $W_0(L_1)$ by adding a small collar to its boundary in S^4. Then add a 1-handle to $W_0(L_1)$ by removing a neighborhood of a disk whose boundary is the curve m_2 shown in Figure 6.28. The neighborhood should be entirely contained in the collar so that the new manifold contains $W_0(L_1)$ in its interior.

Now attach a 2-handle along the curve m_1 that is also shown in Figure 6.28. This can be done ambiently in S^4 because $\ell_1^1 \cup m_1$ is the unlink on ∂B^4: we find a smooth disk in $S^4 \setminus \operatorname{Int} W_0(L_1)$ that has m_1 as its boundary and add a neighborhood of that disk to our manifold. The result is the addition of a 2-handle along m_1 with zero framing.

We claim that the manifold $W_0(L_1) \cup$ (1-handle) $\cup$ (2-handle) is actually homeomorphic to $W_0(L_2)$. In order to see this, we will cancel the (1,2)-handle pair represented by (ℓ_2^1, m_1). Before cancelling, we perform some handle slides to simplify the picture. First take the lower strand of ℓ_1^1 as it passes through ℓ_2^1 and slide it across the 2-disk on $\partial W_0(L_1)$ that ℓ_2^1 bounds. As shown in the center diagram of Figure 6.28, this frees the strand of ℓ_1^1 from m_1 at the expense of adding a full twist to ℓ_1^1.

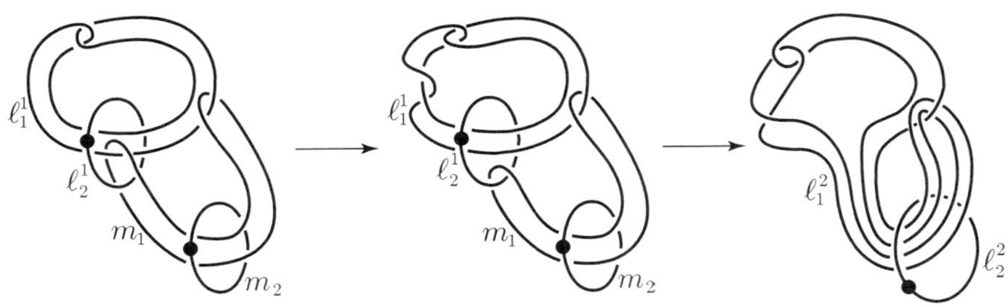

Figure 6.28. First do a handle slide and then cancel a pair of handles

Next slide the 2-handle represented by ℓ_1^1 off the 1-handle represented by ℓ_2^1. This is accomplished by twice sliding the 2-handle attached to ℓ_1^1 over the new 2-handle attached along m_1. We do one handle slide for each of the two strands of ℓ_1^1 that pass through ℓ_2^1. This leaves ℓ_2^1 and m_1 linked to each other, but to nothing else. Thus we can remove the cancelling (1,2)-handle pair that they represent from our picture. The result is shown on the right in Figure 6.28.

Now the new link is isotopic to L_2 so the new manifold actually is $W_0(L_2)$. All the constructions were done ambiently, hence we have $W_0(L_1) \subset W_0(L_2) \subset S^4$. Since ℓ_1^2 is an unknotted curve, the construction can be continued inductively in order to prove the proposition. □

Construction of the example. We can now define the manifold W needed for Example 6.7.1:

$$W = \bigcup_{n=1}^{\infty} W_0(L_n).$$

It follows from Proposition 6.7.4 that $W \subset S^4$ and that W has the homotopy type of S^2. The argument about there being no topological embedding of S^2 into W that is also a homotopy equivalence will occupy the remainder of this section. It will be based on properties of the Alexander polynomial and the Kawauchi invariant developed earlier. The next theorem and its corollary explain the relevance of the Kawauchi invariant to the present situation. For now we will prove only the PL case; later we will generalize the theorem to the topological case.

Theorem 6.7.5. *Suppose $L = \{\ell_1, \ell_2\}$ is a two-component link in S^3 such that $\mathrm{lk}(\ell_1, \ell_2) = 0$ and ℓ_2 is the unknot. If the standard generators of $H_2(W(L); \mathbb{Z})$ can be represented by disjoint PL embedded 2-spheres, then $k(M(L), \gamma_2) = 0$.*

Remark. It is easy to see that each individual element of the standard basis for $H_2(W(L); \mathbb{Z})$ can be represented by a PL embedded 2-sphere. For example, associated with ℓ_i there is a natural PL 2-sphere C_i in $W(L)$; it consists of $B^2 \times \{0\}$, the core of the 2-handle, together with the cone from ℓ_i to the center of B^4. The two 2-spheres intersect (nontransversely) in a single point. Because $\mathrm{lk}(\ell_1, \ell_2) = 0$, the algebraic intersection number of the homology classes, $[C_1] \cdot [C_2]$, is zero. This fact suggests that it should be possible to geometrically separate the two 2-spheres. Despite this algebraic evidence, the 2-spheres cannot in general be separated. The Kawauchi invariant is sensitive enough to detect this.

The following corollary combines Lemma 6.7.3 and Theorem 6.7.5. If one PL 2-sphere can be found in $W_0(L) \subset W(L)$, then the sphere Σ of Lemma 6.7.3 can serve as the second PL 2-sphere in $W(L)$. The corollary will be used to prove that it is not possible to find a PL embedding of the 2-sphere into the manifold W that is a homotopy equivalence.

Corollary 6.7.6. *Suppose $L = \{\ell_1, \ell_2\}$ is a two-component link in S^3 such that $\operatorname{lk}(\ell_1, \ell_2) = 0$ and such that ℓ_2 is the unknot. If the generator of $H_2(W_0(L); \mathbb{Z})$ can be represented by a PL embedded 2-sphere, then $k(M(L), \gamma_2) = 0$.*

Proof of Theorem 6.7.5. Assume Σ_1 and Σ_2 are disjoint PL embedded 2-spheres representing the standard generators of $H_2(W(L); \mathbb{Q})$. We may assume that Σ_1 and Σ_2 are contained in the interior of $W(L)$. Let N_1 and N_2 be disjoint regular neighborhoods of Σ_1 and Σ_2, respectively, in $\operatorname{Int} W(L)$. Define $N = N_1 \cup N_2$ and $Y = \overline{W(L) \smallsetminus N}$.

We claim that $H_j(Y, \partial W(L); \mathbb{Q}) = 0 = H_j(Y, \partial N; \mathbb{Q})$ for $j = 1, 2$. An application of excision yields $H_j(Y, \partial N; \mathbb{Q}) \cong H_j(W(L), N; \mathbb{Q})$, so the fact that $H_j(Y, \partial W(L); \mathbb{Q}) = 0$ for $j = 1, 2$ follows from the exact sequence of the pair $(W(L), N)$. By Alexander duality we have $H_j(Y, \partial W(L); \mathbb{Q}) \cong H^{4-j}(W, N; \mathbb{Q})$; hence the cohomology sequence of the same pair shows that $H_j(Y, \partial W(L); \mathbb{Q}) = 0$ for $j = 1, 2$.

Let $\gamma : \pi_1(\partial W(L)) \to \mathbb{J}$ be the epimorphism γ_2 defined in §6.4. The previous paragraph allows us to uniquely extend γ to $\gamma : \pi_1(Y) \to \mathbb{J}$. By Lemma 6.5.11 we have $k(\partial W(L), \gamma|\partial W(L)) = k(Y, \gamma) = k(\partial N, \gamma|\partial N)$. Thus the proof will be completed by showing that $k(\partial N, \gamma|\partial N) = 0$.

An application of Alexander duality yields $H_1(\overline{S^4 \smallsetminus N_i}; \mathbb{Q}) \cong \mathbb{Q}$, so the Mayer-Vietoris sequence

$$H_2(S^4; \mathbb{Q}) \to H_1(\partial N_i; \mathbb{Q}) \to H_1(N_i; \mathbb{Q}) \oplus H_1(\overline{S^4 \smallsetminus N_i}; \mathbb{Q}) \to H_1(S^4; \mathbb{Q})$$

shows that $H_1(\partial N_i; \mathbb{Q}) \cong \mathbb{Q}$. Proposition 6.5.12 gives $k(\partial N_i, \gamma) = 0$. The fact that $H_i(\partial \widetilde{N}; \mathbb{Q}) \cong H_1(\partial \widetilde{N}_1; \mathbb{Q}) \oplus H_1(\partial \widetilde{N}_2; \mathbb{Q})$ implies $k(\partial N, \gamma) = 0$ as well. □

Remark. The ∂-manifold Y in the last proof does not provide a homology cobordism between $\partial W(L)$ and N because N is not connected. It would be possible to make $N \hookrightarrow Y$ a homology equivalence by connecting N_1 and N_2 with a tube in Y; we choose not to do that, however, because it makes the calculation of $k(\partial N, \gamma)$ more difficult and the proof does not require the full strength of a homology cobordism.

Proof of the PL case of Example 6.7.1. Suppose f were homotopic to a PL embedding $g : S^2 \to W$. By compactness of $g(S^2)$, there would exist an n such that $g(S^2) \subset W_0(L_n)$. Thus the generator of $H_2(W_0(L_n); \mathbb{Z})$ would be represented by a PL embedded 2-sphere. By Corollary 6.7.6, this would mean that $k(M(L_n), \gamma_2) = 0$. But in Example 6.4.8 it was found that $k(M(L_n), \gamma_2) = 2n$. □

Proof of the PL case of the Addendum. Note that $W \subset \mathbb{R}^4$ and $\mathbb{R}^4 \times S^m$ can be embedded in S^{m+4}, so we may assume $W \times S^m \subset S^{m+4}$. Suppose $f \times id$ is homotopic to a PL embedding $g : S^2 \times S^m \to W \times S^m$. By compactness there must exist an n such that $g(S^2 \times S^m) \subset \operatorname{Int} W_0(L_n) \times S^m$. Let N_1 and N_2 be disjoint regular neighborhoods of $g(S^2 \times S^m)$ and the core of $\overline{(W(L_i) \setminus W_0(L_i))} \times S^m \cong S^2 \times S^m \times B^2$, respectively, in $\operatorname{Int} W(L_n) \times S^m$ and define $N = N_1 \cup N_2$. To simplify the notation in the rest of the proof, we use M to denote $W(L_n) \times S^m$ and Y to denote $M \setminus \operatorname{Int} C$. Notice that ∂Y is the disjoint union of ∂M and ∂C.

As in the proof of Theorem 6.7.5, above, we claim that $H_j(Y, \partial M; \mathbb{Z}) = 0$ and $H_j(Y, \partial C; \mathbb{Z}) = 0$ for $j = 1, 2$. Excision, Alexander duality, and the exact sequence of the pair (M, N) combine to yield this conclusion in exactly the same way as in the previous proof. Note that a little extra care is needed in case $m = 2$ since then there are extra terms in $H_2(N; \mathbb{Q})$ and $H_2(M; \mathbb{Q})$.

We now define $\gamma : \pi_1(Y) \to \mathbb{J}$. Since $\partial M = (\partial M(L_i)) \times S^{n-4}$, we may define $\gamma | \partial M$ to be the composition of the homomorphism induced by the projection $\partial M \to \partial M(L_i)$ and the homomorphism $\gamma_2 : \pi_1(\partial M(L_i)) \to \mathbb{J}$ that was defined in §6.4 (and used in the proof of the 4-dimensional case). By the preceding paragraph, $\gamma|\partial M$ extends uniquely to $\gamma : \pi_1(Y) \to \mathbb{J}$. Also using the preceding paragraph, along with Lemma 6.5.11, we see that

$$k(\partial M, \gamma|\partial M) = k(Y, \gamma) = k(\partial N, \gamma|\partial N).$$

But $k(\partial M, \gamma|\partial M) = 2n$ (by Example 6.4.8) and $k(\partial N, \gamma|\partial N) = 0$; this contradiction shows that no PL embedding g can exist. The fact that $k(\partial N, \gamma|\partial N) = 0$ is demonstrated exactly as in the proof of the 4-dimensional case. The fact that $k(\partial M, \gamma|\partial M) = 2n$ follows from the observation that the infinite cyclic cover of ∂M associated with γ is simply the Cartesian product of the cover of $\partial W_0(L_n)$ with S^m. Since $m \geq 2$, the S^m factor does not contribute anything to the first homology. □

The remainder of the section is devoted to a proof of the topological case of Example 6.7.1. The reader who is interested only in the PL case could omit this material. The topological case of Example 6.7.1 follows from the next theorem, which is a topological version of Corollary 6.7.6. It would also be possible to prove a topological version of Theorem 6.7.5, but that would involve even more technical detail and we have no need for the more general theorem. Once Theorem 6.7.7 is established, the proof of the 4-dimensional topological case of Example 6.7.1 is completed in exactly the same way as was the proof of the 4-dimensional PL case.

Theorem 6.7.7. *Suppose $L = \{\ell_1, \ell_2\}$ is a two-component link in S^3 such that $\operatorname{lk}(\ell_1, \ell_2) = 0$ and ℓ_2 is the unknot. If the standard generator of the* *homology group* $H_2(W_0(L);\mathbb{Z})$ *can be represented by a topologically embedded 2-sphere, then* $k(M(L),\gamma_2)=0$.

Suppose $g : S^2 \to W_0(L)$ is a topological embedding representing the standard generator of $H_2(W_0(L);\mathbb{Z})$. Let us use X to denote the set $g(S^2)$. For the remainder of this section, all homology groups are assumed to have coefficients in $\mathbb{Q}$.

Note that X is an ANR. Since X is homeomorphic to S^2 there exists a sequence of connected neighborhoods $N_1, N_2, \dots$ of X in $\operatorname{Int} W_0(L)$ satisfying the following conditions.

(1) $N_{j+1} \subset \operatorname{Int} N_j$ for each $j \geq 1$.

(2) $\cap_{j=1}^{\infty} N_j = X$.

(3) The inclusion induced homomorphism $H_k(N_{j+1}) \to H_k(N_j)$ is zero for $k \neq 0$ or 2.

(4) If $\alpha_j : H_2(N_{j+1}) \to H_2(N_j)$ denotes the inclusion induced homomorphism, then $\operatorname{Im}\alpha_j \cong \mathbb{Q}$.

The fact that the inclusion map $X \hookrightarrow W$ is a homotopy equivalence allows us to impose one more requirement on the neighborhoods.

(5) If $\beta_j : H_2(N_j) \to H_2(W)$ denotes the inclusion induced homomorphism, then $\beta_j|\operatorname{Im}\alpha_j : \operatorname{Im}\alpha_j \to H_2(W)$ is an isomorphism.

We will use α'_j and β'_j to denote the inclusion induced homomorphisms $\alpha'_j : H_1(N_{j+1} \smallsetminus X) \to H_1(N_j \smallsetminus X)$ and $\beta'_j : H_1(N_j \smallsetminus X) \to H_1(W \smallsetminus X)$. Recall that $W_0(L)$ is constructed from B^4 by attaching a 1-handle and a 2-handle. Let us denote the boundary of the cocore of the 2-handle by b_i. Note that b_i is a loop on $\partial W_0(L) \subset W \smallsetminus X$ and that b_i bounds a disk $c_i \subset W_0(L)$. The pair (c_i, b_i) represents a generator of $H_2(W_0(L), \partial W_0(L)) \cong \mathbb{Q}$.

Lemma 6.7.8. *If* α'_j *and* β'_j *are as above, then* $\operatorname{Im}\alpha'_j \cong \mathbb{Q}$ *and* $\beta'_j|\operatorname{Im}\alpha'_j : \operatorname{Im}\alpha'_j \to H_1(W \smallsetminus X)$ *is an isomorphism. Moreover,* b_i *represents a generator of* $H_1(W \smallsetminus X)$.

Proof. Since N_j is a subset of S^4 we see that the inclusion induced homomorphism $H_2(N_j \smallsetminus X) \to H_2(N_j)$ is onto. Thus the long exact sequence of the pair $(N_j, N_j \smallsetminus X)$ shows that $H_2(N_j) \to H_2(N_j, N_j \smallsetminus X)$ is the zero homomorphism. Now consider the following commutative diagram.

$$\begin{array}{ccccccc}
H_2(N_{j+1}) & \xrightarrow{0} & H_2(N_{j+1}, N_{j+1}\setminus X) & \xrightarrow{\partial_{j+1}} & H_1(N_{j+1}\setminus X) & \longrightarrow & H_1(N_{j+1}) \\
\downarrow & & \cong\downarrow & & \downarrow \alpha_j' & & \downarrow 0 \\
H_2(N_j) & \xrightarrow{0} & H_2(N_j, N_j\setminus X) & \xrightarrow{\partial_j} & H_1(N_j\setminus X) & \longrightarrow & H_1(N_j) \\
\downarrow & & \cong\downarrow & & \downarrow \beta_j' & & \downarrow \\
H_2(W) & \xrightarrow{0} & H_2(W, W\setminus X) & \xrightarrow[\cong]{\partial} & H_1(W\setminus X) & \longrightarrow & H_1(W)=0
\end{array}$$

The vertical arrows in the second column are isomorphisms by excision. Each group in the second column is isomorphic to $\check{H}^2(X)\cong\mathbb{Q}$ by Alexander duality. An easy diagram-chasing argument shows that $\operatorname{Im}\alpha_j' = \operatorname{Im}\partial_j \cong \mathbb{Q}$. Essentially the same argument shows that $\operatorname{Im}\beta_j'\circ\alpha_j' = \operatorname{Im}\partial = H_1(W\setminus X)\cong\mathbb{Q}$. Since every onto homomorphism $\mathbb{Q}\to\mathbb{Q}$ is an isomorphism, we have that $\beta_j'|\operatorname{Im}\alpha_j' : \operatorname{Im}\alpha_j' \to H_1(W\setminus X)$ is an isomorphism.

Since $X\hookrightarrow W$ and $W_0(L)\hookrightarrow W$ induce isomorphisms on $\check{H}^2$, $X\hookrightarrow W_0(L)$ does as well. Thus, the horizontal arrow in the lower left corner of the following diagram is an isomorphism.

$$\begin{array}{ccccc}
H_2(W_0(L),\partial W_0(L)) & \longrightarrow & H_2(W_0(L), W_0(L)\setminus X) & \xrightarrow{\cong} & H_2(W, W\setminus X) \\
\cong\downarrow & & \downarrow\cong & & \downarrow\cong \\
H^2(W_0(L)) & \xrightarrow{\cong} & \check{H}^2(X) & = & \check{H}^2(X)
\end{array}$$

The second arrow in the top row is an isomorphism by excision. The vertical arrows are the Alexander and Poincaré duality isomorphisms. The diagram commutes by naturality of duality. Since (c_i, b_i) represents a generator of $H_2(W_0(L),\partial W_0(L))$, it also represents a generator of $H_2(W, W\setminus X)$. But $\partial : H_2(W, W\setminus X)\to H_1(W\setminus X)$ is an isomorphism, so b_i represents a generator of $H_1(W\setminus X)$. □

We now turn our attention to $W(L)$. Let Σ denote the PL 2-sphere which is the core of $\overline{W(L)\setminus W_0(L)}$. Again, in order to simplify notation, we use M to denote $W(L)$. Let A be a PL ray in M that starts at a point of Σ and converges to X. Choose A in such a way that $A\cap\Sigma$ consists of one point and $A\cap X=\emptyset$. Further, choose A so that, for each j, A intersects ∂N_j transversely in exactly one point. In this way we form a compact, connected set $C = X\cup A\cup\Sigma$. By taking the union of N_j with a regular neighborhood of $(A\setminus \operatorname{Int} N_j)\cup\Sigma$ we can form a connected neighborhood P_j of C in M. The sequence of neighborhoods $\{P_j\}$ satisfies the following conditions.

(1) $P_{j+1}\subset \operatorname{Int} P_j$ for each $j\geq 1$.

(2) $\cap_{j=1}^{\infty} P_j = C$.

(3) The inclusion induced homomorphism $H_k(P_{j+1}) \to H_k(P_j)$ is zero for $k \neq 0$ or 2.

(4) If $\phi_j : H_2(P_{j+1}) \to H_2(P_j)$ denotes the inclusion induced homomorphism, then $\operatorname{Im}\phi_j \cong \mathbb{Q} \oplus \mathbb{Q}$.

(5) If $\psi_j : H_2(P_j) \to H_2(M)$ denotes the inclusion induced homomorphism, then $\psi_j|\operatorname{Im}\phi_j : \operatorname{Im}\phi_j \to H_2(M)$ is an isomorphism.

The last three conditions are achieved with the aid of a Mayer-Vietoris sequence.

Lemma 6.7.9. *For every j and for every k, the inclusion induced homomorphism $H_k(M, P_{j+1}) \to H_k(M, P_j)$ is zero.*

Proof. In case $k \neq 2$ or 3, the conclusion follows immediately from the following commutative diagram.

$$\begin{array}{ccccc} 0 = H_k(M) & \longrightarrow & H_k(M, P_{j+1}) & \longrightarrow & \widetilde{H}_{k-1}(P_{j+1}) \\ \| & & \downarrow & & \downarrow 0 \\ 0 = H_k(M) & \longrightarrow & H_k(M, P_j) & \longrightarrow & \widetilde{H}_{k-1}(P_j) \end{array}$$

In case $k = 2$, the argument is nearly the same but we must extend the diagram one place to the left. We use the fact that ψ_j is onto, by Property (5) above.

$$\begin{array}{ccccccc} H_2(P_{j+1}) & \xrightarrow{\psi_{j+1}} & H_2(M) & \longrightarrow & H_2(M, P_{j+1}) & \longrightarrow & H_1(P_{j+1}) \\ \downarrow & & \| & & \downarrow & & \downarrow 0 \\ H_2(P_j) & \xrightarrow{\psi_j} & H_2(M) & \xrightarrow{0} & H_2(M, P_j) & \xrightarrow{\text{1-1}} & H_1(P_j) \end{array}$$

Finally, in case $k = 3$, the argument is a little more delicate. Let θ_j denote the inclusion induced homomorphism $\theta_j : H_3(M, P_{j+1}) \to H_3(M, P_j)$. We have the following commutative diagram:

$$\begin{array}{ccccccc} 0 = H_3(M) & \longrightarrow & H_3(M, P_{j+1}) & \xrightarrow{\partial_{j+1}} & H_2(P_{j+1}) & \xrightarrow{\psi_{j+1}} & H_2(M) \\ \| & & \theta_j \downarrow & & \downarrow \phi_j & & \| \\ 0 = H_3(M) & \longrightarrow & H_3(M, P_j) & \xrightarrow{\partial_j} & H_2(P_j) & \xrightarrow{\psi_j} & H_2(M) \end{array}$$

For any $x \in H_3(M, P_{j+1})$, $\psi_j \partial_j \theta_j(x) = 0$, so $\psi_j \phi_j \partial_{j+1}(x) = 0$. But Property (5) implies that $\psi_j|\operatorname{Im}\phi_j$ is monic. Hence $\phi_j \partial_{j+1}(x) = 0$. By commutativity of the diagram we have $\partial_j \theta_j(x) = 0$. Exactness of the bottom row assures that ∂_j is a monomorphism, so $\theta_j(x) = 0$. $\square$

Lemma 6.7.10. $H_*(M \smallsetminus C, \partial M) = 0$.

Proof. Let M_0 denote the manifold obtained from M by deleting a small open collar on ∂M. Alexander duality gives

$$H_k(M \smallsetminus C, \partial M) \cong H_k(M \smallsetminus C, M \smallsetminus M_0) \cong \check{H}^{4-k}(M_0, C) \cong \check{H}^{4-k}(M, C).$$

Now the definition of Čech cohomology gives

$$\check{H}^{4-k}(M, C) = \lim_{j \to \infty} H^{4-k}(M, P_j).$$

Hence the conclusion follows from Lemma 6.7.9 and the Universal Coefficient Theorem for Cohomology (see (Munkres, 1984, Theorem 53.1), for example). □

Define $\gamma : \pi_1(\partial M) \to \mathbb{J}$ to be the epimorphism γ_2 described in §6.4. Lemma 6.7.10 allows γ to be uniquely extended to $\pi_1(M \smallsetminus C)$. Let $p : \widetilde{M \smallsetminus C} \to M \smallsetminus C$ denote the associated infinite cyclic cover. Recall that for each pair of polyhedra (K, L) in $M \smallsetminus C$, we use the notation $H_k(K, L; \Lambda)$ as a shorthand for the homology group $H_k(p^{-1}(K), p^{-1}(L); \mathbb{Q})$.

Lemma 6.7.11. *For each $j \geq 1$, the inclusion induced homomorphism*

$$a_j : H_1(N_{j+1} \smallsetminus C; \Lambda) \to H_1(N_j \smallsetminus C; \Lambda)$$

satisfies

$$a_j(H_1(N_{j+1} \smallsetminus C; \Lambda)) \subset (t-1)(H_1(N_j \smallsetminus C; \Lambda)).$$

Proof. The only difference between $C \cap N_j$ and $X \cap N_j$ is $A \cap N_j$, which is a 1-dimensional polyhedron, so removing $A \cap N_j$ from the 4-dimensional manifold N_j has no effect on the first homology group. It therefore follows from Lemma 6.7.8 that, for the inclusion induced homomorphism $\alpha''_j : H_1(N_{j+1} \smallsetminus C; \mathbb{Q}) \to H_1(N_j \smallsetminus C; \mathbb{Q})$, $\operatorname{Im} \alpha''_j \cong \mathbb{Q}$. It also follows from the last statement in Lemma 6.7.8 that $p^{-1}(N_j \smallsetminus C)$ is connected for each j, so $H_0(N_j \smallsetminus C; \Lambda) = H_0(p^{-1}(N_j \smallsetminus C); \mathbb{Q}) \cong \mathbb{Q}$. Consider the following commutative diagram in which each row is a portion of a Milnor sequence. (In order to keep the lengths of the rows in the diagram down to a manageable size, we use $N'_j = N_j \smallsetminus C$ and $N'_{j+1} = N_{j+1} \smallsetminus C$.)

$$\begin{array}{ccccccccc}
H_1(N'_{j+1};\Lambda) & \xrightarrow{t-1} & H_1(N'_{j+1};\Lambda) & \xrightarrow{p_{j+1}} & H_1(N'_{j+1};\mathbb{Q}) & \xrightarrow{\delta_{j+1}} & H_0(N'_{j+1};\Lambda) & \xrightarrow[0]{t-1} & H_0(N'_{j+1};\Lambda) \\
{\scriptstyle a_j}\downarrow & & {\scriptstyle a_j}\downarrow & & \downarrow{\scriptstyle \alpha''_j} & & \downarrow{\scriptstyle \cong} & & \downarrow{\scriptstyle \cong} \\
H_1(N'_j;\Lambda) & \xrightarrow{t-1} & H_1(N'_j;\Lambda) & \xrightarrow{p_j} & H_1(N'_j;\mathbb{Q}) & \xrightarrow{\delta_j} & H_0(N'_j;\Lambda) & \xrightarrow[0]{t-1} & H_0(N'_j;\Lambda)
\end{array}$$

Exactness of the top row implies that δ_{j+1} is an epimorphism. Thus $\delta_j \alpha''_j$ is epic and so $\delta_j | \operatorname{Im} \alpha''_j$ is an epimorphism from $\operatorname{Im} \alpha''_j$ to $H_0(N_j \smallsetminus C; \Lambda)$. By the previous paragraph, each of these groups is isomorphic to $\mathbb{Q}$. Since all epimorphisms $\mathbb{Q} \to \mathbb{Q}$ are isomorphisms, we see that $\delta_j | \operatorname{Im} \alpha''_j$ is a monomorphism. Now the composition $\delta_j p_j a_j$ is zero, so $\delta_j \alpha''_j p_{j+1} = 0$. The previous two sentences together imply that $\alpha_j'' p_{j+1} = 0$. Thus $p_j a_j = 0$ and so $\operatorname{Im} a_j \subset \ker p_j = \operatorname{Im}(t-1)$. □

Lemma 6.7.12. *The inclusion induced homomorphisms*

$$b_j : H_1(P_{j+1} \smallsetminus C; \Lambda) \to H_1(P_j \smallsetminus C; \Lambda)$$

and

$$c_j : H_1(M \smallsetminus C, P_{j+1} \smallsetminus C; \Lambda) \to H_1(M \smallsetminus C, P_j \smallsetminus C; \Lambda)$$

satisfy

$$b_j(H_1(P_{j+1} \smallsetminus C; \Lambda)) \subset (t-1)(H_1(P_j \smallsetminus C; \Lambda))$$

and

$$c_j(H_1(M \smallsetminus C, P_{j+1} \smallsetminus C; \Lambda)) \subset (t-1)(H_1(M \smallsetminus C, P_j \smallsetminus C; \Lambda)).$$

Proof. The fact about b_j follows from Lemma 6.7.11 and a Mayer-Vietoris sequence argument. By Lemma 6.7.9 and excision, the inclusion induced homomorphism $H_1(M \smallsetminus C, P_{j+1} \smallsetminus C; \mathbb{Q}) \to H_1(M \smallsetminus C, P_j \smallsetminus C; \mathbb{Q})$ is zero. Thus the second part of the Lemma can be seen from the following commutative diagram in which each row is a Milnor sequence.

$$\begin{array}{ccccc}
H_1(M \smallsetminus C, P_{j+1} \smallsetminus C, \Lambda) & \xrightarrow{t-1} & H_1(M \smallsetminus C, P_{j+1} \smallsetminus C, \Lambda) & \longrightarrow & H_1(M \smallsetminus C, P_{j+1} \smallsetminus C; \mathbb{Q}) \\
\downarrow & & \downarrow{\scriptstyle c_j} & & \downarrow{\scriptstyle 0} \\
H_1(M \smallsetminus C, P_j \smallsetminus C, \Lambda) & \xrightarrow{t-1} & H_1(M \smallsetminus C, P_j \smallsetminus C, \Lambda) & \longrightarrow & H_1(M \smallsetminus C, P_j \smallsetminus C; \mathbb{Q})
\end{array}$$
□

Proof of Theorem 6.7.7. We are assuming that the topological embedding g exists and, from this assumption, wish to derive a contradiction. In view of Lemma 6.7.10, we may apply Lemma 6.5.10 to the pair $(M \smallsetminus C, \partial M)$. Thus $H_1(M \smallsetminus C; \Lambda)_{(t-1)} \cong H_1(\partial M; \Lambda)_{(t-1)}$. In particular, $H_1(M \smallsetminus C; \Lambda)_{(t-1)}$ is finitely generated as a Λ-module. We will derive our contradiction by proving that $H_1(M \smallsetminus C; \Lambda)_{(t-1)} = \{0\}$. By Proposition 6.5.8, it suffices to show that $(t-1) : H_1(M \smallsetminus C; \Lambda) \to H_1(M \smallsetminus C; \Lambda)$ is onto, which will follow from an elementary homological algebra argument involving the following diagram:

$$\begin{array}{ccccccc}
H_1(P_{j+2} \smallsetminus C; \Lambda) & \longrightarrow & H_1(M \smallsetminus C; \Lambda) & \xrightarrow{f_{j+2}} & H_1(M \smallsetminus C, P_{j+2} \smallsetminus C; \Lambda) & \longrightarrow & \widetilde{H}_0(P_{j+2} \smallsetminus C) = 0 \\
\downarrow & & \| & & \downarrow{\scriptstyle c_{j+1}} & & \downarrow \\
H_1(P_{j+1} \smallsetminus C; \Lambda) & \xrightarrow{e_{j+1}} & H_1(M \smallsetminus C; \Lambda) & \xrightarrow{f_{j+1}} & H_1(M \smallsetminus C, P_{j+1} \smallsetminus C; \Lambda) & \longrightarrow & \widetilde{H}_0(P_{j+1} \smallsetminus C) = 0 \\
{\scriptstyle b_j}\downarrow & & \| & & \downarrow & & \downarrow \\
H_1(P_j \smallsetminus C; \Lambda) & \xrightarrow{e_j} & H_1(M \smallsetminus C; \Lambda) & \longrightarrow & H_1(M \smallsetminus C, P_j \smallsetminus C; \Lambda) & \longrightarrow & \widetilde{H}_0(P_j \smallsetminus C) = 0
\end{array}$$

For the sake of completeness we include details. Let f_{j+2}, f_{j+1}, e_{j+1}, and e_j be the indicated homomorphisms in the diagram. Choose $x \in H_1(M \smallsetminus C; \Lambda)$. We must produce $u \in H_1(M \smallsetminus C; \Lambda)$ such that $x = (t-1)u$. By Lemma 6.7.12, there exists $y \in H_1(M \smallsetminus C, P_{j+1} \smallsetminus C; \Lambda)$ such that

$c_{j+1}f_{j+2}(x) = (t-1)y$. Hence $f_{j+1}(x) = (t-1)y$. By exactness, f_{j+1} is onto, so there exists $z \in H_1(M \smallsetminus C; \Lambda)$ for which $f_{j+1}(z) = y$. Now

$$f_{j+1}(x-(t-1)z) = f_{j+1}(x) - (t-1)f_{j+1}(z) = 0,$$

so there is some $w \in H_1(P_{j+1} \smallsetminus C; \Lambda)$ such that $e_{j+1}(w) = x-(t-1)z$. Finally, another application of Lemma 6.7.12 gives $v \in H_1(P_j \smallsetminus C; \Lambda)$ such that $b_j(w) = (t-1)v$. It is simple to check that $u = z + e_j(v)$ satisfies $x = (t-1)u$. □

Historical Notes. The manifold in Example 6.7.1 was originally constructed in (Matsumoto and Venema, 1979), where it is shown that f is not homotopic to any PL embedding. The extension to high dimensions is in (Venema, 1998). The first proof that f is not homotopic to any topological embedding is found in (Liem et al., 1998). Other proofs of the topological case of the 4-dimensional example are found in (Matsumoto, 1979a), (Kojima, 1983), and (Ohkawa, 1982). Theorem 6.7.5 is due to Kawauchi (1980).

Exercise

6.7.1. Prove the topological case of the Addendum to Example 6.7.1.

6.8. Disk bundle neighborhoods and taming

In this final section of the chapter we state two fundamental theorems and point out some of their consequences. We will not prove either of them but include statements in order to round out our study of codimension-two embeddings. The theorems have significant implications for taming and approximation.

The first theorem asserts that locally flat topological embeddings in codimension two have nice neighborhoods.

Definition. Let Q be a k-dimensional manifold topologically embedded in the n-manifold M. A *disk bundle neighborhood* for Q is a closed ∂-manifold neighborhood N of Q in M such that there is a retraction $r : N \to Q$ for which $r^{-1}(x) \cong B^{n-k}$ for each $x \in Q$ and such that N is locally a product.

The last part of the definition means that for each $x \in Q$ there is a neighborhood U of x in Q and a homeomorphism $h_U : r^{-1}(U) \to U \times B^{n-k}$ such that $r|r^{-1}(U) = \pi \circ h_U$, where $\pi : U \times B^{n-k} \to U$ is the projection. As a result, one can assume that $h_U(r^{-1}(U) \cap Q) = U \times \mathbf{0}$.

One of the few ways in which codimension-two embeddings are better behaved than codimension-three embeddings is that locally flat submanifolds always have disk bundle neighborhoods in codimension two. The same is true in codimension one: locally flat, codimension-one embeddings have disk bundle (or, more descriptively, arc bundle) neighborhoods.

Theorem 6.8.1 (Existence of Normal Bundles). *Every locally flat topological $(n-2)$-manifold in an n-manifold has a disk bundle neighborhood.*

This theorem contrasts sharply with known examples in codimension three. For example, (Hirsch, 1968) presents examples of PL embeddings of PL 4-manifolds in S^7 and PL embeddings of S^4 into a PL 7-manifold that have no topological disk bundle neighborhoods.

The next theorem addresses the questions of existence and uniqueness of PL manifold structures on topological manifolds. Although really a codimension-zero theorem, it is included here because of its important consequences for codimension-two embeddings.

Theorem 6.8.2 (PL Structure). *Let M^n be a topological manifold of dimension $n \geq 5$ and let C be a closed subset of M such that some neighborhood of C has a PL structure.*

(1) *If $H^4(M, C; \mathbb{Z}_2) = 0$, then there is a PL structure on M that agrees with the given structure near C.*
(2) *Given one PL structure on M that agrees with the given structure near C, the isotopy classes of PL structures on M that agree with the given one near C are in one-to-one correspondence with the elements of $H^3(M, C; \mathbb{Z}_2)$.*

We present two applications of these theorems to codimension-two embeddings.

Corollary 6.8.3 (Codimension-two Taming). *Suppose $h : Q^{n-2} \to M^n$ is a locally flat topological embedding of a PL $(n-2)$-manifold Q into the PL n-manifold M $(n \geq 5)$. If $H^3(Q; \mathbb{Z}_2) = 0$, then h is isotopic to a PL embedding.*

Proof. Let N be a disk bundle neighborhood of $h(Q)$ in M. The open manifold $\operatorname{Int} N$ has two PL structures: one that is induced from the PL structure on Q and the local product structure on N, and one it inherits as an open subset of M. By Theorem 6.8.2, Part 2, there is an isotopy that pushes the former to the latter. This isotopy pushes h to a PL embedding. □

Corollary 6.8.4 (No locally flat approximation). *The topological embedding g in Example 6.6.1 cannot be approximated by locally flat embeddings.*

Proof. If g could be approximated by locally flat embeddings, then the embedding $g \times \operatorname{Id}$ in the Addendum to Example 6.6.1 could be approximated by locally flat embeddings. But, by Corollary 6.8.3 above, this would make $g \times \operatorname{Id}$ approximable by PL embeddings, contradicting the addendum. □

We close with one final application of the PL Structure Theorem. It does not relate specifically to codimension-two embeddings, but will be important in later chapters.

Theorem 6.8.5 (PL Product Structure). *A topological manifold M^n, $n \geq 5$, has a PL structure if and only if $M \times \mathbb{R}$ has a PL structure. Furthermore, any given PL structure on $M \times \mathbb{R}$ can be realized as the product of some PL structure on M with the obvious structure on $\mathbb{R}$.*

Historical Notes. Theorem 6.8.1 was first stated by Kirby (1970). A corrected proof appears in (Kirby and Siebenmann, 1975). Theorem 6.8.2 is also due to Kirby and Siebenmann; it was first announced in (Kirby and Siebenmann, 1969). The statement quoted here is taken from (Kirby and Siebenmann, 1971). Details of the proof may be found in (Kirby and Siebenmann, 1977). The simply connected case of Theorem 6.8.5 was originally proved by W. Browder (1965).

Chapter 7

Codimension-one Embeddings

The optimal codimension-one results arise in the topological category and, for the most part, involve embeddings of codimension-one manifolds, not complexes, in manifolds. Among the positive aspects, which revolve more around local flatness than around PL approximation or ϵ-tameness, there are three prominent results. The first, developed in §7.3, is a local unknottedness theorem for locally flat approximations to a given embedding: any two sufficiently close, locally flat approximations to a given topological embedding of a compact codimension-one manifold are ambient isotopic, with suitable controls on the isotopy. The second, treated in §7.5 and §7.6, is the characterization of locally flat embeddings of codimension-one manifolds in terms of the 1-LCC condition. The third is the locally flat approximation theorem for manifold embeddings, covered in §7.7.

In addition, §7.1 lays out some elementary separation criteria for codimension-one embedded manifolds. §7.4 presents (a statement of) Edwards's Cell-like Approximation Theorem, and makes preparations for later application of that result. §7.8 touches lightly upon codimension-one analogs of the Kirby-Siebenmann obstruction theory, the codimension-two version of which appears in §6.8. §7.9 presents conditions under which an embedding is 1-LCC. §7.10 treats sewings of crumpled cubes; it gives conditions under which prescribed wildness on two sides of a codimension-one manifold can be welded together in an n-manifold, and along the way it gives some additional examples of wild codimension-one embeddings. §7.11 presents an example of a wildly embedded codimension-one sphere with a manifold mapping cylinder neighborhood, and it establishes that codimension-one embedded manifolds with such mapping cylinder neighborhoods are locally flat if they satisfy an additional freeness condition.

As the chapter progresses, it brings to bear several major theorems whose proofs are beyond the scope of this work. These include: local contractibility of the group of homeomorphisms of a compact manifold in §7.4, the Cell-like Approximation Theorem in §7.4, and the Annulus Theorem in §7.5.

7.1. Codimension-one separation properties

Codimension-one submanifolds locally separate their supermanifolds. Global separation can depend on subtler issues. This section explores how (co)homological data affect global separation.

Proposition 7.1.1. *If M is a connected n-manifold and S is a connected $(n-1)$-manifold embedded in M as a closed subset, then $M \smallsetminus S$ has either one or two components. If S separates M and M' is any connected manifold neighborhood of S in M, then S also separates M'. If, in addition, $H_1(M;\mathbb{Z}_2) \cong 0$, then $M \smallsetminus S$ has two components.*

Proof. The first statement follows from exactness of the sequence

$$H_1(M, M \smallsetminus S; \mathbb{Z}_2) \to \widetilde{H}_0(M \smallsetminus S; \mathbb{Z}_2) \to \widetilde{H}_0(M; \mathbb{Z}_2) \cong 0$$

and the duality-based isomorphism $H_1(M, M \smallsetminus S; \mathbb{Z}_2) \cong H_c^{n-1}(S; \mathbb{Z}_2) \cong \mathbb{Z}_2$. If M' is a connected manifold neighborhood of S in M, then the first vertical arrow in the diagram

$$\begin{array}{ccccc} \mathbb{Z}_2 \cong H_1(M', M' \smallsetminus S; \mathbb{Z}_2) & \longrightarrow & \widetilde{H}_0(M' \smallsetminus S; \mathbb{Z}_2) & \longrightarrow & \widetilde{H}_0(M'; \mathbb{Z}_2) \cong 0 \\ \cong\downarrow & & \downarrow & & \\ \mathbb{Z}_2 \cong H_1(M, M \smallsetminus S; \mathbb{Z}_2) & \longrightarrow & \widetilde{H}_0(M \smallsetminus S; \mathbb{Z}_2) & \longrightarrow & \widetilde{H}_0(M; \mathbb{Z}_2) \cong 0 \end{array}$$

is an isomorphism by excision. When $\widetilde{H}_0(M \smallsetminus S; \mathbb{Z}_2) \cong \mathbb{Z}_2$, commutativity and exactness force $\widetilde{H}_0(M' \smallsetminus S; \mathbb{Z}_2) \cong \mathbb{Z}_2$ as well. In case $H_1(M; \mathbb{Z}_2) \cong 0$, the extended sequence

$$0 \cong H_1(M; \mathbb{Z}_2) \to H_1(M, M \smallsetminus S; \mathbb{Z}_2) \to \widetilde{H}_0(M \smallsetminus S; \mathbb{Z}_2) \to \widetilde{H}_0(M; \mathbb{Z}_2) \cong 0$$

allows us to conclude that $M \smallsetminus S$ has exactly two components. □

Definitions. A connected $(n-1)$-manifold S in an n-manifold M is *two-sided (in M)* if S has a connected neighborhood N_S such that $N_S \smallsetminus S$ is disconnected; otherwise S is *one-sided*. Generally, a disconnected $(n-1)$-manifold in M is *two-sided* there if each of its components is.

Proposition 7.1.1 assures that all compact, codimension-one submanifolds separate S^n and, hence, are two-sided.

Corollary 7.1.2. *Every $(n-1)$-manifold S in an n-manifold M is locally two-sided; that is, each $s \in S$ has arbitrarily small connected neighborhoods N_s such that $N_s \setminus S$ has two components. Hence, if S itself is two-sided and U is one of the sides, then S is 0-LCC in $\overline{U}$.*

Proof. In any coordinate neighborhood W of s, take $N_s \subset W$ as a connected neighborhood of s such that $N_s \cap S$ equals the component of $W \cap S$ containing s. □

Lemma 7.1.3. *Let M denote an orientable n-manifold and let S be a connected $(n-1)$-manifold embedded in M as a closed subset. Then S is two-sided if and only if it is orientable.*

Proof. Consider a connected neighborhood N_S of S, where $N_S \setminus S$ is disconnected if and only if S is two-sided. Produce a smaller neighborhood N' of S that deformation retracts to S in N_S. A look at the Mayer-Vietoris sequence for $N_S = (N_S \setminus S) \cup N'$ reveals that $H_1(N_S \setminus S) \oplus H_1(N') \to H_1(N_S)$ is surjective. Note that the image of $H_1(N')$ in $H_1(N_S)$ coincides with that of $H_1(S)$.

When S is two-sided, $\text{incl}_* : H_1(N_S \setminus S; \mathbb{Z}) \to H_1(N_S; \mathbb{Z})$ is surjective: each loop in S is homotopic in N_S to an approximating loop in $N_S \setminus S$ on either side, by the preceding lemma; when S is one-sided, the square of any loop in S is homotopic to one in $N_S \setminus S$. Thus, the cokernel of incl_* is a torsion group. Read off the (non)separation conclusions from the exact sequence:

$$0 \to \text{Torsion} \to H_1(N_S, N_S \setminus S; \mathbb{Z}) \to \widetilde{H}_0(N_S \setminus S; \mathbb{Z}) \to 0$$

and the duality isomorphism $H_c^{n-1}(S; \mathbb{Z}) \cong H_1(N_S, N_S \setminus S; \mathbb{Z})$. □

Corollary 7.1.4. *Suppose M is a connected, orientable n-manifold and $N \subset M$ is a compact, connected, nonorientable $(n-1)$-manifold. Then $M \setminus N$ is connected.*

Corollary 7.1.5. *Let M be an n-manifold with $H_1(M; \mathbb{Z}_2) \cong 0$. If the $(n-1)$-manifold S embeds in M as a closed subset, then S is orientable.*

Lemma 7.1.6. *Let M be a connected n-manifold and let $S \subset M$ be a connected $(n-1)$-manifold embedded in M as a closed subset. Then S separates M if and only if the inclusion-induced homomorphism $H_c^{n-1}(M; \mathbb{Z}_2) \to H_c^{n-1}(S; \mathbb{Z}_2)$ is trivial.*

Proof. This follows from examination of:

$$\begin{array}{ccccc} H_c^{n-1}(M;\mathbb{Z}_2) & \longrightarrow & H_c^{n-1}(S;\mathbb{Z}_2)\cong\mathbb{Z}_2 & & \\ \downarrow\cong & & \downarrow\cong & & \\ H_1(M;\mathbb{Z}_2) & \longrightarrow & H_1(M,M\smallsetminus S;\mathbb{Z}_2) & \longrightarrow & \widetilde{H}_0(M\smallsetminus S;\mathbb{Z}_2)\to 0. \end{array}$$

□

Corollary 7.1.7. *If M is an n-manifold and S is an $(n-1)$-manifold embedded in M as a closed subset, where $H_1(S;\mathbb{Z}_2)\cong 0$, then S is two-sided in M.*

Proof. Reduce to the case in which S is connected, and let U be a connected neighborhood of S that strong deformation retracts to S in M. Inspection of the diagram ($\mathbb{Z}_2$ coefficients throughout)

$$\begin{array}{ccccc} H_1(U) & \longrightarrow & H_1(U,U\smallsetminus S) & \xrightarrow{\cong} & H_c^{n-1}(S) \\ \downarrow 0 & & \downarrow\cong & & \downarrow \mathrm{Id}_S^* \\ H_1(M) & \longrightarrow & H_1(M,M\smallsetminus S) & \xrightarrow{\cong} & H_c^{n-1}(S) \end{array}$$

yields that $H_1(U)\to H_1(U,U\smallsetminus S)$ is trivial. By duality, $H_c^{n-1}(U)\to H_c^{n-1}(S)$ is trivial, and 7.1.6 applies. □

Corollary 7.1.8. *Given a compact two-sided $(n-1)$-manifold S in the n-manifold M, there exists $\epsilon>0$ such that for any embedding $\lambda:S\to M$ within ϵ of* incl_S, *$\lambda(S)$ is 2-sided in M.*

Proof. Identify a neighborhood U_S of S such that each component of S separates the relevant component of U_S. When $\lambda,\mathrm{incl}_S:S\to U_S$ are homotopic, $H_c^{n-1}(U_S;\mathbb{Z}_2)\to H^{n-1}(\lambda(S);\mathbb{Z}_2)$ can be factored through the trivial homomorphism $H_c^{n-1}(U_S;\mathbb{Z}_2)\to H^{n-1}(S;\mathbb{Z}_2)$. □

Corollary 7.1.9. *Given a compact two-sided $(n-1)$-manifold S in the n-manifold M and a neighborhood U of S, there exists $\epsilon>0$ such that for any two disjoint embeddings $\lambda_0,\lambda_1:S\to M$ within ϵ of* incl_S, *U contains a compact subset C with $\lambda_0(S)\cup\lambda_1(S)$ as its frontier.*

Example 7.1.10. *Let $\theta:S'\to S$ be a 2-1 covering map between compact, connected $(n-1)$-manifolds. Then $W=\mathrm{Map}(\theta)$ is a compact n-dimensional ∂-manifold containing S as a one-sided subset of* $\mathrm{Int}\,W$, *and every embedding $\lambda:S\to\mathrm{Int}\,W$ homotopic to* incl_S *satisfies $\lambda(S)\cap S\neq\emptyset$.*

Proof. Consider any $\lambda:S\to W$ homotopic to $\mathrm{incl}_S:S\to W$; clearly λ induces an isomorphism at the π_1-level. By definition of the mapping cylinder, $\partial W\cong S'$ is a strong deformation retract of $W\smallsetminus S$. Accordingly, if $\lambda(S)$ were disjoint from S, $\lambda_\#:\pi_1(S)\to\pi_1(W)$ would factor through

$\pi_1(\partial W = S') \to \pi_1(W) \cong \pi_1(S)$, an impossibility, as the latter homomorphism fails to be surjective. □

Proposition 7.1.11. *Let M denote a connected n-manifold, S a connected $(n-1)$-manifold embedded in M as a closed and separating subset, and U a component of $M \smallsetminus S$. Then for each $s \in S$ and neighborhood N of s, there exists a neighborhood $N' \subset N$ of s such that*

$$\text{incl}_* : H_k(N' \cap U; \mathbb{Z}) \to H_k(N \cap U : \mathbb{Z})$$

is trivial for all $k > 0$. Furthermore, if S is 1-LCC in $\overline{U}$, then S is k-LCC in $\overline{U}$ for all $k \geq 0$.

Proof. Being an ANR, $\overline{U}$ is locally contractible. Hence, given a neighborhood N of $s \in S$, one can find a smaller neighborhood N' such that $\text{incl}_* : H_k(N' \cap \overline{U}) \to H_k(N \cap \overline{U})$ is trivial for all $k > 0$ ($\mathbb{Z}$ coefficients throughout this argument). We will show that when N is chosen so its intersection with S is contractible, then $\text{incl}_* : H_k(N \cap U) \to H_k(N \cap \overline{U})$ will be an isomorphism ($k > 0$), which will give that $\text{incl}_* : H_k(N' \cap U) \to H_k(N \cap U)$ is trivial. Inspection of the long exact sequence for $(N, N \smallsetminus S)$ and duality yields $H_k(N, N \smallsetminus S) \cong H_c^{n-k}(N \cap S) \cong 0$ for $k > 1$, from which it follows that $\text{incl}_* : H_k(N \smallsetminus S) \to H_k(N)$ is an isomorphism when $k > 1$. Diagram chasing assures the same holds true for $k = 1$, because $H_1(N, N \smallsetminus S) \cong H_c^{n-1}(N \cap S) \cong \mathbb{Z}$, so $H_1(N, N \smallsetminus S) \to \widetilde{H}_0(N \smallsetminus S)$ is an isomorphism. Let V denote the other component of $M \smallsetminus S$. Clearly $H_k(N \smallsetminus S) \cong H_k(N \cap U) \oplus H_k(N \cap V)$. A straightforward Mayer-Vietoris argument gives that

$$H_k(N) \cong H_k(N \cap \overline{U}) \oplus H_k(N \cap \overline{V}) \quad (k > 0).$$

Naturality assures that $\text{incl}_* : H_k(N \cap U) \to H_k(N \cap \overline{U})$ is an isomorphism.

Note that S is 0-LCC in $\overline{U}$ by Corollary 7.1.2. When it is also 1-LCC there, application of the local Hurewicz Theorem 0.8.3 confirms that S is k-LCC in $\overline{U}$ for all $k \geq 2$. □

Exercise

7.1.1. Suppose M is a connected n-manifold and $S \subset M$ is a closed $(n-1)$-manifold such that $M \smallsetminus S$ is connected. For each $\alpha_S \in \pi_1(S)$ there exists $\alpha' \in \pi_1(M \smallsetminus S)$ such that $(\text{incl}_{M \smallsetminus S})_\#(\alpha') = 2 \cdot (\text{incl}_S)_\#(\alpha_S)$.

7.2. The 1-LCC characterization of local flatness for collared embeddings

For a compactum in the trivial dimension range, being 1-LCC embedded implies it admits an ϵ-push into its complement. An analog holds for two-sided

1-LCC embeddings of manifolds in codimension one. This new 1-LCC push-off result warrants close attention, as it presents a pivotal codimension-one technique in relatively simple form. The same technique will reappear with more intricate variations in subsequent sections. As a peripheral benefit, the push-off result quickly leads to the 1-LCC characterization of local flatness for codimension-one submanifolds collared on one side. The full 1-LCC characterization of local flatness is treated in §7.6.

Proposition 7.2.1 (1-LCC push-off). *Suppose M is a connected PL n-manifold, $n \geq 5$, and S is a compact, connected, two-sided $(n-1)$-manifold 1-LCC embedded in M, where $M \smallsetminus S$ has two components, U_+ and U_-. Then for each $\epsilon > 0$ there exists an ϵ-push ψ of (M, S) such that $\psi(S) \subset U_+$.*

Proof. Apply Generalized Controlled Engulfing Theorem 3.3.7 for the given integer n and for $r = n-3$ to obtain $\delta > 0$ corresponding to $\epsilon/3$. After noting that both $\overline{U}_+$ and $\overline{U}_-$ are neighborhood retracts, successively choose open neighborhoods $W^+_{n-2} \supset W^+_{n-3} \supset \cdots \supset W^+_1 \supset W^+_0$ of $\overline{U}_+$ for which there exist strong deformation retractions of W^+_i to $\overline{U}_+$ in W^+_{i+1}, $i < n-2$, that move points less than δ and that never move any point of U_- into U_+. Require, in addition, that $W^+_{n-2} \cap U_- \subset B(S;\epsilon)$. Choose open neighborhoods $W^-_3 \supset W^-_2 \supset W^-_1 \supset W^-_0$ of $\overline{U}_-$ with analogous properties, where $W^-_3 \cap U_+ \subset B(S;\epsilon)$.

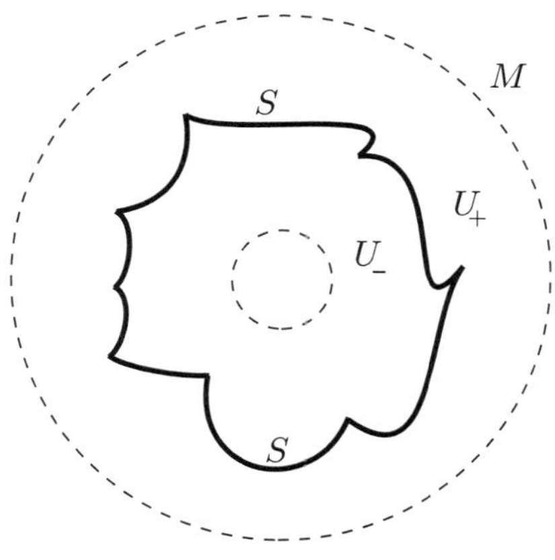

Figure 7.1. The two sides of S in M

Find a compact PL neighborhood P of S in $W^+_0 \cap W^-_0$. Let $C_+ = \mathrm{Cl}(U_+ \smallsetminus P)$ and $C_- = \mathrm{Cl}(U_- \smallsetminus P)$. Subdivide to obtain a triangulation T of P with mesh $T < \epsilon/3$, and let K denote the $(n-3)$-skeleton of T. In T', the first barycentric subdivision of T, let K' denote the simplicial complement of (the subdivided) $K \cup F$, where F denotes the frontier of P. Note that by the special restrictions on the strong deformation retractions of W^+_i to $\overline{U}_+$

spelled out in the preceding paragraph, every relative $(n-3)$-complex in $(W_i^+ \setminus C_+, U_+ \setminus C_+)$ admits a δ-deformation ranging through $W_{i+1}^+ \setminus C_+$, first to $\overline{U}_+ \setminus C_+$, and then into $U_+ \setminus C_+$ by Lemma 3.3.3 (Proposition 7.1.11 assures that S is LCC^n in $\overline{U}_+$).

Now by Theorem 3.3.7 there exists an $(\epsilon/3)$-isotopy ϕ_+ of $M \setminus C_+$ compactly supported in $W_{n-2}^+ \setminus C_+$ such that $\phi_+(U_+ \setminus C_+) \supset K \setminus C_+$; extend via the identity on C_+ to a new push, still denoted as ϕ_+, such that $\phi_+(U_+) \supset K$. Use the same procedure to obtain an $(\epsilon/3)$-isotopy ϕ_- of M supported in $W_3^- \setminus C_-$ such that $\phi_-(U_-) \supset K'$. Stretch across the join structure of T via a third $(\epsilon/3)$-push θ of (M, S) supported in $P \subset W_0^+ \cap W_0^-$ such that

$$\phi_+(U_+) \cup \theta\phi_-(U_-) = C_+ \cup P \cup C_- = M.$$

Apply ϕ_+^{-1} and note that

$$U_+ \cup \phi_+^{-1}\theta\phi_-(U_-) = \phi_+^{-1}(M) = M.$$

As all three pushes are supported in $W_{n-2}^+ \cap W_3^- \subset B(S;\epsilon)$, $\psi = \phi_+^{-1}\theta\phi_-$ is an ϵ-push of (M, S). Most importantly, $\psi(S) \subset U_+$, since obviously $\psi(S)$ is disjoint from $\psi(U_-)$. $\square$

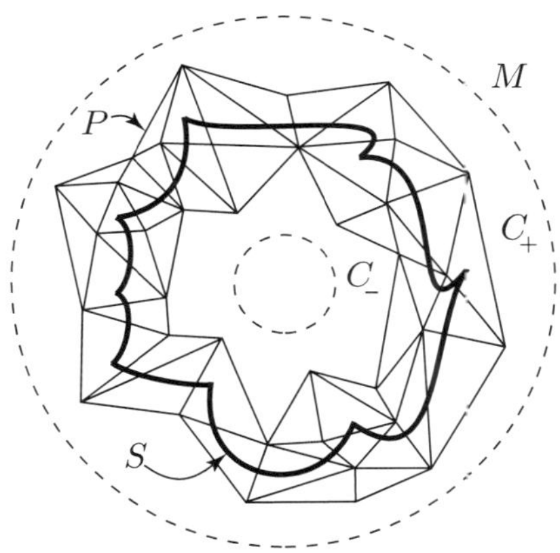

Figure 7.2. The PL neighborhood P of S.

Lemma 7.2.2 (Collar Sliding). *Suppose S is a manifold and $\lambda : S \times [0,1] \to S \times [0,1]$ is an embedding such that $\lambda | S \times 0 = \text{incl}_{S\times 0}$. Then there exists a homeomorphism $h : S \times I \to S \times [0,1] \setminus \lambda(S \times [0, \frac{1}{2}))$ such that $h(s,0) = \lambda(s, \frac{1}{2})$ and $h(s,1) = \langle s, 1\rangle$. Moreover, if each $\lambda(\{s\} \times [0,1])$, $s \in S$, is within $\delta > 0$ of $s \times [0,1]$, then $h(S \times I) \subset B(s \times I; 2\delta)$ for all s.*

Proof. Extend λ to an embedding of $S \times [-1,1]$ in $S \times [-1,1]$ via the Identity on $S \times [-1,0]$. Let ϕ denote the piecewise linear self-homeomorphism of $[-1,1]$ fixing the endpoints, sending 0 to $\frac{1}{2}$ and acting linearly on the complementary subintervals. The self-homeomorphism $\lambda(\mathrm{Id}_S \times \phi)\lambda^{-1}$ defined on $\lambda(S\times[-1,1])$ extends to a self-homeomorphism Φ of $S\times[-1,1]$ via the identity on the complement of $\lambda(S\times[-1,1])$, and Φ restricts to give a homeomorphism $h: S\times I \to S\times[-1,1] \smallsetminus \lambda(S\times[-1,\frac{1}{2}))$. When δ bounds the motion in the S direction under λ, then for any point $\langle s,t\rangle \in S\times I$ moved by h, $\langle s,t\rangle = \lambda(s',t')$ where $d_S(s,s') < \delta$, and

$$h(s,t) \in \lambda(s'\times I) \subset B(s'\times I;\delta) \subset B(s\times I; 2\delta).$$

□

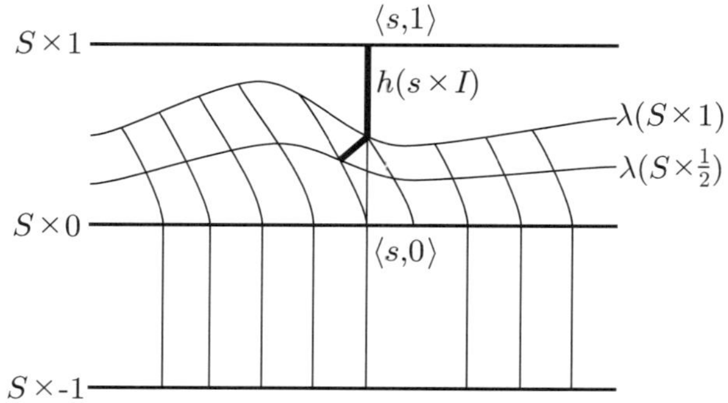

Figure 7.3. A special $S\times I$ product structure

Remark. If $S\times I$ is a PL manifold and λ is a PL collar, then h is a PL embedding.

Theorem 7.2.3. *Suppose M is a PL n-manifold, $n \geq 5$, and S is a compact $(n-1)$-manifold in M such that S is two-sided and 1-LCC embedded. If S has a collar on one side, then S is bicollared.*

Proof. Name a collar $c: S\times I \to M$ on one side of S. Assume both S and M to be connected and M a small enough neighborhood of S that $M \smallsetminus S$ has two components, with U denoting the one missing the image of c. For $i = 1, 2, \ldots$ use Proposition 7.2.1 to obtain a $(1/6i)$-push ψ_i of (M,S) such that $\psi_i(S) \subset U$ and ψ_i fixes $c(S\times[u_i,1])$, where $\{u_i \in (0,1]\}_{i=1}^{\infty}$ is a sequence decreasing to 0 and $\operatorname{diam} c(s\times[0,u_i]) < 1/6i$ for all $s\in S$. Do this so $\psi_{i+1}c(S\times I) \subset \psi_i c(S\times(0,1])$ for all i. Then choose $t_i \in (0,u_i)$ such that $\psi_i c(S\times(t_i,1]) \supset \psi_{i+1}c(S\times I)$, and declare $e_i: S\to U$ to be the embedding sending $s\in S$ to $\psi_i c(s\times t_i)$. For notational simplicity, require $\{t_i\}_{i\geq 1}$ to be a strictly decreasing sequence. According to Lemma 7.2.2, the region R_i bounded by e_i and e_{i+1} is a product $S\times[t_{i+1},t_i]$, the arc fibers of which have diameter less than $1/i$, since (by the proof of the Lemma) these arc fibers live in some $\psi_i c(s \times [t_i, u_i])$ plus the union of intersecting arcs $\psi_{i+1} c(s' \times [t_{i+1}, u_i])$. Thus, $S \cup (\cup_i R_i)$ is a collar on S in $\overline{U}$. □

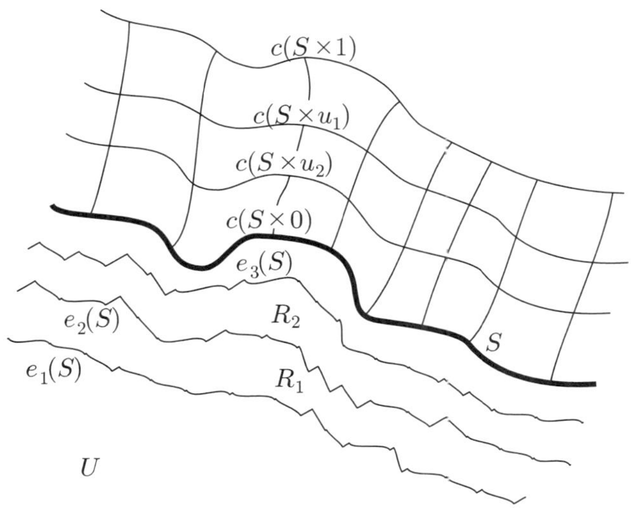

Figure 7.4. The collar on S in $\overline{U}$

Historical Notes. Proposition 7.2.1 and a strengthened Theorem 7.2.3—namely, a 1-LCC local flatness theorem for codimension-one manifolds that can be approximated by locally flat embeddings—were developed in (Seebeck, 1970).

Exercises

7.2.1. Suppose M is a connected PL n-manifold, $n \geq 5$, S is a compact, connected $(n-1)$-manifold that separates M, and U is a component of $M \smallsetminus S$. Then for each $\epsilon > 0$ there exists $\delta > 0$ such that, for any $(n-3)$-complex pair $(K, L) \subset (U \cup B(S; \delta), U)$ and any neighborhood O of S, there is a compactly supported ϵ-push ψ of (M, S) such that $\psi(U \cup O) \supset K$ and $\psi|(U \smallsetminus O) \cup L = \mathrm{Id}|(U \smallsetminus O) \cup L$.

7.2.2. Let $c_0, c_1 : \partial W \times I \to W$ be collars on the boundary of a ∂-manifold W, with $c_0(\partial W \times I) \subset c_1(\partial W \times I)$ and $\operatorname{diam} c_i(w \times I) < \epsilon$ for all $w \in \partial W$, $i = 0, 1$. Then there exists a homeomorphism

$$h : \partial W \times I \to c_1(\partial W \times I) \smallsetminus c_0(\partial W \times [0, \frac{1}{2}))$$

such that $\operatorname{diam} h(w \times I) < 2\epsilon$ for all $w \in \partial W$.

7.2.3. Let W be a ∂-manifold such that ∂W is compact and for every compact subset C of W there is a collar $c : \partial W \times I \to W$ such that

$C \subset c(\partial W \times I)$. Then $W \cong \partial W \times [0, \infty)$. Moreover, if W and the collars are PL, then W is PL homeomorphic to $\partial W \times [0, \infty)$.

7.3. Unknotting close 1-LCC embeddings of manifolds

Throughout §7.3, S will denote a PL $(n-1)$-manifold topologically embedded as a two-sided subset of the PL n-manifold M. The main result, Theorem 7.3.1, assures that any two locally flat approximations to S in M are ambient isotopic under a controlled push of (M, S). It is an exact analog for manifolds of Codimension-three Unknotting Theorem 5.4.2; the codimension-three result cannot be extended to a local unknotting theorem for $(n-1)$-complexes in M, however, since such an extension is known to fail for codimension-two manifolds. En route to establishing 7.3.1, we will show in Theorem 7.3.11 that any two disjoint locally flat approximations cobound an embedded product $S \times I$ with short I-factor.

Theorem 7.3.1 (Local Unknotting for Embeddings of Manifolds). *Let S denote a compact PL $(n-1)$-manifold topologically embedded in a PL n-manifold M^n, $n \geq 5$, as a two-sided subset. Given $\epsilon > 0$ there exists $\delta > 0$ such that, for any two locally flat embeddings λ_0, λ_1 of S in M^n within δ of the inclusion, there exists an ϵ-push θ_t of (M^n, S) such that $\theta_1\lambda_0 = \lambda_1$.*

The proof, which occupies the rest of this section, also depends heavily upon the following result of Edwards and Kirby (1971).

Theorem 7.3.2 (Local Contractibility). *Given a compact manifold S and $\epsilon > 0$, there exists $\delta > 0$ such that if $\Lambda : S \times [-1, 1] \to S \times [-2, 2]$ is an embedding within δ of the inclusion, then there is an isotopy $\Phi_t : S \times [-1, 1] \to S \times [-2, 2]$ such that $\Phi_0 = \Lambda$, $\Phi_1 | S \times \{0\} = \mathrm{incl}_{S\times\{0\}}$, $\rho(\Phi_t, \mathrm{incl}_{S\times[-1,1]}) < \epsilon$ and $\Phi_t | S \times \{\pm 1\} = \Lambda | S \times \{\pm 1\}$ for each $t \in I$.*

As an immediate consequence of 7.3.2, one can define an ambient isotopy Φ'_t on $S \times [-2, 2]$ as $\Phi'_t = \Phi_t \Lambda^{-1}$ on $\Lambda(S \times [-1, 1])$ and as the identity elsewhere. Clearly $\Phi'_0 = \mathrm{Id}$, $\Phi'_1 \Lambda_0 = \mathrm{incl}_{S\times\{0\}}$ and $\rho(\Phi'_t, \mathrm{Id}) < 2\epsilon$. Application of $(\Phi'_1)^{-1}$ to $S \times I \to S \times [-2, 2]$ implies:

Corollary 7.3.3. *Given a compact manifold S and $\epsilon > 0$, there exists $\delta > 0$ such that if $\Lambda : S \times [-1, 1] \to S \times [-2, 2]$ is an embedding within δ of the inclusion, then there is an embedding $\lambda : S \times I \to S \times [-2, 2]$ such that $\lambda(s, 0) = \langle s, -2 \rangle$, $\lambda(s, 1) = \Lambda_0(s)$, and $\lambda(s \times I)$ is within ϵ of $\{s\} \times [-2, 0]$ for each $s \in S$.*

Corollary 7.3.4. *Given a compact manifold S and $\epsilon > 0$, there exists $\delta > 0$ such that if $\Lambda : S \times [-1, 1] \to S \times [-2, 2]$ is an embedding within δ of the inclusion and $\Lambda_0(S) \cap (S \times [0, 2]) = \emptyset$, then there is an embedding*

$h : S \times I \to S \times [-2, 2]$ such that $h(s, 0) = \Lambda_0(s)$, $h(s, 1) = \langle s, 0\rangle$, and $h(s \times I)$ is within ϵ of $\langle s, 0\rangle$ for each $s \in S$.

Proof. This follows from Corollary 7.3.3 and Corollary 7.2.2, and from the consequence of the latter that

$$h(s \times I) \subset B(s \times [-2, 2]; 2\delta) \cap (S \times [-\delta, 0]). \qquad \square$$

Lemma 7.3.5. *Let Y be a locally compact ANR, $C \subset Y$ a compact ANR, and $W \subset Y$ a neighborhood of C. For each $\delta > 0$ there exist $\eta > 0$ and a neighborhood W' of C such that, given any embedding $e : C \to W'$ within η of* incl_C*, W' admits a strong deformation retraction $\mu_t : W' \to W$ to $e(C)$ that moves points less than δ.*

Proof. Determine a compact neighborhood $W' \subset W$ of C that admits a $(\delta/2)$-retraction to C, and find $\eta > 0$ such that any embedding $e : C \to Y$ within η of incl_C is $(\delta/2)$-homotopic to incl_C in W'. An application of the Estimated Homotopy Extension Theorem (Corollary 0.6.5) secures a δ-retraction $R_e : W' \to e(C)$. Properties of ANRs allow prearrangements yielding that $R_e, \text{incl}_{W'} : W' \to W$ are δ-homotopic. $\square$

For the next several lemmas, assume that S and M are manifolds satisfying the hypotheses of Theorem 7.3.1.

Lemma 7.3.6. *Given $\epsilon > 0$, there exists $\eta > 0$ such that for every pair $\lambda_0, \lambda_1 : S \to M$ of locally flat embeddings within η of* incl_S*, there is an ϵ-push ψ of (M, S) such that $\psi(\lambda_0(S)) \cap \lambda_1(S) = \emptyset$.*

Proof. Assume S to be connected and W_S to be a connected neighborhood of S such that $W_S \smallsetminus S$ has two components, U_+ and U_-. Just as in the proof of Proposition 7.2.1, apply Generalized Controlled Engulfing Theorem 3.3.7 for the given integer n and $r = n - 3$ to obtain $\delta > 0$ corresponding to $\epsilon/3$ there. As before, choose open neighborhoods $W^+_{n-2} \supset W^+_{n-3} \supset \cdots \supset W^+_1 \supset W^+_0$ of $\overline{U}_+$ such that not only do there exist strong deformation retractions of W^+_i to $\overline{U}_+$ in W^+_{i+1} $(i = 0, 1, \ldots, n-3)$ moving points less than δ and never moving any point of U_- into U_+, but also (by Lemma 7.3.5) that there exists $\eta_i > 0$ such that for any embedding $\lambda : S \to W^+_i$ within η_i of $\text{incl} : S \to W^+_i$ there is a strong deformation retraction of W^+_i to $\lambda(S)$ in W^+_{i+1} moving points less than δ. Require, in addition, that $W^+_{n-2} \cap U_- \subset B(S; \epsilon) \subset W_S$. Determine open neighborhoods $W^-_3 \supset W^-_2 \supset W^-_1 \supset W^-_0$ of $\overline{U}_-$ and positive numbers $\eta^*_2, \eta^*_1, \eta^*_0$ with analogous properties, where $W^-_3 \cap U_+ \subset B(S; \epsilon)$.

Choose a compact PL neighborhood P of S in $W^+_0 \cap W^-_0$, and impose on P a small mesh triangulation T with specified $(n-3)$-skeleton K and simplicial complement K' of $K \cup F$ (F denoting the frontier of P) in the dual 2-skeleton, as in 7.2.1. Set $\eta = \min\{\eta_i, \eta^*_j, d(S, M \smallsetminus P)\}$. Let $C_+ =$

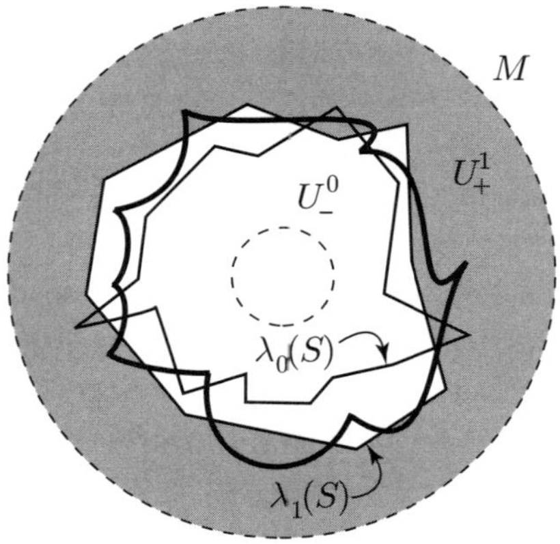

Figure 7.5. The setup for obtaining disjoint approximations.

$\mathrm{Cl}(U_+ \setminus P)$ and $C_- = \mathrm{Cl}(U_- \setminus P)$, as before. Given locally flat embeddings $\lambda_0, \lambda_1 : S \to M$ within η of incl_S, let U_+^1 denote the component of $W_S \setminus \lambda_1(S)$ containing C_+ and U_-^0 the component of $W_S \setminus \lambda_0(S)$ containing C_-. (Use of U_+^1 and U_-^0 represents the most significant change from the proof of Theorem 7.2.1.) Observe that any relative $(n-3)$-complex in $(W_i^+ \setminus C_+, U_+^1 \setminus C_+)$ is δ-homotopic in $(W_{i+1}^+ \setminus C_+, U_+^1 \setminus C_+)$ to a complex mapped into $U_+^1 \setminus C_+$. Similarly, any relative 2-complex in $(W_i^- \setminus C_+, U_-^0 \setminus C_-)$ is δ-homotopic in $(W_{i+1}^- \setminus C_-, U_-^0 \setminus C_-)$ to a complex mapped into U_-^0. Hence, Theorem 3.3.7 promises an $(e/3)$-push ϕ_+ of (M, S) compactly supported in $W_{n-2}^+ \setminus C_+$ such that $\phi_+(U_+^1) \supset K$, and it promises a second $(e/3)$-push ϕ_- of (M, S) compactly supported in $W_3^- \setminus C_-$ such that $\phi_-(U_-^0) \supset K'$. Then there is also a third $(\epsilon/3)$-push θ supported in P—the stretch across the join structure of P—such that

$$\phi_+(U_+^1) \cup \theta\phi_-(U_-^0) \supset C_+ \cup P \cup C_- = W_S.$$

Letting $\psi = \phi_+^{-1}\theta\phi_-$, one sees that ψ is an ϵ-push of (M, S) and $\psi(\lambda_0(S)) \subset U_+^1$, so $\psi(\lambda_0(S)) \cap \lambda_1(S) = \emptyset$, as desired. $\square$

Given two disjoint embeddings λ_0, λ_1 of a codimension-one manifold in a manifold M, we use $[\lambda_0, \lambda_1]$ to denote the unique compact region, if such a region exists, having the union of these images as frontier; if λ_0, λ_1 are disjoint close approximations to a two-sided codimension-one manifold S in M, we use the same symbolism $[\lambda_0, \lambda_1]$ to denote the compact region near S their images cobound; the existence of such a compact region is assured by Corollary 7.1.9.

Lemma 7.3.7. *Given $\epsilon > 0$, there exists $\eta > 0$ such that for every pair $\lambda_0, \lambda_1 : S \to M$ of disjoint embeddings within η of* incl_S, *there are strong ϵ-deformations of $[\lambda_0, \lambda_1]$ onto $\lambda_0(S)$ and $\lambda_1(S)$, respectively.*

Proof. Let W_S be a connected neighborhood of S in M that is separated by S. Reduce W_S so it admits an $(\epsilon/6)$-retraction to S. Apply Corollary 7.1.9 and Lemma 7.3.5 to produce $\delta_1 > 0$ such that, for any embedding $\lambda : S \to M$ within δ_1 of incl_S, $\lambda(S)$ separates W_S and W_S admits an $(\epsilon/3)$-retraction $W_S \to \lambda(S)$. In the presence of two δ_1-approximations $\lambda_0, \lambda_1 : S \to W_S$ to incl_S with disjoint images, we have an $(\epsilon/3)$-retraction $R : W_S \to [\lambda_0, \lambda_1]$ sending one of the components of $W_S \setminus [\lambda_0, \lambda_1]$ to $\lambda_0(S)$ and sending the other to $\lambda_1(S)$. Next, find a compact neighborhood $W' \subset W_S$ of S and $\delta_2 > 0$ such that any two maps $f, f' : W' \to W_S$ δ_2-close to $\text{incl} : W' \to W_S$ are $\epsilon/3$-homotopic in W_S. Set $\delta = \min\{\delta_1, \delta_2\}$. Finally, repeat the initial procedure to find $\eta \in (0, \delta)$ such that, for any embedding $\lambda : S \to M$ within η of incl_S, $\lambda(S) \subset W' \subset W_S$ (so it separates both W' and W_S) and W' admits a δ_2-retraction $W' \to \lambda(S)$. Hence, if $\lambda_0, \lambda_1 : S \to M$ are two disjoint embeddings within η of incl_S, there is a homotopy $\mu_t : [\lambda_0, \lambda_1] \to W_S$ such that $\mu_0 = \text{incl}_{[\lambda_0, \lambda_1]}$, μ_t is constant on, say, $\lambda_0(S)$, μ_1 is a δ-retraction to $\lambda_0(S)$ and μ_t moves points less than $\epsilon/3$. Then $R\mu_t$ functions as a strong ϵ-deformation of $[\lambda_0, \lambda_1]$ to $\lambda_0(S)$. □

Lemma 7.3.8. *Given $\epsilon > 0$, there exists $\delta > 0$ such that for each pair $\lambda_0, \lambda_1 : S \to M$ of disjoint, locally flat embeddings within δ of* incl_S, *each neighborhood U of $[\lambda_0, \lambda_1] \setminus \lambda_1(S)$, and each neighborhood O of $\lambda_1(S)$ there is an ϵ-push ψ of (M, S) fixed on $\lambda_1(S) \cup (M \setminus U)$ such that $\psi([\lambda_0, \lambda_1]) \subset O$.*

Proof. For the given integer n and positive number ϵ apply Theorem 3.3.7 to get $\epsilon' > 0$ such that whenever one is provided with enough ϵ'-homotopies of $(n-3)$-complexes in any PL n-manifold N, then one also has an $(\epsilon/3)$-isotopy of N engulfing such a complex. Then use Lemma 7.3.7 to obtain $\delta > 0$ corresponding to $\epsilon' > 0$ with the properties mentioned there. Assume δ to be sufficiently small that images of δ-approximations to incl_S separate small neighborhoods of S.

Consider any two locally flat δ-approximations λ_0, λ_1 with disjoint images. As an aid, name another locally flat embedding λ_0^* of S into a bicollar on $\lambda_0(S)$ in U, with λ_0^* very close to λ_0 and $[\lambda_0^*, \lambda_1]$ properly containing $[\lambda_0, \lambda_1]$. Let W denote the interior of $[\lambda_0^*, \lambda_1]$. We will produce an ϵ-push of (M, S) moving $[\lambda_0, \lambda_1]$ into O by obtaining an appropriate, compactly supported isotopy of W.

Find a PL ∂-manifold P such that P is closed in W, $P \cup \lambda_1(S) \supset [\lambda_0, \lambda_1]$ and $\partial P \cap \lambda_0(S) = \emptyset$. Impose a triangulation T on P having mesh less than $\epsilon/3$, with diameters of simplices going to 0 as simplices approach

$\lambda_1(S)$, and let K denote the $(n-3)$-skeleton of P. Since W is the union of $[\lambda_0', \lambda_1']$ where λ_0' varies over close approximations to λ_0^*, and λ_1' varies over close approximations to λ_1 in W, Lemma 7.3.7 promises the existence of ϵ'-homotopies deforming $(n-3)$-complexes into $O_W = O \cap W$. Controlled Engulfing Theorem 3.3.7 provides a compactly supported $(\epsilon/3)$-isotopy ϕ_+ of W to itself such that $\phi_+(O_W) \supset K$.

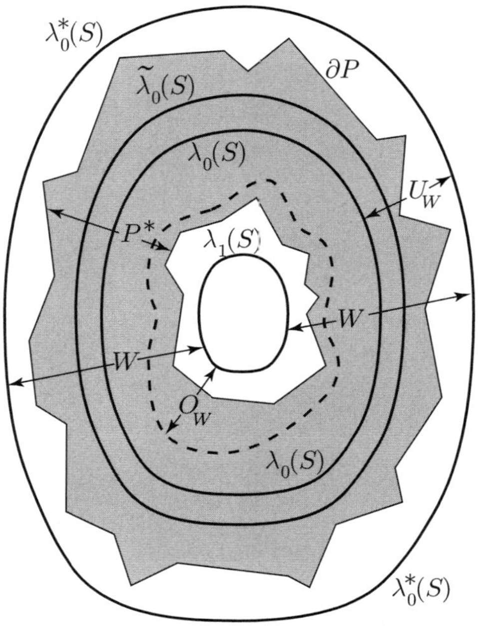

Figure 7.6. Guides for engulfing $[\lambda_0, \lambda_1]$ by O_W.

Let U_W denote $W \smallsetminus [\lambda_0, \lambda_1] = \mathrm{Int}[\lambda_0^*, \lambda_0]$ and let $Sppt_+$ denote the support of ϕ_+. The extra wrinkle in this argument is the observation that, since only a finite part of P extends outside O_W, P contains a finite subcomplex P^* such that $P^* \supset (P \smallsetminus O_W) \cup Sppt_+$. Let K' denote the simplicial complement of $K \cup \overline{P \smallsetminus P^*}$ (subdivided) in the first barycentric subdivision of $T|P^*$. Find another locally flat approximation $\widetilde{\lambda}_0$ to λ_0 in W such that $[\widetilde{\lambda}_0, \lambda_1]$ properly contains $[\lambda_0, \lambda_1]$ and $P \supset [\widetilde{\lambda}_0, \lambda_1] \smallsetminus \lambda_1(S)$. Let $\widetilde{W}$ represent the interior of $[\widetilde{\lambda}_0, \lambda_1]$ and $\widetilde{U}_W = U_W \cap \widetilde{W}$; note that $\widetilde{W} \subset P \subset W$. Again there are ϵ'-homotopies deforming any relative 2-complex in $(\widetilde{W}, \widetilde{U}_W)$ into $\widetilde{U}_W$, so there exists a compactly supported isotopy $\phi_- : \widetilde{W} \to \widetilde{W}$ moving points less than $\epsilon/3$ such that $\phi_-(\widetilde{U}_W) \supset K' \cap \widetilde{W}$. Extend via the identity on $W \smallsetminus \widetilde{W}$ to regard ϕ_- as defined on all of W, with $\phi_-(U_W) \supset K'$ and with ϕ_- fixed on $W \smallsetminus P \subset U_W$. Exploit the usual stretch across the join structure of P^* to produce a third $(\epsilon/3)$-push θ, compactly supported in P^*, such that $\phi_+(O_W) \cup \theta\phi_-(U_W) \supset P^*$. Then

$$W \supset \phi_+(O_W) \cup \theta\phi_-(U_W) \supset (P \setminus P^*) \cup P^* \cup (W \setminus P) = W.$$

Now for $\psi = (\phi_+)^{-1}\theta\phi_-$ we have $O_W \cup \psi(U_W) = W$ and $\psi([\lambda_0, \lambda_1]) \cap \psi(U_W) = \emptyset$. Thus ψ extends via the identity over $M \setminus W$ to an ϵ-push of (M, S) such that $O \supset O_W \supset \psi([\lambda_0, \lambda_1])$. □

As a routine consequence of Lemmas 7.3.6 and 7.3.8 we obtain:

Lemma 7.3.9. *Given $\epsilon > 0$, there exists $\eta > 0$ such that for each pair $\lambda_0, \lambda_1 : S \to M$ of locally flat embeddings within η of* incl_S *and each bicollar $g : S \times [-1, 1] \to M$ on $\lambda_1(S)$, there is an ϵ-push ψ of (M, S) such that $\psi(\lambda_0(S)) \subset g(S \times (0, 1))$.*

Lemma 7.3.10. *Given $n \geq 5$, a compact PL $(n-1)$-manifold S, and $\delta > 0$, there exists $\eta > 0$ such that every locally flat embedding $\lambda : S \times \{0\} \to S \times [-2, 2]$ within η of* $\mathrm{incl}_{S\times\{0\}}$ *extends to an embedding $\Lambda : S \times [-1, 1] \to S \times [-2, 2]$ within δ of* $\mathrm{incl}_{S\times[-1,1]}$.

Proof. Apply Lemma 7.3.8 for the inclusion $S \times \{0\} \hookrightarrow S \times [-2, 2]$ and positive number $\delta/4$ to obtain $\eta' > 0$. Choose a large integer $k > 0$ such that $2/k < \eta'$ and set $\eta = 1/k$.

Consider a locally flat embedding $\lambda : S \times \{0\} \to S \times [-2, 2]$ within η of $\mathrm{incl}_{S\times\{0\}}$. Extend λ to an embedding $g : S \times [-1, 1] \to S \times (-\eta, \eta)$ such that $g(s \times [-1, 1]) \subset B(\langle s, 0\rangle; \eta)$ for all $s \in S$. Arrange the parameterization of the bicollar determined by g so that $g(S \times \{1\})$ separates $g(S \times \{0\})$ from $S \times \{2\}$.

For $i = 0, 1, 2, ..., k$, set $t(i) = (k - i)/k$. The initial choice of η' assures the existence of a controlled $(\delta/4)$-push of $(S \times [-2, 2], S \times \{0\})$ moving $g(S \times \{\pm t(i)\})$ very close to $S \times \{\pm\eta\}$. These pushes will be followed by large moves that change only the $[-2, 2]$ coordinates and effect a precise shuffle repositioning each $g(S \times \{\pm t(i)\})$ very close to $S \times \{\pm t(i)\}$ (respecting $\pm$ signs).

Use $\kappa : S \times \{0\} \to S \times \{\eta\} \subset S \times [-2, 2]$ to denote the obvious embedding, and let g_t denote the embedding sending $\langle s, 0\rangle$ to $g(s, t)$. Note that both κ and g_t are within $2\eta = 2/k < \eta'$ of $\mathrm{incl}_{S\times\{0\}}$. Thus, Lemma 7.3.8 provides a $(\delta/4)$-homeomorphism Ψ'_1 of $S \times [-2, 2]$ to itself fixed outside a small neighborhood of $[g_{t(1)}, \kappa]$—in particular, fixed on $g(S \times [-1, t(2)])$—and moving $g(S \times \{t(1)\})$ into $S \times (\eta/2, \eta)$. Follow Ψ'_1 by another homeomorphism that changes only $[-2, 2]$ coordinates, fixes $\Psi'_1 g(S \times [-1, t(2)]) = g(S \times [-1, t(2)])$ and moves $\Psi'_1 g(S \times \{t(1)\})$ into $S \times [t(1), 1]$. Call the composite Ψ_1. Repeat, obtaining homeomorphisms $\Psi_2, \Psi_3, ..., \Psi_{k-1}$ of $S \times [-2, 2]$ to itself that change S coordinates by less than $\delta/4$ and satisfy $\Psi_i g(S \times \{t(i)\}) \subset$

$S\times[t(i),t(i-1)]$. Restrict supports so that the composite $\Psi = \Psi_{k-1}\cdots\Psi_2\Psi_1$ satisfies $\Psi(g(S\times\{t(i)\}))\subset S\times[t(i),t(i-1)]$. Also require if a point of $g(S\times[-1,1])$ is fixed by $\Psi_1,\dots,\Psi_j$ but moved by Ψ_{j+1}, then its image can be moved by Ψ_{j+2} but cannot be moved by subsequent Ψ_i. Define Ψ by exactly the same process for the other side of the bicollar. One can see that $\Psi g : S\times[-1,1]\to S\times[-2,2]$ changes second coordinates by less than $2/k<\eta'<\delta/4$. Since Ψg changes first coordinates less than $\delta/2$, $\Lambda=\Psi g$ is δ-close to $\mathrm{incl}_{S\times[-1,1]}$. □

Remark. Lemma 7.3.10 is one place in this section where the hypothesis about the codimension-one submanifold S being PL plays a role in the argument, simply by assuring that $S\times[-2,2]$ is PL.

Theorem 7.3.11. *Let S denote a compact PL $(n-1)$-manifold topologically embedded in a PL n-manifold M^n, $n\geq 5$, as a two-sided subset and let ϵ be a positive number. Then there exists $\delta>0$ such that for any two locally flat embeddings λ_0,λ_1 of S in M^n within δ of the inclusion, where $\lambda_0(S)\cap\lambda_1(S)=\emptyset$, there exists an embedding $\Lambda : S\times[0,1]\to M$ such that $\Lambda_0=\lambda_0$, $\Lambda_1=\lambda_1$ and* $\operatorname{diam}\Lambda(s\times[0,1])<\epsilon$ *for all $s\in S$.*

Proof. Once the constraint δ is in place and we get to the locally flat approximations, we will extend λ_1 to an embedding $g : S\times[-2,2]\to M$ for which the fiber arcs $g(s\times[-2,2])$, $s\in S$, have small images. The image bicollar will play the role of $S\times[-2,2]$ in Lemma 7.3.10. The plan is to produce a controlled push ψ of (M,S) that, in spirit, moves $\lambda_0(S)$ into $g(S\times(-1,0))$ extremely close to λ_1. There will be an obvious short product structure on something like $[\lambda_0,\psi\lambda_0]$, Lemma 7.3.10 will provide a short product structure on $[\psi\lambda_0,\lambda_1]$, and these two pieces will fit together as a short product structure on $[\lambda_0,\lambda_1]$.

The crucial issue is size control; it is somewhat delicate due to the need to pass back and forth between the abstract product space $S\times[-2,2]$, where Lemma 7.3.10 applies, and its image under g, where we must operate. To highlight the distinction we use d_M to denote a metric on M and d to denote both the restriction of d_M to S and the product metric on $S\times[-2,2]$. Here are rules for obtaining the required δ. Set $\delta_1=\epsilon/15$. Apply Corollary 7.3.4 for S and $\epsilon/15$ $(=\delta_1)$ to obtain $\delta_2>0$. Take $\eta_1>0$ to be the positive number corresponding to S and δ_2 promised in Lemma 7.3.10. Set $\delta_3=\min\{\delta_1,\eta_1/6\}>0$, and take δ_4 to be a positive number promised by Lemma 7.3.8 with ϵ replaced by $\min\{\epsilon/3,\delta_3\}$. Finally, let $\delta=\min\{\delta_1,\delta_2,\delta_3,\delta_4/3\}$.

Consider disjoint, locally flat embeddings $\lambda_0,\lambda_1 : S\to M$ within δ of incl_S. Assume δ to be small enough that each $\lambda_j(S)$ is two-sided (Corollary 7.1.8). Specify an embedding $g : S\times[-2,2]\to M$ with $g_0=\lambda_1$, diam $g(s \times [-2,2]) < \delta$ for all $s \in S$, and $g(S \times [-2,2]) \cap \lambda_0(S) = \emptyset$. For definiteness, parameterize so $g(S \times \{-2\}) \subset [\lambda_0, \lambda_1]$. Note that, since $\rho(\lambda_1, \text{incl}_S) < \delta \leq \delta_1 = \epsilon/15$,

$$x, y \in S, d(x,y) < \delta_1 \Rightarrow d_M(\lambda_1(x), \lambda_1(y)) < 3\epsilon/15.$$

As a result,

$$x', y' \in S \times [-2,2], d(x', y') < \delta_1 \Rightarrow d_M(g(x'), g(y')) < \epsilon/3.$$

This means that for any embedded $S \times I$ in $S \times [-2,2]$ for which fiber arcs have diameter less than δ_1, the image fiber arcs under g have diameter less than $\epsilon/3$.

Let $c : S \times [0,2] \to M$ be an embedding that determines a collar on $\lambda_0(S)$ in $M \smallsetminus g(S \times [-2,2])$, where $c_0 = \lambda_0$, diam $c(s \times [0,2]) < \delta$ for all s, and c_1 separates the two boundary components of $[\lambda_0, \lambda_1]$. Set $\lambda_0' = c_1$.

The choice of $\delta \leq \delta_3$ ensures that

$$x, y \in \lambda_1(S), d_M(x,y) < 4\delta_3 \Rightarrow d(\lambda_1^{-1}(x), \lambda_1^{-1}(y)) < 6\delta_3 \leq \eta_1.$$

Find a neighborhood $O \subset g(S \times (-1,1))$ of $\lambda_1(S)$ so small that

$$x, y \in O, \ d_M(x,y) < 4\delta_3 \Rightarrow d(g^{-1}(x), g^{-1}(y)) < \eta_1.$$

Note that $\rho_M(\lambda_0', \text{incl}_S) < 2\delta$, yielding $\rho_M(\lambda_0', \lambda_1) < 3\delta \leq \delta_4$. Applying Lemma 7.3.8 to $[\lambda_0', \lambda_1]$, we obtain a δ_3-push ψ of (M, S) which is fixed on $\lambda_0(S) \cup \lambda_1(S)$ and satisfies $\psi([\lambda_0', \lambda_1]) \subset O \smallsetminus g(S \times [0,2])$. Then $\psi c(S \times [0,1])$ provides an $S \times I$ structure on $[\lambda_0, \psi\lambda_0']$ for which the fiber arcs have diameter less than $\delta + 2\delta_3 \leq 3\delta_3 \leq 3\delta_1 < \epsilon/3$. Moreover,

$$\rho_M(\psi\lambda_0', \lambda_1) \leq \rho_M(\psi\lambda_0', \lambda_0') + \rho_M(\lambda_0', \lambda_1) < \delta_3 + 3\delta \leq 4\delta_3.$$

Thus, for $s \in S = S \times \{0\} \subset S \times [-2,2]$, $d(g^{-1}\psi\lambda_0'(s), s) < \eta_1$ by the choice of O. Lemma 7.3.10 promises an embedding

$$\Lambda : S \times [-1,1] \to S \times [-2,2]$$

within δ_2 of the inclusion, where $\Lambda_0 = g^{-1}\psi\lambda_0'$. Here $\Lambda_0(S) \cap (S \times [0,2]) = \emptyset$, so Corollary 7.3.4 assures the existence of an $(\epsilon/15 = \delta_1)$-product structure on $[g^{-1}\psi\lambda_0', \text{incl}_{S\times 0}]$, and its image under g is an $(\epsilon/3)$-product structure on $[\psi\lambda_0', \lambda_1]$, as desired. □

Proof of Theorem 7.3.1. First apply Theorem 7.3.11 with positive number $\epsilon/2$ to obtain $\delta \in (0, \epsilon/2)$, and next apply Lemma 7.3.8 with $\delta/2$ to obtain $\eta \in (0, \delta/2)$. Given two locally flat η-approximations λ_0, λ_1 to incl_S, use 7.3.8 to produce a $(\delta/2)$-push ϕ of (M, S) moving $\lambda_0(S)$ off $\lambda_1(S)$. Then $\phi\lambda_0, \lambda_1$ are disjoint δ-approximations to incl_S, so Theorem 7.3.11 yields an $(\epsilon/2)$-push ψ of (M, S) supported close to $[\phi\lambda_0, \lambda_1]$ and sending $\phi\lambda_0$ to λ_1. □

Corollary 7.3.12. *Suppose M is a connected PL n-manifold, $n \geq 5$, S is a compact, connected, PL $(n-1)$-manifold that separates M, and U is a component of $M \setminus S$. Then S is collared in $\overline{U}$ if and only if S can be pointwise approximated by locally flat embeddings in U.*

Corollary 7.3.13. *Suppose M is a PL n-manifold, $n \geq 5$, and S is a compact PL $(n-1)$-manifold 1-LCC embedded in M as a two-sided subset. Also suppose S can be pointwise approximated by locally flat embeddings. Then S is bicollared.*

Proof. 1-LCC Pushoff Proposition 7.2.1 indicates that S can be pointwise approximated by locally flat embeddings on either side. □

Historical Notes. The results of this section, as well as the entire approach, again are due to (Seebeck, 1970).

Edwards and Kirby were not alone in addressing local contractibility of the manifold homeomorphism group. Černavskiĭ (1969c) had an independent, possibly earlier, proof of the main result.

Exercises

7.3.1. Prove Corollary 7.3.12. [Hint: See the proof of Theorem 7.2.3.]

7.3.2. Suppose M is a PL n-manifold $(n \geq 5)$ and $S \subset M$ a PL $(n-1)$-manifold that is one-sided and 1-LCC embedded in M. Suppose also that S can be pointwise approximated by locally flat embeddings. Then S has an I-bundle neighborhood.

7.3.3. Suppose M is a PL n-manifold, $n \geq 5$, and S is a closed, PL $(n-1)$-manifold topologically embedded in M as a two-sided subset. Then S is ϵ-tame if and only if it is 1-LCC and it can be pointwise approximated by PL embeddings.

7.4. The Cell-like Approximation Theorem

At this juncture we begin to make use of Edwards's Cell-like Approximation Theorem, stated below. Many methods found in its proof have already been employed in this book, and others are completely accessible to all readers. Nevertheless, we omit the rather lengthy argument and refer readers to (Edwards, 1980) or the more complete exposition in (Daverman, 1986).

Theorem 7.4.1 (Cell-like Approximation). *A proper, surjective, cell-like mapping $f : M^n \to X$ defined on an n-manifold M^n, $n \geq 5$ is a near-homeomorphism if and only if X is a finite-dimensional space with the Disjoint Disks Property.*

Definition. A metric space X has the *Disjoint Disks Property*, abbreviated as DDP, if for every pair of maps $f_1, f_2 : I^2 \to X$ and for every $\epsilon > 0$ there exist maps $F_1, F_2 : I^2 \to X$ such that $\rho(F_i, f_i) < \epsilon$ $(i = 1, 2)$ and $F_1(I^2) \cap F_2(I^2) = \emptyset$.

Corollary 7.4.2. *Every cell-like map $f : M^n \to N^n$ between n-manifolds, $n \geq 5$, is a near-homeomorphism. Specifically, if $\epsilon : N^n \to (0, \infty)$ is continuous, then there exists a homeomorphism $g : M^n \to N^n$ such that $\rho(g(x), f(x)) < \epsilon(x)$ for all $x \in M^n$.*

Corollary 7.4.3. *Let $f : M^n \to N^n$ be a cell-like map between n-manifolds, $n \geq 5$, C a closed subset of N^n such that $f|f^{-1}(C)$ is 1-1, and $\epsilon : N^n \to [0, \infty)$ a continuous function such that $\epsilon^{-1}(0) = C$. Then there exists a homeomorphism $g : M^n \to N^n$ satisfying $\rho(g(x), f(x)) < \epsilon(x)$ for all $x \in M^n \smallsetminus f^{-1}(C)$ and $g|f^{-1}(C) = f|f^{-1}(C)$.*

The intent for the remainder of §7.4 is to develop conditions, for later use, under which a cell-like image of a manifold has the DDP. In support of that aim, the immediate issue is to prove that such a cell-like image is an ANR provided it is finite dimensional.

Proposition 7.4.4. *Suppose $p : Y \to X$ is a closed, cell-like mapping defined on a locally compact ANR Y, (K, L) is a pair of finite simplicial complexes, $\mu : K \to X$ is a map, $\nu : L \to Y$ is a map such that $p\nu = \mu|L$ and $\epsilon > 0$. Then there exists a map $\widetilde{\mu} : K \to Y$ with $\widetilde{\mu}|L = \nu$ and there exists a homotopy $H_t : K \to X$ such that $H_0 = p\widetilde{\mu}$, $H_1 = \mu$, $H_t|L = \mu|L$ and $\rho(H_t, \mu) < \epsilon$ for all $t \in I$.*

The preceding proposition supplements Approximate Lifting Proposition 3.2.10. The next lemma serves as the principal tool, and its proof retraces the one given for 3.2.10.

Lemma 7.4.5. *Under the hypothesis of Proposition 7.4.4, there exists $\delta > 0$ such that, for any two maps $\alpha_0, \alpha_1 : K \to Y$ extending ν with $\rho(p\alpha_e, \mu) < \delta$ for $e = 0, 1$, there is a homotopy $h_t : K \to Y$ such that $h_e = \alpha_e$ and, for all $t \in I$, $h_t|L = \nu$ and $\rho(ph_t, \mu) < \epsilon$.*

With Lemma 7.4.5 in hand, the derivation of Proposition 7.4.4 proceeds like the one showing why every non-isolated point in a locally connected complete metric space can be joined via a path to another point nearby. One constructs a sequence of lifts $\alpha_i : K \to Y$ such that not only do the images $p\alpha_i$ converge to μ but also successive images are connected via shorter and shorter homotopies, by Lemma 7.4.5.

Corollary 7.4.6. *Suppose $p : Y \to X$ is a closed, cell-like mapping defined on a locally compact ANR Y, W is an open subset of X, $w \in p^{-1}(W)$ and $i \geq 0$. Then $p_* : \pi_i(p^{-1}(W), w) \to \pi_i(W, p(w))$ is an isomorphism.*

Corollary 7.4.7. *If $p : Y \to X$ is a closed, cell-like mapping defined on a locally compact ANR Y, then X is LC^k for all integers $k \geq 0$.*

In light of Theorem 0.6.1, we also have:

Corollary 7.4.8. *If $p : Y \to X$ is a closed, cell-like mapping from a locally compact ANR to a finite-dimensional metric space X, then X is an ANR.*

Lemma 7.4.9. *If Y is a locally compact ANR satisfying the DDP, then every map $g : N^2 \to Y$ defined on a compact 2-dimensional ∂-manifold N^2 can be approximated by embeddings. Moreover, if $S \subset Y$ is a closed set that has empty interior and is 0-LCC in Y, then each $g : N^2 \to Y$ can be approximated by an embedding $\lambda : N^2 \to Y$ such that $S \cap \lambda(N^2)$ is 0-dimensional.*

Proof. Find a countable collection $\{(D_i, E_i)\}_{i=1}^{\infty}$ of disjoint 2-cell pairs that separate points of N^2—that is to say, for any two points $x, x' \in N^2$, there exists an integer $i \geq 1$ such that $x \in D_i$ and $x' \in E_i$. In the space $C(N^2, Y)$ of continuous functions from N^2 to Y, let

$$O_i = \{f \in C(N^2, Y) \mid f(D_i) \cap f(E_i) = \emptyset\}.$$

Clearly O_i is open in $C(N^2, Y)$. By the DDP and Estimated Homotopy Extension Theorem 0.6.4, each $f \in C(N^2, Y)$ can be approximated by some $f' \in C(N^2, Y)$ such that $f'(D_i) \cap f'(E_i) = \emptyset$; in other words, O_i is dense in $C(N^2, Y)$. The Baire Category Theorem assures that each $f \in C(N^2, Y)$ can be approximated by $\lambda \in \cap_i O_i$, an embedding.

Let S denote a closed 0-LCC subset of Y. Choose triangulations $T_1, T_2, \ldots$ of I^2 with $\text{mesh}T_i < 1/i$. Let L_i denote the 1-skeleton of T_i and

$$O_i' = \{f \in O_i \mid f(L_i) \cap S = \emptyset\}.$$

Since S is 0-LCC in Y, O_i' is an open dense subset of $C(N^2, Y)$. Each $\lambda \in \cap_i O_i'$ is an embedding for which $\lambda(N^2) \cap S \subset \lambda(N^2 \setminus \cup_i L_i)$, a 0-dimensional set. □

A similar argument yields:

Lemma 7.4.10. *Suppose the space X is a union of locally compact ANRs Y_1 and Y_2, each Y_i is a closed subset of X and has the DDP, $S = Y_1 \cap Y_2$ has empty interior and is 0-LCC in Y_i, and any two maps $f_i : I^2 \to Y_i$ can be approximated, arbitrarily closely, by maps $F_i : I^2 \to Y_i$ such that $F_1(I^2) \cap F_2(I^2) = \emptyset$. Then S contains disjoint, 0-dimensional, σ-compact subsets Z_1, Z_2 such that any map from a compact 2-dimensional ∂-manifold N^2 to Y_i can be approximated by a map $g_i : N^2 \to Y_i$ with $g_i(N^2) \cap S \subset Z_i$.*

Lemma 7.4.11. *Let $p : M \to X$ be a closed, cell-like mapping from a connected n-manifold M onto a metric space X that contains a connected $(n-1)$-manifold S as a closed subset, where $X \setminus S$ is disconnected. Then $X \setminus S$ has precisely two components, and S is 0-LCC in the closure of each.*

Proof. Proposition 3.2.9 promises that p induces a cohomology isomorphism $p^* : H_c^{n-1}(S; \mathbb{Z}_2) \cong \mathbb{Z}_2 \to H_c^{n-1}(p^{-1}(S); \mathbb{Z}_2)$. The duality methods of §7.1 assure that $p^{-1}(S)$ separates M into at most two components, so it must separate into exactly two components, for otherwise $X \setminus S$ would be connected. Hence, $X \setminus S$ has exactly two components. Localization yields the 0-LCC conclusion. □

Proposition 7.4.12. *Let $p : M \to X$ be a cell-like map from an n-manifold M onto an ANR X that contains a connected $(n-1)$-manifold S, $n \geq 5$, as a closed subset, where $X \setminus S$ is the disjoint union of components U_1, U_2 such that each $\overline{U}_j$ satisfies the Disjoint Disks Property and, moreover, any two maps $f_j : I^2 \to \overline{U}_j$ can be approximated, arbitrarily closely, by maps $F_j : I^2 \to \overline{U}_j$ such that $F_1(I^2) \cap F_2(I^2) = \emptyset$. Then X has the Disjoint Disks Property and p is a near-homeomorphism.*

Proof. Apply Lemma 7.4.10 to obtain disjoint 0-dimensional σ-compact sets $Z_1, Z_2 \subset S$ such that any map from a compact 2-dimensional ∂-manifold N^2 to $\overline{U}_j$ can be approximated by a map $g : N^2 \to \overline{U}_j$ with $g(N^2) \cap S \subset Z_j$.

Since S is an ANR there exists a neighborhood V of S in X and a retraction $r : V \to S$. Define retractions $r_j : V \cup U_j \to \overline{U}_j$ by $r_j|\overline{U}_j = \text{incl}$ and $r_j|V \setminus U_j = r|V \setminus U_j$. Choose a neighborhood V' of S so that $r|V'$ is homotopic to the inclusion in V. Choose a third neighborhood V'' of S so that the closure of V'' is contained in V'. By the proof of the Estimated Homotopy Extension Theorem (Theorem 0.6.4) there is a map $r' : X \to X$ such that $r'|V'' = r|V''$, $r'|X \setminus V = \text{incl}$, and $r'(V) \subset V$. Define $r'' : X \to X$ by $r''|\overline{U}_j = r_i r'|\overline{U}_j$; then $r''|S = \text{incl}$, $r''(V'') \subset S$, and $r''(\overline{U}_j) = \overline{U}_j$. Observe that r'' can be made arbitrarily close to the identity.

Given maps $f_1, f_2 : I^2 \to X$, choose compact ∂-manifolds $A_i \subset I^2$ such that $f_i^{-1}(S) \subset A_i \subset f_i^{-1}(V'')$. Define $f_i' = r'' f_i$. Then $f_i'(A_i) \subset S$ and each component of $I^2 \setminus A_i$ is mapped by f_i' into either $\overline{U}_1$ or $\overline{U}_2$. Add the components that are mapped to $\overline{U}_1$ to A_i and define B_i to be the union of the closures of the remaining components. Then A_i and B_i are compact boundary submanifolds of I^2 that satisfy the following conditions.

(1) $A_i \cup B_i = I^2$,

(2) $A_i \cap B_i \subset \partial A_i \cap \partial B_i$, and

(3) $f_i'(A_i) \subset \overline{U}_1$ and $f_i'(B_i) \subset \overline{U}_2$.

Since S is a PL manifold, we can make a further adjustment so that

(4) $f_i'(A_i \cap B_i) \subset S \setminus (Z_1 \cup Z_2)$ and

(5) $f_1'(A_1 \cap B_1) \cap f_2'(A_2 \cap B_2) = \emptyset$.

Invoking the hypothesis that U_i has the DDP, one can obtain further approximations f_i'' to f_i' satisfying analogs of conditions (1)–(5), as well as

(6) $f_i''|A_i \cap B_i = f_i'|A_i \cap B_i$,

(7) $f_i''(A_i) \subset \overline{U}_1$ and $f_i''(B_i) \subset \overline{U}_2$, and

(8) $f_1''(A_1) \cap f_2''(A_2) = \emptyset = f_1''(B_1) \cap f_2''(B_2)$.

Then by choice of Z_1, Z_2 one can produce yet another set of approximations F_i satisfying the analogs of (1)–(8), as well as

(9) $S \cap (F_1(\operatorname{Int} A_1) \cup F_2(\operatorname{Int} A_2)) \subset Z_1$ and

(10) $S \cap (F_1(\operatorname{Int} B_1) \cup F_2(\operatorname{Int} B_2)) \subset Z_2$.

It follows from the prearranged (see (4) and (6))

$$Z_1 \cap Z_2 = \emptyset = (Z_1 \cup Z_2) \cap (F_1(A_1 \cap B_1) \cup F_2(A_2 \cap B_2))$$

that $F_1(I^2) \cap F_2(I^2) = \emptyset$. □

Corollary 7.4.13. *Let $p : M \to X$ be a cell-like map defined on an n-manifold M, $n \geq 5$, onto a metric space X containing an $(n-1)$-manifold S embedded in M as a closed, 1-LCC subset, where $X \setminus S$ is an n-manifold. Then X is an n-manifold and p is a near-homeomorphism.*

Proof. Since S locally separates X and the desired conclusion is local, it suffices to consider the case where X and S are connected and $X \setminus S$ has two components, U_1 and U_2. As S is LCC[1] in $\overline{U}_i = S \cup U_i$ $(i \in \{1, 2\})$ and U_i is an n-manifold, $\overline{U}_i$ has the DDP and the hypotheses of 7.4.12 are satisfied. □

Proposition 7.4.14. *Let $p : S^n \to X$ be a cell-like map onto a metric space X that contains an $(n-1)$-sphere S, $n \geq 5$, and $X \setminus S$ is the disjoint union of components U, V where $\overline{U} = A$ embeds in S^n and $\overline{V} = B$ is an n-cell. Then X has the Disjoint Disks Property and p is a near-homeomorphism.*

Proof. As in Proposition 7.4.12, given maps $f_1, f_2 : I^2 \to X$, for $\epsilon > 0$ and $i = 1, 2$ produce ϵ-approximations f_i' to f_i and ∂-manifolds A_i, B_i in I^2 satisfying

(1) $I^2 = A_i \cup B_i$,

(2) $A_i \cap B_i = \partial A_i \cap \partial B_i$,

(3) $f_1'(A_1) \cup f_2'(A_2) \subset A$ and $f_1'(B_1) \cup f_2'(B_2) \subset B$, and

(4) $f_1'(B_1) \cap f_2'(B_2) = \emptyset$ (since B is an n-cell).

Temporarily regard A as a subset of S^n. Identify a neighborhood W of A and retraction $R : W \to A$ such that $R(W \setminus A) \subset \operatorname{Bd} A = S$. Restrict W so

R moves points less than ϵ. Then approximate $f_i'|A_i \to A \subset S^n$ by $g_i : A_i \to W$, with g_i ϵ-close to f_i', $g_i|A_i \cap B_i = f_i'|A_i \cap B_i$, and $g_1(A_1) \cap g(A_2) = \emptyset$.

Once again treating A as a subset of X, exploit the n-cell structure of B to adjust Rg_i to a map $g_i' : A_i \to X$ such that $\rho(g_i', g_i) < \epsilon$, $g_i'|g_i^{-1}(A) = g_i'|g_i^{-1}(A)$ and $g_i'(g_i^{-1}(W \setminus A)) \subset \operatorname{Int} B$. Define $F_i : I^2 \to X$ as $F_i|A_i = g_i'$ and $F_i|B_i = f_i'|B_i$. Note $\rho(F_i, f_i) < 4\epsilon$. Intersections between $F_1(I^2)$ and $F_2(I^2)$ occur at points of $F_1(A_1) \cap F_2(A_2) \cap \operatorname{Int} B$ or of $F_i(A_i) \cap F_j(B_j) \subset \operatorname{Int} B$, $i \neq j$, and can be removed easily by general position. Hence, X has the DDP. □

Historical Notes. R. D. Edwards outlined a proof of the Cell-like Approximation Theorem in his ICM 1978 article (Edwards, 1980). Details of Edwards's proof for $n \geq 6$ are presented in (Daverman, 1986); the 5-dimensional case is treated in (Daverman and Halverson, 2007). Corollary 7.4.2 is originally due to L. C. Siebenmann (1972); its analog in dimension $n = 3$ was done by S. Armentrout (1971) and in dimension $n = 4$ by M. H. Freedman and F. S. Quinn (1990).

Cannon (1978), (1979) introduced the Disjoint Disks Property and early on he conjectured its fundamental role for the Cell-like Approximation Theorem.

The hypothesis about finite-dimensional-image in the Cell-like Approximation Theorem is a necessary one. A. N. Dranishnikov (1989) established the existence of a cell-like map on a 3-dimensional compactum with infinite-dimensional image; this automatically gave a dimension-raising cell-like map defined on S^n, $n \geq 7$. Improving upon Dranishnikov's example slightly, J. Dydak and J. J. Walsh (1993) produced dimension-raising cell-like maps on 2-dimensional compacta and, hence, on S^5. In contrast, work of G. Kozlowski and J. J. Walsh (1983) certifies that cell-like maps defined on 3-manifolds have 3-dimensional images.

Exercise

7.4.1. If Y is a locally compact ANR with the DDP, then each map $f : I^2 \to Y$ can be approximated by a 1-LCC embedding.

7.5. Determining n-cells by embeddings of M_n^{n-1} in S^n

The combined aim of this section and the next is to characterize local flatness of codimension-one manifold embeddings in terms of the 1-LCC condition. Taking a step in that direction, this section establishes (Corollary 7.5.10) that, given an $(n-1)$-sphere $\Sigma \subset S^n$ and component W of $S^n \setminus \Sigma$, $\overline{W}$ is an n-cell if and only if there is an embedded Menger continuum $e(M_n^{n-1})$ in S^n with $\Sigma \subset e(M_n^{n-1}) \subset \overline{W}$. Rounding this out, the next section demonstrates that for any 1-LCC embedded $(n-1)$-sphere $\Sigma \subset S^n$ and complementary domain W, there exists such an embedding $e : M_n^{n-1} \to \overline{W}$.

A secondary goal of the section at hand is a positional characterization of the $(n-1)$-dimensional Menger space in S^n. To that end, we present an *ad hoc* definition of objects called $\mathcal{S}$-curves, which include the standard Menger space M_n^{n-1}, and ultimately (Theorem 7.5.7) we show that any two such $\mathcal{S}$-curves are homeomorphic. This topological analysis of M_n^{n-1} is not essential to the primary purpose: all the lemmas developed in this section lead to 7.5.7 and can be ignored, provided one broadens Theorem 7.5.8 (using the same proof presented here) to detect the n-cell though embeddings of arbitrary $\mathcal{S}$-curves in S^n, not simply through embedded copies of M_n^{n-1}.

The starting point is an elementary result from decomposition theory. A sequence of sets $X_1, X_2, \ldots$ in a metric space is called a *null sequence* if $\operatorname{diam} X_i \to 0$ as $i \to \infty$.

Proposition 7.5.1 (Null sequence decompositions into flat n-cells). *Let $B_1, B_2, \ldots$ be a null sequence of pairwise disjoint, flat n-cells in S^n, U an open subset of S^n containing $\cup_i B_i$, and G the decomposition of S^n having the sets B_i as nondegenerate elements. Then G is shrinkable fixing $S^n \setminus U$. In particular, there exists a surjective map $f : S^n \to S^n$ such that the nondegenerate point preimages of f are the cells $B_1, B_2, \ldots$ and $f|S^n \setminus U =$ Id.*

Proof. Consider any nondegenerate $g_0 \in G$. Since g_0 is flat, regard it as the standard ball of radius 1 centered at O in $\mathbb{R}^n = S^n \setminus \{\infty\}$. Given $\delta > 0$, restrict further, if necessary, so $B(g_0; \delta) \subset U$ and let k denote the smallest positive integer such that $k\delta/3 > 1$. We will produce a homeomorphism $\Theta : S^n \to S^n$ such that

(a) $\Theta|S^n \setminus B(g_0; \delta) = \text{Id}$,

(b) $\operatorname{diam} \Theta(g_0) < \delta$, and

(c) for $g \in G$ either $\operatorname{diam} \Theta(g) < \delta$ or $\Theta(g) = g$.

The homeomorphism Θ will be expressed as a composition $\Theta = \theta_{k-1} \cdots \theta_1$, of $k-1$ homeomorphisms, where each θ_j moves points less than $\delta/3$ and compresses $\theta_{j-1} \cdots \theta_1(g_0)$ radially into the ball of radius $(k-j)\delta/3$.

Using the nullity of the nondegenerate elements, require θ_1 to be the identity outside an open set $U_1 \subset U$ so near g_0 that all $g \in G$ meeting U_1 have diameter less than $\delta/3$. Similarly, after $\theta_{j-1}, \ldots, \theta_1$ have been defined, require θ_j to be the identity outside an open set $U_j \subset U_{j-1}$ so close to $\theta_{j-1} \cdots \theta_1(g_0)$ that, for any other $g \in G$ whose image under $\theta_{j-1} \cdots \theta_1$ meets U_j, $\operatorname{diam} \theta_{j-1} \cdots \theta_1(g) < \delta/3$.

For $i = -1, 0, 1, \ldots, k$, set $\alpha_i = (k-i)/k$. Each θ_j can be defined so as to have similar effect on all rays R emanating from O: there will be a positive number $\xi_j \in (\alpha_{j-1}, \alpha_{j-2})$ such that the segment of R of length ξ_j based at O lies in U_j. There will be four special points on R: the points P_j, Q_j, S_j, T_j at distances $\alpha_{j+1}, \alpha_j, \alpha_{j-1}, \xi_j$, respectively, from O; θ_j will move only those points between P_j and T_j, will send S_j to Q_j and will be linear on the intervals $[P_j, S_j]$ and $[S_j, T_j]$. Thus, each θ_j will move points at most $1/k < \delta/3$ and will compress $\overline{B(O;\alpha_{j-1})}$ to $\overline{B(O;\alpha_j)}$.

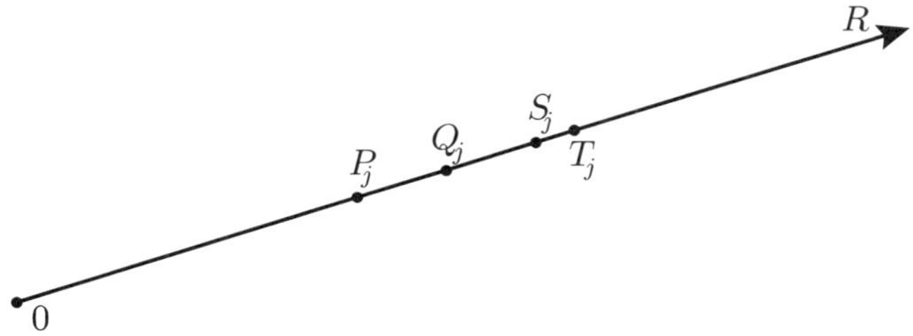

Figure 7.7. The action of θ_j on R

Accordingly, for $g \in G$, $g \neq g_0$, and $j \in \{2, 3, \ldots, k-1\}$, either $\theta_j \cdots \theta_1(g) = \theta_{j-1} \cdots \theta_1(g)$ or $\operatorname{diam} \theta_j \cdots \theta_1(g) < \delta$; moreover, if after the jth compression, $\operatorname{diam} \theta_j \cdots \theta_1(g) \geq \delta/3$, then $\operatorname{diam} \theta_j \cdots \theta_1(g) < \delta$ and $\Theta(g) = \theta_j \cdots \theta_1(g)$. Finally, since $\alpha_{k-1} = 1/k < \delta/3$ and

$$\Theta(g_0) = \theta_{k-1} \cdots \theta_1(g_0) = \overline{B(O;\alpha_{k-1})} \subset B(O;\delta/3),$$

Θ shrinks g_0 to sufficiently small size. By construction Θ moves no point of $S^n \smallsetminus U_1$.

Upon performing this shrinking in pairwise disjoint neighborhoods of each of the finitely many large n-cells in the collection G, we see that G itself is shrinkable fixing $S^n \smallsetminus U$. □

Remark. While the statement and proof of the last result are elementary, they are still quite delicate. For example, a decomposition into points and a null sequence of cellular arcs need not be shrinkable (Daverman and Walsh, 1982).

Corollary 7.5.2. *If $B_1, B_2, \ldots$ is a null sequence of pairwise disjoint, flat n-cells in the interior of B^n and G is the decomposition of B^n having the sets B_i as nondegenerate elements, then the decomposition space B^n/G is an n-cell. Furthermore, if G_m is the decomposition of the ∂-manifold $D_m^n = B^n \smallsetminus \cup_{i=1}^m \operatorname{Int} B_i$ having $B_{m+1}, B_{m+2}, \ldots$ as nondegenerate elements, then the associated decomposition space is homeomorphic to D_m^n.*

Proof. Regard D_m^n as a subset of S^n. Apply Proposition 7.5.1 and Theorem 2.3.4 with $U = \operatorname{Int} D_m^n$. □

We will make use, without proof, of the fundamental Annulus Theorem. This will be discussed more extensively in §8.8.

Theorem 7.5.3 (Annulus Theorem). *Let B' denote a flat n-cell in the interior of an n-cell B ($n > 4$). Then $B \setminus \operatorname{Int} B'$ is homeomorphic to $S^{n-1} \times I$.*

Let $B_1, \ldots, B_k$ be pairwise disjoint, flat n-cells in the interior of an n-cell B. The ∂-manifold $B \setminus \cup_i \operatorname{Int} B_i$ is called an *n-cell with k holes.* The ∂-manifold D^n_m of Corollary 7.5.2 is a relevant example of an n-cell with m holes.

Corollary 7.5.4 (Generalized Annulus Theorem). *If B^* and C^* are n-cells with k holes, $n > 4$, then every homeomorphism h from a component of ∂B^* to a component of ∂C^* extends to a homeomorphism $H : B^* \to C^*$.*

The proof is an exercise.

Definition. An $(n-1)$-*dimensional Sierpiński curve* is a compact metric continuum X which admits an embedding h in S^n such that the components of $S^n \setminus h(X)$ form a null sequence $U_1, U_2, \ldots$ satisfying: (1) each $S^n \setminus U_i$ is an n-cell, (2) $\overline{U}_i \cap \overline{U}_j = \emptyset$ whenever $i \neq j$, and (3) $\overline{\cup_i U_i} = S^n$. For brevity we will say that a compact continuum $X' \subset S^n$ is an $\mathcal{S}$-curve if it is the image of an embedding $h : X \to S^n$, where X satisfies conditions (1)–(3).

The prototypical $\mathcal{S}$-curve is the standard Menger space M^{n-1}_n. The immediate goal is to prove that any two such $\mathcal{S}$-curves are topologically equivalent.

Lemma 7.5.5. *If X is an $(n-1)$-dimensional $\mathcal{S}$-curve in S^n, then for each $\epsilon > 0$ there is an embedding $e : X \to S^n$ such that $\rho(e, \mathrm{incl}_X) < \epsilon$ and the components of $S^n \setminus e(X)$ are bounded by flat $(n-1)$-spheres.*

Proof. For any component U of $S^n \setminus X$, $S^n \setminus U$ is an n-cell which can be re-embedded in its own interior so the image of $\partial(S^n \setminus U)$ is bicollared and, hence, flat. It follows almost automatically that the image of X under this re-embedding is an $\mathcal{S}$-curve. The re-embedding can be controlled to move points only a short distance and to have support very close to $\overline{U}$. Infinite repetition, with increasingly strict motion controls, yields the lemma. □

Definitions. Say that an $(n-1)$-dimensional $\mathcal{S}$-curve $X \subset S^n$ is *special* if each of the components of $S^n \setminus X$ is bounded by a flat sphere. Let X be a special $(n-1)$-dimensional $\mathcal{S}$-curve in S^n and $U_0, U_1, U_2 \ldots$ the components of $S^n \setminus X$. A *subdivision* of X is a division of X into a finite number of such $\mathcal{S}$-curves, brought about by taking a simplicial subdivision T of the compact ∂-manifold R obtained by adding to X all but a finite number $U_0, \ldots, U_m$ of its complementary domains in such a way that the $(n-1)$-skeleton of T lies entirely in X, contains the boundary of R, and does not meet the boundary of any component of $S^n \setminus X$ other than $U_0, \ldots, U_m$. The intersection of the n-cells of T with X gives a collection of $(n-1)$-dimensional $\mathcal{S}$-curves. The subdivision is said to have *mesh less than* ϵ if each n-cell in T has diameter less than ϵ.

Lemma 7.5.6. *Suppose X and Y are special $(n-1)$-dimensional $\mathcal{S}$-curves in S^n $(n \neq 4)$, U and V are components of $S^n \setminus X$ and $S^n \setminus Y$, respectively, h is any homeomorphism of* $\operatorname{Bd} U$ *onto* $\operatorname{Bd} V$, *and $\epsilon > 0$. Then there exist ϵ-subdivisions of X and Y whose $(n-1)$-dimensional skeleta correspond under a homeomorphism h' that extends h.*

Proof. List the components $U_0 = U, U_1, U_2, \ldots$ of $S^n \setminus X$ and, similarly, the components $V_0 = V, V_1, V_2, \ldots$ of $S^n \setminus Y$. Choose an integer $m > 0$ such that all U_i and V_i, $i > m$, have diameter less than ϵ. Form the decomposition space A_X of $S^n \setminus \cup_{i=0}^m U_i$ determined by the nondegenerate elements $\overline{U}_{m+1}, \overline{U}_{m+2}, \ldots$ and, similarly, the decomposition space A_Y of $S^n \setminus \cup_{i=0}^m V_i$ determined by $\overline{V}_{m+1}, \overline{V}_{m+2}, \ldots$. Let $\pi_X : S^n \setminus \cup_{i=0}^m Ui \to A_X$ and $\pi_Y : S^n \setminus \cup_{i=0}^M Vi \to A_Y$ denote the associated decomposition maps. Here A_X and A_Y are n-cells with holes—an equal number of holes, by design. Corollary 7.5.4 assures that the homeomorphism $\pi_Y h(\pi_X)^{-1} : \pi_X(\operatorname{Bd} U) \to \pi_Y(\operatorname{Bd} V)$ extends to a homeomorphism $H : A_X \to A_Y$.

For each $\delta > 0$ there exists a simplicial triangulation T of W' of mesh less than δ whose $(n-1)$-skeleton Σ intersects none of the countably many points having nondegenerate preimages under either $H\pi_X$ or π_Y. Then the sets $K = (H\pi_Z)^{-1}(\Sigma)$ and $K' = (\pi_Y)^{-1}(\Sigma)$ each correspond in 1-1 fashion with Σ. Moreover, when σ is an n-simplex of T, then $(\pi_Y)^{-1}(\partial\sigma) \subset Y$ and $(H\pi_X)^{-1}(\partial\sigma) \subset X$ are flat $(n-1)$-spheres in S^n, by Corollary 7.4.3 to the Cell-like Approximation Theorem; as a result, K and K' effect subdivisions of X and Y, respectively. Since point preimages under $H\pi_X$ and π_Y have diameter less than ϵ, one can choose T of sufficiently small mesh that K and K' have mesh less than ϵ. The desired homeomorphism $h' : K \to K'$ can be defined as the restriction of $(\pi_Y)^{-1}H\pi_X$. □

Theorem 7.5.7. *Any two $(n-1)$-dimensional $\mathcal{S}$-curves in S^n are homeomorphic.*

Proof. Consider any two special $\mathcal{S}$-curves X and Y in S^n. For $i = 1, 2, \ldots$ Lemma 7.5.6 promises an embedding $e_i : X \to B(Y; 1/i)$ such that $Y \subset B(e_i(X); 1/i)$. These embeddings submit to controls ensuring that $\{e_i\}$ forms a Cauchy sequence. Moreover, given any two points $x_1, x_2 \in X$ there exist disjoint n-cells $C_1, C_2 \subset S^n$ such that $e_j(x_1) \in C_1$ and $e_j(x_2) \in C_2$ for sufficiently large j. Hence, the sequence $\{e_i\}$ converges to a homeomorphism $X \to Y$. □

Theorem 7.5.8. *Let e denote an embedding of the $(n-1)$-dimensional Menger space M_n^{n-1} into S^n $(n \geq 5)$ and V a component of $S^n \smallsetminus e(M_n^{n-1})$. Then $S^n \smallsetminus V$ is an n-cell.*

Proof. List the components $U_1, U_2, \ldots$ of $S^n \smallsetminus M_n^{n-1}$ and also the components $V = V_1, V_2 \ldots$ of $S^n \smallsetminus e(M_n^{n-1})$. Choose these indices so that $\operatorname{Bd} V_i = e(\operatorname{Bd} U_i)$ for each i.

Every $\overline{V}_i$ is contractible, since it is a simply connected, homologically trivial (by duality) ANR.

Examine the decompositions G and G' of $S^n \smallsetminus U_1$ and $S^n \smallsetminus V_1$ having $\overline{U}_2, \overline{U}_3, \ldots$ and $\overline{V}_2, \overline{V}_3, \ldots$ as their respective nondegenerate elements. Both G and G' are cell-like, upper semicontinuous decompositions; the nondegenerate elements of G form a null sequence of flat n-cells in the interior of the n-cell $S^n \smallsetminus U_1$. Let $\varphi : S^n \smallsetminus U_1 \to A$ and $\varphi' : S^n \smallsetminus V_1 \to A'$ denote the associated decomposition maps. Obviously there is a unique homeomorphism $e' : A \to A'$ such that the following diagram is commutative:

$$\begin{array}{ccccc} M_n^{n-1} & \xrightarrow{\subset} & S^n \smallsetminus U_1 & \xrightarrow{\varphi} & A \\ \downarrow{\scriptstyle e} & & & & \downarrow{\scriptstyle e'} \\ e(M_n^{n-1}) & \xrightarrow{\subset} & S^n \smallsetminus V_1 & \xrightarrow{\varphi'} & A' \end{array}$$

By Corollary 7.5.2, A is an n-cell, so A' is an n-cell as well. A minor modification of Corollary 7.4.3 provides a homeomorphism $\Phi' : S^n \smallsetminus V_1 \to A'$ that agrees with φ' on $\operatorname{Bd} V_1$. Hence, $S^n \smallsetminus V_1$ is also an n-cell. □

Corollary 7.5.9. *For any embedding $e : M_n^{n-1} \to S^n$ $(n \geq 5)$ of the $(n-1)$-dimensional Menger space, $e(M_n^{n-1})$ is an $\mathcal{S}$-curve.*

Corollary 7.5.10. *Let $\lambda : \partial I^n \to S^n$ $(n \geq 5)$ be an embedding and W a component of $S^n \smallsetminus \lambda(\partial I^n)$. Then $\overline{W}$ is an n-cell if and only if λ can be extended to an embedding $\Lambda : M_n^{n-1} \to \overline{W}$.*

Historical Notes. The positional characterization of Sierpiński curves in S^n presented in Theorem 7.5.7 is due to Cannon (1973b), who based his argument on that given by G. T. Whyburn for the 2-dimensional case.

The proof of the Annulus Conjecture was a sweeping breakthrough, by Kirby (1969); the key idea, usually referred to as the torus trick, had profound implications, including the deep analysis of PL and DIFF structures of manifolds (Kirby and Siebenmann, 1977). More about the momentous importance of tori comes up in Chapter 8.

Proposition 7.5.1, in more general form, was proved by R. J. Bean (1967), who credited Bing for the technique.

Exercises

7.5.1. Let G denote an upper semicontinuous decomposition of an n-dimensional ∂-manifold N such that for each nondegenerate element $g_0 \in G$ and each $\epsilon > 0$ there exists an n-cell B with $g_0 \subset B \subset B(g_0; \epsilon)$, where all nondegenerate elements of G that meet B lie in $\operatorname{Int} B$. Then G is shrinkable.

7.5.2. Show that every homeomorphism between two special $\mathcal{S}$-curves in S^n extends to a homeomorphism of S^n.

7.5.3. Prove Corollary 7.5.4.

7.6. The 1-LCC characterization of local flatness

All the groundwork now has been laid for the foundational characterization, in Theorem 7.6.1 below, of locally flat codimension-one manifold embeddings in terms of the 1-LCC condition. The initial steps reduce the issue to the 1-LCC characterization of flat codimension-one spheres in S^n, which is treated in Theorem 7.6.5; its proof, in turn, capitalizes on the Menger space technology of the preceding section (Corollary 7.5.10).

Theorem 7.6.1. *Every 1-LCC embedding of an $(n-1)$-manifold S in an n-manifold M $(n \geq 5)$ is locally flat.*

No hypothesis about M being PL is needed here; all constructions can be localized to Euclidean patches in M.

The indispensable tool is the following Bubble Lemma. Its proof retraces that of 1-LCC Push-off Proposition 7.2.1, using infinite controlled engulfing.

Lemma 7.6.2 (Bubble Lemma). *Suppose S is an $(n-1)$-manifold in a connected PL n-manifold M $(n \geq 5)$, $D \subset S$ is an $(n-1)$-cell such that S is 1-LCC at each point of* $\operatorname{Int} D$*, U is a component of $M \setminus S$ and $\epsilon : \operatorname{Int} D \to (0,1)$ is a continuous function. Then there exists a 1-LCC embedding $e : \operatorname{Int} D \to U$ such that $d(s, e(s)) < \epsilon(s)$ for all $s \in \operatorname{Int} D$.*

Proof. The embedding e will be $\psi|\operatorname{Int} D$, where ψ is a controlled push of M such that $\psi(\operatorname{Int} D) \subset U$ and $\psi|\partial D = \text{incl}_{\partial D}$. The existence of this ψ stems from an engulfing program establishing that infinite codimension-three complexes near $\operatorname{Int} D$ can be pushed into a preassigned component of $M \setminus S$.

Determine a small connected neighborhood W of $\operatorname{Int} D$ such that W intersects S at $\operatorname{Int} D$, and let W_+ be a component of $W \setminus D$. The claim is that, for $k \leq n-3$ and a sufficiently small neighborhood $W' \subset W$ of $\operatorname{Int} D$, any infinite k-complex in W' admits a controlled push φ into W_+, where $\varphi | M \smallsetminus W = \operatorname{incl} | M \smallsetminus W$. To complete the argument, one works with a PL neighborhood N of $\operatorname{Int} D$ in W', pushes the $(n-3)$-skeleton of N to one side of $W \smallsetminus D$, pushes the dual 2-skeleton of N to the other side of $W \smallsetminus D$, and stretches across the join structure of N to obtain ψ, just as in 7.2.1. Of course, controls on the pushes and, more automatically, on the stretch are necessary to assure that all three adjustments operating in W extend over the rest of M via the Identity.

One way to nail down the engulfing claim is to produce ∂-manifolds A and B whose interiors cover $\operatorname{Int} D$ and whose components are compact, and then to use the usual controlled engulfing methodology, applied component by component, to obtain that complexes near either A or B can be pushed into W_+ with control. Given a k-complex K in W', express it as a union of closed subpolyhedra K_A and K_B, where K_A, K_B are near A, B, respectively. Push K_A into W' with enough control that image of K_B is still near B. Then push that image into W', fixing the image of K_A. Details are left to the reader. □

Lemma 7.6.3. *Suppose the $(n-1)$-sphere $\Sigma \subset S^n$ is the union of two $(n-1)$-cells D and D' such that $\partial D = D \cap D' = \partial D'$, D is 1-LCC in S^n and* $\operatorname{Int} D'$ *is 1-LCC in S^n. Then Σ is 1-LCC in S^n.*

Proof. Focus on $s \in \partial D$; the conclusion is obvious for other points of Σ. Given any neighborhood N_1 of s, find a smaller neighborhood N_2 such that $N_2 \cap (D' \smallsetminus D)$ is simply connected (and connected). Use the hypothesis about D being 1-LCC in S^n to locate another neighborhood $N_3 \subset N_2$ such that all loops in $N_3 \smallsetminus D$ are null-homotopic in $N_2 \smallsetminus D$.

Hence, each loop $f : \partial I^2 \to N_3 \smallsetminus \Sigma$ extends to a map $F_1 : I^2 \to N_2 \smallsetminus D$. Let Z denote the component of $I^2 \smallsetminus (F_1)^{-1}(D')$ containing ∂I^2. Since $D' \cap N_2$ is an ANR, $F_1 | \operatorname{Fr} Z$ extends to a map sending a small neighborhood of $\operatorname{Fr} Z$ in $I^2 \smallsetminus Z$ into $D' \cap N_2$. Thus, there exist a compact, connected ∂-manifold Q, $\partial I^2 \subset Z \subset Q \subset I^2$, and map $F_2 : Q \to N_2 \smallsetminus D$ such that $F_2 | Z = F_1 | Z$ and $F_2(Q \smallsetminus Z) \subset N_2 \cap (D' \smallsetminus D)$. The connectedness of Q implies each component of $I^2 \smallsetminus Q$ is bounded by a simple closed curve, so by the simple-connectedness of $N_2 \cap (D' \smallsetminus D)$, $F_2 | Z$ extends to a map $F_3 : I^2 \to N_2 \smallsetminus D$ with $(F_3)^{-1}(\Sigma) = I^2 \smallsetminus Z$.

As $N_2 \cap D'$ is two-sided in N_2 (Corollary 7.1.7), choose a connected open set $U_{D'}$ such that $N_2 \cap D' \subset U_{D'} \subset N_2$ and $U_{D'} \smallsetminus D'$ has two components; $N_2 \cap D'$ is LCC^1 in the closure (rel N_2) of each of these components, by Proposition 7.1.11. Cover $(F_3)^{-1}(D')$ by another compact ∂-manifold $Q' \subset (F_3)^{-1}(U_{D'}) \cap \operatorname{Int} I^2$.

We claim that the image of each component C of Q' under F_3 meets the closure of only one component of $U_{D'} \setminus D'$. To examine that image, let Z^c denote $I^2 \setminus Z$. By duality,

$$\check{H}^1(Z^c) \cong H_1(\operatorname{Int} I^2, \operatorname{Int} Z) \cong \widetilde{H}_{\mathrm{C}}(\operatorname{Int} Z) \cong 0.$$

Since Z^c splits into the disjoint union of the compact sets $Z^c \cap C$ and $Z^c \setminus C$, $\check{H}^1(Z^c \cap C) \cong 0$. Consequently,

$$0 \cong \check{H}^1(Z^c \cap \operatorname{Int} C) \cong H_1(\operatorname{Int} C, \operatorname{Int} C \setminus Z^c) \to \widetilde{H}_0(\operatorname{Int} C \setminus Z^c) \to 0,$$

so $\operatorname{Int} C \setminus Z^c$ is connected, and its image under F_3 meets exactly one of the components of $U_{D'} \setminus D'$. Applying Lemma 3.3.3 to $F_3|C$ for each C, we obtain a map $F_4 : I^2 \to N_2 \setminus \Sigma$ such that $F_4|I^2 \setminus Q' = F_3|I^2 \setminus Q'$ (and $F_4(C) \subset U_{D'} \setminus D'$); in particular, $F_4|\partial I^2 = f$. □

Corollary 7.6.4. *Suppose S is an $(n-1)$-manifold 1-LCC embedded in an n-manifold M $(n \geq 5)$ and $s \in S$. Then there exist a neighborhood N_s of s in M and a 1-LCC embedded $(n-1)$-sphere $\Sigma \subset N_s$ such that $N_s \approx \mathbb{R}^n$ and $\Sigma \cap S$ contains a neighborhood of s in S.*

As a result, Theorem 7.6.1 reduces to the following:

Theorem 7.6.5. *An $(n-1)$-sphere Σ in S^n $(n \geq 5)$ is flat if and only if it is 1-LCC embedded.*

Proof. Let $\lambda : \partial I^n \to \Sigma$ be a homeomorphism and W a component of $S^n \setminus \Sigma$. The goal will be to extend λ to an embedding $e : M_n^{n-1} \to \overline{W}$ and to apply Corollary 7.5.10.

Let $\kappa = \{k_1, k_2, \dots\}$ be a sequence of integers, $k_i \geq 3$. Associated with κ is an $(n-1)$-dimensional $\mathcal{S}$-curve, X_κ, constructed in a manner modelled on that of M_n^{n-1} in §3.5. Let T_0 be the trivial subdivision of I, just as in that construction. Let T_1 be the subdivision of I into k_1 subintervals of equal length. Assuming T_j to be a subdivision of I into intervals of equal length $1/k_1 \cdots k_j$, let T_{j+1} denote the subdivision obtained by sectioning each interval of T_j into k_{j+1} subintervals of equal length. As a result, T_{j+1} induces a subdivision T_{j+1}^n of I^n into a multitude of isometric subcubes. Set $P_0 = I^n$ and let P_{j+1} denote the union of all n-dimensional subcubes of T_{j+1}^n that lie in P_j and intersect its $(n-1)$-skeleton L_j (as determined by T_j^n). Then $X_\kappa = \cap_j P_j$.

By definition $X_\kappa = M_n^{n-1}$ in the special case $\kappa = \{3, 3, 3, \dots\}$.

For $j = 1, 2, \dots$ there exists a retraction $r_j : P_j \to L_{j-1}$ (since P_j fills no n-cube of T_{j-1}^n), where r_j moves no point more than d_{j-1}, the diameter of the n-cubes from T_{j-1}^n.

The proofs of the next two results are based on routine inverse limit arguments.

Lemma 7.6.6. *The inverse limit of the sequence $\{L_j, r_j|L_j\}$ is X_κ.*

Lemma 7.6.7. *Suppose $\{L_j, r_j|L_j\}$ as above, and suppose $\{\lambda_j : L_j \to S^n\}$ is a sequence of embeddings such that $\lambda_{j+1}|L_j = \lambda_j$ for all $j \geq 1$. Suppose also that for each $\epsilon > 0$ there exists an index m_ϵ such that diam $\lambda_j(\partial\sigma) < \epsilon$ for all $j \geq m_\epsilon$ and all n-cells $\sigma \subset P_j$ in T_j^n. Then there exists an embedding $\Lambda : X_\kappa \to S^n$ such that $\Lambda|L_j = \lambda_j$ for all $j \geq 0$.*

Continuing with the proof of 7.6.5, we choose an integer $k_1 \geq 3$ such that, for the subdivision T_1 of I into k_1 intervals of equal length and T_1^n the associated subdivision of I^n determined by the product of n copies of T_1, we have diam $\lambda(\sigma \cap L_0) < 1/(3^{n-1} \cdot 2) = \delta_1$ for all $\sigma \in T_1^n$. Order the n-cubes $\sigma_1, \sigma_2, \ldots, \sigma_{m(1)}$ of T_1^n in P_1 so that $\partial\sigma_i \cap L_0^{i-1}$ is a PL $(n-1)$-cell E_i, where $L_0^{i-1} = L_0 \cup \bigcup_{t=1}^{i-1} \partial\sigma_t$. We will apply Lemma 7.6.2 recursively to extend λ to an embedding $\lambda : L_0^i \to \overline{W}$ such that diam $\lambda(\partial\sigma_i) < 1/2$ for all σ_i. In addition, we will have assured that $\lambda(E_i)$ is 1-LCC, so $\lambda(\partial\sigma_i)$ will be 1-LCC by Lemma 7.6.3. At the end of the process, when $i = m(1)$, we will have an embedding $\lambda_1 : L_1 \to \overline{W}$ with $\lambda_1(\partial\sigma_i)$ being a 1-LCC embedded sphere and diam $\lambda_1(\partial\sigma_i) < 1/2$ for $i = 1, 2, \ldots, m(1)$.

For the promised assurance that $\lambda(E_i)$ is 1-LCC, we will employ the following variation on Lemma 7.6.3. The proof is an exercise.

Lemma 7.6.8. *Suppose the $(n-1)$-cell $E \subset S^n$ is the union of two $(n-1)$-cells E' and E'' such that $E' \cap E''$ is an $(n-2)$-cell in the boundary of each, E is 1-LCC in S^n and $E \smallsetminus E'$ is 1-LCC in S^n. Then E is 1-LCC in S^n.*

The recursive process runs as follows. Start with $S_0 = \lambda(\partial I^n)$. We produce $(n-1)$-spheres S_i $(i = 1, \ldots, m(1))$ and extensions of λ over L_i so $S_i \subset \lambda(L_i)$ using the Bubble Lemma. It gives approximations $\lambda|\operatorname{Int} E_i$ by new 1-LCC embeddings λ_i' with image very close to $\lambda(E_i)$; since each $\partial\sigma_i \smallsetminus \operatorname{Int} E_i$ is a PL $(n-1)$-cell E_i', $\lambda|\partial E_i'$ extends to a homeomorphism (still called λ) of E_i' to the closure of $\lambda_i'(\operatorname{Int} E_i)$. The sphere S_i is obtained from S_{i-1} by replacing $\lambda(E_i)$ with $\lambda(E_i') = \lambda_i(E_i')$. Let W_j denote the component of $S^n \smallsetminus S_j$ contained in W. In the successive applications of the Bubble Lemma require that $\lambda_i'(E_i) \subset W_i$. Inductively assuming that $\lambda(L_{i-1}) \cap W_{i-1} = \emptyset$, we see that $\lambda|L_i = L_{i-1} \cup E_i'$ is 1-1 and $\lambda(L_i) \cap W_i = \emptyset$, as required.

To control sizes of the $\lambda(\sigma_j)$, one can partition the cubes σ_i into n pairwise disjoint groups. Begin with the cubes from the first group, proceed to those in the second group, and so on. One way to obtain an acceptable grouping is to insist that k_1 be odd and to partition the intervals of T_1 into two pairwise disjoint groups, designated as black and white, with the two intervals containing the endpoints of I being black. The first group of cubes from T_1^n consists of all products of black intervals; generally, the jth group ($j > 1$) consists of all n-cubes from T_1^n expressed as a product involving exactly $j-1$ white intervals. There are exactly n such groups, not $n+1$, since no product determined by n white intervals meets L_0. When using the Bubble Lemma with the first group, insist that $\operatorname{diam}\lambda(\sigma_i) < 1/(3^{n-1}\cdot 2) = \delta_1$. Consequently, when treating cubes from the second group, we see that $\lambda(E_i)$ has diameter at most $3\delta_1$, and we can extend λ over E_i' so $\operatorname{diam}\lambda(\sigma_i) < 3\delta_1$. Each time we progress from one group to the next the diameters of the $\lambda(E_i)$ can triple. Thus, at the end, the extension λ satisfies $\operatorname{diam}\lambda(\sigma_i) < 3^{n-1}\cdot\delta_1 = 1/2$ for all i.

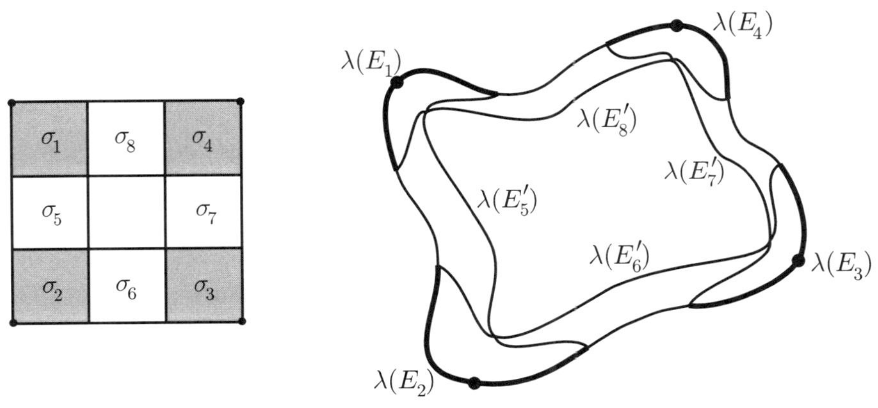

Figure 7.8. P_1 and $\lambda(L_1)$ for $n = 2$, $k_1 = 3$

The next step simply repeats this procedure, except that the role of $\lambda|\partial I^n$ now is taken over successively by $\lambda_1|\partial\sigma_i$, and the role of W by the component of $S^n \smallsetminus \lambda(\partial\sigma_i)$ contained in W. Critical size control is imposed by choosing an integer $k_2 \geq 3$ such that for the subdivision T_2 of I into k_1k_2 intervals of equal length and T_1^n the associated iterated product subdivision of I^n, we have $\operatorname{diam}\lambda_1(\sigma\cap L_1) < 1/(3^{n-1}\cdot 4)$ for all $\sigma \in T_2^n$. Fix an n-cube C_j of T_1^n in P_1 and order the n-cubes $\sigma_1, \sigma_2, \dots, \sigma_{m(2)}$ of T_2^n in C_j so that $\partial\sigma_i \cap C_jL_1^{i-1}$ is an $(n-1)$-cell, where $C_jL_1^{i-1} = \partial C_j \cup \cup_{t=1}^{i-1}\partial\sigma_i$. Apply Lemma 7.6.2 to extend λ_1 to an embedding $\lambda_1 : C_jL_1^i \to \overline{V}$ such that each $\lambda_1(\partial\sigma_i)$ is a 1-LCC embedded $(n-1)$-sphere of diameter less than $1/4$. At the end of the second stage of this process we have an extension of λ_1 to an embedding $\lambda_2 : L_2 \to \overline{V}$ such that $\lambda_2(\partial\sigma)$ is a 1-LCC embedded sphere of diameter less than $1/4$ for all n-cubes $\sigma \subset P_2$ from T_2^n.

Continue in the same way, thereby generating infinite sequences $\kappa = \{k_1, k_2, k_3, \dots\}$ of positive integers, with each $k_i \geq 3$, $\{T_1, T_2, T_3, \dots\}$ of subdivisions of I, with T_i determining $k_1k_2\cdots k_i$ equal length subintervals, and $\{\lambda_i : L_i \to \overline{W}\}$ of embeddings such that $\lambda_{i+1}|L_i = \lambda_i$ and, for each

n-cube $\sigma \subset P_i$ from T_i^n, $\lambda_i(\sigma)$ is a 1-LCC embedded sphere of diameter less than $1/2^i$.

Lemma 7.6.7 furnishes an embedding $\Lambda : X_\kappa \to \overline{W}$ of the associated $\mathcal{S}$-curve X_κ such that $\Lambda|\partial I^n = \lambda|\partial I^n$. By Theorem 7.5.7, X_κ is topologically equivalent to the standard Menger space M_n^{n-1}, and then Corollary 7.5.10 assures that $\overline{W}$ is an n-cell. Finally, since W could be either component of $S^n \setminus \Sigma$, Σ is flat. □

Corollary 7.6.9. *Every 1-LCC embedding of an $(n-1)$-dimensional ∂-manifold S in an n-manifold M $(n \geq 5)$ is locally flat.*

Proof. The Bubble Lemma and Lemma 7.6.3 indicate any sufficiently small $(n-1)$-disk D in S lies on a 1-LCC embedded $(n-1)$-sphere Σ_D that lives in a coordinate chart of M, and D can be assumed to be standardly embedded in Σ_D. □

Corollary 7.6.10. *Every $(n-1)$-cell $E \subset S^n$ as in Lemma 7.6.8 is flat.*

Corollary 7.6.11. *An $(n-1)$-cell E in S^n $(n \geq 5)$ is flat if and only if E is 1-LCC at points of* $\operatorname{Int} E$ *and ∂E is locally homotopically unknotted.*

Historical Notes. The approach to Theorem 7.6.1 presented here was developed by Černavskiĭ (1973). Another argument was given by Daverman (1973b).

The 3-dimensional analogs of the main results of this section were developed by Bing (1961b), and the 4-dimensional analogs were produced by Freedman and Quinn (1990).

On a topic related to Corollary 7.6.10, Černavskiĭ (1967) showed the union of two locally flat $(n-1)$-cells that intersect in an $(n-2)$-cell standardly embedded in the boundary of each to be flat itself, and Kirby (1968b) did the same for $n = 4$. Earlier, P. H. Doyle (1960) established the 3-dimensional version. More recently, Černavskiĭ (2006) provided a new proof of the result.

Exercises

7.6.1. Show that if $\sigma_i \in T_1^n$ in the proof of 7.6.5 belongs to the $(j+1)$st group and contains no point with coordinate 0 or 1, then E_i is congruent to a rescaled version of $[I^j \cup (\partial I^j \times I)] \times I^{n-j}$.

7.6.2. Prove Lemma 7.6.8.

7.6.3. Prove Corollary 7.6.11.

7.7. Locally flat approximations

Theorem 7.7.1 (Locally Flat Approximation). *Let M be an n-manifold $(n \geq 5)$, Q an $(n-1)$-manifold topologically embedded in M as a closed* *subset and $\epsilon : Q \to (0, 1)$ a continuous function. Then there exists a locally flat embedding $\lambda : Q \to M$ such that $\rho(\lambda(q), q) < \epsilon(q)$ for all $q \in Q$.*

To establish this foundational result, we will follow the lead of Ancel and Cannon (1979) who exploited a notion of embedding relation originally introduced in (Cannon, 1975). Embedding relations involve considerable new terminology, which will be laid out in what immediately follows, to provide context for statements of some forthcoming rather technical results. The conclusion of this section contains a brief appendix in which basic properties of embedding relations are developed.

A *relation* $R : X \to Y$ is simply a subset of $X \times Y$ whose projection to the first factor is surjective; in other words, R is a multi-valued function. Its *image* $R(X)$ has the usual meaning, and its *inverse* $R^{-1} : R(X) \to X$ is the relation $\{\langle y, x\rangle \in R(X) \times X \mid \langle x, y\rangle \in R\}$. A relation $R : X \to Y$ is *continuous* if for each closed subset C of Y, $R^{-1}(C)$ is closed in X (expecting inverses of open sets to be open would be unreasonable), and it is an *embedding relation* if any two of its point images are disjoint.

For simplicity, assume that all spaces are locally compact, separable and metrizable. A relation $R : X \to Y$ is *proper* if both R and R^{-1} are continuous with compact point images.

Given a metric space (X, d) and $\epsilon > 0$, (ϵ) denotes the relation $(\epsilon) : X \to X$ defined as $\{\langle x, y\rangle \in X \times X \mid d(x, y) < \epsilon\}$. When such expressions appear as terms in a composition of relations, we occasionally omit the parentheses to shorten the formulae. It should be noted that, if $R : X \to Y$ is a relation, even when the metric on $X \times Y$ is, say, the sum of the metrics on its factors, $(\epsilon) \circ R \subset X \times Y$ is not identical to the ϵ-neighborhood of R but, instead, is a subset. Also, given two functions $f, f' : Y \to X$, $\rho(f, f') < \epsilon$ if and only if $f' \subset (\epsilon) \circ f$ as relations.

A relation $R : X \to Y$ is 1-LCC if for each $x \in X$ and neighborhood U of $R(x)$ in Y there exists a neighborhood $V \subset U$ of $R(x)$ such that loops in $V \setminus \operatorname{Im} R$ are null-homotopic in $U \setminus \operatorname{Im} R$. Quite obviously, when R is an embedding relation, $R(X)$ is closed in Y and $\pi : Y \to Y/R$ is the quotient map determined by the decomposition of Y into singletons from $Y \setminus R(X)$ and the sets $\{R(x) \mid x \in X\}$, then R is 1-LCC if and only if $\pi R(X)$ (which is an embedded copy of X) is a 1-LCC subset of Y/R, in the usual sense.

A *cell-like embedding relation* is a proper embedding relation $R : X \to Y$ from a locally compact metric space X to an ANR Y such that the point images under R are non-empty, disjoint, cell-like sets. By definition, continuous embedding relations between compact ANRs are necessarily proper if point images are cell-like. Our attention will focus on cell-like embedding relations $R : S^{n-1} \to S^n$; it follows easily that then $R(S^{n-1})$ is compact and $R^{-1} : R(S^{n-1}) \to S^{n-1}$ is a genuine cell-like mapping. It is perfectly appropriate to regard a cell-like embedding relation simply as the inverse of a cell-like mapping. Approximation of one embedding relation by another permits the domain of the inverse cell-like map to change, in a controlled way, while preserving the target. In the strategy employed here, given a cell-like embedding relation $R : S^{n-1} \to S^n$, one will find a better approximating relation R'; the associated image $R'(S^{n-1})$ then will be close, in a reasonably rich sense, to $R(S^{n-1})$ and both images will admit cell-like maps (R^{-1} and $(R')^{-1}$) to S^{n-1}. Ultimately, upon passage to a limit, the resulting cell-like embedding relation will admit an approximating 1-LCC embedding of S^{n-1}, which will be flat, by Theorem 7.6.1.

In partial compensation for the introduction of unfamiliar concepts, we will address only the most familiar case of 7.7.1: an embedding of the $(n-1)$-sphere in S^n. Restated in the language of embedding relations, the precise aim of §7.7 is to establish the following 1-LCC variation of Theorem 7.7.1.

Theorem 7.7.2 (1-LCC Approximation of Relations). *Suppose $R : S^{n-1} \to S^n$ $(n \geq 5)$ is a cell-like embedding relation and L is a neighborhood of R in $S^{n-1} \times S^n$. Then L contains a 1-LCC cell-like embedding relation $R' : S^{n-1} \to S^n$.*

Given a cell-like embedding relation $R : S^{n-1} \to S^n$, we will denote by $\pi_R : S^n \to S^n/R$ the quotient map associated with the decomposition of S^n into the sets $R(x)$, $x \in S^{n-1}$, and the singletons of $S^n \setminus R(S^{n-1})$. A central difficulty is that, generally, S^n/R need not be a manifold.

Corollary 7.7.3. *Under the hypotheses of Theorem 7.7.2, L contains a locally flat embedding $\lambda : S^{n-1} \to S^n$.*

Proof. Given $R : S^{n-1} \to S^n$ and L, use 7.7.2 to obtain a 1-LCC embedding relation $R' : S^{n-1} \to S^n$ in L. Find $\epsilon > 0$ such that the ϵ-neighborhood N_ϵ of R' is a subset of L. Proposition 7.4.13 assures that S^n/R' has the DDP and, hence, in view of the Cell-like Approximation Theorem, that the decomposition induced by R' is shrinkable. Choose $\delta > 0$ such that

$$(\pi')^{-1} \circ (\delta) \circ (\pi R') \subset N_\epsilon,$$

where $\pi' : S^n \to S^n/R'$ is the decomposition map, and apply Theorem 2.3.3 to obtain a map $\mu : S^n \to S^n$ realizing that decomposition—in other words, μ satisfies

$$\{\mu^{-1}(s) \mid s \in S^n\} = \{(\pi')^{-1}(x) \mid x \in S^n/R'\}$$

—and require $\rho(\pi'\mu, \pi') < \delta$ as well. It follows that $\lambda = \mu R' \subset N_\epsilon \subset L$ is an embedding and a 1-LCC approximation to R. Hence, $\lambda(S^{n-1})$ is (locally) flat, by Theorem 7.6.1. □

Application of Corollary 7.7.3 to an arbitrary embedding $\lambda' : S^{n-1} \to S^n$ and to the neighborhood $L = (\epsilon) \circ \lambda'$ in $S^{n-1} \times S^n$ immediately yields Theorem 7.7.1 for codimension-one spheres in S^n.

Lemma 7.7.4. *For any cell-like embedding relation $R : S^{n-1} \to S^n$, $\operatorname{Im} R$ separates S^n into two components.*

Proof. By Proposition 3.2.9, $R(S^{n-1})$ has the Čech cohomology of S^{n-1}. Repeat the analysis given in Proposition 7.1.1. □

Let $f : \partial I^2 \to S^n \smallsetminus \operatorname{Im} R$ be a loop, and let $f^* : I^2 \to S^n/R$ be a map extending $\pi_R \circ f$ such that $\operatorname{Im} f^*$ misses one of the complementary domains of the $(n-1)$-sphere $\operatorname{Im}(\pi_R \circ R)$ in S^n/R. Then the relation $\hat{F} = \pi_R^{-1} \circ f^* : I^2 \to S^n$ is called an *R-disk bounded by f*. In the proof of the 1-LCC Approximation Theorem of Relations we shall show that every such loop f near a point image of R bounds a "small" R-disk $\hat{F}$. That notion of smallness is measured as R-diameter, where the *R-diameter* of a set $X \subset S^n$ is defined as

$$R\text{-}\operatorname{diam}(X) = \inf\{\epsilon > 0 \mid \text{for some } s \in S^{n-1}, X \subset \epsilon \circ R \circ \epsilon(s)\};$$

for simplicity, we also define $R\text{-diam}(\hat{F}) = R\text{-diam}(\operatorname{Im} \hat{F})$.

Lemma 7.7.5 (Basic Lemma). *Suppose $R : S^{n-1} \to S^n$ $(n \geq 5)$ is a cell-like embedding relation, $\hat{F} : I^2 \to S^n$ is an R-disk, and L, O are neighborhoods of $R, \hat{F}$, respectively. Then L contains a cell-like embedding relation $R''' : S^{n-1} \to S^n$ and O contains a continuous function $F^* : I^2 \to S^n$ such that $R'''(S^{n-1}) \cap F^*(I^2) = \emptyset$.*

Proof that Basic Lemma 7.7.5 implies Theorem 7.7.2. For purposes of this argument, given two relations $L', L'' : S^{n-1} \to S^n$, we will say that L' is *slice-trivial* in L'' if $L' \subset L''$ and $L'(s)$ is null-homotopic in $L''(s)$ for each $s \in S^{n-1}$.

Let $f_1, f_2, \ldots : S^1 \to S^n$ denote a countable set of embeddings dense in the space of all loops in S^n.

Set $R_0 = R$ and let $L_0 \subset L$ be a compact neighborhood of R_0. Assume inductively that cell-like embedding relations $R_0, \ldots, R_{i-1} : S^{n-1} \to S^n$, compact neighborhoods $L_0 \supset R_0, \cdots, L_{i-1} \supset R_{i-1}$ in $S^{n-1} \times S^n$, and continuous functions $F_1, \ldots, F_{i-1} : I^2 \to S^n$ bounded by $f_1, \ldots, f_{i-1}$ have been determined satisfying the following four conditions for $j = 0, 1, \ldots, i-1$:

(1_j) $R_j \subset \operatorname{Int} L_j \subset L_j \subset (1/j) \circ R_j \circ (1/j)$;

(2_j) L_j is slice-trivial in L_{j-1};

(3_j) $L_j^{-1} \circ L_j \subset (1/j)$ (the $j = 0$ case is vacuous); and

(4_j) if f_j bounds an R_{j-1}-disk and if

$$\epsilon_j = \inf\{R_{j-1}\text{- diam}(\hat{F}) \mid \hat{F} \text{ is an } R_{j-1}\text{-disk bounded by } f_j\},$$

then R_{j-1}- diam(Im F_j) $< 2\epsilon_j$ and $\operatorname{Im} L_j \cap \operatorname{Im} F_j = \emptyset$.

Choose R_i, L_i and F_i as follows. In case f_i bounds an R_{i-1}-disk and ϵ_i is the infimum defined in (4_i), an easy consequence of the Basic Lemma gives a cell-like embedding relation $(R_i : S^{n-1} \to S^n) \subset \operatorname{Int} L_{i-1}$ and a continuous function $F_i : I^2 \to S^n$ bounded by f_i such that $\operatorname{Im} R_i \cap \operatorname{Im} F_i = \emptyset$ and R_{i-1}-diam(Im F_i) $< 2\epsilon_i$. That Condition (4_i) holds is obvious. There is a compact neighborhood L_i of R_i in $\operatorname{Int} L_{i-1} \cap [(1/i) \circ R_i \circ (1/i)]$ by (5) in the Appendix on Continuous Relations; with this choice of L_i, (1_i) will be satisfied. Condition (2_i) can be obtained using (9) of the Appendix. Since $R_i^{-1} \circ R_i = \operatorname{Id} \subset (1/i)$, Condition (3_i) can be obtained by Composition Theorem (6) of the Appendix.

In the other case, where f_i bounds no R_{i-1}-disk, Condition (4_i) is vacuous; one then can take $R_i = R_{i-1}$, F_i arbitrary and L_i satisfying $(1_i) - (3_i)$, in order to complete the inductive step.

Define $R' : S^{n-1} \to S^n$ as $R' = \cap_{i=0}^{\infty} L_i$. We claim that R' is a 1-LCC cell-like embedding relation. Clearly it is contained in $L_0 \subset L$ and clearly $\operatorname{Im} R' \cap \operatorname{Im} F_i = \emptyset$ for all i such that f_i bounds an R_{i-1}-disk.

(i) R' is a proper relation. Being an intersection of compact sets, R' itself is compact and, thus, proper, by (5) in the Appendix.

(ii) R' is cell-like. For each $x \in S^{n-1}$, $R'(x)$ has a neighborhood system $L_0(x) \supset L_1(x) \supset \cdots$, with $L_i(x)$ compact, nonvoid, and contractible in $L_{i-1}(x)$, by (2_i). As a result, $R'(x) = \cap_i L_i(X)$ is cell-like.

(iii) R' is 1-1. Indeed,

$$(R')^{-1} \circ R' \subset \cap_i (L_i^{-1} \circ L_i) \subset \cap_i (1/i) = \operatorname{Id}_{S^{n-1}};$$

hence, point images under R' are disjoint.

Consequently, R' is a cell-like embedding relation. Showing it to be 1-LCC is the only remaining issue.

(iv) R' is 1-LCC. Consider any point $x \in S^{n-1}$ and any neighborhood U of $R'(x)$ in S^n. The task ahead is to find a neighborhood V of $R'(x)$ in $U \subset S^n$ such that loops in $V \setminus \operatorname{Im} R'$ are null-homotopic in $U \setminus \operatorname{Im} R'$.

We rely on (6) of the Appendix again to supply technical estimates. Since

$$U \supset R'(x) = (\operatorname{Id} \circ R' \circ \operatorname{Id}) \circ (\operatorname{Id} \circ R'^{-1} \circ \operatorname{Id}) \circ (\operatorname{Id} \circ R' \circ \operatorname{Id})(x),$$

Composition Theorem (6) assures the existence of an $\alpha > 0$ and an integer $I > 0$ satisfying:

(1) $U \supset (2\alpha \circ L_I \circ 2\alpha) \circ (2\alpha \circ L_I^{-1} \circ 2\alpha) \circ (\alpha \circ R' \circ \alpha)(x)$.

Set $\beta = \alpha/2$ and choose an integer $J > 2/\alpha$. Then $i > J$ implies

$$(2)\ \beta \circ R' \circ \beta(x) \subset (\alpha/2) \circ L_{i-1} \circ (\alpha/2)(x)$$
$$\subset (\alpha/2) \circ [(\alpha/2) \circ R_{i-1} \circ (\alpha/2)] \circ (\alpha/2)(x), \text{ by } (1_{i-1})$$
$$= \alpha \circ R_{i-1} \circ \alpha(x).$$

Having chosen α, β, I, and J, we specify an $(n-1)$-cell neighborhood D of x in $S^{n-1} \cap \beta(x)$, a compact neighborhood V' of $R'(x)$ in $\beta \circ R' \circ \beta(x)$ intersecting $\operatorname{Im} R'$ only in $R'(\operatorname{Int} D)$ and a compact neighborhood V of $R'(x)$ which contracts in V' (recall that $R'(x)$ is cell-like).

We show that each loop $f : S^1 \to S^n$ in $V \smallsetminus \operatorname{Im} R'$ contracts in $U \smallsetminus \operatorname{Im} R'$. Pick $K >$Max$\{I, J\}$ so large that, when $i > K$,

(3) $R_{i-1}(D) \subset \beta \circ R' \circ \beta(x)$,

(4) $R_{i-1}(S^{n-1} \smallsetminus \operatorname{Int} D) \cap V' = \emptyset$, and

(5) $\operatorname{Im} f \cap \operatorname{Im} R_{i-1} = \emptyset$.

Using (5) and the density of $\{f_i\}$, pick $i > K$ such that the loop f_i is homotopic to f in $V \smallsetminus \operatorname{Im}(R' \cup R_{i-1})$. We now explain why the associated extension $F_i : I^2 \to S^n$ given by (4_i) has image in $U \smallsetminus \operatorname{Im} R'$, which will complete the proof.

Since V contracts in V', f admits a continuous extension $g : I^2 \to V'$. Let π denote the decomposition map $S^n \to S^n/R_{i-1}$ associated with R_{i-1}, and identify S^{n-1} with its image under the embedding $\pi \circ R_{i-1}$. The set $\pi \circ f(S^1)$ lies in one of the two components of $(S^n/R_{i-1}) \smallsetminus S^{n-1}$, and the set $\pi \circ g(I^2)$ intersects S^{n-1} only in the $(n-1)$-cell $\pi \circ R_{i-1}(D)$ by (4). By the Tietze Extension Theorem, there is a continuous function $f^* : I^2 \to S^{n-1}/R_{i-1}$ extending $\pi \circ f$, the image of which lies in $\pi \circ g(I^2) \cup \pi \circ R_{i-1}(D)$ and misses one component of $(S^n/R_{i-1}) \smallsetminus S^{n-1}$, namely, the component not containing $f(S^1)$.

The relation $\hat{F} = \pi^{-1} \circ f^* : I^2 \to S^n$ is an R_{i-1}-disk bounded by f. Since $\operatorname{Im} f^* \subset \pi \circ g(I^2) \cup \pi \circ R_{i-1}(D)$, it follows that

$$\operatorname{Im} \hat{F} \subset g(I^2) \cup R_{i-1}(D) \subset \beta \circ R' \circ \beta(x)$$

(by (3) and the choice of $V' \supset g(I^2)$). But $\beta \circ R' \circ \beta(x) \subset \alpha \circ R_{i-1} \circ \alpha(x)$ (by (2)). Hence, by definition of R_i-diam$(\hat{F})$ and rules governing the choice of F_i, F_i is a singular disk in $S^n \smallsetminus L_i$ bounded by f_i and lying, for some $y \in S^{n-1}$, in the set $(2\alpha) \circ R_{i-1} \circ (2\alpha)(y)$, by (4_{i-1}). The only issue remaining is to show that $2\alpha \circ R_{i-1} \circ 2\alpha(y) \subset U$.

Since $\operatorname{Im} F_i \subset \alpha \circ R' \circ \alpha(x) \cap (2\alpha) \circ R_{i-1} \circ (2\alpha)(y)$ and is nonempty, we have $\operatorname{Im} F_i$ contained in

$$(2\alpha) \circ R_{i-1} \circ (2\alpha)(y) \subset [(2\alpha) \circ R_{i-1} \circ (2\alpha)][(2\alpha) \circ R_{i-1}^{-1} \circ (2\alpha)][\alpha \circ R' \circ \alpha](x),$$

and this latter set lies in U, by (1). $\square$

Before proceeding it might be beneficial to review the extensive collection of definitions and notation from §5.5 concerning Štan′ko moves, the template (A, B, C, D, e) in $\hat{I}^2$, the n-dimensional expansions

$$\mathscr{A} = A \times \hat{I}^{n-2}, \quad \mathscr{B} = B \times \hat{I}^{n-2}, \quad \mathscr{C} = C \times [-1, 1] \times \hat{I}^{n-3},$$

the 2-cells $\mathscr{D} = D \times \mathbf{0} \subset \hat{I}^n$ and $e \times \hat{I} = e \times \{0\} \times \hat{I} \times \mathbf{0} \subset \hat{I}^n$, semi-capped surfaces, Delta structures, branching systems, Štan′ko complexes, and the special homeomorphism $\Phi_n : \hat{I}^n \to \hat{I}^n$.

Here is the setting for the Basic Lemma, the proof of which occupies most of the remainder of this section. All this data and notation is presumed to be in place until the completion of that proof.

$R : S^{n-1} \to S^n$, a cell-like embedding relation, with associated cell-like decomposition map: $\pi = \pi_R : S^n \to S^n/R$;

$(\hat{F} = \pi^{-1} \circ f^*) : I^2 \to S^n$, an R-disk with $f^* : I^2 \to S^n/R$ a continuous function and with $\hat{F}|\partial I^2$ an embedding;

L, a neighborhood of R in $S^{n-1} \times S^n$;

O, a neighborhood of $\hat{F}$ in $I^2 \times S^n$.

We identify S^{n-1} with $\operatorname{Im}(\pi \circ R)$ via the homeomorphism $\pi \circ R : S^{n-1} \to \operatorname{Im}(\pi \circ R)$ so that $R = \pi^{-1}|S^{n-1}$, and we also make use of the notation:

W, a component of $S^n/R \smallsetminus S^{n-1}$ containing $f^*(\partial I^2)$;

L_0, a neighborhood of π^{-1} in $S^n/R \times S^n$ whose restriction to S^{n-1} is L.

Anticipating the verifications to be made near the end of the proof of the Basic Lemma, we now impose certain essential controls using the Composition Theorem (6): since

$$[\mathrm{Id} \circ \mathrm{Id} \circ \pi^{-1} \circ \mathrm{Id}_{S^n/R} \circ \pi \circ \mathrm{Id} \circ \mathrm{Id}] \circ \mathrm{Id} \circ \pi^{-1} = \pi^{-1} \subset L_0$$

(where all unsubscripted "Id" denote $\mathrm{Id} : S^n \to S^n$), and since

$$\mathrm{Id} \circ \pi^{-1} \circ \mathrm{Id}_{S^n/R} \circ f^* = \hat{F} \subset O,$$

there is an $\epsilon > 0$ such that

(S†) $(\epsilon \circ \epsilon \circ \pi^{-1} \circ (2\epsilon) \circ \pi \circ \epsilon \circ \epsilon) \circ \epsilon \circ \pi^{-1} \subset L_0$ and

(S‡) $\epsilon \circ \pi^{-1} \circ \epsilon \circ f^* \subset O$.

Theorem 7.7.6 (Štan′ko Complex Mapping). *For each $\epsilon > 0$ there exist a branching system $\Delta : \Delta_0 \to \Delta_1 \to \Delta_2 \to \cdots$, with D_0 of $\Delta_0 = (D_0, E_0, \Gamma_0)$ equal to I^2, and a continuous function $h : C(\Delta) \to S^n/R$ satisfying:*

(1) $h(C(\Delta) \smallsetminus \operatorname{Int} E^*) \subset W \subset S^n/R \smallsetminus S^{n-1}$,

(2) $h(D_i^* \cup D_{i+1}^* \cup D_{i+2}^* \cup \cdots) \subset B(S^{n-1}; \epsilon/2^{i-1})$ *for* $i > 0$,

(3) $\rho(h \circ (^*)|D_0, f^*) < \epsilon$,

(4) $h \circ (^*)|\partial D_0 = f^*|(\partial D_0 = \partial I^2)$,

(5) $\operatorname{diam} h(P_i) < \epsilon/2^i$ *for* $i > 0$ *and each component* P_i *of* $D_i^* \cup E_{i-1}^*$. *In addition, the map* h *may be chosen so there exists a PL injective map* $h' : C(\Delta) \to S^n$ *with* $\pi \circ h' = h$.

Remark. Unlike the output of the related codimension-three Štan′ko Embedding Theorem (5.5.5), h' need not be a homeomorphism: a sequence $\{x_1, x_2, x_3, \dots \mid x_i \in h'(D_i^*)\}$ can accumulate at a point of $h'(E_0^*)$, causing $(h')^{-1}$ to be discontinuous. An example will be provided later, in §7.10, of an actual embedding $R : S^{n-1} \to S^n$ and disjoint simple closed curves J_1 and J_2 in $\pi^{-1}(W)$ such that for every singular disk D_1 in $\mathrm{Cl}(\pi^{-1}(W))$ bounded by J_1 and every singular disk D_2 in S^n bounded by J_2, $D_1 \cap D_2 \neq \emptyset$. If $\hat{F}$ would be an R-disk sending ∂D homeomorphically onto J_1 and if J_2 would bound a component of $h(E_i^*)$, then points of $h'(C(\Delta) \setminus E^*)$ necessarily would accumulate at $h'(\operatorname{Int} E_i^*)$. This possibility accounts for a thickening procedure to be employed later, in Štan′ko Complex Embedding Theorem 7.7.8, and, to a large extent, for the prolonged diversion through the realm of embedding relations. To establish the Embedding Theorem we will employ a carefully constructed cell-like embedding relation $R' : S^n \to S^n$, will replace R by $R' \circ R$, and will obtain a PL embedding of $C(\Delta) \to S^n$ for the modified R.

As in §5.5, the Mapping Theorem is a consequence of iterated applications of the following, which in turn is simply a variation on Lemma 5.5.6 for the slightly more general context to be faced.

Lemma 7.7.7. *Suppose* $g : (D, \partial D) \to (\overline{W}, W)$ *is a map of pairs, where* D *is a disk, and* $\delta > 0$. *Then there exist a Delta structure* $\Delta = (D, E, \Gamma)$ *and a map* $f : D^* \to \overline{W}$ *satisfying:*

(1′) $f(D^* \setminus \operatorname{Int} E^*) \subset W$,

(2′) $f(E^*) \subset B(S^{n-1}; \delta)$,

(3′) $\rho(f \circ (^*)|D, g) < \delta$,

(4′) $f \circ (^*)|\partial D = g|\partial D$, *and*

(5′) $\operatorname{diam} f(P) < \delta$ *for each component* P *of* E^* *or* Γ^*.

Proof. Being the finite-dimensional image of S^n under a cell-like map, S^n/R is an ANR (Corollary 7.4.8). Since $\overline{W}$ is a closed subset of S^n/R bounded by the $(n-1)$-sphere $\pi R(S^{n-1})$, $\overline{W}$ is also an ANR.

Claim: each $s \in \pi R(S^{n-1})$ has arbitrarily small pairs of neighborhoods $N' \subset N$ such that $H_1(N' \cap W; \mathbb{Z}) \to H_1(N \cap W; \mathbb{Z})$ is trivial. Given N with $N \cap \pi R(S^{n-1})$ contractible, choose $N' \subset N$ with $\mathrm{incl} : N' \to N$ homotopically trivial; by Proposition 3.2.9 it suffices to show that $H_1(\pi^{-1}(N' \cap W); \mathbb{Z}) \to H_1(\pi^{-1}(N \cap W); \mathbb{Z})$ is trivial, and that property holds just as in the proof of Proposition 7.1.11.

Hence, there exist positive numbers $\alpha < \beta < \eta < \delta/4$ such that

(η) η-loops in W bound singular $(\delta/4)$-disks in $\overline{W}$,

(β) β-loops in W bound singular, orientable disks with handles of diameter less than η in W (recall Lemma 5.5.3), and

(α) any two points of W within α of each other are joined by $(\beta/2)$-arcs in W.

Triangulate D with mesh so small that the image under g of each simplex has diameter less than α, and let $D^{(0)}, D^{(1)}$ and $D^{(2)}$ denote the successive skeleta of this triangulation. Since $g(\partial D) \subset W$ we may define f on ∂D as $g|\partial D$ and define f for any other vertex v of $D^{(0)}$ as a point in W so close to $g(v)$ that vertices of the same simplex have images in W within α of one another and within α of their image under f.

Apply (α) to extend f over $|D^{(1)}|$ so that the image of each 1-simplex lies in W and has diameter less than $\beta/2$.

Apply (β) for each $\sigma \in D^{(2)}$ to obtain an orientable disk-with-handles Q_σ bounded by $\partial\sigma$ and a continuous extension $f_\sigma : Q_\sigma \to W$ of $f|\partial\sigma$ sending Q_σ to a set of diameter less than η.

In the interior of each Q_σ identify complete sets $\Gamma_\sigma, \Gamma'_\sigma$ of handle curves, as before.

By (η) there exists for each $\sigma \in D^{(2)}$ a finite disjoint union E_σ of disks whose boundaries equal Γ'_σ and whose interior points have no intersection with $\cup_\sigma Q_\sigma$ as well as a continuous extension $f : E_\sigma \to \overline{W}$ taking each component of E_σ to a set of diameter less than $\delta/4$. Define

$$D^* = |D^{(1)}| \cup [\cup_\sigma (Q_\sigma \cup E_\sigma)], \quad E^* = \cup_\sigma E_\sigma, \quad \Gamma^* = \cup_\sigma \Gamma_\sigma.$$

It should be clear that D^* is the semi-capped surface of a Delta structure $\Delta = (D, E, \Gamma)$, where $(^*)|\ |D^{(1)}| = \text{Id}$ and $(^*){:}\sigma \to Q_\sigma \cup E_\sigma$. Clearly f as defined on D^* satisfies $(1'), (3'), (4')$ and $(5')$. If $(2')$ is not already satisfied, it can only stem from the presence of a component of $E^* \cup \Gamma^*$ whose image misses S^{n-1}; the preimage of such a component should simply be deleted from E and Γ. $\square$

Proof of Mapping Theorem 7.7.6. Choose a sequence $\delta_0 > \delta_1 > \cdots$ of positive numbers such that

(i) $4\delta_i < \epsilon/2^i$ for $i \geq 0$ and

(ii) δ_{i+1}-loops in $\overline{W}$ bound singular δ_i-disks in $\overline{W}$.

Lemma 7.7.7 provides a Delta structure $\Delta_0 = (D_0, E_0, \Gamma_0)$, where $D_0 = I^2$, plus a continuous function $h : D_0^* \to \overline{W}$ satisfying the conditions below for $j = 0$, where $f_0 = f|D_0$:

(1_j) $h(D_j^* \setminus \text{Int}\, E_j^*) \subset W$,

(2_j) $h(E_j^*) \subset B(S^{n-1}; \delta_{j+1})$,
(3_j) $\rho(h \circ (^*)|D_j, f_j) < \delta_{j+1}$,
(4_j) $h \circ (^*)|\partial D_j = f_j|\partial D_j$, and
(5_j) $\operatorname{diam} h(P) < \delta_{j+2}$ for each component P of E_j^* or Γ_j^*.

Assume inductively that $\Delta_0 \to \cdots \to \Delta_{i-1}$, $f_j : D_j \to \overline{W}$ and

$$h|D_0^* \cup \cdots \cup D_{i-1}^* : D_0^* \cup \cdots \cup D_{i-1}^* \to \overline{W}$$

have been obtained satisfying (1_j)–(5_j) for each $j \in \{0, \ldots, i-1\}$. For each component γ of Γ_{i-1}^*, let $D(\gamma)$ be a disk with boundary γ. By (5_{i-1}), $\operatorname{diam} h(\gamma) < \delta_{i+1}$, so (ii) assures the existence of a continuous extension $f_i(\gamma) : D(\gamma) \to \overline{W}$ of $h|\gamma$, the image of which has diameter less than δ_i. Set $D_i = \cup_\gamma D(\gamma)$ and define f_i on D_i as $f_i = \cup_\gamma f_i(\gamma)$. By Lemma 7.7.7 there exist a Delta structure (+) $\Delta_i = (D_i, E_i, \Gamma_i)$ and a map $h|D_i^* : D_i^* \to \overline{W}$ satisfying (1_i)–(5_i). This completes the inductive construction of

$$\Delta : \Delta_0 \to \Delta_1 \to \Delta_2 \to \cdots \quad \text{and} \quad h : C(\Delta) \to S^n/R.$$

Conditions (1), (3) and (4) of the Mapping Theorem obviously hold here. For each component $P \cup Q$ of $D_i^* \cup E_{i-1}^*$ ($i \geq 1$), $P \subset D_i^*$, $Q \subset E_{i-1}^*$, we have

$$\operatorname{diam} h(P) \leq 2\rho(h \circ (^*)|P, f_i|P) + \operatorname{diam} f_i(P) < 3\delta_i$$

and $\operatorname{diam} h(Q) < \delta_{i+1}$; thus, $\operatorname{diam} h(P \cup Q) < 4\delta_i < \epsilon/2^i$ and (5) is satisfied. Also, for $i \geq 1$, $d(h(P), S^{n-1}) \leq \epsilon/2^i$, by ($2_{i-1}$), since $h(P) \cap h(E_{i-1}^*) = \emptyset$. But $\operatorname{diam} h(P) < \epsilon/2^i$, so $h(P) \subset B(S^{n-1}; \epsilon/2^{i-1})$, and (2) is satisfied. □

Theorem 7.7.8 (Štan′ko Complex Embedding). *Let Δ, h and h' be as in the conclusion of Mapping Theorem 7.7.6. Then the ϵ-neighborhood of* Id : $S^n \to S^n$ *contains a cell-like embedding relation $R' : S^n \to S^n$ such that there is a PL embedding $h'' : C(\Delta) \to S^n$ with $(R')^{-1} \circ h'' = h'$.*

Proof. In view of Conditions (1) and (2) of the Mapping Theorem, the function $(h')^{-1}|h'(C(\Delta))$ is already continuous except possibly at points of $h'(\operatorname{Int}(E_0^* \cup E_1^* \cup \cdots))$. We will split S^n apart in stages, starting with $h'(\operatorname{Int} E_0^*)$, to provide enough room to isolate $h''(\operatorname{Int} E_0^*) = h'(\operatorname{Int} E_0^*)$ from $h''(C(\Delta) \smallsetminus E_0^*)$. This will make $(h'')^{-1}$ continuous at points of $h''(\operatorname{Int} E_0^*)$. Iteration of the splitting will accomplish the same goal at images of $E_1^*, E_2^*, \ldots$ and will complete the proof of this Embedding Theorem.

The basic splitting move is the inverse of a simple collapsing map. Define $r : \hat{I}^2 \to [0,1]$ as $r(x) = (1/4) \cdot d(x, \partial \hat{I}^2) \in [0, 1/2]$ and set

$$\hat{I}^2 \times_r \hat{I}^{n-2} = \cup\{x \times [r(x) \cdot \hat{I}^{n-2}] \mid x \in \hat{I}^2\} \subset \hat{I}^2 \times \hat{I}^{n-2} \subset \hat{I}^n.$$

Note that $\hat{I}^2 \times_r \hat{I}^{n-2}$ is a closed neighborhood of $\operatorname{Int} \hat{I}^2 \times \mathbf{0}$ in $\hat{I}^n$. Let $\psi : \hat{I}^2 \times_r \hat{I}^{n-2} \to \hat{I}^2 \times \mathbf{0}$ denote projection to the first factor. Define a map

$\Psi : \hat{I}^n \to \hat{I}^n$ extending ψ, fixed on ∂I^n, sending each $x \times \hat{I}^{n-2}$ onto itself, and having as its nondegenerate point preimages precisely the nondegenerate point preimages of ψ. The relation Ψ^{-1} is called the *basic splitting relation.*

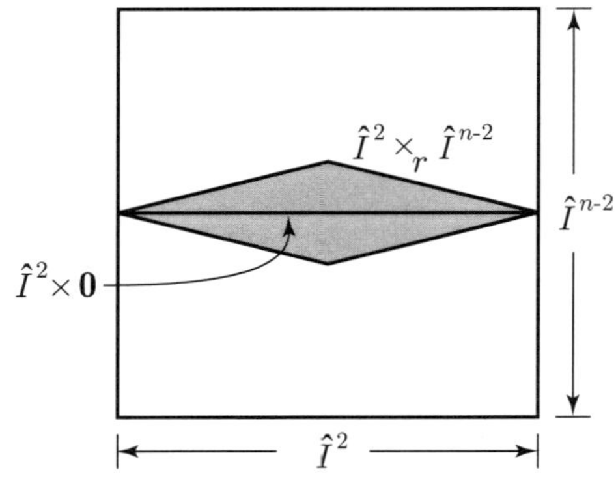

Figure 7.9. The basic splitting relation Ψ^{-1}

For each component E of $h'(E_0^*)$ there is a PL embedding $P_E : \hat{I}^2 \times \hat{I}^{n-2} \to S^n$ taking $\hat{I}^2 \times \mathbf{0}$ onto E and taking each fiber $x \times \hat{I}^{n-2}$ onto a very small set. The embeddings $\{P_E \mid E \subset h'(E_0^*)\}$ should be chosen with disjoint images. Define $R : S^n \to S^n$ splitting S^n along $h'(E_0^*)$ by the formula

$$R_0(x) = \begin{cases} P_E \Psi^{-1} P_E^{-1}(x) & \text{if } x \in P_E(\hat{I}^n) \\ x & \text{if } x \notin \cup_E P_E(\hat{I}^n). \end{cases}$$

The restriction on $\hat{I}^{n-2}$ fiber size assures that R_0 lives in the ϵ-neighborhood (ϵ) of the relation $\text{Id} : S^n \to S^n$. Define $h_0 : C(\Delta) \to S^n$ as

$$h_0(x) = \begin{cases} h'(x) & \text{if} \quad x \in E_0^* \\ R_0 \circ h'(x) & \text{if} \quad x \notin E_0^*. \end{cases}$$

Then h_0 is PL and injective, $R_0^{-1} \circ h_0 = h'$, and $h_0^{-1} | h_0(C(\Delta))$ is continuous at the points of $h_0(E_0^*)$. Choose a compact neighborhood N_0 of R_0 in (ϵ) slice-trivial in (ϵ), with $N_0^{-1} \circ N_1 \subset (1)$.

In the same manner choose $R_1 : S^n \to S^n$ splitting S^n along $h_0(E_1^*)$, fixing $R_0 \circ h'(D_0^*)$, and satisfying $R_1 \circ R_0 \subset \text{Int}\, N_0$. Define $h_1 : C(\Delta) \to S^n$ by

$$h_1(x) = \begin{cases} h_0(x) & \text{if} \quad x \in E_1^* \\ R_1 \circ h_0(x) & \text{if} \quad x \notin E_1^*. \end{cases}$$

Choose a compact neighborhood N_1 of $R_1 \circ R_0$ in $\text{Int}\, N_0$, slice-trivial in $\text{Int}\, N_0$, with $N_1^{-1} \circ N_1 \subset (1/2)$.

In general, have R_i split S^n along $h_{i-1}(E_i^*)$, fixing $R_{i-1} \circ \cdots \circ R_0 \circ h'(D_{i-1}^*)$, and satisfying $R_i \circ R_{i-1} \circ \cdots \circ R_0 \subset \operatorname{Int} N_{i-1}$. Define $h_i : C(\Delta) \to S^n$ as

$$h_i(x) = \begin{cases} h_{i-1}(x) & \text{if} \quad x \in E_i^* \\ R_i \circ h_{i-1}(x) & \text{if} \quad x \notin E_i^*. \end{cases}$$

Again choose a compact neighborhood N_i of $R_i \circ \cdots \circ R_0$ in $\operatorname{Int} N_{i-1}$, slice-trivial in $\operatorname{Int} N_{i-1}$, with $N_i^{-1} \circ N_i \subset (1/(i+1))$.

Finally, define R' as $[R' = \cap_i N_i] : S^n \to S^n$. Just as in the proof of the 1-LCC Approximation Theorem, R' is a cell-like embedding relation. Define $h'' : C(\Delta) \to S^n$ as $h'' = \cup_i (h_i | D_i^*)$. That h'' is the embedding required in this Embedding Theorem is easily confirmed. □

Theorem 7.7.9 (Unknotting). *Suppose $C(\Delta)$ is a Štan′ko complex PL embedded in a PL n-manifold M ($n \geq 5$) and Z is a compact subset of $C(\Delta)$. Then there exist a PL 3-cell Y_Z^3 and a PL embedding $\psi : Y_Z^3 \times I^{n-3} \to M$ such that $Z \subset \psi(Y_Z^3 \times \mathbf{0})$.*

Proof. All but the case $n = 5$ is covered by Lemma 5.5.8. By Lemma 5.5.7 Z is contained in some PL-embedded, collapsible, finite 2-complex in M. According to (Price, 1966), any two homotopic PL embeddings of a collapsible finite k-complex in M are ambient isotopic provided $n \geq 2k+1$, so the remaining $n = 5$ case follows like the others. □

Proof of Basic Lemma 7.7.5. Take Δ, h and h' from the conclusion of Štan′ko Complex Mapping Theorem 7.7.6, and then take the relation R' and the embedding h'' from the conclusion of the Embedding Theorem 7.7.8. Identify $C(\Delta)$ with $h''(C(\Delta))$ via the homeomorphism h''. Recall the combined identification map

$$(^*) : D_0 \sqcup D_1 \sqcup \cdots \to C(\Delta) = D_0^* \cup D_1^* \cup \cdots = h''(C(\Delta)) \subset S^n.$$

For $i \geq 0$ identify that D_i associated with the Delta structure $(+)\Delta_i = (\Delta_i, E_i, \Gamma_i)$ with the D_i from the template $(+)(A_i, B_i, C_i, D_i, e_i)$ in such a manner that $E_i \cup \Gamma_i \subset \operatorname{Int} B_i$ and $(D_i \cap e_i)^* = D_i^* \cap E_{i-1}^* \subset C(\Delta) \subset S^n$. Keep in mind that $\operatorname{Im}(R' \circ R) \cap C(\Delta) \subset \cup_i E_i^* \subset \cup_i B_i^*$.

By Unknotting Theorem 7.7.9 there exist a regular neighborhood N_i of $D_i^* \cup E_{i-1}^*$ in $C(\Delta)$, a PL 3-manifold Y_i and an embedded PL product $Y_i \times \hat{I}^{n-3} \subset S^n$ such that $N_i \subset Y_i = Y_i \times \mathbf{0} \subset Y_i \times \hat{I}^{n-3} \subset S^n$. For $i > 0$ we will use the sets $D_i^* \cup E_{i-1}^* \subset N_i$ and the product structure $\hat{I}^3 \times \hat{I}^{n-3}$ to construct an embedding $\alpha_i : \hat{I}^n = \mathscr{A}_i \cup \mathscr{B}_i \to S^n$ suitable for use in a basic Štan′ko move. The α_i's will be constructed in three steps, which proceed exactly like those of the second proof of Fundamental Lemma 5.5.2, only with X replaced throughout by $\operatorname{Im}(R' \circ R)$. Those steps are not reproduced here, but the properties of these α_i are listed below for easy reference later on in the verification steps:

(1) $\operatorname{Im}(\mathscr{A}_i \cup \mathscr{B}_i) \subset B(D_i^* \cup E_{i-1}^*; \epsilon/i)$ (component by component);

(2) of the sets in the list

$$[D_0^* \cap A_0^*], [B_0^*], [A_1^*, \operatorname{Im} \mathscr{A}_1], [B_1^*, \operatorname{Im} \mathscr{B}_1], [A_2^*, \operatorname{Im} \mathscr{A}_2], [B_2^*, \mathscr{B}_2], \ldots,$$

only the ones in the same or adjacent square brackets can intersect,

(3) $X \cap \alpha_i(\mathscr{A}_i) \subset \alpha_i(\mathscr{C}_i)$;

(4) $\alpha_i(\mathscr{A}_i \cap \Phi_n(\mathscr{C}_i)) \subset S^n \setminus C(\Delta)$.

(5) $\operatorname{Im}(R' \circ R) \cap \alpha_i(\mathscr{B}_i) \subset \alpha_{i+1}(\mathscr{C}_{i+1})$;

(6) $\alpha_i(\mathscr{B}_i) \cap \alpha_{i+1}(\mathscr{A}_{i+1} \cap \Phi_n(\mathscr{C}_{i+1})) = \emptyset$;

(7) $\alpha_i(\mathscr{B}_i \cap \Phi_n(\mathscr{C}_i)) \subset S^n \setminus C(\Delta)$.

With the embeddings α_i all in place, we define the infinite Štan′ko move that leads to the desired cell-like embedding relation and establishes the Basic Lemma. Set

$$R''(x) = \begin{cases} \alpha_i \circ \Phi_n(x) \circ \alpha_i^{-1}(x) & \text{if } x \in \alpha_i(\mathscr{C}_i) \cap \operatorname{Im}(R' \circ R) \\ x & \text{otherwise,} \end{cases}$$

and note that R'' is a function. The cell-like embedding relation $R''' : S^{n-1} \to S^n$ whose existence is posited in the statement of the Basic Lemma is given by $R''' = R'' \circ R' \circ R$, and the continuous function $F^* : I^2 \to S^n \setminus \operatorname{Im} R'''$ named there is given by $F^* = (^*)|(I^2 = D_0)$.

Verification that R''' and F^* have the desired properties is a lengthy process. Each of the verification items labelled (Vi) below opens with a statement of what it confirms. The first of them is a technical calculation ultimately used for showing that $R''' \subset L$, that R''' is proper and that R''' is cell-like.

(V1) If $\langle x, y\rangle \in R' \circ R$ and $\langle x, R''y\rangle \in R''' \setminus \cup_{i<N}[S^{n-1} \times \operatorname{Int} \alpha_i(\mathscr{A}_i \cup \mathscr{B}_i)]$ then

$$\langle x, R''y\rangle \in (\epsilon/N) \circ [R' \circ \pi^{-1} \circ (\epsilon/N) \circ \pi \circ (R')^{-1}] \circ (\epsilon/N) \circ [R' \circ R].$$

It suffices to check the case where $R''(y) \neq y$, and that is done by finding points $y_1, y_2 \in S^n$ such that

$$\langle x, y\rangle \in R' \circ R, \; y_1 \in (\epsilon/N)(y),$$

$$y_2 \in R' \circ \pi^{-1} \circ (\epsilon/N) \circ \pi \circ (R')^{-1}(y_1), \text{ and } R''(y) \in (\epsilon/N)(y_2).$$

By construction of R'' there exist an integer $j \geq N$ and a component P of $\alpha_j(\mathscr{A}_j \cup \mathscr{B}_j)$ containing both y and $R''(y)$. Let Q be the component of $D_j^* \cup E_{j-1}^*$ intersecting P. By (1), $P \subset (\epsilon/j)(Q)$. Thus, there exist points $y_1, y_2 \in Q$ satisfying $y_1 \in (\epsilon/j)(y)$ and $R''(y) \subset (\epsilon/j)(y_2) \subset (\epsilon/N)(y_2)$.

Since $h(Q) = \pi \circ (R')^{-1}(Q)$ has diameter less than ϵ/j by Conclusion (5) of 7.7.6, $y_2 \in R'' \circ \pi^{-1} \circ (\epsilon/N) \circ \pi \circ (R')^{-1}(y_1)$.

(V2) $R''' \subset L$. By (V1) with $N = 1$,

$$R''' \subset \epsilon \circ [R' \circ \pi^{-1} \circ \epsilon \circ \pi \circ (R')^{-1}] \circ \epsilon \circ [R' \circ R]$$

and the latter set lies in L_0, by Condition (S†) of the setting for the Basic Lemma. Since $L_0|S^{n-1} = L, R''' \subset L$.

(V3) R''' is proper. It suffices to show that R''' is compact. Let $\langle x_1, R''y_1\rangle$, $\langle x_2, R''y_2\rangle, \ldots$ be a sequence in R'''. Passing to a subsequence, if necessary, we see that it suffices to address two cases.

Case 1. The points $y_1, y_2, \ldots$ all lie in $\alpha_i\mathscr{A}_i$ for some fixed index i. Since $R''|\ [\alpha_i\mathscr{A}_i \cap \operatorname{Im}(R' \circ R)]$ is continuous, $\langle x_1, R''y_1\rangle, \langle x_2, R''y_2\rangle, \ldots$ all belong to the compact set

$$[(R' \circ R)^{-1}, R''][\alpha_i\mathscr{A}_i \cap \operatorname{Im}(R' \circ R)] \subset [(R' \circ R)^{-1}, R''] \operatorname{Im}(R' \circ R) \subset R'''.$$

Hence, the points cluster in R'''.

Case 2. For each integer $N > 0$, only finitely many of the points $y_1, y_2, \ldots$ lie in $\cup_{i<N} \operatorname{Int} \alpha_i\mathscr{A}_i$. Then the intersection of the sequence $\{\langle x_i, R''y_i\rangle\}_{i=N}^{\infty}$ with the set $Z_N = R''' - \cup_{i<N}[S^{n-1} \times \operatorname{Int} \alpha_i(\mathscr{A}_i \cup \mathscr{B}_i)]$ is contained in the portion of

$$(\epsilon/N) \circ [R' \circ \pi^{-1} \circ (\epsilon/N) \circ \pi \circ (R')^{-1}] \circ (\epsilon/N) \circ [R' \circ R|$$

outside $\cup_{i<N}[S^{n-1} \times \operatorname{Int} \alpha_i(\mathscr{A}_i \cup \mathscr{B}_i)]$ and is nonempty for each such N. The intersection of the Z_n (over all $N \geq 1$) equals

$$R' \circ R \setminus \cup_{i=1}^{\infty}[S^{n-1} \times \operatorname{Int} \alpha_i(\mathscr{A}_i \cup \mathscr{B}_i)],$$

which is a compact subset of R'''. It follows easily that the points cluster at a point of R'''. We conclude that R''' is compact and, thus, proper.

(V4) R''' is injective. Since $R' \circ R$ is injective, it suffices to show that R'' is injective. We have

$$\operatorname{Im}(R' \circ R) = [\operatorname{Im}(R' \circ R) \setminus \cup_i \operatorname{Im} \alpha_i] \cup [\operatorname{Im}(R' \circ R) \cap (\alpha_1\mathscr{C}_1 \cup \alpha_2\mathscr{C}_2 \cup \cdots)]$$

by conditions (3) and (5). The set $R''[\operatorname{Im}(R' \circ R) \cap \alpha_i\mathscr{C}_i]$ lies in $\alpha_i\Phi_n(\mathscr{C}_i)$. The sets $\alpha_i\Phi_n(\mathscr{C}_i)$ and $\alpha_j\Phi_n(\mathscr{C}_j)$ miss one another if $j \neq i-1, i, i+1$, by (2). Intersections of the form $\alpha_i\Phi_n(\mathscr{C}_i) \cap \alpha_{i+1}\Phi_n(\mathscr{C}_{i+1})$ lie in

$$\alpha_i(\mathscr{B}_i \cap \Phi_n\mathscr{C}_i) \cap \alpha_{i+1}(\mathscr{A}_{i+1} \cap \Phi_n\mathscr{C}_{i+1})$$

by (2) and the latter intersection is empty by (6). Thus R''' is injective.

(V5) R''' has nonempty point images. This follows immediately, since R'' is a function and $R' \circ R$ has nonempty point images.

(V6) $F' \subset O$. We have $F^* = (^*)|I^2 = h''(^*)|(I^2 = D_0)$, and

$$h''|D_0^* \subset R' \circ \pi^{-1} \circ h \subset R' \circ \pi^{-1} \circ \epsilon \circ f^* \subset \epsilon \circ \pi^{-1} \circ \epsilon \circ f^* \subset O$$

by definition of h'', limits on the distance between h and f^*, limits on the motion of R', and Condition (S‡) of the Setting.

(V7) $\operatorname{Im} F^* \cap \operatorname{Im} R'' = \emptyset$. Here we have

$$\operatorname{Im} F^* \cap \operatorname{Im}(R' \circ R) \subset E_0^* \subset \alpha_1 \mathscr{C}_1;$$

$$C(\Delta) \cap R''(\alpha_1 \mathscr{C}_1 \cap \operatorname{Im}(R' \circ R)) = \emptyset; \text{ and}$$

$$R''(\alpha_i \mathscr{C}_i \cap \operatorname{Im}(R' \circ R)) \cap \operatorname{Im} F^* = \emptyset \text{ for } i > 1,$$

since $\alpha_i(\mathscr{A}_i \cup \mathscr{B}_i) \cap \operatorname{Im} F^* = \emptyset$ for $i > 1$. Disjointness of the two images follows from these three observations.

(V8) R''' is cell-like. Consider any $x \in S^{n-1}$ and $X = R' \circ R(x)$. If X meets only finitely many of the sets $\alpha_i \mathscr{A}_i$, then $R''|X$ is a re-embedding of X, so $R''(X)$ is also cell-like. Assume then that X meets infinitely many of the $\alpha_i \mathscr{A}_i$. Let U_i denote the union of the components of $\alpha_i \mathscr{A}_i$ intersecting X, V_i the union of the components of $\alpha_i(\mathscr{A}_i \cup \mathscr{B}_i)$ intersecting U_i, and $X_i = X \cap \alpha_i \mathscr{A}_i$. There clearly exist homotopies $H_i : U_i \times [0,1] \to S^n$ starting at the inclusion, fixing FrU_i and having images in V_i such that $H_i(-,1)|X_i = R''|X_i$. By the same calculation performed in (V1), $V_N \cup V_{N+1} \cup \cdots$ is contained in

$$(\epsilon/N) \circ R' \circ \pi^{-1} \circ (\epsilon/N) \circ \pi \circ (R')^{-1} \circ (\epsilon/N) \circ R' \circ R(x) \smallsetminus \cup_{i<N} [S^{n-1} \times \operatorname{Int} \alpha_i(\mathscr{A}_i \cup \mathscr{B}_i)].$$

It follows from the Composition Theorem that, as $N \to \infty$, the various sets named immediately above converge uniformly to the compact set

$$X \smallsetminus \cup_{i=1}^{\infty} \operatorname{Int} \alpha_i(\mathscr{A}_i \cup \mathscr{B}_i) \subset R''(X).$$

Thus $X, U_1, U_2, \ldots, H_1, H_2 \ldots$ satisfy the hypotheses of Lemma 7.7.10 below, application of which will establish that $R''(X)$ is cell-like and will complete the proof of the Basic Lemma. □

Suppose given the following: a cell-like continuum X in S^n; pairwise disjoint compact subsets $U_1, U_2, \ldots$ in S^n; $X^- = X \smallsetminus \cup_{i=1}^{\infty} \operatorname{Int} U_i$; a sequence $\{\epsilon_i\}$ of positive numbers, with $\epsilon_i \to 0$; and a collection of homotopies $H_i : U_i \times [0,1] \to S^n$ starting at the identity, moving only points of $\operatorname{Int} U_i$ and having image in $B(X^-; \epsilon_i)$. Define a function $f : X \to S^n$ as $f|X^- = \text{incl}_{X^-}$ and $f|X \cap U_i = H_i(-,1)$.

Lemma 7.7.10. *If f is injective, then $f(X)$ is cell-like.*

Remark. Even if it is injective, f need not be a homeomorphism.

Proof. Consider $x_1, x_2, \ldots \in X$. If infinitely many x_i belong to one of the sets $A_i = (X \cap U_i) \cup X^-$, then the sequence $f(x_1), f(x_2), \ldots$ clusters at some point of $f(A_i)$, since $f|A_i$ is continuous. On the other hand, if none of the A_i contains infinitely many points of $\{x_1, x_2, \ldots\}$, then we can assume that $x_i \in \operatorname{Int} U_{j(i)}$, where $j(1) < j(2) < \cdots$. Hence $f(x_i) \in B(X^-; \epsilon_{j(i)})$ for each

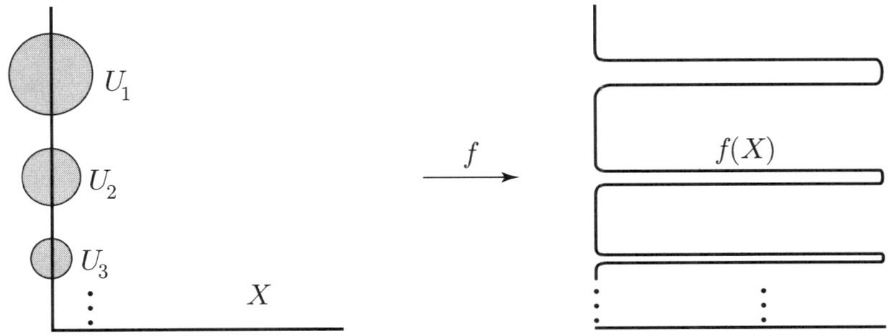

Figure 7.10. An injective f which is not a homeomorphism

i, and so $f(x_1), f(x_2), \ldots$ clusters at some point of $X^- \subset f(X)$. This yields that $f(X)$ is compact.

As an aid to showing $f(X)$ is cell-like, for each positive integer N define a map $\mu_N : S^n \to S^n$ as the inclusion on $S^n \setminus \cup_{i=1}^N \operatorname{Int} U_i$ and as $H_i(-,1)$ on $\operatorname{Int} U_i$, $i = 1, \ldots, N$. Similarly, define another collection of maps $\hat{\mu}_N : f(X) \to S^n$ as inclusion on $f(X) \setminus f(\cup_{i=1}^N \operatorname{Int} U_i)$ and as $[H_i(-,1)|X \cap U_i]^{-1}$ on $f(X \cap U_i)$. The injectivity of f assures that each $\hat{\mu}_N$ is well-defined and continuous.

Given an arbitrary neighborhood U of $f(X)$, choose an integer N so large that $i > N$ implies $H_i(U_i \times [0,1]) \subset U$. As a consequence,

$$\mu_N(X \cap U_i) = H(X \cap U_i, 0) \subset H_i(U_i, 0) \subset U \text{ for all } i > N,$$

and

$$\mu_N(X \cap U_i) = H(X \cap U_i) = f(X \cap U_i) \subset U \text{ for } i = 1, 2, \ldots, N.$$

Hence, $\mu_N(X) \subset U$. The map $(\operatorname{Id}, \mu_N) : S^n \to S^n \times S^n$ sends X into $S^n \times U$. Since $(\operatorname{Id}, \mu_N)(X)$ is cell-like, being homeomorphic to X, there is a neighborhood V of $(\operatorname{Id}, \mu_N)(X)$ that is contractible in $S^n \times U$. In turn, there is another neighborhood W of X in S^n such that $(\operatorname{Id}, \mu_N)(W) \subset V$.

Choose an integer $M > N$ so large that $i > M$ implies $H_i(U_i \times [0,1]) \subset W$. As above, $\hat{\mu}_M(f(X)) \subset W$.

Using $\pi_2 : S^n \times S^n \to S^n$ to denote projection to the second factor, consider the continuous function

$$\mu_N \circ \hat{\mu}_M = \pi_2 \circ (\operatorname{Id}, \mu_N) \circ \hat{\mu}_M : f(X) \to S^n.$$

It is null homotopic in U, since $(\operatorname{Id}, \mu_N) \circ \hat{\mu}_M(f(X)) \subset (\operatorname{Id}, \mu_N)(W) \subset V$ and V is null homotopic in $\pi_2^{-1}(U)$. But $\operatorname{incl}_{f(X)}$ and $\mu_N \circ \hat{\mu}_M$ are homotopic in U via the homotopy that fixes $f(X \cap U_i)$ for $i \notin \{N+1, \ldots, M\}$ and that moves $f(X \cap U_i)$ to $X \cap U_i$ by the reverse of the homotopy $H_i(-,t)|(X \cap U_i)$

in the other finite set of cases. Thus, $f(X)$ is cell-like, as it is null homotopic in an arbitrary neighborhood U. □

Theorem 7.7.11. *Let $R : S^{n-1} \to S^n$ be a cell-like embedding relation and $\pi : S^n \to S^n/R$ the associated decomposition map. Let C denote the closure of one of the components of $S^n/R \setminus \pi \circ R(S^{n-1})$, $T : C \to S^n$ a cell-like embedding relation, and L a neighborhood of T. Then L contains a 1-LCC cell-like embedding relation $T''' : C \to S^n$.*

Proof. The argument is a fairly straightforward repetition of the one just given for Theorem 7.7.2, except that in this setting the focus rests on loops lying toward only one specific side of the sequence of cell-like embedding relations. Here $\pi^{-1} : S^n/R \to S^n$ restricts to a cell-like embedding relation $\hat{R} : C \to S^n$. One needs an analog of Basic Lemma 7.7.5 for $\hat{R}$-disks that meet $\operatorname{Im} \hat{R}$ only at points of $\hat{R}(\operatorname{Bd} C)$. Given a countable dense collection $\{f_i : S^1 \to S^n\}$ of embedded loops in S^n, as in the proof that 7.7.5 implies 7.7.2, the adapted Basic Lemma gives rise to a sequence of cell-like embedding relations $\hat{R}_i : C \to S^n$, neighborhoods L_i of $\hat{R}_i$ and controlled mappings $F_i : I^2 \to S^n$ such that $\operatorname{Im} \hat{R}_i \cap \operatorname{Im} F_i = \emptyset$ as before, unless $f_i(S^1) \cap \hat{R}_{i-1}(\operatorname{Int} C) \neq \emptyset$, in which case $\hat{R}_i = \hat{R}_{i-1}$ and F_i, which is essentially irrelevant, is chosen arbitrarily. □

Corollary 7.7.12. *Let $R : S^{n-1} \to S^n$ be a cell-like embedding relation and let $\pi : S^n \to S^n/R$ be the associated decomposition map. Suppose C is the closure of one of the components of $S^n/R \setminus \pi \circ R(S^{n-1})$ and $\pi \circ R(S^{n-1})$ is 1-LCC in C. Then C is an n-cell.*

Proof. Let $T''' : C \to S^n$ denote the 1-LCC cell-like embedding relation promised by Theorem 7.7.11, and use π''' to denote the quotient map for the decomposition of S^n whose elements are $\{T'''(x) \mid x \in C\}$ and the singletons from $S^n \setminus \operatorname{Im} T'''$. Here the frontier of $\pi'''T'''(C)$ in S^n/T''' is the 1-LCC embedded $(n-1)$-sphere $\pi'''T'''(\operatorname{Bd} C)$, so Proposition 7.4.13 assures that S^n/T''' is the n-sphere. Furthermore, being 1-LCC, $\pi'''T'''(\operatorname{Bd} C)$ is flatly embedded, so $\pi'''T'''(C) \cong C$ is an n-cell. □

Corollary 7.7.13. *If $\lambda : S^{n-1} \to S^n$ is an embedding, $n \geq 5$, and C is the closure of one of the components of $S^n \setminus \lambda(S^{n-1})$ such that $\lambda(S^{n-1})$ is 1-LCC in C, then C is an n-cell.*

Theorem 7.7.14. *Let $\lambda : S^{n-1} \to S^n$ be an embedding, $n \geq 5$, C the closure of one of the components of $S^n \setminus \lambda(S^{n-1})$, and $\epsilon > 0$. Then there exists an embedding $\lambda' : C \to S^n$ such that $\rho(\lambda', \operatorname{incl}_C) < \epsilon$ and $\operatorname{Cl}(S^n \setminus \lambda'(C))$ is an n-cell.*

Proof. Apply Theorem 7.7.11 to obtain a 1-LCC embedding relation $T''' : C \to S^n$ in the ϵ-neighborhood of $\operatorname{incl}_C : C \to S^n$. Form the associated decomposition $\mathscr{G}$ into the sets $\{T'''(x) \mid x \in C\}$ and the singletons from $S^n \setminus \operatorname{Im} T'''$, and let C' denote the closure of the component of $S^n/\mathscr{G}$ not containing the image of $T'''(\operatorname{Int} C)$. Corollary 7.7.12 indicates that C' is an n-cell. Proposition 7.4.14 yields that $S^n/\mathscr{G}$ is topologically S^n. Hence, $S^n \cong S^n/\mathscr{G}$ is expressed as the union of a copy of C and the n-cell C', where $C \cap C' = \partial C'$. Obtaining this reembedded copy of C pointwise close to C itself comes about by a controlled shrinking of $\mathscr{G}$, as in the proof of Corollary 7.7.3. □

The Locally Flat Approximation Theorem (7.7.1) combines with Corollary 7.6.11 to give

Corollary 7.7.15. *A locally homotopically unknotted $(n-2)$-sphere Σ in S^n $(n \geq 5)$ is flat if and only if Σ bounds an $(n-1)$-cell $E \subset S^n$.*

As another consequence of 7.7.1, local issues about codimension-one embeddings involving manifolds reduce to problems about embeddings of $(n-1)$-spheres in S^n.

Theorem 7.7.16. *Suppose $n \geq 5$, Q is an $(n-1)$-manifold embedded in an n-manifold M as a closed subset, and $q \in Q$. Then there exist an $(n-1)$-sphere Σ in S^n, a neighborhood N_q of q in M, and an embedding $e : N_q \to S^n$ such that $e(N_q \cap Q) \subset \Sigma$.*

Proof. Exploiting coordinate charts in M, we can simplify the setting so that $q \in Q \subset \mathbb{R}^n \subset S^n$. Fix an $(n-1)$-cell B with $q \in \operatorname{Int} B \subset B \subset Q$, and choose a neighborhood U of q in S^n for which $Q \cap \overline{U} \subset \operatorname{Int} B$. According to Theorem 7.7.1 we can assume that Q is locally flat at each point of $Q \setminus \overline{U}$. Here ∂B is flat (Corollary 7.7.15). Thus we can assume that ∂B is the standard $(n-2)$-sphere in S^n and that a collar C on ∂B in B lies in the standard copy of $S^{n-1} \subset S^n$.

Consider the universal cover $p : E \to S^n \setminus \partial B$; E is homeomorphic to $\operatorname{Int} B^{n-1} \times \mathbb{R}$ in such a way that $p(\operatorname{Int} B \times \{0\}) \subset S^{n-1}$. Lift $\operatorname{Int} B$ to the universal cover E via $\lambda : \operatorname{Int} B \to E$ with the collar C on ∂B mentioned earlier going into, say, $\operatorname{Int} B^{n-1} \times \{0\}$, and extend λ to a lift $\lambda' : N \to E$ defined on some neighborhood N of $\operatorname{Int} B$ for which $N \cap Q = \operatorname{Int} B$. Find an embedding h of the universal cover into a neighborhood of $p(\operatorname{Int} B^{n-1} \times \{0\})$ such that $p\lambda(x) = p\lambda'(x) = x$ for all $x \in C$ and $h(E) \cap S^{n-1} = p(\operatorname{Int} B \times \{0\})$. Then

$$\Sigma = h\lambda'(B) \cup (S^{n-1} \setminus p(\operatorname{Int} B \times \{0\}))$$

is an $(n-1)$-sphere in S^n and $e = h\lambda'$ embeds the neighborhood N of q in the required way. □

Appendix on Embedding Relations

Standard results about continuous relations that carry over from the function setting are:

(1) If $R : X \to Y$ and $S : Y \to Z$ are continuous, then $S \circ R : X \to Z$ is continuous.

(2) If $R : X \to Y$ is continuous with compact point images and $C \subset X$ is compact, then $R(C)$ is compact.

(3) If $R : X \to Y$ is continuous with nonempty, connected point images and $C \subset X$ is connected, then $R(C)$ is connected.

For simplicity we assume all spaces under consideration to be locally compact, separable metric spaces. The relations most useful in geometric topology are the proper relations—recall that a relation $R : X \to Y$ is *proper* if both R and R^{-1} are continuous with compact point images. The equivalent but asymmetric defining property usually ascribed is the following:

(4) A relation $R : X \to Y$ is proper provided R is continuous with compact point images and the inverse of each compact subset of Y is compact.

The basic results on proper relations are:

(5) A proper relation $R : X \to Y$ is a closed subset of $X \times Y$. Each neighborhood of R in $X \times Y$ contains a proper neighborhood $N : X \to Y$ of R in $X \times Y$. Each closed subset of a proper relation is a proper relation.

(6) Composition Theorem. Suppose $R : X \to Y$ and $S : Y \to Z$ are relations where R^{-1} and S both are continuous with compact point images and U is a neighborhood of $S \circ R$ in $X \times Z$. Then there exist neighborhoods V of R in $X \times Y$ and W of S in $Y \times Z$ such that $W \circ V \subset U$.

Proof. Fix $y \in Y$ and observe that the (possibly empty) compact set $R^{-1}(y) \times S(y)$ lies in U, since

$$U \supset S \circ R = S \circ \mathrm{Id}_Y \circ R = (R^{-1} \times S)(\mathrm{Id}_Y).$$

There exist open neighborhoods B_y of $R^{-1}(y)$ in X and C_y of $S(y)$ in Z such that $B_y \times C_y \subset U$. Continuity of R^{-1} and S gives an open neighborhood A_y of y such that $R^{-1}(A_y) \subset B_y$ and $S(A_y) \subset C_y$.

The paracompactness of Y assures the existence of a precise, locally finite, open refinement $\{A'_y \mid y \in Y\}$ of the open cover $\{A_y \mid y \in A\}$ that covers Y and satisfies $\mathrm{Cl}(A'_y) \subset A_y$ for each $y \in Y$. Define $V : X \to Y$ and $W : Y \to Z$ by the formulae $V^{-1}(y) = \cap\{B_{y(0)} \mid y \in \mathrm{Cl}(A'_{y(0)})\}$ and $W(y) = \cap\{C_{y(0)} \mid y \in \mathrm{Cl}(A'_{y(0)})\}$. □

(7) COROLLARY. If $R : X \to Y$ and $S : Y \to Z$ are continuous with compact point images and U is a neighborhood of $S \circ R$ in $X \times Z$, then there is a neighborhood V of R in $X \times Y$ such that $S \circ V \subset U$.

Proof. Choose compact sets $X_1, X_2, \ldots$ whose interiors cover X. The relations $R|X_i : X_i \to Y$ are proper by (4). Accordingly, by the Composition Theorem, there exist neighborhoods V_i of $R|X_i$ in $X_i \times Y$ with $S \circ V_i \subset U$, and $V = \cup_i V_i$ satisfies the requirements of the Corollary. □

The characteristic feature of cell-like embedding relations is that, by and large, they can be approximated by continuous functions.

(8) CONTINUOUS APPROXIMATION THEOREM. Suppose $R : X \to Y$ is a continuous cell-like embedding relation from a finite-dimensional space X to an ANR Y. Then each neighborhood of R in $X \times Y$ contains a continuous function from X to Y.

See (Cannon, 1975) for a proof. The Continuous Approximation Theorem is not used in this book.

(9) SLICE TRIVIALITY THEOREM. Suppose $R : X \to Y$ is a continuous relation with cell-like point images and $L'' : X \to Y$ is a neighborhood of R. Then there exists another neighborhood L' of R such that $L' \subset L''$ and $x \in X$ implies $L'(x)$ contracts in $L''(x)$.

Proof. For each $x \in X$ find an open set $W \supset R(x)$ in Y and a neighborhood V of x such that $V \times W \subset L''$. Cell-likeness of $R(x)$ leads to another $W_x \supset R(x)$ that contracts in W. Moreover, x has a neighborhood $U_x \subset V$, such that $R(x') \subset W_x$ for all $x' \in U_x$. In case X is compact, corresponding to the open cover $\mathcal{U} = \{U_x \times W_x \mid x \in X\}$ of R is a $\delta > 0$, a kind of Lebesgue number for $\mathcal{U}$, such that for the δ-neighborhood L' of $R \subset X \times Y$ and for arbitrary $x \in X$, some $U_z \times W_z \in \mathcal{U}$ contains $\{x\} \times L'(x)$. Hence, $L'(x) \subset W_z$ contracts in $L''(x)$. The general case (X locally compact) is left to the reader. □

Historical Notes. There are other approaches to the Locally Flat Approximation Theorem. Cannon, Bryant and Lacher (1979) performed multiple grope replacements in M, changing it to ANR homology n-manifold Y equipped with a cell-like map $p : Y \to M$ and a 1-LCC embedding $\lambda : Q \to Y$ of the given $(n-1)$-manifold Q such that $Y \smallsetminus \lambda(S)$ is an n-manifold and $p\lambda = \mathrm{incl}_Q$. Either work of S. Ferry (1979) or application of Quinn's Index Theorem (see §8.5) yields that Y actually is an n-manifold. Upon approximating p by a homeomorphism (Corollary 7.4.2), one obtains a 1-LCC approximation $h\lambda(Q)$ to Q.

Ferry (1992) provided another argument for the Locally Flat Approximation Theorem which combines surgery below the middle dimension with an application of his α-Approximation Theorem, a topic treated briefly in Chapter 8.

As an alternative, in case Q of Theorem 7.7.1 is two-sided, let $C+, C-$ denote the closures of the components of $M \setminus Q$. Form a new space M^* from $C+ \sqcup (Q \times [-1,1]) \sqcup C-$ by identifying each $\langle q, -1 \rangle$ with $q \in C-$ and, similarly, identifying each $\langle q, +1 \rangle$ with $q \in C+$. By (Quinn, 1987) M^* is the cell-like image of an n-manifold, and quite obviously it satisfies the Disjoint Disks Property, so M^* itself is an n-manifold. Moreover, there is an evident cell-like map $p : M^* \to M$ that sends each arc $q \times [-1,1]$ to q; approximating p by a homeomorphism, one notices that the image of $Q \times \{0\} \subset Q \times [-1,1] \subset M^*$ is a bicollared approximation to the original Q.

L. L. Lininger (1965) and N. Hosay (1963) independently proved the 3-dimensional version of Theorem 7.7.11 for embeddings $T : C \to S^n$; Daverman (1977), (1987) did the same in higher dimensions.

Exercises

7.7.1. Suppose $\Sigma^{n-2} \subset S^n$ is a locally flat $(n-2)$-sphere that bounds a topologically embedded $(n-1)$-cell $B \subset S^n$, $n \geq 5$. Then Σ^{n-2} is flat.

7.7.2. If $R : X \to S^n$ is a 1-LCC cell-like embedding relation defined on the compact, k-dimensional space X, where $k \leq n-3$ and $n \geq 5$, then the decomposition of S^n into points of $S^n \setminus \operatorname{Im} R$ and the sets $\{R(x) \mid x \in X\}$ is shrinkable.

7.8. Kirby-Siebenmann obstruction theory

The PL Structure Theorem 6.8.2 of Kirby and Siebenmann has crucially important consequences in codimension one, just as it does in codimension two.

Say that an embedding $\varphi : Q^{n-1} \to M^n$ of one PL manifold in another is *PL locally flat* if φ is PL and if all link pairs in $(M^n, \varphi(Q^{n-1}))$ are PL standard pairs.

Corollary 7.8.1 (Codimension-One Taming). *Suppose Q^{n-1} is a PL $(n-1)$-manifold such that $H^3(Q^{n-1}; \mathbb{Z}_2) = 0$, $h : Q^{n-1} \to M^n$ is a locally flat topological embedding of Q^{n-1} into a PL n-manifold M^n, $n \geq 5$, and $\epsilon > 0$. Then h is ambient isotopic to a PL locally flat embedding via an ϵ-isotopy of M^n.*

The proof coincides with that for the codimension-two taming result (Corollary 6.8.3).

Corollary 7.8.2. *If Q^{n-1} is a PL $(n-1)$-manifold, $n \geq 5$, such that $H^3(Q^{n-1};\mathbb{Z}_2) \neq 0$, then there exist a PL n-manifold M^n and a locally flat embedding $h : Q^{n-1} \to M^n$ that is not ambient isotopic to a PL locally flat embedding.*

Proof. The Product Structure Theorem promises a PL structure on $M^n = Q^{n-1} \times \mathbb{R}$ not compatible with the obvious product structure on $Q^{n-1} \times \mathbb{R}$. If there were a PL locally flat embedding $Q^{n-1} \to M^n$ isotopic to $Q^{n-1} \to Q^{n-1} \times \{0\}$, then an open PL bicollar on the image could be expanded to produce a PL homeomorphism $Q^{n-1} \times \mathbb{R} \to M^n$. □

Locally Flat Approximation Theorem 7.7.1 then leads to a PL Approximation Theorem for the PL manifolds with trivial $\mathbb{Z}_2$-cohomology in dimension 3.

Corollary 7.8.3 (Codimension-One PL Approximation). *Suppose Q^{n-1} is a PL $(n-1)$-manifold such that $H^3(Q^{n-1};\mathbb{Z}_2) = 0$, and suppose $h : Q^{n-1} \to M^n$ is a topological embedding of Q^{n-1} into a PL n-manifold M^n, $n \geq 5$. Then h can be approximated, arbitrarily closely, by PL locally flat embeddings.*

7.9. Detecting 1-LCC embeddings

§4.6 presents conditions for detecting 1-LCC embeddings of codimension-three compacta. This section does the same for embeddings of codimension-one manifolds. Among the conditions covered are local flatness modulo certain twice-flat subsets (Theorem 7.9.2), singular regular neighborhoods (Theorem 7.9.8), and a local spanning property (Theorem 7.9.15).

Here is a slight strengthening of an observation appearing earlier in the proof that 7.7.5 implies 7.7.2.

Lemma 7.9.1. *Let β denote an $(n-1)$-cell in an n-manifold M^n, $f : I^2 \to M^n$ a map such that $f(\partial I^2) \cap \beta = \emptyset$, and Z the component of $I^2 \smallsetminus f^{-1}(\beta)$ containing ∂I^2. Then there exists a map $g : I^2 \to M^n$ such that $g|Z = f|Z$ and $g(I^2 \smallsetminus Z) \subset \beta$; moreover, g can be obtained so that $g(I^2 \smallsetminus Z) \cap \partial\beta \subset f(I^2) \cap \partial\beta$.*

Proof. This follows using the Tietze Extension Theorem to extend

$$f|\overline{Z} \cap (I^2 \smallsetminus Z) : \overline{Z} \cap (I^2 \smallsetminus Z) \to \beta$$

to a map $I^2 \smallsetminus Z \to \beta$. For the strengthened conclusion, simply apply the result using an $(n-1)$-cell $\beta' \subset \beta$ such that $\beta' \supset f(I^2) \cap \beta$ and $\beta' \cap \partial\beta = f(I^2) \cap \partial\beta$. □

Theorem 7.9.2. *Suppose Σ^{n-1} is an $(n-1)$-manifold embedded as a closed subset of an n-manifold M^n and X is a closed subset of Σ^{n-1} such that Σ^{n-1} is 1-LCC in M^n at each point of $\Sigma^{n-1} \setminus X$ and X is LCC^1 in both Σ^{n-1} and M^n. Then Σ^{n-1} is 1-LCC in M^n.*

Proof. Start with $s \in \Sigma^{n-1}$ and $\epsilon > 0$. Identify an $(n-1)$-cell β in Σ with $s \in \operatorname{Int}\beta \subset \beta \subset B(s;\epsilon)$, and then choose a simply connected neighborhood $W_s \subset B(s;\epsilon)$ with $W_s \cap \Sigma^{n-1} \subset \beta$. Given any loop $f : \partial I^2 \to W_s \setminus \Sigma$, extend f to $F : I^2 \to W_s$. Since X is LCC^1 in M^n, Lemma 3.3.3 assures that F can be approximated by a map $F^* : I^2 \to W_s \setminus X$ such that $F^*|\partial I^2 = F|\partial I^2 = f$. Let H denote the component of $I^2 \setminus (F^*)^{-1}(\Sigma)$ containing ∂I^2, and note that $F^*(\overline{H}) \cap X = \emptyset$. Apply Lemma 7.9.1 to obtain a map $g : I^2 \to F^*(H) \cup \beta$ such that $g|H = F^*|H$ and $g(I^2 \setminus H) \subset \beta$. Since X is also LCC^1 in Σ^{n-1}, a mild extension of Lemma 3.3.3 assures that g can be approximated by a map $g^* : I^2 \to F^*(H) \cup \beta$ such that $g^*|H = g|H = F^*|H$ and $g^*(I^2 \setminus H) \subset \beta \setminus X$. Note that $g^*(I^2) \subset (F^*(H) \cup \beta) \setminus X \subset B(s;\epsilon)$. Now g^* can be approximated by a map $g : I^2 \to B(S;\epsilon) \setminus \Sigma^{n-1}$ with $g|\partial I^2 = g^*|\partial I^2 = f$, since Σ^{n-1} is 1-LCC at points of $\Sigma^{n-1} \cap g^*(I^2)$. □

Corollary 7.9.3. *If the $(n-1)$-sphere $\Sigma^{n-1} \subset S^n$, $n \geq 5$, is locally flat modulo a finite set, then Σ^{n-1} is flat.*

Corollary 7.9.4. *If the $(n-1)$-sphere $\Sigma^{n-1} \subset S^n \geq 5$, is locally flat modulo a twice-flat k-cell or k-sphere, $k \leq n-4$, then Σ^{n-1} is flat.*

Remark. Suspensions of examples like the Fox-Artin 2-sphere in S^3 (§2.8.3) indicate that Corollary 7.9.4 fails when $k = n-3$.

Definition. Let Σ^{n-1} denote an $(n-1)$-sphere topologically embedded in S^n and $C \subset \Sigma^{n-1}$ a Cantor set. Extending a previous definition for cells and spheres in Σ^{n-1}, we say that C is *twice flat* if it is flat as a subset of both Σ^{n-1} and S^n.

Corollary 7.9.5. *If the $(n-1)$-sphere $\Sigma^{n-1} \subset S^n$, $n \geq 5$, is locally flat modulo a twice-flat Cantor set, then Σ^{n-1} is flat.*

Definition. Let Σ denote a connected $(n-1)$-manifold topologically embedded in an n-manifold M as a closed, separating subset, and let U denote a component of $M \setminus \Sigma$. We say that Σ can be *homeomorphically approximated from U* if for each $\epsilon > 0$ there exists an embedding $\lambda_\epsilon : \Sigma \to U \subset M$ such that $\rho(\lambda_\epsilon, \text{incl}_\Sigma) < \epsilon$.

Theorem 7.9.6. *Let Σ denote a connected $(n-1)$-manifold topologically embedded in an n-manifold M as a closed, separating subset, and let U denote a component of $M \setminus \Sigma$ such that Σ can be homeomorphically approximated from U. Then Σ is 1-LCC embedded in $\overline{U}$.*

The proof is an exercise. Here there is no need to restrict n.

Corollary 7.9.7. *Let Σ denote a connected $(n-1)$-manifold topologically embedded in an n-manifold M as a closed, separating subset. Then Σ is locally flat in M if and only if it can be homeomorphically approximated from each component of $M \smallsetminus \Sigma$.*

Definition. Let Q denote a connected $(n-1)$-manifold topologically embedded in an n-manifold M as a closed, separating subset, and let U denote a component of $M \smallsetminus Q$. We say that Q is *singularly collared in* $\overline{U}$ if there exists a map $\mu : Q \times [0,1] \to \overline{U}$ such that $\mu_0 = \mathrm{incl}_Q$ and $\mu(Q \times (0,1]) \subset U$.

Theorem 7.9.8. *Suppose Q is a connected $(n-1)$-manifold topologically embedded in a connected n-manifold M as a closed, separating subset, and suppose U is a component of $M \smallsetminus Q$ such that Q is singularly collared in $\overline{U}$. Then Q is 1-LCC in $\overline{U}$.*

The argument for 7.9.8 hangs on the notion of degree for maps between (orientable) manifolds, on a result about degree being locally determined relative to the target, and on another result that degree-one maps induce epimorphisms of fundamental groups.

Definitions. An *orientation* of a connected, orientable n-manifold U is a choice of generator of $H^n_c(U;\mathbb{Z})$. Given connected n-manifolds U, V equipped with orientations γ_U, γ_V, respectively, the *degree* of a proper map $f : U \to V$ is the integer d such that $f^*(\gamma_V) = d \cdot \gamma_U$. Functorial properties assure that the degree of a composite (of proper mappings) is the product of degrees.

Lemma 7.9.9. *Suppose $f : M \to N$ is a proper map between connected, oriented n-manifolds and V is a connected open subset of N such that $U = f^{-1}(V)$ is connected. Then the degree of f equals the degree of $f|U$, provided U, V are oriented with the orientations obtained by restriction from M, N, respectively.*

Proof. Let $\rho_U : H^n_c(U;\mathbb{Z}) \to H^n_c(M;\mathbb{Z})$ and $\rho_V : H^n_c(V;\mathbb{Z}) \to H^n_c(N;\mathbb{Z})$ denote the isomorphisms induced by extension. Equality of the degrees follows immediately from the commutativity of

$$\begin{array}{ccc} H^n_c(U;\mathbb{Z}) & \xleftarrow{(f|U)^*} & H^n_c(V;\mathbb{Z}) \\ \cong \big\downarrow \rho_U & & \cong \big\downarrow \rho_V \\ H^n_c(M;\mathbb{Z}) & \xleftarrow{f^*} & H^n_c(N;\mathbb{Z}), \end{array}$$

together with the prescription that the vertical isomorphisms preserve preferred orientations. □

Corollary 7.9.10. *Every proper, non-surjective mapping $f : M \to N$ between orientable n-manifolds has degree* 0.

Proof. Properness implies that $f(M)$ is closed in N. Applying the analysis of 7.9.9 to a connected open subset V of $N \setminus f(M)$, observe that f^* factors through the trivial group $H^n_c(\emptyset; \mathbb{Z})$. □

Lemma 7.9.11. *Suppose $f : M \to N$ is a proper map between connected, oriented n-manifolds, V is a connected open subset of N, $U_1, U_2 \dots$ are the components of $f^{-1}(V)$, and $f|U_i : U_i \to V$ has degree d_i ($i = 1, 2, \dots$). Then $d_i = 0$ for all but finitely many values and the degree of f equals $\Sigma_i d_i$.*

Proof. We re-employ the notation from the preceding lemma and examine the diagram:

$$\begin{array}{ccc} H^n_c(f^{-1}(V); \mathbb{Z}) & \xleftarrow{(f|)^*} & H^n_c(V; \mathbb{Z}) \\ \downarrow \rho & & \cong \downarrow \rho_V \\ H^n_c(M; \mathbb{Z}) & \xleftarrow{f^*} & H^n_c(N; \mathbb{Z}). \end{array}$$

As above, choices of orientations for M, N give rise to preferred generators γ_V of $H^n_c(V; \mathbb{Z})$ and γ_i for $H^n_c(U_i; \mathbb{Z})$. Here, by hypothesis $(f|U_i)^*$ sends γ_V to $d_i \cdot \gamma_i \in H^n_c(U_i; \mathbb{Z})$. For $v \in V$, at most finitely many components of $f^{-1}(V)$ meet $f^{-1}(v)$, by properness, and Corollary 7.9.10 attests that $d_i = 0$ for those U_i that do not surject to V. Moreover, $H^n_c(f^{-1}(V); \mathbb{Z}) \cong H^n_c(\cup U_i; \mathbb{Z}) \cong \oplus_i H^n_c(U_i; \mathbb{Z})$, so the image of γ_V under the homomorphism in the upper row is $\xi = \Sigma_i(d_i \cdot \gamma_i)$. Under the extension $\rho : H^n_c(\cup_i U_i; \mathbb{Z}) \to H^n_c(M)$, we see that $\rho(\xi) = (\Sigma_i d_i) \cdot \gamma_M$, since each γ_i is sent to the preferred orientation class $\gamma_M \in H^n_c(M; \mathbb{Z})$ via the extension ρ. □

Lemma 7.9.12. *Let p denote a positive integer. If $\theta : \widetilde{M} \to M$ is a p-fold covering map between oriented PL n-manifolds, then the degree of θ is $\pm p$.*

Proof. This widely known result typically is based on other definitions of orientability. In the context at hand, the usual diagram

$$\begin{array}{ccc} H^n_c(V; \mathbb{Z}) & \xleftarrow{(h|V)^*} & H^n_c(V; \mathbb{Z}) \\ \cong \downarrow \rho_V & & \cong \downarrow \rho_V \\ H^n_c(M; \mathbb{Z}) & \xleftarrow{h^* = \mathrm{Id}^*} & H^n_c(M; \mathbb{Z}) \end{array}$$

reveals that, given a homeomorphism $h : M \to M$ properly homotopic to Id_M and open subset V such that $h(V) = V$, $h|V$ must be orientation preserving. Let $V \subset M$ denote the interior of a PL n-cell evenly covered by θ, $U_1, \ldots, U_p$ the components of $\theta^{-1}(V)$, and ϵ_i the degree of $\theta|U_i : U_i \to V$. Since $\theta|U_i$ is a homeomorphism, $\epsilon_i = \pm 1$. Should there be a pair $\{\epsilon_i, \epsilon_j\}$ with $\epsilon_i = -\epsilon_j$, then just as in (Rourke and Sanderson, 1972, p. 44ff) an isotopy $\widetilde{M} \to \widetilde{M}$ carrying U_i to U_j could be used to produce a homeomorphism $h : M \to M$ properly isotopic to Id_M whose restriction to V reverses orientations. Consequently, Lemma 7.9.11 implies that θ has degree $p \cdot \epsilon_1 = \pm p$. □

Lemma 7.9.13. *Any proper map $f : M \to N$ between connected, oriented, PL n-manifolds of degree ± 1 induces an epimorphism $f_* : \pi_1(M) \to \pi_1(N)$.*

Proof. Suppose to the contrary that $f_*(\pi_1(M)) \neq \pi_1(N)$. Construct the covering space $\theta : \widetilde{N} \to N$ corresponding to $f_*(\pi_1(M))$. Then f lifts to a map $\widetilde{f} : M \to \widetilde{N}$ such that $f = \theta\widetilde{f}$. Localization as in Lemma 7.9.9 assures that f is surjective, for otherwise it would have degree 0.

It can be easily shown that $\widetilde{f}$ is a proper mapping. When θ is a p-fold covering, $p < \infty$, both θ and $\widetilde{f}$ are proper. By Lemma 7.9.12 $|\text{degree}(\widetilde{f})| = [\pi_1(N) : f_*(\pi_1(M))] > 1$. But this is impossible, as it would yield $1 = |\text{degree}(f)| = |\text{degree}(\theta)| \cdot |\text{degree}(\widetilde{f})| > 1$.

We conclude by explaining why θ must be a finite-sheeted cover. Adjusting f via a proper homotopy, we can assume that for some small open set V in N, $f|f^{-1}(V)$ is PL. Let V' denote the interior of an n-simplex σ in V whose preimage under f consists of finitely many n-simplices, each mapped homeomorphically to σ. If θ were infinite-sheeted, $\widetilde{f}$ would have degree 0, as it could not be onto. Let $U_1, \ldots, U_k$ denote the components of $f^{-1}(V')$ and ϵ_i the degree of $f|U_i : U_i \to V'$. The collection $\{U_1, \ldots, U_k\}$ is partitioned into finitely many subcollections corresponding to the preimages under $\widetilde{f}$ of the various components V^* of $\theta^{-1}(V')$. Over each subcollection the various degrees of the restricted $\widetilde{f}$ sum to 0, by Lemma 7.9.11. Upon composing θ and $\widetilde{f}$, we obtain $\epsilon_1 + \cdots + \epsilon_k = 0$, yielding degree$(f) = 0$, contrary to hypothesis. □

Proof of Theorem 7.9.8. Focus on a point $q \in Q$. In light of Corollary 7.7.16, we can transfer to the setting in which there are an embedding $\lambda : N_q \to S^n$ defined on some neighborhood N_q of q in M and an $(n-1)$-sphere Σ in S^n with $\lambda(N_q \cap Q) \subset \Sigma$. Let U' denote the component of $S^n \smallsetminus \Sigma$ containing points of $\lambda(N_q \cap U)$ arbitrarily close to $s = \lambda(q)$. It will suffice to prove that U' is 1-LC at s.

Fix $\epsilon > 0$. As Q is singularly collared in $\overline{U}$, there exist a small $(n-1)$-cell D on Σ with $s \in \operatorname{Int} D$ and a map $\mu : D \times [0,1] \to B(s;\epsilon) \cap \overline{U'}$ such that $\mu_0 = \mathrm{incl}_D$ and $\mu(D \times (0,1]) \subset U'$.

Lemma 7.9.14. *In the setting above, $\overline{U'}$ contains an open subset W such that* $\operatorname{Int} D \subset W \subset f(D \times [0,1])$.

Proof. Use μ to produce a map h of $\Sigma \times [0,1]$ to $\Sigma \cup \mu(D \times [0,1]) \subset S^n$ such that $h(z,0) = z$ for all $z \in \Sigma$, $h(z',t) = z'$ for all $z' \in \Sigma \setminus \operatorname{Int} D$, and $h(\operatorname{Int} D \times \{t\}) \cap \Sigma = \emptyset$ for all $t > 0$. Define W as the intersection of $\overline{U'}$ and the component of $S^n \setminus h_1(\Sigma)$ that contains $\operatorname{Int} D$. We show that $W \subset \mu(D \times [0,1])$.

Suppose to the contrary that there exists

$$w \in W \setminus \mu(D \times [0,1]) \subset W \setminus h(\Sigma \times [0,1]).$$

Choose $z_0 \in N_s \setminus \overline{U'}$ and regard $S^n \setminus \{z_0\}$ as $\mathbb{R}^n$. As in (Hurewicz and Wallman, 1948), for $x \neq z_0$ let π_x denote the radial map of $\mathbb{R}^n \setminus \{x\} = S^n \setminus \{x, z_0\}$ onto the unit $(n-1)$-sphere centered at x. Then $\pi_w h_0$ and $\pi_w h_1$ are homotopic maps of Σ to S^{n-1}. For any $y \in S^n \setminus \overline{U'}$ we see that w and y belong to different components of $S^n \setminus h(\Sigma \times \{0\})$ and belong to the same component of $S^n \setminus h(\Sigma \times \{1\})$. This leads to the desired contradiction, because according to Theorem VI.10 of (Hurewicz and Wallman, 1948), $\pi_w h_0$ is an essential map and $\pi_w h_1$ is an inessential map. □

Continuing with the proof of Theorem 7.9.8, apply Lemma 7.9.14 to obtain the promised connected open subset W of $\overline{U'}$ such that

$$\operatorname{Int} D \subset W \subset \mu(D \times [0,1]) \subset B(s;\epsilon).$$

Identify the component Y of $\mu^{-1}(W)$ containing $\operatorname{Int} D \times \{0\}$. Set $W_U = W \cap U'$, $Y_U = Y \cap \mu^{-1}(W_U)$, and $\mu_U = \mu | Y_U$. It follows immediately that $\mu_U : Y_U \to W_U$ is a proper map between (orientable) n-manifolds, which implies that the degree of μ_U is defined. We shall prove that μ_U has degree ± 1.

By Theorem 7.7.14 we can assume that U' is embedded in S^n with an n-cell as its complement. Hence, μ_U extends to a map

$$\widetilde{\mu} : Y_U \cup (\operatorname{Int} D \times (-1,0]) \to S^n$$

with $\widetilde{\mu} | \operatorname{Int} D \times (-1,0]$ an embedding into $S^n \setminus U'$ such that

$$\widetilde{\mu}(\operatorname{Int} D \times (-1,0]) \cap \Sigma = \widetilde{\mu}(\operatorname{Int} D \times \{0\}).$$

It follows that $\widetilde{\mu}$ is a proper mapping between orientable n-manifolds. Since the image obviously contains the connected open subset $\widetilde{\mu}(\operatorname{Int} D \times (-1,0))$ over which $\widetilde{\mu}$ is a homeomorphism, Lemma 7.9.9 assures that $\widetilde{\mu}$ and μ_U have degree ± 1.

As a result, μ_U induces an epimorphism at the fundamental group level (Lemma 7.9.13). To each loop α in W_U corresponds a loop α' in Y_U with $(\mu_U)_*([\alpha']) = [\alpha]$. Since α' is contractible in $\operatorname{Int} D \times (0,1]$, α is contractible in $\mu(\text{Int}\, D \times (0,1]) \subset B(s;\epsilon) \cap U'$. Hence, U' is 1-LC at s, and Q is 1-LCC in $\overline{U}$. □

Definition. Suppose Q is an $(n-1)$-manifold topologically embedded in an n-manifold M as a closed subset, and suppose U is a component of $M \setminus Q$. We say that Q can be *locally spanned from* U if, for each neighborhood N_q of each $q \in Q$, there exist $(n-1)$-cells $B \subset Q$ and $D \subset \overline{U}$ such that $s \in \text{Int}\, B$, $D \cap Q = \partial B$, and $B \cup D \subset N_q$.

Theorem 7.9.15. *Suppose Σ is a connected $(n-1)$-manifold topologically embedded in a connected n-manifold M as a closed, separating subset, and suppose U is a component of $M \setminus \Sigma$ such that Σ can be locally spanned from U. Then Σ is 1-LCC in $\overline{U}$.*

Proof. Consider a map f of ∂I^2 into a small subset of U. Extend f to a map F of I^2 into a small subset of $\overline{U}$. In view of the locally spanned condition, we determine a finite collection of very small $(n-1)$-cells $B_1, ..., B_k$ in Σ whose interiors cover $\Sigma \cap F(I^2)$, and corresponding $(n-1)$-cells $D_1, ..., D_k$ in $\overline{U}$ such that $D_i \cap \Sigma = \partial B_i$ and $D_i \cap f(\partial I^2) = \emptyset$ for each i. Application of Lemma 7.9.1 yields a new map $F_1 : I^2 \to U$ with small image contained in $F(I^2) \cup D_1 \setminus \text{Int}\, B_1$ and with $F_1|\partial I^2 = F|\partial I^2 = f$, where in particular $F_1(I^2) \cap \partial B_1 \subset F(I^2) \cap \partial B_1$. Consequently, $F_1(I^2) \cap \Sigma \subset F(I^2) \cap (\Sigma \setminus \text{Int}\, B_1)$. Another $k-1$ applications of Lemma 7.9.1 yield a new map $F_k : I^2 \to U$ with small image in $(F(I^2) \cup (\cup_i D_i)) \setminus \Sigma$ and $F_k|\partial I^2 = F|\partial I^2 = f$. The key is that, once a point of Σ is removed from the image of F_i, it reappears in none of the succeeding images. □

Corollary 7.9.16. *An $(n-1)$-sphere Σ in S^n is 1-LCC if and only if Σ is locally spanned from each component of $S^n \setminus \Sigma$.*

Corollary 7.9.17. *An $(n-1)$-sphere Σ in S^n, $n \geq 5$, is flat if and only if Σ is locally spanned from each component of $S^n \setminus \Sigma$.*

Historical Notes. The distinction between the cases $n = 3$ and $n > 3$ of codimension-one embeddings was first displayed by J. C. Cantrell's doctoral dissertation, ultimately refined into the result about $(n-1)$-spheres in S^n, $n > 3$, necessarily being flat if they are locally flat modulo a single point. Key ideas appeared in the treatment of Theorem 2.9.3. Corollary 7.9.3 is a mild generalization. Kirby (1968a) provided an elegant geometric construction to establish Corollary 7.9.5 for all $n > 3$.

For embeddings in 3-manifolds flatness results emerged in an order rather opposite to those in high dimensions. Bing's original flattening theorem (1959b) was the low-dimensional version of Corollary 7.9.7. Later he used it to show that a surface in a 3-manifold is locally flat if it is 1-LCC embedded (Bing, 1961b).

J. Hempel (1964) showed that a compact 2-manifold in S^3 which is singularly collared on both sides is 1-LCC embedded; Daverman (1976) developed the argument used here for Theorem 7.9.8 that works in all ambient dimensions; Bryant and Lacher (1975) independently obtained the same result and generalizations to other codimensions as well.

P. Olum (1953) proved that maps of degree 1 between connected oriented n-manifolds induce epimorphisms at the π_1 level. D. B. A. Epstein (1966) introduced a valuable geometric notion of degree; he showed that a proper map : $M \to N$ between connected, oriented n-manifolds has degree p if and only if p is the minimum integer such that, for some map g properly homotopic to f and small open n-cell $V \subset N$, $g^{-1}(V)$ has exactly p components, on each of which g restricts to a homeomorphism.

A singular regular neighborhood can be regarded as the image of a homotopy of a codimension-one manifold S that instantly deforms S into its complement. Instead of homotopies, one might consider a sequence of approximations to the inclusion; a codimension-one sphere Σ in S^n is said to be *free* if, for each $\epsilon > 0$ and each component U of $S^n \setminus \Sigma$, there is a map $f_\epsilon : \Sigma \to U$ that moves points less than ϵ. As of this writing, whether free $(n-1)$-spheres in S^n, $n \geq 3$, must be 1-LCC embedded is still unknown; a partial result about the implications of freeness toward flatness appears in §7.11.

C. E. Burgess (1965) introduced the locally spanned concept and proved Theorem 7.9.15 in the 3-dimensional setting; his proof immediately applies in all dimensions.

Exercises

7.9.1. Suppose $\Sigma^{n-1} \subset S^n$, $n \geq 5$, is a wildly embedded $(n-1)$-sphere with complementary domains U, V for which there is a homeomorphism $\psi : \overline{U} \to \overline{V}$ with $\psi|\Sigma = \text{incl}_\Sigma$. Also suppose Σ is locally flat modulo a k-cell, $k \leq n-3$, or a Cantor set C. Show that C is wildly embedded in S^n.

7.9.2. Prove Theorem 7.9.6.

7.9.3. Show that most embeddings of S^{n-1} in S^n, $n \geq 5$, are locally flat (i.e., show that the locally flat embeddings form a dense, G_δ-subset of $\text{Emb}(S^{n-1}, S^n)$).

7.10. Sewings of crumpled n-cubes

Definitions. Let Σ be an $(n-1)$-sphere topologically embedded in S^n and U one of the components of $S^n \setminus \Sigma$. Then $\overline{U}$ is a *crumpled n-cube*, and the sphere Σ is called the *boundary* of $\overline{U}$, denoted $\mathrm{Bd}\,\overline{U}$. A crumpled n-cube C in S^n is a *closed n-cell complement* if $\mathrm{Cl}(S^n \setminus C)$ is an n-cell.

Restated in this terminology, Theorem 7.7.14 certifies that every crumpled n-cube ($n \geq 5$) admits an embedding in S^n as a closed n-cell complement. This signals that wildness on one side of an embedded codimension-one manifold forces no corresponding wildness on the other side. Our attention next turns to a complementary issue: what sorts of wildness on one side can be matched with wildness on the other side?

Definitions. A *sewing* of crumpled n-cubes C_1 and C_2 is a homeomorphism $h : \mathrm{Bd}\, C_1 \to \mathrm{Bd}\, C_2$. Associated with any such sewing h is the *sewing space* $C_1 \cup_h C_2$, namely, the quotient space obtained from the disjoint union $C_1 \sqcup C_2$ after identification of each $x \in \mathrm{Bd}\, C_1$ with $h(x) \in \mathrm{Bd}\, C_2$.

Sewings of crumpled cubes enhance the proliferation of wildness. The crucial question about a sewing of crumpled n-cubes is whether the sewing space is an n-manifold. When it is, the manifold necessarily is S^n (see Corollary 7.10.3), and then S^n contains a separating $(n-1)$-sphere Σ bounding copies of the two crumpled cubes, which are matched up along Σ exactly as prescribed by the sewing.

In light of Theorem 7.7.14, the sewing will always yield S^n when no wild point in one crumpled cube is matched with a wild point in the other. This section presents several examples indicating that an arbitrary sewing of crumpled cubes need not have a manifold as its sewing space and it probes conditions under which a sewing space is S^n, despite possible overlapping of the wildness. It also introduces an inflation technique for producing wildness, which it exploits to fabricate wild spheres in $\mathbb{R}^n$ that are locally flat modulo subspheres flatly embedded in $\mathbb{R}^n$. The most elaborate example—Example 7.10.14, which relies upon the construction of ramified wild Cantor sets from §4.8—is a crumpled cube C such that $C \cup_{\mathrm{Id}} C$ does not yield S^n. This C contains two embedded loops such that any singular disk in C bounded by the first meets every singular disk in S^n bounded by the second. That is precisely the feature necessitating the elaborate blow-up procedure of §7.7 used to establish the 1-LCC Approximation Theorem.

Here is an elementary consequence of Theorem 7.7.14.

Proposition 7.10.1. *If C is any crumpled n-cube in S^n, $n \geq 5$, and $h : \mathrm{Bd}\, C \to \partial B^n$ any sewing to the boundary of an n-cell, then $C \cup_h B^n \cong S^n$.*

Proposition 7.10.2. *For any sewing $h : \mathrm{Bd}\, C_1 \to \mathrm{Bd}\, C_2$ of crumpled n-cubes, $n \geq 5$, there exists a cell-like mapping $S^n \to C_1 \cup_h C_2$.*

Proof. Apply Theorem 7.4.14 or the preceding proposition to regard C_2 as embedded in S^n so that $\mathrm{Cl}(S^n \setminus C_2)$ is an n-cell B. Specify a collar

$\lambda : \partial B \times [0,1] \to B$ on $\partial B = \operatorname{Bd} C_2$ (with $\lambda_0 = \text{incl}_{\partial B}$), set

$$B' = C_2 \cup \lambda(\partial B \times [0,1]),$$

and note that, by Corollary 2.4.12 to the Generalized Schönflies Theorem, B' is also an n-cell (see Figure 7.11). Define a sewing $h' : \operatorname{Bd} C_1 \to \partial B'$ as $h'(x) = \lambda(h(x), 1)$. Then $C_1 \cup_{h'} B' \cong S^n$, by Proposition 7.10.1, and the decomposition of $S^n = C_1 \cup_{h'} B'$ into points and the arcs

$$\{\lambda(\{b\} \times [0,1]) \subset B' \mid b \in \partial B\}$$

gives rise to a cell-like mapping $S^n \to C_1 \cup_h C_2$. □

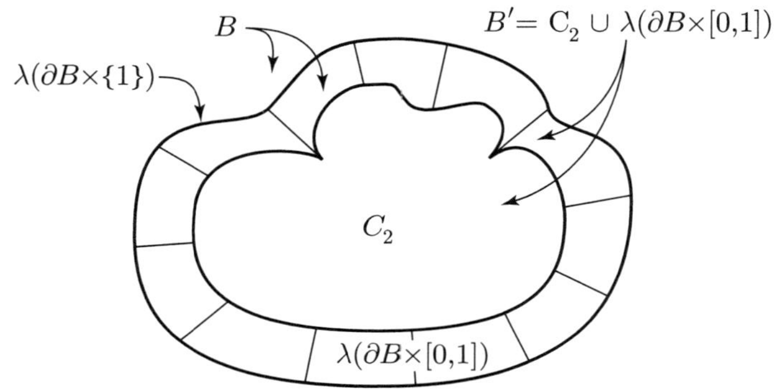

Figure 7.11. The domain of the cell-like map $S^n \to C_1 \cup_h C_2$

With an application of Corollary 7.4.2 we obtain:

Corollary 7.10.3. *If a sewing $h : \operatorname{Bd} C_1 \to \operatorname{Bd} C_2$ of crumpled n-cubes ($n \geq 5$) yields a manifold, then $C_1 \cup_h C_2 \approx S^n$.*

Proposition 7.10.4. *Let C_1 and C_2 be closed n-cell complements in S^n, $n \geq 5$, and $h : \operatorname{Bd} C_1 \to \operatorname{Bd} C_2$ a sewing. Then a necessary condition for $C_1 \cup_h C_2$ to be S^n is that any two maps $f_i : I^2 \to C_i$, $i \in \{1,2\}$, can be approximated, arbitrarily closely, by maps $F_i : I^2 \to S^n$ such that*

$$F_2(I^2) \cap h(\operatorname{Bd} C_1 \cap F_1(I^2)) = \emptyset.$$

Proof. Let $\lambda_i : C_i \to S^n, i \in \{1,2\}$, be embeddings such that

$$\lambda_1(C_1) \cap \lambda_2(C_2) = \lambda_1(\operatorname{Bd} C_1) = \lambda_2(\operatorname{Bd} C_2),$$

$\lambda_1(C_1) \cup \lambda_2(C_2) = S^n$, and $\lambda_1 | \operatorname{Bd} C_1 = \lambda_2 h$. Find a small neighborhood W_i of $\lambda_i(C_i)$ and a retraction $R_i : W_i \to \lambda_i(C_i)$ close to the identity on W_i, with $R_i(W_i \smallsetminus \lambda_i(C_i)) \subset \lambda_i(\operatorname{Bd} C_i)$. Approximate the maps $\lambda_i f_i : I^2 \to \lambda_i(C_i)$ by maps $g_i : I^2 \to W_i$ with disjoint images and with $\lambda_i^{-1} R_i g_i$ close to f_i. Set $U_i = g_i^{-1}(W_i \smallsetminus \lambda_i(C_i))$, observe that $\lambda_i^{-1} R_i g_i(U_i) \subset \operatorname{Bd} C_i$, and use the fact that each C_i is a closed n-cell complement in S^n to adjust each $\lambda_i^{-1} R_i g_i$ to a new map $F_i : I^2 \to S^n$ such that $F_i|I^2 \smallsetminus U_i = \lambda_i^{-1} R_i g_i|I^2 \smallsetminus U_i = \lambda_i^{-1} g_i|I^2 \smallsetminus U_i$ and $F_i(U_i) \subset S^n \smallsetminus C_i$. Set $Y_i = (F_i)^{-1}(\operatorname{Bd} C_i)$. It follows that

$$\begin{aligned} F_2(I^2) \cap h(\operatorname{Bd} C_1 \cap F_1(I^2)) &\subset \lambda_2^{-1} g_2(Y_2) \cap h\lambda_1^{-1} g_1(Y_1) \\ &\subset \lambda_2^{-1}(g_2(Y_2) \cap g_1(Y_1)) = \emptyset. \end{aligned}$$ □

Theorem 7.10.5. *Let C_1 and C_2 be crumpled n-cubes ($n \geq 5$) satisfying the Disjoint Disks Property and let $h : \operatorname{Bd} C_1 \to \operatorname{Bd} C_2$ be a sewing such that any two maps $f_i : I^2 \to C_i$, $i \in \{1, 2\}$, can be approximated, arbitrarily closely, by maps $F_i : I^2 \to C_i$ such that*

$$F_2(I^2) \cap h(\operatorname{Bd} C_1 \cap F_1(I^2)) = \emptyset.$$

Then $C_1 \cup_h C_2$ is topologically S^n.

Proof. This is mainly a rephrasing of Proposition 7.4.12. Its hypotheses hold by Proposition 7.10.2 and Corollary 7.4.8. □

Remark. Regardless of whether C_1 and C_2 satisfy the Disjoint Disks Property, a sewing $h : \operatorname{Bd} C_1 \to \operatorname{Bd} C_2$ yields S^n if h satisfies the mismatch property of Theorem 7.10.5 (Cannon and Daverman, 1981).

Theorem 7.10.6. *For any crumpled n-cube C, $n \geq 5$, $C \cup_{\mathrm{Id}} C \cong S^n$ if and only if C satisfies the Disjoint Disks Property.*

Proof. If C satisfies the Disjoint Disks Property, then the identity sewing satisfies the mismatch property of Theorem 7.10.5 and $C \cup_{\mathrm{Id}} C \cong S^n$.

For the other implication, in case $C \cup_{\mathrm{Id}} C \cong S^n$, there exists a retraction $r : S^n \to C$ that is 1-1 over $\operatorname{Bd} C$; r simply folds one of the copies of C over onto the other. As in 7.10.4, let $\lambda_i : C \to S^n$, $i \in \{1, 2\}$, be embeddings such that $\lambda_1(C) \cup \lambda_2(C) = S^n, \lambda_1(C) \cap \lambda_2(C) = \lambda_1(\operatorname{Bd} C) = \lambda_2(\operatorname{Bd} C)$ and $r\lambda_1 = r\lambda_2 = \mathrm{Id}_C$. Given maps $\mu_1, \mu_2 : I^2 \to C$, approximate $\lambda_1\mu_1, \lambda_2\mu_2$ by maps $\mu_1', \mu_2' : I^2 \to S^n$ with disjoint images. Then $r\mu_1', r\mu_2'$ are maps of I^2 to C such that

$$r\mu_1'(I^2) \cap r\mu_2'(I^2) \cap \operatorname{Bd} C = \emptyset.$$

Make a further (general position) approximation over the n-manifold $\operatorname{Int} C$ to obtain maps $\mu_1'', \mu_2'' : I^2 \to C$ with disjoint images. □

Corollary 7.10.7. *If C is a crumpled n-cube, $n \geq 5$, satisfying the Disjoint Disks Property, then there exists an involution $u : S^n \to S^n$ with orbit space homeomorphic to C. In particular, the fixed point set of u is a wild $(n-1)$-sphere (provided C is not an n-cell).*

Definition. The *orbit space* of an involution $u : S^n \to S^n$ is the quotient of S^n obtained by identifying each $x \in S^n$ with $u(x)$.

To obtain quick applications of Theorem 7.10.5 it is advantageous to develop conditions under which a crumpled cube C has the Disjoint Disks Property.

Lemma 7.10.8. *If the boundary of a crumpled n-cube $C \subset S^n$ is locally flat modulo a Cantor set, k-cell or k-sphere Z that is flat in $\operatorname{Bd} C$, where $n \geq 5$ and $1 \leq k \leq n-4$, then C has the Disjoint Disks Property.*

Proof. As before, identify a neighborhood W of C in S^n and retraction $R : W \to C$ for which $R(W \smallsetminus C) \subset \operatorname{Bd} C$. Given maps $f_1, f_2 : I^2 \to C$, produce approximations $f_1', f_2' : I^2 \to W \subset S^n$ such that $f_1'(I^2) \cap f_2'(I^2) = \emptyset$. Restrict W to assure that each Rf_i' is a close approximation to f_i. Let $U_i = (f_i')^{-1}(W \smallsetminus C)$. Invoke the hypothesis about Z being flat in $\operatorname{Bd} C$ to determine maps $F_1, F_2 : I^2 \to C$ such that $F_i | I^2 \smallsetminus U_i = Rf_i' | I^2 \smallsetminus U_i = f_i' | I^2 \smallsetminus U_i$, $F_i(U_i) \subset \operatorname{Bd} C \smallsetminus Z$, and F_i is close to Rf_i'. Since $\operatorname{Bd} C$ is locally collared in C at points of $F_i(U_i)$, the maps F_i can be further adjusted, fixing $I^2 \smallsetminus U_i$ while pushing points of $F_i(U_i)$ away from $\operatorname{Bd} C$, thereby yielding maps $F_i' : I^2 \to C$ such that

$$F_1'(I^2) \cap F_2'(I^2) \cap \operatorname{Bd} C \subset f_1'(I^2 \smallsetminus U_1) \cap f_2'(I^2 \smallsetminus U_2) \cap \operatorname{Bd} C = \emptyset.$$

Finally, they can be adjusted once more over $\operatorname{Int} C$ so as to have disjoint images. □

Example 7.10.9. *There exists a wild $(n-1)$-sphere Σ in $\mathbb{R}^n$ that is locally flat modulo a k-sphere flatly embedded in $\mathbb{R}^n$ $(1 \leq k \leq n-2)$.*

Such an example arises by inflating any crumpled $(n-1)$-cube C in $\mathbb{R}^{n-1}$ whose double $C \cup_{\mathrm{Id}} C$ is S^{n-1} (or, equivalently, C has the Disjoint Disks Property). Think of C as a subset of $\mathbb{R}^{n-1}$—for definiteness, assume C to be collared in $\mathbb{R}^{n-1} \smallsetminus \operatorname{Int} C$—and let $\nu : C \to [0,1]$ be a map such that $\nu^{-1}(0) = \operatorname{Bd} C$. By an *inflation of* C we mean

$$\operatorname{Infl}(C, \nu) = \{\langle x, t\rangle \in \mathbb{R}^{n-1} \times \mathbb{R}^1 = \mathbb{R}^n \mid x \in C \text{ and } |t| \leq \nu(x)\}.$$

The frontier Σ of $\operatorname{Infl}(C, \nu)$ is the union of the two copies of C, the graphs of $\pm\nu$, sewn together via the Identity map along their boundaries and so, by hypothesis, is an $(n-1)$-sphere. Clearly, the topological type of $\operatorname{Infl}(C, \nu)$ does not depend on the choice of map ν, so from here on out we shall refer to such a construction as an *inflation of* C, denoted $\operatorname{Infl}(C)$, without reference to any specific ν.

The $(n-1)$-sphere $\Sigma = \operatorname{Bd}\operatorname{Infl}(C)$ is locally flat modulo the $(n-2)$-sphere $\Sigma \cap (\mathbb{R}^{n-1} \times \{0\}) \cong \operatorname{Bd} C$, which is flat in $\mathbb{R}^n$ (see Exercise 6.3.2 and Theorem 6.3.6). Moreover, Σ is wildly embedded (assuming C is not a cell) because $C \times \{0\}$ is a strong deformation retract of $\operatorname{Infl}(C)$ via a deformation that moves points vertically—in the $\mathbb{R}^1$ direction—and preserves interiors;

hence, the interior of $\mathrm{Infl}(C)$ is 1-LCC at $\langle x, 0\rangle \in \mathrm{Bd}\, C \times \{0\}$ if and only if $\mathrm{Int}\, C$ is 1-LCC at x.

Take C to be a crumpled $(n-1)$-cube in $\mathbb{R}^{n-1}$, $n > 5$, whose frontier is locally flat modulo a Cantor set X standardly embedded in $\mathrm{Bd}\, C$. Theorem 7.10.6 and Lemma 7.10.8 assure that $C \cup_{\mathrm{Id}} C \cong S^{n-1}$. For $k = 1, \ldots, n-2$ let D_k denote a k-cell in $\mathrm{Bd}\, C$ containing X. Then Σ is locally flat modulo the k-cell $D_k \times \{0\} \subset \mathrm{Bd}\, C \times \{0\} \subset \Sigma$, which is flat in $\mathbb{R}^n$ by Corollary 4.6.10 when $k < n-2$ and by Theorem 6.3.6 when $k = n-2$. For that matter, Σ is locally flat modulo the Cantor set $X \times \{0\}$, which also is flat in $\mathbb{R}^n$.

Lemma 7.10.10. *If the boundary of a crumpled n-cube C is locally flat modulo a Cantor set, k-cell or k-sphere that is flat in S^n, where $n \geq 5$ and $1 \leq k \leq n-3$, then C has the Disjoint Disks Property.*

The proof, which is similar to that of 7.10.8, is left as an exercise.

Definition. A group G of homeomorphisms on a space X is said to act *semifreely* on X if there exists a subset Z of X such that for every $g \in G$, $g(z) = z$ for all $z \in Z$ and $g(x) \neq x$ for all $x \in X \setminus Z$ and all $g \neq \mathrm{Id}_X$.

Theorem 7.10.11. *Let C be a crumpled n-cube, $n \geq 5$, such that $C \cup_{\mathrm{Id}} C \cong S^n$. Then there exists a semifree S^1-action on S^{n+1} having an $(n-1)$-sphere as its fixed point set and having orbit space homeomorphic to C.*

Proof. There is a map $p : S^1 \times C \to \mathrm{Infl}(C) \cup_{\mathrm{Id}} \mathrm{Infl}(C)$ which is 1-1 on $S^1 \times \mathrm{Int}\, C$ and which behaves like projection to the second factor on $S^1 \times \mathrm{Bd}\, C$. The obvious free S^1-action on $S^1 \times C$ (trivial on the C factor) descends under p to a semifree action on $\mathrm{Infl}(C) \cup_{\mathrm{Id}} \mathrm{Infl}(C)$, and the latter is topologically S^n, since $\mathrm{Infl}(C)$ has the Disjoint Disks Property (another exercise). □

Example 7.10.12. *There exist a crumpled n-cube C having the Disjoint Disks Property and a homeomorphism $h : \mathrm{Bd}\, C \to \mathrm{Bd}\, C$ such that $C \cup_h C$ fails to be a manifold.*

Proof. Start with a wild Cantor set X_n in S^n equipped with a special geometric defining sequence and an embedded loop $e_n(\partial I^2) \subset S^n \setminus X_n$ as in Lemma 4.8.5. Construct an n-cell B in $S^n \setminus e_n(\partial I^2)$ containing X_n as a standardly embedded Cantor set in ∂B, and set $C = S^n \setminus \mathrm{Int}\, B$. Apply Mixing Lemma 4.8.1 to obtain a homeomorphism $\tau : X_n \to X_n$ mixing the admissible subsets of X_n. Since X_n is flat in $\mathrm{Bd}\, C = \partial B$, τ extends to a homeomorphism $h : \mathrm{Bd}\, C \to \mathrm{Bd}\, C$. Now by Proposition 7.10.4 $C \cup_h C$ cannot be an n-manifold: there is a special Cantor set X, the image of X_n in the sewing space, and disjoint loops away from X, one in each copy of C, which ought to bound essentially disjoint disks in the n-manifold (at least for $n \geq 5$), but any two such singular disks must meet somewhere in X, by the mixing property. □

Example 7.10.13. *There exist a crumpled n-cube $C^* \subset S^n$ and embedded loops $e_n(\partial I^2)$, $e'(\partial I^2)$ in* $\operatorname{Int} C^*$ *such that the image of every singular disk in C^* bounded by e_n intersects the image of every singular disk in S^n bounded by e'.*

For $k = n - 1, n$, apply Lemma 4.8.7 to produce wild Cantor sets X_k in S^k equipped with compatible special geometric defining sequences, each with the strong interior inessential property. Identity loops $e_k(\partial I^2) \subset S^k \setminus X_k$ such that every singular disk in S^k bounded by $e_k(\partial I^2)$ contains an admissible subset of X_k (Corollary 4.8.6). Build k-cells $B_k \subset S^k \setminus e_k(\partial I^2)$ with X_k standardly embedded in ∂B_k and ∂B_k locally flat modulo X_k. Form the complementary crumpled k-cube $C_k = S^k \setminus \operatorname{Int} B_k$.

Note that C_k has the Disjoint Disks Property by Lemma 7.10.8 (provided $k \geq 5$) . Inflate C_{n-1} to a crumpled n-cube $C' = \operatorname{Infl}(C_{n-1})$. There is a natural embedding $e' : \partial I^2 \to C'$ for which every singular disk $F'(I^2)$ in C' bounded by $e'(\partial I^2)$ contains an admissible subset of X_{n-1}. Construct an $(n-1)$-cell $\beta' \subset \operatorname{Bd} C'$ containing X_{n-1} in its boundary as a standardly embedded subset, with β' locally flatly embedded in $\operatorname{Bd} C'$ modulo X_{n-1}, and construct a similar $(n-1)$-cell $\beta \subset \operatorname{Bd} C_n$ with X_n standardly embedded in $\operatorname{Bd} \beta$. Produce a homeomorphism $h : \beta' \to \beta$ such that $h|X_{n-1} : X_{n-1} \to X_n$ mixes the admissible subsets of the Cantor sets.

Let $C^* = C' \sqcup_h C_n$ denote the object obtained from the disjoint union of C' and C_n by gluing $\beta' \subset \operatorname{Bd} C'$ to $\beta \subset \operatorname{Bd} C_n$ via h. Regard C' as embedded in S^n so $S^n \setminus \operatorname{Int} C'$ is an n-cell B'. Thicken β' to an n-cell $D' \subset B'$ locally flat modulo $X_{n-1} \subset \partial D'$, where β' and X_{n-1} are flatly embedded in $\partial D'$. Then D' is flat in S^n, so $S^n \setminus \operatorname{Int} D'$ is an n-cell $D \supset C'$. The homeomorphism $h : \beta' \to \beta$ extends to a homeomorphism $H : \partial D \to \operatorname{Bd} C_n$. Then $C^* = C' \cup_h C_n$ has an obvious embedding in $D \cup_H C_n \cong S^n$.

This object C^* contains two noteworthy loops: $e_n : \partial I^2 \to C_n \subset C^*$ and $e' : \partial I^2 \to C' \subset C^*$. Consider singular disks $F_n(I^2), F'(I^2)$ in S^n, C' bounded by e_n, e', respectively. Then $F_n(I^2)$ contains an admissible subset of $j(X_n) \subset j(C_n)$. To see that $F'(I^2)$ contains an admissible subset of $j'(X_{n-1}) \subset j'(C')$, one can modify F' using Lemma 7.9.1 to obtain another map $F^* : I^2 \to C'$ such that $F^*(I^2) \subset F'(I^2) \cup \beta^*$, where β^* denotes the image of $\beta = h(\beta')$ in C^*, and where $F^*(I^2) \cap \partial\beta^* = F'(I^2) \cap \partial\beta^*$. Hence, $F'(I^2) \cap j'(X_{n-1}) = F^*(I^2) \cap j(X_{n-1})$ contains the image under j' of an admissible subset of X_{n-1}. This implies that $F'(I^2)$ and $F_n(I^2)$ intersect.

Clearly C^* cannot have the Disjoint Disks Property. As a result, it also serves as:

Example 7.10.14. *There exists a crumpled n-cube C^* such that $C^* \cup_{\mathrm{Id}} C^*$ fails to be a manifold.*

Historical Notes. Bing was the first to produce periodic homeomorphisms of $\mathbb{R}^n$ and S^n having wild fixed point sets; in perhaps his most widely known example of this type, he demonstrated (Bing, 1952) that $AH \cup_{\mathrm{Id}} AH \cong S^3$, where AH denotes the crumpled 3-cube bounded by Alexander's horned sphere, so S^3 admits an involution fixing a wild 2-sphere. R. J. Daverman and W. T. Eaton (1969) proved that an arbitrary sewing $h : \mathrm{Bd}\, C_1 \to \mathrm{Bd}\, C_2$ of crumpled 3-cubes can be approximated by another sewing h' such that $C_1 \cup_{h'} C_2 \cong S^3$; nothing comparable is known about arbitrary sewings of crumpled n-cubes, $n \geq 5$. Eaton (1972) showed the mismatch property of Theorem 7.10.5 to be a necessary and sufficient condition for a sewing of two crumpled 3-cubes to yield S^3; Cannon and Daverman (1981) showed it be a sufficient condition for a sewing of crumpled n-cubes to yield S^n, $n \geq 4$. Daverman (see comments in (1981)) introduced the inflation process as a method of constructing wild codimension-one embeddings. He also (2007) provided various mismatch properties under which a sewing of crumpled cubes yields S^n.

Exercises

7.10.1. If the inflation $\mathrm{Infl}(C)$ of a crumpled $(n-1)$-cube C is bounded by a sphere Σ, $n \geq 5$, then Σ is collared from $\mathrm{Cl}(\mathbb{R}^n \setminus \mathrm{Infl}(C))$.

7.10.2. Prove Lemma 7.10.10.

7.10.3. If the crumpled cube C satisfies the Disjoint Disks Property, then so does $\mathrm{Infl}(C)$.

7.10.4. For $n > 3$ the suspension of any crumpled n-cube C satisfies the Disjoint Disks Property.

7.10.5. If the crumpled cube C satisfies the Disjoint Disks Property, then each map $f : I^2 \to C$ can be approximated by an embedding $F : I^2 \to C$ such that $F(I^2) \cap \mathrm{Bd}\, C$ is 0-dimensional.

7.11. Wild examples and mapping cylinder neighborhoods

The presence of mapping cylinder neighborhoods imposes considerable regularity on an embedding, but not enough regularity to ensure local flatness. At the heart of §7.11 is a construction in Example 7.11.2 of a codimension-one sphere wildly embedded in S^n despite possessing a mapping cylinder neighborhood. Complementing the example is an initial result indicating that the combination of mapping cylinder neighborhood and freeness implies local flatness for codimension-one manifold embeddings.

The methods for producing Example 7.11.2 lead to other applications, including the construction here of a wild Cantor set whose embedding satisfies a strong homogeneity property.

Theorem 7.11.1. *Suppose Σ^{n-1} is a connected, two-sided $(n-1)$-manifold in an n-manifold M, $n \geq 5$, such that Σ^{n-1} has a mapping cylinder neighborhood and is free. Then Σ^{n-1} is bicollared.*

Proof. Specify a component U of $M \smallsetminus \Sigma^{n-1}$. It suffices to show that Σ^{n-1} is 1-LCC in $\overline{U}$. Apply the hypothesis to obtain a proper map $\psi : N^{n-1} \to \Sigma^{n-1}$ defined on an $(n-1)$-manifold N^{n-1} such that Σ^{n-1} has a closed neighborhood in $\overline{U}$ naturally homeomorphic to $\mathrm{Map}(\psi)$, the mapping cylinder of ψ.

Consider $s \in \Sigma^{n-1}$ and $\epsilon > 0$ such that $B(s;\epsilon)$ lies in a Euclidean patch in M. Identify a small $(n-1)$-cell $D \subset \Sigma^{n-1} \cap B(s;\epsilon)$ with $s \in \operatorname{Int} D$. Pushing down the mapping cylinder structure of $\mathrm{Map}(\psi)$, if necessary, we assume the part W of the mapping cylinder determined by $\psi^{-1}(D)$ lies in $B(s;\epsilon)$. Let D_1, D_2, D_3 be additional $(n-1)$-cells with

$$s \in \operatorname{Int} D_{i+1} \subset D_{i+1} \subset \operatorname{Int} D_i \subset D_i \subset \operatorname{Int} D \subset D = D_0,$$

and then let W_i denote the portion of $\mathrm{Map}(\psi)$ determined by $\psi^{-1}(\operatorname{Int} D_i)$ $(i = 0, 1, 2, 3)$. Delete all points of N^{n-1} from W_i to form W_i^* $(i = 1, 2, 3)$.

We claim that any loop α in $W_3^* \smallsetminus \Sigma$ is null-homotopic in $B(s;\epsilon) \cap U$. Find $\gamma \in (0,1)$ so close to 1 that the image $W_3^- \subset W^*$ of $\psi^{-1}(\operatorname{Int} D_3) \times (0,\gamma)$ in $\mathrm{Map}(\psi)$ contains α. Use freeness of Σ^{n-1} in U to obtain a map $g : D \to U$ so close to $\mathrm{incl}_D : D \to \overline{U}$ in the ANR $\overline{U}$ to allow a homotopy $\mu : D \times [0,1/3] \to \overline{U}$ between $\mu_0 = \mathrm{incl}_D$ and $\mu_{1/3} = g$; do this so the image of μ lies in the portion of $\mathrm{Map}(\psi)$ corresponding to $N^{n-1} \times [\gamma, 1]$. The mapping cylinder structure offers the means to push the image of g out to the frontier $\operatorname{Fr}\mathrm{Map}(\psi)$ (relative to $\overline{U}$); push first through the image of $N^{n-1} \times [\gamma, 1)$ to the level corresponding to γ, and then through the image of $N^{n-1} \times [0,\gamma]$ out to the frontier. This gives an extension of μ to a map $\mu : D \times [0,1] \to \overline{U}$ satisfying

$\mu(D \times [1/3, 1]) \subset U$,

$\mu_1(D) \subset \operatorname{Fr}\mathrm{Map}(\psi)$ (relative to $\overline{U}$),

$\mu(D \times [1/3, 2/3]) \subset N^{n-1} \times [\gamma, 1) \subset \mathrm{Map}(\psi) \cap U$,

$\mu(D \times [2/3, 1]) \subset N^{n-1} \times [0, \gamma] \subset \mathrm{Map}(\psi) \cap U$, and

$\mu(D \times [0, 1)) \cap \operatorname{Fr}\mathrm{Map}(\psi) = \emptyset$.

Furthermore, this can be arranged so that

$\mu(D \times [0, 1)) \subset B(s;\epsilon)$,

$\mu(\partial D_0 \times [0, 1]) \cap W_1 = \emptyset$,

$\mu((D_0 \setminus \operatorname{Int} D_i) \times [0,1]) \cap W_{i+1} = \emptyset$ $(i = 1, 2)$, and
$\mu(D_i \times [0,1]) \subset W_{i-1}$ $(i = 1, 2, 3)$.

By the argument given for Lemma 7.9.14, $\mu(D_0 \times [0,1])$ contains points of U very close to each point of $\operatorname{Int} D_1$; in view of the connections along mapping cylinder lines and away from $\mu(\partial D_0 \times [0,1])$, the same argument gives that $\mu(D_0 \times [0,1]) \supset W_1$. In like fashion, $\mu(D_i \times [0,1]) \supset W_{i+1}$ $(i = 1, 2)$.

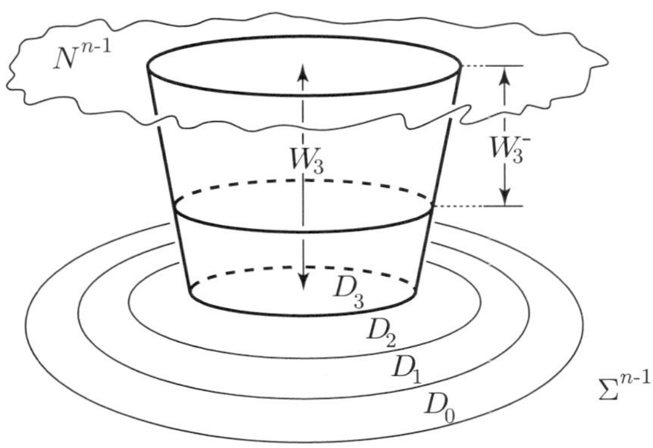

Figure 7.12. The neighborhood W_3 and other structures near s

Let Y_1^* denote the component of $\mu^{-1}(W_1^*)$ containing $\operatorname{Int} D_1 \times \{0\}$. As in the proof of Theorem 7.9.8, one can append a collar $\operatorname{Int} D_1 \times (-1, 0]$ to W_1^* and extend $\mu | Y_1^* : Y_1^* \to W_1^*$ to a map

$$\nu : Y_1^* \cup (\operatorname{Int} D_1 \times (-1, 0]) \to W_1^* \cup (\operatorname{Int} D_1 \times (-1, 0])$$

in the obvious way. Here ν has degree ± 1, since it is a homeomorphism over the appended collar (Lemma 7.9.9). Note that $Y_1^* \supset \operatorname{Int} D_2 \times (0, 1)$.

Form $Y_3^- = \mu^{-1}(W_3^-)$. Properties of μ force $Y_3^- \subset \operatorname{Int} D_2 \times (2/3, 1) \subset Y_1^*$. List the components $U_1, U_2, \dots$ of Y_3^- and use d_i to denote the degree of $\mu | U_i : U_i \to W_3^-$. Then $\Sigma_i d_i = \pm 1$, by Lemma 7.9.11.

Let W_1^-, like W_3^-, denote the portion of W_1 corresponding to the image of $N^{n-1} \times (0, \gamma)$ and let Y_1^- denote the component of $\mu^{-1}(W_1^-)$ containing $\operatorname{Int} D_2 \times (2/3, 1)$. Now we have $Y_3^- \subset Y_1^- \subset D_0 \times (2/3, 1)$. Another appeal to Lemma 7.9.11 assures that $\mu | Y_1^- : Y_1^- \to W_1^-$ is a degree ± 1 map between connected manifolds. Thus, Lemma 7.9.13 promises a loop $\alpha' \subset Y_1^-$ such that $\mu(\alpha')$ is homotopic to α in W_1^-. As α' is null homotopic in $D \times (2/3, 1)$, α is null homotopic in $\mu(D \times (2/3, 1)) \subset B(s; \epsilon) \cap U$. □

Example 7.11.2. *For $n \geq 6$, S^n contains a wildly embedded $(n-1)$-sphere Σ with a mapping cylinder neighborhood.*

This example introduces a remarkably useful and direct new method for producing wildness. It involves decompositions into acyclic sets. Typically the methodology brings about an associated decomposition space S that contains an object which obviously is "wild", in the sense of failing to be 1-LCC embedded; however, it can be far from obvious that S is a manifold. Although the given decomposition is only acyclic, not cell-like, in many instances there does exist a cell-like map from another manifold onto S, in which event the Cell-like Approximation Theorem possibly could be exploited to detect that S is a manifold.

For the specific issue at hand, Example 7.11.2, the decomposition space S associated with an acyclic decomposition of S^n contains an object Σ related to the $(n-1)$-sphere, and the latter obviously has a neighborhood P with the structure of a mapping cylinder. Moreover, the sphere-like subspace Σ has wildness features, by virtue of containing a 1-sphere that fails to be 1-LCC in S.

Construction of the Example. Assume $n \geq 7$; something similar can be done for $n = 6$, but we will ignore that special case. Fix a finite, acyclic 2-complex A that is PL embedded in S^{n-2} with contractible complement (see Example 0.10.3). Take a regular neighborhood $N(A)$ of the embedded A and then spin an $(n-2)$-ball B, $N(A) \subset \operatorname{Int} B \subset B \subset S^{n-2}$, to produce a PL embedding of $N(A) \times S^1$ in S^{n-1}. Treat S^{n-1} as an equatorial sphere in S^n. Form the decomposition of S^n having the sets $\{A \times \{s\} \mid s \in S^1\}$ as nondegenerate elements, and let $p : S^n \to S$ denote the map to the associated decomposition space S. We will show that both S and $p(S^{n-1})$ are spheres. The image under p of an annular neighborhood of S^{n-1} in S^n will be a mapping cylinder neighborhood of $p(S^{n-1})$. The 1-sphere $p(A \times S^1)$ will be wildly embedded in S (and in $p(S^{n-1})$ as well) because it fails to be 1-LCC.

Lemma 7.11.3. *The quotient space $N(A)/A$ is the cell-like image of a ∂-manifold W under a cell-like map that restricts to a homeomorphism on a neighborhood of ∂W.*

Proof. The complex A is embedded in S^{n-2} with contractible complement. Therefore, the closed complement C' of a collar on $\partial N(A)$ in $S^{n-2} \setminus \operatorname{Int} N(A)$ is contractible, and $N(A)/A \cong (S^{n-2} \setminus \operatorname{Int} N(A))/C'$, since each is a cone over $\partial N(A)$. □

Lemma 7.11.4. *Suppose M is an m-manifold and $p : M \times S^1 \to X$ is a closed, surjective mapping for which there exist an $(m-2)$-dimensional, compact ANR Z in M and a closed subset C of S^1 such that the nondegenerate point preimages under p are the sets $\{Z \times \{s\} \mid s \in C\}$. Let D be a* *dense subset of C. Then each map $f : I^2 \to X$ can be approximated by a map $F : I^2 \to X$ such that $F(I^2) \cap p(Z \times C) \subset p(Z \times D)$.*

Proof. Given $f : I^2 \to X$, choose a triangulation T on I^2 so images of its simplices under f have small diameter. Let L denote the subcomplex of T containing the 1-skeleton plus all 2-simplices $\tau \in T$ with $f(\tau) \cap p(Z \times C) = \emptyset$. Here f can be approximately lifted to a map $g : L \to M$: for each 2-simplex $\tau \in L$ set $g|\tau = p^{-1} f|\tau$; for vertices and 1-simplices of L not contained in any 2-simplex there, the existence of such an approximate lift g follows readily from the dimension restriction on Z. The desired map F will coincide with pg on L. For each 2-simplex $\sigma \in T \smallsetminus L$, $g|\partial\sigma$ is homotopic in $(U \smallsetminus Z) \times J$ to a map sending $\partial\sigma$ into $U \times \{d\}$, where U is a small neighborhood of Z in M, J is a small subset of S^1, and $d \in D$. Thus, $F|\partial\sigma = pg|\partial\sigma$ extends over σ to a map $F : \sigma \to p((U \smallsetminus Z) \times J) \cup (U \times \{d\})$, which gives the desired approximation to f. □

Corollary 7.11.5. *The space $(N(A)/A) \times S^1$ satisfies the DDP.*

Proof. Specify disjoint dense subsets D_1, D_2 of S^1. Let $p : N(A) \times S^1 \to X = (N(A)/A) \times S^1$ denote the decomposition map. Given two maps $f_i : I^2 \to (N(A)/A) \times S^1$, approximate by maps $F_i : I^2 \to (N(A)/A) \times S^1$ with $F(I^2) \cap p(A \times S^1) \subset p(A \times D_i)$ $(i = 1, 2)$. A general position adjustment near points of $F_1(I^2) \cap F_2(I^2)$ in the ∂-manifold $p((N(A) \smallsetminus A) \times S^1)$ yields disjoint approximations. □

It follows from the Cell-like Approximation Theorem that $(N(A)/A) \times S^1$ is a ∂-manifold, and hence $p(S^{n-1})$ is a manifold. The latter is a sphere since it is a simply-connected homology sphere (by the Vietoris-Begle Theorem). Similarly, $S = p(S^n)$ is an n-sphere. Finally, $p(A \times S^1)$ is wildly embedded in S since it fails to be 1-LCC: it has a compact neighborhood P of the form $(c * \partial(N(A) \times [-1, 1])) \times S^1$ and its complement in P deformation retracts to $\partial(N(A) \times [-1, 1]) \times S^1$. □

Definition. A subset X of a space S is *strongly homogeneously embedded* in S if every homeomorphism $h : X \to X$ extends to a homeomorphism $H : S \to S$.

Example 7.11.6. *For $n \geq 6$, S^n contains a wild, strongly homogeneously embedded Cantor set.*

Proof. Again we use acyclic decompositions to build a space S containing a Cantor set X that is both strongly homogeneously embedded and wild, in the sense of failing to be 1-LCC embedded. The real work involves showing that S is a manifold.

As before, consider the acyclic 2-complex A PL embedded in S^{n-2} with contractible complement. Let C be a Cantor set,

$$C \subset \operatorname{Int} I = \operatorname{Int} I \times \{1/2\} \subset \operatorname{Int} I^2.$$

Consider $S^{n-2} \times I^2$ as a PL subset of S^n.

Let $p : S^n \to S$ be the decomposition map associated with the decomposition of S^n into points and the sets $\{A \times \{c\} \mid c \in C\}$. The Cantor set of interest is $X = p(A \times C)$. Obviously it is strongly homogeneously embedded in S because for any homeomorphism $h : X \to X$ there exists a homeomorphism $g : C \to C$ rendering the following diagram commutative:

$$\begin{array}{ccc} A \times C & \xrightarrow{\operatorname{Id}\times g} & A \times C \\ \big\downarrow{\scriptstyle p|_{A\times C}} & & \big\downarrow{\scriptstyle p|_{A\times C}} \\ X & \xrightarrow{\quad h \quad} & X \end{array}$$

This g extends to a homeomorphism $G : I^2 \to I^2$ that reduces to the identity on ∂I^2, and

$$\operatorname{Id} \times G : S^{n-2} \times I^2 \to S^{n-2} \times I^2$$

extends to $H : S^n \to S^n$ via the Identity off $S^{n-2} \times I^2$. Then H, in turn, induces a homeomorphism $\hat{H} : S \to S$ as $\hat{H} = pHp^{-1}$, and $\hat{H}|X = h$.

By Corollary 7.11.5, S has the DDP.

Finally, we explain why S is the cell-like image of a manifold. Let N_1 be a regular neighborhood of A in $\operatorname{Int} N(A)$ and let E_1 be the union of a pair of disjoint 2-cells in $\operatorname{Int} I^2$, with $\operatorname{Int} E_1 \supset C$. As $N(A) \times I^2 \subset S^n$ is a regular neighborhood of a copy of A, $\partial(N(A) \times I^2)$ bounds a compact, contractible n-manifold Q_0. We claim that $\operatorname{Int} Q_0$ contains a pair of disjoint copies of Q_0 whose union Q_1 satisfies $Q_0 \setminus \operatorname{Int} Q_1 \cong N(A) \times (I^2 \setminus \operatorname{Int}(N_1 \times E_1))$. Form a second compact contractible n-manifold Q' by removing $\operatorname{Int}(N_1 \times E_1)$ from $N(A) \times I^2$ and attaching in a copy of Q_0 to what remains along each component of $\partial(N_1 \times E_1)$. Remarkably, Q' is PL homeomorphic to Q_0, by the Relative h-Cobordism Theorem (Rourke and Sanderson, 1972, p. 87): the union of Q_0, Q' and a collar joining ∂Q_0 to $\partial Q'$ is an n-sphere, hence that union bounds an $(n+1)$-cell W, the resulting triple (W, Q_0, Q') is a relative h-cobordism, so $W \cong Q_0 \times [0,1]$ with Q_0 corresponding to $Q_0 \times \{0\}$ and Q' to $Q_0 \times \{1\}$.

Let $N_1 \supset N_2 \supset \cdots N_k \supset \cdots$ be regular neighborhoods of A in S^{n-2} such that $N_k \subset \operatorname{Int} N_{k-1}$ and $\cap_k N_k = A$. For $k \geq 2$ let E_k be a union of 2^k pairwise disjoint 2-cells in $\operatorname{Int} E_{k-1}$, where $\cap_k E_k = C$. Each component of $N_{k-1} \times E_{k-1} \setminus \operatorname{Int}(N_k \times E_k)$ is homeomorphic to $N(A) \times I^2 \setminus \operatorname{Int}(N_1 \times E_1)$. Now $S^{n-2} \times I^2$ contains a sequence $Q_0 \supset Q_1 \supset \cdots Q_{k-1} \supset Q_k \supset \cdots$, where for $k > 0$ Q_k consists of 2^k pairwise disjoint copies of Q_0, $Q_k \subset \operatorname{Int} Q_{k-1}$ and

$$Q_{k-1} \smallsetminus \operatorname{Int} Q_k \cong N_{k-1} \times E_{k-1} \smallsetminus \operatorname{Int}(N_k \times E_k).$$

As a result, there is a natural surjective map $q : Q_0 \to p(N(A) \times I^2) \subset S$ sending distinct components of $\cap_k Q_k$ to distinct points of X and being 1-1 on $Q_0 \smallsetminus \cap_k Q_k$. Rather obviously, the components of $\cap Q_i$ are cell-like sets, so q is a cell-like mapping and therefore a near-homeomorphism. It follows that S is a manifold. □

Constructions like those of Example 7.11.2 or §2.6 give wild but strongly homogeneously embedded 1-spheres in $S^5 \times S^1$ and $S^3 \times S^1$, respectively. Exactly the same methods, with S^k in place of S^1, lead to wild but strongly homogeneously embedded k-spheres in codimensions 5 and 3. Whether there is a wild and strongly homogeneously embedded (or even just homogenously embedded) codimension-one manifold example remains an open question.

Historical Notes. Theorem 7.11.1 is due to Bryant and Lacher (1975), who did considerably more, showing there that the combination of mapping cylinder neighborhood and a generalized concept of freeness implies local flatness for embedded manifolds of all other codimensions. In low dimensions freeness is not a necessary ingredient: V. Nicholson (1969) proved that complexes in 3-manifolds with mapping cylinder neighborhoods are tame, and Lacher and A. Wright (1970) showed that 3-manifolds with mapping cylinder neighborhoods in 4-manifolds are locally flat.

S. Ferry and E. Pederson (1991) produced a catalogue of wildly embedded circles in S^n ($n \geq 7$) similar to the wild circles of Example 7.11.2. Theirs are indexed by Wall's finiteness obstruction.

M. A. Kervaire (1969) proved that every PL homology n-sphere, $n \geq 5$, bounds a compact, contractible, PL $(n+1)$-manifold; Kervaire derived the same result in the smooth category, provided one allows modification of the homology sphere by taking its connected sum with a (unique) smooth homotopy sphere.

Alternate ways of getting a cell-like mapping from a manifold onto spaces like these acyclic decomposition spaces are treated in Chapter 8.

The methods arising in the development of Example 7.11.2 are those used to settle the Double Suspension Problem, discussed later here in §8.10. The connection is exposed in Exercise 7.11.2 below.

The strongly homogeneous but wildly embedded Cantor set is due to Daverman (1979).

Exercises

7.11.1. Suppose Σ^{n-1} is a connected, two-sided $(n-1)$-manifold in a connected n-manifold M ($n \geq 5$), $\psi : \Sigma^{n-1} \to \Sigma^{n-1}$ is a cell-like mapping, and U is a component of $M \smallsetminus \Sigma^{n-1}$ such that Σ^{n-1} has a closed neighborhood in $\overline{U}$ naturally homeomorphic to the mapping cylinder of ψ. Then Σ^{n-1} is collared from $\overline{U}$.

7.11.2. Show that the join of S^1 and $\partial N(A)$, where $N(A)$ is the acyclic, $(n-2)$-dimensional ∂-manifold described in Example 7.11.2, is topologically S^n.

7.11.3. Show that S^n contains a wild, homogenously embedded $(n-2)$-torus ($n > 2$). [Hint: spin the Bing sling of Subsection 2.8.5.]

Chapter 8

Codimension-zero Embeddings

This concluding chapter organizes a whirlwind tour past some codimension-zero results. Offering few proofs, it highlights major developments and presents historical perspectives.

8.1. Manifold characterizations

We begin with a reminder of two important tools, both of which are treated in (Rourke and Sanderson, 1972).

An *h-cobordism* is a triple (W^n, M_0, M_1) that consists of a compact n-dimensional ∂-manifold W^n with disjoint boundary components M_0 and M_1 such that each inclusion $M_i \hookrightarrow W^n$ is a homotopy equivalence.

Theorem 8.1.1 (h-Cobordism). *If (W^n, M_0, M_1) is an h-cobordism, where W^n is a simply connected PL ∂-manifold and $n \geq 6$, then W^n is PL homeomorphic to $M_0 \times [0,1]$.*

When combined with G. Perelman's recent solution to the 3-dimensional Poincaré Conjecture, the h-Cobordism Theorem yields the following corollary.

Corollary 8.1.2 (PL Poincaré Theorem). *Any PL n-manifold homotopy equivalent to S^n is PL homeomorphic to S^n, $n \neq 4$.*

For an arbitrary h-cobordism (W^n, M_0, M_1), W^n simply connected or not, there is an element $\tau(W, M_0)$ of the Whitehead group of $\pi_1(W^n)$ that determines whether W^n is a product. This element is called the *torsion* of the cobordism.

Theorem 8.1.3 (s-Cobordism). *Let (W^n, M_0, M_1) be an h-cobordism, where W^n is a compact PL n-dimensional ∂-manifold, $n \geq 6$. There exists a well-defined element $\tau(W^n, M_0) \in \mathrm{Wh}(\pi_1(W^n))$ such that W^n is PL homeomorphic to $M_0 \times [0,1]$ if and only if $\tau(W^n, M_0)$ is trivial. Conversely, for every finitely presented group G, $\tau_0 \in \mathrm{Wh}(G)$ and $n \geq 6$, there exists a PL h-cobordism (W^n, M_0, M_1) such that $\pi_1(W^n) \cong G$ and $\tau(W^n, M_0) = \tau_0$.*

Noncompact manifolds deserve some consideration as well.

Theorem 8.1.4. *A contractible PL manifold W^n, $n \neq 4$, is PL homeomorphic to $\mathbb{R}^n$ if and only if W^n is simply connected at infinity.*

See Exercise 3.1.2 for the cases $n \geq 5$.

Corollary 8.1.5. *If V^m, W^n are contractible PL manifolds of dimensions $m, n \geq 1$, where $m + n \geq 5$, then $V^m \times W^n$ is PL homeomorphic to $\mathbb{R}^{m+n}$.*

A locally compact space X is *relatively 2-connected at infinity* if for each compact $C \subset X$ there exists another compact $D \subset X$, $D \supset C$, such that for every map $f : (I^2, \partial I^2) \to (X, X \smallsetminus D)$ there is a homotopy $\mu_t : I^2 \to X$ between f and a map into $X \smallsetminus C$ that sends ∂I^2 into $X \smallsetminus C$ for all $t \in I$.

Theorem 8.1.6. *Suppose W^n is an n-dimensional PL ∂-manifold, $n \geq 5$, with ∂W^n compact. Then W^n is PL homeomorphic to $\partial W^n \times [0,1)$ if and only if $\mathrm{incl} : \partial W^n \to W^n$ is a homotopy equivalence and W^n is relatively 2-connected at infinity.*

Engulfing technology assures that every compact subset C of W^n can be PL engulfed by a collar on ∂W^n, and then collar sliding or monotone union methods lead to a covering of W^n by copies of $\partial W^n \times [1 - 2^{-k}, 1 - 2^{-(k+1)}]$ that fit together perfectly.

Historical Notes. The h-Cobordism Theorem is due to Smale (1962), while the s-Cobordism Theorem was proved independently by D. Barden (1963), Mazur (1963a), (1963b), and Stallings. The 3-dimensional Poincaré Theorem was recently established by Perelman; a detailed exposition of his proof may be found in (Morgan and Tian, 2007). Freedman (1982) proved a 4-dimensional topological Poincaré Theorem.

Theorem 8.1.4 was first done by Stallings (1962b). McMillan handled the 3-dimensional case of the same result, modulo the now-established 3-dimensional Poincaré Conjecture. Freedman (1982) did the 4-dimensional case for topological manifolds and topological homeomorphisms. Outside the topological category, dimension 4 is strikingly unusual: there are a multitude of different PL and smooth structures on $\mathbb{R}^4$ (Gompf, 1985), (Taubes, 1987).

8.2. The α- and β-Approximation Theorems

When can a map $f : M \to N$ between a general pair of n-manifolds be adjusted to a homeomorphism? Siebenmann showed that, if $\dim M = \dim N \geq 5$, then any cell-like map $M \to N$ can be approximated arbitrarily closely by homeomorphisms. Siebenmann's result has some noteworthy improvements.

Let α denote an open cover of N. Define an *α-equivalence (over N)* to be a proper map $f : M \to N$ from another n-manifold M to N which has a homotopy inverse g for which there exist homotopies $fg \simeq \mathrm{Id}_N$ and $gf \simeq \mathrm{Id}_M$ limited by α and $f^{-1}(\alpha)$, respectively. A homotopy $\Phi : X \times I \to Y$ is *limited by a collection $\mathcal{U}$* of subsets of Y if, for each $x \in X$, some $U_x \in \mathcal{U}$ contains $\Phi(\{x\} \times I)$.

Theorem 8.2.1 (α-Approximation). *Let N be an n-manifold, $n \geq 6$. For every open cover α of N there is another open cover β of N such that if M is any n-manifold and $f : M \to N$ any β-equivalence, then f can be α-approximated by a homeomorphism.*

This α-Approximation Theorem posits a control over the target manifold, independent of the domain, and concludes that a map submitting to that control is reasonably close to a homeomorphism. There is a complementary result involving even simpler control exclusively on the source. Given an open cover β of space Y, say that a continuous function $g : Y \to X$ is a *β-map* provided the collection $\{g^{-1}(x) \mid x \in X\}$ refines β.

Theorem 8.2.2 (β-Approximation). *Let α be an open cover of an n-manifold M, $n \geq 6$. There exists another open cover β of M such that any proper β-map $g : M \to N$ onto another n-manifold N is homotopic through α-maps to a homeomorphism.*

Historical Notes. Chapman and Ferry (1979) proved α-Approximation Theorem 8.2.1. Exploiting 8.2.1, Ferry (1979) developed the β-Approximation Theorem 8.2.2 shortly thereafter.

8.3. Ends of manifolds

Throughout this section W will denote a non-compact, n-dimensional, PL ∂-manifold ($n > 5$) having compact boundary (empty boundary allowed). The question under consideration is the conditions needed to guarantee that W can be PL embedded in a compact PL ∂-manifold W' such that $W' \setminus W$

consists entirely of components of $\partial W'$. When that occurs, one says that W *admits a (compact) boundary* and calls W' a *completion* of W.

An *end* of W is a function ε that assigns to each compact subset C of W a nonempty component U of $W \smallsetminus C$ in such a way that if $C \subset C'$, then $U \supset U'$. The open set U is called a *neighborhood* of the end. An end ε is *isolated* if there exists a neighborhood U of ε such that for any compact $K \subset W$ only one component of $U \cap (W \smallsetminus K)$ has noncompact closure.

Given a closed subset C of a compact manifold M, the ends of $W = M \smallsetminus C$ correspond to components K of C: any neighborhood U_K of such a component K contains a neighborhood of the corresponding end of W; that end is isolated precisely when K is an open subset of C.

A *collar for an end* ε is the closure N of a neighborhood of ε such that N is a connected ∂-manifold PL embedded in W so that $N \cong \partial N \times [0,1)$. Obviously W admits a compact boundary if and only if it has only a finite number of ends, each of which has a collar.

The strategy for devising a collar on an isolated end ε is to produce more and more highly connected neighborhoods of ε. According to Theorem 8.1.6, it suffices to find a closed neighborhood N of ε such that $\partial N \hookrightarrow N$ is a homotopy equivalence and N is relatively 2-connected at ∞. Ultimately, under the right conditions on ε, one can secure such a neighborhood N.

A 0-*neighborhood* of an isolated end ε is a closed connected neighborhood V of ε such that V is a PL ∂-manifold in W and ∂V is connected. Every isolated end of W has arbitrarily small 0-neighborhoods: if the connected, PL ∂-manifold V is a closed neighborhood of ε, one can drill tunnels through V to join up the various boundary components and to create thereby a 0-neighborhood $V' \subset V$.

Consider an inverse sequence of groups and homomorphisms

$$G_1 \xleftarrow{\varphi_1} G_2 \xleftarrow{\varphi_2} G_3 \xleftarrow{\varphi_3} G_4 \longleftarrow \cdots.$$

Its *inverse limit* is the subgroup

$$\{\langle g_1, g_2, \ldots, \rangle \in \Pi_{i=1}^{\infty} G_i \mid \varphi_{i+1}(g_{i+1}) = g_i \text{ for all } i\}.$$

A *subsequence* of this inverse sequence is a sequence

$$G_{n(1)} \xleftarrow{\varphi_1'} G_{n(2)} \xleftarrow{\varphi_2'} G_{n(3)} \xleftarrow{\varphi_3'} G_{n(4)} \longleftarrow \cdots,$$

where $0 < n(1) < n(2) < \cdots$ and φ_i' is the composite $\varphi_{n(i)} \circ \cdots \circ \varphi_{n(i+1)-1}$.

An inverse sequence of groups is *stable* if there exists a subsequence such that the induced sequence

$$\operatorname{Im} \varphi_1' \xleftarrow{\varphi_1'|} \operatorname{Im} \varphi_2' \xleftarrow{\varphi_2'|} \operatorname{Im} \varphi_3' \xleftarrow{\varphi_3'|} \operatorname{Im} \varphi_4' \longleftarrow \cdots$$

of image groups and restricted homomorphisms has isomorphisms everywhere. In that spirit, say that π_1 is *stable* at an isolated end ε if there exists a sequence $N_1 \supset N_2 \supset \cdots$ of path-connected neighborhoods of ε and (for some base points $x_i \in N_i$ and for φ_i equal to the inclusion-induced $\pi_1(N_{i+1}, x_{i+1}) \to \pi_i(N_i, x_{i+1})$ followed by a change of base point automorphism) the sequence

$$\pi_1(N_1, x_1) \xleftarrow{\varphi_1} \pi_1(X_2, x_2) \xleftarrow{\varphi_2} \pi_1(N_3, x_3) \longleftarrow \cdots$$

is stable. When that occurs, define $\pi_1(\varepsilon)$, *the fundamental group of the end* ε, as $\operatorname{Im}\varphi_1' \cong \operatorname{Im}\varphi_i'$. The stability of π_1 at ε is independent of all the relevant choices.

Now suppose ε is an isolated end of W and π_1 is stable at ε. Then a 1-*neighborhood* of ε is a closed 0-neighborhood N of ε such that the natural inclusion-induced homomorphisms $\pi_1(\partial N) \to \pi_1(N)$ and $\pi_1(\varepsilon) \to \pi_1(N)$ are isomorphisms.

Lemma 8.3.1. *If π_1 is stable at ε and $\pi_1(\varepsilon)$ is finitely presented, then ε has arbitrarily small 1-neighborhoods.*

A space X is *dominated by a finite complex* K if there exist maps $d : K \to X$ and $u : X \to K$ such that $du \simeq \mathrm{Id}_X$. Let $\mathcal{D}$ denote the class of spaces that have the homotopy type of a (not necessarily finite) CW complex and are dominated by a finite complex.

Given an isolated end ε of W such that π_1 is stable at ε and $\pi_1(\varepsilon)$ is finitely presented, call ε a *tame end* if, in addition, every sufficiently small 1-neighborhood of ε belongs to $\mathcal{D}$. The presence of tame ends is clarified by the next result.

Lemma 8.3.2. *Suppose ε is an isolated end of W such that π_1 is stable at ε and $\pi_1(\varepsilon)$ is finitely presented. The following statements are equivalent:*

(1) There exists a connected neighborhood U of ε such that $U \in \mathcal{D}$ and the inclusion-induced $\pi_1(\varepsilon) \to \pi_1(U)$ has a left inverse;

(2) Some 1-neighborhood of ε belongs to $\mathcal{D}$;

(3) Every sufficiently small 1-neighborhood of ε belongs to $\mathcal{D}$;

(4) Every sufficiently small 0-neighborhood of ε belongs to $\mathcal{D}$.

A k-neighborhood V of ε, $k \geq 2$, is a 1-neighborhood such that $\pi_i(V, \partial V) \cong 0$ ($i = 0, 1, \ldots, k$). When ε has arbitrarily small k-neighborhoods, then with handle-swapping and handle-sliding moves, as in the proof of the s-Cobordism Theorem, one can improve them to obtain arbitrarily small $(k+1)$-neighborhoods for $k = 1, 2, \ldots, n-4$. Success in securing the collar on ε is at hand if one can take the next step.

Theorem 8.3.3. *If V is a connected PL ∂-manifold of dimension $n \geq 5$ with one end ε such that V is an $(n-2)$-neighborhood of ε, π_1 is stable at ε and the natural inclusion $\pi_1(\varepsilon) \to \pi_1(V)$ is an isomorphism, then V is PL homeomorphic to $\partial V \times [0,1)$.*

At this point in the lengthy process, when confronted with the need to improve $(n-3)$-neighborhoods to $(n-2)$-neighborhoods, one encounters an unavoidable new obstacle. Let R be a ring. A *projective* R-module is one that is isomorphic to a direct summand of a free R-module. Two R-modules M_1, M_2 are *stably isomorphic* if there exists a finitely generated, free R-module F such that $M_1 \oplus F, M_2 \oplus F$ are R-isomorphic. A finitely generated R-module is *stably free* if it is stably isomorphic to a free R-module. The stable isomorphism classes of finitely generated, projective R-modules form an abelian group (with direct sum as the group operation), called the *projective class group of R*, and written $\widetilde{K}_0(R)$. The class of stably free R-modules corresponds to the zero element of this group. Finally, given a group G, $\widetilde{K}_0(G)$ denotes $\widetilde{K}_0(\mathbb{Z}[G])$, where $\mathbb{Z}[G]$ is the integral group ring determined by G. To make the final improvement, one wants and, it turns out, needs $H_{n-2}(\widetilde{V}, \partial\widetilde{V})$ to be stably free as a $\mathbb{Z}[\pi_1(\varepsilon)]$-module (for arbitrarily small $(n-3)$-neighborhoods V of ε).

Proposition 8.3.4. *Suppose ε is an isolated end of W that has arbitrarily small $(n-3)$-neighborhoods V. Then, for any such V, $H_{n-2}(\widetilde{V}, \partial\widetilde{V}; \mathbb{Z}[\pi_1(\varepsilon)])$ is a projective $\mathbb{Z}[\pi_1(\varepsilon)]$-module. Furthermore, if ε is a tame end, then $H_{n-2}(\widetilde{V}, \partial\widetilde{V}; \mathbb{Z}[\pi_1(\varepsilon)])$ is a finitely generated projective $\mathbb{Z}[\pi_1(\varepsilon)]$-module.*

Theorem 8.3.5 (Collaring). *A tame end ε of W is collared in W if and only if an obstruction $\sigma(\varepsilon) \in \widetilde{K}_0(\pi_1(\varepsilon))$ is trivial. Here $\sigma(\varepsilon)$ is the stable isomorphism class of $H_{n-2}(\widetilde{V}, \partial\widetilde{V}; \mathbb{Z}[\pi_1(\varepsilon)])$, for any $(n-3)$-neighborhood V of ε.*

Remarks. Up to sign, the obstruction $\sigma(\varepsilon)$ is Wall's finiteness obstruction $\sigma(V)$, for any closed $(n-3)$-neighborhood V of ε.

All obstructions can be realized. Given $n \geq 6$, a finitely presented group G and $\sigma_0 \in \widetilde{K}_0(G)$, there exists a PL n-manifold W having a single (necessarily isolated) end ε such that ε is tame, $\pi_1(\epsilon) \cong G$ and $\sigma(\varepsilon) = \sigma_0 \in \widetilde{K}_0(G)$. (cf. (Wall, 1965a), (1966).)

Corollary 8.3.6. *Let W be a connected PL n-manifold, $n \geq 6$, having a single end ε, where ε is tame and $0 \cong \sigma(W) \in \widetilde{K}_0(\pi_1(\varepsilon))$. Then W is homeomorphic to the interior of a compact ∂-manifold.*

As all submodules of a finitely generated $\mathbb{Z}$-module are free, $\widetilde{K}_0(G) \cong 0$ when G is the trivial group. More generally, the projective class group

$\widetilde{K}_0(G)$ is also known to be trivial for the fundamental group G of many compact, aspherical, Riemannian manifolds. For tori (i.e., for free Abelian groups), this is a result of H. Bass (1968); see also (Swan, 1978).

Corollary 8.3.7. *A PL manifold W^n is the interior of a compact ∂-manifold with 1-connected boundary if and only if it is simply-connected at ∞ and $H_*(W;\mathbb{Z})$ is finitely generated.*

Theorem 8.3.8. *Let W be a manifold with a single isolated end ε and let Q be a compact connected manifold with $\chi(Q) = 0$. Then ε is tame if and only if the end of $W \times Q$ has a collar.*

Corollary 8.3.9. *Let W be a connected n-manifold, $n \geq 5$, having a single end ε such that $\pi_1(\varepsilon)$ is stable and ε is tame. Then $W \times S^1$ is homeomorphic to the interior of a compact ∂-manifold.*

M. W. Davis (1983) has an example of a contractible, one-ended (necessarily) manifold W^n, $n \geq 4$, for which the end has arbitrarily small closed ∂-manifold neighborhoods V such that $\partial V \hookrightarrow V$ is a homotopy equivalence but W has no completion. Despite the fact that the neighborhoods V mentioned have the homotopy type of a finite complex, the end is not tame in the strict sense employed in this discussion, because π_1 fails to be stable at the end.

Historical Notes. In his Ph.D. thesis, Siebenmann (1965) developed this detailed analysis of conditions under which a manifold W admits a compact boundary. Theorem 8.3.3 is also his work (1969).

Earlier, W. Browder, J. Levine and G. R. Livesay (1965) established Corollary 8.3.7 for simply connected ends.

L. S. Husch and T. M. Price (1970) showed that, modulo the now-established 3-dimensional Poincaré Conjecture, a 3-dimensional ∂-manifold M is the interior of a compact ∂-manifold if ∂M is compact, M has single isolated end ε, π_1 is stable at ε, and $\pi_1(\varepsilon) \neq \mathbb{Z}/2$.

The obstruction $\sigma(W)$ was developed by C. T. C. Wall (1965a), (1966) to provide a negative answer to Whitehead's problem asking whether a finitely dominated CW complex must be homotopy equivalent to a finite complex. In other words, Collaring Theorem 8.3.5 promises that a tame end ε is collared if and only if ε has a neighborhood homotopy equivalent to a finite complex.

Theorem 8.3.8 was derived by Siebenmann (1965) and S. M. Gersten (1966) independently, as part of a more general product formula. Corollary 8.3.9 was originally proved by M. Mather (1965).

8.4. Ends of maps

Like manifolds, maps themselves occasionally admit completions. Given a ∂-manifold W and map $e : W \to X$, a *completion* of e consists of an embedding of $W \subset W'$ into a ∂-manifold W' with $W' \smallsetminus W \subset \partial W'$ and an extension of e to a proper map $e' : W' \to X$. Unlike with manifolds, however, one does not distinguish various distinct ends for a map; instead, the target space X parameterizes the end.

The conditions under which a completion of a map exists are rather elaborate. A *neighborhood* of the end of e is an open subset U of W such that $e|W \smallsetminus U$ is proper. The end is *onto* if each neighborhood of the end surjects to X via e. The end is 0-LC if for each neighborhood U of the end, each $x \in X$, and each neighborhood V_x of x, there exist a neighborhood $U' \subset U$ of the end and a neighborhood $V_x' \subset V$ of x such that any two points in $U' \cap e^{-1}(V_x')$ can be joined by paths in $U \cap e^{-1}(V_x)$. The end is 1-LC if U' and V_x' can be obtained so that, in addition to the 0-LC condition, loops in $U' \cap e^{-1}(V_x')$ are null-homotopic in $U \cap e^{-1}(V_x)$.

A 0-LC map $e : W \to X$ has *locally constant fundamental group at the end* if for each $x \in X$ there are neighborhoods U of the end and V_x of x such that $U \cap e^{-1}(V_x)$ has a regular covering space whose end is 1-LC over V_x. The local fundamental group of the end is the group of covering transformations of this cover. Well-defined up to isomorphism, it determines a twisted coefficient system on X.

Given a metric space X and continuous $\varepsilon : X \to (0, \infty)$, say that a homotopy $h : Y \times I \to X$ has *diameter* less than ε *over* X if the diameter of each path $h(\{y\} \times I)$ is less than the minimum of ε on that path.

The end is *tame* if, for each neighborhood U of the end and each map $\varepsilon : X \to (0, \infty)$, there exist a neighborhood V of the end, $V \subset U$, and a homotopy h_t of W such that $h_0 = \mathrm{Id}_W$, $h_1(W) \subset W \smallsetminus V$, $h_t|W \smallsetminus U = \mathrm{Id}_{W \smallsetminus U}$ and $eh : M \times I \to X$ has diameter less than ε over X. Tameness of the end is a necessary condition: if e has a completion, then the homotopies required for tameness can be readily obtained by pushing along collar lines in a neighborhood of the added boundary.

This homotopical notion of tameness has a homological analog. Suppose $e : W \to X$ is 0-LC and has locally constant fundamental group π at the end. The end of e is *homologically tame* if π is finitely presented and for every neighborhood of the end and every continuous $\varepsilon : X \to (0, \infty)$ there is a neighborhood U' of the end such that for every $K \subset X$ the homomorphism

$$H_*(e^{-1}(K), e^{-1}(K) \smallsetminus U; \mathbb{Z}\pi) \to H_*(e^{-1}(K^\varepsilon), e^{-1}(K^\varepsilon) \smallsetminus U'; \mathbb{Z}\pi)$$

is trivial. In this formula K^ε is just shorthand for $B(K; \varepsilon)$.

Lemma 8.4.1. *Suppose the end of the map $e : W \to X$ from a manifold W to a locally compact, locally 1-connected metric space X is onto, 0-LC and has locally constant, finitely presented fundamental group π. Then the end is tame if and only if it is homologically tame.*

With the preliminaries now completed, here is a statement of the End Theorem.

Theorem 8.4.2 (End). *Suppose $e : W \to X$ is a map from an n-manifold, $n \geq 6$, to a locally compact, locally 1-connected space X. Suppose also that the end of e is tame, onto, and has locally constant fundamental group π such that the Whitehead groups $Wh(\pi \times \mathbb{Z}^k)$ are trivial for all $k \geq 0$. Then e has a completion $e' : W' \to X$.*

Siebenmann's Collaring Theorem in §8.3, an unparameterized version of this End Theorem, amounts to producing a completion for the constant map $e : W \to \{\text{point}\}$.

Corollary 8.4.3. *Let W be a PL n-manifold and $Y \subset W$ be a compact ANR that is 1-LCC embedded in N, where $n \geq 6$ and $n - \dim Y \geq 3$. Then Y has a mapping cylinder neighborhood in W.*

The idea behind the proof of the Corollary is to identify a closed neighborhood $N \subset W$ of Y, with N an n-dimensional PL ∂-manifold equipped with a retraction $r : N \to Y$. Examine the restriction

$$r' = r|N \smallsetminus Y : N \smallsetminus Y \to Y.$$

Its end is 0-LC, due to the codimension restriction on Y. The 1-LCC hypothesis assures that the end is 1-LC; this means, of course, that r' has locally constant fundamental group π at the end, where π is the trivial group. One then shows that the end of r' is tame by showing it is homologically tame.

Hence, Theorem 8.4.2 applies. In the resulting completion $r^* : N^* \to Y$, a collar on $\partial N^* = N^* \smallsetminus (N \smallsetminus Y)$ projects in a natural way to a manifold mapping cylinder neighborhood of $Y \subset W$.

At the heart of the End Theorem is a result—the Thin h-Cobordism Theorem—offering controls on collar images as one approaches the end. Suppose M is a compact ∂-manifold, with ∂M the disjoint union of $(n-1)$-manifolds N_0, N_1, $e : M \to X$ is continuous and $\delta > 0$. Say that (M, N_0) is a *(δ, h)-cobordism over X* if for $i = 0, 1$ there is a homotopy $M \times [0, 1] \to M$ starting at the identity, ending with a map into N_i and having diameter $< \delta$ over X. In addition, say that (M, N_0) has a *δ-product structure over X* if there is a homeomorphism $h : N_0 \times [0, 1] \to M$ such that h_0 is the Identity on N_0 and, when considered as a homotopy, h has diameter less than δ over X.

Theorem 8.4.4 (Thin h-Cobordism). *Suppose X is a compact, locally 1-connected metric space, $n \geq 5$ and $\epsilon > 0$. Then there exists $\delta > 0$ such that if M is a compact, n-dimensional ∂-manifold, (M, N_0) is a (δ, h)-cobordism over X, and M has a regular covering $\widetilde{M} \to M$ so that the composite $\widetilde{M} \to X$ is $(\delta, 1)$-connected over X, where the group of covering translations π satisfies $Wh(\pi \times \mathbb{Z}^k) \cong 0$ for all k, then (M, N_0) has an ϵ-product structure over X.*

Remark. There is a much more general version of this Thin h-Cobordism Theorem for non-compact X and proper maps $e : M \to X$.

A *cell-like resolution* of a space X is a pair (M, f) consisting of an n-dimensional topological manifold M and a proper, cell-like, surjective mapping $f : M \to X$; X is said to be *resolvable* if it has a cell-like resolution.

Corollary 8.4.5. *Let X be an n-dimensional metric space, $n \geq 5$. The following statements are equivalent:*

a) X is resolvable;

b) $X \times \mathbb{R}^k$ is resolvable for some integer $k \geq 0$;

c) $X \times \mathbb{R}^k$ is resolvable for all $k \geq 0$.

Proof. Obviously a) $\Longrightarrow$ c) $\Longrightarrow$ b). To see that b) $\Longrightarrow$ a), name a cell-like map $f : M \to X \times \mathbb{R}^k$ and consider $e = \text{proj} \circ f : M \to X \times \mathbb{R}^{k-1}$, where proj: $X \times \mathbb{R}^k \to X \times \mathbb{R}^{k-1}$ is the projection. End Theorem 8.4.2 applies, providing a completion $e' : M' \to X \times \mathbb{R}^{k-1}$. Restrict e' to (a component of) $\partial M' \setminus M$; it is easily seen that this restricted map provides a resolution of $X \times \mathbb{R}^{k-1}$. Induction gives a resolution of X itself. □

Corollary 8.4.6. *For any n-dimensional metric space X, the following statements are equivalent:*

a) X is resolvable;

b) $X \times \mathbb{R}^k$ is a manifold for some integer $k \geq 0$;

c) $X \times \mathbb{R}^2$ is a manifold.

Under the standard hypotheses, completions are unique in a strongly controlled sense.

Theorem 8.4.7. *Suppose $W \to X$ satisfies the conditions of Theorem 8.4.2, and suppose $e' : W' \to X$ and $e'' : W'' \to X$ are two completions of e. Then for every $\varepsilon : X \to (0, \infty)$ there is an isotopy of W that takes a collar neighborhood of $W' \setminus W$ to a collar neighborhood of $W'' \setminus W$ and whose image under e has diameter less than ε.*

Historical Notes. The program outlined in this section is due to F. Quinn (1979), (1982).

8.5. Quinn's obstruction and the topological characterization of manifolds

Loosely put, homology manifolds are spaces possessing the local homology properties of manifolds. Formally, an *n-dimensional homology manifold* is a locally compact, finite-dimensional, metric ANR X such that at each point $x \in X$,

$$H_r(X, X \setminus \{x\}) = \begin{cases} \mathbb{Z} & \text{if } r = n \\ 0 & \text{otherwise.} \end{cases}$$

Clearly topological manifolds are themselves examples of homology manifolds. Work of R. L. Wilder (1979) assures that all homology 2-manifolds are topological 2-manifolds, but that classical result does not generalize to any dimensions above 2. Numerous nonmanifold examples arise from cell-like decompositions. The quotient of S^3 obtained by identifying the Whitehead continuum to a point constitutes a minimal 3-dimensional example; likewise, the quotient of S^n obtained using a non-cellular arc in place of the Whitehead continuum gives minimal higher-dimensional examples.

Corollary 7.4.7 and either the Vietoris-Begle Theorem (0.4.1) or Proposition 3.2.9 lead to:

Proposition 8.5.1. *Every n-dimensional resolvable space X is an n-dimensional homology manifold.*

Quinn (1987) discovered an obstruction to the existence of cell-like resolutions. The obstruction has the following properties:

Theorem 8.5.2 (Index). *Let X be a connected, homology n-manifold, $n > 3$. Then there is an integer $i(X)$, called the index of X, such that $i(X) = 1$ if and only if X has a cell-like resolution. Moreover, the index function i satisfies:*

(1) $i(X) \equiv 1 \bmod 8$,

(2) $i(X) = i(U)$ for all connected open subsets $U \subset X$,

(3) $i(X \times Y) = i(X) \cdot i(Y)$, and

(4) i is invariant under cell-like, surjective mappings between homology n-manifolds.

The important business of realizing the possible Quinn obstructions is treated in the next section.

The combination of Quinn's Index Theorem 8.5.2 and Edwards's Cell-like Approximation Theorem 7.4.1 gives the following notable characterization of topological manifolds:

Theorem 8.5.3. *A connected homology n-manifold X, $n \geq 5$, is a topological n-manifold if and only if $i(X) = 1$ and X has the DDP.*

Corollary 8.5.4. *A connected homology n-manifold X, $n \geq 5$, is a topological n-manifold provided some open subset of X is an n-manifold and X has the DDP.*

Corollary 8.5.5. *Let X be a connected homology n-manifold, $n \geq 4$. Then $i(X) = 1$ if and only if $X \times \mathbb{R}^2$ is a manifold.*

Finally, here is a characterization of ∂-manifolds.

Theorem 8.5.6. *Suppose X is a locally compact ANR containing a closed subset C such that $X \smallsetminus C$ is an n-manifold, $n \geq 5$, C is an $(n-1)$-manifold, and $H_*(X, X \smallsetminus \{c\}; \mathbb{Z}) \cong 0$ for all $c \in C$. Then X is a ∂-manifold with $\partial X = C$ if and only if C is 1-LCC in X.*

Proof. Attach an open collar $C \times (-1, 0]$ to X along $C = C \times \{0\}$. An elementary Mayer-Vietoris computation confirms that $X^* = X \cup (C \times (-1, 0])$ is a homology n-manifold. Quinn's Index Theorem assures the existence of a cell-like resolution for X^*. Then X^* is an n-manifold, as it has the DDP by Corollary 7.4.13. Either Theorem 7.2.3 or Theorem 7.6.1 yields that $C \times \{0\}$ is bicollared in X^*. In other words, C is collared in X, so X is a ∂-manifold. □

End Theorem 8.4.2 can be used to prove an analogous result in codimensions ≥ 2. This allows the codimension-three and codimension-two flatness theorems (Corollary 5.7.3 and Theorem 6.3.6) to be generalized to manifold factors in homology manifolds. In particular, it gives a proof of Theorem 6.3.6. A space X is a *manifold factor* if $X \times \mathbb{R}^2$ is a manifold (cf. Corollary 8.4.6). A subset X of a space Y is *locally flat* at $x \in X$ if there exist a neighborhood U of x in Y and an integer $k \geq 1$ such that $(U, U \cap X) \cong ((U \cap X) \times \mathbb{R}^k, (U \cap X) \times \{0\})$.

Theorem 8.5.7. *Assume M is a homology n-manifold, $n \geq 6$, X is a closed subset of M, X is a manifold factor, and $M \smallsetminus X$ is an n-manifold.*

(1) *If $\dim X \leq n-2$ and X is locally flat in M, then M is a manifold.*
(2) *If $\dim X \leq n-3$, then X is locally flat in M if and only if X is 1-LCC in M.*
(3) *If $\dim X = n-2$, then X is locally flat in M if and only if X is locally homotopically unknotted in M.*

Historical Notes. Ferry (1979) published the first proof of Theorem 8.5.6. The same result also was claimed by Černavskiĭ and Seebeck. Černavskiĭ's claim was later retracted. Although Seebeck circulated a manuscript, his work never appeared in print. Theorem 8.5.7 is from (Quinn, 1979).

8.6. Exotic homology manifolds

A pressing question asks: does every homology manifold admit a resolution? Could it be that Quinn's index is always trivial? The answer is negative. J. Bryant, S. Ferry, W. Mio and S. Weinberger (1996) showed that all possible Quinn obstructions can be realized:

Example 8.6.1. *For every compact simply-connected n-dimensional topological manifold M, $n \geq 6$, and every integer $k \equiv 1 \bmod 8$ there exists a compact n-dimensional ANR homology manifold X homotopy equivalent to M with $i(X) = k$.*

Nonresolvable homology manifolds are called *exotic.* The exotic homology manifolds are important new objects in high-dimensional manifold theory. They provide valuable periodicity in the theory of manifold structure sets. An *n-dimensional homology ∂-manifold* is an ANR pair (X, Z), such that Z is closed in X, $X \setminus Z$ is a homology n-manifold, and $H_*(X, X \setminus \{z\}; \mathbb{Z}) \cong 0$ for all $z \in Z$. The (simple-homotopy) structure set $S(X)$ associated with such a pair (X, Z) is the collection of all simple homotopy equivalences $\psi : Y \to X$ defined on another n-dimensional homology ∂-manifold (Y, Z') such that $\phi|Z' : (Z' = \psi^{-1}(Z)) \to Z$ is a homeomorphism, modulo the relation given by s-cobordism involving homology ∂-manifolds.

Theorem 8.6.2. *Let X be an n-dimensional homology ∂-manifold, $n \geq 5$. Then $S(X)$ is isomorphic to $S(X \times B^4)$.*

The latter admits the structure of an abelian group, more or less like that of higher homotopy groups. Such group structures can be imposed on topological manifold structure sets, but periodicity as in 8.6.2 fails when restricted to genuine manifolds.

The impossibility of having a codimension-one embedding of one homology manifold in another if they have different indices is easy to see. In codimension two there can be no locally homotopically unknotted embedding of an exotic homology manifold in a genuine one—Quinn's End Theorem would give a manifold mapping cylinder neighborhood N of the embedded homology manifold X, and a second application of the End Theorem to the map $\widetilde{\partial N} \to X$ on the universal cover would give that X is resolvable. Inequality of indices is no barrier to embeddability when the codimension is greater than two.

Theorem 8.6.3. *Let X be a compact homology n-manifold, V^s a compact s-dimensional ∂-manifold, $s-n \geq 3$, and $f : X \to V^s$ a homotopy equivalence. Then f is homotopic to an embedding.*

Theorem 8.6.4. *For any compact homology n-manifold X, $n \geq 5$, there exists an embedding of X in some topological $(n+3)$-manifold.*

Theorem 8.6.5. *If X is a homology n-manifold X, $n \geq 5$, then there exist a homology n-manifold X' satisfying the DDP and a cell-like, surjective mapping $f : X' \to X$.*

Thus, there are homology manifolds with the DDP representing every possible homotopy type. These objects carry some of the useful properties of genuine manifolds.

Proposition 8.6.6. *Let X be a homology n-manifold with the DDP, $n \geq 5$, and let P and Q be polyhedra of dimensions p and q, respectively. Then all maps $f : P \to X$ and $g : Q \to X$ can be approximated by maps f' and g' such that $\dim[f'(P) \cap g'(Q)] \leq p+q-n$; moreover, if $p+q-n \leq n-3$, then f', g' can be obtained such that $f'(P) \cap g'(Q)$ is 1-LCC embedded in X.*

Corollary 8.6.7. *Suppose X is a homology n-manifold with the DDP and K is a finite k-complex, $2k+1 \leq n$. Then every map $f : K \to X$ can be approximated by a 1-LCC embedding.*

Theorem 8.6.8. *Suppose X^n is a homology n-manifold, $n \geq 5$, having the DDP, M^m is a closed PL m-manifold, $3m \leq 2n-2$, and $f\colon M^m \to X^n$ is a $(2m-n+2)$-connected map. Then f is homotopic to a 1-LCC embedding.*

Corollary 8.6.9. *Let $\lambda : M^m \to X$ be an embedding of a closed, PL m-manifold in a homology n-manifold X with the DDP, $3m \leq 2n-2$. Then λ can be approximated by a 1-LCC embedding.*

Historical Notes. Theorems 8.6.3 and 8.6.4 were proved by (Bryant and Mio, 1999) and independently by H. Johnston (1999). Theorem 8.6.5 was treated in (Bryant et al., 2007). Proposition 8.6.6 was handled in (Bryant, 1986). Theorem 8.6.8 and the Corollary (to its proof) were done by (Bryant and Mio, 2000). In addition, a transversality result for 1-LCC embeddings of two metastable range manifolds in a homology n-manifold with the DDP was developed in both (Bryant and Mio, 2000) and (Johnston, 1999).

Exercise

8.6.1. Show that every homology n-manifold X homotopy equivalent to T^n is resolvable. [Hint: find a connected homology n-manifold Y such that both $\mathbb{R}^n \setminus B^n$ and some open subset U of X embed in Y, by lifting the homotopy equivalence to a controlled map $\widetilde{X} \to \mathbb{R}^n$ and using Proposition 8.8.2.]

8.7. Homotopy tori are tori

Let T^n denote the n-torus $S^1 \times \cdots \times S^1$ (n factors of S^1). Consider pairs (M, f) where M denotes a closed PL n-manifold and $f : M \to T^n$ a homotopy equivalence. Define an equivalence relation on this collection by declaring $(M, f) \sim (M', f)$ if there exists a PL homeomorphism $H : M \to M'$ with $f \simeq f'H$. The set of equivalence classes, denoted as $\mathscr{S}(T^n)$, is called the homotopy structure set for T^n. It turns out that $\mathscr{S}(T^n)$ is a finite set with remarkable properties:

Theorem 8.7.1. *For $n \geq 5$ there is a bijection $H^3(T^n; \mathbb{Z}_2) \to \mathscr{S}(T^n)$ that is natural with respect to covering maps $T^n \to T^n$.*

Corollary 8.7.2. *For every $(M, f) \in \mathscr{S}(T^n)$, $n \geq 5$, there exists a finite covering map $\theta : \widetilde{T}^n \to T^n$ for which the pull-back $\widetilde{f} : \widetilde{M} \to \widetilde{T}^n$ is homotopic to a PL homeomorphism.*

In the topological category, the related structure set is simple.

Theorem 8.7.3. *For $n \geq 5$ every n-manifold homotopy equivalent to T^n is homeomorphic to T^n.*

Consequently, PL structures on the n-torus are not unique—the Hauptvermutung fails even for compact PL manifolds.

Historical Notes. Theorem 8.7.1 was established by W. C. Hsiang and Shaneson (1970) and by Wall (1969), independently; A. Casson is also known to have obtained the result, recast in a more geometric form. Theorem 8.7.3 was proved in (Hsiang and Wall, 1969).

8.8. Approximating stable homeomorphisms of $\mathbb{R}^n$ by PL homeomorphisms

A homeomorphism $h : S \to S$ is *stable* if it can be expressed as a composition $h = h_k \circ \cdots \circ h_1$ of homeomorphisms $h_i : S \to S$ such that $h_i | U(i) = \mathrm{incl}_{U(i)}$ on some nonempty open subset $U(i)$ of S, for $i = 1, 2, \ldots, k$.

Back in the 1960s stable homeomorphisms were a hot topic, due to close ties with the then-unsettled annulus conjecture. The n-dimensional stable homeomorphism conjecture propounded that all orientation-preserving homeomorphisms of S^n are stable. M. Brown and H. Gluck (1964) proved that the n-dimensional stable homeomorphism conjecture implies the n-dimensional annulus conjecture, and that the annulus conjecture in all dimensions $\leq n$ implies the stable homeomorphism conjecture in dimensions $\leq n$.

Elementary PL theory assures that, on any connected, PL manifold, the orientation-preserving, PL self-homeomorphisms are stable. With controlled engulfing methods, guided by homotopies moving points along rays emanating from the origin (and their images under a given homeomorphism), Connell (1963) proved a near-converse.

Theorem 8.8.1. *Every stable homeomorphism $h : \mathbb{R}^n \to \mathbb{R}^n$ $(n \geq 5)$ can be approximated by a PL homeomorphism.*

The following clever trick has proved to be useful in other contexts.

Proposition 8.8.2. *Every bounded homeomorphism $h : \mathbb{R}^n \to \mathbb{R}^n$ is stable.*

Proof. Define $q : \mathbb{R}^n \to \operatorname{Int} B^n$ as $q(x) = x/(1 + \|x\|)$ and $h' : R^n \to \mathbb{R}^n$ as

$$h'(x) = \begin{cases} qhq^{-1}(x) & \text{for } x \in \operatorname{Int} B^n \\ x & \text{for } x \in \mathbb{R}^n \smallsetminus \operatorname{Int} B^n. \end{cases}$$

The boundedness of h implies that h' is a homeomorphism.

Let $D^n \subset B^n$ denote the ball or radius $1/2$ centered at the origin. Rescale $\mathbb{R}^n$ so there exists an open set $V \subset \mathbb{R}^n$ with $V \cup h(V) \subset D^n$. Modify q so $q|D^n$ is the inclusion $D^n \to \operatorname{Int} B^n$ and $q(x) = x/(1 + \|x\|)$ for points x outside a compact neighborhood of D^n in $\operatorname{Int} B^n$. For h' defined as before, h' is the identity on $\mathbb{R}^n \smallsetminus B^n$, $h|V = h'|V$, and $h = [h(h')^{-1}] \circ h'$. Consequently, h is stable. □

Corollary 8.8.3. *Suppose the homeomorphism $g : T^n \to T^n$ is homotopic to the Identity. Then every lift $\widetilde{g} : \mathbb{R}^n \to \mathbb{R}^n$ of g to the universal cover is bounded and, hence, stable.*

Corollary 8.8.4. *All orientation-preserving homeomorphisms $g : T^n \to T^n$ are stable.*

Proof. Find a PL homeomorphism $h : T^n \to T^n$ such that $h \circ g$ is homotopic to Id. Then $h \circ g$ and $g = h^{-1}(h \circ g)$ are stable. □

Kirby (1969) settled the stable homeomorphism conjecture and, with it, the annulus conjecture. The crux of the proof involved showing:

Theorem 8.8.5. *Orientation-preserving homeomorphisms $g : \mathbb{R}^n \to \mathbb{R}^n$ $(n \geq 5)$ are stable.*

Corollary 8.8.6. *Every homeomorphism $\mathbb{R}^n \to \mathbb{R}^n$ $(n \geq 5)$ can be approximated by a PL homeomorphism.*

To make sense of what comes next, it pays to say that a homeomorphism h between open subsets of $\mathbb{R}^n$ is *stable* if each point of the domain has a neighborhood W such that $h|W$ extends to a stable homeomorphism of $\mathbb{R}^n$. This is a condition about atlases of coordinate charts; using it, one can define *stable manifolds* in the usual way, as well as *stable homeomorphisms* between stable manifolds. Clearly then a homeomorphism between stable manifolds is stable if its restriction to some open set is stable.

The key to proving Theorem 8.8.5 then is to put a patch of the action on the n-torus. Consider a homeomorphism g of $\mathbb{R}^n$ to itself. Find an immersion α of $T^n \setminus D^n$ in $\mathbb{R}^n$ (where D^n is a PL n-ball in T^n). Let $\widetilde{T^n \setminus D^n}$ denote $T^n \setminus D^n$ with the PL structure induced by $g \circ \alpha$. This leads to the following commutative diagram:

$$\begin{array}{ccc} T^n \setminus D^n & \xrightarrow{\mathrm{Id}} & \widetilde{T^n \setminus D^n} \\ \alpha \downarrow & & \downarrow g\alpha \\ \mathbb{R}^n & \xrightarrow{g} & \mathbb{R}^n \end{array}$$

where α and $g\alpha$ are PL and hence stable. Hence, the diagram reveals that g is stable if and only if Id is stable.

The end of $\widetilde{T^n \setminus D^n}$ obviously has a PL neighborhood V topologically homeomorphic to $S^{n-1} \times \mathbb{R}$. By attaching a PL collar and ball to V, using either Corollary 8.3.7 and the PL Poincaré Theorem for the $(n-1)$-sphere $(n > 5)$ or a result of (Wall, 1967) that a PL end topologically homeomorphic to $S^4 \times \mathbb{R}$ is actually PL homeomorphic to it, we see that $\widetilde{T^n \setminus D^n}$ can be extended to a PL structure on a PL manifold τ^n, which is a topological n-torus.

Let $2D^n$ denote a larger n-disk in T^n containing D^n. The topological Generalized Schönflies Theorem assures that $\mathrm{Id}|T^n \setminus 2D^n$ extends to a homeomorphism f of T^n onto τ^n. Since the extended homeomorphism is stable, so is $\mathrm{Id} : T^n \setminus 2D^n \to \widetilde{T^n \setminus 2D^n}$, which suffices.

Historical Notes. M. W. Hirsch (1961) showed the existence of smooth (and PL) immersions of the punctured n-torus in $\mathbb{R}^n$. Later Ferry (1974) gave an explicit formula for them.

With the techniques outlined here, Kirby (1969) proved that the group of homeomorphisms of $\mathbb{R}^n$ (with the compact-open topology) is locally contractible. This was generalized to compact manifolds and to interiors of compact ∂-manifolds (Edwards and Kirby, 1971). Černavskiĭ (1969c) had an independent proof of this same result. Applications of local contractibility have already occurred here, most notably in §7.3.

From the analysis of PL structures on the n-torus emerged the landmark *Foundational Essays on Topological Manifolds, Smoothings, and Triangulations* of (Kirby and Siebenmann, 1977). Included in that momentous work was the existence of manifolds with no PL structure, the PL Structure Theorem given as Theorem 6.8.2 here, the PL Product Structure Theorem given as Theorem 6.8.5 here, the result that concordant PL structures on a given manifold are PL equivalent in an isotopic sense, and the classification of PL or differentiable structures on a manifold in terms of a stable reduction of the topological tangent bundle to a PL or differentiable bundle. Applications of these fundamental codimension-zero results to codimension two and one appear in Chapters 6 and 7.

Exercises

8.8.1. Every stable homeomorphism $h : S^n \to S^n$ is isotopic to the identity.

8.8.2. Suppose $f, g : \mathbb{R}^n \to \mathbb{R}^n$ are homeomorphisms for which there exists $B > 0$ such that $d(f(x), g(x)) < B$ for all $x \in R^n$. Then f is stable if and only if g is stable.

8.9. Rigidity: Homotopy equivalence implies homeomorphism

A closed manifold N is said to be *topologically rigid* if every homotopy equivalence $f : M \to N$ from another manifold M is homotopic to a homeomorphism. In other words, N is topologically rigid if the homotopy structure set $\mathscr{S}(N)$, discussed in §8.7, consists of a single element. Spheres are familiar examples of topologically rigid manifolds; as mentioned in the preceding section, tori also are topologically rigid (for $n > 4$).

The Borel Conjecture posits that all closed aspherical manifolds are topologically rigid.

Riemannian manifolds provide fertile ground for testing the Borel Conjecture, since all complete Riemannian manifolds with non-positive sectional curvature are aspherical, by the Cartan-Hadamard Theorem. In support of the Borel Conjecture itself, L. Bieberbach (1912) proved that every homotopy equivalence between compact, flat Riemannian manifolds is homotopic to an affine diffeomorphism. G. D. Mostow (1968) showed that every homotopy equivalence between compact hyperbolic manifolds is homotopic to an isometry, and a few years later he proved that a homotopy equivalence between compact, non-positively curved locally symmetric spaces is homotopic to diffeomorphism, provided that the target has no 1- or 2-dimensional geodesic subspaces which locally are direct Cartesian factors.

The results of Bieberbach and Mostow offer evidence in favor of rigidity, without quite attaining that level, due to assumptions about Riemannian structure on both domain and range. F. T. Farrell, W. C. Hsiang and

L. E. Jones have been the leaders of prolonged, far-reaching efforts concerning rigidity. Their work involves a vast array of techniques, including methods from controlled topology, like those Quinn used in attacking the End Theorem, but extending across surgery theory, K-theory, and geometric analysis. Here is a far-reaching result, due to Farrell and Jones (1993), which encapsulates progress at a juncture a few years back.

Theorem 8.9.1. *All closed non-positively curved Riemannian n-manifolds, $n > 4$, are topologically rigid.*

For additional developments in this direction, consult the survey put together by C. W. Stark (2002).

8.10. Simplicial triangulations

Shortly after Kirby and Siebenmann established the existence of compact manifolds with no PL structure, the Double Suspension Problem was resolved in the affirmative. The problem asked whether there was a PL non-simply connected homology $(n-2)$-sphere Σ—namely, a $(n-2)$-manifold with the homology of S^{n-2}—whose double suspension is topologically S^n. The affirmative solution implied that even simple manifolds like spheres could admit non-PL simplicial triangulations. (A manifold M is said to admit a *simplicial triangulation* if it is homeomorphic to a simplicial complex.)

Example 8.10.1. *For $n \geq 5$, S^n admits a non-PL, simplicial triangulation.*

Proof. If Σ is a non-simply connected PL homology $(n-2)$-sphere such that $S^0 * S^0 * \Sigma = S^1 * \Sigma$ an n-manifold, then in the obvious triangulation T determined by the join structure $S^1 * \Sigma$, the link of any 1-simplex from the S^1 factor is the homology sphere Σ, which is not even topologically equivalent, let alone PL equivalent, to S^{n-2}. If a manifold, $S^0 * S^0 * \Sigma$ must be S^n, by Proposition 2.4.9. □

An example revealing how the double suspension of a homology sphere can be a manifold appears here in Exercise 7.11.2.

A result of Cannon (1979) assures that the double suspension of every PL homology $n-2$-sphere is S^n. This gave rise to a characterization of the manifolds within the class of simplicial complexes.

Theorem 8.10.2. *A simplicial complex K is the underlying set of a topological n-manifold if and only if, for each simplex $\sigma \in K, \mathrm{lk}(\sigma, K)$ has the homology of S^{n-k-1} and, for each vertex $v \in K, \mathrm{lk}(v, K)$ is simply connected.*

In this setting, $\mathrm{lk}(\sigma, K)$ must be a $(k-1)$-sphere for each codimension-k simplex $\sigma \in K$, $k \leq 3$, and thus $\mathrm{lk}(\sigma, K)$ must be a 3-manifold for each

4-simplex σ. The double suspension result implies that $|K \smallsetminus K^{(0)}|$ is an n-manifold. Simple connectedness of vertex links yields that the open star of each vertex is topologically equivalent to $\mathbb{R}^n$ (see Corollary 8.3.7; this was also known for $n = 4$ even before Perelman's work on the Poincaré Conjecture).

The existence of non-PL simplicial triangulations then caused people to wonder whether all manifolds would admit, at the very least, a simplicial triangulation.

Theorem 8.10.3. *Every topological n-manifold M, $n \geq 5$, admits a simplicial triangulation if and only if there exists a homology 3-sphere Σ^3 such that Σ^3 has Rochlin invariant 1 and $\Sigma^3 \# \Sigma^3$ bounds an acyclic 4-dimensional PL ∂-manifold.*

Whether such a homology 3-sphere Σ^3 exists remains unsettled. The connected sum mentioned in Theorem 8.10.3 is the oriented version. For each oriented homology 3-sphere Σ^3, unlike for $\Sigma^3 \# (-\Sigma^3)$, which bounds the acyclic ∂-manifold $(\Sigma^3 \smallsetminus \operatorname{Int} 3\text{-cell}) \times I$, the requirement that $\Sigma^3 \# \Sigma^3$ bounds an acyclic manifold has content. Each homology 3-sphere Σ^3 is the boundary of a smooth, compact ∂-manifold W with trivial second Stiefel-Whitney class $w_2(W)$. The signature of W—namely, the signature of the intersection form on $H^2(W)$—is divisible by 8, and a theorem of Rochlin (1952) shows that, modulo 16, this value does not depend on the choice of W. The *Rochlin invariant* of Σ then is the value of signature$(W)/8$, modulo 2.

Historical Notes. In unpublished work, Edwards produced the first example of a non-simply connected homology sphere whose double suspension is a manifold and, therefore, a topological sphere. He also showed, among other things, that the triple suspension of any homology n-sphere is S^{n+3}.

Theorem 8.10.3 is due to D. Galewski and R. Stern (1980) and, independently, T. Matumoto (1978).

Bibliography

Akbulut, S. 1977. *On 2-dimensional homology classes of 4-manifolds*, Math. Proc. Camb. Phil. Soc. **82**, 99–106.

______. 1978. *On representing homology classes of 4-manifolds*, Invent. Math. **49**, 193–198.

______. 1984. *A fake 4-manifold*, Four-manifold Theory (Durham, N.H., 1982), Contemp. Math., vol. 35, Amer. Math. Soc., Providence, R.I., pp. 75–141.

______. 1991a. *A solution to a conjecture of Zeeman*, Topology **30**, 513–515.

______. 1991b. *A fake compact contractible 4-manifold*, J. Differential Geom. **33**, 335–356.

Akbulut, S. and R. C. Kirby. 1979. *Mazur manifolds*, Michigan Math. J. **26**, 259–284.

Akin, E. 1969. *Manifold phenomena in the theory of polyhedra*, Trans. Amer. Math. Soc. **143**, 413–473.

Alexander, J. W. 1924a. *On the subdivision of 3-space by a polyhedron*, Proc. Nat. Acad. Sci. U.S.A. **10**, 6–8.

______. 1924b. *An example of a simply connected space bounding a region which is not simply connected*, Proc. Nat. Acad. Sci. U.S.A. **10**, 8–10.

______. 1924c. *Remarks on a point set constructed by Antoine*, Proc. Nat. Acad. Sci. U.S.A. **10**, 10–12.

______. 1928. *Topological invariants of knots and links*, Trans. Amer. Math. Soc. **30**, 275–306.

______. 1930. *The combinatorial theory of complexes*, Ann. of Math. (2) **31**, 292–320.

Alford, W. R. 1962. *Some "nice" wild 2-spheres in E^3*, Topology of 3-manifolds and Related Topics (M. K. Fort, Jr., ed.), Prentice-Hall, Englewood Cliffs, N.J., pp. 29–33.

Ancel, F. D. 1984. *Approximating cell-like maps of S^4 by homeomorphisms*, Four-manifold Theory (Durham, N.H., 1982), Contemp. Math., vol. 35, Amer. Math. Soc., Providence, R.I., pp. 143–164.

Ancel, F. D. and J. W. Cannon. 1979. *The locally flat approximation of cell-like embedding relations*, Ann. of Math. (2) **109**, 61–86.

Ancel, F. D. and C. R. Guilbault. 1995. *Compact contractible n-manifolds have arc spines* ($n \geq 5$), Pacific J. Math. **168**, 1–10.

Andrews, J. J. and M. L. Curtis. 1962. *n-space modulo an arc*, Ann. of Math. (2) **75**, 1–7.

Antoine, L. 1921. *Sur l'homeomorphie de deux figures et de leurs voisinages*, J. Math. Pures Appl. **86**, 221–325.

Artin, E. 1925. *Zur Isotopie zweidimensionaler Flächen im R^4*, Hamburgh Abh. **4**, 174–177.

Armentrout, S. A. 1971. *Cellular decompositions of 3-manifolds that yield 3-manifolds*, Memoirs Amer. Math. Soc., vol. 107, Amer. Math. Soc., Providence, R.I.

Barden, D. 1963. *The structure of manifolds*, Doctoral Dissertation, Cambridge University.

Barge, J. 1980. *An algebraic proof of a theorem of J. Milnor*, Topology Symposium, Siegen 1979 (Proc. Sympos., Univ. Siegen, Siegen, 1979), Lecture Notes in Math., vol. 788, Springer, Berlin, pp. 396–398.

Bass, H. 1968. *Algebraic K-theory*, W. A. Benjamin, Inc., New York-Amsterdam.

Bean, R. J. 1967. *Decompositions of E^3 with a null sequence of starlike equivalent nondegenerate elements are E^3*, Illinois J. Math **19**, 21–23.

Berkowitz, H. W. 1971. *A counterexample to a proof of Homma*, Yokohama Math. J. **19**, 1–3.

Berkowitz, H. W. and J. Dancis. 1970a. PL *approximations and embeddings of manifolds in the $\frac{3}{4}$-range*, Topology of Manifolds (Proc. Inst., Univ. of Georgia, Athens, Ga., 1969), Markham Publ. Co., Chicago, pp. 335–340.

———. 1970b. *PL approximations of embeddings and isotopies of polyhedra in the metastable range*, Topology of Manifolds (Proc. Inst., Univ. of Georgia, Athens, Ga., 1969), Markham Publ. Co., Chicago, pp. 341–352.

Bieberbach, L. 1912. *Über die Bewegungsgruppen der Euklidischen Räume II. Die Gruppen mit einem endlichen Fundamentalbereich*, Math. Ann. **72**, 400–412.

Bing, R. H. 1952. *A homeomorphism between the 3-sphere and the sum of two solid horned spheres*, Ann. of Math. (2) **56**, 354–362.

———. 1954. *Locally tame sets are tame*, Ann. of Math. (2) **59**, 145–158.

———. 1956. *A simple closed curve that pierces no disk*, J. Math. Pures Appl. (9) **35**, 337–343.

———. 1957a. *Upper semicontinuous decompositions of E^3*, Ann. of Math. (2) **65**, 363–374.

———. 1957b. *A decomposition of E^3 into points and tame arcs such that the decomposition space is topologically different from E^3*, Ann. of Math. (2) **65**, 484–500.

———. 1959a. *The Cartesian product of a certain nonmanifold and a line is E^4*, Ann. of Math. (2) **70**, 399–412.

———. 1959b. *Conditions under which a surface in E^3 is tame*, Fund. Math. **47**, 105–135.

———. 1961a. *A wild surface each of whose arcs is tame*, Duke Math. J. **28**, 1–15.

———. 1961b. *A surface is tame if its complement is 1-ULC*, Trans. Amer. Math. Soc. **101**, 294–305.

———. 1962a. *Pointlike decompositions of E^3*, Fund. Math. **50**, 431–453.

———. 1962b. *Each disk in E^3 contains a tame arc*, Amer. J. Math. **84**, 583–590.

———. 1963. *Stable homeomorphisms on E^5 can be approximated by piecewise linear ones*, Notices Amer. Math. Soc. **10**, 666, Abstract 607-16.

———. 1964. *Some aspects of the topology of 3-manifolds related to the Poincaré conjecture*, Lectures on Modern Mathematics, Vol. II, Wiley, New York, pp. 93–128.

———. 1968. *Radial engulfing*, Conference on Topology of Manifolds (Michigan State Univ., E. Lansing, Mich., 1967), Prindle, Weber & Schmidt, Boston, pp. 1–18.

———. 1975. *Vertical general position*, Geometric Topology (L. C. Glaser and T. B. Rushing, ed.), Lecture Notes in Math., vol. 438, Springer-Verlag, Berlin and New York, pp. 16–41.

———. 1983. *The Geometric Topology of 3-Manifolds*, Amer. Math. Soc. Colloq. Publ., vol. 40, Amer. Math. Soc., Providence, R.I.

Bing, R. H. and J. M. Kister. 1964. *Taming complexes in hyperplanes*, Duke Math. J. **31**, 491–511.

Blanchfield, R. C. 1957. *Intersection theory of manifolds with operators with applications to knot theory*, Ann. of Math. (2) **65**, 340–356.

Blankinship, W. A. 1951. *Generalization of a construction of Antoine*, Ann. of Math. (2) **53**, 276–291.

Bothe, H. 1964. *Ein eindimensionales Kompaktum in E^3, die sich nicht lagetrau in die Mengersche Universalkurve einbetten lässt*, Fund. Math. **54**, 251–258.

Boyd, W. S. and A. H. Wright. 1973. *A 1-ALG simple closed curve in E^3 is tame*, Canad. J. Math. **25**, 646–656.

Browder, W. 1965. *Structures on $M \times R$*, Proc. Cambridge Philos. Soc. **61**, 337–345.

Browder, W., J. Levine, and G. R. Livesay. 1965. *Finding a boundary for an open manifold*, Amer. J. Math. **87**, 1017–1028.

Brown, M. 1960. *A proof of the generalized Schoenflies theorem*, Bull. Amer. Math. Soc. **66**, 74–76.

______. 1961. *The monotone union of open n-cells is an open n-cell*, Proc. Amer. Math. Soc. **12**, 812–814.

______. 1962. *Locally flat imbeddings of topological manifolds*, Ann. of Math. (2) **75**, 331–341.

______. 1967. *Wild cells and spheres in high dimensions*, Michigan Math. J. **14**, 219–224.

Brown, M. and H. Gluck. 1964. *Stable structures on manifolds: I. Homeomorphisms of S^n*, Ann. of Math. (2) **79**, 1–17.

Bryant, J. L. 1966a. *Taming polyhedra in the trivial range*, Topology Seminar (Wisconsin, 1965), Ann. of Math. Studies, No. 60, Princeton Univ. Press, Princeton, N.J., pp. 153–158.

______. 1966b. *Taming polyhedra in the trivial range*, Michigan Math. J. **13**, 377–384.

______. 1968. *Euclidean space modulo a cell*, Fund. Math. **63**, 43–51.

______. 1969. *On embeddings of compacta in Euclidean space*, Proc. Amer. Math. Soc. **23**, 46–51.

______. 1971a. *On embeddings of 1-dimensional compacta in E^5*, Duke Math. J. **38**, 265–270.

______. 1971b. *On embeddings with locally nice cross-sections*, Trans. Amer. Math. Soc. **155**, 327–332.

______. 1972. *Approximating embeddings of polyhedra in codimension three*, Trans. Amer. Math. Soc. **170**, 85–95.

______. 1973. *Euclidean n-space modulo an $(n-1)$-cell*, Trans. Amer. Math. Soc. **179**, 181–192.

______. 1986. *General position theorems for generalized manifolds*, Proc. Amer. Math. Soc. **98**, 667–670.

______. 1990. *Triangulation and general position of PL diagrams*, Topology Appl. **34**, 211–233.

______. 2002. *Piecewise linear topology*, Handbook of Geometric Topology, North-Holland, Amsterdam, pp. 219–259.

Bryant, J. L., S. Ferry, W. Mio, and S. Weinberger. 1996. *Topology of homology manifolds*, Ann. of Math. (2) **143**, 435–467.

______. 2007. *Desingularizing homology manifolds*, Geom. Topol. **11**, 1289–1314.

Bryant, J. L. and R. C. Lacher. 1975. *Embeddings with mapping cylinder neighborhoods*, Topology **14**, 191–201.

______. 1977. *Blowing up homology manifolds*, J. London Math. Soc. (2) **16**, no. 2, 372–376.

Bryant, J. L., R. C. Lacher, and B. J. Smith. 1975. *Free spheres with mapping cylinder neighborhoods*, Geometric Topology (L. C. Glaser and T. B. Rushing, ed.), Lecture Notes in Math., vol. 438, Springer-Verlag, Berlin and New York, pp. 58–65.

Bryant, J. L. and W. Mio. 1999. *Embeddings of homology manifolds in codimensions ≥ 3*, Topology **38**, 811–821.

______. 2000. *Embeddings in generalized manifolds*, Trans. Amer. Math. Soc. **352**, 1131–1147.

Bryant, J. L. and C. L. Seebeck III. 1968a. *Locally nice embeddings in codimension three*, Bull. Amer. Math. Soc. **74**, 378–380.

———. 1968b. *Locally nice embeddings of polyhedra*, Quart. J. Math. Oxford Ser. (2) **19**, 257–274.

———. 1968c. *Locally nice embeddings of polyhedra*, Conference on the Topology of Manifolds (Michigan State Univ., E. Lansing, Mich., 1967), Prindle, Weber & Schmidt, Boston, Mass., pp. 23–28.

———. 1969. *An equivalence theorem for embeddings of compact absolute neighborhood retracts*, Proc. Amer. Math. Soc. **20**, 256–258.

———. 1970. *Locally nice embeddings in codimension three*, Quart J. Math. Oxford (2) **21**, 265–272.

Burde, G. and H. Zieschang. 1985. *Knots*, Walter de Gruyter, Berlin and New York.

Burgess, C. E. 1965. *Characterizations of tame surfaces in E^3*, Trans. Amer. Math. Soc. **114**, 80–97.

Burgess, C. E. and J. W. Cannon. 1971. *Embeddings of surfaces in E^3*, Rocky Mountain J. Math. **1**, 259–344.

Cannon, J. W. 1973a. *ULC properties in neighborhoods of embedded surfaces and curves in E^3*, Canad. J. Math. **25**, 31–73.

———. 1973b. *A positional characterization of the (n-1)-dimensional Sierpiński curve in S^n*, Fund. Math. **79**, 107–112.

———. 1975. *Taming cell-like embedding relations*, Geometric Topology (L. C. Glaser and T. B. Rushing, ed.), Lecture Notes in Math. vol. 438, Springer-Verlag, Berlin and New York, pp. 66–118.

———. 1978. *The recognition problem: what is a topological manifold?*, Bull. Amer. Math. Soc. **84**, 832–866.

———. 1979. *Shrinking cell-like decompositions of manifolds. Codimension three*, Ann. of Math **110**, 83–112.

Cannon, J. W., J. L. Bryant, and R. C. Lacher. 1979. *The structure of generalized manifolds having nonmanifold set of trivial dimension*, Geometric Topology, Academic Press, New York. edited by J. C. Cantrell, pp. 261–300.

Cannon, J. W. and G. Conner. 2000. *The combinatorial structure of the Hawaiian earring group*, Topology Appl. **106**, 225–271.

Cannon, J. W. and R. J. Daverman. 1981. *Cell-like decompositions arising from mismatched sewings: applications to 4-manifolds*, Fund. Math. **111**, 211–233.

Cantrell, J. C. 1963a. *Almost locally flat embeddings of S^{n-1} in S^n*, Bull. Amer. Math. Soc. **69**, 716–718.

———. 1963b. *Separation of the n-sphere by an $(n-1)$-sphere*, Trans. Amer. Math. Soc. **108**, 185–194.

———. 1964. *n-frames in euclidean k-space*, Proc. Amer. Math. Soc. **15**, 574–578.

Cantrell, J. C. and C. H. Edwards Jr. 1963. *Almost locally polyhedral curves in Euclidean n-space*, Trans. Amer. Math. Soc. **107**, 451–457.

Cappell, S. and J. Shaneson. 1974. *The codimension two placement problem and homology equivalent manifolds*, Ann. of Math. (2) **99**, 277–348.

———. 1976. *Piecewise linear embeddings and their singularities*, Ann. of Math. (2) **103**, 163–228.

———. 1977. *Totally spineless manifolds*, Illinois J. Math. **21**, 231–239.

Černavskiĭ, A. V. 1965. *On topological imbeddings of polyhedra in euclidean spaces*, Dokl. Akad. Nauk SSSR **165**, 1257–1260 (Russian). Also in Soviet Math. Dokl. **6** (1965), 1606–1610.

———. 1967. *On singular points of topological imbeddings of manifolds and the union of locally flat cells*, Dokl. Akad. Nauk SSSR **165**, 528–530 (Russian). Also in Soviet Math. Dokl. **7** (1966), 433–436.

______. 1969a. *Piecewise linear approximations of embeddings of cells and spheres in codimensions higher than two*, Mat. Sb. (N.S.) **80**, 339–364 (Russian). Also in Math. USSR-Sb. **9** (1969), 321–343.

______. 1969b. *Topological embeddings of manifolds*, Dokl. Akad. Nauk SSSR **187**, 1247–1250. Also in Soviet Math. Dokl. **10** (1969), 1037–1040.

______. 1969c. *Local contractibility of the group of homeomorphisms of a manifold*, Mat. Sb. (N.S.) **79**, 307–356 (Russian). Also in Math. USSR-Sb. **8** (1969), 288–333.

______. 1970. *Piece-wise linear approximation of embeddings of manifolds in codimensions greater than two. A supplement to the article "Piece-wise linear approximation of embeddings of cells and spheres in codimensions greater than two"*, Mat. Sb. (N.S.) **82 (124)**, 499–500 (Russian). Also in Math. USSR-Sb. **11** (1970), 465–466.

______. 1973. *Coincidence of local flatness and local simple-connectedness for imbeddings of (n-1)-dimensional manifolds into n-dimensional manifolds when* $n > 4$, Mat. Sb. (N.S.) **93 (133)**, 279–286 (Russian). Also in Math. USSR-Sb. **20** (1973), 297–304.

______. 2006. *Theorem on the union of two topologically flat cells of codimension 1 in* $\mathbb{R}^n$, Abstr. Appl. Anal., 1–9. Art. ID 84602.

Chapman, T. A. 1979. *Locally homotopically unknotted embeddings of manifolds in codimension two are locally flat*, Topology **18**, 339–348.

______. 1981. *Approximation results in topological manifolds*, Memoirs Amer. Math. Soc., vol. 251, Amer. Math. Soc., Providence, R.I.

Chapman, T. A. and S. Ferry. 1979. *Approximating homotopy equivalences by homeomorphisms*, Amer. J. Math. **101**, 583–607.

Cobb, J. I. 1968. *Taming almost PL embeddings*, Notices Amer. Math. Soc. **15**, 371. Abstract #68T-241.

Cohen, M. M. 1967. *Simplicial structures and transverse cellularity*, Ann. of Math. (2) **85**, 218–245.

______. 1969. *A general theory of relative regular neighborhoods*, Trans. Amer. Math. Soc. **136**, 189–229.

______. 1973. *A course in simple-homotopy theory*, Graduate Texts in Math., vol. 10, Springer-Verlag, New York.

Connell, E. H. 1963. *Approximating stable homeomorphisms by piecewise linear ones*, Ann. of Math. (2) **78**, 326–338.

______. 1967. *A topological h-cobordism theorem*, Illinois J. Math. **11**, 300–309.

Connelly, R. 1970a. *Unknotting close polyhedra in codimension three*, Topology of Manifolds (Proc. Inst., Univ. of Georgia, Athens, Ga., 1969), Markham Publishing Co., Chicago, pp. 384–388.

______. 1970b. *Unknotting close embeddings of polyhedra in codimension greater than three*, Ph.D. dissertation, University of Michigan.

______. 1971. *A new proof of Brown's collaring theorem*, Proc. Amer. Math. Soc. **27**, 180–182.

Crowell, R. H. and R. H. Fox. 1977. *Introduction to Knot Theory*, Graduate Texts in Math., vol. 57, Springer-Verlag, New York, Heidelberg, Berlin.

Curtis, C. L., M. H. Freedman, W. C. Hsiang, and R. Stong. 1996. *A decomposition theorem for h-cobordant smooth, simply-connected compact 4-manifolds*, Invent. Math. **123**, 343–348.

Dancis, J. 1966. *Some nice embeddings in the trivial range*, Topology Seminar (Wisconsin, 1965), Ann. of Math. Studies, No. 60, Princeton Univ. Press, Princeton, N.J., pp. 159–170.

______. 1970. *An introduction to embedding spaces for geometrical topologists*, Topology of Manifolds (Proc. Inst., Univ. of Georgia, Athens, Ga., 1969), Markham Publ. Co., Chicago, pp. 389–393.

Daverman, R. J. 1973a. *Factored codimension one cells in Euclidean n-space*, Pacific J. Math. **46**, 37–43.

———. 1973b. *Locally nice codimension one manifolds are locally flat*, Bull. Amer. Math. Soc. **79**, 410–413.

———. 1973c. *On the scarcity of tame disks in certain wild cells*, Fund. Math. **79**, 63–77.

———. 1975. *On the absence of tame disks in certain wild cells*, Geometric Topology (L. C. Glaser and T. B. Rushing, ed.), Lecture Notes in Math. vol. 438, Springer-Verlag, Berlin, pp. 142–155.

———. 1976. *Singular regular neighborhoods and local flatness in codimension one*, Proc. Amer. Math. Soc. **57**, 357–362.

———. 1977. *Every crumpled n-cube is a closed n-cell complement*, Michigan Math. J. **24**, 225–241.

———. 1978. *Embeddings of $(n-1)$-spheres in Euclidean n-space*, Bull. Amer. Math. Soc. **84**, 377–405.

———. 1979. *Embedding phenomena based upon decomposition theory: wild Cantor sets satisfying strong homogeneity properties*, Proc. Amer. Math. Soc. **75**, 177–182.

———. 1981. *Infinite inflations of crumpled cubes*, Topology Appl. **12**, 35–43.

———. 1986. *Decompositions of Manifolds*, Academic Press, Orlando, Florida.

———. 1987. *Every crumpled 4-cube is a closed 4-cell complement*, Topology Appl. **26**, 107–113.

———. 2007. *Sewings of crumpled cubes revisited*, Topology Appl. **155**, 105–111.

Daverman, R. J. and W. T. Eaton. 1969. *A dense set of sewings of two crumpled cubes yields S^3*, Fund. Math. **59**, 51–60.

Daverman, R. J. and D. Halverson. 2007. *The cell-like approximation theorem: dimension 5*, Fund. Math. **197**, 81–121.

Daverman, R. J. and F. C. Tinsley. 1986. *Laminations, finitely generated perfect groups, and acyclic mappings*, Michigan Math. J. **33**, 343–351.

———. 1994. *A controlled plus construction for crumpled laminations*, Trans. Amer. Math. Soc. **342**, 807–826.

Daverman, R. J. and J. J. Walsh. 1982. *A nonshrinkable decomposition of S^n involving a null sequence of cellular arcs*, Trans. Amer. Math. Soc. **272**, 771–784.

Davis, J. F. and P. Kirk. 2001. *Lecture Notes in Algebraic Topology*, Graduate Studies in Math., vol. 35, Amer. Math. Soc., Providence, R.I.

Davis, M. W. 1983. *Groups generated by reflections and aspherical manifolds not covered by Euclidean space*, Ann. of Math. (2) **117**, 293–394.

Doyle, P. H. 1960. *Unions of cell pairs in E^3*, Pacific. J. Math. **10**, 521–524.

Doyle, P. H. and J. G. Hocking. 1960. *Some results on tame disks and spheres in E^3*, Proc. Amer. Math. Soc. **11**, 832–836.

Dranishnikov, A. N. 1989. *On a problem of P. S. Alexandrov*, Math. USSR-Sb. **63**, 539–545.

Duvall, P. and L. S. Husch. 1980. *A continuum of dimension n which does not embed up to shape in $2n$-space*, Proc. Internat. Conf. on Geometric Topology, Polish Scientific Publ., Warsaw, pp. 113–119.

Dydak, J. and J. J. Walsh. 1993. *Infinite dimensional compacta with cohomological dimension two: an application of the Sullivan Conjecture*, Topology **32**, 93–104.

Eaton, W. T. 1972. *The sum of solid spheres*, Michigan Math. J. **19**, 193–207.

———. 1977. *Notes for the University of Texas topology seminar*. Unpublished manuscript, January, 1977.

Eaton, W. T., C. P. Pixley, and G. A. Venema. 1977. *A topological embedding which cannot be approximated by PL embeddings*, Notices Amer. Math. Soc. **24**, A-302.

Edgar, G. A. 1990. *Measure, Topology, and Fractal Geometry*, Undergraduate Texts in Mathematics, Springer-Verlag, New York.

Edwards, R. D. 1975a. *The equivalence of close piecewise linear embeddings*, Gen. Topology and its Appl. **5**, 147–180.

———. 1975b. *Demension Theory, I*, Geometric Topology (L. C. Glaser and T. B. Rushing, Jr., ed.), Lecture Notes in Math., vol. 438, Springer-Verlag, Berlin, pp. 195–211.

———. 1980. *The topology of manifolds and cell-like maps*, Proc. Internat. Congr. Mathematicians, Helsinki, 1978, Acad. Sci. Fenn., Helsinki, pp. 111–127.

———. 1986. *Approximating codimension* ≥ 3 σ*-compacta with locally homotopically unknotted embeddings*, Topology and its Appl. **24**, 95–122.

———. 1982. *Homotopically tame implies geometrically tame*. Unpublished notes.

Edwards, R. D. and R. C. Kirby. 1971. *Deformations of spaces of embeddings*, Ann. of Math. (2) **83**, 63–88.

Epstein, D. B. A. 1966. *The degree of a map*, Proc. London Math. Soc. **16**, 369–383.

Engelking, R. 1978. *Dimension Theory*, North-Holland Publishing Co., Amsterdam.

Farrell, F. T. and L. E. Jones. 1993. *Topological rigidity for compact nonpositively curved manifolds*, Differential Geometry, Proc. Symp. Pure. Math., Vol. 53, Part 3, Amer. Math. Soc., Providence, R.I., pp. 229–274.

Federer, H. 1969. *Geometric measure theory*, Die Grundlehren der mathematischen Wissenschaften, Band 153, Springer-Verlag, New York.

Fenn, R. 1971. *On Dehn's lemma in 4 dimensions*, Bull. London Math. Soc. **3**, 79–81.

———. 1983. *Techniques of Geometric Topology*, London Math. Soc. Lecture Note Series, vol. 57, Cambridge Univ. Press, Cambridge.

Ferry, S. 1974. *An immersion of* $T^n \setminus D^n$ *into* $\mathbb{R}^n$, Enseignement Math. (2) **20**, 177–178.

———. 1979. *Homotoping* ϵ*-maps to homeomorphisms*, Amer. J. Math. **101**, 567–582.

———. 1992. *On the Ancel-Cannon Theorem*, Topology Proc. **17**, 41–58.

Ferry, S. and E. Pedersen. 1991. *Some mildly wild circles in* S^n *arising from algebraic* K*-theory*, K-Theory **4**, 479–499.

Flores, A. 1934. *Über n-dimensionale Komplexe, die im* E^{2n+1} *absolut Selbstverschlungen sind*, Ergeb. Math. Colloq. **6**, 4–7.

Fox, R. H. 1950. *On the total curvature of some tame knots*, Ann. of Math. (2) **52**, 258–260.

———. 1962. *A quick trip through knot theory*, Topology of 3-manifolds and Related Topics (M. K. Fort, Jr., ed.), Prentice-Hall, Englewood Cliffs, N.J., pp. 120–167.

Fox, R. H. and E. Artin. 1948. *Some wild cells and spheres in three-dimensional space*, Ann. of Math. (2) **49**, 978–990.

Fox, R. H. and O. G. Harrold. 1962. *The Wilder arcs*, Topology of 3-manifolds and Related Topics (M. K. Fort, Jr., ed.), Prentice-Hall, Englewood Cliffs, N.J., pp. 184–187.

Freedman, M. H. 1982. *The topology of four-dimensional manifolds*, J. Differential Geom. **17**, 352–453.

Freedman, M. H. and F. Quinn. 1990. *Topology of 4-manifolds*, Princeton Univ. Press, Princeton, N.J.

Galewski, D. E. and R. J. Stern. 1980. *Classification of simplicial triangulations of topological manifolds*, Ann. of Math. (2) **111**, 1–34.

Gersten, S. M. 1966. *A product formula for Wall's obstruction*, Amer. J. Math. **88**, 337–346.

Giffen, C. H. 1977. *Disciplining dunce hats in 4-manifolds*. Unpublished manuscript.

Gillman, D. 1964. *Note concerning a wild sphere of Bing*, Duke Math. J. **31**, 247–254.

———. 1967. *The spinning and twisting of a complex in a hyperplane*, Ann. of Math. (2) **85**, 32–41.

Gluck, H. 1963. *Embeddings in the trivial range*, Bull. Amer. Math. Soc. **69**, 824–831.

Gompf, R. E. 1985. *An infinite set of exotic* R^4*'s*, J. Differential Geom. **21**, 283–300.

Gompf, R. E. and A. I. Stipsicz. 1999. *4-manifolds and Kirby Calculus*, Graduate Studies in Math., vol. 20, Amer. Math. Soc., Providence, R.I.

Greathouse, C. 1963. *Locally flat, locally tame, and tame embeddings*, Bull. Amer. Math. Soc. **69**, 820–823.

Grünbaum, B. 1970. *Imbeddings of simplicial complexes*, Comment. Math. Helv. **44**, 502–513.

———. 2003. *Convex Polytopes*, Second Edition, Graduate Texts in Math., vol. 221, Springer-Verlag, New York and Berlin.

Gugenheim, V. K. A. M. 1953. *Piecewise linear isotopy and embeddings of elements and spheres (I)*, Proc. London Math. Soc. **3**, 29–53.

Guilbault, C. R. 1992. *An open collar theorem for 4-manifolds*, Trans. Amer. Math. Soc. **331**, no. 1, 227–245.

———. 1995. *Some compact contractible manifolds containing disjoint spines*, Topology **34**, 99–108.

Haefliger, A. 1962. *Knotted $(4k-1)$-spheres in $6k$-space*, Ann. of Math. (2) **75**, 452–466.

Hatcher, A. 2002. *Algebraic Topology*, Cambridge Univ. Press, Cambridge.

Hausdorff, F. 1919. *Dimension und äußeres Maß*, Math. Annalen **79**, 157–179.

———. 1944. *Mengenlehre*, Dover Publications, New York.

Haver, W. E. 1975. *Mappings between ANRs that are fine homotopy equivalences*, Pacific J. Math. **58**, 457–461.

Hempel, J. 1964. *A surface in S^3 is tame if it can be deformed into each complementary domain*, Trans. Amer. Math. Soc. **111**, 273–287.

Hilton, P. J. and U. Stammbach. 1971. *A course in homological algebra*, Graduate Texts in Math., vol. 4, Springer-Verlag, New York and Berlin.

Hirsch, M. W. 1961. *On embedding differentiable manifolds in euclidean space*, Ann. of Math. (2) **73**, 556–571.

———. 1966a. *On tubular neighbourhoods of manifolds I*, Proc. Cambridge Philos. Soc. **62**, 177–181.

———. 1966b. *On tubular neighbourhoods of manifolds II*, Proc. Cambridge Philos. Soc. **62**, 183–185.

———. 1968. *On tubular neighborhoods of piecewise linear and topological manifolds*, Conference on the Topology of Manifolds (Michigan State Univ., E. Lansing, Mich., 1967), Prindle, Weber & Schmidt, Boston, Mass., pp. 63–80.

Hirschhorn, P. S. and J. G. Ratcliffe. 1980. *A simple proof of the algebraic unknotting of spheres in codimension two*, Amer. J. Math. **102**, 489–491.

Hocking, J. G. and G. S. Young. 1988. *Topology*, 2nd ed., Dover Publications Inc., New York.

Homma, T. 1962. *On the embedding of polyhedra in manifolds*, Yokohama Math. J. **10**, 5–10.

———. 1966. *A theorem of piecewise linear approximation*, Yokohama Math. J. **14**, 47–54.

———. 1968. *A theorem of piecewise linear approximation, II*, Yokohama Math. J. **16**, 107–124.

Hosay, N. 1963. *The sum of a real cube and a crumpled cube is S^3*, Notices Amer. Math. Soc. **9**, 666.

Hosokawa, F. 1958. *On ∇-polynomials of links*, Osaka Math. J. **10**, 273–282.

Hsiang, W. C. and J. Shaneson. 1970. *Fake tori*, Topology of Manifolds (Proc. Inst., Univ. of Georgia, Athens, Ga., 1969), Markham Publ. Co., Chicago, pp. 18–51.

Hsiang, W. C. and C. T. C. Wall. 1969. *On homotopy tori II*, Bull. London Math. Soc. **1**, 341–342.

Hu, S.-T. 1959. *Homotopy Theory*, Pure and Applied Mathematics, vol. 8, Academic Press, New York and London.

———. 1965. *Theory of Retracts*, Wayne State University Press, Detroit.

Hudson, J. F. P. 1966a. *Concordance and isotopy of PL embeddings*, Bull. Amer. Math. Soc. **72**, 534–535.

______. 1966b. *Piecewise linear embeddings and isotopies*, Bull. Amer. Math. Soc. **72**, 536–537.

______. 1967. *Piecewise linear embeddings*, Ann. of Math. (2) **85**, 1–31.

______. 1969a. *Piecewise Linear Topology*, Mathematical Lecture Notes Series, W. A. Benjamin, New York.

______. 1969b. *On sunny collapsing*, Proc. Cambridge Philos. Soc. **66**, 247–250.

______. 1970. *Concordance, isotopy, and diffeotopy*, Ann. of Math. (2) **91**, 425–448.

Hungerford, T. W. 1974. *Algebra*, Graduate Texts in Math., vol. 73, Springer-Verlag, New York, Heidelberg, Berlin.

Hurewicz, W. and H. Wallman. 1948. *Dimension Theory*, Princeton Univ. Press, Princeton, N.J.

Husch, L. S. and I. Ivanšić. 1981. *Embeddings of compacta up to shape*, Shape Theory and Geometric Topology (S. Mardešić and J. Segal, ed.), Lecture Notes in Math., vol. 870, Springer-Verlag, Berlin, pp. 119–134.

Husch, L. S. and T. M. Price. 1970. *Finding a boundary for a 3-manifold*, Ann. of Math. (2) **91**, 223–235.

Irwin, M. C. 1962. *Combinatorial embeddings of manifolds*, Bull. Amer. Math. Soc. **68**, 25–27.

______. 1965. *Embeddings of polyhedral manifolds*, Ann. of Math. (2) **82**, 1–14.

______. 1966. *Polyhedral immersions*, Proc. Cambridge Philos. Soc. **62**, 45–50.

Ivanšić, I., R. B. Sher, and G. A. Venema. 1981. *Complement theorems beyond the trivial range*, Illinois J. Math. **25**, 209–220.

Johnston, H. 1999. *Transversality for homology manifolds*, Topology **38**, 673–697.

Kato, M. 1970. *Embedding spheres and balls in codimension* ≤ 2, Invent. Math. **10**, 89–107.

Kato, M. and Y. Matsumoto. 1972. *Simply connected surgery of submanifolds in codimension two. I*, J. Math. Soc. Japan **24**, 586–608.

Kawauchi, A. 1974a. *On partial Poincaré duality and higher-dimensional knots with* $\pi_1 = \mathbb{Z}$, Masters Thesis, Kobe University.

______. 1974b. *A partial Poincaré duality theorem for infinite cyclic covers*, Quart. J. Math. Oxford Ser. (2) **26**, 437–458.

______. 1975. *Three dimensional homology handles and circles*, Osaka J. Math. **12**, 565–581.

______. 1977. *On quadratic forms of 3-manifolds*, Invent. Math. **43**, 177–198.

______. 1978. *On the Alexander polynomials of cobordant links*, Osaka J. Math. **15**, 151–159.

______. 1980. *On a 4-manifold homology equivalent to a bouquet of surfaces*, Trans. Amer. Math. Soc. **262**, 95–112.

______. 1996. *A Survey of Knot Theory*, Birkhäuser, Basel.

Keesling, J. E. 1986. *Hausdorff dimension*, Topology Proc. **11**, 349–383.

Kervaire, M. A. 1962. *La théorème de Barden-Mazur-Stallings*, Comment. Math. Helv. **40**, 31–32.

______. 1965a. *On higher dimensional knots*, Differential and Combinatorial Topology (S. S. Cairns, ed.), Princeton Univ. Press, Princeton, N.J., pp. 105–119.

______. 1965b. *Les nœuds de dimensions supérieures*, Bull. Soc. Math. France **93**, 225–271.

______. 1969. *Smooth homology spheres and their fundamental groups*, Trans. Amer. Math. Soc. **144**, 67–72.

Kervaire, M. A. and J. W. Milnor. 1961. *On 2-spheres in 4-manifolds*, Proc. Nat. Acad. Sci. **47**, 1651–1657.

Kervaire, M. A. and C. Weber. 1978. *A survey of multidimensional knots*, Knot Theory (J.-C. Hausmann, ed.), Springer-Verlag, New York, pp. 61–143.

Kirby, R. C. 1968a. *On the set of nonlocally flat points of a submanifold of codimension one*, Ann. of Math. (2) **88**, 281–290.

———. 1968b. *The union of flat $(n-1)$-balls is flat in $\mathbb{R}^n$*, Bull. Amer. Math. Soc. **74**, 614–617.

———. 1969. *Stable homeomorphisms and the annulus conjecture*, Ann. of Math. (2) **89**, 575–586.

———. 1970. *Codimension-two locally flat embeddings have normal bundles*, Topology of Manifolds (Proc. Inst., Univ. of Georgia, Athens, Ga., 1969), Markham Publishing Co., Chicago, pp. 416–423.

———. 1978. *A calculus for framed links in $\mathbb{S}^3$*, Invent. Math. **45**, 35–56.

———. 1989. *The Topology of 4-manifolds*, Lecture Notes in Mathematics, vol. 1374, Springer-Verlag, Berlin and New York.

———. 1997. *Problems in low-dimensional topology*, Geometric Topology, Part II (William H. Kazez, ed.), AMS/IP Studies in Advanced Mathematics, Volume 2, Amer. Math. Soc. and International Press, Providence, R.I., pp. 35–473.

Kirby, R. C. and L. C. Siebenmann. 1969. *On the triangulation of manifolds and the Hauptvermutung*, Bull. Amer. Math. Soc. **75**, 742–749.

———. 1971. *Some theorems on topological manifolds*, Manifolds–Amsterdam 1970 (Proc. Nuffic Summer School), Springer-Verlag, Berlin.

———. 1975. *Normal bundles for codimension 2 locally flat imbeddings*, Geometric Topology (L. C. Glaser and T. B. Rushing, ed.), Lecture Notes in Mathematics, vol. 438, Springer-Verlag, Berlin and New York, pp. 310–324.

———. 1977. *Foundational Essays on Topological Manifolds, Smoothings, and Triangulations*, Ann. of Math. Studies, vol. 88, Princeton Univ. Press, Princeton, N.J.

Kister, J. M. 1963. *Microbundles are fibre bundles*, Bull. Amer. Math. Soc. **69**, 854–857.

———. 1964. *Microbundles are fibre bundles*, Ann. of Math. (2) **80**, 190–199.

Klee, V. 1955. *Some topological properties of convex sets*, Trans. Amer. Math. Soc. **78**, 30–45.

Kojima, S. 1983. *Milnor's $\bar{\mu}$-invariants, Massey products, and Whitney's trick in 4 dimensions*, Topology Appl. **16**, 43–60.

Kozlowski, G. *Images of ANRs.* Famous unpublished manuscript.

Kozlowski, G. and J. J. Walsh. 1983. *Cell-like mappings on 3-manifolds*, Topology **22**, 147–151.

Kuga, K. 1984. *Representing homology classes of $S^2 \times S^2$*, Topology **23**, 133–137.

Lacher, R. C. 1969. *Cell-like spaces*, Proc. Amer. Math. Soc. **20**, 598–602.

———. 1977. *Cell-like mappings and their generalizations*, Bull. Amer. Math. Soc. **83**, no. 4, 495–552.

Lacher, R. C. and A. H. Wright. 1970. *Mapping cylinders and 4-manifolds*, Topology of Manifolds (Proc. Inst., Univ. of Georgia, Athens, Ga., 1969), Markham Publ. Co., Chicago, pp. 424–427.

Lefschetz, S. 1931. *On compact spaces*, Ann. of Math. (2) **32**, 521–538.

Levine, J. 1965. *Unknotting spheres in codimension 2*, Topology **4**, 9–16.

———. 1966. *Polynomial invariants of knots in codimension two*, Ann. of Math. (2) **84**, 537–554.

———. 1969. *Knot cobordism groups in codimension two*, Comment. Math. Helv. **44**, 229–244.

Lickorish, W. B. R. 1965. *The piecewise linear unknotting of cones*, Topology **4**, 67–91.

———. 1997. *An Introduction to Knot Theory*, Graduate Texts in Math., vol. 175, Springer-Verlag, New York, Heidelberg, Berlin.

———. 2003. *Knotted contractible 4-manifolds in S^4*, Pacific J. Math. **208**, 283–290.

Lickorish, W. B. R. and L. C. Siebenmann. 1969. *Regular neighbourhoods and the stable range*, Trans. Amer. Math. Soc. **139**, 207–230.

Liem, V. T., Y. Matsumoto, and G. A. Venema. 1998. *A homotopy equivalence that is not homotopic to an embedding*, Topology Appl. **90**, 211–222.

Liem, V. T. and G. A. Venema. 1991. *Characterization of knot complements in the 4-sphere*, Topology Appl. **42**, 231–245.

______. 1993. *On the asphericity of knot complements*, Can. J. Math. **45**, 340–356.

Lininger, L. L. 1965. *Some results on crumpled cubes*, Trans. Amer. Math. Soc. **118**, 534–549.

Luukkainen, J. and J. Väisälä. 1977. *Elements of Lipschitz topology*, Ann. Acad. Sci. Fenn. Ser AI **3**, 85–122.

Mandelbrot, B. B. 1982. *The Fractal Geometry of Nature*, W. H. Freeman and Company, New York.

Mardešić, S. and J. Segal. 1982. *Shape Theory*, North-Holland Mathematical Library, vol. 26, North-Holland Publishing Co., Amsterdam, NL.

Massey, W. S. 1967. *Algebraic Topology: An Introduction*, Harcourt, Brace & World, Inc., New York.

______. 1977. *Algebraic Topology: An Introduction*, Graduate Texts in Math., Vol. 56, Springer-Verlag, New York. Reprint of the 1967 edition.

______. 1991. *A Basic Course in Algebraic Topology*, Graduate Texts in Math., vol. 127, Springer-Verlag, New York.

Mather, M. 1965. *Counting homotopy types of manifolds*, Topology **4**, 93–94.

Matsumoto, Y. 1973. *Knot cobordism groups and surgery in codimension two*, J. Faculty Sci. Univ. Tokyo **20**, 253–317.

Matoušek, J. 2003. *Using the Borsuk-Ulam Theorem*, Universitext, Springer-Verlag, Berlin and New York.

Matsumoto, Y. 1975a. *A 4-manifold which admits no spine*, Bull. Amer. Math. Soc. **81**, 467–470.

______. 1975b. *Some counterexamples in the theory of embedding manifolds in codimension two*, Sci. Papers College Gen. Ed. Univ. Tokyo **25**, 49–57.

______. 1979a. *An extension of Kawauchi's results to topological embeddings.* Handwritten notes (unpublished).

______. 1979b. *Wild embeddings of piecewise linear manifolds in codimension two*, Geometric Topology (J. C. Cantrell, ed.), Academic Press, New York, pp. 393–428.

Matsumoto, Y. and G. A. Venema. 1979. *Failure of the Dehn lemma on contractible 4-manifolds*, Invent. Math. **51**, 205–218.

Matumoto, T. 1978. *Triangulation of manifolds*, Algebraic and Geometric Topology (Proc. Sympos. Pure Math., Stanford Univ., Stanford, CA, 1976), Part 2, Amer. Math. Soc., Providence, R.I., pp. 3–6.

Mazur, B. C. 1959. *On embeddings of spheres*, Bull. Amer. Math. Soc. **65**, 59–65.

______. 1961a. *A note on some contractible manifolds*, Ann of Math. (2) **73**, 221–228.

______. 1961b. *On embeddings of spheres*, Acta Math. **105**, 1–17.

______. 1963a. *Differential topology from the point of view of simple homotopy theory*, Inst. Hautes Études Sci. Publ. Math. **15**, 1–93.

______. 1963b. *Relative neighborhoods and the theorems of Smale*, Ann. of Math. (2) **77**, 232–249.

______. 1964a. *Corrections to: "Differential topology from the point of view of simple homotopy theory" and further remarks*, Inst. Hautes Études Sci. Publ. Math. **22**, 81–91.

______. 1964b. *The method of infinite repetition in pure topology. I*, Ann. of Math. (2) **80**, 201–226.

______. 1966. *The method of infinite repetition in pure topology. II. Stable applications*, Ann. of Math. (2) **83**, 387–401.

McMillan, D. R., Jr. 1964. *A criterion for cellularity in a manifold*, Ann. of Math. (2) **79**, 327–337.

———. 1978. *An arc in a PL n-manifold with no neighborhood that embeds in $S^n, n \geq 4$*, Michigan Math. J. **25**, 29–43.

McMillan, D. R., Jr. and H. Row. 1969. *Tangled embeddings of one-dimensional continua*, Proc. Amer. Math. Soc. **22**, 378–385.

Menger, K. 1926. *Über umfassendste n-dimensionale Mengen*, Proc. Akad. Wetenschappen, Amsterdam **29**, 1125–1128.

Miller, R. T. 1970. *Close isotopies on piecewise-linear manifolds*, Trans. Amer. Math. Soc. **151**, 597–628.

———. 1972. *Approximating codimension three embeddings*, Ann. of Math. (2) **95**, 406–416.

Milnor, J. W. 1961. *Two complexes which are homeomorphic but combinatorially distinct*, Ann. of Math. (2) **74**, 575–590.

———. 1964. *Most knots are wild*, Fund. Math. **54**, 335–338.

———. 1968. *Infinite cyclic coverings*, Conference on the Topology of Manifolds (Michigan State Univ., E. Lansing, Mich., 1967), Prindle, Weber and Schmidt, Boston, pp. 115–133.

Milnor, J. W. and J. D. Stasheff. 1974. *Characteristic Classes*, Ann. of Math. Studies No. 76, Princeton Univ. Press, Princeton, N.J.

Mislin, G. 1995. *Wall's finiteness obstruction*, Handbook of Algebraic Topology, North Holland, Amsterdam, pp. 1259–1291.

Mitchell, W. J. R. and D. Repovš. 1988. *The topology of cell-like mappings. Conference on Differential Geometry and Topology (Sardinia, 1988)*, Rend. Sem. Fac. Sci. Univ. Cagliari **58**, 265–300.

Moise, E. E. 1977. *Geometric Topology in Dimensions 2 and 3*, Graduate Texts in Math., vol. 47, Springer-Verlag, New York, Heidelberg, Berlin.

Moore, R. L. 1925. *Concerning upper semi-continuous collections of continua*, Trans. Amer. Math. Soc. **27**, no. 4, 416–428.

Morgan, John and Gang Tian. 2007. *Ricci flow and the Poincaré conjecture*, Clay Mathematics Monographs, vol. 3, American Mathematical Society, Providence, RI.

Morse, M. 1960. *A reduction of the Schönflies extension problem*, Bull. Amer. Math. Soc. **66**, 113–117.

Mostow, G. D. 1968. *Quasi-conformal mappings in n-space and the rigidity of hyperbolic space forms*, Inst. Hautes Études Sci. Publ. Math. **34**, 53–104.

Munkres, J. R. 1975. *Topology: A First Course*, Prentice-Hall, Englewood Cliffs, N.J.

———. 1984. *Elements of Algebraic Topology*, Addison-Wesley Publ. Co., Menlo Park, CA.

———. 2000. *Topology*, Second Edition, Prentice-Hall, Englewood Cliffs, N.J.

Newman, M. H. A. 1926. *On the theory of combinatory analysis situs*, Proc. Roy. Acad. Amsterdam **29**, 610–641.

———. 1948. *Boundaries of ULC sets in Euclidean n-space*, Proc. Acad. Sci. USA **34**, 193–196.

———. 1966. *The engulfing theorem for topological manifolds*, Ann. of Math. (2) **84**, 555–571.

Nicholson, V. 1969. *Mapping cylinder neighborhoods*, Trans. Amer. Math. Soc. **143**, 259–268.

Ohkawa, T. 1982. *Homological separation of 2-spheres in a 4-manifold*, Topology **21**, 297–313.

Olum, P. 1953. *Mappings of manifolds and the notion of degree*, Ann. of Math. (2) **58**, 453–480.

Papakyriakopoulos, C. D. 1957. *Dehn's lemma and the asphericity of knots*, Ann. of Math. (2) **66**, 1–26.

Penrose, R., J. H. C. Whitehead, and E. C. Zeeman. 1961. *Imbeddings of manifolds in Euclidean space*, Ann. of Math. (2) **75**, 613–623.

Price, T. M. 1966. *Equivalence of embeddings of k-complexes in E^n for $n \leq 2k+1$*, Michigan Math. J. **13**, 65–69.

______. 1975. *On codimension two embeddings*, Geometric Topology (Proc. Conf., Park City, Utah, 1974), Springer, Berlin, pp. 365–370. Lecture Notes in Math., Vol. 438.

Price, T. M. and C. L. Seebeck III. 1974. *Somewhere locally flat codimension one manifolds with 1-ULC complements are locally flat*, Trans. Amer. Math. Soc. **193**, 111–122.

______. 1975. *A codimension two taming theorem*, Geometric Topology (Proc. Conf., Park City, Utah, 1974), Springer, Berlin, pp. 371–394. Lecture Notes in Math., Vol. 438.

Quinn, F. 1979. *Ends of maps, I*, Ann. of Math. (2) **110**, 275–331.

______. 1982. *Ends of maps, III: Dimension 4 and 5*, J. Differential Geometry **17**, 503–521.

______. 1983. *Resolutions of manifolds, and the topological characterization of manifolds*, Invent. Math. **72**, 267–284.

______. 1987. *An obstruction to the resolution of homology manifolds*, Michigan Math. J. **34**, 285–291.

Rochlin, V. A. 1952. *New results in the theory of four-dimensional manifolds*, Doklady Akad. Nauk. SSSR (N.S) **84**, 221–224 (Russian).

Rolfsen, D. 1990. *Knots and Links*, Publish or Perish, Inc., Houston, Texas. Second printing.

Rourke, C. P. 1967. *Covering the track of an isotopy*, Proc. Amer. Math. Soc. **18**, 320–324.

______. 1970. *Embedded handle theory, concordance and isotopy*, Topology of Manifolds (Proc. Inst., Univ. of Georgia, Athens, Ga., 1969), Markham Publ. Co., Chicago, pp. 431–438.

Rourke, C. P. and B. J. Sanderson. 1972. *Introduction to Piecewise-Linear Topology*, Ergebnisse der Mathematik und ihrer Grenzgebiete, Band 69, Springer-Verlag, Berlin.

Rushing, T. B. 1969. *Taming embeddings of certain polyhedra in codimension three*, Trans. Amer. Math. Soc. **145**, 87–103.

______. 1972. *Topological Embeddings*, Academic Press, New York.

______. 1982. *The Mazur-Morse method*, Memorias Seminario Especial de Topologia **IV**, 71–79.

______. 1992. *Hausdorff dimension of wild fractals*, Trans. Amer. Math. Soc. **334**, 597–613.

Schönflies, A. 1908. *Die Entwicklung der Lehre von den Punktmannigfaltigkeiten*, Vol. II, Teubner, Leipzig.

Seebeck, C. L., III. 1970. *Collaring an $(n-1)$-manifold in an n-manifold*, Trans. Amer. Math. Soc. **148**, 63–68.

______. 1971. *Tame arcs on wild cells*, Proc. Amer. Math. Soc. **29**, 197–201.

______. 1975. *Locally homotopically unknotted embeddings of manifolds in codimension two*, Geometric Topology (Proc. Conf., Park City, Utah, 1974), Springer, Berlin, pp. 427–430. Lecture Notes in Math., Vol. 438.

Shaneson, J. L. 1968. *Embeddings with codimension two of spheres in spheres and H-cobordisms of $S^1 \times S^3$*, Bull. Amer. Math. Soc. **74**, 972–974.

Shapiro, A. 1957. *Obstructions to the imbedding of a complex in a Euclidean space*, Ann. of Math. (2) **66**, 256–269.

Sher, R. B. 1968. *Concerning wild Cantor sets in E^3*, Proc. Amer. Math. Soc. **19**, 1195–1200.

______. 1971. *Tame polyhedra in wild cells and spheres*, Proc. Amer. Math. Soc. **30**, 160–174.

Siebenmann, L. C. 1965. *The obstruction to finding a boundary for an open manifold of dimension greater than five*, Ph.D. dissertation, Princeton University.

______. 1968. *On detecting Euclidean space homotopically among topological manifolds*, Invent. Math **6**, 245–261.

______. 1969. *On detecting open collars*, Trans. Amer. Math. Soc. **142**, 201–227.

______. 1972. *Approximating cellular maps by homeomorphisms*, Topology **11**, 271–294.

Siebenmann, L. C. and J. Sondow. 1966/1967. *Some homeomorphic sphere pairs that are combinatorially distinct*, Comment. Math. Helv. **41**, 261–272.

Smale, S. 1961. *Generalized Poincaré's conjecture in dimensions greater than four*, Ann. of Math. (2) **74**, 391–406.

______. 1962. *On the structure of manifolds*, Amer. J. Math. **84**, 387–399.

Spanier, E. H. 1966. *Algebraic Topology*, McGraw-Hill Book Co., New York.

Stallings, J. R. 1960. *Polyhedral homotopy-spheres*, Bull. Amer. Math. Soc. **66**, 485–488.

______. 1962a. *On topologically unknotted spheres*, Ann. of Math. (2) **77**, 490–503.

______. 1962b. *The piecewise-linear structure of Euclidean space*, Proc. Cambridge Phil. Soc. **58**, 481–488.

______. 1963. *On topologically unknotted spheres*, Ann. of Math **77**, 490–503.

______. 1965a. *The embedding of homotopy types into manifolds*. Mimeographed notes, Princeton University.

______. 1965b. *On infinite processes leading to differentiability in the complement of a point*, Differential and Combinatorial Topology (A Symposium in Honor of Marston Morse), Princeton Univ. Press, Princeton, N.J., pp. 245–254.

Štan′ko, M. A. 1969. *The imbedding of compacta in euclidean space*, Dokl. Akad. Nauk SSSR **186**, 1269–1272 (Russian). Also in Soviet Math. Dokl. **10** (1969), 758–761.

______. 1970. *Imbedding of compacta in euclidean space*, Mat. Sb. (N.S.) **83 (125)**, 234–255 (Russian).

______. 1971a. *Approximation of imbeddings of compacta in codimensions greater than two*, Dokl. Akad. Nauk. SSSR **198**, 783–786 (Russian). Also in Soviet Math. Dokl. **12** (1971), 906–909.

______. 1971b. *Solution of Menger's problem in the class of compacta*, Dokl. Acad. Nauk. SSSR **201**, 1299–1302 (Russian). Also in Soviet Math. Dokl. **12** (1971), 1846–1849.

______. 1973. *Approximation of compacta in E^n in codimension greater than two*, Mat. Sb. (N.S.) **90 (132)**, 625–636, 640 (Russian). Also in Math. USSR-Sb. **19**, (1973) 615–626.

Stark, C. W. 2002. *Topological rigidity theorems*, Handbook of Geometric Topology, North Holland, Amsterdam, pp. 1045–1084.

Stillwell, J. 1980. *Classical Topology and Combinatorial Group Theory*, Graduate Texts in Math., vol. 72, Springer-Verlag, New York, Heidelberg, Berlin.

Sumners, D. W. 1966. *Higher-dimensional slice knots*, Bull. Amer. Math. Soc. **72**, 894–897.

______. 1970. *Homotopy torsion in codimension two knots*, Proc. Amer. Math. Soc. **24**, 229–240.

Swan, R. G. 1978. *Projective modules over Laurent polynomial rings*, Trans. Amer. Math. Soc. **237**, 111–120.

Swarup, G. A. 1975. *An unknotting criterion*, J. Pure Appl. Algebra **6**, 291–296.

Taubes, C. G. 1987. *Gauge theory on aymptotically periodic 4-manifolds*, J. Differential Geom. **25**, 363–430.

Torres, G. 1953. *On the Alexander polynomial*, Ann. of Math. (2) **57**, 57–89.

Torres, G. and R. H. Fox. 1954. *Dual presentations of the group of a knot*, Ann. of Math. (2) **59**, 211–218.

Väisälä, J. 1979. *Demension and measure*, Proc. Amer. Math. Soc. **76**, 167–168.

van Kampen, E. R. 1932. *Komplexe in euklidischen Räumen*, Abh. math. Sem. Hamburg **9**, 72–78, 152–153.

van Mill, J. 1989. *Infinite-Dimensional Topology*, North-Holland Mathematical Library, No. 43, North-Holland Publishing Co., Amsterdam.

Venema, G. A. 1985. *Embeddings of locally connected compacta*, Trans. Amer. Math. Soc. **292**, 613–627.

______. 1987. *Approximating codimension two embeddings of cells*, Pacific J. Math. **126**, 165–195.

______. 1995. *Duality on noncompact manifolds and complements of topological knots*, Proc. Amer. Math. Soc. **123**, 3251–3262.

______. 1997. *Local homotopy properties of topological embeddings in codimension two*, Geometric Topology, Part I (William H. Kazez, ed.), AMS/IP Studies in Advanced Mathematics, Volume 2, Amer. Math. Soc. and International Press, Providence, R.I., pp. 388–405.

______. 1998. *A manifold that does not contain a compact core*, Topology Appl. **90**, 197–210.

Wall, C. T. C. 1965a. *Finiteness conditions for CW-complexes*, Ann. of Math. (2) **81**, 56–69.

______. 1965b. *Unknotting tori in codimension one and spheres in codimension two*, Proc. Cambridge Philos. Soc. **61**, 659–664.

______. 1966. *Finiteness conditions for* CW *complexes. II*, Proc. Roy. Soc. Ser. A **295**, 129–139.

______. 1967. *On bundles over a sphere with a fibre euclidean space*, Fund. Math. **61**, 57–72.

______. 1969. *On homotopy tori and the annulus theorem*, Bull. London Math. Soc. **1**, 95–97.

______. 1970. *Surgery on compact manifolds*, London Math. Soc. Monographs, vol. 1, Academic Press, London.

______. 1999. *Surgery on Compact Manifolds*, Second Edition, Mathematical Surveys and Monographs, vol. 69, Amer. Math. Soc., Providence, R.I.

Weber, C. 1967. *Plongements de polyhèdres dans le domaine métastable*, Comment. Math. Helv. **42**, 1–27.

West, J. E. 1975. *Compact ANR's have finite type*, Bull. Amer. Math. Soc. **81**, 163–165.

Whitehead, J. H. C. 1935. *A certain open manifold whose group is unity*, Quart. J. Math. (2) **6**, 268–279.

Whitney, H. 1944a. *The self-intersection of a smooth n-manifold in $2n$-space*, Ann. of Math. (2) **45**, 220–246.

______. 1944b. *The singularities of a smooth n-manifold in $(2n-1)$-space*, Ann. of Math. (2) **45**, 247–293.

Wilder, R. L. 1979. *Topology of Manifolds*, Amer. Math. Soc. Colloq. Publ., vol. 32, Amer. Math. Soc., Providence, R.I. Reprint of the 1963 edition.

Wright, D. G. 1977. *Sticky arcs in E^n $n \geq 4$*, Proc. Amer. Math. Soc. **66**, 181–182.

Zassenhaus, H. J. 1958. *The Theory of Groups*, Second Edition, Chelsea Publishing Co., New York.

Zeeman, E. C. 1960. *Unknotting spheres*, Ann. of Math. (2) **72**, 350–361.

______. 1962a. *Isotopies and knots in manifolds*, Topology of 3-manifolds and Related Topics (M. K. Fort, Jr., ed.), Prentice-Hall, Englewood Cliffs, N.J., pp. 187–193.

______. 1962b. *The Poincaré conjecture for $n \geq 5$*, Topology of 3-manifolds and Related Topics (M. K. Fort, Jr., ed.), Prentice-Hall, Englewood Cliffs, N.J., pp. 198–204.

______. 1963a. *Seminar on Combinatorial Topology*, Inst. des Hautes Études Sci., Paris. Mimeographed notes, revised 1965.

______. 1963b. *Unknotting combinatorial balls*, Ann. of Math. (2) **78**, 501–526.

______. 1964. *On the dunce hat*, Topology **2**, 341–358.

______. 1965. *Twisting spun knots*, Trans. Amer. Math. Soc. **115**, 471–495.

Selected Symbols and Abbreviations

$\searrow$, 100
$\doteq$, 299
$\vee$, 1

$A(X,\gamma;t)$, 301
ANR, 12
AR, 12

$B(A;\epsilon)$, 115
B^n, xiv
∂M, 2

$C(X,Y)$, 53
$\mathrm{Cell}(M,Y)$, 60
$C_{\mathrm{PL}}(X,Y)$, 53

$d(x,y)$, 53
dem, 125
$\dim_{\mathrm{h}}$, 137
DDP, 367
D^n, 1

$E(L)$, 302
$\mathrm{Emb}(X,Y)$, 53
$\mathrm{Emb}_{\mathrm{PL}}(X,Y)$, 53
(ϵ), 383

Fr, 418

$\mathrm{Homeo}(X,Y)$, 53
$\mathrm{Homeo}_{\mathrm{PL}}(X,Y)$, 53
$H_*(X;\mathbb{Z}[\mathbb{J}])$, 275
$\check{H}_c^p(A,B;G)$, 6

$\mathrm{Infl}(C,\nu)$, 414
$\mathrm{Int}\, M$, 2
I^n, 1

$\mathbb{J}$, 275

$k(X,\gamma)$, 314
k-LC, 10
k-LCC, 31
K^ε, 432
$K_{p,q}$, 25

Λ, 299
LC^k, 10
LCC^k, 31

$M(L)$, 302
$\mathrm{Map}(f)$, 123
$\mathrm{Map}(f,K_0)$, 186
M_n^k, 131
$\widetilde{M}_n^k$, 132
$M\!B$, 8
m_p, 137

N_f, 58

$\mathcal{O}(B)$, 300

π_f, 186
$\pi_n(X,x_0)$, 9
$\pi_n(X,A,x_0)$, 10

R- $\mathrm{diam}(X)$, 385

$\rho(f, g)$, 53
$R[\mathbb{J}]$, 275
$\mathbb{R}^n$, xiii
$\mathbb{R}^n_+$, 2

$S(f)$, 100
$\mathscr{S}(T^n)$, 439
S^n, xiv
$\mathrm{Spin}^k(A)$, 38
$\mathrm{Surj}(X, Y)$, 53
$\mathrm{Surj}_{\mathrm{PL}}(X, Y)$, 53
$\mathrm{Susp}(K)$, 35

1-UV, 168

$W(L)$, 335

$\mathbb{Z}$, 275
$\mathbb{Z}[G]$, 275, 430

Index

absolute
  neighborhood retract, 12
  retract, 12
  self-intersector, 182
abstract regular neighborhood, 213, 219
acyclic, 17
Alexander
  3-cell, 49
  Cantor set, 51, 52
  horned sphere, 47, 50
  invariant, 274
  isotopy, 202
  matrix, 305
  polynomial, 301
Alexander-Newman Theorem, 27
Alexandroff Theorem, 14
Alford
  arc, 89
  sphere, 88
algebraic unknotting, 279, 295
α-equivalence, 427
ambient
  dimension, xvi
  space, xiv
Annulus Theorem, 65, 374
ANR, 12
antipodal
  map, 183
  points, 183
Antoine
  3-cell, 46, 171
  Cantor set, 46
  necklace, 42, 125, 169, 172, 175
  sphere, 46
Approximation Problem, xvi
Approximation Theorems, 162, 227, 243, 263, 325, 347, 366, 370, 382, 384, 398, 403, 438, 440
AR, 12

ball, xiv
β-map, 427
β-statement, 168
bicollar, 62, 356
  pinched, 93
binary icosahedral group, 18, 281
Bing
  dogbone space, 60
  hooked rug, 83
  house with two rooms, 189
  shrinking criterion, 56
  sling, 80, 424
Blankinship Cantor set, 169, 172, 175
Borel Conjecture, 442
Borromean rings, 311
boundary
  manifold (∂-manifold), 2
  of a $\mathcal{E}$-manifold, 2
  operator for homotopy groups, 11
∂-manifold, 2
bouquet of circles, 1
branching system, 236, 388
Bryant-Seebeck engulfing, 115, 127
Bubble Lemma, 377, 382

Cantor set, 42
cell, xiv
  with holes, 374
cell-like
  embedding relation, 383
  mapping, 109

resolution, 434
set, 107
Cell-like Approximation Theorem, 366
cellular, 59
map, 60
move, 146
set, 59
cellularity criterion, 73, 106
center line of a solid torus, 44
closed
n-cell complement, 410
manifold, 2
cocore of a handle, 329
codimension, xvii
collapse, 100
$(j, \epsilon)-$, 254
collar
for an end, 428
of a ball pair, 285
of a subset, 64
pinched, 92
singular, 405
complementary domains, 26
completion
of a manifold, 428
of map, 432
complex
CW, xiv
simplicial, xiv
concordance, 198
concordant, 199
Cone Unknotting Theorem, 201
Relative, 215
connected
k-connected pair, 11
k-connected space, 10
locally k-co-connected, 31
locally k-connected, 10
continuous relation, 383
contractible retraction, 213
controlled engulfing, 115
generalized, 119, 213, 262, 266, 354, 359, 362
coordinate neighborhood, xiv
core of a handle, 329
covering isotopy, 209
Theorem
Relative, 216
theorem, 210
covering transformations, 298
cube, 1, 131
crumpled, 410
with handles, 84
CW complex, xiv

DDP, 367
deck transformations, 298
decomposition space, 55
deficiency of a group, 283
defining sequence
abstract, 172
compatible, 176
geometric, 173
degree, 405
of a Laurent polynomial, 312
Dehn's Theorem, 282
Delta structure, 236, 388
(δ, h)-cobordism over X, 433
δ-homotopy, 118
δ-product structure over X, 433
demension, 125
diameter of a homotopy, 432
dimension
of a topological space, 14
of embedding, 125
theorem, 8
disjoint disks property, 367
disk, 1
bundle neighborhood, 346
dominated by a finite complex, 429
double Fox-Artin arc, 77
Double Suspension Theorem, 423, 443
dual
face of a cube, 133
skeleton of a cube, 133
skeleton of a triangulation, 98
Duality Theorem, 6, 8, 317
dunce hat, 189

embedded space, xiv
embedding, xiv
dimension, 125, 160
relation, 383
cell-like, 383
Embedding Existence Theorems, 162, 182, 185, 190, 191, 194, 217, 237, 274, 290, 325, 327, 335, 391, 438
end
of a manifold, 428
of a map, 432
End Theorem, 433
engulfing, 97
Bryant-Seebeck, 115
controlled, 115
generalized, 119
radial, 123
Stallings, 99
Zeeman, 105
ϵ-mapping, 14
ϵ-mapping cylinder neighborhood, 129
ϵ-push, 115
ϵ-regular neighborhood, 124
ϵ-tame, 157
equivalent

embeddings, xiv
Euclidean space, xiii
Eventual Hurewicz Theorem, 16
everywhere wild, 73
Existence Problem, xvi
exotic homology manifold, 437
expansion, 185
exterior of a link, 302
eyebolt, 84

face of a cube, 131
faithful
embedding of boundary manifolds, 194
map of ∂-manifolds, 38
map of pairs, 201
fiber
of an embedding of D^j, 249
preserving, 209, 249
fine homotopy equivalence, 110
flat, xv
cell, 26, 27, 66, 164, 264, 382
locally, 201
locally flat modulo A, 164
sphere, 26, 27, 62, 64, 93, 164, 264, 278, 284, 289, 399, 404
twice, 92
Flatness Problem, xv
Flattening Theorems, 26, 27, 29, 62, 64, 66, 93, 201, 264, 284, 289, 293, 295, 356, 366, 382, 399, 404, 418
Fox-Artin
arc, 75
double, 77
sphere, 79
fractal, 137
framing of a handle, 329
full
subcomplex, 98
subset of Emb(X, M), 158
fundamental group of the end, 429

Generalized Schönflies Theorem, 62, 64

h-Cobordism Theorem, 425
Thin, 434
Weak, 105
handle theory, 276
handle-like decomposition, 220
Hauptvermutung, 283, 439
Hausdorff
dimension, 137
measure, 137
Hawaiian earring, 5
homeomorphically approximated, 404
homogeneously embedded, 83
strongly, 421
homologically tame, 432
homology
∂-manifold, 437
cobordant, 313
manifold, 435
homotopy group
boundary operator, 11
definition, 9
induced homorphism, 10
relative, 10
homotopy sequence of a pair, 11
Hopf link, 28, 211
Hurewicz Isomorphism Theorem
global, 16
Local (or Eventual), 16
Relative, 17

icosahedron, 18
index of a homology manifold, 435
Index Theorem, 435
induced homomorphism, 10
infinite cyclic cover, 298
inflation, 414
interior of a ∂-manifold, 2
Invariance of Domain, 2
inverse limit of groups, 428
inverse set, 61
isolated end, 428
isotopy
covering, 209
extension, 209

(j, ϵ)-collapse, 254
Jordan
Curve Theorem, 19
Separation Theorem, 8

k-connected
pair, 11
space, 10
Kawauchi invariant, 314
Kirby calculus, 329
Klee trick, 66
knotted, xv

Laurent polynomial, 299
degree, 312
LC^k, 10
Levine's Unknotting Theorem, 282
limited by a collection, 427
link calculus, 329
linking number, 302
local
coefficients, 275
contractibility, 358
locally
1-alg, 31
co-connected, 31

collared, 64
connected in dimension k, 10
constant fundamental group, 432
flat, xvi, 32, 201, 436
flat modulo A, 164
homotopically unknotted, 31
solvable, 158
spanned, 409
tame, 33
longitude, 28, 302, 329

Main Problem, xiv
manifold, 2
factor, 436
topological, 162
manifold-with-boundary, 2
map
r-connected, 185
1-UV, 168
proper, 58
mapping
cylinder, 123
neighborhood, 129, 417
reduced, 185
retraction, 186
simplicial, 124, 185
torus, 326
Matsumoto
link, 307
manifold, 333
Mazur swindle, 95
meets boundary regularly, 201
Menger
continuum, 138, 141, 233, 374
modified, 132
standard, 131
sponge, 131
meridian, 28, 302, 329
mesh of a cover, 137
metastable range, 192, 193
mildly wild arc, 79
Milnor
duality, 317
sequence, 276, 299
mismatch property, 413
modified Menger continuum, 132
Monotone Union Theorem, 285
Relative, 285

near-homeomorphism, 56
neighborhood, 2
of an end of a manifold, 428
of an end of a map, 432
Newman contractible manifold, 19
nice immersion, 228
nondegeneracy set, 58
null sequence, 372

one-point union, 1
one-sided, 350
orbit space, 413
order
ideal, 300
of a module, 300
ordinary polynomial, 312
organization of this book, xvi
orientable, 6
orientation, 405
orthogonal complement, 323

pillbox, 47
pinched
bicollar, 93
collar, 92
PL, xiv, xv
almost, 265
locally flat, 279, 402
modulo a point, 92
on an open subset, 265
Schönflies Conjecture, 26
Structure Theorem, 347
plug, 84
Poincaré Theorem
PL, 425
Weak, 99
Poincaré-Lefschetz duality, 6
polyhedron, xiv
presentation matrix, 300
preserve fibers, 209
primary
decomposition, 315
submodule, 314
Product Neighborhood Theorem, 201
Relative, 215
projective
class group, 430
module, 430
proper
mapping, 58
relation, 383
property 1-UV, 168
pseudo-isotopy, 154, 162

quotient map, 55

R-diameter, 385
R-disk, 385
radial engulfing, 123
ratio-changing general position, 151
regular neighborhood
ϵ-, 124
abstract, 213, 219
meets boundary regularly, 201
regular value, 225
relation, 383

cell-like embedding, 383
continuous, 383
Relative
Covering Isotopy Theorem, 216
Hurewicz Isomorphism Theorem, 17
Monotone Union Theorem, 285
Product Neighborhood Theorem, 215
relative homotopy group, 10
relatively 2-connected at infinity, 426
removing a 2-handle, 329
resolvable, 434
retract
absolute, 12
neighborhood, 12
retraction
contractible, 213
Rochlin invariant, 444

s-Cobordism Theorem, 426
saturated set, 55
Schönflies
Conjecture, 26
Theorem
Generalized, 62, 64
PL, 19
topological, 19
Seifert
matrix, 305
surface, 302
Seifert-van Kampen Theorem, 3, 21
semi-capped surface, 236, 388
semifree action, 415
sewing crumpled cubes, 411
shadow, 100, 217
shrinkability criterion, 56
shrinkable
decomposition, 56
fixing C, 58
strongly, 58
Sierpiński
carpet, 131
curve, 374
simple closed curve, 19
simple-homotopy equivalence, 185
simplicial
complement, 98
complex, xiv
mapping cylinder, 124, 185
triangulation, 443
simply connected at infinity, 106
$\sin(1/x)$-continuum, 59
singular collar, 405
singular set
of a map, 100
skeleton
dual, 98
of a cube, 131
of a triangulation, 98
slice knot, 278
slice-trivial, 385
sliced map, 217
Slicing Lemma, 217
solid torus, 42
solvable, 152
locally, 158, 161
sphere, xiv
spinning, 36, 281
Stability of Embedding Theorem, 217
stable
fundamental group at an end, 429
homeomorphism, 439, 441
inverse sequence, 428
manifold, 441
stably
free, 430
isomorphic, 430
Stallings engulfing, 99, 108, 194, 289
modified, 104, 108, 289
standard
n-cube, 130
Menger continuum, 132
mistake, 125, 130
Štan′ko
complex, 236, 388
move, 235, 388
strongly
homogeneously embedded, 421
shrinkable, 58
sunny collapsing, 211
supporting isotopy, 115
surjection, 53
suspension, 35
symmetrically situated, 182

tame
ϵ-tame, 157
as a subspace, xv
Cantor set, 42
compactum, 227
embedding, xv
end of a manifold, 429
end of a map, 432
locally, 33
Taming Problem, xv
Taming Theorems, 158, 160, 161, 261, 263, 265, 347, 402
template, 235, 388
Thin h-Cobordism Theorem, 434
topological
embedding, xiv
manifold, xiv, 162
topologically rigid, 442
torsion, 426
submodule, 301

torus
  knot, 25, 28
  solid, 42
trivial range, xvii, 145
twice flat, 92, 404
two-sided, 350

unknotted, xv
Unknotting Problem, xv, 196
Unknotting Theorems, 146, 165, 196, 199, 201, 212, 213, 215, 220, 240, 261, 264, 278, 279, 281, 358, 393
unlink, 28
unraveled at infinity, 284
upper half space, 1

vertically simple, 217
Vietoris-Begle Theorem, 9, 109, 435

Wang sequence, 276
Weak
  h-Cobordism Theorem, 105
  Poincaré Theorem, 99
weakly flat, 78
wedge, 1
weight of a group, 283
Whitehead
  continuum, 111
  group, 280, 281, 426, 433
  link, 112, 211, 335
  manifold, 112
  Theorem, 17
Whitney trick, 195
wild
  at a point, 30, 32
  Cantor set, 42, 169, 174, 421
  embedding, xv
  mildly, 79
  point, 51
  set, 51
  subspace, xv

Zeeman engulfing, 105
Zeeman's Unknotting Theorem, 210

Titles in This Series

108 **Enrique Outerelo and Jesús M. Ruiz,** Mapping degree theory, 2009
107 **Jeffrey M. Lee,** Manifolds and differential geometry, 2009
106 **Robert J. Daverman and Gerard A. Venema,** Embeddings in manifolds, 2009
105 **Giovanni Leoni,** A first course in Sobolev spaces, 2009
104 **Paolo Aluffi,** Algebra: Chapter 0, 2009
103 **Branko Grünbaum,** Configurations of points and lines, 2009
102 **Mark A. Pinsky,** Introduction to Fourier analysis and wavelets, 2009
101 **Ward Cheney and Will Light,** A course in approximation theory, 2009
100 **I. Martin Isaacs,** Algebra: A graduate course, 2009
99 **Gerald Teschl,** Mathematical methods in quantum mechanics: With applications to Schrödinger operators, 2009
98 **Alexander I. Bobenko and Yuri B. Suris,** Discrete differential geometry: Integrable structure, 2008
97 **David C. Ullrich,** Complex made simple, 2008
96 **N. V. Krylov,** Lectures on elliptic and parabolic equations in Sobolev spaces, 2008
95 **Leon A. Takhtajan,** Quantum mechanics for mathematicians, 2008
94 **James E. Humphreys,** Representations of semisimple Lie algebras in the BGG category $\mathcal{O}$, 2008
93 **Peter W. Michor,** Topics in differential geometry, 2008
92 **I. Martin Isaacs,** Finite group theory, 2008
91 **Louis Halle Rowen,** Graduate algebra: Noncommutative view, 2008
90 **Larry J. Gerstein,** Basic quadratic forms, 2008
89 **Anthony Bonato,** A course on the web graph, 2008
88 **Nathanial P. Brown and Narutaka Ozawa,** C^*-algebras and finite-dimensional approximations, 2008
87 **Srikanth B. Iyengar, Graham J. Leuschke, Anton Leykin, Claudia Miller, Ezra Miller, Anurag K. Singh, and Uli Walther,** Twenty-four hours of local cohomology, 2007
86 **Yulij Ilyashenko and Sergei Yakovenko,** Lectures on analytic differential equations, 2007
85 **John M. Alongi and Gail S. Nelson,** Recurrence and topology, 2007
84 **Charalambos D. Aliprantis and Rabee Tourky,** Cones and duality, 2007
83 **Wolfgang Ebeling,** Functions of several complex variables and their singularities (translated by Philip G. Spain), 2007
82 **Serge Alinhac and Patrick Gérard,** Pseudo-differential operators and the Nash–Moser theorem (translated by Stephen S. Wilson), 2007
81 **V. V. Prasolov,** Elements of homology theory, 2007
80 **Davar Khoshnevisan,** Probability, 2007
79 **William Stein,** Modular forms, a computational approach (with an appendix by Paul E. Gunnells), 2007
78 **Harry Dym,** Linear algebra in action, 2007
77 **Bennett Chow, Peng Lu, and Lei Ni,** Hamilton's Ricci flow, 2006
76 **Michael E. Taylor,** Measure theory and integration, 2006
75 **Peter D. Miller,** Applied asymptotic analysis, 2006
74 **V. V. Prasolov,** Elements of combinatorial and differential topology, 2006
73 **Louis Halle Rowen,** Graduate algebra: Commutative view, 2006
72 **R. J. Williams,** Introduction the the mathematics of finance, 2006
71 **S. P. Novikov and I. A. Taimanov,** Modern geometric structures and fields, 2006

70 **Seán Dineen,** Probability theory in finance, 2005
69 **Sebastián Montiel and Antonio Ros,** Curves and surfaces, 2005
68 **Luis Caffarelli and Sandro Salsa,** A geometric approach to free boundary problems, 2005
67 **T.Y. Lam,** Introduction to quadratic forms over fields, 2004
66 **Yuli Eidelman, Vitali Milman, and Antonis Tsolomitis,** Functional analysis, An introduction, 2004
65 **S. Ramanan,** Global calculus, 2004
64 **A. A. Kirillov,** Lectures on the orbit method, 2004
63 **Steven Dale Cutkosky,** Resolution of singularities, 2004
62 **T. W. Körner,** A companion to analysis: A second first and first second course in analysis, 2004
61 **Thomas A. Ivey and J. M. Landsberg,** Cartan for beginners: Differential geometry via moving frames and exterior differential systems, 2003
60 **Alberto Candel and Lawrence Conlon,** Foliations II, 2003
59 **Steven H. Weintraub,** Representation theory of finite groups: algebra and arithmetic, 2003
58 **Cédric Villani,** Topics in optimal transportation, 2003
57 **Robert Plato,** Concise numerical mathematics, 2003
56 **E. B. Vinberg,** A course in algebra, 2003
55 **C. Herbert Clemens,** A scrapbook of complex curve theory, second edition, 2003
54 **Alexander Barvinok,** A course in convexity, 2002
53 **Henryk Iwaniec,** Spectral methods of automorphic forms, 2002
52 **Ilka Agricola and Thomas Friedrich,** Global analysis: Differential forms in analysis, geometry and physics, 2002
51 **Y. A. Abramovich and C. D. Aliprantis,** Problems in operator theory, 2002
50 **Y. A. Abramovich and C. D. Aliprantis,** An invitation to operator theory, 2002
49 **John R. Harper,** Secondary cohomology operations, 2002
48 **Y. Eliashberg and N. Mishachev,** Introduction to the h-principle, 2002
47 **A. Yu. Kitaev, A. H. Shen, and M. N. Vyalyi,** Classical and quantum computation, 2002
46 **Joseph L. Taylor,** Several complex variables with connections to algebraic geometry and Lie groups, 2002
45 **Inder K. Rana,** An introduction to measure and integration, second edition, 2002
44 **Jim Agler and John E. McCarthy,** Pick interpolation and Hilbert function spaces, 2002
43 **N. V. Krylov,** Introduction to the theory of random processes, 2002
42 **Jin Hong and Seok-Jin Kang,** Introduction to quantum groups and crystal bases, 2002
41 **Georgi V. Smirnov,** Introduction to the theory of differential inclusions, 2002
40 **Robert E. Greene and Steven G. Krantz,** Function theory of one complex variable, third edition, 2006
39 **Larry C. Grove,** Classical groups and geometric algebra, 2002
38 **Elton P. Hsu,** Stochastic analysis on manifolds, 2002
37 **Hershel M. Farkas and Irwin Kra,** Theta constants, Riemann surfaces and the modular group, 2001